HERPETOLOGY

F. HARVEY POUGH
Arizona State University West

ROBIN M. ANDREWS
Virginia Polytechnic Institute and State University

JOHN E. CADLE
Harvard University

MARTHA L. CRUMP
Northern Arizona University

ALAN H. SAVITZKY
Old Dominion University

KENTWOOD D. WELLS
University of Connecticut

Prentice Hall
Upper Saddle River, NJ 07458

Library of Congress Cataloging-in-Publication Data
Herpetology / F. Harvey Pough . . . [et al.].
          p.   cm.
     Includes bibliographical references and index.
     ISBN 0-13-850876-3
     1. Herpetology.   I. Pough, F. Harvey.
     QL641.H47   1998
     597.9—dc21   97-19347
                                             CIP

Executive Editor: Sheri L. Snavely
Editor in Chief: Paul F. Corey
Editorial Director: Tim Bozik
Assistant Vice President of Production and Manufacturing: David W. Riccardi
Executive Managing Editor: Kathleen Schiaparelli
Project Management: J. Carey Publishing Service
Marketing Manager: Jennifer Welchans
Manufacturing Manager: Trudy Pisciotti
Buyer: Ben Smith
Creative Director: Paula Maylahn
Art Director: Heather Scott
Art Manager: Gus Vibal
Art Editor: Karen Branson
Text Designer: Anne Flanagan
Cover Designer: Heather Scott
Cover Photograph: Jean-Paul Ferrero, Auscape International Pty. Ltd.
Editorial Assistants: Lisa Tarabokjia, Nancy Gross
Photo Editor: Lori Morris-Nantz
Photo Research Administrator: Melinda Reo
Photo Researcher: Cindy Lee Overton

© 1998 by Prentice-Hall, Inc.
Simon & Schuster/A Viacom Company
Upper Saddle River, New Jersey 07458

Printed in the United States of America

10 9 8 7 6 5 4 3 2 1

ISBN 0-13-850876-3

Prentice-Hall International (UK) Limited, *London*
Prentice-Hall of Australia Pty. Limited, *Sydney*
Prentice-Hall of Canada, Inc., *Toronto*
Prentice-Hall Hispanoamericana, S. A., *Mexico*
Prentice-Hall of India Private Limited, *New Delhi*
Prentice-Hall of Japan, Inc., *Tokyo*
Simon & Schuster Asia Pte. Ltd., *Singapore*
Editora Prentice-Hall do Brasil, Ltda., *Rio de Janeiro*

Brief Contents

Contents

P A R T   I V

What Are Their Prospects
for Survival?

Preface

Amphibians and reptiles are successful organisms that represent an approach to terrestrial vertebrate life quite different from that adopted by birds and mammals. The internal processes of amphibians and reptiles differ in many respects from the corresponding processes in birds and mammals, and amphibians and reptiles function differently from birds and mammals in communities and ecosystems. Understanding how and why amphibians and reptiles differ from birds and mammals enriches a biological education, and the study of herpetology is a great deal more than just the study of amphibians and reptiles.

In our view, understanding amphibians and reptiles as organisms requires a perspective that integrates their morphology, physiology, behavior, and ecology and places that information in a phylogenetic context. This book does that—it presents the biology of amphibians and reptiles as the product of phylogenetic history and environmental influences acting in both ecological and evolutionary time. We emphasize how amphibians and reptiles function in the broadest sense. For example, ectothermal temperature regulation is reflected in nearly every aspect of the biology of amphibians and reptiles, from their body shapes (extremely small body size and elongate body shape are feasible only for ectotherms) to their role in ecosystems (low energy flow and high conversion efficiency are the result of ectothermy).

In this book we have emphasized the integration of information from different biological specialties to produce a picture of amphibians and reptiles as animals that do remarkable things and play important roles in modern ecosystems. Evolution provides the context in which the distinctive characteristics of amphibians and reptiles must be evaluated, and both ancestral and derived features are central to an understanding of the biology of amphibians and reptiles.

Collaboration by the six authors—whose research specializations include autecology, synecology, evolution, morphology, physiology, and behavior—has produced a treatment that interweaves these specialties. We find the interrelationships among different levels of biological organization fascinating and have tried to build students' understanding of these relationships from chapter to chapter. In the case of lizards, for example, one or more aspects of the intricate correlations among phylogeny, foraging mode, diet, morphology, exercise physiology, predator avoidance, social system, and reproductive mode is discussed in nearly every chapter. We have used this technique of building topics chapter by chapter in the hope that students will find the complex relationships that emerge intellectually stimulating.

Our goal has been to write a textbook, not a reference book, and that intention is reflected in many features of this volume. The level of treatment assumes that readers are starting with the background provided by a course in Vertebrate Zoology. We have avoided the jargon of biological specialties wherever possible, believing that encountering an unfamiliar technical term breaks a reader's concentration, even if the term has been defined previously. Space has limited the number of citations we could include, and we have placed more emphasis on recent review papers and other citations that will help students to enter the literature of a topic than on acknowledging the first work in a field.

Above all, this book is the product of the lifelong fascination each of us has felt for the animals we study. We hope we will succeed in conveying our sense of excitement to readers.

Acknowledgments

This book has benefited from assistance from many quarters. Gregory W. Payne originally assembled a group of authors to write a herpetology textbook. David Hillis (University of Texas) was an early member of the group, and he has generously allowed us to include material that he developed. The book would never have been completed without the enthusiasm, perseverance, and professionalism of our editor, Sheri Snavely. Her patience in working with six authors was monumental, the stuff of which legends are made, and we are enormously grateful to her. The skill of our production manager, Jennifer Carey, added greatly to the quality of the book, and her command of the intricacies of publishing was always reassuring. Heather Scott's skill in design has produced an aesthetically pleasing format that allows the eye to move smoothly across the page. Ellen Smith's work as developmental editor was instrumental in establishing a logical progression of information, and she compiled the index and glossary that make topics and terminology accessible to readers. Laura Schuett's superb illustrations reveal anatomical details while capturing the essence of the animals we find so fascinating. All of us have been aided by facilities provided by our home institutions, without which this book could not have been written.

Colleagues all over the world have responded generously to our questions and requests for manuscript reviews, photographs, and original data. Academic schedules being what they are, our opportunities to work on the book often coincided with our colleagues' departures for field work, and we are especially grateful for the promptness of their replies at such inconvenient times.

Michael Angilletta, *University of Pennsylvania*
David L. Auth, *University of Florida*
John M. R. Baker, *The Open University*
Edmund D. Brodie, Jr. *Utah State University*
Edmund D. Brodie III, *University of Kentucky*
Gordon Burghardt, *University of Tennessee*
Russell L. Burke, *University of Michigan*
Mark A. Chappell, *University of California, Riverside*
Justin Congdon, *Savannah River Ecology Laboratory*
D. Andrew Crain, *University of Florida*
Alison Cree, *University of Otago*
David Crews, *University of Texas*

David Cundall, *Lehigh University*
Kenneth C. Dodd, Jr., *National Biological Service*
Samantha K. Dozier, *Old Dominion University*
William A. Dunson, *Pennsylvania State University*
Arthur C. Echternacht, *University of Tennessee*
Sharon B. Emerson, *University of Utah*
Peter Feinsinger, *Northern Arizona University*
Lee A. Fitzgerald, *Texas A&M University*
M. J. Fouquette, Jr., *Arizona State University, Tempe*
Margaret H. Fusari, *University of California, Santa Cruz*
Erik W. A. Gergus, *Arizona State University West*
J. Whitfield Gibbons, *University of Georgia*
Mac F. Given, *Neumann College*
David M. Green, *McGill University*
Harry W. Greene, *University of California, Berkeley*
John D. Groves, *The North Carolina Zoo*
Craig Guyer, *Auburn University*
Tim R. Halliday, *The Open University*
James Hanken, *University of Colorado*
Lindesay Harkness, *Chapel Hill, NC*
Robert W. Henderson, *Milwaukee Public Museum*
Frank R. Hensley, *University of North Carolina, Greensboro*
Paul E. Hertz, *Barnard College*
W. Ronald Heyer, *National Museum of Natural History*
John B. Iverson, *Earlham College*
Robert G. Jaeger, *University of Southwestern Louisiana*
Farish A. Jenkins, Jr., *Harvard University*
Lisa Kammerlocher, *Arizona State University West*
Jeffrey W. Lang, *University of North Dakota*
Harvey B. Lillywhite, *University of Florida*
Keith B. Malmos, *Arizona State University West*
John M. Matter, *Clemson University*
Grace McLaughlin, *University of Florida*
Roy McDiarmid, *National Museum of Natural History*
F. M. Anne McNab, *Virginia Polytechnic Institute and State University*
Raymond A. Mendez, *Portal, AZ*
Charles W. Myers, *The American Museum of Natural History*
Kristian Omland, *University of Connecticut*
David Owens, *Texas A & M University*
Daniel M. Pavuk, *Bowling Green State University*
Joseph H. K. Pechmann, *Savannah River Ecology Lab*
Christopher E. Petersen, *Old Dominion University*

Marion R. Preest, *University of Miami*

Carl Qualls, *Virginia Polytechnic Institute and State University*

Fiona Qualls, *Virginia Polytechnic Institute and State University*

Alan Richmond, *University of Massachusetts*

Kathleen A. Roberts, *Old Dominion University*

James Perran Ross, *Crocodile Specialist Group, IUCN*

Rodolfo Ruibal, *University of California, Riverside*

Anthony P. Russell, *University of Calgary*

Barbara A. Savitzky, *Christopher Newport University*

Ivan Sazima, *Universidad Estudual Campinas*

Gordon W. Schuett, *Arizona State University West*

Richard A. Seigel, *Southeastern Louisiana University*

Kurt Schwenk, *University of Connecticut*

Vaughan H. Shoemaker, *University of California, Riverside*

Wade C. Sherbrooke, *Southwestern Research Station, The American Museum of Natural History*

Mary Jane Spring, *University of Connecticut*

Howard Suzuki, *Gainesville, FL*

John B. Thorbjarnarson, *Wildlife Conservation Society*

Bruce J. Turner, *Virginia Polytechnic Institute and State University*

R. Wayne Van Devender, *Appalachian State University*

Paul Verrell, *Washington State University*

Marvalee H. Wake, *University of California at Berkeley*

Richard J. Wassersug, *Dalhousie University*

Romulus Whitaker, *Madras Crocodile Bank Trust*

Henry Wilbur, *University of Virginia*

What Are Amphibians and Reptiles?

CHAPTER

1

Herpetology as a Field of Study

The word herpetology is based on the Greek *herpes*, meaning a creeping thing. That may not sound like an enthusiastic way to describe an animal, but the ancient world held some reptiles in high regard. Alexander the Great, a Macedonian who united many of the Greek city-states around 500 B.C., derived some of his power from a legendary serpent that protected him in its coils when he was a baby. Some ancient legends are preserved in our current world. The staff entwined by a serpent that was carried by Æsclepius, the Roman god of healing, appears as the caduceus (a winged staff entwined by two snakes) of modern medicine. The Romans built shrines to Æsculapius throughout their empire and released Æsculapian snakes (*Elaphe longissima*) at the shrines. This species is native to southern Europe, including Italy and Asia Minor. It also has isolated populations in central Europe, far to the north of other occurrences of the species. These northern populations may be descendants of snakes released 2,000 years ago at shrines to Æsculapius.

The great eighteenth-century Scandinavian biologist Carl von Linné had a less favorable opinion of the creeping animals. Writing under the Latinized version of his name, Carolus Linnaeus, he initiated the hierarchical method of naming organisms that we are familiar with as the binominal classification system. His work, *Systema Naturae* (*The System of Nature*), assembled organisms in groups. Linnaeus did not distinguish amphibians from reptiles, referring to both of them as amphibians and characterizing them as "foul and loathsome." He noted that for this reason,

the Lord had not exerted himself to create many of them.

Later generations of biologists would contend that Linnaeus was sadly mistaken in both those statements. Herpetologists have found amphibians and reptiles to be fascinating subjects for study, and they have identified a large number of species to study — about 4,600 amphibians and 6,000 reptiles. (For comparison, there are about 4,000 species of mammals and 9,000 birds.) Thus the study of herpetology covers more species of animals than does either ornithology or mammalogy, and it includes a great range of body forms, behaviors, and life history patterns.

Studies of amphibians and reptiles have played key roles in biological specializations as diverse as developmental biology, ecology, and medicine. Many of these contributions are the result of unique characteristics that make a species of amphibian or reptile suitable for a particular technique. For example, the large eggs of many frogs and salamanders allow embryonic development to be observed under a microscope. Much of our understanding of the way cells move during gastrulation (when an egg changes from a hollow ball of cells to a structure with distinct layers of endoderm, mesoderm, and ectoderm) resulted from work in which embryologists marked individual cells of frog and salamander embryos with dye and observed their movements. In a similar vein, the diurnal (daytime) activity patterns of many lizards and their use of color and movement in social behavior have made these animals central figures in studies of behavioral ecology.

Herpetological studies have also contributed to advances in molecular biology and medicine. A terrestrial salamander can be exposed to a rapid increase in temperature when the sun shines on the rock it's hiding under, and salamanders have biochemical mechanisms that prevent heat damage in this situation. Some of the earliest descriptions of the protective action of heat shock proteins (also known as chaperonins) were based on studies of their action in salamanders. Subsequently heat shock proteins have been found to occur in nearly all organisms, and they have recently been employed in biomedical contexts, including a treatment for cancer.

The Diversity of Amphibians and Reptiles

Many people are familiar with the major groups of amphibians and reptiles — salamanders, frogs, turtles, crocodilians, and lizards and snakes — from visits to zoos or from televised nature programs. This chapter provides an overview of some basic characteristics of these familiar groups. Detailed descriptions of amphibians and reptiles and their evolutionary relationships are presented in Chapters 2, 3, and 4.

Amphibians

The extant (currently living) species of amphibians include three groups — the salamanders (Urodela), frogs (Anura), and caecilians (Gymnophiona). The more than 4,600 species of amphibians encompass enormous diversity in body form, size, ecology, and behavior (see Duellman and Trueb 1986).

Modern amphibians are grouped as the Lissamphibia. The prefix *liss-* means smooth, and refers to their scaleless skin. The lack of a protective scaly covering and other characteristics of the skin shape many aspects of the lives of amphibians, including social behavior (hedonic glands in the skin produce pheromones used in courtship) and defense (poison from granular glands in the skin makes many amphibians unpalatable, and at least one species is lethally toxic to predators as large as humans). Most important is the permeability of the skin to water (Chapter 5). The skin is a site for gas exchange (uptake of oxygen and release of carbon dioxide) by amphibians, and in biological systems permeability to gases is inseparable from permeability to water. Water evaporates rapidly from the skins of most species of amphibians, and an amphibian can dehydrate and die in a few hours if it does not have access to water.

The high rate at which water evaporates from the skin of most amphibians limits their activity in time and space. That is, an amphibian can be active only during periods when high humidity and low wind speed reduce the evaporative stress. Thus, amphibians are typically active at night (especially rainy nights), and amphibian faunas are most diverse in moist environments.

Skin permeability has another aspect, however, that allows amphibians to live in unexpectedly dry places. Amphibians do not drink water; they absorb water through the skin, and they can absorb water from moist soil. Frogs, toads, and salamanders that live in desert habitats spend many months in burrows underground, drawing water from the soil around them to maintain their water balance. Deserts do not provide water for amphibians to drink and, paradoxically, amphibians would not be able to exist in arid environments if their skins were not permeable.

Urodeles (Salamanders).   Salamanders have elongate trunks and tails. Most salamanders have four legs, although the limbs of some aquatic species are greatly reduced. Two trends are prominent in the evolution of salamanders — loss of lungs and the widespread occurrence of paedomorphosis (the occurrence of larval characteristics in adults).

It seems surprising that a terrestrial animal can live without lungs, but the largest family of salamanders, the Plethodontidae, is characterized by lunglessness, and extremely small lungs are found in some other lineages as well. Lunglessness is possible for salamanders because the skin is a major site of gas exchange. Several explanations have been proposed for the advantage of lunglessness among plethodontid salamanders. The most comprehensive of these hypotheses relates the evolutionary loss of lungs to the specialization of the hyoid apparatus in the throat as a mechanism for projecting the tongue to capture prey. The hyoid cartilages form part of the breathing apparatus of salamanders with lungs. Muscular compression of the buccal region of the mouth and throat forces air into the lungs, and the hyoid cartilages of lunged salamanders are robust. With loss of the lungs (and hence no longer needing to use the hyoid apparatus to ventilate the lungs), plethodontid salamanders have developed the hyoid cartilages as a mechanism for tongue projection during prey capture. The most specialized plethodontids, a group known as bolitoglossines, have tongues that can be projected farther than the head and trunk length of the salamander combined, and aimed well enough to capture tiny, moving prey items (see Chapter 9) (Figure 1–1).

Paedomorphosis is an example of an evolutionary phenomenon known as heterochrony (discussed in Chapter 7). Heterochrony refers to alterations in the timing and rate of developmental processes (primarily during embryonic life) that change the body form of adults. Paedomorphic adults are aquatic and have larval characteristics, such as the presence of external gills and lateral line systems and the absence of eyelids and adult tooth patterns. Paedomorphosis is characteristic of some entire families of aquatic salamanders, such as the Proteidae (*Necturus*, the North American mudpuppy, and *Proteus*, the European cave salamander). In other families, such as the Ambystomatidae, some species are paedomorphic whereas others metamorphose into terrestrial adults. Paedomorphosis is facultative in some species of salamanders; certain populations — or even only certain individuals within a population — retain larval characteristics, whereas other populations or individuals metamorphose. These variations probably represent evolutionary responses to different ecological situations.

Anurans (Frogs and Toads).   The anurans, with about 4,000 species worldwide, form the largest group of amphibians, and they are the most ecologically diverse. In fact, the variety of anurans in the world greatly exceeds the number of names that can be used to distinguish them, and we will use the word frog to refer to any anuran, and toad, tree frog, and similar variants when they are part of the common name of a species. The immediately distinctive characteristic of all frogs is the absence of a tail, and the name Anura is formed from two Greek words meaning without a tail.

Most anurans have short bodies, large heads, and four well-developed limbs. The relative lengths of the forelimbs and hind limbs help to sort anurans into locomotor categories (Figure 1–2). Species with short hind limbs are generally runners, hoppers, or walkers, whereas those with long hind limbs are swimmers or jumpers (moving 10 body lengths or more in a single leap). Among jumping frogs, large hind limb muscles relative to total body mass identify good jumpers. Long hind limbs are often associated with climbing frogs (Emerson 1988b).

Anurans have an enormous variety of reproductive modes, extending from aquatic larvae (tadpoles) to direct development (eggs that hatch into tiny frogs without a free-living tadpole stage) and even to viviparity (birth of tiny frogs). Free-living aquatic larvae

Figure 1–1 **Tongue projection by a bolitoglossine salamander.** *(Photograph by G. Roth.)*

Figure 1–2 **Anuran body form.** Most frogs that jump and swim well have long hind limbs. Walkers and climbers have long forelimbs and hind limbs, whereas hoppers and burrowers have short limbs. *(Source: Pough et al. 1992.)*

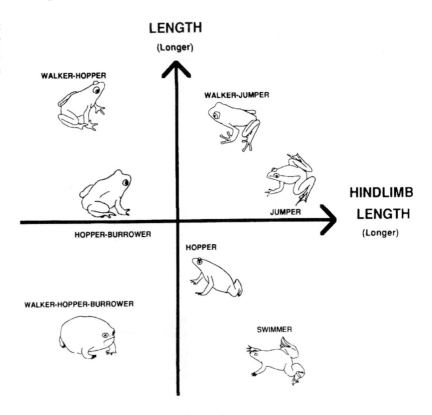

that metamorphose into adults are the most common reproductive mode. Tadpoles are very different animals from frogs, with complex specializations associated with feeding (Chapter 9). Most tadpoles are herbivores, and they are extremely efficient filter feeders. A tadpole is basically a swimming sieve attached to a gut, and enormous anatomical changes occur at metamorphosis when a tadpole changes to a frog.

Because tadpoles and frogs are such different animals, the selective forces acting on them are different and they may follow separate evolutionary paths. A striking example of separate evolutionary trajectories is the difference in tadpole and adult specializations seen in a group of leptodactylid frogs from South America. *Lepidobatrachus*, *Ceratophrys*, and *Chacophrys* form a closely related group of strange-looking frogs with short legs and enormous heads and mouths. (*Ceratophrys* is sold in the pet trade under the name Pac-Man frogs.) The adults of all three genera are ground-dwellers that eat other frogs, in some cases luring their prey within reach by twitching their toes to simulate the movements of an insect. Because adults of all three genera eat other frogs, this dietary specialization is presumed to have been present in their common ancestor.

The tadpole of the common ancestor was a filter feeder, and tadpoles of the three genera display three different feeding specializations (Figure 1–3). *Chacophrys* tadpoles are filter feeders and have a typical tadpole body form. Tadpoles of *Ceratophrys* and *Lepidobatrachus* both prey on other tadpoles, but they look different. *Ceratophrys* tadpoles are much like the filter-feeding tadpoles of *Chacophrys*, but tadpoles of *Lepidobatrachus* have enormous heads and jaws. This specialization of *Lepidobatrachus* tadpoles appears to be the result of heterochrony. That is, a large head and jaws develop only at metamorphosis in *Ceratophrys* and *Chacophrys*, but these characters develop during the larval period of *Lepidobatrachus* (Hanken 1993). Thus, the feeding specializations of tadpoles of the three genera have followed different evolutionary paths, even though adults of all three genera share the same feeding specialization.

Gymnophiona (Caecilians). The least-known group of amphibians is the caecilians — elongate, legless, burrowing and aquatic animals found in tropical habitats around the world (see Nussbaum 1992). Most of these animals spend their entire lives underground

(a)

(b)

(c)   (d)

Figure 1–3 **Specializations of adult frogs and their tadpoles.** An adult *Lepidobatrachus* (a) and a tadpole (c). Note the enormous development of the head and jaws in both the adult and the tapole. In contrast, the adult *Ceratophrys* (b) has a large head and jaws, but the tadpole (d) lacks these specializations, even though it is a predator on other tadpoles. *(Photographs by (a) Lawrence E. Naylor/Photo Researchers Inc., (b) Stephen Dalton/Photo Researchers, Inc., (c) Rodolfo Ruibal, (d) R. Wayne Van Devender.)*

or in the water, and thus are difficult to study. Relatively little is known about their natural history, but work with captive animals has revealed elaborate reproductive specializations, including species in which the embryos develop within the mother and feed on a lipid-rich substance they scrape from the walls of the oviducts with specialized fetal teeth (Chapter 7).

Reptiles

The extant reptiles, in the sense covered by herpetology, include turtles, crocodilians, squamates (snakes and lizards), and tuatara. The qualifier "in the sense covered by herpetology" is needed because the taxa of animals listed do not include all the forms descended from a common ancestor. Crocodilians are closely related to dinosaurs and birds, and a complete list of the exant reptiles would have to include birds. Birds are so different from the other groups of reptiles, however, that they are excluded from herpetology, and we will use the term reptiles to mean all extant reptiles except birds. Chapter 2 describes the relationships of living and extinct amphibians and reptiles (including dinosaurs and birds) in more detail.

All reptiles (in the herpetological sense) except crocodilians have a heart with a single ventricle, as do amphibians and fishes. (Crocodilians, birds, and mammals have a solid septum dividing the ventricle into left and right compartments.) Oxygenated blood from the lungs and deoxygenated blood from the veins both enter the single ventricle of a reptile. Surprisingly, the two bloodstreams are normally kept separate in the heart despite the lack of a septum. Oxygenated blood is pumped to the head and body via the aortic arches, and deoxygenated blood is sent to the lungs via the pulmonary artery (see Chapter 6).

While it is true that an undivided ventricle is a primitive character of amphibians and reptiles (i.e., one that was present in the ancestor of the groups) that is retained in reptiles, that does not mean that a reptilian heart is inferior to a mammalian heart. The undivided ventricle has an important functional benefit for reptiles. Because there is no ventricular septum, a reptile can adjust the proportion of blood flowing from the heart that goes to the body versus the lungs. This phenomenon is called an intracardiac blood shunt, and reptiles use the ability to shift blood between the body and the lungs to accelerate heating and slow cooling. Therefore, this ancestral characteristic — the absence of a ventricular septum — is ad-

vantageous in the context of reptilian biology. In fact, crocodilians, which do have a ventricular septum, use a different method of creating an intracardiac shunt when they are adjusting blood flow to accelerate warming. (Intracardiac shunts are discussed further in Chapter 6.)

Testudinea (Turtles). Just about everybody can recognize a turtle. The shell, which is a distinguishing feature of turtles, is a remarkable structure that encloses the entire animal in a bony case with openings only at the front and rear. The shell also has limited the morphological diversity of turtles — there are aquatic and terrestrial turtles, but no arboreal or gliding species.

The habits of a turtle can usually be deduced from its appearance. Terrestrial turtles like the tortoises (*Geochelone* and several other genera) and North American box turtles (*Terrapene*) have high, domed shells and stout limbs. Some tortoises, especially the gopher tortoises of North America (the genus *Gopherus*), use their spadelike front feet to dig burrows that may extend for 20 meters or more.

Aquatic turtles usually have webbed feet and relatively flat shells that offer less resistance to movement in water than do the domed shells of terrestrial species. Some aquatic turtles, such as musk turtles (*Sternotherus*), mud turtles (*Kinosternon*), and snapping turtles (*Chelydra*), spend more time walking on the bottom than they do swimming, and these species (especially musk turtles) have a more distinctly dome shape to their shells than do turtles that swim quickly to capture prey or escape predators. Some aquatic turtles are still more specialized: the shells of softshelled turtles (*Apalone* and about 14 other genera) have flat shells that lack a bony layer and an external covering of scales. The leatherback sea turtle (*Dermochelys*) is another streamlined species in which the dermal bones are greatly reduced and the stiff dermal scales have been replaced by a flexible covering of skin.

Squamata (Lizards and Snakes). Squamates are the largest group of reptiles. Snakes and lizards are part of the same evolutionary lineage. That is, in an evolutionary sense, the animals known as snakes are specialized lizards, and there is no correct name for a group that includes only the animals popularly called lizards. The term squamates can be used for some topics because the phenomenon being discussed is common to both kinds of animals. In other cases, however, we will refer to lizards or snakes to make distinctions between the two kinds of squamates.

Squamates are an enormously diverse group, with species that live in habitats extending from below-ground to the treetops, from deserts to the ocean, and from the equator to the Arctic Circle (Bauer 1992). Many lizards are diurnal, brightly colored, and use conspicuous visual displays in their social behavior. These characters have made lizards familiar elements of the fauna and important subjects for behavioral and ecological studies (Vitt and Pianka 1994). Snakes are often secretive and rely on scent rather than vision in their predatory and social behavior. As a result, snakes are usually a less conspicuous part of the fauna than are lizards. Nonetheless, snakes are important components of ecosystems in many parts of the world and display a broad range of specializations (Seigel et al. 1987, Shine 1991, Seigel and Collins 1993).

Leglessness has evolved repeatedly among squamates. All snakes are legless, and several families of lizards include species that are functionally or actually legless. Legless squamates must be elongate because they form curves along the length of the body to push against the substrate during locomotion (Chapter 8). Snakes have elongated trunks with very high numbers of presacral vertebrae (i.e., anterior to where the hip would be if snakes had hips) and relatively few post-sacral vertebrae. Legless lizards, in contrast, have relatively short trunks and very long tails. Many legless squamates are surface dwellers. Their slim bodies allow them to move easily through dense vegetation or leaf litter. Other legless species are fossorial (burrowing). Some of these animals construct open tunnels in compact soils, whereas others move through loose soil by a process known as sand swimming. Amphisbaenians are an evolutionary lineage of specialized burrowing squamates that occur in tropical habitats around the world (Gans 1992). Only one of the 135 species of amphisbaenians retains limbs.

Arboreal lizards are often flattened from side to side, and many species have crests on the head that may help to obscure their outlines and make them less conspicuous to predators that see them silhouetted against the sky. African chameleons (*Chamaeleo* and other genera) provide good examples of crests, and similar structures can be found on other arboreal lizards such as the casque-headed lizard of Central America (*Corytophanes cristatus*).

Specialized arboreal snakes, such as the Central American vine snake *Imantodes cenchoa*, hunt lizards, frogs, and birds that perch at the tips of branches. Their extremely elongate bodies and tails spread their weight over a large area and allow them to crawl across leaves and twigs (Figure 1–4). Vine snakes represent an extreme development of elongation among snakes, but other arboreal snakes, such as boas, pythons, and vipers, are slimmer than their terrestrial relatives.

Not all snakes are long and thin. In fact, being short and fat has some definite advantages. Snakes swallow their prey whole, and the cartoon image of a snake with a lump representing a large meal in its stomach is based on fact. A stout body allows snakes to swallow large prey, and stout bodies are one of several specializations of vipers. The stoutest vipers are members of the African genus *Bitis*, such as the puff adder and Gaboon viper (*B. arietans* and *B. gabonica*). The Gaboon viper grows to a length of only 1.2 meters, but there is a record of a Gaboon viper eating an antelope.

Many lizards enter water to escape from predators, but only a few species actually forage underwater. The best-known aquatic species is the Galápagos marine iguana (*Amblyrhynchus cristatus*), which feeds on marine algae that it scrapes from rocks, diving as deep as 10 meters in the process. The distinction between aquatic and terrestrial species is blurred among snakes. Many genera, including familiar forms such as water snakes (*Nerodia* and *Natrix*) and some garter snakes (*Thamnophis*), forage both in and out of water. More specialized aquatic snakes, such as the homalopsines of Asia and Australia, have nostrils with valves that exclude water. The most specialized aquatic snakes, the acrochordids (the Indo-Australian wart snakes) and hydrophiines (sea snakes), lack enlarged ventral scales and never emerge from the water.

Crocodylia (Crocodilians). Only 21 species of crocodilians survive today, and most of them are listed as threatened or endangered. The largest living reptile is probably the Australian saltwater crocodile (*Crocodylus porosus*). Estimating the maximum size of crocodilians is difficult because individuals continue to grow slowly long after they reach maturity. Thus, the oldest crocodilians are the largest ones, but in a world where humans are their major predators, few crocodilians live

Figure 1–4 **Extremes of body form among snakes.** (a) An elongate arboreal colubrid snake from Central America (*Imantodes cenchoa*). (b) A heavy-bodied terrestrial viper, the African puff adder (*Bitis arietans*). (*Photographs by (a) Michael and Patricia Fogeden/Bruce Coleman, (b) F. Harvey Pough.*)

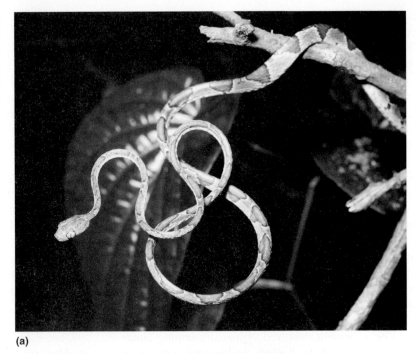

(a)

(b)

long enough to grow to their maximum size. The largest living saltwater crocodiles are about 7 meters long and weigh more than 1,000 kilograms.

The most conspicuous morphological differences among crocodilians involve the shape of the snout. Broad-snouted species, including the American alligator (*Alligator mississippiensis*), are generalized feeders that eat hard-bodied prey such as turtles, snails, and armored fishes. Long-snouted species, of which the Indian gharial (*Gavialis gangeticus*) is the prime example, are fish-eating specialists that capture fast-moving prey with a rapid sideways movement of the head. Some crocodiles (*Crocodylus*) with broad snouts prey on mammals, seizing them when they come to the water to drink, and large crocodiles have been known to attack and kill humans.

Crocodilians are members of the archosaurian lineage, which includes dinosaurs and birds (Chapter 2). As such, they provide a basis for understanding the ecology, behavior, and physiology of dinosaurs (Sues 1989). Crocodilians provide extensive parental care to their young, and evidence is emerging of social structures among crocodilians that may involve many individuals (Lang 1989). These observations support the hypothesis that similar kinds of parental care and social behavior were characteristic of dinosaurs.

Rhynchocephalia, Sphenodontida (Tuatara). Commonly known by their Maori name, tuatara, the Sphenodontia contains only two species, both in the genus *Sphenodon* (Figure 1–5). Indeed, until the redis-

Figure 1–5 **Henry, a tuatara in the Southland Museum at Invercargill, New Zealand.** Henry was the model for the tuatara that appears on the New Zealand 5-cent coin. *(Photographs by F. Harvey Pough.)*

covery of the second species was reported in 1990, only a single species of tuatara was recognized (Daugherty et al. 1990). Modern tuatara (the Maori language does not add an *s* to form the plural) are confined to offshore islands of New Zealand. Both species of tuatara are lizard-like in appearance but are distinguished from lizards by several primitive features of their internal anatomy. Tuatara eggs take 11 to 16 months to hatch, and the hatchlings of *S. punctatus* require 11 to 13 years to mature and 25 years or more to grow to the adult size of 50 centimeters and about 2 kilograms. Tuatara are active primarily at night, when they emerge from their burrows and feed on invertebrates and small vertebrates.

The more abundant species, *Sphenodon punctatus*, is found on 29 islands off the shore of New Zealand, whereas *S. guntheri* occurs only on North Brother Island and has a total population of about 300 adults. Both species are long-lived — a maximum age of at least 60 years has been recorded for a female *S. punctatus* on Stephens Island — and have low reproductive rates. The major threats to populations of tuatara come from introduced mammals, especially ship and Norway rats, which prey on eggs, juveniles, and adult tuatara (Cree et al. 1995). Several of the populations of *S. punctatus* consist of fewer than 20 individuals, and tuatara do not appear to be reproducing on islands with large numbers of rats. The recovery plan for tuatara adopted by the New Zealand Department of Conservation calls for establishing breeding colonies of tuatara in captivity to produce young animals to restock some of these islands after the rats have been exterminated (Cree and Butler 1993).

Shared Characters of Amphibians and Reptiles

By now you should be convinced that the more than 10,000 species of amphibians and reptiles display an enormous diversity of behavioral and morphological features. But if that's so, why is there an area of biological specialization called herpetology that includes both amphibians and reptiles? After all, reptiles are no more closely related to amphibians than they are to mammals — why should one taxonomic discipline study two such distantly related groups as amphibians and reptiles?

Historical accident is part of the reason for the existence of the field of herpetology. Remember that in the eighteenth century Linnaeus did not distinguish amphibians from reptiles — he lumped them together with other vertebrates that were not bony fishes, birds, or mammals. But historical inertia is not the major reason why herpetologists have continued to study both amphibians and reptiles. These two groups share an ancestral characteristic that makes their biology uniquely different from that of any other group of terrestrial vertebrates. Amphibians and reptiles are ectotherms — that is, they obtain the energy they need to raise their body temperatures to the levels that permit normal activity from the sun, either directly (by basking in the sunlight) or indirectly (by resting on a warm surface such as a rock that was previously heated by the sun). In contrast, endotherms (such as birds and mammals) produce heat chemically by metabolizing carbohydrate, lipid, and protein from the food they eat.

Ectothermy has consequences that are apparent in nearly all aspects of an animal's life. Amphibians and reptiles, despite the enormous differences between the two groups and the variation seen within each group, are more similar to each other in many aspects of their biology than either is to birds or mammals, which are endotherms. Understanding the significance of ectothermy is a key to understanding how and why the ecology, behavior, morphology, and physiology of amphibians and reptiles are different from those of birds and mammals, and why ectotherms and endotherms play different roles in ecosystems.

The mechanics of ectothermal temperature regulation are complex, and its details are discussed in Chapter 5, but a brief explanation of some features of ectothermy shows why ectotherms are so different from endotherms.

Ectothermal Thermoregulation

Many ectotherms control their body temperatures at high levels and within narrow limits during their periods of activity. Lizards, especially species that live in open, sunny habitats, provide the best examples of how effective ectothermal thermoregulation can be. Many species of lizard maintain their body temperatures between 35 and 42°C while they are thermoregulating. That is as warm as the body temperatures of most birds and mammals.

The desert iguana (*Dipsosaurus dorsalis*), a lizard found in the deserts of the southwestern United States and northern Mexico, has been particularly well studied (Figure 1–6). A desert iguana spends the night in its burrow, plugging the opening with soil. During the night the lizard's body temperature falls to the temperature of the burrow, which is about 20°C in summer. Desert iguanas are late risers, and it is not until some hours after sunrise that a lizard begins to move toward the mouth of its burrow. After opening the entrance, the lizard pokes its head out. If the burrow mouth is in the sun, the lizard's body temperature begins to rise as the lizard is heated by solar radiation. By the time the lizard whose thermoregulatory behabior is shown in Figure 1–6 emerged from its burrow, its body temperature had risen by 12°C from a nighttime low of 23°C to 35°C.

Outside the burrow the lizard moves through a mosaic of sunny and shaded areas. Its body temperature rises to about 37°C and stabilizes at that level,

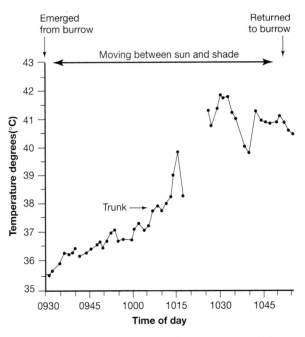

Figure 1–6   **Daily pattern of thermoregulatory behavior for a desert iguana (*Dipsosaurus dorsalis*).** In the morning the lizard waits near the entrance of its burrow until it has warmed up. After it emerges it moves back and forth between sun and shade to maintain its body temperature between 37 and 42°C. In late morning it returns to its burrow. *(Source: Based on data in Dewitt 1967.)*

rising and falling slightly from minute to minute as the lizard moves from sun to shade. By late morning, the sun is nearly overhead and the desert is too hot. The lizard can no longer maintain a safe body temperature, and it ceases activity and retreats to its burrow. In the burrow, the lizard cools again to a body temperature near 20°C, and remains at that temperature until it emerges the following morning. That is a typical pattern of daily temperature change for an ectotherm — warm when it is active and cool when it is in its retreat site.

Costs and Benefits of Ectothermy and Endothermy

A lizard that behaves like the one in the example has two important energy-saving features compared to a mammal such as the antelope ground squirrel (*Ammospermophilus leucurus*), which lives in the same habitat. The first advantage is that the desert iguana uses sunlight to maintain a high body temperature, whereas the squirrel relies on metabolic heat production. As a result of the difference in the metabolic requirements of ectotherms and endotherms, the rates of energy use by ectotherms are one-seventh to one-tenth those of an endotherm of the same body size (Figure 1–7). That difference means that the desert iguana uses far less energy per day than the squirrel.

The total difference in daily energy requirement for the lizard and the squirrel is more than the 10-fold difference in their metabolic rates because of the daily cycle of body temperature the lizard experiences. At night when its body temperature is low, the lizard's energy use falls to about one third of its daytime rate. The squirrel also shows a change in energy consumption at night, but in the opposite direction from the lizard. The squirrel is producing heat by metabolism to replace the heat it loses to the environment. When the environment cools at night, the squirrel loses heat faster and must increase its metabolic rate to produce the additional heat it needs. Thus the difference between energy use by the lizard and the mammal is greater at night than it is during the day.

A third factor enters the equation — activity. The squirrel is more active than the lizard, and activity requires energy. The combined effects of the three factors — the use of solar energy rather than metabolic energy by the lizard to keep itself warm, the reduction in the lizard's body temperature at night, and the sedentary behavior of the lizard compared to the mammal — are dramatic. A lizard uses only about 3 percent as much energy in a day as a mammal of the same body size. That lower metabolic requirement of the lizard translates into lower food demands (and hence the ability to live in environments with low biological productivity) and into greater efficiency at transforming the energy in food into body tissue.

Of course, ectothermy has disadvantages as well as benefits. Ectotherms cannot maintain their ideal body temperatures under all environmental conditions. The activities of amphibians and reptiles are limited in time and space by environmental conditions to a greater extent than are those of birds and mammals. Nocturnal activity and cold climates present especially difficult conditions for ectotherms, and species in temperate regions are usually active only during the day and in warm seasons.

If ectotherms do not completely cease activity when thermoregulation is difficult, they may have body temperatures that depart substantially from the levels they would normally maintain. Variation in body temperature affects all aspects of organismal function, including the speed at which nerve impulses

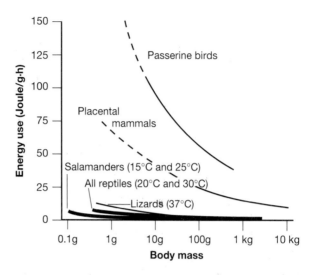

Figure 1–7 **Mass-specific energy use.** Metabolic rates (in Joules·[g·h]$^{-1}$) of ectothermal and endothermal vertebrates are shown as a function of body size (measured as body mass and shown on a logarithmic scale). The metabolic rates of endotherms (birds and mammals) are 7 to 10 times those of ectotherms (salamanders, all reptiles, lizards) of the same body size. The mass-specific metabolic rates of both endotherms and ectotherms increase at small body sizes, and the rates for small ectotherms are substantially lower than those of small endotherms. *(Source: Pough 1980.)*

travel and the force with which muscles contract. Everything an animal does is affected by changes in its body temperature, including activities that are critical to its survival, such as sprinting to escape a predator or striking at prey (Figure 1–8). Because they are ectotherms, amphibians and reptiles often operate at body temperatures that do not permit maximum levels of performance. In contrast, the endothermal thermoregulation of birds and mammals ensures that these animals (with the exception of species that become torpid or hibernate) are always very close to the body temperature that yields maximum performance.

Body Size and Shape

A striking feature of amphibians and reptiles is how small most of them are compared to birds and mammals. For example, about 80 percent of lizard species and 90 percent of frogs and salamanders weigh less than 20 grams as adults. In contrast, most mammals weigh more than 20 grams. Thus a typical frog, sala-

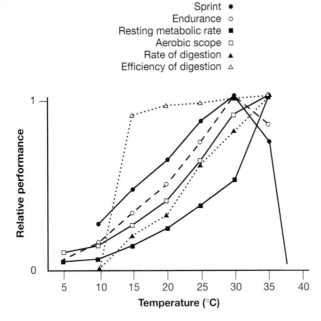

Figure 1–8   **Body temperature affects virtually every-thing an ectotherm does.** This example shows that the snake *Natrix maura* can crawl faster and farther as body temperature increases. Its resting metabolic rate increases with temperature, as does its capacity to increase oxygen consumption (aerobic scope). The rate and efficiency of di-gestion also increase with increasing temperature. *(Source: Hailey and Davies 1988.)*

mander, or lizard is very much smaller than a bird or mammal. In fact, it's only among turtles, snakes, and crocodilians that you find many species that weigh more than 500 grams (i.e., the size of a large white rat).

The smallest body sizes of amphibians and reptiles are also smaller than those of birds and mammals. Mice, bats, chickadees, and sparrows weigh 10 to 20 grams. Very few birds and mammals weigh less than 5 grams, but that is a common body size for amphibians and reptiles. In fact, adults of many small species of amphibians and reptiles weigh less than half a gram.

What is it about amphibians and reptiles that al-lows them to be so much smaller than birds and mam-mals? The answer lies in the energetic cost of being a very small animal. You will notice that the lines repre-senting energy use in Figure 1–7 rise toward the left axis of the graph — that is, at small body sizes. This pattern is characteristic of both endotherms and ec-totherms.

To put the relationship between body size and en-ergy into words, the energy requirement of a gram of tissue (called the mass-specific energy requirement) is greater for a small animal than for a larger one of the same kind. For example, at a body temperature of 30°C, a lizard weighing 100 grams has a mass-specific energy requirement of $1.98 \, \text{J} \cdot (\text{g} \cdot \text{h})^{-1}$ whereas for a 10-gram lizard the value is $3.14 \, \text{J} \cdot (\text{g} \cdot \text{h})^{-1}$. (The symbol J indicates joule, a unit of energy equal to 0.239 calo-ries.) The mass-specific energy requirement becomes progressively greater as body size gets smaller.

This mass-specific energy requirement increases at small body sizes for both endotherms and ectotherms, but the energy requirements of endotherms are 7 to 10 times larger than those of ectotherms at all body sizes. Because of that difference in energy needs, a very small mammal or bird must eat much more food every day than does a small frog or lizard. The rapid increase in mass-specific metabolic rates at small body sizes, com-bined with the high energy requirements of en-dotherms, means that being a small endotherm is ener-getically expensive. In fact, it is so expensive to be a very small endotherm that only a very few species do it (e.g., some hummingbirds), and these small en-dotherms lower body temperature to save energy at night and sometimes during the day as well.

In contrast, being a very small ectotherm is rela-tively inexpensive, and many species of amphibians and reptiles live in a body size range that is free of birds and mammals. The primary competitive and predatory interactions of these small amphibians and reptiles are with each other and with invertebrates such as spiders, scorpions, and centipedes.

Relatively few amphibians and reptiles grow as large as most birds and mammals, and the few evolu-tionary lineages that do include large species are char-acterized by morphological or ecological specializa-tions. (Snakes and turtles, for example, have body forms that do not occur among birds and mammals.) Perhaps endothermy is generally more effective than ectothermy for large terrestrial vertebrates, although no biological basis has been demonstrated for such a relationship.

Ectothermy and Efficiency

One more consequence of ectothermy should be con-sidered here because it leads to a major difference in the way amphibians and reptiles function in ecosystems compared to birds and mammals: Ectothermy is an efficient way of life in terms of how an organism uses the energy in the food it eats.

Ecologists and organismal biologists are concerned with the way an organism partitions the energy it obtains in its food. That is, what proportion of the total energy goes to maintenance (processes like respiration, circulation, transporting molecules in and out of cells, and so on) and how much goes to production (growth of an individual or development of eggs and embryos). It is in the balance between the two major categories of energy use, maintenance and production, that ectotherms have a major advantage over endotherms.

Because endotherms rely on heat generated by metabolic processes, they devote a large proportion of the food they consume to keeping themselves warm. In fact, about 98 percent of the energy a bird or mammal obtains from its food is used to generate heat, and only about 2 percent is available for production of new tissue. In contrast, ectotherms get heat from the sun without having to use energy from their food. As a result of their ectothermy, the proportion of the energy amphibians and reptiles consume that is con-

verted to new animal tissue (the conversion efficiency) is close to 50 percent, about 25 times higher than the conversion efficiency of birds and mammals. Table 1–1 illustrates this difference with examples of the conversion efficiencies of several species of ectotherms and endotherms. The average conversion efficiency for 19 species of birds and mammals was 1.4 percent, whereas for 12 species of amphibians and reptiles the average conversion efficiency was 50 percent.

Amphibians and Reptiles in Terrestrial Ecosystems

The difference in conversion efficiency between amphibians and reptiles, on one hand, and birds and mammals on the other has important implications for energy flow through ecosystems (Chapter 14). An example will illustrate this principle. The Hubbard Brook Experimental Forest in New Hampshire has been the site of ecosystem studies since the 1960s. The numbers of different kinds of animals have been counted, the annual production of young and growth

Table 1–1 **Efficiency of biomass conversion by ectotherms and endotherms.***

Ectotherms		Endotherms	
Species	**Efficiency**	**Species**	**Efficiency**
Red-backed salamander *Plethodon cinereus*	48	Kangaroo rat *Dipodomys merriami*	0.8
Mountain salamander *Desmognathus ochrophaeus*	76–98	Field mouse *Peromyscus polionotus*	1.8
Panamanian anole *Anolis limifrons*	23–28	Meadow vole *Microtus pennsylvanicus*	3.0
Side-blotched lizard *Uta stansburiana*	18–25	Red squirrel *Tamiasciurus hudsonicus*	1.3
Hognose snake *Heterodon contortrix*	81	Least weasel *Mustela rixosa*	2.3
Python *Python curtus*	6–33	Savanna sparrow *Passerculus sandwichensis*	1.1
Adder *Vipera berus*	49	Marsh wren *Telmatodytes palustris*	0.5
Average of 12 species	50	Average of 19 species	1.4

*These are net conversion efficiencies calculated as (energy converted/energy assimilated) × 100.
Source: Pough 1980.

of adults have been measured, and the amount of food the animals consume and the amount of energy they use for maintenance have been calculated. The values of all these variables can be compared by converting them to energy per unit area of forest (i.e., kilojoules per hectare).

Thomas Burton studied the role of salamanders in energy flow through the Hubbard Brook ecosystem (Burton and Likens 1975). One species of salamander, the red-backed salamander (*Plethodon cinereus*), makes up about 90 percent of the total salamander community in the Hubbard Brook forest. Burton concentrated on this species, and for comparison he used information about birds and mammals that had been gathered by other ecologists working at Hubbard Brook.

The salamanders at Hubbard Brook consumed a small amount of food compared to what was consumed by birds: only 46,000 kJ·hectare^{-1} annually for salamanders, compared to 209,000 kJ·hectare^{-1} for birds. From that perspective, the salamanders don't seem very important.

The picture changes, however, when the annual production of new tissue by salamanders and birds is compared. The conversion efficiency of the salamanders is so high (about 60 percent) that they are producing more new animal mass every year than the birds or mammals. The question that interests a predator is "What is there to eat?" And the answer is "Salamanders!"

This analysis shows that salamanders provide more energy (kilojoules) to the food chain than do either birds or mammals. That is, salamanders are quantitatively important in the forest ecosystem — the salamander community represents a lot of energy.

In addition, the role salamanders play in the ecosystem also differs qualitatively from that of birds and mammals. Red-backed salamanders, like so many other species of amphibians and reptiles, are small. Adult red-backed salamanders weigh about 1 gram — they are much smaller than the birds and small mammals at Hubbard Brook. These small salamanders feed on tiny invertebrates that are much too small for a bird or mammal to eat. The salamanders, however, capture these small invertebrates and convert them to salamander biomass very efficiently. Thus, in the context of the Hubbard Brook ecosystem the salamanders are harvesting the energy in prey that is not directly available to birds and mammals because it comes in small packages, and converting the energy into sala-

mander-size packages that the birds and mammals can consume. Both the small body size of the salamanders and their high conversion efficiencies are direct results of ectothermy (Pough 1980, 1983).

Development of Herpetology as a Field of Study

The historical development of herpetology from the eighteenth century to the present traces a path of increasingly integrative and synthetic studies. During the eighteenth and nineteenth centuries, herpetology was largely descriptive. Knowledge of the diversity of amphibians and reptiles on a worldwide basis grew as the colonial powers dispatched military and commercial expeditions across the globe. Many of these expeditions included a naturalist, who was often the medical officer. In other cases the naturalists were civilian volunteers, like Charles Darwin, who sailed with the HMS *Beagle* from 1831 to 1836 as it surveyed the coast of South America for the British Admiralty. The observations he made on this voyage formed the conceptual basis for his great work, *On the Origin of Species by Means of Natural Selection*, which was published in 1859. Darwin collected herpetological specimens on this voyage that can still be studied today at the British Museum (Natural History) (Figure 1–9).

Amphibians and reptiles from all over the world were deposited in museums in Europe, where they

Figure 1–9 **A marine iguana (*Amblyrhynchus cristatus*).** The label indicates that this mounted specimen in the collection of the British Museum (Natural History) was collected by Captain R. FitzRoy of the HMS *Beagle*. (*Photograph by Alan Savitzky.*)

formed the basis for increasingly complete catalogues of the world's fauna (Adler 1989). In France from the middle of the eighteenth century to the middle of the nineteenth century, George de Buffon, Bernard Lacépède, François Daudin, Constante Duméril and his son Auguste, and Gabriel Bibron published multi-volume works on natural history that described minerals, plants, and animals (including amphibians and reptiles). England moved onto the herpetological stage in the second half of the nineteenth century with the work of John Edward Gray, Keeper of Zoology at the British Museum (Natural History), and his successors Albert Gunther and George Boulenger. Boulenger's catalogues of the amphibians and reptiles in the collection of the British museum, published in nine volumes between 1882 and 1896, superseded those of Gray and Gunther and summarized knowledge of herpetological diversity on a worldwide basis. Boulenger's classifications of amphibians and reptiles have influenced herpetology to the present, and his catalogues are still indispensable references.

In North America this exploratory phase of herpetology looked mainly to the vast and little-known western part of the continent. As expeditions and individuals explored the West, they collected specimens of the flora and fauna. The first monograph on North American herpetology was published in three volumes between 1836 and 1838 by John Holbrook, a physician practicing in Charleston, South Carolina. The Academy of Natural Sciences of Philadelphia and the National Museum of Natural History (Smithsonian Institution) in Washington, D.C., were the repositories for many of the specimens collected by these expeditions. Spencer Baird and Charles Girard at the National Museum of Natural History, and Edward Drinker Cope, who held positions at Haverford College and the University of Pennsylvania and was associated with the Academy of Natural Sciences, described many of the specimens collected from western North America. Leonhard Stejneger joined the staff of the National Museum in 1889 and, with Thomas Barbour, Director of the Museum of Comparative Zoology at Harvard University, published a total of five editions of the *Check List of North American Amphibians and Reptiles* in the first half of the twentieth century. A sixth edition was published by Karl Schmidt of the Field Museum of Natural History in Chicago in 1953. These checklists summarized the North American herpetofauna that had been described since the work of Holbrook, more than a century earlier.

The herpetological literature has expanded enormously in the twentieth century, and a comprehensive list of major topics and authors is far beyond the scope of this summary. The paragraphs that follow illustrate a few major trends and identify just a small fraction of the workers who have contributed to our current state of knowledge. Additional examples are provided in subsequent chapters.

Studies of herpetological systematics have increasingly shifted from the description of species to studies of the relationships among species and genera in families. A few of the many landmark works in this area include studies of lizards (Charles Camp 1923, Arnold Kluge 1967, Richard Estes and Gregory Pregill 1988), turtles (Arthur Loveridge and Ernest Williams 1957), and anurans (Robert Inger 1967, Arnold Kluge and J. S. Farris 1969, Frank Blair 1972).

Many of these studies of systematics included a substantial amount of information about natural history, and studies specifically focused on the ecology and behavior of individual species or of communities of amphibians and reptiles became a prominent part of the herpetological literature in the twentieth century. G. Kingsley Noble's (1931) *The Biology of the Amphibia* integrated systematics, natural history, and behavior. It is still a valuable reference. More recent and more taxonomically restricted works include classic studies such as Ernest Williams's (1983) analysis of *Anolis* lizards in the Caribbean. Each island has a distinct group of species, but on island after island *Anolis* lizards partition the structural habitat in the same way. In forests, for example, there is typically a robust species that lives on the ground and lower part of tree trunks, another robust species that lives higher on the trunk and in the crown of the tree near the trunk, a slender species that lives in the twigs at the edge of the crown, and a giant species that lives in the crown and preys on the smaller species (Figure 1–10).

Studies of streamside salamanders in the Appalachian Mountains by Nelson Hairston Sr. (1987) have revealed that species occur in a predictable sequence along an axis extending from the edge of a stream. Competition for resources and predation by large salamanders on smaller species have been proposed as explanations for these patterns, which are repeated in valley after valley.

In contrast to the two previous examples, which used natural experiments (i.e., comparisons of different Caribbean islands or different Appalachian moun-

Figure 1–10 **Habitats of *Anolis* lizards.** *Anolis* lizards divide the habitat in terms of body size of the lizard (size axis), perch height on the tree (perch axis), and exposure to sun or shade (climate axis). The size axis is particularly evident for species that live in the crown of the tree: a small species lives on twigs at the outer edge of the canopy, larger species are closer to the trunk, and a giant species lives in the crown of the canopy, where it preys on the other species. The perch axis reflects differences in perch height on the trunk; the climate axis reflects variation in the proportion of sun and shade and extends from species that live in full sun to those living in perpetual shade beneath the forest canopy. *(Source: Williams 1983.)*

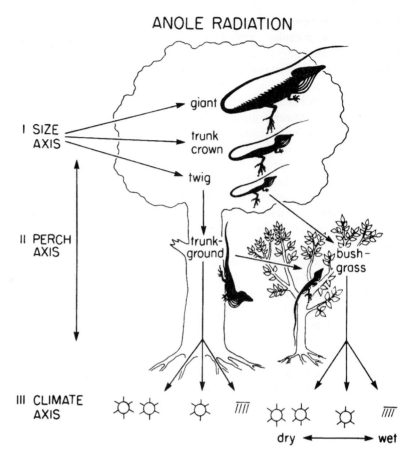

tain valleys), Henry Wilbur (Wilbur et al. 1983) has created artificial communities of amphibian larvae to examine competition and predation. The results of these experimental manipulations show that the sequence in which the eggs of different species of amphibians hatch has an enormous influence on the results of competitive and predatory interactions among the larvae.

Early naturalists often made incidental observations of the behavior of amphibians and reptiles, but serious study of the behavior of these animals did not begin until the twentieth century, with the work of G. Kingsley Noble. He conducted pioneering studies of calling and mating behavior of frogs (Noble and Noble 1923) and sexual selection in lizards (Noble and Bradley 1933). Noble also founded the Department of Experimental Biology at the American Museum of Natural History, which later became the Department of Animal Behavior and for some time was one of the few centers for behavioral research in this country. By the 1950s, behavioral studies in herpetol-

ogy were focusing on two major areas. The structure and function of frog calls were investigated by Frank Blair and his students at the University of Texas (Blair 1963) and Charles M. Bogert at the American Museum of Natural History (Bogert 1960). The cacophony at a breeding pond where males of several species of anurans call simultaneously is bewildering to a herpetologist — how can a female frog find her way to a male of the correct species? These studies showed that a female anuran can use several cues to find a male. Not only do the vocalizations differ among species (trills, whistles, whines, grunts, and more other sounds than there are words to describe them), but males of each species call from particular places — on the bank, in shallow water, in deep water, from bushes overhanging the pond, and so on. By going to the right calling site for her species, a female will encounter calling males.

Studies of the displays of lizards by Charles Carpenter at the University of Oklahoma (Carpenter 1967) and A. Stanley Rand at the Smithsonian Tropical Research Institute in Panama (Rand 1967) revealed similarly complex communication systems. Many lizards bob their heads, do push-ups, and expand a colorful gular fan that lies beneath the chin during social interactions. The cycle of up-and-down movements of the head and body, the tempo of expansion and contraction of the gular fan, and its color and pattern provide redundant information about the species and sex of a displaying male. A female can identify a male of her species even if she can see only some parts of the display.

The modern era of behavioral studies, with a focus on communication, sexual selection, and mating systems, began in the mid-1970s with the publication of E. O. Wilson's *Sociobiology* (1975), which emphasized evolutionary interpretations of animal behavior. This book was soon followed by the publication of important empirical studies and reviews of the behavior of frogs (Kentwood Wells 1977a), salamanders (Stevan Arnold 1977, Tim Halliday 1977), and lizards (Judy Stamps 1977), and a major symposium on the social behavior of reptiles (Neil Greenberg and David Crews 1977). The themes developed in these publications focus on the effects of specific behaviors on the Darwinian fitness of individual animals. For example, when a female frog has moved to the area where males of her species are calling, how does she decide which of the males to mate with, and will her choice of a mate determine the fitness of her offspring? Questions of this sort remain the principal focus of behavioral studies today.

Beginning in the middle of the twentieth century, studies of temperature regulation by reptiles (A. Sergeyev 1939, Raymond Cowles and Charles Bogert 1944, Roger Avery 1976, Donald Bradshaw 1986) laid the foundations for a new biological specialty, physiological ecology, that examines the way living organisms interact with their physical environment. Because amphibians and reptiles are ectotherms, they respond with behavioral adjustments to minute-to-minute changes in physical variables such as the intensity of sunlight and the speed of air movement. These behaviors are easy to observe and interpret, and herpetological studies have led the way in many areas of physiological ecology (Carl Gans and F. Harvey Pough 1982). Increasingly these studies have focused on performance — how well does an animal carry out an activity such as escaping from a predator or defending a territory that is important for its Darwinian fitness?

The evolutionary significance of variation in performance capacity among individuals is particularly important — does an individual that runs faster or fights better than another have a greater chance of surviving and of passing its genes on the next generation (Albert Bennett 1987)? It has even been possible to use experimental techniques to alter the body size of hatchling lizards so as to compare the rates of survival of larger- and smaller-than-normal individuals with those of normal size (Barry Sinervo 1993). Manipulations of this sort might be used in conservation and management of endangered species by producing hatchlings designed to flourish in the environment where they will be released (Pough 1993, Sinervo 1994).

In the past decade, the widespread adoption of cladistic methods for studies of evolutionary relationships (see Chapter 2) has added a new dimension to studies of organismal function. Knowing the order in which evolutionary lineages have separated and anatomical characters have appeared allows one to ask questions about the evolution of characters that do not fossilize, such as physiology and behavior (Raymond Huey and Albert Bennett 1986). Evolutionary changes in characters such as preferred body temperature can be traced, and cause-and-effect hypotheses can be tested (e.g., Benjamin Dial and Lee Grismer 1992). This methodology brings to studies of evolution a degree of rigor that was never before possible, and will provide important insights into how and why organisms work the way they do.

Anatomical studies have always played a major role in herpetology, and this field of study illustrates the broad influence of herpetology in modern biology. Herpetological examples abound in reviews of organismal and evolutionary biology such as *Predator–Prey Relationships* (edited by Martin Feder and George Lauder 1986), *Complex Organismal Functions: Integration and Evolution in Vertebrates* (edited by David Wake and Gerhard Roth 1989), *Quantitative Genetic Studies of Behavioral Evolution* (edited by Christine Boake 1994), and *Ecological Morphology: Integrative Organismal Biology* (edited by Peter Wainwright and Stephen Reilly 1994). The chapters that follow cite examples in many other areas of biology in which studies of amphibians and reptiles have led to conceptual and empirical advances with broad significance for biology as a whole.

An enormous amount of information about amphibians and reptiles is reviewed in three multivolume series — *Biology of the Reptilia* (Carl Gans, 1969–1992), *Physiology of the Amphibia* (John Moore editor, volume 1; Brian Lofts editor, volumes 2 and 3), and *Amphibian Biology* (Harold Heatwole, senior editor). In addition, two herpetological publications were so successful that related volumes have appeared, rather like the sequels that follow a blockbuster Hollywood movie. *Lizard Ecology: A Symposium* (William Milstead 1967) was succeeded by *Lizard Ecology: Studies of a Model Organism* (Raymond Huey, Eric Pianka, and Thomas Schoener 1983), and *Lizard Ecology: Historical and Experimental Perspectives* (Laurie Vitt and Eric Pianka 1994). *Snakes: Ecology and Evolutionary Biology* (Richard Seigel, Joseph Collins, and Susan Novak 1987) was followed by *Snakes: Ecology and Behavior* (Seigel and Collins 1993).

The current work in herpetology described in these books integrates the characteristics of amphibians and reptiles with the way they function, both as organisms and as members of ecological communities and ecosystems. The following chapters will continue that perspective, presenting amphibians and reptiles as animals that represent a successful approach to terrestrial vertebrate life that differs substantially from the way that birds and mammals live.

Summary

Many of the features that determine how modern amphibians and reptiles work are ancestral characters in an evolutionary sense. The permeable skins of amphibians, the three-chambered hearts of reptiles, and the ectothermy of both groups were inherited essentially unchanged from the ancestors of the modern groups. Despite their primitive (in the biological sense) status, the characteristics of amphibians and reptiles are in no sense inferior to the derived conditions seen among birds and mammals. On the contrary, these ancestral characters have been retained in the modern lineages of amphibians and reptiles because they work so well.

Our presentation of herpetology will emphasize integration of diverse features of amphibians and reptiles in this functional context. Understanding how and why amphibians and reptiles differ from birds and mammals and how they work requires an appreciation of the consequences of ectothermy and endothermy in terms of morphology, physiology, ecology, and behavior, and an understanding of the evolutionary history and phylogenetic relationships of the living groups. That is the theme of this book.

2

The Place of Amphibians and Reptiles in Vertebrate Evolution

Herpetology is concerned with an enormously diverse collection of animals, and it is appropriate to begin a survey of their diversity by providing a phylogenetic (evolutionary) context for their radiation. A phylogenetic perspective helps one visualize the place of the organisms in vertebrate evolution and provides a historical framework for interpreting patterns of physiological, morphological, or behavioral evolution. Phylogeny helps scientists formulate testable hypotheses about the sequence in which changes have occurred, and thus can lead to inferences about the mechanisms of evolution.

Largely for historical reasons, herpetology concerns an array of organisms for which the historical (phylogenetic) context often is not apparent. Why would anyone study organisms as different as frogs and crocodiles? In fact, most herpetologists are not so versatile, but the sheer phylogenetic diversity of amphibians and reptiles means that it is especially important to keep their relationships in mind when reading a book such as this. We owe the current scope of herpetology as a discipline partly to the Swedish naturalist, Carl von Linné, who wrote under the Latinized form of his name, Carolus Linnaeus. In a classification published between 1735 and 1758, Linnaeus gave a name to every known species of plant and animal. He placed all vertebrates that were not mammals, birds, or bony fishes into the class Amphibia (which thus included amphibians and reptiles, but also sharks and sundry other vertebrates such as holocephalans and lampreys). Only through the work of many naturalists through the first half of the nineteenth century were reptiles finally accorded a distinct status. Although herpetology has divested itself of the fish groups, it still concerns itself with all tetrapods other than mammals and birds. It thus seems appropriate to provide an overview of the evolutionary relationships of tetrapods at the outset. We also consider major problems concerning the relationships among the major groups within amphibians and reptiles and review hypotheses concerning the relationship of tetrapods to nontetrapod vertebrates. To set the stage, we will briefly review some elements of phylogenetic systematics that are central to our discussion of amphibians and reptiles.

Phylogenetic Systematics

Our system of naming organisms is based on the work of Linnaeus, which antedated Darwin by a century. The Linnaean system uses the familiar binominal nomenclature (genus and species) to identify species and groups those species into higher categories (families and orders). These named entities are often referred to as taxa (singular, taxon), from the Greek *taxo* (arrangement).

The appearance in 1966 of an English translation of the work of Willi Hennig, a German biologist, began a revolution in the way evolutionary relationships are analyzed. Hennig's method, which is known as phylogenetic systematics or cladistics, emphasizes the importance of a monophyletic (from the Greek *mono* [single] and *phylo* [tribe]) evolutionary origin. A monophyletic evolutionary lineage (also called a clade) includes an ancestor and all of its descendants. Only monophyletic evolutionary lineages are recognized in phylogenetic systematics. Thus, birds are included in Reptilia because they are descendants of the common ancestor that give rise to all the other taxa usually included in Reptilia. There is no way to give a name to a monophyletic group consisting of all reptiles except birds. As we noted in Chapter 1, a discussion of the biology of birds is beyond the scope of this book, and we are using the term reptiles as a shorthand to mean all reptiles except birds, but this is a convenience that is not technically correct.

Phylogenetic systematics uses derived characters to identify monophyletic lineages and to discover the order of branching of evolutionary lineages. A derived character is simply a character that differs from the ancestral form. Form example, a lineage of salamanders (the Plethodontidae) have grooves that extend from the nostrils to the upper lip. These nasolabial grooves, as they are called, are unique to plethodontid salamanders—they do not occur in any other lineage of salamanders. Thus, nasolabial grooves are a derived character of plethodontids. Characters used in phylogenetic analyses may be morphological, as in the case of nasolabial grooves, but can equally well be behavioral, physiological, chromosomal, or molecular.

A derived character is called an apomorphy, from the Greek *apo* (away from) and *morph* (form). In other words, an apomorphy is a structure that has moved away from the ancestral form. A shared derived character is found in two or more taxa; an example is nasolabial grooves, which are found in all species of plethodontid salamanders. A shared derived character is called a synapomorphy, from the Greek *syn* (together). Possession of the shared derived character of nasolabial grooves is part of the evidence that plethodontid salamanders are a monophyletic lineage. Not all derived characters are shared, of course, and a derived character that appears in only one taxon is an autapomorphy, from the Greek *auto* (self). Some derived characters have evolved independently in different groups and are therefore convergences.

Ancestral characters are called plesiomorphies (from the Greek *plesio*, close to). Ancestral characters do not provide any information about evolutionary relationships. That is not to say that ancestral characters are unimportant. On the contrary, ancestral char-

acters can be profoundly important in how an animal lives. Ectothermy, for example, is an ancestral character of amphibians and reptiles that has ramifications in many aspects of their ecology and behavior. It is essential to understand the mechanisms and implications of ectothermy to understand the biology of lizards and turtles, but the fact that lizards and turtles are ectotherms does not provide any information about their evolutionary relationship.

By analyzing the occurrence of shared derived characters, it is possible to determine the evolutionary sequence of branching that has produced the monophyletic lineages we know today. These phylogenetic diagrams (often called cladograms) are hypotheses about the evolutionary relationships of the groups included. Cladograms are constructed by postulating the sequence of changes from ancestral to derived characters that will produce the distribution of derived characters seen in the taxa being studied. Because many such sequences are usually possible for a large data set, some criterion must be employed to choose among the alternatives. Usually the cladogram that requires the smallest number of changes from ancestral to derived conditions, called the most parsimonious cladogram, is preferred. Two adjacent branches are called sister lineages. For example, in Figure 2–1 Reptilia and Mammalia are sister lineages among extant amniotes. Lineages outside the clade under consideration are called outgroups. For example, Mammalia and Lissamphibia are both outgroups to Reptilia.

Like any scientific hypothesis, the hypothesis of evolutionary relationships represented by a cladogram is always subject to falsification by new evidence or a better analysis of existing evidence. Alternative hypotheses about evolutionary relationships are very common, and we will identify many of those as we discuss the phylogenetic relationships of the extant amphibians and reptiles. In some cases we have groups that are clearly monophyletic lineages that can be defined by shared derived characters, but it has not yet been possible to determine the order in which they separated. That situation is called an unresolved polytomy when the branching sequence of three or more lineages cannot be determined. The neobatrachian frogs shown in Figure 3–17 are an example of a polytomy: 12 lineages have been identified, but the sequence in which the lineages branched is unknown.

Some taxa recognized in current taxonomy cannot be distinguished by shared derived characters. These

Figure 2–1 **Phylogeny of major groups of tetrapods and their relatives.** *(Source: Adapted from Gauthier et al. 1989, Ahlberg and Milner 1994, and Laurin and Reisz 1995.)*

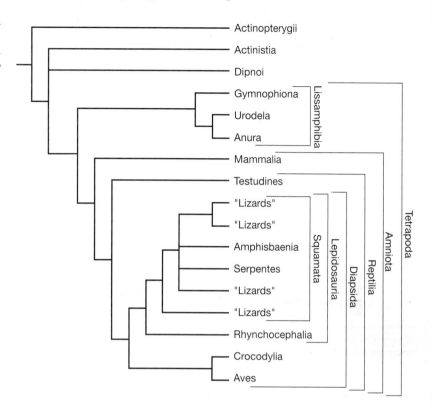

are sometimes groups that were recognized before shared derived characters began to be used as a basis for group recognition. Often they are characterized by lacking the derived features of other taxa. For example, "batagurid" turtles (discussed in Chapter 4) lack the shared derived features of testudinids, but no unequivocal derived characters unite all "batagurids." We suspect that testudinids are included among "batagurids," but we lack sufficient knowledge of character distributions to discern exactly how. Groups of this sort are called nonmonophyletic or paraphyletic (from the Greek *para*, beside or beyond). We will use quotation marks to identify assemblages that are probably nonmonophyletic.

Tetrapod Phylogeny

In the next two chapters we describe the extant amphibians and reptiles and provide evidence for their evolutionary origins. Here we discuss relationships among extant tetrapods and evidence for the monophyly of major tetrapod groups. The major extant clades of tetrapods and their relatives are outlined in

Figure 2–1. The remainder of this chapter will discuss evidence favoring the major hypotheses of tetrapod relationships.

Extensive discussions of evidence for the monophyly of these groups and their interrelationships can be found in Benton (1990) and Schultze and Trueb (1991). If we consider only the extant tetrapod groups, the fundamental divergence within Tetrapoda results in two clades, the Lissamphibia (frogs, salamanders, and caecilians) and Amniota (all other tetrapods) (Figure 2–1). A guide to the major groups of vertebrates we will refer to throughout this book is given in the following paragraphs.

Actinopterygii. The ray-finned fishes. Actinopterygii is the sister group of Sarcopterygii (lobe-finned fishes.) In cladistic classification, all of the following groups are lobe-finned fishes.

Actinistia. Represented by a single extant species, *Latimeria chalumnae*, Actinistia is a diverse group of lobe-finned fishes extending back into the Paleozoic.

Dipnoi. Dipnoi includes three species of extant lungfish in Africa, South America, and Australia, as well as diverse fossil species extending well back into the Paleozoic.

Tetrapoda. Tetrapoda includes all land vertebrates with weight-bearing limbs having distinct carpal (forefoot) and tarsal (hind foot) regions and distal phalanges (fingers and toes). The extant groups of tetrapods are the Lissamphibia and the Amniota. Fossil groups of concern to us here are *Ichthyostega*, *Acanthostega*, and *Tulerpeton*. (Note that some workers [e.g., Gauthier et al. 1989] restrict the name Tetrapoda to Lissamphibia + Amniota and fossils more closely related to them than to other vertebrates.)

Lissamphibia. Lissamphibia is a clade that includes the last common ancestor of frogs (Anura), salamanders (Urodela), and caecilians (Gymnophiona), and all of the descendants of this last common ancestor. We informally refer to these organisms as amphibians throughout this book. The fossil taxa more closely related to lissamphibians than to amniotes are referred to as temnospondyls, but there is no agreement yet as to whether temnospondyls are a monophyletic group, and following Milner (1993), we use this term informally. Some workers have proposed using the term *Amphibia* as the formal name for this clade (e.g., Cannatella and Hillis 1993). Both terms, Lissamphibia and Amphibia, are widely used in the current literature.

Amniota. Amniota includes the last common ancestor of Mammalia and Aves and all of its descendants. The extant groups included are mammals (Mammalia), turtles (Testudines), snakes, lizards, and amphisbaenians (Squamata), tuatara (Rhynchocephalia), crocodilians (Crocodylia), and birds (Aves).

Reptilia. Reptilia includes the last common ancestor of Testudines and Aves and all of its descendants. We explicitly include birds in Reptilia because they are cladistically part of that clade. Historically birds have not been included in the discipline of herpetology, and we will refer to them only in passing.

The Transition from Fishes to Tetrapods

For more than a century the debate over the origin of tetrapods has focused on the relationships between the earliest tetrapods and several groups of lobe-finned fishes (sarcopterygians), including lungfishes (Dipnoi), Actinistia, and Paleozoic fishes such as osteolepiforms (e.g., *Eusthenopteron*), porolepiforms, and Panderichthyida. Some theories asserted that Tetrapoda was diphyletic, with salamanders (Urodela) derived from sarcopterygians independently from other tetrapods (see Schultze 1994). However, there is now consensus that tetrapods are monophyletic, and attention has shifted to the problems of the interrelationships of the extant groups Tetrapoda, Dipnoi, and Actinistia, and of these to the Paleozoic sarcopterygians; and to the morphological transition from fish-like creatures to tetrapods (Ahlberg and Milner 1994). There is general agreement that all three Paleozoic sarcopterygian fish groups (Osteolepiformes, Porolepiformes, Panderichthyida) are more closely related to Tetrapoda than are the extant sarcopterygians, *Latimeria* and Dipnoi (Ahlberg and Milner 1994, Schultze 1994). Thus, we will first consider the biology and relationships of the best-known Paleozoic tetrapods and their relationship to these fish groups, and then discuss the relationships of tetrapods to the extant sarcopterygians.

Three very distinct genera of early tetrapods (*Ichthyostega*, *Acanthostega*, and *Tulerpeton*) appear in the Upper Devonian, 360 MaBP (million years before the present), of Greenland and Russia (Figure 2–2). *Ichthyostega* and *Acanthostega* are the best known. *Tulerpeton* is known mainly from cranial elements. These were carnivorous animals 50 to 120 centimeters long. Curiously, all of these animals were polydactylous, having from six to eight digits (Coates and Clack 1990). These extra digits are consistent with predictions from theoretical and developmental studies of the origin of tetrapod limbs (Shubin and Alberch 1986). All early tetrapods, including later Paleozoic groups such as temnospondyls and anthracosaurs, are characterized by a very distinctive form of the teeth in cross section. The dentine is folded into a complex labyrinthine pattern, which prompted the name Labyrinthodontia, seen in the older literature for some of these Paleozoic tetrapod groups. Labyrinthodontia is now known to be paraphyletic. Nevertheless, it was the unique structure of

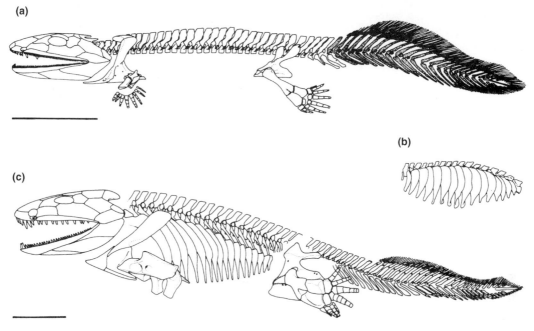

Figure 2–2   **The earliest tetrapods, *Acanthostega* and *Ichthyostega*.** Reconstructed skeletons of *Acanthostega* (a) and *Ichthyostega* (b). The expanded thoracolumbar ribs of an extant mammal, the two-toed anteater (c), are shown for comparison with *Ichthyostega*. *(Source: Courtesy of M. I. Coates and J. A. Clack.)*

the teeth that gave an initial clue to the fish relatives of tetrapods, for the same tooth structure is found in *Ichthyostega*.

Although osteolepiform fishes (represented by *Eusthenopteron*) have historically been the model lobefins from which early tetrapod evolution was viewed, recently discovered well-preserved fossils from Canada and Russia of another group of Paleozoic lobefins known as Panderichthyida (365 MaBP) have put us one step closer to the earliest tetrapods (Vorobyeva and Schultze 1991, Ahlberg and Milner 1994). In fact, some panderichthyid fossils have been confused with tetrapods. Panderichthyids possess an unusual combination of characteristics, making them truly intermediate between osteolepiforms and true tetrapods. In essence, they look like tetrapods with paired fins (Figure 2–3). The eyes are dorsally placed on rather crocodile-like skulls. They have laterally compressed bodies without dorsal or anal fins and, like tetrapods but unlike other fishes, they have identifiable frontal bones in the skull, a derived character shared with tetrapods. These features suggest that the tetrapod-like body and other features of panderichthyids appeared before limbs. None of these features shared between panderichthyids and tetrapods are seen in *Eusthenopteron*, which is a much more fishlike creature. Like other lobefins that have been postulated to be among the closest relatives of tetrapods, panderichthyids have the peculiar labyrinthine structure seen in the teeth of the earliest tetrapods. They were probably shallow-water predators.

The Ecological Transition in Tetrapod Origins

The morphological and physiological transition from an aquatic to a terrestrial mode of life required changes in many organ systems. The most important of these changes were in the support and locomotor systems. Strong weight-bearing elements and the associated musculature are critical to the ability to move on land in comparison to water. The skull of panderichthyids, like that of other fishes, was relatively immobile and firmly attached to the pectoral girdle. The earliest tetrapods are characterized by pectoral girdles that are capable of movement independent of the skull.

Figure 2–3 **Panderichthys, a sarcopterygian sister group to tetrapods.** *(Source: Vorobyeva and Schultze 1991.)*

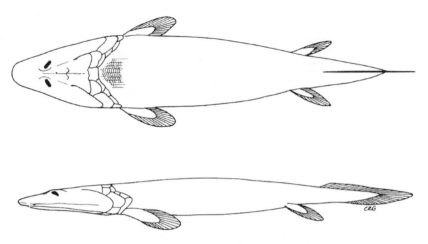

The pelvic girdle and its articulation with a specially differentiated portion of the vertebral column (sacrum and modified sacral ribs) are derived characters of tetrapods. The evolution of limbs capable of terrestrial support in the earliest tetrapods modified and added to the basic structure seen in panderichthyids. The proximal limb elements (humerus, radius, and ulna in the forelimbs; femur, tibia, and fibula in the hind limbs) of *Ichthyostega* and *Acanthostega* are more robust than those of panderichthyids, but the corresponding elements are homologous in all three groups.

Critically, the phalanges (bony elements of the digits) of tetrapods, homologous to the postaxial skeletal elements of panderichthyids (Coates and Clack 1990), acquire new organization and function with the evolution of tetrapods. The proximal limb elements of early tetrapods retain an organizaton similar to that of panderichthyids, but the distal elements are organized more like hands and feet than like fins. Nevertheless, functional interpretations suggest that the limbs of both *Acanthostega* and *Ichthyostega* were probably adapted more for aquatic than for terrestrial locomotion (Coates and Clack 1990).

Changes in the body form and proportions of early tetrapods are coincident with changes in the skeleton and reflect increasing support for terrestriality. The skull became relatively flat with a more elongate, broader snout region, and the eyes became more dorsal. These changes in skull shape may be associated with the acquisition of a buccal pump respiratory mechanism for filling the lungs (see Chapter 5), and with changes in the feeding system required by terrestriality (see Chapter 9). Fishes use primarily suction feeding, whereby food is drawn into the mouth by the sudden creation of negative pressure in the buccal cavity using their highly mobile jaws and branchial apparatus. Tetrapods feeding on land use either the tongue or jaws to seize food. Thus, we can assume that changes in the feeding system were early innovations of tetrapods.

Interestingly, some modern tetrapods (e.g., some salamanders) migrate annually between a terrestrial and aquatic medium, using the tongue for food acquisition on land and suction feeding in water. Experiments show that the mechanics of this transition in feeding mode are quite simple. Basically, terrestrial adult salamanders retain the structural and functional components of their larval feeding system and simply add components, such as a tongue, for feeding on land (Lauder and Reilly 1988). Both feeding modes are possible for adult salamanders that went through an aquatic larval stage. By analogy, the evolution of terrestrial feeding modes in the earliest tetrapods need not have involved radical reorganization or remolding of the ancestral feeding apparatus, but only the addition of components associated with terrestriality.

Although the soft anatomy of the respiratory system is not preserved in the fossil record, aspects of the anatomy of the palate and stapes in *Acanthostega* suggest that it still retained a fishlike breathing mechanism, including an open spiracle and a fishlike branchial skeleton and opercular chamber (Coates and Clack 1991; Clack 1992, 1994). The stapes of *Acanthostega* buttressed the palate against the braincase and apparently had no role in hearing (the stapes of *Ichthyostega* is unknown). This observation agrees with other evidence suggesting that these early tetrapods were not fully terrestrial.

Many other functional and anatomical changes required for the evolution of terrestriality have left no evidence in the fossil record. Sensory systems, in particular the eyes and ears, would have changed to accommodate differences in the transmission of sensory signals through air and water. The evolution of terrestrial hearing, including a stapes associated with a tympanum, seems to occur later in tetrapod evolution than *Ichthyostega* and *Acanthostega*. By the time temnospondyls appear, some 30 million years later, the structure of the hearing apparatus approaches that of modern amphibians, especially frogs. In fact, the detailed structure of the hearing apparatus has provided strong evidence supporting the relationship of temnospondyls to the Lissamphibia (Lombard and Bolt 1979, Bolt and Lombard 1985).

Monophyly of Lissamphibia

All morphological and molecular phylogenetic studies suggest that lissamphibians are monophyletic relative to other extant tetrapods. However, controversy still exists as to whether Lissamphibia is monophyletic relative to several groups of Paleozoic temnospondyl amphibians that are potential close relatives of any of the three modern amphibian orders.

Nearly 30 million years elapsed between the appearance of the earliest tetrapods, *Acanthostega*, *Ichthyostega*, and *Tulerpeton*, in the fossil record and the explosion of tetrapod diversity in the Late Devonian. This period is critical for understanding early amphibian and amniote evolution, for it is at this time that several tetrapod groups, including microsaurs, nectrideans, anthracosaurs, temnospondyls, and the earliest Amniota, appear in the fossil record (Figure 2–4). Thus, lissamphibians in whole or in part could potentially be most closely related to any of these groups. We consider this problem after first establishing a case for the monophyly of Lissamphibia.

Lissamphibia is now recognized as a clade, but that understanding took a long time to achieve. Parker (1956) and Parsons and Williams (1963) established clear evidence of lissamphibian monophyly. Their results have been extended by numerous workers (Milner 1988, 1993). The following list includes some of the unique or nearly unique derived features shared (with a few exceptions) by extant amphibians:

1. The teeth of modern amphibians are pedicellate and bicuspid (Figure 2–5). Each tooth crown sits on a base (pedicel), from which the crown is separated by a fibrous connection. Moreover, the teeth have two cusps, lingual (inner side of the jaw) and labial (outer side). Such a tooth structure is unique to Lissamphibia and some of their temnospondyl relatives.

2. The sound-conducting apparatus of the middle ear of extant amphibians consists of two elements, the stapes (columella), which is the usual element in tetrapods, and the operculum, a structure unique to lissamphibians that is connected via muscles to the suprascapula in the pectoral girdle (Figure 2–5). Caecilians lack opercular muscles because they do not have limbs, and interpretations vary as to whether the operculum itself is fused into their highly derived skull or is absent altogether. In the inner ear, all modern amphibians have two sensory epithelial patches, the papilla basilaris, found in other tetrapods, and the papilla amphibiorum, unique to lissamphibians. The papilla basilaris receives relatively high-frequency sound input via the stapes. The papilla amphibiorum receives relatively low-frequency input via the opercular apparatus.

3. The stapes is directed dorsolaterally from the fenestra ovalis, a character shared by lissamphibians and some of their presumed Paleozoic relatives. The stapes of all other tetrapods and their sarcopterygian relatives is directed ventrolaterally.

4. The fat bodies of modern amphibians develop from the germinal ridge (which also gives rise to the gonads), a developmental origin unique among tetrapods.

5. The skin contains both mucous and poison (granular) glands that are broadly similar in structure among extant amphibians.

6. Specialized receptor cells in the retina, called green rods, are present in frogs and salamanders. (Greed rods apparently are absent from caecilians, perhaps because of their very reduced eyes.)

7. All living amphibians have a sheet of muscle underlying the eyes, the levator bulbi muscle, which permits them to elevate the eye. This ability is unique to amphibians.

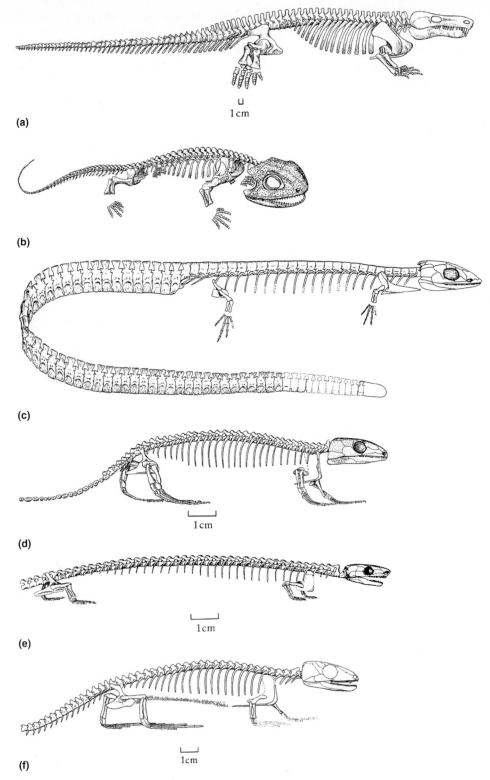

Figure 2–4 **Diversity of late Paleozoic tetrapods.** (a) *Limnoscelis* and (b) *Discosauriscus*, both Anthracosauria. (c) *Ptyonius*, Nectridea. (d) *Tuditanus* and (e) *Rhynchokonkos*, both Microsauria. (f) *Hylonomus*, Amniota: Reptilia. *(Source: Williston 1912, Spinar 1952, Bossy 1976, Carroll and Gaskill 1978, Carroll and Baird 1972.)*

8. All extant amphibians have unusual respiratory mechanisms, including major reliance on cutaneous respiration and the use of a buccopharyngeal force pump to get air into the mouth and lungs.

9. Extant amphibians have short, straight ribs that do not encircle the body. The ribs of Paleozoic tetrapods (other than some temnospondyls thought to be closely related to lissamphibians) are long, robust, and encircle the viscera.

10. Modern amphibians have two occipital condyles with which the skull articulates with two cotyles on the atlas. Most other tetrapods have a single occipital condyle.

11. Lissamphibians share similar reductions in skull bones and fenestration patterns compared to Paleozoic tetrapods (Figure 2–6). These shared derived characters include loss of the supratemporals, intertemporals, tabular, postparietals, jugals, and postorbitals. Other elements, such as the pterygoid and parasphenoid bones in the palate, are reduced to produce a similar configuration of bones among the three modern amphibian groups. Nonetheless, the skull morphology of caecilians is highly unusual compared to that of frogs and salamanders.

Characters 1 through 11 are derived characters unique, or virtually unique, to Lissamphibia and provide strong evidence for the monophyly of the group relative to all other extant tetrapods. Many other-derived characters of Lissamphibia have evolved independently in other tetrapods. For example, Trueb and Cloutier (1991) identified 29 shared derived characters of Lissamphibia, of which 17 were unique to them and 12 others evolved independently in other tetrapods. Several presumed derived characters of Lissamphibia (e.g., characters in No. 2 above) cannot be evaluated for caecilians because of their highly derived morphology. Thus, the actual level at which these characters are synapomorphies (i.e., for all Lissamphibia or only for Anura + Urodela) cannot be determined. Other characters, such as the numerous losses of skull bones tabulated in No. 10, are perhaps the result of a general pattern of paedomorphosis among lissamphibians, and hence are subject to high levels of convergence. These are general problems that have plagued the establishment of lissamphibian monophyly. Nevertheless, the numerous shared unique features strongly support that hypothesis.

Molecular evidence also supports the monophyly of Lissamphibia relative to Amniota and bony fishes (*Latimeria*, Dipnoi, and Actinopterygians). Relation-

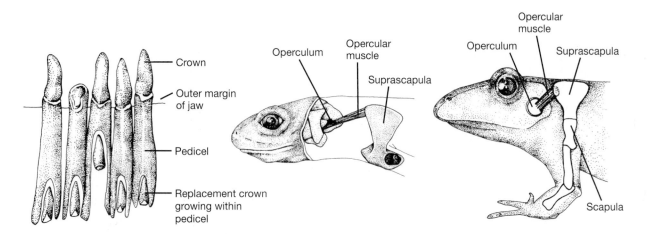

Figure 2–5 **Shared derived characters of Lissamphibia.** Pedicellate teeth (a), opercular apparatus of salamanders and frogs (b). (*Source: (a) Based on Parson and Williams 1963, (b) modified from Hetherington 1992.*)

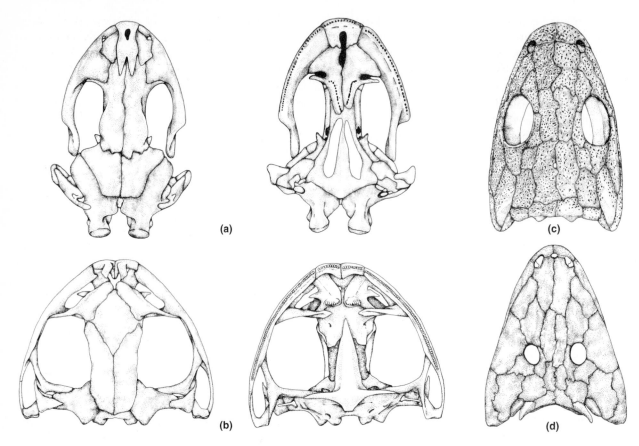

Figure 2–6   **Skulls of lissamphibians and Paleozoic tetrapods.** (a) *Phaeognathus hubrichti*, Urodela: Plethodontidae; (b) *Gastrotheca walkeri*, Anura: Hylidae; (c) *Dendrepeton*, Edopoid temnospondyl; (d) *Seymouria*, Antharacosauria. *(Source: Redrawn from Duellman and Trueb 1986 and Carroll 1988.)*

ships among these groups have been investigated using sequences of ribosomal genes (Hillis 1991, Hedges and Maxson 1993). The results clearly support the monophyly of Lissamphibia relative to the other groups, and support Amniota as the sister group to Lissamphibia. Of course, the molecular data cannot address questions concerning relationships among extant groups and those represented only by fossils. Therein lie the major remaining controversies concerning the relationships and monophyly of frogs, salamanders, and caecilians. In other words, if we accept the monophyly of Lissamphibia relative to other extant tetrapod groups, do they form a clade relative to the known Paleozoic tetrapod groups?

Relationships Between Lissamphibians and Paleozoic Amphibians

The explosive radiation of early tetrapods 300 to 335 MaBP produced numerous groups that are possible early relatives of the Lissamphibia. The earliest fossil that can clearly be assigned to an extant lissamphibian clade is *Triadobatrachus*, the sister taxon to Anura, from the Lower Triassic (245 MaBP). Thus, the closest extinct relatives of Lissamphibia must have existed in the late Paleozoic, roughly between 300 and 250 MaBP. Although several tetrapod groups existed during this period, most current hypotheses consider two Paleozoic tetrapod groups, the temnospondyls and microsaurs, as possible close relatives of Lissamphibia or parts thereof (Milner 1993) (Figure 2–7). We consider this problem in more detail because it bears on the monophyly of Lissamphibia relative to Paleozoic tetrapod groups. Little controversy surrounds the relationship of temnospondyls to Lissamphibia specifically or to the Anura-Caudata clade. The relationship of microsaurs to the lissamphibian groups is less clear.

Temnospondyls are represented by about 170 genera known from the Mississipian to the Lower Jurassic (approximately 335 to 210 MaBP). They ranged in length from only a few centimeters to more than a meter. Attention has focused on a paraphyletic assemblage of temnospondyls referred to as dissorophoids, of which the best known forms are *Amphibamus, Doleserpeton,* and Branchiosauridae (Figure 2–8). These are among the smallest temnospondyls (skull lengths approximately 12 mm). Moreover, progressive modifications of the palate, dentition, ear, pectoral girdle, and humerus toward the lissamphibian conditions are observed within temnospondyls and especially within dissorophoids, but not in any other Paleozoic groups (Bolt and Lombard 1985, Bolt 1991, Milner 1993). It is now recognized that dissorophoids and other temnospondyls form successive outgroups to Lissamphibia, and that many lissamphibian characters can be viewed as end points of general trends occurring within temnospondyls (Figure 2–9).

The morphology of caecilians is the most difficult to reconcile with the evolutionary trends within temnospondyls. Moreover, caecilians share some characters with the Microsauria, which are possibly their closest relatives. Microsaurs are a diverse group of elongate, rather small (less than 75 centimeters long) Pennsylvanian–Lower Permian amphibians (Carroll and Gaskill 1978). Based on similarities in palatal structure, temporal fenestration, dentition, body elongation, reduced orbits, and braincase structure, Carroll and Currie (1975) argued that microsaurs were the closest relatives of caecilians among late Paleozoic tetrapods. Many of these shared characters (e.g., reduced orbits, body elongation) are characters often associated with burrowing in tetrapods, and others possibly reflect the small size associated with paedomorphosis (Bolt 1991, Milner 1993). Thus, if temnospondyls gave rise to caecilians as well as to frogs and salamanders, then the burrowing-associated characters shared by caecilians and microsaurs must be convergent. This idea is also supported by developmental studies of skull ossification in caecilians (Wake and Hanken 1982). On the other hand, if the burrowing-associated characters are evidence of a caecilian–microsaur relationship, then many characters shared among caecilians, frogs, and salamanders (e.g., pedicellate bicuspid teeth, absence of jugals and ectopterygoids) must have arisen convergently because these are not present in microsaurs.

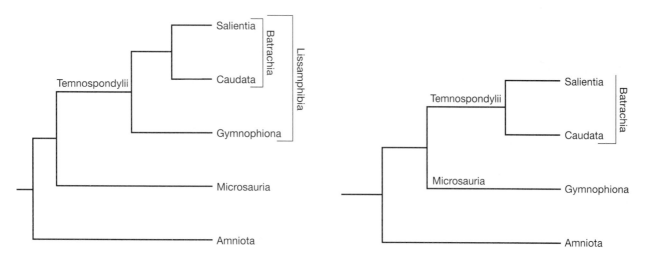

Figure 2–7 **Alternative hypotheses of interrelationships among lissamphibians, temnospondyls, and microsaurs.** *(Source: Milner 1993.)*

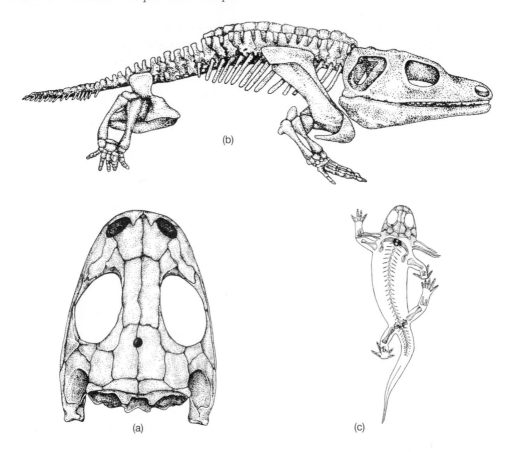

Figure 2–8   **Dissorophoid temnospondyls.** Skull of *Doleserpeton annectens* (a);   *Cacops* (b), a relatively large dissorhophid with a total length of 40 centimeters; *Branchiosaurus* (c), a branchiosaur retaining evidence of external gills. *(Source: Redrawn from (a) Bolt 1977, (b) Carroll 1988, (c) Boy 1971.)*

Although a preponderance of evidence supports a sister group relationship of microsaurs to all lissamphibians plus temnospondyls, that hypothesis is only marginally better supported than an alternative in which microsaurs are viewed as the closest relatives of caecilians (Milner 1993). Thus, although the monophyly of Lissamphibia relative to other extant tetrapods seems established beyond question, the interrelationships of temnospondyls, microsaurs, and lissamphibians must still be resolved.

Lissamphibians as Paedomorphic Temnospondyls

Accepting temnospondyls as the closest relatives of Lissamphibia (or at least of frogs and salamanders), we can consider the processes leading to the highly derived skeletal morphology of lissamphibians in comparison to Paleozoic temnospondyls, and in particular the dissorophoids. Miniaturization and heterochrony (see Chapter 7) may have been important evolutionary processes in the origin of Lissamphibia. Persuasive arguments suggest that many of the derived features of lissamphibians evolved through these processes (Bolt 1977, 1979).

One common result of heterochrony is the morphologically juvenile state referred to as paedomorphosis (*paedo* = child, *morph* = form). Paedomorphosis refers to the retention of juvenile characteristics in adult stages of an organism relative to the ancestral developmental sequence for that organism (see Chapter 7). For example, as adults, some salamanders retain gills (a juvenile condition), and we infer that these salamanders are derived from transforming ancestors. In essence, they have arrested the transfor-

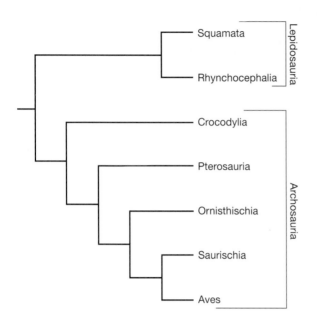

Figure 2–9   **Major clades of Diapsida.** *(Source: Adapted from Gauthier et al. 1989 and Laurin and Reisz 1995.)*

mation process and retain some juvenile features, despite the sexual maturation of their gonads. Paedomorphosis has been enormously important in producing morphological diversity in extant amphibians and reptiles. At the same time it can frustrate efforts to reconstruct phylogeny because arresting development in diverse lineages can result in the convergent appearance of many morphological characters, such as the presence of gills in adults of diverse lineages of salamanders.

Extant amphibians can be thought of as paedomorphic temnospondyls, and this perspective explains many of their unusual shared morphological characteristics. One common result of paedomorphosis is size reduction (juveniles are smaller than adults), and modern amphibians are very small compared to Paleozoic tetrapods. For example, some labyrinthodonts are estimated to have been 7 meters long and to have had skulls close to 2 meters long. Turning to the fossil record of temnospondyls, we see a striking evolutionary trend toward size reduction, of which dissorophoids and Lissamphibia are simply the end point. In other words, lissamphibians are miniaturized temnospondyls.

The heterochronic process producing small size in lissamphibians leaves many other imprints on their morphology. In fact, some of the most characteristic

features of Lissamphibia can be interpreted as paedomorphic features. We give just three examples, made possible by the remarkable preservation of developmental sequences, including larvae, juveniles, and adults, of some Paleozoic temnospondyls known as branchiosaurs (Lower Permian of Germany) (Figure 2–8). Branchiosaurs are probably the sister group to Lissamphibia (Trueb and Cloutier 1991, Milner 1993). Ontogenetic series of branchiosaurs are so well preserved that it is possible to examine the sequence in which the bones of the skull ossified during development.

1. Bones such as the supratemporals, postfrontals, prefrontals, jugals, postorbitals, and ectopterygoids are the last to appear during development of branchiosaurs. It is precisely these bones that are absent from lissamphibian skulls, suggesting that frogs, caecilians, and salamanders have arrested their development at a stage before these bones form. All of the bones appearing early in the development of branchiosaurs (nasals, frontals, parietal, lacrimals, etc.) are present in lissamphibians.

2. The orbits of lissamphibians are large relative to Paleozoic forms. Dissorophoids are similar to lissamphibians in having relatively large eyes. Sensory organs such as the eyes form relatively early in development and appear relatively large in early developmental stages. As a result of paedomorphosis, lissamphibians and derived dissorophoids have relatively large eyes compared to many other Paleozoic temnospondyls.

3. Finally, one of the most characteristic features of Lissamphibia, bicuspid pedicellate teeth, is attributable to retention of a juvenile condition observed in dissorophoids. The development of teeth in Paleozoic dissorophoid amphibians (as well as in extant lissamphibians) went through a sequence in which the larvae had nonpedicellate, monocuspid teeth. At metamorphosis, these teeth were replaced by teeth that were bicuspid and pedicellate. In dissorhophoids but not lissamphibians, these bicuspid pedicellate teeth were gradually replaced by adult teeth that were monocuspid and had the characteristic labyrinthine structure. Thus, the adult lissamphibian tooth condition is precisely that shown by juvenile Paleozoic temnospondyls in being pedicellate, bicuspid, and lacking labyrin-

thine structure. In other words, adult lissamphibians retain the juvenile condition shown by ancestral temnospondyls.

Many peculiar aspects of lissamphibian morphology are comprehensible when lissamphibians are viewed as paedomorphic temnospondyls. Furthermore, understanding the process giving rise to these peculiarities sheds light on a fundamental evolutionary process governing morphological evolution in many tetrapods. We shall see other examples of heterochrony affecting morphological evolution in amphibians and reptiles later in this book.

Relationships Among Extant Orders of Lissamphibia

The relationships among Anura, Urodela, and Gymnophiona are not yet conclusively resolved despite an enormous amount of research in this area. No one hypothesis of relationships among the three orders is overwhelmingly supported when all data sources are considered, although particular data sets support some alternatives over others. For example, Trueb and Cloutier (1991) found that soft anatomical characters grouped salamanders and caecilians together, whereas osteological characters grouped salamanders and frogs together. Combined analysis of all morphological characters support a frog + salamander clade. Ribosomal DNA sequences group salamanders and caecilians (Larson and Wilson 1989), but that solution is only marginally better than the alternative grouping of salamanders and frogs (Hillis 1991, Hay et al. 1995). A combined analysis of molecular and morphological data sets marginally supports the frog + salamander clade (Hillis 1991). Considering all data sources, the two viable alternatives are a frog + salamander clade or a salamander + caecilian clade. The third possibility, a frog + caecilian clade, has never been seriously considered. Most evidence favors a frog + salamander clade (Milner 1988, Trueb and Cloutier 1991).

Characters supporting a frog + salamander (i.e., Anura + Urodela) clade include the opercular apparatus and associated ear structures. As noted, the opercular apparatus is lost in caecilians, perhaps as a result of limb loss. However, the apparatus is also reduced within salamanders by loss of one or more components in various groups. True dermal scales are absent in frogs and salamanders (present in caecilians and in Paleozoic tetrapods), and ectopterygoids and postfrontals are absent from their skulls. Finally, two developmental characters, absence of segmentation of the sclerotome and reduction or loss of male Müllerian ducts, are shared by frogs and salamanders but not caecilians. On the other hand, the limbs of the earliest known caecilian, *Eocaecilia* (Lower Jurassic), share some unusual features with those of extant salamanders (Jenkins and Walsh 1993). In particular, the structure of the femurs and the presence of an interglenoid tubercle are interpreted as strongly supporting a caecilian + salamander relationship, and the tubercle is shared with microsaurs alone among Paleozoic tetrapods (Jenkins and Walsh 1993). Clearly, the uncertainty concerning relationships among the modern amphibian groups will ultimately bear on hypotheses relating these to Paleozoic temnospondyls and microsaurs.

The Radiation of Amniotes

We have now traced the phylogeny of tetrapods from their origins to the basic split among extant groups, Lissamphibia and Amniota (Figure 2–1). We also considered the evolutionary relationships of taxa associated with the amphibian clade. Finally we turn to the Amniota and examine the relationships among its clades.

The Major Clades

All of the modern groups of amniotes can be traced to the Triassic in the fossil record, and most even into the late Paleozoic. A series of extraembryonic membranes (amnion, chorion, allantois, and yolk sac) is a shared derived character of amniotes and gives the group its name. The phylogeny of Amniota has been extensively studied, and broad agreement exists about the relationships among major groups (Figure 2–9). Among extant groups, Mammalia is the sister group to Reptilia, which includes turtles (Testudines), tuatara (Rhynchocephalia), lizards and snakes (Squamata), crocodilians (Crocodylia), and birds (Aves). The monophyly of Amniota is supported by derived characters of the skull, pectoral girdle, and appendicular skeleton (Laurin and Reisz 1995). Other aspects of soft anatomy may be derived characters of amniotes that have been lost in certain groups (Gauthier et al. 1988). For example, a single penis with erectile tissue is found in crocodilians, birds, mammals, and turtles, but has been lost in Lepidosauria (Rhynchocephalia + Squamata) as well as most birds. Despite these losses, a single penis is considered a shared derived character of Amniota.

Within Reptilia a fundamental split gives rise to two clades, one represented by turtles and their extinct relatives (Anapsida) and the other including all other reptiles (Diapsida). These names refer to the pattern of temporal fenestration in the skulls, with the Anapsida lacking fenestration and the Diapsida having a pair of bony bars on either side delimiting two temporal openings. As indicated in Figure 2–9, extant diapsids are represented by two major clades. Lepidosauromorpha encompasses the extant clade Lepidosauria (Squamata + Rhynchocephalia) and several fossil groups (e.g., mosasaurs), whereas Archosauromorpha encompasses the extant groups Crocodylia + Aves and fossil groups, including dinosaurs and pterosaurs.

Reptilian monophyly is supported by characters of the skull. Most early lepidosauromorphs were small and probably insectivorous. Hence, they did not fossilize well, and consequently the relationships of early lepidosauromorphs are poorly understood. Early archosauromorphs, on the other hand, were larger and more robust and they are more completely understood. Their early radiation occurred in the late Paleozoic and eventually gave rise to the clade Archosauria, whose two primary divisions, Ornithosuchia and Pseudosuchia, led to dinosaurs and birds, on the one hand, and crocodilians on the other (Figure 2–9). The extant clades of Diapsida are Lepidosauria and Archosauria. Both Lepidosauria and Archosauria are strongly supported by many synapomorphies of skeletal and soft anatomy (Gauthier et al. 1988, 1989). Birds are cladistically part of Archosauria and are the closest relatives of crocodilians among extant taxa. Fossil groups such as dinosaurs and pterosaurs are more closely related to birds than are crocodilians.

The monophyly of Lepidosauromorpha is supported by several shared derived characters of the ribs and appendicular skeleton, but because early lepidosauromorphs are poorly known, the support for this clade is somewhat weak (Gauthier et al. 1988). On the other hand, Lepidosauria is characterized by many derived characters of the skeletal and soft anatomy. The latter characters include a transverse cloacal slit (versus an anteroposterior orientation in other tetrapods), loss of a single penis and evolution of paired penes (hemipenes) residing in the tail base, and regular episodes of ecdysis in which the outer layer of the epidermis is shed.

The radiation of Archosauria was marked by two notable trends (Gauthier et al. 1989). First, even early members of the clade show derived cranial modifications associated with increased predation efficiency, including elaborated cranial musculature and sharp, thecodont dentition (i.e., teeth set in sockets in the jaw bones). These and other modifications reach their culmination in some dinosaurs, which add additional features such as raptorial forelimbs suitable for grabbing prey. Many features of birds that are associated with flight, such as long forelimbs and birdlike wrists, fused clavicles (furcula), a fused bony sternum, hollow bones, and long forelimbs, evolved earlier in the archosaur radiation in association with predatory habits (Gauthier and Padian 1985). Second, modifications in the postcranial skeleton of archosaurs permitted an erect stance and a narrow-track gait, and the ability to breathe while running (Parrish 1986). Many successive modifications associated with locomotion occur within Archosauria.

Controversial Aspects of Amniote Phylogeny

Figures 2–1 and 2–9 are based on analyses including both extant and fossil taxa (Gauthier et al. 1988, Laurin and Reisz 1995). However, an important conclusion of the analyses of Gauthier et al. (1988) concerned the effect that inclusion of fossils had on phylogenetic hypotheses of extinct and extant amniotes. When data only from extant taxa were considered, the topology of the resulting tree was substantially different from Figure 2–1 in postulating a sister group relationship between mammals and archosaurs (crocodilians + birds). Indeed, similar hypotheses have been proposed by others based on less comprehensive character analyses including extant and fossil taxa (e.g., Gardiner 1993; see also Benton 1990). Moreover, molecular sequence data (18s and 28s ribosomal genes) also support a close bird–mammal relationship (Hedges et al. 1990). The different conclusions concerning amniote relationships that are reached when these different taxon and character sets are used underscore the critical importance that character distributions have on tree topologies, and demonstrate that fossils can significantly influence those topologies (Donoghue et al. 1989). When molecular and morphological data are analyzed together, the topology is identical to that shown in Figures 2–1 and 2–9 (Eernisse and Kluge 1993). Thus, the relationships depicted in Figure 2–1 are considered the best currently available hypothesis of relationships among amniotes.

The position of turtles (Testudines) within Amniota remains unclear. Although all recent comprehensive analyses of amniote phylogeny have placed turtles firmly within a clade containing other reptiles relative to Synapsida, the position of turtles within that clade varies somewhat (Gauthier et al. 1988, Laurin and Reisz 1995, Lee 1995, Rieppel and deBraga 1996). All of these authors except Rieppel and deBraga (1996) place turtles as a sister group to Diapsida. Lee (1993) and Laurin and Reisz (1995) demonstrated that the relationships of turtles were with several fossil groups, including pareiosaurs and procolophonids, a clade referred to as Parareptilia (Figure 2–1). Rieppel and deBraga (1996) revived an old idea that the anapsid skulls of turtles were secondarily derived, and not homologous to those of Paleozoic fossil groups having similar skull morphology. Their results place turtles among the Diapsida (archosaurs and lepidosaurs). Thus, the phylogenetic position of turtles, in recent years residing at the base of the reptilian clade, is once again challenged. We follow the results of Lee (1993) and Laurin and Reisz (1995) in

their more traditional placement, but recognize that the placement of turtles is presently insecure.

Summary

Phylogenetic systematics (also known as cladistics) emphasizes the importance of derived characters (characters that have changed from the ancestral condition) in defining monophyletic evolutionary lineages (clades). The results of phylogenetic analyses may vary depending on the characters and organisms included. Robust hypotheses about phylogenetic relationships are supported by several different lines of evidence.

The monophyletic origin of tetrapods is clearly established, and the extant groups form two clades, the Lissamphibia (frogs, salamanders, and caecilians) and Amniota (all other tetrapods). Most analyses agree in placing frogs and salamanders as each other's sister group. The exact relationships among lissamphibians and two fossil groups, temnospondyls and microsaurs, are not certain. Mammalia and Reptilia (including birds) are two extant sister taxa of amniotes. Within Reptilia, the position of turtles is somewhat controversial, but most studies agree in supporting their sister group status to other reptiles.

By focusing on the sequence of appearance of derived characters, phylogenetic systematics casts light on evolutionary mechanisms and provides a basis for hypotheses about cause-and-effect relationships in evolution. These methods can be applied to the evolution of physiology and behavior as well as morphology.

3

Classification and Diversity of Extant Amphibians

Living amphibians—the Urodela, Gymnophiona, and Anura—comprise more than 4,600 described species—approximately equivalent to the number of extant mammal species. Most amphibians are tropical organisms, and many new ones are discovered each year. The new species are primarily frogs, which are far more diverse than the other two groups combined: more than 4,100 species of extant amphibians are frogs, compared to about 415 species of salamanders and 165 species of caecilians (Frost 1985, Duellman 1993). Diversity is not evenly distributed within families. For example, more than half of the extant salamanders are plethodontids, and more than half of all frogs are in just three families: "Leptodactylidae," Hylidae, and "Ranidae." We introduce the systematics, morphology, and natural history of amphibians in this chapter. References to the systematics literature are provided primarily as an introduction to the literature on each group and are not comprehensive.

Salamanders (Urodela)

Typical salamanders are four-limbed, short-bodied animals with a long tail. The limbs of some elongate aquatic and terrestrial species such as *Amphiuma*, *Siren*, *Pseudobranchus*, and *Oedipina* are reduced or lost entirely. The skulls of salamanders are reduced through absence of many bones. Sizes in salamanders vary tremendously, from adult sizes of 30 millimeters total length to nearly 2 meters in length. Most species are terrestrial as adults but return to water for breed-

ing. Other species are completely terrestrial or aquatic, and many tropical plethodontids are arboreal.

Salamanders lack middle ear cavities and external ears. Primitively, the hearing apparatus consists of two components, collectively referred to as the opercular apparatus (see Figure 2–5). It consists of the columella (= stapes), which is found in other tetrapods, and the operculum and associated opercularis muscle, which are unique to amphibians. Both the columella and the operculum are bony or cartilaginous elements that are associated with the fenestra ovalis of the inner ear. The columella receives relatively high-frequency airborne sound. The operculum, which is connected via the opercularis muscle to the suprascapula of the pectoral girdle, receives low-frequency sound from the air or substrate. One or more components of the opercular apparatus are absent in various salamander families.

Salamanders have the largest genomes of any tetrapods, with nuclear DNA content reaching more than 100 picograms per diploid nucleus in *Onychodactylus* (Hynobiidae) and cryptobranchids and nearly 200 picograms in *Amphiuma* (Amphiumidae) (Morescalchi 1979). Although the large genomes are interesting from a genetic perspective, the correspondingly large cell nuclei can also influence evolutionary patterns and morphological processes at the organismal level (Sessions and Larson 1987, Wake 1991b).

Fertilization is external in the Cryptobranchoidea and probably in sirenids. It is internal via spermatophores in all other salamanders. Ancestrally, development involves a larval stage, but many species (e.g., many plethodontids) have direct development in which eggs hatch directly into juveniles. *Salamandra salamandra* and species of *Mertensiella* give birth to advanced larvae in water, and *Salamandra atra* and *Mertensiella luschani antalyana* bear fully metamorphosed young. Larvae, when present, have true teeth on the jaws, gill slits, and external gills.

Salamander larvae are essentially similar in body form to juveniles and adults, with the exception of having external gills and other features associated with an aquatic existence, such as tail fins. Balancers, ectodermal projections on the side of the head, develop in pond-dwelling larvae of some families. These structures may aid in maintaining balance until the limbs develop or may help prevent the larvae from sinking into bottom sediment. Larvae, in contrast to adults, generally lack eyelids, have a histologically different skin morphology, and lack maxillae. These features change into the adult conditions at metamorphosis. The palate of salamanders is totally remodeled during metamorphosis, a unique characteristic of the group.

Some salamanders (e.g., Cryptobranchidae, Proteidae, Sirenidae, *Ambystoma mexicanum*, and some Plethodontidae and Dicamptodontidae) never completely metamorphose under natural conditions and reproduce in a larval state. In other salamanders, metamorphosis may or may not take place in response to environmental cues, a phenomenon called facultative metamorphosis. Facultative metamorphosis may occur at either the individual or population level. Virtually all salamanders retain some larval or juvenile characteristics as adults (paedomorphosis), a factor contributing to the difficulty of reconstructing salamander phylogeny on the basis of morphological traits (see below).

The Urodela is characterized by a suite of derived skeletal and muscular features, primarily of the skull, hyoid, and cranial muscles (Milner 1988). The terms Urodela and Caudata have been used virtually interchangeably for the extant salamanders, but, as pointed out by Evans and Milner (1996), this usage creates problems when fossils are considered. Evans and Milner (1996) provide persuasive arguments for standardizing use of the term Urodela for all extant salamanders, and the term Caudata for a more inclusive grouping including all known fossils and extant forms. We follow this usage here, but alternative views have been expressed (e.g., Cannatella and Hillis 1993).

Content and Distribution

Sixty-one genera and approximately 415 species are recognized. Extant salamanders occur principally in North America and northern Eurasia. However, in addition to substantial North American diversity, one clade of the Plethodontidae has an extensive radiation in Central and South America.

Fossil Record

Salamanders appear in the Middle Jurassic (170 to 159 millions of years before the present [MaBP]) of England and Kirghizstan (Evans et al. 1988, Nessov 1988). The extensive salamander fossil record was reviewed by Estes (1981); Evans and Milner (1996) reviewed newly discovered Mesozoic forms. The only articulated Jurassic salamander is *Karaurus sharovi* from Russia (Ivachnenko 1978, Carroll 1988). All salamander fossils are from the Holarctic region with the exception of a sirenid from the Cretaceous of Sudan (Evans et al. 1996) and an enigmatic form from Israel (Nevo and Estes 1969).

Systematics. Edwards (1976), Milner (1983), Green and Sessions (1991), Larson and Dimmick (1993).

Systematics and Phylogeny of Extant Salamanders

Evidence for the monophyly of Urodela comes from the characters of the jaw adductor musculature, the ossification sequence of the skull bones, and the late appearance of the maxillae (Milner 1988). Paedomorphosis produces convergent evolution of many morphological characteristics (Wake 1991b), and its pervasive occurrence in salamanders has made it difficult to achieve robust phylogenies for the major groups using only morphological characters. Phylogenetic analyses of molecular data alone differ from those based on morphology (Hillis 1991, Larson 1991, Hay et al. 1995), but recent attempts to integrate the two types of data have been moderately successful (Larson and Dimmick 1993).

A phylogeny for the salamander families based on morphology and ribosomal gene sequences is shown in Figure 3–1. Sirenidae is the sister group to all other salamanders. The remaining salamanders (Neocaudata) are grouped into two clades: the Cryptobranchoidea (Cryptobranchidae and Hynobiidae) and the

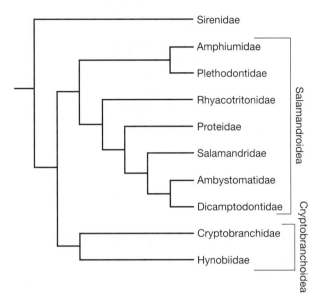

Figure 3-1 **Phylogenetic relationships among sala-mander families.** This tree is based on a combined analysis of ribosomal RNA sequences and morphological characters. *(Source: Modified from Larson and Dimmick 1993.)*

Salamandroidea (all remaining salamanders). The latter clade includes all salamanders known to have internal fertilization.

Relationships within the Salamandroidea are still controversial. Nevertheless, four groups appear in most analyses (Evans and Milner 1996): (1) Plethodontidae + Amphiumidae; (2) Rhyacotritonidae; (3) Proteidae; (4) Ambystomatidae + Dicamptodontidae + Salamandridae. The placement of Proteidae within Salamandroidea is the most problematic relationship at present.

Sirenidae

Sirenids are long, slender, eel-like salamanders that lack pelvic girdles and hind limbs (Figure 3–2). A keratinized beak is present, and they have many other unusual features. Paedomorphosis accounts for many of these features, such as the lack of eyelids, the presence of external gills, nonpedicellate teeth, the absence (*Pseudobranchus*) or reduction (*Siren*) of maxillae, a reduction in the number of digits on the forelimbs, and skin that is histologically similar to larval skin.

Siren lacertina reaches nearly a meter in total length, but other species are much smaller. *Pseudo-*

branchus individuals are only 150 to 200 millimeters long as adults. Sirenids are fully aquatic and inhabit swamps, lakes, and marshes with slow-moving water. They prey on invertebrates such as crayfish and other crustaceans, insects, and worms. *Siren intermedia,* and probably other sirenids, burrows into mud at the bottom of drying ponds and aestivates for up to a year within a cocoon to avoid desiccation (Gehlbach et al. 1973).

Eggs of sirenids are attached to vegetation in water or in rudimentary nests submerged in vegetation (Godley 1983). Females of *Siren intermedia* are known to attend the clutches (Godley 1983). Several lines of evidence suggest that fertilization is external, although courtship has not been observed. Cloacal glands, which produce spermatophores in male salamanders and function in sperm storage in females, are absent, and sperm are not present in the oviducts of females collected during the oviposition season (Sever 1991, Sever et al. 1996).

Content and Distribution. 2 genera (*Siren, Pseudobranchus*), 4 species. Coastal plain and Mississippi Valley of the southeastern United States and extreme northeastern Mexico (Figure 3–3).

Systematics. Martof (1974), Moler and Kezer (1993).

Cryptobranchidae

Cryptobranchids undergo incomplete metamorphosis. Adults lack eyelids and retain one pair of gill slits, which are closed in *Andrias*. The bodies are dorsoventrally compressed and the heads are extremely flattened. These completely aquatic salamanders exhibit a highly unusual mode of asymmetrical suction feeding in which the bilateral elements of the mandibles and hyoid move independently (Chapter 9) (Cundall et al. 1987, Elwood and Cundall 1994). Cryptobranchids also have unusual specializations associated with cutaneous respiration (Chapter 6). Cryptobranchids are the largest extant salamanders, with adult sizes reaching 1.5 to 1.8 meters for *Andrias* and 750 millimeters for *Cryptobranchus*. All species inhabit cold mountain streams. Fertilization is external and eggs are laid in strings under rocks in streams. Males of *Cryptobranchus* construct nests into which several females may deposit eggs. Consequently, a nest may contain 1,000 to 2,000 eggs (Nickerson and Mays 1973). Males guard their nests at least through early larval stages.

Figure 3-2 **Salamander diversity.** (a) *Siren lacertina* (Sirenidae, southeastern United States). (b) *Onychodactylus fischeri* (Hynobiidae, Asia). (c) Fire salamander with young (*Salamandra salamandra*, Salamandridae). (d) *Proteus anguinus* (Proteidae, southern Europe). (e) *Ambystoma maculatum* (Ambystomatidae, United States). (f) *Dicamptodon ensatus* (Dicamptodontidae, western United States). *(Photographs by (a, d, f) R. W. Van Devender, (b) Harry Greene, (c) Hans Reinhard/Bruce Coleman Ltd, (e) F. Harvey Pough.*

Figure 3-3 **Distribution of salamander families Sirenidae, Hynobiidae, and Salamandridae.**

Sirenidae ○
Hynobiidae ●

Salamandridae ○

Content and Distribution. *Andrias* (2 species; Japan and central China), *Cryptobranchus* (1 species; eastern North America) (Figure 3–4).

Systematics. Dundee (1971), Sessions et al. (1982)., Routman et al. (1994).

Hynobiidae

Hynobiids are relatively small salamanders (100 to 250 millimeters total length). They undergo complete metamorphosis and therefore have eyelids and lack gill slits as adults (Figure 3–2). Lungs are reduced in *Ranodon* and absent in *Onychodactylus*, which is fully aquatic. Fertilization is external, either directly or, in the case of *Ranodon sibericus*, the female deposits eggs on top of a spermatophore previously deposited by the male. Eggs are laid in gelatinous masses attached to rocks or vegetation in ponds or streams. Several species (e.g., *Onychodactylus* spp. and *Batrachuperus mustersi*) inhabit high mountain streams, even as adults. In most other species, adults are terrestrial and migrate to water for

breeding. The larval period in the high-altitude species *Batrachuperus mustersi* is 2 years (Reilly 1983).

Content and Distribution. 7 genera, about 36 species. Disjunct distribution in Asia from the Ural Mountains to the Pacific Ocean and south to China and Japan, Afghanistan, and Iran (Figure 3–3).

Systematics. Dunn (1923), Thorn (1968), Zhao et al. (1988).

Amphiumidae

Like sirenids, amphiumids are elongate, paedomorphic, aquatic salamanders that lack eyelids. However, unlike sirenids, amphiumids retain both pairs of limbs and girdles, lack external gills (but retain one pair of gill slits), and their teeth are pedicellate.

Two species of *Amphiuma* (*A. tridactylum* and *A. means*) attain lengths of approximately 1.1 meters, but maximum size in *A. pholeter* is only 35 centimeters. Amphiumids inhabit sluggish streams and rivers of swamps, occasionally moving overland. Fertilization

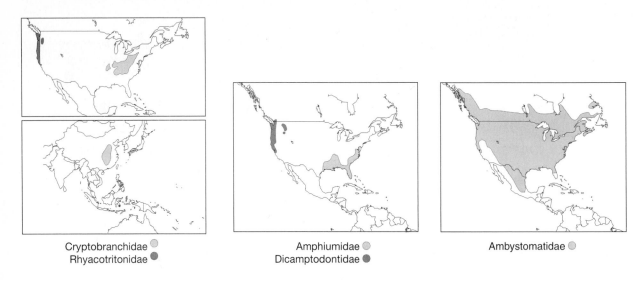

Figure 3-4   **Distribution of salamander families Cryptobranchidae, Rhyacotritonidae, Amphiumidae, Dicamptodontidae, and Ambystomatidae.**

is internal, with males depositing spermatophores directly into the cloaca of females during courtship. Eggs are laid on mud near water and are attended by the female. *Amphiuma* preys actively on a wide variety of vertebrates and invertebrates, including insects, crayfish, snails, amphibians, reptiles, and fish.

Content and Distribution. 1 genus (*Amphiuma*), 3 species. Coastal plain and lower Mississippi Valley of the southeastern United States (Figure 3–4).

Systematics. Salthe (1973).

Plethodontidae

The plethodontids are the most diverse and species-rich group of salamanders and are the only salamanders that have extensively radiated in the tropics (Figure 3–5). All species are lungless and are characterized by a nasolabial groove that aids in chemoreception. Plethodontids occupy subterranean, aquatic, terrestrial, and arboreal habitats. Much of the spectacular radiation of plethodontids is marked by adaptive transitions in locomotory and feeding structures (Wake and Larson 1987). Equally important, life history traits have evolved from an ancestral life cycle involving aquatic larvae (most Desmognathinae and Hemidactyliini), toward increased terrestriality and direct development, on one hand (Plethodontini and Bolitoglossini), or permanently aquatic forms that reproduce as larvae on the other (some Hemidactyliini).

However, within desmognathine salamanders the reverse trend, from terrestriality toward longer larval periods and increasing use of aquatic habitats, seems to have occurred (Titus and Larson 1996).

The origin of lunglessness among plethodontids is most often attributed to adaptation for a mountain stream habitat, the presumed ancestral environment of plethodontids (Ruben and Boucot 1989, Beachy and Bruce 1992, Ruben et al. 1993). Lunglessness evolved independently in *Onychodactylus* (Hynobiidae) and lungs are reduced in *Ranodon* (Hynobiidae) and *Rhyacotriton* (Rhyacotritonidae), all of which also live in fast-flowing streams.

Plethodontids include the smallest (*Thorius;* 30 millimeters total adult length) and some of the largest (*Pseudoeurycea belli;* 320 millimeters total length) terrestrial salamanders. Body forms vary greatly, from relatively robust to elongate, slender species such as *Batrachoseps, Thorius,* and *Oedipina.* Many arboreal species have webbed feet and prehensile tails. Some forms of *Ensatina eschscholtzii klauberi* have bright antipredator warning (aposematic) coloration, and several species of plethodontids are involved in mimicry complexes (e.g., *Plethodon cinereus* mimics the efts of the salamandrid, *Notophthalmus,* and *Desmognathus imitator* mimics *Plethodon jordani;* Brodie and Howard 1973, Brodie and Brodie 1980).

Major clades within the plethodontids are shown in Figure 3–6. The Desmognathinae (*Phaeognathus,*

Figure 3-5 **Diversity within Plethodontidae.** (a) *Typhlomolge rathbuni*, a permanently larval hemidactyliine from Texas. (b) *Ensatina eschscholtzii klauberi*, an aposematically colored plethodonine from California. (c) *Hydromantes platycephalus*, a cave and crevice-dweller from California. (d) *Oedipina ignea*, a fossorial bolitoglossine from Mexico. (e) *Pseudoeurycea belli*, a large terrestrial bolitoglossine from Mexico. (f) *Bolitoglossa occidentalis*, a small arboreal bolitoglossine from Central America. *(Photographs by (a) Charles E. Mohr/Photo Researchers, (b) Harry Greene, (c) John Cadle, (d, e, f) R. Wayne Van Devender.)*

Desmognathus) are characterized by having four larval gill slits, a unique ligament extending from the atlas to the lower jaw (atlantomandibular ligament), and a suite of derived features associated with peculiar burrowing and feeding modes (Wake 1966, Schwenk and Wake 1993). The Plethodontinae have the alternative states of a normal mouth-opening mechanism and three larval gill slits. Within the Plethodontinae, the hemidactyliines (*Eurycea, Gyrinophilus, Haideotriton, Hemidactylium, Pseudotriton, Stereochilus, Typhlomolge, Typhlotriton*) have an aquatic larval stage, whereas the plethodonines (*Aneides, Plethodon, Ensatina*) and bolitoglossines (*Batrachoseps, Hydromantes,* and 12 exclusively Neotropical genera) have direct development. The latter two clades are distinguished by the presence (plethodonines) or absence (bolitoglossines) of second basibranchials in the hyoid. *Haideotriton, Ty-phlomolge,* some species of *Eurycea,* and one species of *Gyrinophilus* are permanently aquatic.

The Plethodontidae contains over half of extant salamanders, with major centers of diversity in eastern North America, in the highlands of southern Mexico and Guatemala, and in the highlands of eastern Costa Rica and western Panama. One tribe of the Plethodontidae, the primarily Neotropical Bolitoglossini, includes nearly half of all extant species of salamanders, and many new species are still being discovered. Sadly, their rain forest habitats are also being destroyed, and species are likely becoming extinct before discovery and description. Plethodontids have formed a paradigm for studies of patterns and mechanisms of evolution in amphibians, especially those focusing on the role of heterochrony in generating morphological novelties (Chapter 7; Wake 1992).

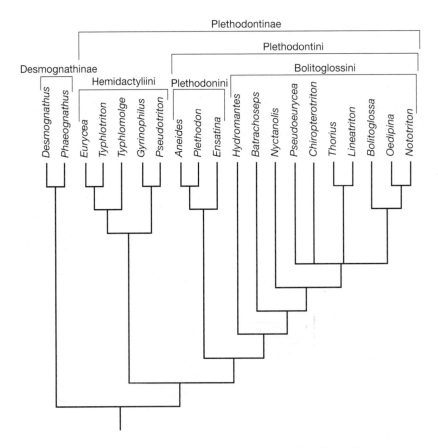

Figure 3-6   **Phylogeny of Plethodontidae.** The relationships among the subfamilies, tribes, and selected genera based on an analysis of morphological characters are shown. (*Source: Modified from Wake 1966, Larson et al. 1981, and Wake and Elias 1983.*)

Figure 3-7 **Distribution of the sala-mander families Plethodontidae and Proteidae.**

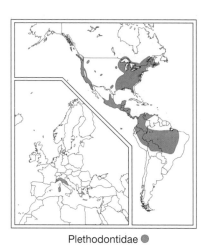

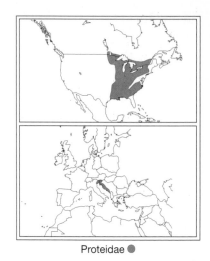

Plethodontidae ●

Proteidae ●

Figure 3-7 **Distribution of the sala-mander families Plethodontidae and Proteidae.**

Content and Distribution. 28 genera, approximately 266 species. Distribution disjunct (Figure 3–7): eastern North America from southern Canada to Florida (Desmog-nathinae, Hemidactyliini, Plethodontini); extreme western North America from southeastern Alaska to Baja California (Plethodontini and Bolitoglossini); scattered localities in the central United States and Rocky Mountains (Plethodon-tini); Mexico, Central America, and South America to Bo-livia and eastern Brazil (Bolitoglossini); southern Europe and Sardinia (*Hydromantes* only, a bolitoglossine that also occurs in western North America).

Systematics. Dunn (1926), D. B. Wake (1966, 1993). Tilley and Mahoney (1996), Titus and Larson (1996). Highton (1991), Highton et al. (1989, *Plethodon*). Larson et al. (1981; Plethodontinae). Jackman and Wake (1994; *En-satina*). Wake and Elias (1983), Wake (1987; Bolitoglossi-nae). Good and Wake (1993; *Nototriton*), Hanken and Wake (1994; *Thorius*), Darda (1994; *Chiropterotriton*), Larson (1983; *Bolitoglossa*), Wake et al. (1978; *Hydromantes*).

Rhyacotritonidae

Rhyacotritonids, represented by a single genus, *Rhya-cotriton*, are characterized by unique squared-off glands posterior to the vent in adult males (Sever 1988). They exhibit a unique combination of mor-phological traits that distinguishes them from other families, including the absence of an operculum and opercularis muscle (otherwise lost in only a few hyno-biids), and greatly reduced lungs and associated struc-tures (also seen in plethodontids and some hynobiids). Both larvae and adults of *Rhyacotriton* inhabit cold, fast-flowing, well-shaded seepages and streams in humid old-growth conifer forests. Maximum adult

sizes are up to about 60 millimeters snout–vent length, but body size varies among populations. Lar-val periods are 3 to 5 years (Nussbaum and Tait 1977). The species of *Rhyacotriton* are morphologically simi-lar but extremely differentiated genetically, suggesting a very old radiation.

Content and Distribution. 1 genus (*Rhyacotriton*), 4 species. Coastal Pacific Northwest of the United States from northern California to Washington; Cascade Range of Oregon and Washington (Figure 3–4).

Systematics. Good and Wake (1992).

Proteidae

Proteids are aquatic, paedomorphic salamanders. They have large gills and caudal fins. Proteids are un-usual among salamanders in lacking maxillary bones (also absent in the sirenid *Pseudobranchus*), in having two pairs of larval gill slits, and in having a diploid chromosome count of 38 (compared to fewer than 30 in other Salamandroidea).

Proteids attain adult sizes up to about 45 centime-ters total length. Species of *Necturus* inhabit lakes or streams of North America, whereas *Proteus* lives in limestone caves in southern Europe. *Proteus* is similar to other cave-dwelling salamanders such as *Typhlo-molge* (Plethodontidae) in having a slender body and limbs, whitish skin, and reduced eyes.

Content and Distribution. 2 genera, 6 species. *Proteus anguinus:* alps of Italy and Yugoslavia. *Necturus* (5 species): eastern North America (Figure 3–7).

Systematics. Hecht and Edwards (1976), Maxson et al. (1988).

Salamandridae

Salamandrids are small to moderate (200 millimeters) in size and have smooth to extremely rugose skin. *Salamandra*, *Mertensiella*, and *Chioglossa* are terrestrial as adults, whereas the remaining genera (newts) spend prolonged periods annually or their entire adult lives in water. The aquatic species have well-developed tail fins. Many salamandrids have conspicuous skin glands that produce highly toxic secretions (tetrodotoxins). Many also have aposematic coloration and elaborate defensive displays (Brodie 1977, Daly et al. 1987).

Life histories of salamandrids are highly varied. Courtship behaviors are often elaborate (Arnold 1977, Halliday 1977), usually involving prolonged interaction between partners. Fertilization is internal via spermatophores. Females of the European salamandrids *Salamandra* and *Mertensiella* retain developing eggs within the body, either depositing advanced larvae into water (*S. salamandra* and *M. caucasica*) or giving birth to fully metamorphosed young (*S. atra* and *M. luschani antalyana*). Eggs of oviparous salamandrids are deposited in ponds or streams. Adults of many North American and Eurasian salamandrids are terrestrial and migrate to ponds for breeding (e.g., *Taricha* and *Triturus*). *Notophthalmus* (eastern North America) has a more complex life cycle. Adults are permanently aquatic and ontogeny may involve a larval stage only, or the larvae may metamorphose into an immature eft stage that is terrestrial for 1 to 14 years, eventually returning to ponds and transforming into the adult stage (Healy 1974, Gill 1978).

Content and Distribution. 15 genera, approximately 55 species. Eastern and western North America (*Notophthalmus*, *Taricha*), Europe, northwest Africa, and western Asia (8 genera), and eastern Asia (eastern India to Japan; 5 genera) (Figure 3–3).

Systematics. Titus and Larson (1995), Halliday and Arano (1991), Tan and Wake (1995).

Ambystomatidae

Ambystomatids are robust salamanders of moderate size (adults are 100 to 300 millimeters in length). They share many osteological characters with plethodontids, but in general have many characters that are ancestral within Salamandroidea. Duellman and Trueb (1986) pointed out that ambystomatids share a derived pattern of spinal nerves with plethodontids, but otherwise have no uniquely derived characters and are therefore of questionable monophyly. However, molecular studies suggest phylogenetic unity of the group (Shaffer et al. 1991).

Metamorphosis is either facultative or obligate in some species of ambystomatids. The Mexican axolotl, *Ambystoma mexicanum*, is a permanently aquatic species that has been used in a wide variety of research in developmental and experimental biology (Shaffer 1993). Adults of transforming species are terrestrial. Most ambystomatids breed in early spring, and eggs are usually deposited in ponds or slow-moving streams. However, several species (e.g., *Ambystoma opacum*) breed during the fall and deposit eggs on land near water. The nest sites are flooded and larvae develop in water.

Hybridization among species of the *Ambystoma jeffersonianum-laterale* complex has produced several triploid all-female species, of which the best studied are *Ambystoma platineum* and *A. tremblayi* (Kraus 1989, Hedges et al. 1992, Spolsky et al. 1992). Reproduction in these triploids occurs through gynogenesis. In this mode of unisexual reproduction, eggs develop without a genetic contribution from males, although sperm penetration of the eggs is required to initiate cleavage. In the case of *A. platineum* and *A. tremblayi*, mating takes place with males of one of the sympatric parental species (*A. jeffersonianum* or *A. laterale*) and sperm penetrate, but do not fertilize, the eggs of the triploids.

Phylogenetic studies of species of *Ambystoma* suggest great antiquity for lineages within the genus (Shaffer et al. 1991, Shaffer 1993), even though species of the *Ambystoma tigrinum* complex are less than 5 million years old (Shaffer and McKnight 1996).

Content and Distribution. 1 genus (*Ambystoma*), about 32 species. North America from southern Canada to the southern edge of the Mexican Plateau (Figure 3–4).

Systematics. Shaffer et al. (1991), Shaffer (1993), Jones et al. (1993), Reilly and Brandon (1994).

Dicamptodontidae

Dicamptodontids are moderate to large salamanders (to 17 centimeters snout–vent length in *Dicamptodon ensatus*). *Dicamptodon copei* is permanently aquatic and retains external gills. Metamorphosis is facultative in individuals or populations of the other species. *Dicamptodon* inhabits damp coniferous forests with

Figure 3-8 **Diversity of caecilians.** (a) *Ichtyophis bannanicus* (Ichthyophiidae, Asia), (b) *Dermophis mexicanus* (Caeciliidae, Central America, terrestrial) (c) *Potamotyphlus kaupii* (Caeciliidae, South America, aquatic). *(Photographs by (a) Harry Greene, (b) R. Wayne Van Devender, (c) John Cadle.)*

(a)

(b)

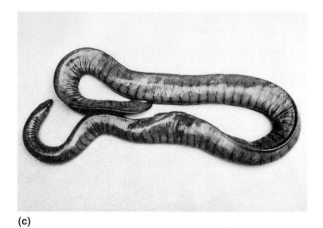

(c)

cold streams or cold mountain lakes. Metamorphosed adults are terrestrial. Eggs are deposited in cold streams, and there is a lengthy larval period (2 to 5 years). Species of *Dicamptodon* are virtually indistinguishable morphologically but extremely differentiated genetically and in some life history characters.

Content and Distribution. 1 genus (*Dicamptodon*), 4 species. Pacific northwest from northern California to southern Canada; Rocky Mountains of central Idaho and adjacent Montana (Figure 3–4).

Systematics. Good (1989), Nussbaum (1976).

Caecilians (Gymnophiona)

Caecilians are elongate terrestrial, burrowing, or aquatic amphibians with distinctly annulated bodies (Figure 3–8). Many morphological structures are reduced in caecilians, as in many other elongate or burrowing tetrapods. Tails are greatly reduced or absent altogether. Limbs and their girdles are absent in all extant species but are present in the earliest known fossil form, *Eocaecilia*. The eyes are reduced in extant species and are covered with skin or bone. The left lung may be reduced or absent, and in one aquatic species lungs are absent (see discussion under Caeciliidae). Dermal scales are present in some species of caecilians, lying within the annular grooves that delimit the body segments (annuli). Fertilization is internal in all species, and is effected by a protrusible copulatory organ (phallodeum) formed by a portion of the cloacal wall in males. Some species have aquatic eggs and larvae, whereas others lay eggs in terrestrial sites and undergo direct development. About 75 percent of all species for which reproductive mode is known are viviparous, and the developing young are nourished by secretions of cells in the oviductal epithelium (Wake 1977). In general, very little is known of the natural history of caecilians because of their secretive nature. Many species are known only from one or a few museum specimens.

The annuli ringing the body of caecilians are one of the group's most characteristic features. All caecilians have a series of primary annuli that are probably homologous with the costal grooves of salamanders. Each primary annulus is associated with one vertebra along most of the trunk. In many species of caecilians, primary annuli may become subdivided by secondary annuli, which in turn may become subdivided by tertiary annuli. In each case the appearance of the annuli follows a definite ontogenetic pattern: primaries appear first, followed by secondaries, which form in a wave beginning posteriorly and spreading anteriorly. Finally, the tertiaries form in the same way as the secondaries (Nussbaum and Wilkinson 1989).

The scales of caecilians are composed of layers of collagenous fibers topped with mineralized nodules (squamulae) (Zylberberg and Wake 1990). Several scales are aligned, much like an oblique stack of coins, in a pouch located in the dermis below the annular grooves. Scale pouches partly or completely encircle the body along each groove and are nestled among the mucous and poison glands of the dermis. Scales are unknown in extant amphibians other than caecilians, but mineralized osteoderms are present in many fossil amphibians.

Some of the more distinctive features of caecilians are found in the structure of the skull, which is always well ossified. The skulls of most caecilians are stegokrotaphic, that is, they are completely roofed except for sensory openings (for the eyes, nares, and tentacles). Rhinatrematids, scolecomorphids, and some caeciliids have a partially open temporal region that is slightly kinetic (zygokrotaphy). Fusion of some skull bones in caecilians results in several compound elements, including a maxillopalatine (maxilla + palatine) and a large element termed the os basale (otic and occipital bones + parasphenoid) that forms most of the posteroventral and posterior portions of the skull (Figure 3–9). Teeth are present on the maxil-

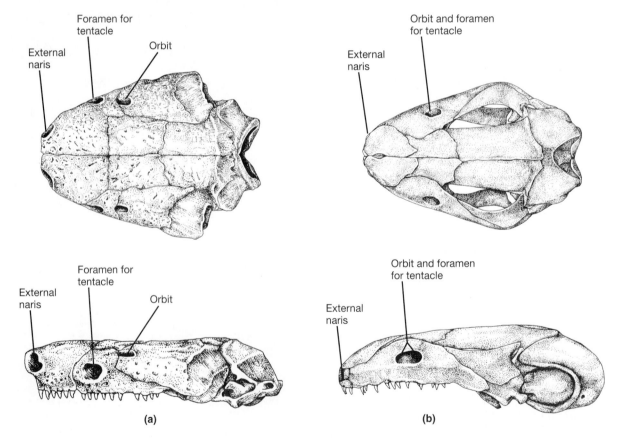

Figure 3-9   **Skull morphology of caecilians.** (a) Stegokrotaphic skull of *Dermophis mexicanus* (Caeciliidae). (b) Zygokrotaphic skull of *Epicrionops bicolor* (Rhinatrematidae). *(Source: Modified from M. Wake and Hanken 1982, Nussbaum 1977.)*

lopalatines, premaxillae, vomers and pseudodentaries. The lower jaws bear a long retroarticular process where the muscles involved in jaw movement insert. Caecilians have evolved a dual jaw adduction mechanism, unique among tetrapods, consisting of joint action of the mandibular adductors (ancestral component among tetrapods) and the interhyoideus muscles (a component unique to caecilians) (Bemis et al. 1983, Nussbaum 1983).

All caecilians have a specialized sensory organ, the tentacle, that opens to the surface of the head through an aperture located between the eyes and nostrils. The position of the aperture varies considerably and is useful in identifying species. A portion of the tentacular structures is protrusible through the aperture to varying degrees in different taxa. The tentacle itself is a complex of structures (muscles, glands, ducts, and so forth) that develops in close association with the eyes and Jacobson's organs (Billo and Wake 1987). It functions in chemoreception.

Content and Distribution

Thirty-four genera and approximately 165 species are recognized. Caecilians are pantropical except for Madagascar and land east of Wallace's Line (Papuan–Australian region). They have not been discovered in central tropical Africa, although they are present in East and West Africa.

Fossil Record

Caecilians are known from the Paleocene of Brazil, the Late Cretaceous of Bolivia and Sudan, and the Early Jurassic of Arizona, U.S.A. (ca. 190 MaBP; Jenkins and Walsh 1993, Evans et al. 1996). With the exception of the last record, all fossils are isolated vertebrae and shed little light on caecilian evolution. The Jurassic specimens, however, are spectacularly preserved skulls and postcrania of an unusual form known as *Eocaecilia micropodia*. Aside from being the earliest and best-preserved caecilian fossil, *Eocaecilia* retains well-developed, although slightly reduced, limbs and girdles, has well-developed eyes, and has a completely stegokrotaphic skull. The last feature is significant in suggesting that the reduced skull roof present in the rhinatrematids may be a derived rather than ancestral condition for modern caecilians (see Nussbaum 1983).

Systematics. Taylor (1968), Nussbaum and Wilkinson (1989).

Systematics and Phylogeny of Caecilians

Evidence for the monophyly of Gymnophiona includes most of the unusual caecilian features already described. The phylogeny of caecilians was most recently reviewed by Nussbaum and Wilkinson (1989), Hass et al. (1993), and Hedges et al. (1993). Phylogenetic inferences have included data sets based on morphology (Nussbaum 1979b), immunological comparisons (Case and Wake 1977, Hass et al. 1993), chromosomal morphology (Nussbaum 1991), and molecular sequences from ribosomal genes (Hedges et al. 1993). Five families of caecilians are currently recognized. Following Hedges et al. (1993), the aquatic typhlonectids are included in the Caeciliidae because both molecular and morphological evidence suggests a close relationship between typhlonectids and some caeciliids.

Molecular and morphological analyses are remarkably concordant concerning relationships among the caecilian families (Figure 3–10). Rhinatrematidae is the sister group to the remaining extant families.

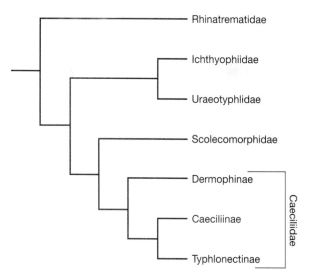

Figure 3-10   **Phylogenetic relationships among caecilian families.** This tree is based on analyses of morphological and molecular characters. Three clades (subfamilies) within Caeciliidae are shown to emphasize the relationship of the aquatic Typhlonectinae to more terrestrial caeciliids, but the monophyly of the other two subfamilies is uncertain (see text). (*Source: Modified from Hillis 1991 and Wilkinson and Nussbaum 1996.*)

Successive clades are the Ichthyophiidae, Uraeoty-phlidae, Scolecomorphidae, and Caeciliidae. The major uncertainty concerns the relationships of the Uraeotyphlidae, which may form a clade with the Ichthyophiidae (Wilkinson and Nussbaum 1996), or may be more closely related to the Scolecomorphi-dae–Caeciliidae clade.

Rhinatrematidae

Rhinatrematids retain several relatively primitive characteristics among extant caecilians. They have a true tail containing vertebrae, muscles, and skin an-nuli. The mouth is at the tip of the snout, unlike most other caecilians in which the snout projects over the mouth. The tentacular opening is adjacent to the an-terior edge of the eye, which is considered to be an ancestral condition relative to its more anterior posi-tion in other caecilians (Nussbaum 1977). The skulls are zygokrotaphic. Numerous scales are present in the annuli of rhinatrematids.

Very little is known of the natural history of rhina-trematids, although the zygokrotaphic skulls and ter-minal mouths do not suggest great specializations for burrowing. They may be cryptic surface forms. They are relatively small caecilians, up to about 330 mil-limeters in length. Rhinatrematids are oviparous, and larvae are known for some species of *Epicrionops*. Lar-vae of *E. petersi* have been found in mud at the edge of streams (Duellman and Trueb 1986).

Content and Distribution. 2 genera (*Epicrionops, Rhi-natrema*), 9 species. *Epicrionops*: western Amazonia and northwestern South America (Colombia, Ecuador, Peru, Venezuela). *Rhinatrema*: eastern Amazonia and the Guianan region (Brazil, French Guiana, Surinam, Guyana) (Figure 3–11).

Systematics. Nussbaum (1977).

Ichthyophiidae

Like rhinatrematids, ichthyophiids have a true tail, but their skulls are more solidly roofed (stegokro-taphic). The mouth may be nearly terminal or subter-minal. Scales are present in the body annuli. The ten-tacle is between the eye and nostril, but closer to the eye. Ichthyophiids attain lengths up to about 500 mil-limeters. Females of *Ichthyophis glutinosus* lay clusters of eggs in moist soil or a burrow near water and at-tend the clutches. Larvae are aquatic.

Content and Distribution. 2 genera (*Caudacaecilia, Ichthyophis*), about 36 species. India, Sri Lanka, Southeast Asia, the Philippines, mainland Malaysia, Sumatra, and Bor-neo (Figure 3–11).

Systematics. Taylor (1968).

Uraeotyphlidae

Uraeotyphlids have a true tail and somewhat ste-gokrotaphic skulls. The mouth may be subterminal or more recessed. The tentacular opening is far forward in the skull, underneath the nostril. Scales are present. Uraeotyphlids are small caecilians, up to about 300 millimeters in length. The species are probably oviparous and possibly have direct development, as larvae have never been found.

Content and Distribution. 1 genus (*Uraeotyphlus*), 4 species. Extreme southern India (Figure 3–11).

Systematics. Nussbaum (1979b).

Scolecomorphidae

Scolecomorphids are characterized by several highly unusual aspects of morphology. The zygokrotaphic skulls lack several bony elements, including post-frontals, pterygoids, and stapes. Orbits are absent and the eyes are vestigial. They are attached to the tenta-cles and may be protruded from the tentacular open-ing when the tentacles are extended. Calcified spines are present on the phallodea in some species of *Scole-comorphus*. These are otherwise unknown among cae-cilians. Only primary annuli are present, and scales are unknown in this family. The largest species of scolecomorphids attain lengths of about 450 millime-ters. *Crotaphotrema* is probably oviparous, whereas species of *Scolecomorphus* are probably viviparous (Nussbaum and Wilkinson 1989).

Content and Distribution. 2 genera (*Crotaphotrema, Scolecomorphus*), 5 species. Disjunct distribution: Cameroon, Malawi, and Tanzania (Figure 3–11).

Systematics. Nussbaum (1985).

Caeciliidae

With the exception of the aquatic typhlonectines, cae-ciliids are terrestrial burrowers. No tail is present, and the mouth is recessed underneath the snout. The po-sition of the tentacular opening and the presence of scales vary. The skulls are completely stegokrotaphic. Primary annuli may or may not be divided by sec-

Figure 3-11   **Distribution of caecilian families Rhinatrematidae, Ichthyophiidae, Uraeotyphlidae, Scolecomorphidae, and Caeciliidae.**

Rhinatrematidae ⦿
Ichthyophiidae ●

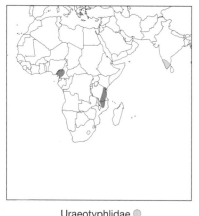

Uraeotyphlidae ⦿
Scolecomorphidae ●

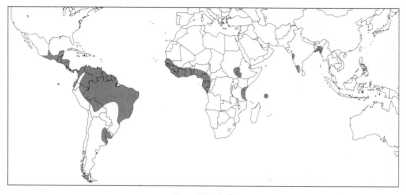

Caeciliidae ●

ondary annuli, but tertiary annuli are invariably absent.

The Caeciliidae contains among the smallest (*Idiocranium russeli* and *Grandisonia brevis;* approximately 100 millimeters in length) and the largest caecilians (some species of *Caecilia* and *Oscaecilia* attain 1.5 meters in length). Life histories are equally varied and include species with direct development (South America, Africa, India, and the Seychelles), with aquatic larvae (*Sylvacaecilia,* some *Grandisonia*), and viviparous species (most New World and some African species). All of the typhlonectines are viviparous and have aquatic larvae. Females of some oviparous and viviparous species are known to give parental care. For example, Sanderson (1937) found females of *Idiocranium russeli* coiled about their eggs on a small mound within a nest cavity under a dense grass mat. He also reported females of the viviparous species *Ge-*

otrypetes seraphini coiled around neonates within a nest chamber.

Most caeciliids appear to be burrowers, although some are surface-active, particularly after heavy rains. However, the typhlonectines are semiaquatic to fully aquatic. The aquatic species have compressed bodies and well-developed dorsal fins on the posterior body. Some species of caeciliids have bright patterns including orange, yellow, pink, or bluish colors.

A remarkable species of typhlonectine, *Atretochoana eiselti,* known only from a single specimen from an unknown South American locality, is the largest known lungless tetrapod (725 millimeters in length; Nussbaum and Wilkinson 1996). Additionally, it is the only known choanate (lungfishes + tetrapods) in which the internal nares are sealed off from the buccal cavity by fusion of the choanal valves. Although other unusual features occur in its skull, cranial musculature, and circulatory system, *Atretochoana* is the apparent sister taxon to *Potamotyphlus*, a rather typical typhlonectine.

Content and Distribution. 28 genera, approximately 100 species. The internal systematics is unsettled and a resolved phylogeny has been difficult to achieve. 3 subfamilies: Typhlonectinae (5 genera, 12 species; South America), Caeciliinae (4 genera, ca. 45 species; Panama to northern South America), Dermophinae (19 genera, ca. 43 species; Central and South America, Africa, the Seychelles, India) (Figure 3–11). Typhlonectinae is monophyletic, but monophyly of the other two subfamilies is uncertain.

Systematics. Wilkinson (1989, 1996), Savage and Wake (1972; *Dermophis, Gymnopis*), Lahanas and Savage (1992; *Oscaecilia*), Nussbaum and Hinkel (1994; *Boulengerula*). Hedges et al. (1993) placed Typhlonectidae within Caeciliidae.

Frogs (Anura)

Living frogs (Anura) and their extinct sister taxon *Triadobatrachus* form a monophyletic group (Salientia) characterized by numerous derived characters. The skulls are extremely reduced, lacking many elements present ancestrally. The frontals and parietal bone are fused to form a frontoparietal. Teeth are absent from the dentary except in one extant species. Tails are short (*Triadobatrachus*) or absent (Anura), and the caudal vertebrae are fused into a rod (urostyle) in Anura. Vertebral numbers are reduced, numbering no more than nine presacral vertebrae in Anura. The ilia and proximal tarsal elements are elongate.

The Anura are further distinguished from *Triadobatrachus* in having greatly elongated hind limbs and feet. The tibiale and fibulare (astragalus and calcaneum) of the ankle are elongate and at least partially fused. In the forelimb the radius and ulna are fused, and in the hind limb the tibia and fibula are fused. These features are possibly associated with jumping, a major mode of locomotion for most frogs. The elements of the hyoid apparatus are fused into a hyoid plate, and a number of elements are lost from the skulls. The tongue is attached in the front of the mouth in most frogs, and has a free posterior edge. Protrusion involves flipping the tongue out of the mouth using the attached anterior end as a pivot point.

Anuran larvae, if present, lack true teeth, but most have keratinous jaw sheaths and denticles. Most larvae have a large branchial basket and internal gills. The larvae (tadpoles) are very different morphologically from juvenile and adult frogs, and metamorphosis involves a radical reorganization of the body plan. Most tadpoles are filter-feeding or herbivorous organisms, and metamorphosis results in enormous changes in food procurement and digestion, as well as all other systems.

Reproductive diversity in frogs is extremely varied. Most frogs have external fertilization, effected by close contact between the sexes in an embrace called amplexus. Most species of frogs have no parental care, but other species brood eggs, larvae, or young on the limbs or back, in specialized pouches or depressions on the back, in the vocal sacs of males, or in the stomach. Many species attach egg masses to vegetation above water or deposit them in aerial bodies of water, such as in epiphytes. In many of these cases, the clutches or tadpoles are guarded or otherwise cared for by a parent. Internal fertilization occurs in at least three species, *Ascaphus truei* (Ascaphidae) and two species of *Eleutherodactylus* (Leptodactylidae). Direct development occurs in many families, and viviparity is documented for several species in two families (*Nectophrynoides* and its relatives in the Bufonidae, and *Eleutherodactylus jasperi* in the Leptodactylidae).

Characters defining the major groups of frogs include various aspects of skeletal structure (especially the skull, vertebrae, and pectoral girdles), muscles of the limbs, presence and morphology of intercalary elements in the digits, morphology of the tadpoles, and behavior (amplexus type). Salient features of these systems that are important in frog systematics are reviewed here.

Axial Skeletal Elements

Compared to caecilians and salamanders, frog skulls are relatively broad and the number of bones is reduced. The elements most frequently lost are the palatines, vomers, quadratojugals, and columellae, and the pattern of loss can provide important clues to relationships. The loss of skull elements in general gives frog skulls the appearance of being rather open and lightly built compared to the skulls of other amphibians. Dentition is also reduced and, with the exception of *Gastrotheca guentheri* (Hylidae), is always absent from the mandible. Teeth are present on the maxillae, premaxillae, and often on the vomers. The dermal bones of the skull roof in frogs may be co-os-

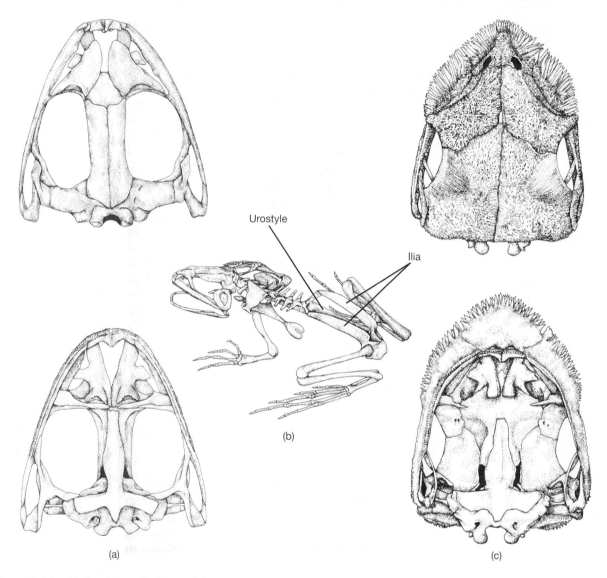

Figure 3-12   **Skeletal morphology of frogs.** (a) Dorsal and ventral views of the skull of *Leptodactylus bolivianus* (Leptodactylidae). (b) Frog skeleton with major characteristics of Anura indicated. (c) Dorsal and ventral views of the skull of *Triprion spatulatus* (Hylidae), showing extensive casquing. In *Triprion* the dermal bones of the skull are co-ossified with the overlying skin. *(Source: Modified from Trueb 1973 and Duellman 1970.)*

sified (fused) with the overlying skin. In addition, in some frogs the dermal bones of the skull are elaborated into heavily ossified and projecting casques, such as seen in the hylids *Hemiphractus, Triprion,* and *Anotheca* (Figure 3–12).

Several vertebral characters are used in frog classification. The number of presacral vertebrae varies from five to nine, with higher numbers considered ancestral. The form of the vertebral centrum, which in early development encloses the notochord, is also of importance. Although the specific developmental patterns are different, mature vertebrae fall into one of several patterns, of which three are common: amphicoelous vertebrae, in which the anterior and posterior ends of the centra are rounded; opisthocoelous vertebrae, in which the anterior ends of the centra are rounded and the posterior ends are concave; and the procoelous condition, in which the centra are anteriorly concave and posteriorly rounded.

Pectoral Girdle and Limbs

Pectoral girdles anchor the forelimbs to the body (Figure 3–13). The proximal limb element (humerus) articulates with the pectoral girdle at the glenoid fossa. Frog pectoral girdles consist of three elements above the glenoid fossa: the scapula, suprascapula, and cleithrum. Ventromedially below the glenoid fossa the gir-

dle elements include various sternal components (omosternum and sternum) along the midline, and the clavicles and coracoids extending from the midline to articulate with the scapula and forelimb at the glenoid fossa. A series of cartilaginous elements, the epicoracoid and procoracoid cartilages, lies between the clavicles and coracoids on either side. Two general types of anuran pectoral girdles, first distinguished by the great American anatomist and evolutionist E. D. Cope, have been used in a systematic context. In most arciferal girdles the epicoracoids are fused anteriorly between the clavicles and they overlap posterior to this region. Arcifery is the most widespread condition of pectoral girdles in anura and is probably the ancestral pattern. Firmisternal girdles, observed in Ranoidea, Microhylidae, and occasionally in bufonids, pipids, and leptodactylids, are characterized by complete fusion of the epicoracoid horns medially and fusion of the sternum to the pectoral arch. Conditions intermediate between the two general types of pectoral girdles are known, thus complicating systematic use of this feature. In fact, frog pectoral girdles are a complex of characters that may or may not covary with one another, and their evolution will be understood more fully once the nature of this covariation is better characterized (Ford 1993). The structure of the girdles also has important functional consequences for locomotion (Emerson 1983, 1984, 1988b) that have yet to be fully integrated into systematic analyses.

Characters of the external and internal morphology of the limbs have also been used in anuran systematics. Externally, the presence, distribution, and development of various tubercles can have systematic importance (e.g., the well-developed outer metatarsal tubercle that is modified into a digging spade in pelobatids), as can the amount and distribution of webbing. The pattern of limb musculature is useful in characterizing major groups. The astragalus and calcaneum (tibiale and fibulare) may be partially or completely fused, and an additional cartilaginous or bony element, the intercalary element, is present between the penultimate and ultimate phalanx of the digits of some taxa. The shape of the bony terminal phalanx is diagnostic for several groups (e.g., Hylidae).

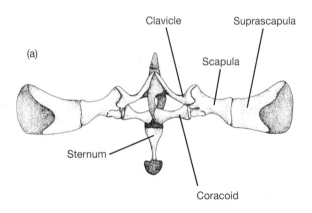

(a)

Clavicle   Suprascapula

Scapula

Sternum

Coracoid

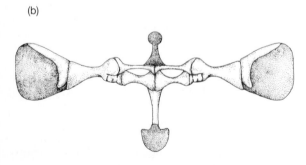

(b)

Figure 3-13   **Morphology of frog pectoral girdles.** (a) Ventral view of the arciferal pectoral girdle of *Leptodactylus bolivianus* (Leptodactylidae). (b) Ventral view of the firmisternal pectoral girdle of *Rana pipiens* (Ranidae). *(Source: Modified from Trueb 1973.)*

Larvae

The use of larval characters in frog systematics was pioneered by Orton (1953, 1957) and further ex-

Figure 3-14 **Four morphological types of anuran tadpoles.** For each tadpole type (Orton 1953, 1957), a ventral view of the oral disc and of the head-body are shown. The dashed line shows the path of water flow. (a) Type 1. (b) Type 2. (c) Type 3. (d) Type 4. *(Source: Modified from Orton 1953, 1957; Starrett 1973.)*

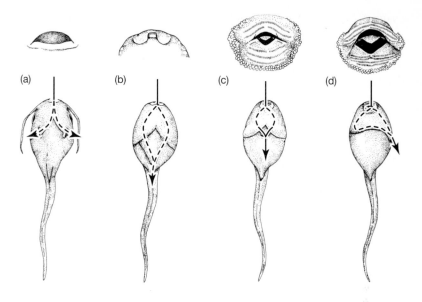

tended by Starrett (1973), Sokol (1975), and Wassersug (1980). Larval characters primarily used in anuran systematics include the structure of the mouthparts (presence and arrangement of papillae, horny jaw sheaths, and denticles), the structure of the branchial chambers, and the position of the spiracle, through which water leaves the chambers (Figure 3–14). Characters of the chondrocranium and the relative positions of cranial nerve ganglia are also used.

Orton's (1953, 1957) Type 1 larvae, characteristic of the pipid *Xenopus* and *Rhinophrynus*, are characterized by wide, slitlike mouths without keratinized mouthparts, and by paired spiracular openings. Type 2 larvae (Microhylidae) have a single midventral spiracle and more complex mouthparts than Type 1, but also lack keratinized mouthparts. Types 1 and 2 larvae are filter-feeding. Type 3 larvae have keratinized mouthparts and midventral spiracles. They are characteristic of *Ascaphus*, Bombinatoridae, and Discoglossidae. Type 4 larvae, characteristic of all other frogs, differ from Type 3 in having sinistral spiracles; that is, the spiracle exits on the left side of the body. Of course, the variety of developmental modes found in frogs means that larval stages may become highly modified. As a result, larval type is not informative about relationships in many groups. Current consen-

sus (Lynch 1973, Sokol 1975) favors the hypothesis that the reduced mouthparts of Types 1 and 2 are secondarily derived rather than ancestral.

Tadpoles differ not only in aspects of general morphology but also in aspects of developmental ecology (Altig and Johnston 1989). Some tadpoles have an extended period as free-swimming and feeding larvae, whereas the eggs of some frogs hatch into nonfeeding tadpoles that are supplied with abundant yolk. This type of tadpole generally has a short larval life and quickly metamorphoses into a small juvenile frog. Because they are nonfeeding, these tadpoles usually have very reduced denticles and jaw sheaths.

Amplexus

Three general classes of amplectic behavior can be recognized (Lynch 1973, Nussbaum 1980). In inguinal amplexus the male grasps the female around the inguinal region, whereas in axillary amplexus the female is grasped immediately behind her forelimbs. In cephalic amplexus the male is positioned far forward on the dorsum of the female and clasps her head. In all cases, amplexus is designed to bring the cloacae of males and females into close proximity so that fertilization can be easily achieved. However, in a variant of cephalic amplexus exhibited by some Malagasy ranids (Mantellinae), sperm are simply shed onto the female's dorsum and trickle over the eggs being shed from her cloaca (Blommers-Schlosser 1975). Inguinal amplexus is characteristic of most relatively primitive frogs such as *Ascaphus, Leiopelma,* Discoglossidae, Pipidae, Rhinophrynidae, Pelobatidae, and Pelodytidae, but also of some Neobatrachia. Axillary amplexus characterizes most Neobatrachia, but the occurrence of inguinal amplexus in some neobatrachians (e.g., some bufonids, Myobatrachidae, leptodactylids, *Brachycephalus, Heleophryne*) makes the direction of evolutionary change in this character within Neobatrachia difficult to ascertain (Ford and Cannatella 1993). Cephalic amplexus is known in some dendrobatids.

Content and Distribution

Approximately 310 genera and about 4,100 extant species of frogs are known, and many new species are described each year. For example, between 1985 and 1992, 484 species were recognized as new (including newly described as well as resurrected species; Duellman 1993). This amounts to approximately 70 species per year, most of which are new tropical species. Frogs are cosmopolitan except where limited by cold or xeric climates, and are absent from many oceanic islands. They are most diverse and are poorly known in tropical regions.

Fossil Record

Triadobatrachus massinoti (Figure 3–15) from the Lower Triassic of Madagascar (230 MaBP) is the earliest salientian (Rage and Rocek 1989). As indicated before, it lacks several derived features of anurans. Several Jurassic fossils are either anurans or pre-anuran salientians. The earliest of these is *Prosalirus bitis* (Early Jurassic, ca. 190 MaBP; Shubin and Jenkins 1995), which occurs in the same North American deposits as the earliest known caecilian. Two others occur later in the Jurassic (*Notobatrachus* and *Vieraella* from Argentina; Báez and Basso 1996), but fossils assigned to the extant families Discoglossidae and Pelobatidae also appear in the mid- to late Jurassic (Evans et al. 1990, Evans and Milner 1993). Fossil pipid tadpoles are known from the Lower Cretaceous (Estes et

al. 1978) and well-preserved pelobatid tadpoles are known from the Miocene (Wassersug and Wake 1995). In general, the fossil record of Anura is poor. It has been reviewed by Sanchiz and Rocek (1996), Rocek and Nessov (1993), Tyler (1991), and Báez (1996).

Systematics. Ford and Cannatella (1993), Hedges and Maxson (1993), Hillis et al. (1993).

Systematics and Phylogeny of Frogs

The monophyly of Anura is supported by many features, including nine or fewer presacral vertebrae, fused caudal vertebrae (urostyle), compound radioulna and tibiofibula, elongate tarsal elements, and peculiar larvae. Other features, such as fusion of the frontal and parietals, edentulous dentaries, and an elongate, anteriorly directed ilium, are found in Anura + *Triadobatrachus*, and are shared derived characters of Salientia.

Lynch (1973) and Ford and Cannatella (1993) reviewed hypotheses of the relationships of frogs (see also Inger 1967, Kluge and Farris 1969). The current understanding of frog phylogeny is summarized in Figure 3–16. Early splits within frogs are well-resolved relative to groups within the clade Neobatrachia, which includes the most species-rich families and by far the greatest diversity of extant frogs. All frogs other than Neobatrachia are conveniently referred to as archaeobatrachians, although such an assemblage may be nonmonophyletic.

Neobatrachia is a clade defined by several shared derived characters of skeletal structure (loss of parahyoid bones, fusion of the third distal carpals to other carpals) and limb muscles (separation of the sartorius and semitendinosus muscles of the hind limbs). Within Neobatrachia, several families compose a clade (Ranoidea) characterized by complete fusion of the epicoracoid cartilages. Ford and Cannatella (1993) identified no shared derived characters for those neobatrachians that are not part of Ranoidea. These other families are collectively referred to as "Bufonoidea," and it is given formal systematic status in much of the traditional literature. Lee and Jamieson (1992, 1993) pointed out that several "bufonoid" families (Myobatrachidae, Leptodactylidae, Hylidae, Bufonidae, Rhinodermatidae) share a derived character in the shape of the perfatorium of the

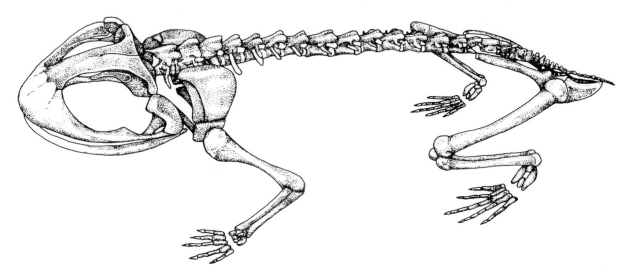

Figure 3-15   ***Triadobatrachus massinoti.*** *Triadobatrachus* (Lower Triassic, Madagascar) is the earliest known salientian and sister taxon to Anura. Total length is about 10 cm. *(Source: Modified from Rage and Rocek 1989.)*

sperm cells, that portion penetrating the egg envelope. If confirmed for the other bufonoid groups (*Allophryne*, Brachycephalidae, *Heleophryne*, Centrolenidae, Pseudidae), this character would corroborate the monophyly of "Bufonoidea" as the sister group to Ranoidea within Neobatrachia.

Several of the major clades recognized on the basis of morphological characters are also supported by molecular data (Pipanura, Mesobatrachia), but others (Ranoidea) are not (Hillis et al. 1993). In addition, "Bufonoidea" is supported as monophyletic by molecular sequence data but not by morphological characters. A general problem with our current understanding of frog phylogeny is that many clades are supported by few morphological characters, and molecular data sets sample very few taxa within higher groups, usually only one species per family. Although this situation will improve, statistical support for most phylogenies currently available is weak.

Ascaphidae

Ascaphidae, represented by a single extant species, *Ascaphus truei* (25 to 50 millimeters in length), is unique among frogs in having an intromittent organ (the tail, hence giving rise to the common name of this frog, the tailed frog). The tail is actually a highly vascularized extension of the cloaca that is supported by cartilaginous rods, the postpubis or Nobleian rods, that are attached to the ventral part of the pelvic girdle. Although these rods are also present in female *Ascaphus*, as well as in *Leiopelma* (Leiopelmatidae) and *Xenopus* (Pipidae), they do not support a tail in these frogs. During amplexus, which is inguinal in *Ascaphus*, the tail is bent forward by contraction of the rectus abdominis muscles and inserted into the cloaca of the female. There are nine presacral vertebrae (a character shared with *Leiopelma*; eight or fewer in other extant frogs), which have amphicoelous centra. The pectoral girdle is arciferal. The larvae are Type 3. *Ascaphus* is currently thought to be the sister taxon to all other frogs (Figure 3–16).

Tailed frogs inhabit cold, torrential streams and are highly aquatic, even as adults. Tympana are absent, the frogs apparently do not call, and amplexus occurs underwater. *Ascaphus* produces a few large eggs that develop very slowly and are laid under rocks in water (Brown 1989, Adams 1993). The tadpoles live

Figure 3-16 **Phylogenetic relation-ships among frog families.** Some other named clades are indicated. Quotation marks around a n̄ame indicate probable paraphyly of the named group. *(Source: Ford and Cannatella 1993.)*

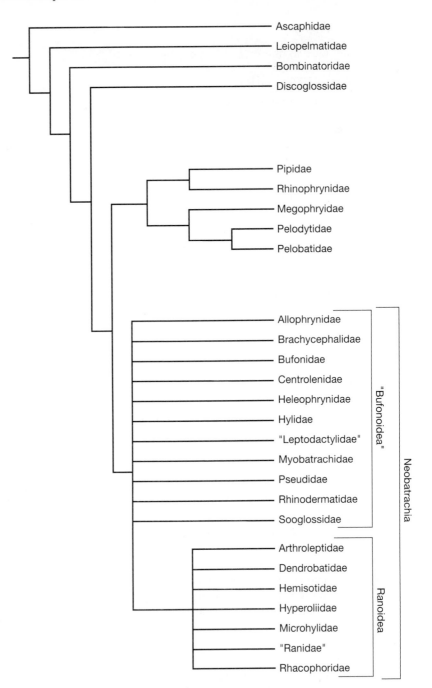

in fast-flowing water and have well-developed suctorial oral discs and reduced tail fins. They have been observed to climb out of streams in spray zones using their oral discs. Internal fertilization, absence of calls and external ears, and tadpole morphology are all probably adaptations for the highly turbulent aquatic environments where *Ascaphus* lives.

Content and Distribution. 1 species, *Ascaphus truei.* Northwestern United States and adjacent southern Canada (Figure 3–17).

Systematics. Metter (1968).

Leiopelmatidae

Leiopelma is unique among frogs in having ossified inscriptional ribs embedded in the ventral body musculature (Green and Cannatella 1993). *Leiopelma* lacks tympana and certain middle ear structures. The pectoral girdle is arciferal. A highly unusual system of heteromorphic sex chromosomes occurs in *Leiopelma*, and up to 15 supernumerary chromosomes are found in *L. hochstetteri* (Green 1988a,b).

The three extant species of *Leiopelma* are small, nocturnal frogs up to 50 millimeters in length (Bell 1978). Two species are terrestrial and the other inhabits streamside habitats. Amplexus is inguinal in two of the species and unknown in the third. Small clutches (20 to 70 eggs) are laid in moist terrestrial situations (Stephenson and Stephenson 1957). Development is direct and rapid, with the tadpoles developing neither jaw sheaths nor denticles (Bell 1978). Males of two species are known to attend clutches and to carry neonates on their backs after hatching.

Content and Distribution. 1 genus (*Leiopelma*), 3 species. New Zealand (Figure 3–18).

Systematics. Daugherty et al. (1982), Worthy (1987a,b), Green and Sharbel (1988).

Bombinatoridae

Bombina and *Barbourula* form a clade characterized by derived features of the skull and hyoid (Ford and Cannatella 1993). The pectoral girdle is arciferal and the presacral vertebrae are opisthocoelous. The tadpoles are Type 3. Species of *Bombina* are small and toadlike, whereas *Barbourula* is large and fully aquatic. Amplexus is inguinal. *Bombina* has bright orange or yellow aposematic coloration on the belly and toxic skin secretions (Bachmeyer et al. 1967). It adopts a defensive posture, the Unken reflex, to display these colors when threatened. *Bombina* lays single eggs in ponds, but the reproductive biology and tadpoles of *Barbourula* are unknown. Two species, *Bombina bombina* and *B. variegata*, have been the subject of extensive studies of hybrid zone dynamics (Szymura 1993, Nürnberger et al. 1995).

Content and Distribution. 2 genera: *Bombina* (6 species), *Barbourula* (2 species). Europe and Asia to China, Korea, and Vietnam (Figure 3–18). Often included in Discoglossidae.

Systematics. Lang (1988), Clarke (1987), Gollman et al. (1993).

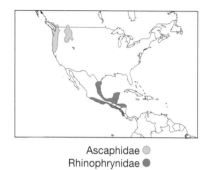

Ascaphidae ●
Rhinophrynidae ●

Figure 3-17   **Distribution of frog families Ascaphidae and Rhinophrynidae.**

Discoglossidae

Discoglossids have arciferal pectoral girdles and aquatic Type 3 larvae. The vertebrae are opisthocoelous and the pectoral girdles are arciferal. These are small to moderate-sized frogs (30 to 70 millimeters snout-vent length). Discoglossids have aquatic eggs and tadpoles. In *Alytes*, amplexus takes place on land and the male maneuvers the fertilized eggs onto his back and hind limbs. Here they are carried and moistened when necessary until near hatching, at which time they are deposited in water. *Discoglossus* lays its eggs directly in water.

Content and Distribution. 2 genera (*Alytes*, *Discoglossus*), 5 species. Western Europe to the Middle East; northwestern Africa (Figure 3–19). Molecular divergence in albumins between *Alytes* and *Discoglossus* suggests an extremely ancient separation of these genera (Maxson and Szymura 1984).

Systematics. Arntzen and Szymura (1984), Maxson and Szymura (1984), Busack (1986), Márquez and Bosch (1995).

Pelobatidae

Pelobatine pelobatids have a well-developed keratinous, spadelike metatarsal tubercle, internally supported by a well-ossified prehallux, on the hind feet. The vertebrae are procoelous and the pectoral girdle is arciferal. Amplexus is inguinal and the larvae are Type 4.

Many pelobatids have glandular, tuberculate skin, including enlarged parotoid glands on the dorsum. *Scaphiopus holbrooki* and several Asian megophryines

Figure 3-18 **Distribution of frog families Bombinatoridae, Leiopelmatidae, Hyperoliidae, Pelodytidae, Rhacophoridae, and Myobatrachidae.**

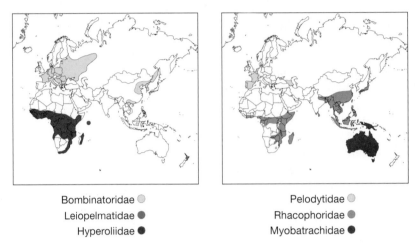

Bombinatoridae ○
Leiopelmatidae ●
Hyperoliidae ●

Pelodytidae ○
Rhacophoridae ●
Myobatrachidae ●

are unusual among frogs in having large pectoral glands similar in structure to the dorsal parotoid glands, but the significance of having these glands on the venter is unknown (Jacobs et al. 1985). Species of *Pelobates*, *Spea*, and *Scaphiopus*, commonly called spade-foot toads, are fossorial frogs that emerge infrequently except during heavy rains. They dig their own burrows or use burrows of other animals. *Spea* and *Scaphiopus* are explosive breeders that breed in ephemeral ponds. Some species of desert regions have exceedingly rapid development times, with as little as 8 days from egg laying to metamorphosis in *Scaphiopus couchi* (Newman 1992).

Content and Distribution. 10 genera, approximately 95 species. North America (*Spea, Scaphiopus*); western Eurasia and northwestern Africa (*Pelobates*); Pakistan and northern India to China, southeast Asia, Philippines, islands of the Sunda Shelf (megophryines) (Figure 3–22). 2 subfami-

lies: Pelobatinae: *Pelobates*, *Spea*, and *Scaphiopus;* Megophryinae: *Atympanophrys*, *Brachytarsophrys*, *Leptobrachella*, *Leptobrachium*, *Leptolalax*, *Megophrys*, and *Scutiger*. Megophryidae (Ford and Cannatella 1993) is included within Pelobatidae following Henrici (1994).

Systematics. Sage et al. (1982), Wiens and Titus (1991), Henrici (1994).

Pelodytidae

The vertebrae are procoelous and the pectoral girdle is arciferal. Amplexus is inguinal and the tadpoles are Type 4. *Pelodytes* is a nocturnal, terrestrial frog except during the breeding season, when it becomes conspicuously diurnal. Eggs are laid in ponds, usually attached to vegetation. Pelodytidae has traditionally been included as a subfamily within Pelobatidae, but the astragalus and calcaneum are entirely fused

Figure 3-19 **Distribution of frog families Discoglossidae, Hemisotidae, Sooglossidae, and Arthroleptidae.**

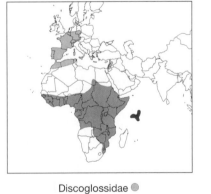

Discoglossidae ○
Hemisotidae ●
Sooglossidae ●

Arthroleptidae ○

in *Pelodytes*, unlike in any pelobatids (Ford and Cannatella 1993). Henrici (1994) concluded that *Pelodytes* was the sister taxon to Pelobatidae + Megophryidae.

Content and Distribution. 2 species: *Pelodytes punctatus* and *P. caucasicus*. Distribution disjunct: western Europe and the Caucasus Mountains of western Asia (Figure 3–18).

Systematics. Henrici (1994).

Rhinophrynidae

Rhinophrynus has opisthocoelous vertebrae and an arciferal pectoral girdle. This frog is highly modified for fossoriality (Trueb and Gans 1983). The Type 1 tadpoles are filter-feeders, lack jaw sheaths and denticles, and exhibit schooling behavior. The limbs are short and powerful, and a well-developed spade is present on the inner metatarsal tubercle and first toe. The body is robust; the head is pointed and has thickened, cornified skin at its tip. The eyes are very small and the tympanum is absent. The skull is reinforced and the pectoral girdle overlaps it. The entire feeding apparatus, including buccal, pharyngeal, esophageal, and hyoid structures, is highly modified to feed on ants and termites underground. Tongue function is unique among frogs and the tongue is protruded straight from a groove in the front of the mouth, rather than being flipped out of the mouth as in other frogs. *Rhinophrynus* inhabits subhumid areas and is highly fossorial, feeding under ground on ants and termites. Surface activity occurs only after heavy rains. Amplexus is inguinal, and eggs are laid in temporary pools.

Content and Distribution. 1 species, *Rhinophrynus dorsalis*. Extreme southern Texas to Costa Rica (Figure 3–17).

Systematics. Fouquette (1969).

Pipidae

Pipids are unique among frogs in lacking tongues. The pectoral girdle is a modified firmisternal type. The vertebrae are opisthocoelous. *Xenopus* and some species of *Pipa* are unusual in having nonpedicellate teeth on the maxillae and premaxillae. Amplexus is inguinal and the tadpoles are Type 1. The size range of pipids is 40 to 170 millimeters snout–vent length (Duellman and Trueb 1986). *Xenopus* is one of the most widely used animals in experimental and developmental biology, and a considerable amount is known concerning its genetics, development, molecular biology, and neurobiology

(Tinsley and Kobel 1996). Most species of *Xenopus* are polyploids (tetraploids to hexaploids). The fossil record of pipids is more extensive than for other frogs (Lower Cretaceous–Miocene of Israel, Africa, and South America; Báez 1996).

African pipids deposit eggs in water and the tadpoles are filter-feeders with the exception of tadpoles of *Hymenochirus*, which are carnivorous. Species of *Pipa*, *Hymenochirus*, and *Pseudhymenochirus* undergo an elaborate courtship ritual involving midwater or surface turnovers by amplectic pairs. In *Pipa* this results in the sticky fertilized eggs adhering to the back of the female, where they are pressed in by the male (Rabb and Rabb 1963a,b). Subsequently, each egg becomes enveloped by swelling of the skin on the dorsum of the female, and further development takes place within these depressions.

Some unusual features of pipids are associated with their totally aquatic existence (Trueb 1996). The bodies are dorsoventrally compressed and the limbs are splayed laterally. Lateral line systems are present except in *Hymenochirus* and *Pseudhymenochirus*. The feet are large and fully webbed. Hands bear long fingers and are unwebbed except in *Hymenochirus* and *Pseudhymenochirus*. In all genera except *Pipa*, the toes have distinct keratinous, clawlike tips (hence the name clawed frogs). Species of *Xenopus* call underwater and have highly modified ears and laryngeal apparatuses to accommodate the properties of sound production and transduction through water (Yager 1992a,b, Elepfandt 1996).

Content and Distribution. 5 genera (*Pipa*, *Silurana*, *Hymenochirus*, *Pseudhymenochirus*, *Xenopus*), approximately 30 species. Tropical South America (*Pipa*), sub-Saharan Africa (all other genera) (Figure 3–22).

Systematics. Trueb and Cannatella (1986), Carr et al. (1987), Cannatella and Trueb (1988a,b), de Sá and Hillis (1990), Cannatella and De Sá (1993), Graf (1996).

Allophrynidae

Allophryne ruthveni is a small frog, 20 to 30 millimeters in snout–vent length, and teeth are entirely lacking. The vertebrae are procoelous, the pectoral girdle is arciferal, and intercalary elements are present in the digits. The skull is unusual in having a strongly ossified cranial roof but reduced maxillae, pterygoids, and squamosals, and lacks palatines altogether. Little is known of the natural history of this rare frog. Males call from vegetation over ponds within rain forests, and eggs are deposited in the ponds (Duellman 1975).

Figure 3-20 **Diversity of frogs.** (a) *Bufo arenarum* (Bufonidae, South America). (b) *Atelopus spumarius* (Bufonidae, South America). (c) *Hyla vasta* (Hylidae, South America). (d) *Centrolene euknemos* (Centrolenidae, South America). (e) *Phyllobates terribilis* (Dendrobatidae, South America). (f) *Hemisus guttatum* (Hemisotidae, Africa) *(Photographs by (a, f) Martha Crump, (b) John Cadle, (c) Art Wolfe, (d) David Dennis/Tom Stack & Associates, (e) Charles W. Myers.)*

Its relationships within Neobatrachia are obscure (Figure 3–16).

Content and Distribution. 1 species, *Allophryne ruthveni*. Guianan region of South America (Figure 3–21).

Systematics. Hoogmoed (1969).

Brachycephalidae

Brachycephalids are tiny frogs (snout–vent lengths of less than 16 millimeters) and have undergone digital reduction (two digits on the hands, three on the feet). Bony dermal ossifications occur dorsal to the procoelous vertebrae in *Brachycephalus*. The pectoral girdle is firmisternal and lacks a sternum. Amplexus is initially inguinal during courtship, later shifting to a more axillary position as eggs are laid (Pombal et al. 1994). Eggs are laid terrestrially and direct development occurs in *Brachycephalus ephippium* and probably in the other species. After eggs are deposited, females coat them with soil particles, which may serve both as camouflage and to prevent desiccation.

Brachycephalus ephippium is a diurnal inhabitant of leaf litter of humid forests. It is bright orange and has skin glands that secrete tetrodotoxin. The other two species of brachycephalids are cryptically colored. *Psyllophryne didactyla* is one of the smallest tetrapods known, approximately 10 in millimeters in snout–vent length.

Content and Distribution. 2 genera (*Brachycephalus*, *Psyllophryne*), 3 species. Southeastern Brazil (Figure 3–21).

Systematics. Izecksohn (1971), McDiarmid (1971).

Bufonidae

Bufonids are unique among anurans in having a Bidder's organ, a rudimentary ovary that develops on the anterior end of the larval testes of males (Roessler et al. 1990). The persistence of Bidder's organs in many adult bufonids is considered a paedomorphic trait. Teeth are also entirely absent, a rare condition among frogs. Presacral vertebrae are procoelous. The pectoral girdle is arciferal or a modified firmisternal type. Aquatic tadpoles are Type 4. The skulls are heavily ossified and usually co-ossified with the overlying skin.

Prominent cutaneous glands, such as the conspicuous parotoid glands located on the posterodorsal portion of the head, are characteristic of many species of bufonids (Figure 3–20). Species of the diverse Neotropical genus *Atelopus* have bright aposematic colors and potent skin toxins. The skin toxins of most bufonids are primarily peptides, but tetrodotoxin, a water-soluble alkaloid, is characteristic of *Atelopus*. Lipophilic alkaloids similar to those in dendrobatids have been found in the bufonid *Melanophryniscus*.

Most species of bufonids are terrestrial, but some (e.g., *Ansonia*) are semiaquatic stream frogs and a few (*Pedostibes* from Southeast Asia) are arboreal. Bufonids use axillary amplexus, and most species deposit strings of eggs in ponds or streams that hatch into Type 4 larvae. However, some bufonids, such as the Philippine *Pelophryne*, deposit eggs in leaf axils several meters aboveground. The tadpoles of *Stephopaedes* and *Mertensophryne* have an unusual fleshy crown completely encircling the eyes and nostrils (Channing 1978, Grandison 1980). The crown probably facilitates respiration at the surface film in their arboreal sites. The tadpoles of *Ansonia*, *Atelopus*, and *Bufo veraguensis* occur in torrential streams and have well-developed suckers on the belly, which they use to attach themselves to the substrate. This modification has evolved convergently in other tadpoles occurring in torrents, for example in the ranid *Staurois*.

Reproductive modes within bufonids span the range observed within frogs as a whole. Within a clade of ten African bufonid genera reproductive modes vary from oviparity with free-swimming larvae to direct terrestrial development and viviparity (*Nectophrynoides*, *Nimbaphrynoides*) (M. H. Wake 1993, Graybeal and Cannatella 1995).

Content and Distribution. 33 genera, approximately 380 species. Cosmopolitan in temperate and tropical regions except east of Wallace's Line (Australopapuan region), Madagascar, and oceanic islands (Figure 3–22). One species, *Bufo marinus*, has been widely introduced into many areas, including Florida, Australia, New Guinea, and many islands, where it has become a serious pest.

Systematics. McDiarmid (1971), Blair (1972), Graybeal (1993), Maxson (1984), Graybeal and Cannatella (1995), Hass et al. (1995).

Heleophrynidae

Heleophryne has amphicoelous vertebrae and an arciferal pectoral girdle. These frogs live along torrential mountain streams. They have fully webbed feet and expanded toe discs. About 100 to 200 eggs are laid under rocks in shallow pools (Branch 1991). Amplexus is inguinal. The tadpoles (Type 4) are unique in having a large suctorial oral disc that has many rows of denticles but lacks jaw

Allophrynidae ○
Brachycephalidae ●
Rhinodermatidae ●

"Leptodactylidae" ○

Pseudidae ○

Centrolenidae ○

Dendrobatidae ○

Figure 3-21   **Distributions of frog families Allophrynidae, Brachycephalidae, Rhinodermatidae, "Leptodactylidae," Pseudidae, Centrolenidae, and Dendrobatidae.**

sheaths. Metamorphosis occurs in 1 to 2 years. Two of the five species are threatened by urban development.

Content and Distribution. 1 genus (*Heleophryne*), 5 species. High mountains of extreme southern South Africa (Figure 3–22). The relationships of *Heleophryne* are uncertain.

Systematics. Poynton (1964), Boycott (1988).

"Leptodactylidae"

The pectoral girdle is arciferal or of a modified firmisternal condition. Vertebrae are procoelous. Amplexus is axillary and larvae are of Type 4. Leptodactylids vary considerably in morphology, habits, and life history. The size range is large, from small *Eleutherodactylus iberia*, only 10 millimeters in snout–vent length, to *Tel-*

matobius culeus, up to about 250 millimeters. Reproduction and life histories are equally varied in leptodactylids. Many telmatobiines have aquatic eggs and larvae, whereas most leptodactylines construct foam nests and have aquatic tadpoles. Others lay eggs on the ground and may have truncated free larval stages. Several genera, including more than 500 species of *Eleutherodactylus*, do not lay their eggs in water and have direct development. Two species of *Eleutherodactylus*, *E. coqui* and *E. jasperi*, and possibly many more, have internal fertilization. *Eleutherodactylus jasperi* is known to be viviparous, with three to five young per brood that are maintained in the fused posterior portions of the oviducts (M. H. Wake 1993).

Telmatobius culeus, an aquatic leptodactylid inhabiting Lake Titicaca on the border of Peru and Bolivia at

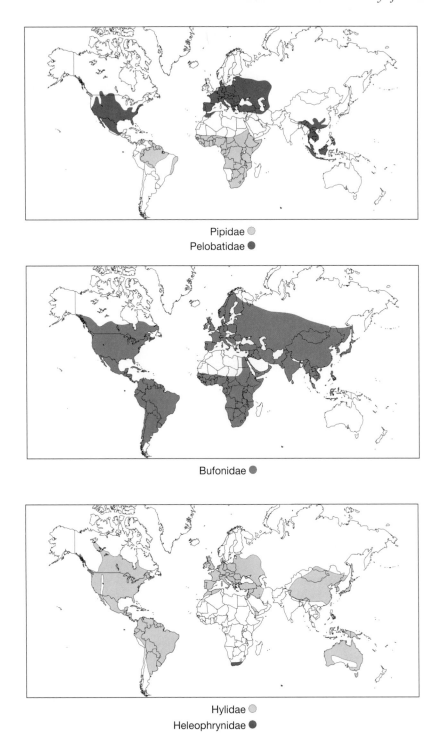

Pipidae ○
Pelobatidae ●

Bufonidae ●

Hylidae ○
Heleophrynidae ●

Figure 3-22 **Distribution of the frog family Pipidae, Pelobatidae, Bufonidae, Hylidae, Heleophrynidae, Microhylidae, and "Ranidae."** *(continued)*

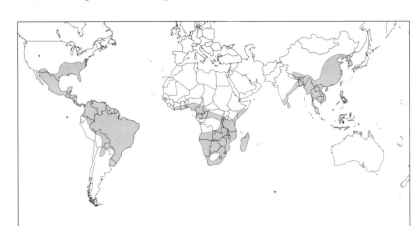

Microhylidae ⬤

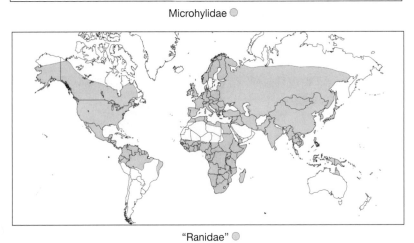

"Ranidae" ⬤

Figure 3-22 *(continued)* **Distribution of the frog family Pipidae, Pelobatidae, Bufonidae, Hylidae, Heleophrynidae, Microhylidae, and "Ranidae."**

nearly 4,000 meters in the Andes, is among the largest frogs in the world. It has many morphological and physiological specializations for an aquatic existence. These frogs live in deep, cold lake waters. Webbing is extensive, as it is in most fully aquatic frogs, and the body is equipped with large, baggy dermal flaps similar to those seen in the urodele *Cryptobranchus* (see Figure 6–1). These flaps increase the surface area for integumentary respiration. As in *Cryptobranchus*, capillaries penetrate to the epidermis in *T. culeus* and the lungs are reduced. These two species also exhibit the same swaying motion to break the water–skin boundary layer (Hutchison et al. 1976).

In contrast to *Telmatobius culeus*, species of *Ceratophrys* are large, nocturnal, terrestrial predators of lowland forests in South America (see Figure 1–3). These frogs are cryptic sit-and-wait predators and consume a wide variety of organisms, but are unusual

in that more than 50 percent of the prey by volume are vertebrates (frogs, snakes, lizards, and rodents) (Duellman and Lizana 1994). Most leptodactylids are small insectivores that inhabit leaf litter of the forest floor, but other habitats include rock crevices, riparian situations, and arboreal sites.

Content and Distribution. 49 genera, 890+ species. Probably nonmonophyletic. Extreme south-central United States; nearly all of Mexico, Central America, and South America (Figure 3–21). 4 subfamilies (Lynch 1971, Duellman and Trueb 1986): Telmatobiinae, the largest and most species-rich group (including *Eleutherodactylus*, *Telmatobius*, and 30 other genera); Ceratophryinae (*Ceratophrys*, *Chacophrys*, *Lepidobatrachus*); Hylodinae (*Crossodactylus*, *Hylodes*, *Megaelosia*); and Leptodactylinae (11 genera of frogs that construct foam nests, including *Leptodactylus*, *Edalorhina*, *Adenomera*, and *Physalaemus*). *Eleutherodactylus*, with well over 500 recognized species, is the most specise-rich genus of vertebrates but will probably be partitioned once the relationships of leptodactylids are better understood. Molecular studies indicate extremely ancient divergences within Leptodactylidae.

Systematics. Heyer (1975, 1994), Lynch (1971, 1978, 1982, 1986, 1993), Heyer and Maxson (1983), Maxson and Heyer (1982, 1988), Wiens (1993).

Myobatrachidae

The pectoral girdle is arciferal. Intercalary elements are absent and digital discs are usually small or absent. The notochord persists in adults. Amplexus is inguinal. Free-swimming and feeding tadpoles are of Type 4, but many myobatrachids have direct development or nonfeeding larvae.

Myobatrachids are extremely varied in size (20 millimeters to greater than 110 millimeters in snout–vent length) and life history. Most species are terrestrial, but some are fossorial, and species of *Taudactylus* and *Rheobatrachus* live along or in torrential mountain streams. Eggs of many species are deposited in water and have typical aquatic tadpoles. However, eggs may be laid on land and undergo either direct development or have aquatic tadpoles, which may be feeding or nonfeeding. Foam nests are constructed by some limnodynastines.

Two unusual forms of egg brooding occur in myobatrachids. Eggs and tadpoles are brooded in a pair of inguinal pouches in males of *Assa darlingtoni* (Ingram et al. 1975). The eggs are laid on the ground and attended by the male. At hatching the tadpoles wriggle up into the pouches, where development through metamorphosis takes place. Two remarkable frogs in the genus *Rheobatrachus* have unique adaptations for brooding their eggs in the stomach (Tyler 1983). *Rheobatrachus* is an aquatic frog that inhabits mountain streams of southern Queensland, Australia. After egg laying and external fertilization, the eggs are swallowed by the female parent, and development and metamorphosis occur in the stomach. The relationships of *Rheobatrachus* within Myobatrachidae are unclear. Discovered only in 1973, no specimens of *Rheobatrachus* have been seen in the field since 1982. They may be extinct, perhaps among the first victims of an epidemic disease spreading among Australian anurans during the early 1980s (Laurance et al. 1996).

Lee and Jamieson (1992) pointed out that all myobatrachids share two characters of sperm ultrastructure, thus corroborating the monophyly of the group. They also suggested that Myobatrachidae was the sister group to other bufonoids (Figure 3–17), whereas previously Myobatrachidae was considered a clade

within "Leptodactylidae." Microcomplement fixation data on albumins (Maxson 1992) show that the divergence between the Limnodynastinae and Myobatrachinae, and some generic divergences within these groups, are extremely ancient (Cretaceous).

Content and Distribution. 21 genera, approximately 110 species. Australia, New Guinea, and Tasmania (Figure 3–18).

Systematics. Blake (1973), Heyer and Liem (1976), and Farris et al. (1982).

Rhinodermatidae

Rhinoderma has procoelous vertebrae. The pectoral girdle is arciferal, but the epicoracoids are completely ossified. Rhinodermatids are small frogs, up to 30 millimeters in body length, that are characterized by a fleshy proboscis at the tip of the snout (hence the name of the family and genus). They inhabit riparian habitats along cold streams. Amplexus is axillary. Eggs are laid on land and attended by males. Development within the eggs is rapid, and the tadpoles (Type 4) are picked up by the males and either carried to pools of water and released (*Rhinoderma rufum*) or carried in the vocal pouches of males through their entire development (*Rhinoderma darwini*).

Content and Distribution. 2 species: *Rhinoderma darwini* and *R. rufum*. Wet temperate southern beech (*Nothofagus*) forests of southern Argentina and Chile (Figure 3–21).

Systematics. Cei (1962, 1980).

Hylidae

Hylids are characterized by claw-shaped terminal phalanges (also found in some hyperoliids). Intercalary elements are present in the digits. The pectoral girdle is arciferal, and the vertebrae are procoelous. Free-swimming tadpoles are of Type 4.

Most hylids are arboreal and have well-developed toe discs (Figure 3–20). However, some are fossorial (*Pternohyla* and *Cyclorana*) or live in tree holes (*Triprion*). Like some other fossorial frogs, fossorial hylids often live in rather arid areas and form cocoons to protect themselves from desiccation during unfavorable periods. Unusual modifications of skull morphology, including extensive co-ossification and casquing, are characteristic of some hylids. These modifications are often, but not invariably, observed in frogs inhabiting subhumid to arid environments.

Most hylids are strong jumpers, but members of the Phyllomedusinae are primarily arboreal walkers and make slow, deliberate movements along tree limbs using an opposable first digit in the hands and feet.

Although extensive glandular development is not generally characteristic of hylids, many phyllomedusines have conspicuous glandular swellings on the dorsum. These glands produce peptides that, in one instance, are used in hunting magic rituals by aboriginal populations (Daly et al. 1992). The Indians introduce secretions of *Phyllomedusa bicolor* into burns in their skin, thereby inducing violent physiological reactions, including vomiting, incontinence, and rapid heart rate, followed by euphoria, which is thought to improve their hunting skills. Several species of *Phyllomedusa* inhabiting arid areas of southern South America are unique among amphibians in having lipid glands in the skin, the secretions of which the frogs methodically spread over their skin with wiping behavior to prevent desiccation (Figure 5–4). Some canopy-dwelling hylids (e.g., some *Phyllomedusa bicolor* and some *Gastrotheca*) have osteoderms in the dorsal skin, which may reduce evaporative water loss.

Reproductive modes are diverse in hylids. Amplexus is axillary. Most hylines typically lay eggs in water and have aquatic feeding tadpoles. Phyllomedusines are largely arboreal and lay their eggs on vegetation over pools or streams, and the tadpoles complete their development in water after hatching.

Reproductive specializations are particularly well-developed in hemiphractines (Del Pino 1989, Weygoldt et al. 1991). In this group, eggs, tadpoles, or young frogs, depending on the species, are carried on the backs of the females, either exposed on the dorsum (*Cryptobatrachus*, *Hemiphractus*, *Stefania*), or within a special dorsal pouch (*Flectonotus*, *Fritziana*, *Gastrotheca*) that may be relatively open or completely closed except for a small aperture. Pouch development is induced by estrogens at sexual maturity in females and then results in permanent structures. After fertilization the eggs are maneuvered onto the back or into the pouch by the female, using her hind limbs. The pouches are highly vascularized, and intimate contact between the embryonic and maternal circulation is established via the pouch lining and peculiar bell-shaped embryonic gills. Development within the pouch of some *Gastrotheca* takes several months, but development in other hemiphractines is much more rapid. Direct development occurs in *Cryptobatrachus*, *Stefania*, *Hemiphractus*, and most *Gastrotheca*. *Flectono-*

(a)

(b)

Figure 3-23   **Diversity of frogs.** (a) *Platypelis grandis* (Microhylidae, Madagascar). (b) *Plethodontohyla inguinalis* (Microhylidae, Madagascar). (c) *Rana palmipes* ("Ranidae," South America). (d) *Mantella baroni* ("Ranidae," Madagascar). (e) *Boophis guibei* (Rhacophoridae, Madagascar). (f) *Heterixalus alboguttatus* (Hyperoliidae, Madagascar). *(Photographs by (a, b, d, e, f) John Cadle, (c) Martha Crump.)*

tus and *Fritziana* produce advanced, nonfeeding tadpoles that complete their development in water, and some species of *Gastrotheca* produce tadpoles that have a lengthy period of development in water after being brooded in the pouch. Phylogenetic studies (Duellman et al. 1988) suggest that direct development is the ancestral mode within hemiphractines,

Figure 3-23   *(continued)*

and that free-living feeding tadpoles have evolved multiple times within the group.

Content and Distribution. 38 genera, approximately 740 species. North, Central, and South America; West Indies; Australopapuan region. A single genus, *Hyla*, ranges throughout southern temperate Eurasia, Japan, and extreme northern Africa (*Hyla* is also found throughout the Americas) (Figure 3–22). Most of the diversity is in the Americas, with more than 500 species. A secondary center with ap-proximately 150 species is in Australia and New Guinea. 4 subfamilies: Pelodryadinae (Australopapuan region), Phyllomedusinae (tropical Central and South America), Hemiphractinae (Panama and South America), and Hylinae (North, Central, and South America; Eurasia and Africa).

Systematics. Maxson et al. (1975), Maxson and Wilson (1975), Maxson (1976), Duellman et al. (1988), Hedges (1986), Cocroft (1994). The numerous morphological studies are reviewed by Duellman and Trueb (1986).

Pseudidae

Pseudids are characterized by elongate ossified intercalary elements. The pectoral girdle is arciferal and the vertebrae are procoelous. Pseudids are aquatic frogs, attaining lengths of up to 70 millimeters in snout–vent length in *Pseudis paradoxa*, but species of *Lysapsus* are much smaller (20 to 40 millimeters). They have fully webbed feet and robust hind limbs. Amplexus is axillary, and the tadpoles are Type 4. Eggs are laid among aquatic vegetation in shallow water. The tadpoles of *Pseudis paradoxa* are the largest of any frog, reaching 250 millimeters in length (Emerson 1988a). They metamorphose into relatively small juveniles.

Content and Distribution. 2 genera (*Lysapsus* and *Pseudis*), 4 species. South America east of the Andes; Magdalena River valley in Colombia (Figure 3–21).

Systematics. Gallardo (1961).

Centrolenidae

The presence of a medial process on the third metacarpal is diagnostic for Centrolenidae. The astragalus and calcaneum are fused and the terminal phalanges are T-shaped. The pectoral girdle is arciferal. Vertebrae are procoelous. Intercalary cartilages and expanded toe discs are present.

Most centrolenids are small (<30 mm), but some *Centrolene* reach nearly 80 millimeters in snout–vent length. In small species the skin on the venter is often transparent, giving rise to the common name glass frogs. Amplexus is axillary. Small clutches of eggs are attached to vegetation or rocks above flowing water and are frequently, perhaps universally, attended by the males (e.g., Lynch et al. 1983, Villa 1984). At hatching, the tadpoles drop into the water below. The tadpoles (Type 4) are elongate and burrow into mud, gravel, or detritus on the bottom of streams. Tadpoles living in oxygen-poor substrates are often bright red as a result of blood flowing close to the surface of their unpigmented skin. This is probably a respiratory adaptation for these environments (Villa and Valerio 1982). Males of *Centrolene* have a prominent bony process on the humerus that is used in intraspecific aggressive interactions. Centrolenids are most diverse in wet mountain forests, and tremendous numbers of new species are being discovered in the Andes.

Content and Distribution. 3 genera (*Centrolene, Cochranella, Hyalinobatrachium*), 120+ species. Southern Mexico to Bolivia and Argentina, and in southeastern Brazil (Figure 3–21).

Systematics. Ruiz-Carranza and Lynch (1991).

Microhylidae

Frogs in the family Microhylidae are characterized by two or three palatal folds in adults, and a suite of larval features such as the absence of cornified denticles, imperforate nares, and a glottis fully exposed on the floor of the mouth (Ford and Cannatella 1993). Vertebrae are procoelous and the pectoral girdles are firmisternal. Amplexus is axillary, but in the African genus *Breviceps* the bodies are so rotund and the limbs so short that the male cannot grasp the female. Instead, he becomes glued to her back via secretions of glands on his ventral surface. Free-swimming tadpoles are Type 2 except in scaphiophrynines, which have a mosaic of characters of Types 2 and 4.

Most microhylids are relatively small frogs, and some are tiny (e.g., species of *Stumpffia* from Madagascar that are 10 to 15 millimeters in snout–vent length as adults). Others may be quite large (e.g., males of *Plethodontohyla inguinalis* may reach 100 millimeters in snout–vent length). Microhylids may be fossorial, terrestrial, or arboreal, and body forms are highly variable (Figure 3–23). For example, expanded toe discs are present in arboreal species, and fossorial species often have depressed bodies and pointed snouts. Microhylids live in habitats ranging from arid deserts to extremely wet rain forests.

The life histories of microhylids are extremely varied. Most microhylines and phrynomerines lay eggs in ponds and have free-swimming, feeding tadpoles. Many scaphiophrynines (e.g., *Platypelis, Plethodontohyla*) lay eggs in tree holes, and the males of some, such as *Plethodontohyla*, are known to guard the clutches. Other scaphiophrynines (*Paradoxophyla* and *Scaphiophryne*) are explosive breeders that lay eggs in ponds. Direct development occurs in asterophrynines, genyophrynines, and some microhylines. *Breviceps* and *Hoplophryne* construct nests within which nonfeeding tadpoles develop. The tadpole of *Otophryne robusta* (northern South America; Microhylinae) is highly unusual in having a sinistral spiracle at the tip of a long siphon projecting from the body wall. These tadpoles burrow into the substrate at the bottom of streams (Wassersug and Pyburn 1987).

Two species of microhylids (*Chiasmocleis ventrimaculatus* and *Gastrophryne olivacea*) form facultative commensal associations with large theraphosid spiders, using the same burrows and foraging areas as the spiders (Cocroft and Hambler 1989). The frogs are apparently recognized by the spiders through chemosensory cues and are not attacked, even though other frogs within the foraging territories are readily eaten by the spiders, as are small vertebrates that prey on frogs, such as snakes.

Content and Distribution. 65 genera, approximately 315 species. Eastern United States throughout Central and South America; sub-Saharan Africa; India and southeast Asia through the Philippines and Sunda Shelf; New Guinea and extreme northern Australia (Figure 3–22). Major radiations, accounting for most of the diversity of microhylids, have occurred in New Guinea and Madagascar. 9 subfamilies: Scaphiophryninae and Cophylinae (Madagascar); Asterophryinae and Genyophryninae (Australopapuan region and Philippines); Brevicipitinae and Phrynomerinae (sub-Saharan Africa); Microhylinae (New World, southeast Asia); Dyscophinae (Madagascar, southeast Asia); Melanobatrachinae (Tanzania and southern India). Scaphiophryninae is the apparent sister taxon to the remaining subfamilies, which are united by derived features of the tadpoles (Type 2 of Orton).

Systematics. Parker (1934), Zweifel (1962, 1972, 1986), Donnelly et al. (1990), Wild (1995).

Dendrobatidae

The pectoral girdle is firmisternal. The fingers bear a pair of dermal scutes on the dorsal surfaces, a character that otherwise appears only in some leptodactylids and myobatrachids. The vertebrae are procoelous.

Amplexus is cephalic where known in dendrobatids, but many species do not amplex. *Aromobates* is nocturnal and seems to be fully aquatic, so it is likely that egg laying takes place in water. However, all other dendrobatids are diurnal and terrestrial, and small clutches of eggs are deposited in terrestrial or arboreal locations and attended by a parent. The tadpoles (Type 4) adhere to the backs of the parent and are carried for a variable period of time before being deposited in water. Females of some species of *Dendrobates* deposit their tadpoles individually in arboreal sites, returning occasionally to deposit unfertilized eggs, which serve as a food source for the tadpoles.

Dendrobatids are famous in the popular literature as dart poison frogs because the toxic skin secretions of *Phyllobates* are used to poison blowgun darts by several South American Indian groups. Indeed, their skin contains some of the most potent naturally occurring alkaloids, which act irreversibly on synaptic and neuromuscular junctions. The alkaloids of dendrobatids are lipophilic (soluble in lipids), as contrasted with water-soluble alkaloids such as tetrodotoxin found in the skin of some other amphibians (e.g., salamandrids). The species of dendrobatids that harbor these compounds have bright aposematic colorations (Figure 3–20). The most toxic frog known is *Phyllobates terribilis* of northwestern Colombia, which can contain nearly 2 milligrams of batrachotoxins per frog, whereas only 0.2 milligram entering directly into the bloodstream constitutes a lethal human dose. During initial fieldwork when the frog was described, researchers used rubber gloves for handling the frogs, and several chickens and dogs died when they ingested refuse from the fieldwork.

Not all dendrobatids are highly toxic, however. Ancestrally within Dendrobatidae, skin alkaloids and aposematic coloration are absent (*Aromobates, Colostethus*). Both characters are present in *Epipedobates, Minyobates, Phyllobates,* and *Dendrobates*. Outside the clade of aposematic dendrobatids, lipophilic alkaloids and associated aposematic coloration have evolved convergently only in Malagasy ranid frogs of the genus *Mantella* (Daly et al. 1996). One myobatrachid (*Pseudophryne*) and bufonid (*Melanophryniscus*) have lipophilic skin alkaloids but are not aposematically colored. Recent experiments and circumstantial evidence point to a dietary origin for some, perhaps most, dendrobatid alkaloids, with certain beetles, ants, and perhaps millipedes being likely sources (Daly, Garraffo, et al. 1994, Daly, Secunda, et al. 1994, Daly 1995, Caldwell 1996).

A remarkable species of dendrobatid, *Aromobates nocturnus*, discovered in 1981 in cloud forests of northern Venezuela, is thought to be the sister group to all other dendrobatids (Myers et al. 1991). It is unusual in being far larger than any other species of dendrobatid (62 millimeters in body length, 25 to 40 percent larger than other dendrobatids) and is nocturnal and fully aquatic, unlike all other species in the family. Although having a foul odor, which gives the frog its generic name, *Aromobates* is not highly poisonous.

The relationship of dendrobatids within Neobatrachia has been especially controversial (Ford 1993), and relationships to either "Leptodactylidae" or "Ranidae" have been suggested. A recent study of tadpole morphology (Haas 1995) indicated that dendrobatid larvae do not share unequivocal derived characters with Ranoidea, but do with several "bufonoid" families. Ribosomal DNA sequence data generally placed Dendrobatidae with "bufonoid" families as well (Hillis et al. 1993).

Content and Distribution. 6 genera (*Aromobates, Colostethus, Dendrobates, Epipedobates, Minyobates,* and *Phyllobates*), about 175 species. Nicaragua to southeastern Brazil (Figure 3–21). The highest diversity is in northwestern South America (Colombia, Ecuador).

Systematics. Myers and Daly (1976), Maxson and Myers (1985), Myers (1987), Myers et al. (1991). The skin toxins of dendrobatids are reviewed by Daly et al. (1987) and Daly (1995).

Hemisotidae

Shared derived characters of Hemisotidae include the absence of a sternum (also observed in *Rhinophrynus* and Brachycephalidae) and a skull highly modified for headfirst burrowing. The pectoral girdle is firmisternal and the vertebrae are procoelous. Species of *Hemisus* are smooth-skinned burrowers of savanna regions. The heads are small, pointed, and posteriorly delimited by a transverse skin fold. These are unlike nearly all other frogs in that they use the snout to dig headfirst into soil (Emerson 1976). Amplexus is axillary. *Hemisus* lays small clutches of large eggs in a cavity dug into the side of temporary pools formed during seasonal rains. Females attend the clutches and may assist in carrying the tadpoles (Type 4) to open water, or in digging a passageway to water from the nest chamber (Branch 1991).

Content and Distribution. 1 genus (*Hemisus*), 8 species. Savanna regions of sub-Saharan Africa (Figure 3–19). *Hemisus* is often placed within its own subfamily of "Ranidae," but shares some features with some Hyperoliidae and "Ranidae" (Ford and Cannatella 1993).

Systematics. Laurent (1972).

Arthroleptidae

The vertebrae are procoelous in arthroleptids and the pectoral girdles are firmisternal. Amplexus is axillary. *Arthroleptis* is a terrestrial inhabitant of African rain forests. Small clutches of large eggs are laid in leaf litter or in nest chambers in the ground, and development is direct (Branch 1991). The astylosternine *Trichobatrachus robustus* lays eggs in streams. During the breeding season, males develop long, hairlike, highly vascularized projections on the posterior flanks and thighs, giving them a shaggy appearance (see Figure 6–2). They are commonly called hairy frogs. Males sit on their clutches of eggs and the hairs are thought to aid in cutaneous respiration, thereby allowing the frogs to remain submerged for longer periods while attending their eggs (Noble 1925, Perret 1966).

Content and Distribution. 7 genera, approximately 75 species. Sub-Saharan Africa (Figure 3–39). 2 subfamilies: Arthroleptinae (*Arthroleptis, Cardioglossa*) and Astylosterninae (*Astylosternus, Leptodactylon, Nyctibates, Scotobleps, Trichobatrachus*). Arthroleptidae is of uncertain monophyly, and its relationships within Ranoidea are unclear (Figure 3–19).

Systematics. Poynton and Broadley (1985).

Sooglossidae

In sooglossids the pectoral girdles are arciferal, but with partial fusion of the epicoracoid cartilages. The vertebrae are amphicoelous. These are small terrestrial frogs (to 40 millimeters in snout–vent length) that inhabit moss forests. *Sooglossus gardineri* is one of the smallest frogs, reaching only 10 to 12 millimeters in snout–vent length as adults. Amplexus is inguinal and eggs are laid terrestrially. Direct development occurs in one species of *Sooglossus*, whereas in the other the eggs hatch into nonfeeding tadpoles and are carried on the back of the female. The relationships of sooglossids have been controversial. Inguinal amplexus is a relatively primitive characteristic shared with many other "bufonoids," but the karyotypes of sooglossids are most similar to those of myobatrachids

Content and Distribution. 2 genera (*Nesomantis, Sooglossus*), 3 species. Seychelles Islands (Figure 3–19).

Systematics. (Nussbaum 1979a, 1980).

"Ranidae"

No unequivocal shared derived characters of "Ranidae" are known. The pectoral girdle is firmisternal and the vertebrae are procoelous. Amplexus is axillary in most taxa. Most species lay eggs in water and have feeding tadpoles (Type 4), but some African petropedetines (*Anhydrophryne*) and Asian ranines (e.g., *Ceratobatrachus, Platymantis*) have direct development. The tadpoles of some species (*Amolops*) live in swift mountain streams and have suctorial discs on the belly, as seen in many tadpoles living in similar environments (see discussion under Bufonidae).

Size and body form are extremely varied in ranids. Although many species are small, the family includes the largest known frog, the west African *Conraua goliath* (300 millimeters in snout–vent length). Some species have robust bodies and are somewhat fossorial (*Pyxicephalus*, *Tomopterna*), whereas others are strong swimmers that inhabit banks of mountain streams (e.g., *Amolops*, *Staurois*). Frogs of the genus *Mantella* (Madagascar) have convergently evolved bright aposematic colorations and lipophilic skin alkaloids like those of Neotropical dendrobatid frogs.

One European ranid, *Rana esculenta*, is a hybridogenetic frog produced by hybridization between *R. lessonae* and *R. ridibunda*. These frogs are unusual in that both sexes are represented in most populations (compare the unisexual parthenogenetic ambystomatid salamanders and *Cnemidophorus* lizards). Moreover, some populations of the *Rana esculenta* complex undergo a modified form of cyclical parthenogenesis, in which sexual generations alternate with asexual ones, similar to processes in some invertebrates such as rotifers and aphids (Graf and Polls Pelaz 1989, Schmidt 1993).

Very unusual mating behavior with minimal contact between partners takes place in the Malagasy mantellines, genus *Mantidactylus* (Blommers-Schlosser 1975). Many *Mantidactylus* attach their eggs to leaves above water. During mating the male places his thighs directly over the head and shoulders of the female, who immediately begins shedding eggs without receiving the assistance of the male in expelling them from her body. Sperm are shed onto the female's back and trickle over the extruded eggs, thus effecting fertilization. Males may leave the female before the completion of egg laying. The stimulus that induces egg laying in *Mantidactylus* is unknown but may be mediated through secretions of the large femoral glands present on the ventral surface of the thighs in males of these ranids. The mating posture would bring these glands in direct contact with the head and body of the female.

Content and Distribution. 46 genera, 700+ species. Cosmopolitan except extreme southern South America; West Indies; most of Australia; ranids do not occur on most oceanic islands (Figure 3–22). Relationships among major groups of ranids are poorly understood.

Systematics. Clarke (1981), Dubois (1981).

Hyperoliidae

The pectoral girdle is firmisternal and intercalary elements are present in the digits. The vertebrae are procoelous. These are small to medium-sized frogs (15 to 80 millimeters in snout–vent lengths), many of which are arboreal, have toe discs, and are brightly colored. However, *Kassina* and a few other species are terrestrial. Amplexus is axillary in most species. Many hyperoliids deposit eggs in ponds, but some attach clutches to vegetation above water. *Afrixalus* lays its eggs on a leaf and then folds the edges of the leaf together, gluing them with oviductal secretions. A few species use tree holes for egg deposition, and *Leptopelis* and *Tachycnemis* lay terrestrial eggs, which hatch into aquatic larvae. Tadpoles of hyperoliids are free-swimming and feeding, having cornified jaw sheaths and denticles (Type 4). *Afrixalus fornasinii* is the only terrestrial frog known to prey on the egg clutches of other species of anurans, although this behavior is also reported in aquatic pipids (Drewes and Altig 1996).

Content and Distribution. 19 genera, 230 species. Most genera in sub-Saharan Africa; *Tachycnemis* and *Heterixalus* endemic to the Seychelles and to Madagascar, respectively (Figure 3–18).

Systematics. Drewes (1984), Channing (1989).

Rhacophoridae

The pectoral girdle is firmisternal in rhacophorids. Intercalary elements are present in the digits. The vertebrae are procoelous and the pectoral girdles are firmisternal. Amplexus is axillary and larvae are Type 4. Body size varies greatly in rhacophorids, from less than 20 millimeters to over 120 millimeters in snout–vent length. Most rhacophorids are arboreal and have enlarged toe discs. Hence, they are sometimes referred to as Old World tree frogs. However, some species are decidedly terrestrial and lack toe discs (e.g., *Aglyptodactylus* from Madagascar). Some species lay eggs in water and have aquatic tadpoles, whereas others (e.g., *Chiromantis*, *Polypedates*, *Rhacophorus*) construct foam nests either in water or on vegetation above water. *Chiromantis* builds its nests on tree branches and a single nest may be communally constructed by many individuals. The foam hardens, thus protecting the eggs from desiccation until hatching, whereupon the larvae drop into the water below. Still other species of rhacophorids place their eggs in tree holes and have nonfeeding or feeding tadpoles

(Wassersug et al. 1981). Some species of *Philautus* lay small clutches in arboreal sites and undergo direct development (Alcala and Brown 1982). Extensive webbing is characteristic of many rhacophorids. *Rhacophorus nigromaculatus* of Southeast Asia uses the extensive webbing and dorsoventrally flattened body to locomote by parachuting (Emerson and Koehl 1990, Emerson et al. 1990).

Content and Distribution. 10 genera, approximately 212 species. Tropical Africa and Madagascar; southern India and Sri Lanka; southern China and Japan to the Philippines and islands of the Sunda Shelf (Figure 3–18).

Systematics. Liem (1970), Dubois (1981).

These three groups are monophyletic relative to other extant tetrapods. Among salamanders, the externally fertilizing groups (Sirenidae and Cryptobranchoidea) are sister groups to the remaining families (Salamandroidea), but relationships within Salamandroidea are not clearly resolved. On the other hand, general agreement exists concerning the relationships among families of caecilians. Although many amphibians are linked for a portion of their life cycle to aquatic habitats because they have an aquatic larval stage, all groups have species that have circumvented a larval stage by various forms of direct development. These include species that lay their eggs in terrestrial environments, those that carry their eggs or young on or in the body, and viviparous species.

Summary

Extant amphibians are mostly frogs (Anura) (88 percent of the species), followed by salamanders (Urodela) (9 percent) and caecilians (Gymnophiona) (3 percent).

4

Classification and Diversity of Extant Reptiles

Extant reptiles, including Testudines (Chelonia), Rhynchocephalia, Squamata, Crocodylia, and Aves, comprise more than 15,800 species. Of these, approximately 8,700 species are birds (Aves), which are not treated in this book. Most of the remaining reptiles (approximately 7,150 species) are squamates (lizards, snakes, and amphisbaenians; 6,850 species), followed by turtles (260 species), crocodilians (22 species), and tuatara (2 species). Nearly two thirds of squamates are lizards and amphisbaenians. The large families of lizards include the Scincidae (1,070 species), Gekkonidae (835 species), and perhaps Iguanidae in the broad sense (more than 900 species). Two thirds of extant snakes (1,800 species) belong to the family Colubridae. The major characteristics and natural history of these groups are introduced in this chapter. References to the systematics literature are provided primarily as an introduction to the literature on each group, but are not comprehensive.

Turtles (Testudines or Chelonia)

Turtles are the most distinctive and instantly recognizable of all organisms, and the shell that entirely encases the body is unique. The shell is composed of dermal ossifications incorporating the ribs, vertebrae, and portions of the pectoral girdle. Because the vertebrae, except for those of the neck and tail, are completely fused into the shell, no movement occurs in the axial skeleton of the body. Likewise, the ribs are fused into the lateral portion of the carapace. Moreover, the pectoral and pelvic girdles are medial to the rib cage, which is an arrangement seen in no other vertebrate (Figure 4–1). This morphology results from two developmental processes: (1) differential growth between the lateral and dorsoventral axes of the trunk, producing a more flattened body, and (2) incorporation of the ribs into the dermis and their deflection dorsally, rather than growth medially and ventrally as in other vertebrates (Burke 1989, 1991). The complete enclosure of the internal organs of turtles within a rigid shell means that the animals cannot breathe by expanding and contracting the rib cage (see Chapter 6).

The turtle shell has two parts, a dorsal carapace and a ventral plastron, which are joined laterally by a bridge (Figure 4–2). The shell is composed of dermal bony elements that are covered externally by keratinous scutes or, in a few instances, leathery skin (Trionychidae, Dermochelyidae, Carettochelyidae). The scutes do not have the same pattern as the underlying bony elements, and the misalignment of sutures in the bony and keratinous portions of the shell adds strength to the structure. The limbs, tail, neck, and head protrude from anterior and posterior openings. The plastron of turtles may be a solid structure or it may have a set of loosely articulated bony elements, depending on the group. The overall shell shape varies greatly, with relatively high domes being characteristic of many terrestrial turtles, and relatively flat, streamlined forms characteristic of aquatic and marine turtles.

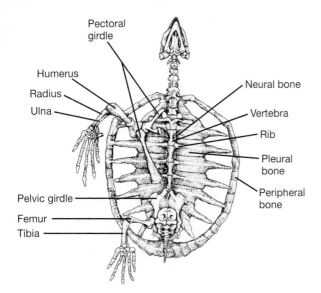

Figure 4–1   **Relationship between the axial and appendicular skeletons of a turtle.** The pectoral and pelvic girdles are enclosed within the rib cage (pleural bones of the carapace). All vertebrae except those of the neck and tail are fused into the carapace. *(Source: Modified from Bellairs 1970.)*

In most turtles both the carapace and the plastron are rigid structures, but several lineages have independently evolved the ability to close the body within the shell. Usually this occurs by the development of a hinge on the front or rear lobe of the plastron, as seen in North American box turtles (*Terrapene*, Emydidae) and mud turtles (*Kinosternon*, Kinosternidae). However, African testudinids of the genus *Kinixys* draw the posterior portion of the carapace downward to protect the rear end, and trionychids of the genus *Lissemys* have anterior and posterior carapace hinges to close both ends. Shell reduction has occurred in some turtles, perhaps most peculiarly in the African testudinid *Malacochersus*. This turtle, unlike most land tortoises, has a very flat shell with the carapace reduced to a series of bony rings surrounding large vacuities. Both the carapace and the plastron are very soft. This turtle inhabits rocky outcrops and wedges itself into crevices by inflating its body with air.

Turtle skulls differ from those of other extant reptiles in lacking temporal fenestrae, and hence are anapsid (Figure 4–3). However, the skulls of many turtles are greatly emarginated from the posterior end, resulting in rather open temporal and posterior regions of the skull. Teeth are completely absent in extant turtles (present in some fossils), and are replaced by a tough keratinous beak. Moreover, in turtles the orientation of the jaw adductor muscles to the braincase is unique among tetrapods, although modified in two distinctive ways among extant turtles (Gaffney 1975). In each case, the muscles pass over a prominence, the trochlea, in passing from the lower jaw to the skull, thus forming a sharp angle in the orientation of the muscle fibers. In other tetrapods the orientation of the jaw adductors is relatively straight, there being no trochlea to reorient the muscle fibers.

The limb structure of turtles is highly variable, reflecting the environment and locomotory modes of different species. Marine species and the freshwater *Carettochelys* have well-developed flippers with extremely elongated digits that are not independently movable. Most aquatic species have well-developed webbing between the digits, which still retain some independent mobility. Terrestrial turtles generally have stout, clublike limbs capable of lifting heavy bodies off the substrate. Their digits are often reduced, and the feet are equipped with thickened pads.

All turtles are oviparous and lay eggs within a nest dug in the ground or lay their eggs on the surface with no nest (e.g., the testudinid *Geochelone denticulata*). In general, no parental care occurs. Male turtles have a penis that is an outgrowth of the cloacal wall (Figure 4–4). It contains erectile tissue and a

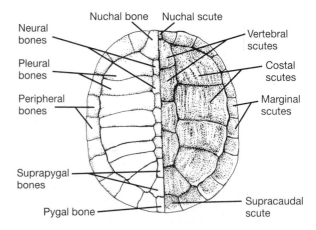

Figure 4–2   **Components of the carapace of a turtle.** Bony components are shown on the left. The overlying epidermal scutes are shown on the right side. *(Source: Modified from Pritchard and Trebbau 1984.)*

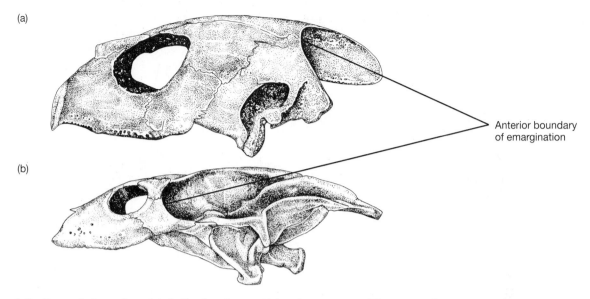

(a)

(b)

Anterior boundary
of emargination

Figure 4–3   **Lateral view of turtle skulls showing variation in structure.** The temporal region of the skull of *Lepidochelys* (a) is only shallowly emarginate, whereas that of *Trionyx* (b) is considerably emarginate. *(Source: Modified from Gaffney 1979.)*

groove in the dorsal surface through which sperm flow during copulation. Ornamentation of the glans of the penis is important in turtle systematics. The courtship of turtles can be quite elaborate, involving prolonged interaction between the sexes before copulation. Turtles are the longest-lived vertebrates, with documented captive records of more than 50 years for some species (Bowler 1977), although longevity in natural populations is probably less (e.g., Gibbons and Semlitsch 1982). Their population biology poses numerous problems for ecologists as well as those setting conservation policy (Congdon et al. 1993, 1994).

Chromosomal evolution and sex-determining mechanisms have been extensively studied in turtles. Temperature-dependent sex determination (see Chapter 7) is probably the only mode of sex determination in the Testudinidae, Chelydridae, Kinosternidae (with the single exception of *Staurotypus*), Dermatemydidae, Cheloniidae, Dermochelyidae, Carettochelyidae, and Pelomedusidae (Janzen and Paukstis 1991). Temperature-dependent sex determination also occurs in most emydids and batagurids, but a variety of patterns of genotypic sex determination are also seen in these families. Genotypic sex determination is probably the rule in Trionychidae and Chelidae.

Content and Distribution

The 12 extant families contain approximately 260 species. They are cosmopolitan in terrestrial, freshwater, and marine habitats except at extremely high latitudes. Regions of exceptionally high diversity are the southeastern United States (primarily Emydidae) and Southeast Asia (primarily "Bataguridae").

Fossil Record

Turtles have an extensive fossil record, owing in part to their heavy, bony skeletons and aquatic or marine habits. The earliest known turtles are all Late Triassic (210 million years before the present [MaBP]): *Proganochelys* from Germany, *Palaeochersis* from South America, and *Proterochersis* from Europe (Gaffney 1990, Rougier et al. 1995). *Proganochelys* is the sister taxon to other known turtles, with *Palaeochersis*, *Proterochersis*, and *Australochelys* (Early Jurassic of Africa) forming successive sister groups to modern turtles (Cryptodira + Pleurodira) (Gaffney and Kitching 1994). Cryptodires are known from the early Jurassic Kayenta Formation of Arizona (Gaffney et al. 1987), which has also yielded the earliest known caecilian and anuran (see Chapter 3). Pleurodires are known from

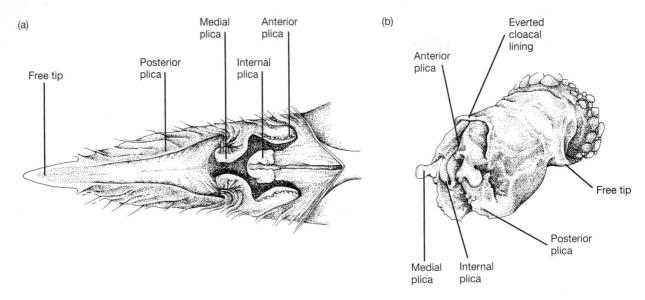

Figure 4–4   **Morphology of the penis of a turtle (*Emydura australis*, Chelidae),** protruded (left) and unprotruded (right) views. The unprotruded view shows the position of the penis in the cloaca as seen in a ventral dissection. *(Source: McDowell 1983.)*

the Upper Jurassic (Rougier et al. 1995). Thus, both major extant clades are known from relatively early in the recorded history of Testudines. The fossil record shows that most families were much more widespread than their present distributions (see Carroll 1988).

The origin of turtles and their peculiar body plan has long been shrouded in mystery, but new fossil discoveries and reinterpretations of amniote relationships have suggested two alternative views. Gauthier et al. (1988), Lee (1993, 1995, 1996), and Laurin and Reisz (1995) include turtles and several Paleozoic fossil anapsid groups together in a monophyletic clade known as Parareptilia. An alternative view, argued by Rieppel and deBraga (1996), places turtles within Diapsida as the sister group to Sauropterygia (nothosaurs and plesiosaurs).

Systematics. Bickham and Carr (1983), Gaffney (1984), Gaffney and Meyland (1988), King and Burke 1989, Gaffney et al. (1991), Iverson 1992.

Systematics and Phylogeny of Turtles

Many of the peculiar features of turtles are shared derived characters of the clade Testudines (Gaffney and Meylan 1988, Gaffney 1990, Lee 1995). These features include the shell and its association with vertebrae, ribs, and girdles; the medial position of the girdles relative to the ribs; the absence of teeth and their replacement by a keratinous beak (palatine and vomerine teeth present in some early fossils); and the trochlea and the unusual configuration of the jaw adductor musculature. Many other shared derived characters of the turtle skeleton, such as loss of the pineal foramen, loss of certain skull bones, and other skeletal details, are more subtle. Because of their many unique structural features, the monophyly of turtles has never been challenged.

The term Testudines is often used for all turtles, including the common ancestor of the fossil *Proganochelys* and the two clades with extant members (Cryptodira and Pleurodira), and all of its descendants. The alternative term Chelonia is widely used for the extant clades only. Lee (1995) provides slightly different definitions for the two terms, but the alternative usage depends on the inclusion of fossil forms only and will not concern us here.

All extant turtles belong to one of two clades, the Pleurodira and Cryptodira (Figure 4–5). These two groups have traditionally been distinguished on the basis of the mode of neck retraction—that is, whether the neck is bent to the side (Pleurodira, the side-

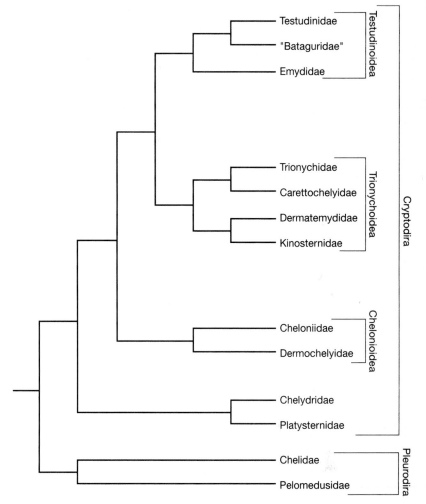

Figure 4–5 **Phylogenetic relationships among turtle families based on morphology.** *(Source: Gaffney and Meylan 1988.)*

necked turtles) or retracted in a vertical plane (Cryptodira). However, early fossil members of both of these clades lack the retraction mechanisms typical of later members of the clades (Gaffney and Meylan 1988). Nevertheless, numerous shared derived characters exist in the details of vertebral and skull anatomy.

The mode of neck retraction is useful for distinguishing extant turtles, and it is based on the peculiar form of the cervical vertebrae that permits bending of the vertebral column in either the horizontal or vertical plane (Williams 1950). Many cryptodires can fully retract the neck and head within the shell, especially if the shell also has hinges to allow closure. Pleurodires, on the other hand, generally bring the head and neck close to the side of the body in the gap between the carapace and plastron. Some cryptodires, such as sea

turtles (Cheloniidae and Dermochelyidae) and snapping turtles (Chelydridae), have lost the ability to completely retract their necks. Pleurodira includes the extant families Chelidae and Pelomedusidae, both of which include only freshwater turtles. Cryptodira includes all other turtles, which may be freshwater, marine, or terrestrial.

Within Cryptodira, four major clades are recognized (Figure 4–5). In order of branching, these are (1) Chelydridae, the sister group to the other families; (2) Chelonioidea (Cheloniidae, Dermochelyidae); (3) Trionychoidea (Kinosternidae, Dermatemydidae, Carettochelyidae, Trionychidae); and (4) Testudinoidea (Emydidae, Testudinidae, and the paraphyletic group "Bataguridae"). The major problematic area in the phylogeny of cryptodires is the relationship of the Testudinidae (land tortoises) to the "Bataguridae." Although there is general agreement that testudinids are monophyletic and are derived from within the batagurids, the relationships among genera of batagurids are so unclear at present that we retain this paraphyletic group until the systematics is better understood (e.g., Gaffney and Meylan 1988). Alternatively, the batagurids are sometimes included within the Testudinidae, but that arrangement obscures the generally accepted monophyly of testudinids relative to batagurids. Similarly, the relationships of *Platysternon megacephalum* (Platysternidae) are controversial, with morphological studies (e.g., Gaffney and Meylan 1988) placing this taxon within the Chelydridae, whereas chromosome structure and some vertebral and shell characters support a relationship closer to the emydids.

Most shared derived characters of turtle families are details of skull and shell structure, the pattern of cranial arteries, and, to lesser extents, chromosomal and penial morphology. There has as yet been no comprehensive study of turtle phylogeny using molecular characters.

Chelidae

Chelids are characterized by unusually extensive emargination of the cheekbones so that only a parietal-squamosal bar remains. Quadratojugals and mesoplastra are absent, distinguishing chelids from pelomedusids. Chelids are aquatic turtles that range in size from about 15 centimeters in carapace length (*Pseudemydura umbrina*) to nearly 50 centimeters (*Chelodina expansa*) (Figure 4–6). Most species inhabit slow-moving freshwater or swamps, although *Chelodina siebenrocki* also occurs in brackish water. Some populations of *Platemys platycephala* exhibit an unusual form of triploidy in which individual cells are diploid or triploid within an individual (Bickham et al. 1985).

The South American chelid, *Chelus fimbriatus* (the matamata), is one of the most unusual and familiar turtles. Its shell is broad and flat, with three keels produced by the strongly protuberant costal and vertebral plates. The shell is often camouflaged by a thick coating of algae. Like *Chelodina* and *Hydromedusa*, *Chelus* has a very long neck, and the skin of the head and neck has many cutaneous flaps that may serve a tactile function may or be sensitive to water currents. The snout has a fleshy proboscis. The mandibles are reduced to slender bony struts and fail to meet at the midline. *Chelus* uses suction feeding to capture fish

and aquatic invertebrates (Formanowicz et al. 1989) (see Chapter 9).

Content and Distribution. 10 genera, approximately 40 species. South America, Australia, and New Guinea (Figure 4–7).

Systematics. Gaffney (1977), Georges and Adams (1992; Australian species), Derr et al. (1987; *Platemys*).

Pelomedusidae

All extant pelomedusids inhabit lakes, rivers, swamps, and marshes, but some fossils apparently were marine. Plastrons are hinged in *Pelusios* but rigid in all other species. Although many pelomedusids are carnivores or omnivores, species of *Podocnemis* are mainly herbivorous. *Podocnemis expansa* forages in rivers and flooded forests, and its diet is principally the fruits and flowers of forest trees that fall into the water (Soini 1984).

The largest extant pelomedusid is *Podocnemis expansa* of the Amazon and Orinoco river systems of South America. Females may reach nearly 90 centimeters in carapace length (Pritchard and Trebbau 1984). However, the fossil pelomedusid *Stupendemys* from the late Tertiary of Venezuela reached at least 2.3 meters in carapace length and is possibly the largest turtle ever to have lived, even larger than the largest extant turtle, *Dermochelys* (Dermochelyidae) (Wood 1976). The smallest pelomedusid is the African *Pelusios nanus*, which attains a carapace length of 12 centimeters.

The nesting behavior of *Podocnemis expansa* is similar in many ways to that of sea turtles. This large riverine pelomedusid lays its eggs on islands within the large rivers it inhabits. Nests are constructed at night and consist of a body pit, within which the nest hole itself is dug. Eggs of several species of *Podocnemis*, particularly *P. expansa* and *P. unifilis*, have been harvested for food for more than a century, as reported by Henry Walter Bates (1876). Management of these species has become of special concern in several countries because of overexploitation (see Chapter 15).

Content and Distribution. 5 genera (*Podocnemis, Peltocephalus, Erymnochelys, Pelomedusa, Pelusios*), about 25 species. Northern South America, Africa, Madagascar, and the Seychelles (Figure 4–7). *Peltocephalus* and *Podocnemis* are endemic to northern South America, *Pelomedusa* is in sub-Saharan Africa and Madagascar, *Erymnochelys* is endemic to Madagascar, and *Pelusios* is found in Africa, Madagascar, and the Seychelles.

Systematics. Gaffney and Meylan (1988).

Figure 4–6   **Diversity of turtles.** (a) *Chelodina longicollis* (Chelidae, Australia). (b) *Pelusios subniger* (Pelomedusidae, Africa). (c) *Chelonia mydas* (Cheloniidae, worldwide). (d) *Apalone spinifera* (Trionychidae, North America). (e) *Terrapene carolina* (Emydidae, North America). (f) *Geochelone denticulata* (Testudinidae, South America). *(Photographs by (a) Jean-Paul Ferrero/AUSCAPE International, (b, e) Harvey Pough, (c, d) R. Wayne Van Devender, (f) John Cadle.)*

Carettochelyidae

Carettochelys reaches about 70 centimeters in carapace length. It inhabits freshwaters of rivers, water holes, and lagoons, and also enters brackish estuaries. It is omnivorous. Eggs are laid at night on high sandbanks during the dry season. In Australia, fully developed embryos become dormant within the eggs until the nests are flooded by rising river waters. In *Carettochelys* the bony carapace is covered with soft skin that, unlike the skin of trionychids, does not have embedded bony scutes. As in sea turtles, the limbs are elongate paddles, and they are clawed. A fleshy proboscis is at the tip of the snout, a derived character shared with trionychids. *Carettochelys* propels itself through water using simultaneous movements of its forelimbs, much as sea turtles do. Other freshwater turtles use alternate limb movements.

Content and Distribution. 1 species, *Carettochelys insculpta*. Southern New Guinea and extreme northern Australia (Figure 4–8). It is the apparent sister taxon to Trionychidae among extant turtles (Figure 4–5).

Systematics. Gaffney and Meylan (1988).

Cheloniidae and Dermochelyidae

These two families, collectively referred to as sea turtles, are treated together because there is general agreement that the two families are monophyletic relative to other turtles and many aspects of their biology are similar. Sea turtles are the only turtles whose forelimbs are more strongly developed than the hind limbs. In both families the limbs are modified into elongate and fully webbed paddles, and, except for females that must come ashore to lay eggs, they are totally marine. The limbs cannot support the body off the substrate when the animals move on land. Although many sea turtles are wide-ranging throughout temperate and tropical seas outside the breeding season, egg laying for all species occurs only on tropical or (*Caretta* only) warm temperate beaches. Individual sea turtle females reproduce on a multiyear cycle, although several large clutches of eggs are usually laid within one breeding season. Nesting is nocturnal in most species, but it is diurnal in *Lepidochelys*. All sea turtles undergo extensive migrations, navigating by using cues from Earth's electromagnetic field (Lohmann 1992). They are also extremely faithful to particular breeding areas (Bowen et al. 1992, Bowan, Avise et al. 1993).

Cheloniids range in size from about 70 centimeters in carapace length (*Lepidochelys*) to about 1.5 meters (*Chelonia mydas* and *Caretta caretta*). *Chelonia mydas* is herbivorous, but other cheloniids are carnivores or omnivores. *Eretmochelys* is largely a spongivore, whereas *Caretta* consumes primarily molluscs and crustaceans such as oysters and crabs.

Dermochelys is unique among extant turtles in that the carapace is composed of thousands of polygonal osteoderms embedded in the leathery skin, giving rise to its common name, leatherback or leathery sea turtle. Rows of enlarged osteoderms form seven longitudinal keels on the carapace and four more on the plastron. Claws are absent from the tips of the flippers. There are many anatomical peculiarities in the skele-

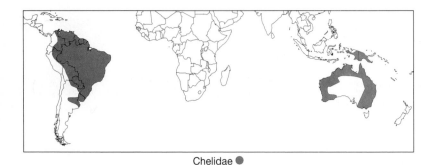
Chelidae ●

Pelomedusidae ●

Figure 4–7 **Distribution of the turtle families Chelidae and Pelomedusidae.**

Figure 4–8 **Distribution of the turtle families Carettochelyidae, "Bataguridae," Chelydridae, and Platysternidae.**

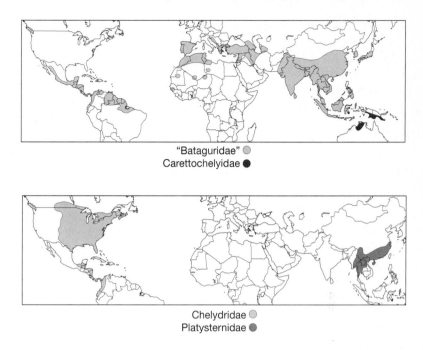

"Bataguridae" ○
Carettochelyidae ●

Chelydridae ○
Platysternidae ●

ton. This is by far the largest extant turtle, but well-documented size records are scarce. Carr (1952) validated a carapace length of 2.4 meters, but the largest documented by Pritchard and Trebbau (1984) was 1.9 meters. Virtually all cheloniids feed on organisms (plants or animals) attached to the substrate, but *Dermochelys* is totally pelagic and feeds primarily in the water column on jellyfishes (Cnidaria). Its jaws are weak and incapable of managing hard-bodied prey. *Dermochelys* also has several physiological specializations for maintaining its body temperature well above ambient sea tempeature (see Chapter 5), which may explain in part its greater northward extension into cold ocean than any other sea turtle.

Content and Distribution. Cheloniidae: 5 genera (*Chelonia, Lepidochelys, Natator, Eretmochelys, Caretta*), 7 species. Dermochelyidae: 1 species, *Dermochelys coriacea*. Both families occur worldwide in temperate and tropical

oceans (Figure 4–9). Phylogenetic analyses of both molecular and morphological data support the sister group relationship of *Dermochelys* to other sea turtles (Cheloniidae). Within Cheloniidae, *Lepidochelys* and *Caretta* are sister taxa relative to *Chelonia*, but the positions of *Eretmochelys* and *Natator* are poorly resolved.

Systematics. Gaffney and Meylan (1988), Bowen, Nelson et al. (1993).

Chelydridae

Chelydridae are large-headed freshwater turtles with powerful jaws and an aggressive nature. The head, neck, and limbs are so large that they cannot be retracted fully within the shell. These turtles are not strong swimmers, and they move mainly by walking on the bottom. *Chelydra* inhabits a wide variety of freshwater and brackish habitats, but usually with abundant aquatic vegetation. It is omnivorous, seemingly eating almost anything it can swallow. Eggs are laid in nests that are often constructed some distance from water. *Chelydra* exhibits an unusual defensive posture (shared with the chelid *Chelus*) in which the head and anterior body are lowered while the hind limbs are extended maximally, thus exposing the dorsal surface of the shell toward a potential threat (Dodd and Brodie 1975).

Macroclemys is almost completely aquatic and is probably the world's heaviest freshwater turtle, growing to nearly 70 centimeters in carapace length and more than 80 kilograms. It inhabits deep-water rivers, lakes, and swamps where there is abundant vegetation. *Macroclemys* has a bifurcate fleshy projection on the tongue that is used to lure fish into the open mouth when the turtles are resting on the bottom (Figure 13–9). However, they also actively forage and are probably fairly catholic carnivores. The diet includes frogs, snakes, molluscs, crustaceans, aquatic plants, and even other turtles.

Content and Distribution. 2 species: *Chelydra serpentina* (eastern North America from southern Canada, south through Central America to western Ecuador), *Macroclemys temminckii* (southeastern United States) (Figure 4–8).

Systematics. Lovich (1993).

Dermatemydidae

Dermatemys reaches a carapace length of about 65 centimeters and is essentially totally aquatic. It is a vegetarian inhabiting large rivers, lakes, and temporary pools where there is aquatic vegetation. It also enters brackish water. Only rarely surfacing, *Dermate-*

mys apparently uses the buccopharyngeal lining for gas exchange. Captive individuals continually take water in through the mouth and expel it from the nostrils (Ernst and Barbour 1989).

Content and Distribution. 1 species, *Dermatemys mawii*. Southern Mexico to northern Honduras (Figure 4–10).

Systematics. Gaffney and Meylan (1988).

Emydidae

Most emydids are freshwater or semiaquatic turtles (Figure 4–6). One species, *Malaclemys terrapin*, inhabits brackish marshes and coastal marine habitats, and species of *Terrapene* are terrestrial, with the exception of the aquatic box turtle, *T. coahuila*. Sizes range from relatively small turtles, such as *Clemmys guttata* and *C. muhlenbergi* (12 centimeters in carapace length) to relatively large *Trachemys scripta* (to 60 centimeters in carapace length). Most are omnivorous as adults, but *Emydoidea* and *Deirochelys* are primarily carnivorous. The plastrons of *Emys*, *Emydoidea*, and *Terrapene* are hinged. All other emydids have solid plastrons. Because of their relative abundance and ease of study, several emydids have been the subjects of long-term ecological studies (e.g., Congdon et al. 1983, Gibbons 1990).

Content and Distribution. 10 genera, 35 species (*Clemmys, Emydoidea, Emys, Terrapene, Chrysemys, Deirochelys, Graptemys, Pseudemys, Trachemys, Malaclemys*). All genera are restricted to North America (to northern Mexico) except *Trachemys* (North, Central, and South America and the West Indies) and *Emys* (Europe, western Asia, and northwest Africa) (Figure 4–10). Our Emydidae corresponds to the

Figure 4–9 **Distribution of the turtle families Cheloniidae and Dermochelyidae.**

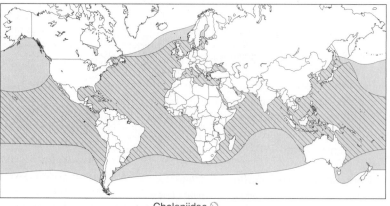

Cheloniidae ⬭
Dermochelyidae ⬭ + ⬤

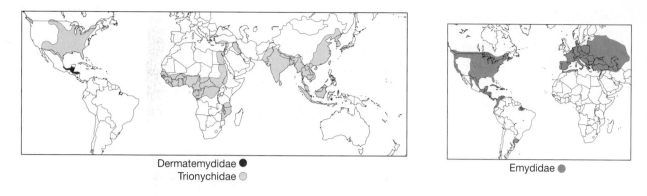

Dermatemydidae ●
Trionychidae ◐

Emydidae ●

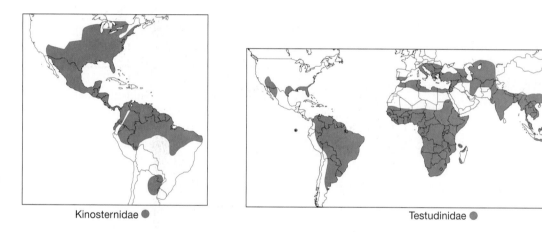

Kinosternidae ●

Testudinidae ●

Figure 4–10 **Distribution of the turtle families Dermatemydidae, Trionychidae, Emydidae, Kinosternidae, and Testudinidae.**

Emydinae of some authors, who also include the "Bataguridae" within Emydidae (e.g., Iverson 1992).

Systematics. McDowell (1964), Bramble (1974), Lovich and McCoy (1992), Vogt (1993), and Lamb et al. (1994), (*Graptemys*); Milstead (1969; *Terrapene*); Seidel (1988; West Indian species).

Kinosternidae

This family includes the smallest turtles of North America (11 to 15 centimeters in carapace length), commonly called mud and musk turtles. The tropical species of *Staurotypus* and *Kinosternon* are larger, reaching 25 centimeters to nearly 40 centimeters. The shells are elongate. The plastron may be very re-

duced, and it is singly or doubly hinged in some species (e.g., *Kinosternon subrubrum* and *K. minor*). These turtles release a foul-smelling musk from glands in the cloaca, hence one of their common names. Like chelydrids, these are mostly bottom-walkers rather than strong swimmers and live in slow-moving water, often with much vegetation. All are carnivorous. *Staurotypus*, uniquely among kinosternids, has sex chromosomes (XX-XY, with males heterogametic and females homogametic).

Content and Distribution. 3 genera (*Kinosternon*, *Staurotypus*, *Claudius*), 22 species (some workers recognize *Sternotherus* as distinct from *Kinosternon*, but that renders *Kinosternon* paraphyletic). Eastern North America from southern Canada, south through Mexico, Central America, and South America (Figure 4–10).

Systematics. Hutchison (1991), Iverson (1991).

Platysternidae

The head of *Platysternon* is so enormous that it cannot be withdrawn into the shell. The tail is exceptionally long, nearly the length of the carapace (to 18 centimeters). The carapace is flattened dorsoventrally and the plastron is well developed. *Platysternon* inhabits cool mountainous rocky streams. These turtles are apparently nocturnal carnivores and seldom bask, although they may forage short distances from water. They are superb climbers in vegetation. Only one or two eggs are laid per clutch.

Content and Distribution. 1 species, *Platysternon megacephalum*. Southeastern China to Burma and Thailand (Figure 4–8). The relationships of this species within Cryptodira are controversial.

Systematics. Whetstone (1978), Haiduk and Bickham (1982), Gaffney and Meylan (1988).

"Bataguridae"

Batagurids are in many respects the Old World ecological equivalents of their close relatives, the emydids. Most species are aquatic to semiaquatic, inhabiting both fresh and brackish water. *Callagur borneoensis* even nests on sea beaches along with nesting sea turtles. *Pyxidea mouhotii* and two species of *Heosemys* are terrestrial. *Heosemys silvatica* uses burrows in the ground. Many species of batagurids are herbivorous or omnivorous, but *Malayemys subtrijuga* eats predominantly snails, and species of *Mauremys* are highly carnivorous. Sizes of batagurids range from quite small (carapace length, 12 to 14 centimeters in several species of *Cuora* and in *Heosemys silvatica*) to relatively large (carapace length, 50 to 60 centimeters in *Hieremys*, *Kachuga*, *Callagur*, and *Orlitia*). Both *Callagur borneoensis* and *Orlitia borneensis* have been reported to reach 75 to 80 centimeters in carapace length.

Content and Distribution. About 23 genera, 60 species. A single New World genus, *Rhinoclemmys* (9 species): northwestern Mexico to northern South America. Remaining genera: northwest Africa, and from Europe to western Asia and the Middle East, across southern Asia to China, Japan, the Philippines, and islands of the Sunda Shelf (Figure 4–8). The greatest diversity is in southern Asia. "Bataguridae" is paraphyletic without the inclusion of Testudinidae (Gaffney and Meylan 1988). If the two groups are merged, the name Testudinidae has priority because it was applied to the group first.

Systematics. McDowell (1964), Hirayama (1984), Sites et al. (1984), Carr and Bickham (1986).

Testudinidae

Commonly called tortoises, testudinids are land turtles and usually have high domed shells (Figure 4–6). The limbs are stout for supporting the heavy bodies, the feet are unwebbed, and the digits contain no more than two phalanges. The head and limbs can be fully withdrawn into the shell, with the heavily scaled limbs forming an effective protective barrier to the outside. Plastrons are hinged in *Testudo* and *Pyxis*. Uniquely among extant land turtles, tortoises in the genus *Kinixys* have a hinged carapace that allows it to be lowered over the hindquarters. Unlike other tortoises, *Malacochersus* has a dorsoventrally compressed shell with a reduced bony component. Terrestriality has evolved repeatedly and independently many times in the history of turtles. Testudinids are merely a recent and diverse example. Box turtles of the genus *Terrapene* (Emydidae) are another recent example, and have converged on testudinids in some aspects of limb and shell structure.

The high domed shells of testudinids are thought to be partly adaptations against predation because the shell diameter exceeds the jaw gape of most potential predators, but small individuals are vulnerable to an array of predators. South American *Geochelone denticulata* individuals up to at least 37 centimeters long are eaten by jaguars, which break open the tops of the carapaces with their canines and consume the body within (Emmons 1989). However, many individuals of *Geochelone* have tooth marking indicating unsuccessful attacks by these large cats. The cats are apparently incapable of extracting a tortoise from either end of the shell. Jaguars can break the relatively flat shells of even large *Podocnemis* (Pelomedusidae) individuals, and this observation gives some credence to the idea that the domed shell of tortoises is at least a partly effective defense against some predators.

Extant testudinids range in size from about 10 to 130 centimeters in carapace length. Gigantism has evolved independently in many oceanic island populations of testudinids, with those of the Galápagos and Aldabra islands (both *Geochelone*) being the most famous. Many such island populations became extinct with the arrival of Europeans (Arnold 1979). Extant continental species are smaller than island species, but some fossil giant tortoises are known from continental areas (see Arnold 1979).

Most testudinids are herbivores or omnivores. They inhabit environments ranging from extremely arid deserts to wet rain forests. Species of *Gopherus* construct deep burrows for refuge using spadelike front feet. Most tortoises have relatively small clutch sizes (in many species only one or two eggs). The largest clutch sizes are reported for the Asian testudinid, *Manouria emys*, which can lay up to about 50 eggs. The eggs of testudinids may be laid in holes dug in the ground, as in many other turtles, but *Manouria emys* constructs a large mound nest of decaying vegetation and may remain with the nest for several days. *Geochelone denticulata* of South America often does not construct a nest at all but deposits its eggs either singly or in small clusters in the leaf litter of the forest floor.

Content and Distribution. Approximately 11 genera, about 40 species. Essentially worldwide in temperate and tropical terrestrial regions, including oceanic islands such as the Galapagos and Aldabra islands; absent from Australia (Figure 4–10).

Systematics. Crumly (1982, 1994), Lamb et al. (1994), Lamb and Lydeard (1994).

Trionychidae

Trionychids have extremely dorsoventrally flattened bodies and reduced bony portions of the shell (Figure 4–6). The carapace lacks the peripheral series of bones (except *Lissemys*, which has them only posteriorly), and the costal bones, those forming the major portion of the carapace (Figure 4–2), are reduced distally in most species. This results in free distal ends of the ribs, which is a feature seen in juvenile turtles as well as adults of several other families. The bones of the plastron are not fused at the midline, similar to the pattern in sea turtles. The bony plates of the shell are covered with leathery skin that has no keratinous scutes. The snout bears a fleshy proboscis and the jaws are covered with fleshy lips rather than keratinous beaks. Trionychids have long necks and their limbs are extensively webbed. Genera of the Cyclanorbinae have cutaneous femoral flaps on the plastron that cover the hind limbs when they are withdrawn into the shell.

Trionychids are fully aquatic turtles, strong swimmers, and primarily carnivorous. Sizes range from relatively small (25 centimeters in *Lissemys punctata* and *Pelodiscus sinensis*) to large (130 centimeters in *Pelochelys bibroni*). Many species bury themselves in the substrate and extend their long necks to project the snout above the water surface to breathe.

Content and Distribution. 14 genera, about 25 species. North America (except the extreme West) from southern Canada to northern Mexico; sub-Saharan Africa and the Nile Valley to the Middle East; India and Southeast Asia to China and Japan; and islands of the Sunda Shelf (Figure 4–10). 2 subfamilies: Cyclanorbinae (*Lissemys*, *Cycloderma*, *Cyclanorbis*), Africa, the Indian subcontinent, and Burma; Trionychinae (all other genera), North America, Africa, the Middle East, Southeast Asia, and islands of the Sunda Shelf.

Systematics. Meylan (1987), Gaffney and Meylan (1988).

Lepidosauria

Among extant reptiles, *Sphenodon*, lizards, and snakes make up the Lepidosauria, the sister taxon to Archosauria (birds + crocodilians) (see Figure 2–9). The lepidosaurs are characterized by a transverse cloacal slit (longitudinal in other tetrapods), regular skin shedding in which the outer layers of the epidermis are shed at one time, the development of breakage planes in the tail (see caudal autotomy in the discussion of Squamata), and many other anatomical features (Gauthier et al. 1988).

Tuatara (Rhynchocephalia)

Among lepidosaurs, Rhynchocephalia is the sister taxon to Squamata (lizards and snakes) (Figure 4–11). Rhynchocephalia includes several extinct taxa and the extant *Sphenodon*. *Sphenodon* (family Sphenodontidae) and its closest fossil relatives are referred to as the Sphenodontida.

Sphenodontida is characterized by numerous shared derived characters of skeletal morphology. The teeth are relatively large, acrodont, and have a distinctive pattern of regionalization (heterodonty) in the jaws. Caniniform teeth are present at the anterior end of the maxillae and dentaries. Teeth are also present on the palatine bones, a feature absent in lizards. During ontogeny the premaxillary teeth are replaced by bony downgrowths of the premaxilla, forming a pair of chisel-like structures in adults. Tuatara have shallow paired outpocketings of the posterior wall of the cloaca that have been interpreted as precursors to the

Figure 4–11 **Phylogenetic relationships among lepidosaurs.** All terminal tips are families except for Serpentes (see Figure 4–24). *(Source: Estes et al. 1988.)*

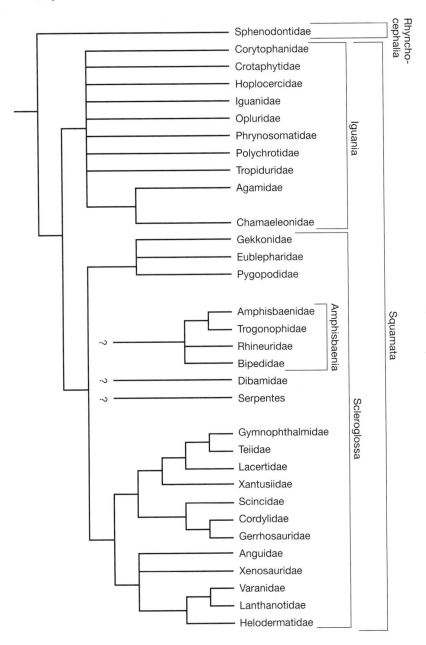

hemipenes of squamates (Arnold 1984). Unlike lizards, *Sphenodon* retains the lower temporal bar in its diapsid skull and hence does not have a streptostylic quadrate as in lizards (see Chapter 9). Osteoderms are absent. As in many lizards, the tail of *Sphenodon* has autotomic planes, which allow the tail to break at specific points (see discussion under Squamata).

Species of *Sphenodon* are lizardlike in body form and reach about 300 millimeters in body length (snout to vent). Their common name, tuatara, comes

from Maori words meaning "spines on the back." The two extant species are morphologically similar but are differentiated genetically, as ascertained by allozyme analysis.

Tuatara are terrestrial and construct their own burrows or use those of ground-nesting seabirds. They bask in the sun at the entrances to their burrows, but are most active at night in cool ambient temperature. They are primarily insectivorous, but earthworms, snails, bird eggs, and nestling birds have also been recorded in the diet. Eggs are laid in shallow holes and development time is long, up to more than a year. *Sphenodon* is apparently long-lived, as captive individuals in zoos have lived for more than 50 years. Many populations of tuatara are urgently threatened owing to introduced rodents, rabbits, birds, and domestic animals.

Fossil Record. A single fossil species, *Sphenodon diversum*, has been described from New Zealand, although sphenodontids are represented by several fossil taxa (Gauthier et al. 1988). Sphenodontida is represented by fossils in the Lower Triassic.

 Content and Distribution. 2 species: *Sphenodon punctatus* and *S. guntheri*. Small islands off New Zealand (Figure 4–12).

 Systematics. Daugherty et al. (1990).

Squamata: Lizards and Snakes

Lizards and snakes together compose the Squamata, the sister group to *Sphenodon* among extant lepidosaurs (Figure 4–13). Snakes are cladistically nested within lizards, and the traditional classification of these two groups as coequal suborders of Squamata (Lacertilia and Serpentes) does not accurately reflect their interrelationships. In other words, Lacertilia is paraphyletic with respect to Serpentes. Nevertheless, because of extensive differences in physiology, behavior, and functional morphology, we consider lizards and snakes separately. This treatment should cause no

confusion as long as the relationships among squamates are kept in mind.

 Squamata can be diagnosed by more than 70 anatomical shared derived characters, primarily of the musculoskeletal system (Estes et al. 1988). The skulls are highly kinetic compared to the skulls of crocodilians, turtles, and *Sphenodon* (see Chapter 9). One of the most characteristic features of squamates are well-developed, paired copulatory organs called hemipenes (singular: hemipenis). These are outpocketings of the posterior wall of the cloaca that reside in the base of the tail of males (Figure 4–14) (Arnold 1984b). Penes of all other amniotes are single and, if internal, reside in the cloaca. A groove in the surface of each hemipenis, the sulcus spermaticus, transports sperm flowing from the cloaca during copulation. These organs are often supplied with spines, ridges, and other ornamentation, and the pattern of ornaments is widely used in squamate systematics. Varanid lizards have calcified structures (hemibacula) that add rigidity to their hemipenes (Card and Kluge 1995). Hemipenes are everted during copulation and retracted by their own intrinsic musculature. Only a single hemipenis is used during copulation. Female squamates have tiny homologues of hemipenes, dubbed hemiclitori (singular: hemiclitoris), housed within their tail bases (Arnold 1984b, Böhme 1995).

 Limb reduction has been a major evolutionary trend within Squamata and loss of limbs has occurred in multiple lineages, at least 62 separate times by one estimate (Greer 1991). In fact, limb reduction, meaning actual loss of bony elements from the ancestral

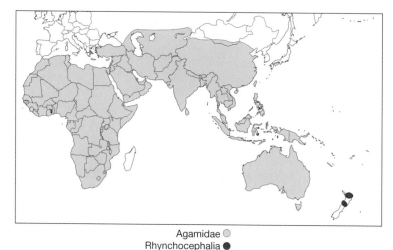

Agamidae ○
Rhynchocephalia ●
(Sphenodontidae)

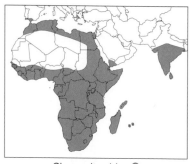

Chamaeleonidae ●

Figure 4–12 **Distribution of Rhynchocephalia (Sphenodontidae), Agamidae, and Chamaeleonidae.**

Figure 4–13 **Diversity of lizards.** (a) *Moloch horridus* (Agamidae, Australia). (b) *Chamaeleo oshaughnesseyi* (Chamaeleonidae, Madagascar). (c) *Brookesia nasus* (Chamaeleonidae, Madagascar). (d) *Diplodactylus spinigerus* (Gekkonidae, Australia). (e) *Lialis burtonis* (Pygopodidae, Australia). (f) *Amphisbaena fuliginosa* (Amphisbaenidae, South America). *(Photographs by (a,d,e) F. Harvey Pough, (b,c,f) John Cadle.)*

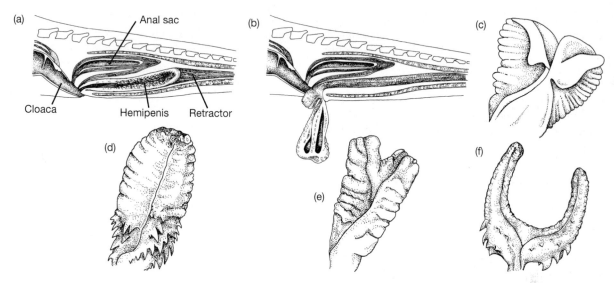

Figure 4–14 **Structure of hemipenes of squamates.** (a) and (b) Parasagittal sections through the base of a snake tail showing a hemipenis inverted (a) and everted (b). (c–f) Everted hemipenes of a lizard and several snakes showing diversity of external ornamentation: (c) *Lacerta agilis*, (d) *Spalerosophis diadema*, (e) *Epicrates angulifer*, (f) *Agkistrodon contortrix*. *(Source: Dowling and Savage 1960.)*

squamate condition, has occurred in every major group except gerrhonotine anguids, Lacertidae, Xenosauridae, Xantusiidae, Varanidae, and iguanians other than chameleons and agamids. Most reductions involve only a few phalanges. Limb reduction is estimated to have occurred only once or a few times in most squamate families, but limb reduction has evolved independently at least 25 times within the family Scincidae. Even many genera of skinks show extreme variation in limb development (Greer 1990). Snakes and many lizards (e.g., amphisbaenians and anguids such as *Ophisaurus* and *Anniella*) have taken this to the extreme and have lost most or all bony elements in the limbs. In virtually all cases in which more than a few bony elements have been lost, body elongation, as seen in snakes and amphisbaenians, has occurred (Gans 1975). Some lizards, such as *Ophisaurus*, have extremely elongate tails as well. The loss of limbs entails other structural changes associated with novel locomotory modes (see Chapter 8), and limbless squamates are often fossorial. The evolutionary and developmental mechanisms underlying limb reduction are as yet very poorly understood.

The skin of squamates is covered with scales of varying shapes and sizes, and their arrangement is important in systematics. Osteoderms often lie underneath the scales of lizards, but osteoderms are absent from snakes. In lizards the skin is sometimes supplied with discrete mite pockets, which vary from shallow depressions scarcely differentiated from the surrounding skin to very deep pockets often covered by an overhanging flap of skin (see Chapter 13). The skin of squamates is generally not richly supplied with glands, but conspicuous pores are often present in the preanal and ventral femoral region of lizards (preanal and femoral pores, respectively). These pores are openings to exocrine glands that produce chemicals used for marking territories and perhaps other functions. The prominence and number of these glands are often sexually dimorphic (males having larger and more numerous pores than females), and the number and location of pores are used in squamate systematics.

Caudal autotomy refers to the ability to lose the tail voluntarily. Although it is a shared derived character of Squamata, the ability to autotomize the tail has been lost many times (Arnold 1984b). Autotomy is observed in *Sphenodon*, many lizards, and several species of snakes. With few exceptions (see discussion under Agamidae), tail loss by these animals is facilitated by the development of fracture planes within specific vertebrae, and by the arrangement of caudal muscle bundles and connective tissue that permits easy separation (Figure 4–15). In some cases there is only a single fracture plane, but more commonly there are several or many fracture planes along the length of the tail. Autotomy is clearly an effective antipredator device. Tail regrowth occurs in most lizards after breakage, although the regenerated tail is shorter and is internally supported by a solid rod of cartilage rather than discrete bony vertebrae. Tail regeneration does not occur in the few colubrid snakes in which the tails autotomize (e.g., *Pliocercus, Scaphiodontophis*), and in snakes separation occurs between vertebrae.

Reproductive and sex-determining mechanisms are extremely diverse in squamates, and even casual inspection suggests that most patterns have evolved numerous times. Oviparity and viviparity occur widely in both lizards and snakes, and egg retention time may vary considerably even within a species. By one estimate, viviparity has arisen at least 45 times just within lizards (Blackburn 1982), and nearly half of these instances are within the family Scincidae. About 20 percent of all lizards are viviparous. Parthenogenesis occurs in the lizard families Agamidae, Chamaeleonidae, Gekkonidae, Teiidae, Lacertidae, and Xantusiidae and is known to occur in one snake, *Rhamphotyphlops braminus* (Typhlopidae). Sex-determining mechanisms have been investigated in relatively few squamates but, as with reproductive mode, tremendous diversity prevails. About a dozen species of lacertids and gekkonids are known to have temperature-dependent sex determination, but genetic sex

determination is also known in these families and in many others. Male heterogamety, female heterogamety, or homomorphic sex chromosomes may occur in both snakes and lizards. Temperature-dependent sex determination is not known in snakes.

Content and Distribution

Squamata includes 24 families of lizards, four families of amphisbaenians, and Serpentes. Lizards and amphisbaenians comprise approximately 420 genera and more than 3,300 species. By far the most species-rich family of lizards is Scincidae, with more than 1,000 species. Many new species of polychrotids, tropidurids, gekkonids, and scincids are discovered each year. Lizards are cosmopolitan in distribution except in extremely high latitudes.

Fossil Record

The earliest known lizards occur in middle Jurassic deposits of England and Scotland, although the presence of sphenodontians by the mid-Triassic suggests that squamates must have been present by then (Evans 1995). Lizards have a spotty fossil record until the Cretaceous, and then a fairly extensive one throughout the late Mesozoic and Cenozoic. Both scincomorph and anguimorph lizards are known from the mid-Jurassic. Gekkotans appear in the late Jurassic, and amphisbaenians and iguanians in the late Cretaceous. All of the Jurassic lizards occur on northern

Figure 4–15 **Morphology of tail autotomy in a lizard.** A fractured vertebra and the fracture planes in the centra of two adjacent vertebrae are shown. The arrangement of muscle bundles is shown around one centrum. When the tail is autotomized, the frayed ends of the muscle bundles collapse to seal the end of the broken tail. (*Source: Sheppard and Bellairs 1972.*)

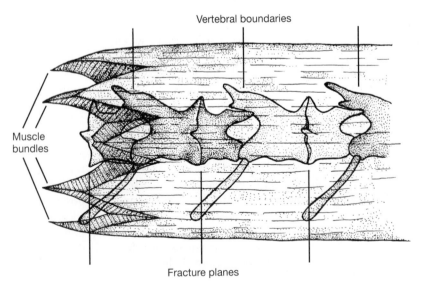

(Laurasian) land masses, and their early history in Gondwana is obscure. Overviews of the fossil record of lizards are given by Estes (1983), Evans (1993, 1995), Gao and Hou (1996), Albino (1996).

   Systematics. Estes and Pregill (1988).

Systematics and Phylogeny of Lizards

The foundation of modern squamate systematics is Camp's (1923) classic monograph. Despite many more characters and refined methods of analysis, the broad outlines of lizard relationships established by him are still largely accepted today. The most comprehensive estimate of squamate phylogeny is based on numerous anatomical characters (Estes et al. 1988). Two clades, the Iguania and Scleroglossa, represent the basal split in the squamate phylogenetic tree (Figure 4–11). Within the Iguania, Chamaeleonidae and Agamidae are thought to be sister taxa, based, among other characters, on sharing a pattern of tooth implantation referred to as the acrodont pattern (Figure 4–16). Acrodont teeth are attached to the apex of the dentaries, maxillae, and premaxillae, whereas the ancestral squamate pattern is a pleurodont dentition in which the teeth are cemented to the medial surface of the tooth-bearing

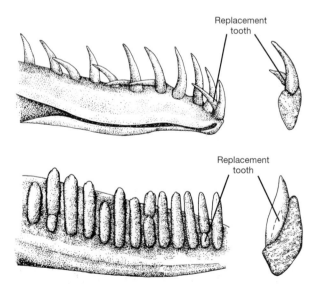

Figure 4–16 **Modes of tooth implantation among squamates.** (a) Acrodont, as seen in chameleons, agamids, and snakes. (b) pleurodont, as seen in other lizards. (*Source: Modified from Kardong 1995.*)

bones. Agamidae is possibly paraphyletic with respect to Chamaeleonidae (Frost and Etheridge 1989). An important point is that Amphisbaenia (amphisbaenians) and Serpentes (snakes) are nested within the Scleroglossa, whereas these two groups are often separated as taxonomic groups distinct from lizards. These evolutionary relationships have been recognized for many years (including Camp's treatise). Successive clades within Scleroglossa are the Gekkota, Scincomorpha, and Anguimorpha, with relationships as shown in Figure 4–11.

   The relationships of Dibamidae, Amphisbaenia, and Serpentes have proved difficult to estimate. In part this difficulty results from loss of limbs in these groups, because many of the informative characters within Scleroglossa are characters of limb morphology. In other respects the difficulty arises from the extreme and often peculiar modifications that obscure their relationships to other squamates. Each of these groups has been associated with several clades of Scleroglossa. For example, in the case of Serpentes, most evidence supports a position within Varanoidea (Figure 4–11) (McDowell and Bogert 1954), and Wu (1996) suggested that the relationships of Amphisbaenia are with Scincomorpha.

   To readers familiar with classical lizard taxonomy, the greatest difference in this account will be the recognition of several families for the traditional family Iguanidae (Frost and Etheridge 1989). The change is not as dramatic as it might appear, however. These new families of iguanians were informally recognized clades for many years and have simply been given formal taxonomic status because Frost and Etheridge (1989) identified no derived characters uniting all of them. However, a recent molecular study (Macey et al. 1997) supports the monophyly of Iguanidae in the traditional sense and recommends a return to previous nomenclature.

   Controversy surrounds some details of lizard phylogeny. A case in point is the classification of Gekkota (Gekkonidae + Pygopodidae). Kluge's (1987) analysis of gekkonoid relationships revealed one shared derived character shared by pygopodids and diplodactyline geckos, which he included in the Pygopodidae. Others (Estes et al. 1988) advocated retaining the more traditional arrangement until stronger support for the alternative arrangement is produced. Similarly, the recognition of Gymnophthalmidae (microteiids) as distinct from Teiidae (macroteiids) is advocated by some but not all workers. An extended treatment of squamate phylogeny, concentrating on lizards, is given in Estes and Pregill (1988). Readers interested in lists of derived characters for major lineages will find them there.

Agamidae

Agamids are moderate-sized to large (*Hydrosaurus* is more than 1 meter in total length), diurnal, primarily terrestrial lizards. Limbs are well developed, and there are no trends toward reduction or loss. The scales are often modified to form extensive crests, frills (e.g., in *Chlamydosaurus*), or spines (*Moloch*). In many agamids these features are sexually dimorphic and are used in intraspecific interactions. Southeast Asian agamids of the genus *Draco* are the only lizards capable of true gliding flight, which they accomplish by extending elongate and highly mobile ribs that support a thin flight membrane made of skin. The Australian *Moloch* is similar in overall appearance to the North American *Phrynosoma* (Phrynosomatidae) in having a tanklike, heavily spined body and short tail (Figure 4–13). Like most species of *Phrynosoma*, it feeds primarily on ants. With the exception of one genus of live-bearing agamids, *Phrynocephalus*, agamids are oviparous. Many species of *Agama* live in dense colonies with well-developed social hierarchies and territories. Several agamids, including *Agama*, *Physignathus*, and *Laudakia*, have evolved intervertebral autotomy, an unusual contrast to the intravertebral autotomy characteristic of other lizards.

Content and Distribution. Approximately 45 genera, 300 species. Africa, southern and Central Asia, and the Indoaustralian Archipelago to Australia, New Guinea, and the Solomon Islands (Figure 4–12). Agamidae is of uncertain monophyly and has traditionally included those lizards with acrodont dentition that were not in Chamaeleonidae. These two groups are often referred to as the monophyletic group Acrodonta (Figure 4–11).

Systematics. Borsuk-Bialynicka and Moody (1984), Joger (1991).

Chamaeleonidae

Chameleons are among the most easily recognizable lizards because of the extensive development of casques, horns, and crests on the head in most species (Figure 4–13). They are well known for their ability to change color. Their feet are zygodactylous, with adjacent digits fused on each hand and foot, forming opposable grasping pads. The tails are prehensile in arboreal species. Their tongues are extremely elongate, involving many peculiar modifications of the hyoid apparatus (Bell 1989, Wainwright et al. 1991). Chameleon eyes can move independently and have many structural modifications that are unique among vertebrates. Unusually, chameleons use accommodation (focusing depth) to measure the distance of objects, whereas virtually all other vertebrates use triangulation. Chameleons have well-known abilities to accurately fire their long tongues to hit distant targets (Figure 9–23).

Most chameleons have laterally compressed bodies. The size range is from less than 25 millimeters total length to more than 500 millimeters. Chameleons are exclusively diurnal and primarily insectivorous. However, birds have been recorded from the diets of the largest species (*Chamaeleo melleri* and *C. oustaleti*). Most chameleons are highly arboreal, but the dwarf chameleons of the genus *Brookesia* (Madagascar) are exceptions to many of the generalizations usually made about chameleons. They are small, drab lizards (usually brown to gray) with little ability to change colors. They have short, nonprehensile tails and are terrestrial inhabitants of rain forests, although they retain the zygodactylous feet characteristic of the family. Most chameleons are oviparous, but species of *Bradypodion* (Africa) and some species of *Chamaeleo* are viviparous. Dwarf forms of chameleons have evolved in both Madagascar (*Brookesia*) and Africa (*Bradypodion* and *Rhampholeon*).

Content and Distribution. 4–6 genera, approximately 130 species. Africa and Madagascar, the Middle East, India and Sri Lanka, and southern Spain. More than half the species are endemic to Madagascar (Figure 4–12). Several phylogenies and systematic arrangements have been proposed, but they disagree substantially.

Systematics. Hillenius (1986, 1988), Klaver and Böhme (1986), Rieppel (1987), Hofman et al. (1991), Raxworthy and Nussbaum (1995).

Iguanidae

Iguanids are moderate-sized (14 centimeters in snout–vent length in *Dipsosaurus*) to large (more than 70 centimeters in *Cyclura*) lizards that may be terrestrial (*Dipsosaurus*, *Cyclura*), rock-dwelling (*Sauromalus*, *Ctenosaura*), or arboreal (*Iguana*, *Brachylophus*). *Amblyrhynchus* has salt glands, freely enters the ocean, and feeds on marine algae. Iguanids are primarily herbivores as adults and consume leaves, fruits, and flowers. In all genera except *Amblyrhynchus* the colon is provided with partitions (Figure 9–42) that probably serve several functions associated with digestion of plants, including slowing the passage of food through the gut, increasing the surface area available for absorption, and providing a microhabitat for an extensive nematode fauna involved in the breakdown of cellulose (Iverson 1980).

Content and Distribution. 8 genera, approximately 34 species. *Amblyrhynchus* and *Conolophus*, Galápagos Islands; *Cyclura*, West Indies; *Brachylophus*, Fiji; *Iguana*, *Ctenosaura*, *Dipsosaurus*, and *Sauromalus*, southwestern United States through tropical South America (Figure 4–17).

Systematics. De Queiroz (1987), Etheridge and De Queiroz (1988).

Opluridae

Oplurids are terrestrial (*Chalarodon*) to rock-dwelling or arboreal (*Oplurus*) lizards that primarily occur in subhumid to arid areas. *Chalarodon* is a rather small lizard (200 millimeters in total length), whereas *Oplurus cuvieri* reaches nearly 400 millimeters. Several species have moderately to greatly spinose tails. These oviparous lizards may be brightly colored during the breeding season, and males actively defend territories with body postures and displays.

Content and Distribution. *Oplurus* (6 species) and the monotypic *Chalarodon madagascariensis*. Madagascar (Figure 4–23).

Systematics. Titus and Frost (1996).

Phrynosomatidae

Phrynosomatidae includes lizards with a broad array of morphological and ecological types. Species of *Phrynosoma* are small, extremely spiny lizards with flat, short bodies and tails. They are solitary and cryptic desert species. *Phrynosoma* exhibits a peculiar defensive be-

Figure 4–17 **Distribution of the lizard families Iguanidae, Phrynosomatidae, Hoplocercidae, Tropiduridae, Polychrotidae, Crotaphytidae, and Corytophanidae.**

Iguanidae ○

Tropiduridae ●

Phrynosomatidae ○
Hoplocercidae ●

Polychrotidae ○

Corytophanidae ●
Crotaphytidae ○

havior in which blood is squirted from sinuses in the orbits (Figure 13–16). This defense is elicited most frequently in response to predation by canids and may have evolved in response to predation by coyotes and foxes (Middendorf and Sherbrooke 1992). *Sceloporus*, the phrynosomatid genus containing the largest number of species, includes terrestrial, arboreal, and rock-dwelling forms, in habitats from tropical rain forests to high mountain scrub. *Phrynosoma douglassi* and some species of *Sceloporus* are viviparous. Species of *Uma* are sand-diving lizards adapted to areas of loose sand. Many aspects of their morphology are modified for such habitats, including the development of fringes on the toes, countersunk lower jaws, and modifications of the eyelids and nostrils (Arnold 1995).

Because of their accessibility to North American investigators, short life spans, and often high population densities, phrynosomatids have become model systems for research in the ecology and cytogenetics of natural vertebrate populations. In particular, *Sceloporus* has figured prominently in the development of life history evolution theory and in the integration of physiological and population ecology (e.g., Tinkle and Dunham 1986, Dunham et al. 1989). Similarly, complex chromosomal polymorphisms and hybridization among species of the *Sceloporus grammicus* complex have provided fruitful systems for understanding hybrid zones and chromosome evolution (e.g., Reed et al. 1995a,b).

Content and Distribution. 10 genera, approximately 125 species. Southern Canada to Panama (Figure 4–17). *Sceloporus* (80 species) and *Phrynosoma* (13 species) account for most of the diversity.

Systematics. Reeder and Wiens (1996), Sites et al. (1992; *Sceloporus*).

Tropiduridae

Tropidurids inhabit an extreme range of habitats, including lowland rain forests or dry forests (*Stenocercus*, *Tropidurus*, and *Uracentron*), savannas (*Tropidurus*), deserts (*Ctenoblepharys* and *Tropidurus*), and high mountain grasslands above tree line (*Liolaemus* occurs at nearly 5,000 meters altitude in the Andes). All species are diurnal. Body form and natural history are variable in this group. Species may be sand-dwellers (*Ctenoblepharys* and some *Liolaemus*), rock-dwelling (some *Stenocercus*, *Tropidurus*, and *Liolaemus*), terrestrial (*Tropidurus*, *Liolaemus*), or arboreal (*Uracentron*). Most species are oviparous, but some *Liolaemus* species are viviparous. Most tropidurids are relatively

small lizards and are primarily insectivorous. *Uracentron azureum* and *Strobilurus torquatus* are arboreal species of Amazonian and Atlantic forests of South America. Both have very spiny tails and are ant specialists. A few *Liolaemus* species are herbivorous, even though they are only 50 to 60 millimeters in snout–vent length.

Content and Distribution. 9–12 genera, approximately 270 species. South America, West Indies (*Leiocephalus*), Galápagos Islands (*Microlophus*) (Figure 4–17). The taxonomy of *Liolaemus* and *Stenocercus* is poorly understood, and many new species in these genera are being discovered. Frost (1992) synonymized *Uracentron*, *Plica*, and *Strobilurus* with *Tropidurus*. The family is of uncertain monophyly (Titus and Frost 1996).

Systematics. Frost (1992), Pregill (1992), Etheridge (1995).

Polychrotidae

Polychrotids include some of the most familiar (*Anolis*) and least familiar (*Urostrophus*, *Anisolepis*) lizards to North American students. Since *Anolis* has been the subject of innumerable studies in behavior, physiology, and ecology that will appear in other chapters, this account concentrates on the biology of lesser known polychrotids, although we begin with *Anolis* and its relatives.

Anolis is a huge genus (nearly 400 recognized species and many more undescribed) that has been well studied in the West Indies (Williams 1983). It is even more diverse in Central and South America, but those species are poorly known. Some workers divide *Anolis* into several genera. *Anolis* and its close relatives, *Chamaeolis* (= *Chamaeleolis*), *Chamaelinorops*, and *Phenacosaurus*, are referred to as anoles. All anoles are characterized by the presence of subdigital lamellae bearing setae similar to those of geckos (Ruibal and Ernst 1965). The setae of anoles are simple and unbranched, and are much smaller than those of geckos (see discussion under Gekkonidae). *Anolis* species are equipped with brightly colored dewlaps that are used in intraspecific communication. *Phenacosaurus* includes ten species from relatively high elevations in the Andes of northern South America and also from the high, flat-topped tablelands known as tepuis in the Guianan region of South America. They are characterized by well-developed casquing on the head, and inhabit grass- and bushland or mountainous rain forests. The other anole genera, *Chamaeleolis* and *Chamaelinorops*, are restricted to the Greater Antilles.

Chamaeolis is a large arboreal anole with extensive casquing, whereas *Chamaelinorops barbouri* is a small, terrestrial, shade-loving species of mountain forests.

Polychrus (five species) are moderately large, long-legged arboreal lizards of wet and dry forests. They move very slowly and have long, prehensile, nonauto-tomic tails. The other polychrotids are less well known. Species of *Urostrophus* and *Anisolepis grilli* have relatively long tails and are mainly arboreal lizards of southern South America. Other species of *Anisolepis* occur in seasonally flooded grasslands. Little is known of the natural history of *Enyalius*, but the species are known from Atlantic and Amazonian forests of east-ern Brazil. *Enyalius leachii* has a color pattern match-ing dead leaves and becomes immobile when dis-turbed. *Pristidactylus* species are known from both *Nothofagus* (southern beech) forests and open forma-tions of southern South America. They are terrestrial or rock-dwelling.

Content and Distribution. 11 genera, >440 species. Southeastern United States through Central America and most of South America; West Indies (Figure 4–17).

Systematics. Williams (1976a,b, 1988, 1989), Etheridge and Williams (1985, 1991), Guyer and Savage (1986, 1992), Cannatella and de Queiroz (1989), Williams et al. (1996).

Hoplocercidae

Hoplocercids are moderate-sized lizards up to about 160 millimeters in snout–vent length. *Enyalioides* and *Morunasaurus* are rain forest lizards, whereas *Hoplocer-cus* is found primarily in the dry forests known as *cer-rados* in Brazil. These are primarily terrestrial lizards with spiny tails (extremely so in *Hoplocercus*), and sev-eral species are known to use burrows in the ground.

Content and Distribution. 3 genera, 10 species: *Enyalioides* (Panama, northwestern South America, Amazo-nia); *Hoplocercus* (cerrados of Brazil; *Morunasaurus* (disjunct: eastern Peru and Ecuador, Panama) (Figure 4–17). The sys-tematics of this group is poorly understood, and both gen-era are possibly paraphyletic.

Systematics. Avila-Pires (1995), Etheridge and de Queiroz (1988).

Crotaphytidae

Crotaphytids are moderately large (approximately 100 to 145 millimeters in snout–vent length) lizards of mesic to arid areas of North America. Species of *Gambelia* and *Crotaphytus reticulatus* occur in flatland deserts, using rocky outcrops occasionally, whereas other species of *Crotaphytus* are exclusively rock-dwelling. In addition to insects, most crotaphytids consume some proportion of vertebrates, especially other lizards, but also rodents and snakes. Vertebrates are major dietary items for *Gambelia wislizenii* and *G. copei*. Unlike all other iguanians except polychrotids, crotaphytids use squealing vocalizations when stressed. Autotomic fracture planes are present in the tails of *Gambelia*, whereas they are absent from *Crota-phytus*. *Crotaphytus* uses an unusual form of saltatorial bipedalism in moving among boulders.

Content and Distribution. 2 genera (*Crotaphytus, Gambelia*), 12 species. South central United States to north-ern Mexico (Figure 4–17).

Systematics. McGuire (1996).

Corytophanidae

These lizards are recognizable by their well-devel-oped head crests and casques. The crests are sexually dimorphic in *Basiliscus* (only males have crests), whereas both sexes have them in *Corytophanes* and *Laemanctus*. *Corytophanes* uses the crest in a defensive display in which the lateral aspect of the body is ori-ented toward a predator, thus making the lizard ap-pear larger than it is. *Basiliscus* species are well known for their ability to run bipedally across the surface of water, and their toes are equipped with enlarged squarish scales associated with this behavior (see Chapter 8). They even occasionally seek refuge un-derwater, lodging themselves among boulders on the bottom for short periods.

Content and Distribution. 3 genera (*Basiliscus, Coryto-phanes, Laemanctus*), 9 species. Central Mexico to north-western South America (Figure 4–17).

Systematics. Lang (1989).

Eublepharidae and Gekkonidae

Members of these two families are commonly called geckos and are sometimes placed together in one fam-ily. Eublepharids are unusual geckos in that most are terrestrial and lack the subdigital setae that give gekkonids exceptional climbing ability. Eublepharids are also the only geckos that retain eyelids. In Gekkonidae the eyes are covered by an immovable spectacle. Most eublepharids are inhabitants of arid or subhumid environments, and some (e.g., *Hemitheconyx* and *Holodactylus*) are highly modified for life in sandy habitats. *Hemitheconyx taylori* is even semifossorial. Within *Coleonyx*, the species of Central America inhabit subhumid tropical forests, whereas the northern species inhabit deserts.

The setae of geckos are projections of the highly modified scales on the ventral surface of the digits (subdigital lamellae or toe pads). Each seta is a complexly branching, hairlike projection 60 to 90 microns in length, with each tip bearing a spatulate expansion. The setae of geckos are much larger and more complex in structure than those of *Anolis* and other polychrotids (see Chapter 8). The elaborate arrangements of lamellae and details of setal structure are important characters in the systematics of gekkonids. Not surprisingly, given the elaborate development of digital modifications, no trends toward limb reduction are seen in gekkonids, although digital reduction has occurred in some cases.

Most geckos are nocturnal, although some gekkonids of tropical forests are diurnal (e.g., *Gonatodes* and *Sphaerodactylus* of the Neotropics and *Lygodactylus* of Africa and Madagascar), as are some species of desert regions (e.g., *Rhotropus* and *Quedenfeldtia* of Africa). Species of *Phelsuma* (Africa, Madagascar, the Indian Ocean islands), commonly called day geckos, are all diurnal and heliotropic. Few geckos are colorful, but many species of *Phelsuma* have bright colors of green, blue, yellow, and red.

Although the digital modifications of geckos are primarily for life on vertical surfaces such as rock faces or in trees, many species are secondarily terrestrial and show a reduction in the elaboration of digital pads or other specializations for terrestriality. For example, *Palmatogecko* (Gekkonidae) of southern Africa lives in areas of shifting sand and has developed extensive webbing between the digits. Similar modifications have occurred repeatedly in gekkonid evolution in response to similar environmental demands.

Vocal communication is rare among squamates, but many geckos other than eublepharids and sphaerodactylines emit clicks or chirps for intraspecific communication, including territorial calls, intersexual communication, and distress calls (see Chapter 12).

Geckos are oviparous with the exception of the diplodactyline genera *Naultinus*, *Heteropholis*, *Hoplodactylus*, and *Rhacodactylus*, which are viviparous. There are only one or two eggs per clutch in oviparous species. The eggshells of eublepharid and diplodactyline geckos are leathery, like those of virtually all squamates. However, the eggshells of gekkonines and sphaerodactylines harden on exposure to air. Calcium in the form of calcium carbonate is stored in expansions of the endolymphatic system onto the neck of gekkonines and sphaerodactylines, and is used for eggshell strengthening and for rapid bone growth. Parthenogenesis is known to occur in a number of gekkonids, including *Heteronotia binoei*, several species of *Hemidactylus*, *Lepidodactylus lugubris*, and *Nactus pelagicus*. As in other parthenogenetic squamates, most of these are species complexes of diploid and polyploid forms, and the parthenogens result from hybridization. Temperature-dependent sex determination is known to occur in *Eublepharis macularius* and species of *Gekko*.

Defensive mechanisms are especially well developed among gekkonids. Tails of all species are highly autotomic. The skin of many species is especially fragile and tears easily, making it difficult for predators to grab them. *Geckolepis* of Madagascar loses the entire skin if it is seized. The skin regenerates within weeks. In this case, skin loss is facilitated by a splitting zone within the connective tissue underlying the epidermis (Schubert et al. 1990).

Content and Distribution. Eublepharidae: 5 genera, about 25 species (extremely disjunct: southwestern United States and Central America; East and West Africa; southern Asia; extreme Southeast Asia and islands of the Sunda Shelf) (Figure 4–18). Gekkonidae: 82 genera, about 870 species (cosmopolitan except northern North America and Eurasia) 3 subfamilies of Gekkonidae: Gekkoninae (65 genera; cosmopolitan in tropical and subtropical regions), Sphaerodactylinae (5 genera; Central and South America, West Indies), Diplodactylinae (12 genera; Australian region, New Zealand). Kluge (1987) included the Diplodactylinae in the Pygopodidae because of a shared derived character of the muscle encircling the external ear opening.

Systematics. Grismer (1988; Eublepharidae), Kluge (1967, 1987), Harris (1982), Harris and Kluge (1984; Sphaerodactylinae).

Pygopodidae

Pygopodids are elongate, snakelike lizards in which forelimbs are absent and the hind limbs are reduced to a scaly flap at the level of the vent (Figure 4–13). The eyes are covered by an immovable spectacle, as in snakes, and an external ear opening may be absent. The tails are very long and autotomic. Pygopodids lay clutches of one to three eggs. Some species are burrowers, whereas others are terrestrial, hiding in leaf litter or under surface objects, and some climb in low vegetation. A few, such as *Aclys*, *Ophidiocephalus*, and *Pletholax*, are highly modified sand-swimmers. *Delma* and some species of *Pygopus* are nocturnal, whereas

Figure 4–18 **Distribution of the lizard families Gekkonidae and Eublepharidae.**

Gekkonidae

Eublepharidae

other species of *Pygopus* are diurnal, and *Lialis* has been reported as being both nocturnal and diurnal. Most pygopodids consume arthropods (insects and spiders), but *Lialis burtoni* and *L. jicari* eat relatively large scincid lizards. Predation on skinks by *Lialis* is facilitated by unusual anatomical modifications, including pointed, recurved, hinged teeth and highly mobile kinetic joints in the elongate skulls (Patchell and Shine 1986a,b).

Content and Distribution. 8 genera, about 36 species (Australia and New Guinea) (Figure 4–19).

Systematics. Kluge (1974, 1976).

Amphisbaenia

The Amphisbaenia are so different in many respects from other squamates that we depart from our separate family accounts to discuss their unusual features as a group. Amphisbaenians are elongate squamates and, with the exception of *Bipes*, are limbless (Figure 4–13). Their pelvic and pectoral girdles are variably reduced or absent in different genera. Most amphisbaenians have very short tails and their bodies are distinctly annulated. Sizes range from adult body lengths of about 10 centimeters in some species of African *Chirindia* to more than 70 centimeters in the South American *Amphisbaena alba*.

Amphisbaenian skulls are heavily ossified and modified for digging (see Chapter 8). Uniquely among squamates, the brain is entirely surrounded by the frontal bones. In most cases the skin can move entirely independently of the underlying trunk. This peculiar modification is associated with their use of rectilinear locomotion within subterranean tunnels: they are able to use virtually any point along the body as a fixed point to anchor against tunnel walls, and the trunk can move forward within the skin during bur-

Figure 4–19 **Distribution of the lizard families Pygopodidae, Lacertidae, Gerrhosauridae, Teiidae, Varanidae, Lanthanotidae, and Helodermatidae.**

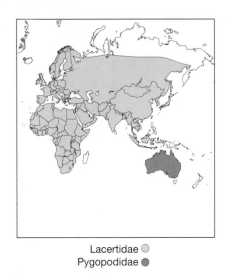

Lacertidae ○
Pygopodidae ●

Gerrhosauridae ●
Teiidae ○

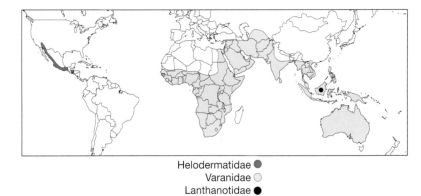

Helodermatidae ●
Varanidae ○
Lanthanotidae ●

rowing. This anatomical feature reduces the drag created by friction of the skin against the tunnel walls. Such rectilinear locomotion is unknown in squamates other than amphisbaenians and some snakes.

Unlike the primary annuli of caecilians and the ventral scales of snakes, which have a one-to-one correspondence to the vertebrae, there are two body annuli per vertebra in all amphisbaenians other than *Blanus*. Some species of amphisbaenians have autotomic tails, but regeneration does not occur if the tail is lost. Head shape varies considerably in amphisbaenians, generally in accordance with the type of burrowing particular species use. Most species have relatively rounded heads, but parallel evolution of keel-headed and spade-shaped heads has occurred on all continents (Gans 1987).

Most elongate vertebrates, including caecilians, snakes, and limbless lizards, undergo reduction in one

of the lungs, and in all of these groups the left lung is reduced. In amphisbaenians the right lung is reduced.

Amphisbaenians occupy environments ranging from lowland rain forests through deciduous subtropical forests to extremely arid deserts. All species burrow, but some occasionally venture onto the surface or can be found under cover objects. All amphisbaenians appear to be oviparous except for several species of *Trogonophis* that are viviparous.

Given the specialized environmental space used by amphisbaenians, it is perhaps not surprising that predators specializing on them have also evolved. In southern Africa a small colubrid snake, *Cryptolycus nanus*, is extremely modified for a burrowing existence and is often associated with the amphisbaenid *Chirindia swynnertoni*. Dietary studies suggest that *Cryptolycus* eats nothing but *Chirindia*. The presumed closest relatives of *Cryptolycus*, terrestrial snakes of the genus *Lycophidion*, are skink specialists (Broadley and Gans 1978).

Amphisbaenians are classified in four families. The Amphisbaenidae is the largest, most widespread, and shows the greatest diversity in burrowing specializations within the group. The Mediterranean genus *Blanus* is usually placed in the Amphisbaenidae, but it has a number of unique features (e.g., 1:1 ratio of annuli to vertebrae) and shares some unusual features with the other families. The single extant species of Rhineuridae, *Rhineura floridana* of central Florida, is unique in having a spatulate digging shield and in some aspects of its internal anatomy.

Species of Trogonophidae are unusual amphisbaenians in many respects. They lack caudal autotomy, have acrodont teeth (pleurodont in other amphisbaenians), and have relatively shorter trunks than other amphisbaenians. The bodies of trogonophids are triangular in cross section, and they use oscillating rather than rectilinear movements in digging. Unlike other amphisbaenians, trogonophids use their tail to apply force during burrowing. In these amphisbaenians, the skull actually rotates on the vertebral column during penetration of the soil.

Species of *Bipes* (Bipedidae) are remarkable among squamates in retaining only the forelimbs. Moreover, the hands have supernumerary phalanges, a feature otherwise very rare in tetrapods. The hands are used while burrowing into soil from the surface, but not for locomotion within subterranean tunnels. *Bipes* individuals have very elongate bodies and live in arid scrub lands or deserts. The short tails are autotomic and have only a single fracture plane.

Content and Distribution. Amphisbaenidae: approximately 19 genera, 133 species; West Indies, South America, sub-Saharan Africa, and disjunct circum-Mediterranean areas. Rhineuridae: 1 species (*Rhineura floridana*); northern Florida. Trogonophidae: 4 genera, 6 species; disjunct distribution in northern Africa and the Middle East. Bipedidae: 1 genus (*Bipes*), 3 species; southern Baja California and the mainland Mexican states of Guerrero and Michoacan (Figure 4–20).

Systematics. Gans (1978, 1990), Papenfuss (1982; *Bipes*).

Dibamidae

These are small (250 millimeters in maximum total length), attenuated burrowers having vestigial eyes covered by a scale. They lack external ear openings. Both ends of the body are blunt. Males have small, flaplike hind limbs similar to those of pygopodids, and females are limbless. Dibamids are oviparous, having a clutch size of one egg in known cases. *Anelytropsis* is found in rather dry habitats in Mexico ranging from forests to deciduous scrubland and pine–oak forests. *Dibamus* species are known from rain forests and secondary forest in Asia. Although the two genera of dibamids have a disjunct distribution, they share a long list of unusual morphological attributes that corroborates their relationship.

Content and Distribution. 2 genera: *Anelytropsis* (1 species; northeastern Mexico), and *Dibamus* (9 species; extreme southeast Asian mainland [Vietnam and Malaysia] east through the southern Philippine Islands, islands of the Sunda Shelf to extreme western New Guinea) (Figure 4–21).

Systematics. Greer (1985).

Teiidae and Gymnophthalmidae

Teiids and gymnophthalmids are often informally referred to as macroteiids and microteiids, respectively, based on a crude distinction in modal body size between the two. However, sizes overlap considerably when the extremes are considered. The smallest species of gymnophthalmids are scarcely 10 centimeters in total length, whereas the teiids *Tupinambis* and *Dracaena* may reach a total length of about 1 meter and 1.3 meters, respectively. These two families are sister groups within Scincomorpha and are sometimes considered a single family, Teiidae (Figure 4–11)

Except for secretive and burrowing species (mostly microteiids), these lizards are active, diurnal species (Figure 4–22). Their habitats range from extremely arid deserts to tropical rain forests to high-altitude *paramos* in the Andes. Species in three genera, *Neusticurus*, *Dracaena*, and *Crocodilurus*, are semiaquatic and will freely enter water for refuge and foraging. *Dracaena*, a large lizard of Amazonian rain forests, is primarily arboreal and has molariform teeth which it uses to crush large snails, its principal food. It may be seen in trees over watercourses or swimming about in flooded forests. All teiids and gymnophthalmids are oviparous, and communal nesting has been reported for species of both families. Parthenogenesis is prevalent among teiids (*Cnemidophorus*, *Kentropyx*, and *Teius*) and gymnophthalmids (*Gymnophthalmus*) (see Chapter 7).

Limb reduction does not occur in macroteiids but has occurred multiple times within microteiids, and in many of these it is accompanied by body elongation.

Figure 4–20 **Distribution of lizard families Amphisbaenidae, Trogoniphidae, Bipedidae, and Rhineuridae.**

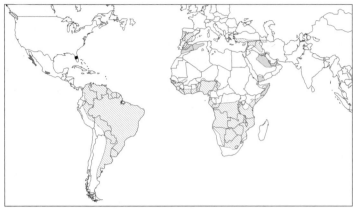

Bipedidae ●
Amphisbaenidae ◌
Trogonophidae ○
Rhineuridae ●

In *Bachia*, the limbs are reduced to tiny flaps that are useless for locomotion, and in *Calyptommatus* limbs are entirely lacking. These lizards are often burrowers in tropical forests (*Bachia*) or occupy specialized habitats such as loose sand (*Calyptommatus*).

Content and Distribution. Teiidae (macroteiids; 9 genera, 105 species). Northern United States through Central America and most of South America; West Indies (Figure 4–19). Gymnophthalmidae (microteiids; 31 genera, 140 species). Southern Mexico to Argentina (Figure 4–21).

Systematics. Dixon (1973; *Bachia*), Presch (1974, 1980), Rieppel (1980), Harris (1985, 1994; *Ptychoglossus*), Kizirian (1996; *Pholidobolus*).

Lacertidae

Lacertids are small to moderate-sized lizards with well-developed limbs. Dorsal scales are usually small and granular. Most species are active terrestrial or rock-dwelling lizards, and all are diurnal. They are primarily insectivores, although some are partially herbivorous. Lacertids are oviparous except for most populations of *Lacerta vivipara* (Europe). Parthenogenesis, resulting from interspecific hybridization, occurs in five species of *Lacerta* in the region of the Caucasus Mountains of southwestern Asia (Uzzell and Darevsky 1975).

Outside Europe, lacertids have diversified into many kinds of habitats. Several species of *Meroles* in southern Africa are adapted to habitats of windblown sand and have developed fringes on the feet to permit walking on these loose surfaces. Their heads are pointed and have countersunk lowerjaws to facilitate their sand-diving habits (Arnold 1995). Juveniles of

Heliobolus lugubris of the Kalahari Desert mimics, in size, behavior, and color pattern, a species of noxious beetle living sympatrically, thus avoiding predation by birds and mammals (Huey and Pianka 1977). The coloration becomes cryptic as individuals grow larger than the beetle models (Huey and Pianka 1977).

Content and Distribution. 29 genera, approximately 215 species. Africa, Eurasia, and islands of the Sunda Shelf (Figure 4–19).

Systematics. Boulenger (1921), Arnold (1989a,b, 1991).

Xantusiidae

Xantusiids are relatively small lizards that occur in humid to arid habitats (Figure 4–22). Several species are habitat specialists (Bezy 1988, 1989a). For example, *Xantusia vigilis* is often associated with the Joshua tree formations of the Mojave desert, and *X. henshawi* occurs only where there are exfoliating granitic outcrops. Species of *Lepidophyma* occur in habitats ranging from rain forests and dry forests to mountainous conifer forests. Several species are associated with limestone outcrops or boulder jumbles and live in caves and crevices of such habitats. Xantusiids are long-lived for such small animals, up to 10 years in the case of *X. vigilis* (10 centimeters in total length). Xantusiids have relatively flat bodies and heads, lack movable eyelids, are secretive, often nocturnal, and relatively sedentary lizards. Several species of *Lepidophyma* reproduce parthenogenetically but, unlike the situation in parthenogenetic teiid lizards, parthenogens of *Lepidophyma* do not appear to result from interspecific hybridization. Moreover, some parthe-

Figure 4–21 **Distribution of the lizard families Dibamidae, Xenosauridae, Gymnophthalmidae, and Xantusiidae.**

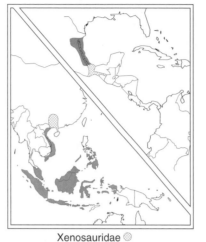

Xenosauridae ⊗
Dibamidae ●

Gymnophthalmidae ●

Xantusiidae ●

nogenetic *Lepidophyma* species are bisexual and others are unisexual (all-female). Within the species *L. flavimaculatum*, populations from southern Central America are all-female, whereas populations from Guatemala and Honduras contain both males and females (Bezy 1989b). All xantusiids are live-bearers.

Content and Distribution. 3 genera, 17 species: *Cricosaura* (eastern Cuba), *Lepidophyma* (Mexico to Panama), *Xantusia* (southwestern United States to northern Mexico) (Figure 4–21).

Systematics. Bezy (1989a), Crother and Presch (1992), Hedges and Bezy (1993).

Scincidae

Most skinks are characterized by smooth, shiny cycloid scales that are underlain by osteoderms, thus giving a very hard exterior to the body. The osteo-derms of skinks are unusual in that each is composed of a mosaic of smaller bones rather than a single bone as in most other lizards. Also unusually among squamates, many skinks have a well-developed secondary palate.

Body forms are extremely variable, from rather typical lizards with robust limbs and short bodies to limbless species with very elongate, snakelike bodies (Figure 4–22). Consequently, habitat diversity is also relatively great, although terrestrial and fossorial skinks are more prevalent than arboreal and aquatic species. Many species in desert regions are sand-swimmers. One Malagasy skink, *Amphiglossus astrolabi*, is aquatic and can be caught using fish traps. *Cryptoblepharus boutonii* is a small skink native to intertidal regions of southern Africa. It feeds on copepods (small invertebrates known as sand fleas), and its high tolerance for saline environments has facilitated its

Figure 4–22 **Diversity of lizards.** (a) *Tupinambis nigropunctatus* (Teiidae, South America). (b) *Proctoporus ventrimaculatus* (Gymnopthalmidae, South America). (c) *Xantusia riversiana* (Xantusiidae, California). (d) *Ctenotus fallens* (Scincidae, Australia). (e) *Shinisaurus crocodilurus* (Xenosauridae, China). (f) *Gerrhonotus multicarinatus* (Anguidae, western United States). *(Photographs by (a,c,d,e,f) F. Harvey Pough, (b) John Cadle.)*

colonization of islands throughout the Indo-Pacific region. The size range of extant skinks is great, from small species of less than 5 centimeters in snout–vent length to *Tiliqua scincoides* (Australia) and *Macroscincus coctei* (Cape Verde Islands), at more than 32 centimeters in snout–vent length.

Most skinks are diurnal, although some are nocturnal or crepuscular. Lygosomines are numerically dominant among skinks and reach their highest diversity in the Australian region. They have colonized many oceanic islands of the Pacific. A single scincine genus, *Eumeces*, is found throughout North and Central America and is also represented in north Africa, Southwest Asia, and Southeast Asia. Species of Feyliniinae and Acontiinae are small groups of limbless African skinks. Nearly 45 percent of skink species are viviparous.

Limb reduction and loss has occurred many times within Scincidae, and intrageneric variation in its occurrence is prevalent. For example, the Australian genus *Lerista* (>60 species) includes species with fully developed limbs and others with some degree of digit and limb reduction. Some species of *Lerista* have totally lost the forelimbs, but at least very rudimentary hind limbs are always retained. In another Australian genus, *Anomalopus*, both limbs may be totally absent. Similar patterns of limb reduction and loss are repeated in African skinks, where some of the most specialized limbless burrowing skinks occur (*Typhlosaurus* and *Acontias*).

Because of their hard, rounded bodies, skinks are difficult lizards for many predators to handle. However, snakes in several different lineages (e.g., *Liophidium*, *Scaphiodontophis*, *Psammodynastes*) have evolved specializations, such as hinged teeth or the development of a gap (diastema) within the tooth row, to facilitate capture of these lizards (see Chapter 9).

Content and Distribution. Approximately 100 genera, 1090 species. Cosmopolitan (Figure 4–23). Four subfamilies (Greer 1970): Feyliniinae (central tropical Africa); Acontiinae (southern Africa and Kenya); Scincinae (North and Central America, Africa and Madagascar, southern Europe and Asia to Japan and the Philippines); Lygosominae (cosmopolitan except northern Eurasia and northern and western North America).

Systematics. Greer (1970).

Cordylidae

Cordylids are characterized by scales arranged in transverse circles around the body, and they often have strongly keeled or very spiny tails. The body is heavily armored with osteoderms. All species except *Cordylus giganteus* are rock-dwelling and use body inflation to wedge themselves into rock crevices for defense. *Cordylus giganteus*, the largest cordylid (>30 centimeters in total length), is terrestrial and digs long burrows in the soil. It has a battery of large spines projecting from the rear of the head and also many rows along the tail. If pursued into its burrow, *C. giganteus* can defend itself by backing toward the intruder, swinging its strong tail from side to side. As a last resort it anchors itself inside its burrow by hooking the head spines into the roof of the tunnel. Another cordylid, *Cordylus cataphractus*, has an unusual defense for a lizard: it rolls into a tight ball, holding the tail base in its mouth and exposing only the hard spiny scales to the exterior. Many cordylids are solitary and territorial, but species of *Platysaurus* live in dense colonies and are the only oviparous cordylids (all others are viviparous). As in many other squamate lineages, limb reduction has occurred in cordylids of the genus *Chamaesaura*, which have long, snakelike bodies and extremely reduced limbs.

Content and Distribution. 4 genera (*Chamaesaura*, *Cordylus*, *Pseudocordylus*, *Platysaurus*), approximately 42 species. Eastern and southern Africa (Figure 4–23).

Systematics. Lang (1991).

Gerrhosauridae

The scales of gerrhosaurids are arranged in transverse rows and are underlain by osteoderms. They do not develop spines to the extent seen in cordylids, however. Gerrhosaurids have a prominent lateral fold along the body, a feature also seen in some anguids. These are small (total length, 15 centimeters in *Cordylosaurus*) to large lizards (total length, 70 centimeters in *Gerrhosaurus validus* and *Zonosaurus maximus*). Reduced-limbed gerrhosaurids (*Angolosaurus* and *Tetradactylus*) are associated with shifting-sand environments or grasslands. The Malagasy species of *Tracheloptychus* and *Zonosaurus* are terrestrial lizards of either rain forests or more open formations, and many species dig burrows in the ground. *Zonosaurus maximus* is semiaquatic and takes to water to escape disturbance, hiding under bottom debris for long periods.

Content and Distribution. 6 genera, approximately 30 species. *Tracheloptychus*, *Zonosaurus* (Madagascar); *Angolosaurus*, *Cordylosaurus*, *Gerrhosaurus*, *Tetradactylus* (sub-Saharan Africa) (Figure 4–19). Gerrhosaurids are often included in the Cordylidae, their sister group.

Systematics. Lang (1990, 1991).

Figure 4–23 **Distribution of the lizard families Opluridae, Cordylidae, Scincidae, and Anguidae.**

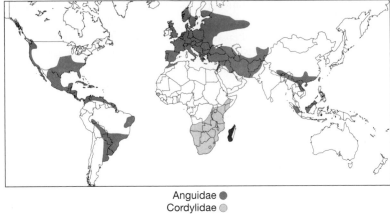

Anguidae ●
Cordylidae ○
Opluridae ●

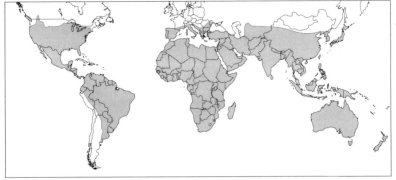

Scincidae ○

Anguidae

Anguidae is a widespread family in tropical and temperate regions. Most species are terrestrial, but *Anniella* is fossorial, and species of *Abronia* are entirely arboreal (Figure 4–22). Habitats include open grasslands (*Ophiodes, Ophisaurus*), sand dunes (*Anniella*), pine–oak and cloud forests (*Abronia*), lowland rain forest (some *Diploglossus*), and chaparral (*Elgaria*). Most anguids appear to be diurnal and to prefer relatively cool and humid environments. Sizes range from total lengths of less than 70 millimeters to more than 1.3 meters (*Ophisaurus apodus*, which feeds on vertebrates). The tails of anguids autotomize readily. Limb reduction and loss occurs in many anguids and has evolved several times within the family. Species of *Ophisaurus* and *Anniella* are entirely limbless. *Ophiodes* has lost the forelimbs but retains tiny hind limbs, and many species of *Diploglossus* have relatively short limbs. Several genera are viviparous (*Barisia, Abronia, Mesaspis, Anniella, Celestus, Anguis,* and *Ophiodes*). *Diploglossus* and *Elgaria* have both oviparous and viviparous species, and *Ophisaurus* and *Gerrhonotus* are

oviparous. Parental care of eggs has been observed in species of *Ophisaurus* and *Gerrhonotus*. Most species of the Central American genus *Abronia* are threatened with extinction because of the destruction of their cloud forest habitats (Campbell and Frost 1993).

Content and Distribution. 15 genera, approximately 102 species. Disjunct distribution in North, Central, and South America; West Indies; western Eurasia and northwest Africa; Southeast Asia and islands of the Sunda Shelf (Figure 4–23). Four subfamilies: Anguinae (North America, Eurasia), Anniellinae (extreme western North America), Diploglossinae (Central and South America, West Indies), and Gerrhonotinae (western North America, Central America).

Systematics. Gauthier (1982), Good (1987a,b, 1988, 1994), Campbell and Frost (1993; *Abronia*), Campbell and Camarillo (1994; *Diploglossus*), Savage and Lips (1993).

Xenosauridae

Species of *Xenosaurus* inhabit lowland and lower mountain humid forests with the exception of the recently discovered *X. rectocollaris*, which occurs in a high-elevation, semiarid habitat. All species are rock-dwellers. The best-known species, *X. grandis*, lives solitarily in crevices in limestone outcrops. Xenosaurids are diurnal, consume primarily arthropods, and have somewhat spiny, nonautotomic tails. *Shinisaurus crocodilurus* (Figure 4–22) is diurnal and semiaquatic, and is known to feed on tadpoles and fish. It been observed basking on branches overhanging streams. All xenosaurids are viviparous.

Content and Distribution. 2 genera, 5 species. *Xenosaurus* (4 species): southern Mexico and Guatemala. *Shinisaurus crocodilurus*: Guangxi Province, China (Figure 4–21).

Systematics. Haas (1960), King and Thompson (1968).

Varanidae and Lanthanotidae

Varanids range in size from small (25 centimeters in total length) to very large—*Varanus komodoensis*, at 3 meters in total length, is the largest extant lizard. One extinct species from the late Quaternary of Australia is estimated to have been 6 meters long. Varanids are active, fast-moving foragers that feed on a variety of relatively small invertebrates and vertebrates (King and Green 1993). Lanthanotidae includes only the monotypic *Lanthanotus borneenesis*, which is a burrowing and semiaquatic lizard with short limbs and a somewhat prehensile tail. Observations on its natural history are scant, but it is oviparous, like the varanids. Unlike *Varanus*, *Lanthanotus* seems to be nocturnal. *Varanus niloticus* is also semiaquatic and has a vertically compressed tail with a dorsal crest for swimming. Crabs and crocodile eggs are its dietary staples.

Content and Distribution. Varanidae: 1 genus, *Varanus* (about 40 species). Africa, across southern Asia to China, and through the Indoaustralian Archipelago to Australia. Two thirds of the species are in Australia. Lanthanotidae, 1 species, *Lanthanotus borneensis*. Borneo (Figure 4–19).

Systematics. Baverstock et al. (1993).

Helodermatidae

Helodermatids are somewhat stout lizards with short, blunt tails that are used for fat storage. The two species are the only venomous lizards in the world. Unlike in snakes, the venom glands of *Heloderma* are nonmuscularized and are located in the tissue alongside the mandibles. The venom flows passively into the mouth through many ducts, where sharp, grooved teeth convey it into prey. The diets of *Heloderma* consist mostly of vertebrates, including mammals, birds and their eggs, lizards, occasional insects, and even a

turtle (*Kinosternon*). *Heloderma* are widely foraging diurnal or nocturnal predators. They prefer cooler temperatures than many lizards, and in this respect they are similar to their anguid relatives. Their tails are somewhat prehensile, and both species climb readily, ascending to considerable heights in order to reach birds' nests. However, they also use ground burrows.

Content and Distribution. 2 species, *Heloderma suspectum* and *H. horridum*. Southwestern United States to Guatemala (Figure 4–19).

Systematics. Bogert and del Campo (1956), Pregill et al. (1986).

Serpentes

Serpentes is characterized by numerous derived characters, but few of these are, in fact, restricted to the group. Most of these characters also appear in other squamates, especially those that are limbless or burrowing, such as amphisbaenians, some skinks, pygopodids, and dibamids. Some of these characters are directly associated with a reduction of limbs and girdles, which are derived features of snakes just as in other limbless tetrapods. The left lung is reduced in snakes, as it in most elongate squamates (except amphisbaenians, in which the right lung is reduced). Autotomic planes are absent from the caudal vertebrae and scleral ossicles are absent from the eyeball. As in xantusiids, pygopodids, eublepharids, and some skinks and lacertids, snakes have a transparent spectacle covering the eye. Many other derived features of snakes involve losses of skull or hyoid elements (lacrimal, jugal, epipterygoid, and squamosal bones in the skull, and several elements of the hyoid). These losses also occur convergently in other squamates. In snakes the lower jaw comprises the dentary anteriorly and a compound bone posteriorly that is formed from the fused articular and prearticular bones. Tropidophiids and Caenophidia have lost the coronoid bones.

The unique features of snakes are observed in many organ systems. The supraoccipital bone is excluded from the margin of the foramen magnum by the exoccipitals. Alone among squamates, snakes have 120 or more precloacal vertebrae (the number ranges upward to more than 400). Some limbless lizards have more than 200 total vertebrae, but most of these are in their long tails. In snakes the ophthalmic branch of the trigeminal nerve is enclosed within the braincase by downgrowths of the parietal bone and enters the orbit through the optic foramen. In other squamates the nerve is not enclosed and enters the orbit posteriorly. Snakes lack muscles in the ciliary body of the eye, which accounts for their peculiar mode of accommodation. Finally, the left systemic (arterial) arch is larger than the right one in snakes, the reverse of the usual condition in tetrapods.

Many of the morphological features that we associate with snakes, as contrasted with lizards, are based on a somewhat misleading comparison of relatively derived snakes (e.g., the clade Macrostomata, of which familiar examples are colubrids and vipers) with relatively generalized lizards (e.g., iguanids). These include many features associated with the evolution of the feeding system of snakes, including characters responsible for the increased gape and mobility of the jaws (see Chapter 9). Some specific examples include the elongation of the quadrates and pterygoids, further loosening of the mandibular symphysis, a general reduction in articulations between bones of the palatomaxillary arch and the skull, and a similar reduction or elimination of articulations between bones of the palate. In fact, these features show considerable evolutionary change within snakes, and relatively primitive snakes (e.g., *Anilius* and some uropeltines) do not exhibit the derived conditions of many of these characters.

The skulls of snakes are more highly kinetic than the skulls of most lizards, but in a rather different way. Whereas most lizards have lost only the lower temporal arch (exceptions are Gekkota and varanids, among others), snakes have lost both temporal arches and the quadrate attaches to the skull via the supratemporal bone. In contrast to most lizards, the brain in snakes is enclosed in a rigid box formed by downgrowths of the frontals and parietal, which have an extensive articulation with the sphenoid below the brain. Thus, the meso- and metakinetic joints of the skull roof, which are present in lizards, are lost in snakes (see Chapter 9). A new and complex prokinetic joint is formed between the frontals and the nasal region, with the nasals, frontals, and septomaxillae involved in the articulation. The snout kinesis of some snakes is also aided by mobility of the prefrontal relative to the skull roof, and of the maxilla relative to the prefrontal (as seen, for an extreme example, in the fang erection mechanism in Viperidae).

Snakes may be oviparous or viviparous, and even closely related species may differ in reproductive mode. In other cases the reproductive mode characterizes large clades, despite otherwise variable life histories. For example, all Boinae and New World na-

tricine colubrids (Thamnophiini) are viviparous, whereas oviparity is characteristic of Pythoninae and most Eurasian natricines. Genetic sex determination occurs in all species in which the mechanism has been determined. Sex chromosome heteromorphism is the rule in Colubroidea (where females are the heterogametic sex) but is unknown in other snakes (Gorman 1973). One parthenogenetic species of snake is known, *Rhamphotyphlops braminus* (Typhlopidae). This species is triploid (Wynn et al. 1987), and thus possibly a product of interspecific hybridization, as are nearly all other squamate parthenogens. However, the parental species have not been identified. Parthenogenesis has permitted wide dispersal of *Rhamphotyphlops braminus*. The species was probably native to the southwest Pacific region, but it has become established on oceanic islands such as Hawaii and Madagascar, mainland areas of the circum-Pacific and Indian oceans (Asia, Central and South America, Africa, and India), and Florida. Recent work suggests that a form of parthenogenesis that produces only diploid males may be widespread among snakes and other squamates (Dubach et al. 1997, Schuett et al. 1997).

Sexual dimorphism among snakes is generally more subtle than in many lizards, owing in large part to the lack of specific structures used for intraspecific territorial and sexual displays. Either males or females may be larger, depending on the species, but females are larger in most species. Similarly, males usually have proportionally longer tails than females. This reaches an extreme in some species of the Malagasy colubrid genus *Liopholidophis*, in which males may have a tail up to 55 percent of total length, whereas tails of females of the same species are only 35 percent of total length. Males of many species of snakes have keels on the supra-anal scales or tubercles on the chin scales. These keels are either absent or smaller in females. In those snakes retaining pelvic elements, these are usually larger in males than in females and are used during courtship to stimulate the females. Sexual differences in coloration are less common in snakes than in lizards but are known in some species, such as the boomslang, *Dispholidus typus* (Colubridae), and many vipers (Shine 1993). Snakes of the Malagasy colubrid genus *Langaha* have peculiar scaly protrusions from the snout. In females these structures are broad and blunt, whereas they are narrow and pointed in males. Their function is entirely unknown.

The body proportions and some details of morphology reflect the microenvironment used by a snake (Cadle and Greene 1993). Suites of correlated morphological features have evolved repeatedly in different snake lineages in response to these environmental demands. Thus, it is often easy to infer the macrohabitat (arboreal, fossorial, aquatic, etc.) of an unfamiliar snake even with a relatively superficial examination. Some examples of these macrohabitats and their associated characters are given in Table 4–1.

Dentition and Venoms

Typically, snake teeth are long, slender, and slightly curved. They are attached to the jaws in a somewhat modified pleurodont manner, in which each individual tooth is set within a shallow socket. Some scientists describe the condition as thecodont. In alethinophidians, teeth are universally present on the maxillae, palatines, and pterygoids in the upper jaw and on the dentary in the lower jaw. The premaxilla bears teeth in some relatively primitive alethinophidians. Scolecophidians have extremely reduced dentition and the pattern of reduction is characteristic within families (see below). Ancestrally within alethinophidians, all of the teeth are roughly of the same form (homodont), but in colubroid snakes various forms of heterodonty develop, in which teeth on different parts of the tooth-bearing bones differ in size. Only within colubroids are individual teeth modified as grooved or hollow fangs. The number of teeth on the tooth-bearing bones varies considerably. For example, only two or three maxillary teeth are present in colubrids such as *Dasypeltis* (African egg-eating snakes) and *Tomodon* (a South American slug eater), whereas the number in some species closely related to these genera is more than 30.

Various classification schemes have been proposed for the maxillary dentition patterns observed within colubroids, but no scheme can adequately reflect the true variation within this large clade. The simplest scheme recognizes four general categories: aglyph for homodont maxillary dentition; opisthoglyph, in which the posterior pair of teeth on each maxilla is enlarged, a condition often referred to as rear-fanged; proteroglyph, in which each maxilla is relatively long, bears a single hollow fang on its anterior end (and usually additional teeth behind the fang), and the fang is not erected by extensive rotation of the maxilla around the prefrontal bone; and solenoglyph, in which the maxilla is extremely reduced, never bears teeth other than a hollow fang, and the fang is erected by rotation

Table 4–1 **Aspects of body form and structure of snakes that are associated with specific habitats.**

Arboreal
Small body mass (high length/mass ratio)
Compressed body
Relatively long tail (possibly prehensile)
Relatively large eye
Enlarged vertebral scale row
Center of gravity shifted posteriorly

Fossorial
Small body size (length)
Reduced head width
Scale reductions
Small eye
Inferior mouth
Skull reinforcements
Narrow snout

Aquatic
Dorsal and terminal displacement of eyes and nostrils
Valvular nostrils

Cryptozoic
Small size

of the maxilla on the prefrontal bone. These terms are general descriptors, but actual patterns within each class are variable. Aglyph and opisthoglyph patterns are characteristic of colubrids, whereas the proteroglyph and solenoglyph patterns are characteristic of elapids and vipers, respectively. The enlarged posterior maxillary teeth of colubrids are often, but not universally, separated from more anterior teeth by a gap, or diastema. Some workers restrict the term opisthoglyph to snakes having enlarged and grooved rear maxillary teeth. But in fact the variation in both tooth size and extent of grooving is virtually continuous within colubrids (Young and Kardong 1996), making such distinctions arbitrary. Even very close relatives display alternative conditions. Grooving, although usually most distinct and associated with the posterior maxillary teeth, may appear on virtually any of the maxillary teeth, and faint grooving is present on palatine, pterygoid, and dentary teeth of some species (Young and Kardong 1996).

The front fangs of vipers and elapids are hollow, with the venom canal a hollow tube distinct from the pulp cavity of the tooth. The details of structure of the fangs and their placement on the maxillary bones suggest that the front fangs of vipers and elapids are homologous with the posterior maxillary fangs of colubrids (Anthony 1955, Jackson and Fritts 1995).

Associated with the fangs of colubroid snakes are glands that produce venoms used in subduing prey (Figure 9–39). These structures are important in the systematics of Colubroidea (Kochva et al. 1967, Kochva and Wollberg 1970). Snake venoms are complex mixtures of proteins and other molecules. Any particular venom may contain several hundred components, which accounts for some of the difficulty in treating of snakebite and also in understanding the evolution of venoms. The toxins of snake venoms are proteins that range from small peptides with only a few amino acids to complex enzymes and nonenzymatic proteins that attain high molecular weights. They are classified by their physiological actions and chemical structures (Dufton 1984, Breckenridge and Dufton 1987). For example, hemorrhagins and hemolysins destroy blood vessel linings and red blood cells, respectively. Myotoxins destroy skeletal muscle, whereas neurotoxins act at synaptic or neuromuscular junctions. In general, elapid venoms are characterized by primarily neurotoxic properties, whereas viperid venoms characteristically have more hemolytic and cytolytic properties.

Snake venom toxins are thought to have arisen evolutionarily from digestive enzymes by processes of gene duplication and divergence (Strydom 1979). These processes are common mechanisms of molecular evolution and have given rise, for example, to the multiple forms of hemoglobin in vertebrates (see Li and Graur 1991). An example from snake venoms are the phospholipases A_2 found in both elapid and viperid venoms (Harris 1997, Moura-da-Silva et al. 1997). Phospholipases A_2 are digestive enzymes involved in vertebrate lipid catabolism. Multiple copies of the phospholipase A_2 genes are found in the venom glands of snakes, but the products of these genes differ somewhat in chemical structure and often fundamentally in physiological action. In viperids, the neurotoxic components of the venoms of *Crotalus durissus terrificus* and *Crotalus scutulatus* are phospholipases A_2. Other forms of phospholipases act directly on the lipid portion of the cell membranes of skeletal muscle and are responsible for the myotoxic effects and necrosis characteristic of many venoms. In elapids the toxic phospholipases act on the presynaptic portions of synaptic junctions, and are responsible for the primarily neurotoxic effects of these venoms. Such divergence in structure and function of toxic compounds is characteristic of the molecular evolution of snake venoms. Some recent advances in our understanding of the evolution of these compounds have been made possible by methods for cloning of toxin genes (e.g., Smith et al. 1995, Middlebrook 1992).

In comparison to venoms of the elapids and viperids, the venoms of colubrids are very poorly understood (Hiestand and Hiestand 1979, Rosenberg et al. 1985). Like elapid and viperid venoms, colubrid venoms are complex mixtures and include phospholipases A as toxic components. Although most attention has been devoted to Duvernoy's gland secretions in colubrids, some studies have shown toxic properties for secretions from some of the other head glands. For example, Laporta-Ferreira and Salomo (1991) reported that proteolytic activity was present not only in the secretions of Duvernoy's gland, but also in the infralabial gland of the South American colubrid *Sibynomorphus neuwiedii*. This is one of many Neotropical snakes that feed on snails by extracting the snails' bodies from their shells using the dentary teeth. The secretion of the infralabial glands is introduced into the snail by the dentary teeth and causes rapid paralysis, thereby allowing the snake to extract the snail from its shell (see Chapter 9). It is very likely that many secretions from the oral glands of colubrids have similar effects on prey, but the natural history of venoms in general is an enormously understudied area.

Sensory Systems

Several senses, especially smell and sight, are well developed among snakes. Specialized receptors for infrared radiation have evolved independently several times within boids and viperids. Although snakes lack external and middle ears, they are able to pick up airborne and substrate-borne sound.

The sense of smell is used extensively in the foraging behavior of snakes and also in intraspecific communication. Noble (1937) pioneered the study of pheromones involved in the courtship behavior of several natricine colubrids (*Thamnophis*, *Nerodia*, *Storeria*), and recent field and laboratory studies of other species have affirmed the importance of chemical communication generally (e.g., Andrén et al. 1997). Pheromones are probably generally important in snake reproductive ecology and are produced in the skin and possibly the scent glands in the tail (Noble 1937, Weldon and Leto 1995).

Vision is important to snakes except for the many burrowing species that have reduced eyes. However, the eyes of snakes are unusual. All tetrapods other than snakes focus the eye by changing the lens curvature using muscles within the ciliary body, which supports the lens peripherally. The absence of ciliary muscles, a shared derived character of Serpentes, means that accommodation must occur by another mechanism. In snakes, the entire lens is moved with respect to the retina by means of muscles within the iris. The loss of ciliary muscles and the peculiar mode of accommodation in snakes led some scientists to suggest that modern snakes underwent degeneration of the eyes during a long period of burrowing ancestry, and that the eyes of modern snakes are reconstituted from degenerate parts (Bellairs and Underwood 1951, Rieppel 1988).

Specialized infrared receptors have evolved independently several times within Boidae (de Cock Buning 1985, Kluge 1993) and once within Colubroidea (Crotalinae). In the Boidae the receptors are located in pits within the upper and lower labial and/or rostral scales, whereas in Crotalinae a single pit is present between the eye and nostril on either side of the head (Figure 9–41). All infrared receptors are innervated by the trigeminal nerve. Experiments show that many snakes that lack specialized pit receptors can nevertheless sense infrared radiation, albeit less efficiently than those with pits (Barrett 1970, Breidenbach 1990). The structure of the pits is such that very precise directionality and distance of the infrared source can be measured. Moreover, the thermal cues are integrated with visual ones in the optic region of the brain, thus giving the snakes superimposed visual and infrared images of their environment (Hartline et al. 1978, Newman and Hartline 1982).

Content and Distribution

Approximately 12 families are generally recognized (more if Colubridae is divided), with more than 1,800 species. Snakes are cosmopolitan except in extremely high latitudes.

Fossil Record

The fossil record of snakes is rather poor compared to that of other squamate groups. Skulls are extremely rare, and most fossils consist of isolated vertebrae. Furthermore, because the systematics of extant snakes relies very little on vertebral characters, correlating the results of systematic research on extant and fossil groups has been problematic (Cadle 1987, Kluge 1993). Thus, the assignment of many fossils to extant families, particularly those of Mesozoic and early Tertiary age, is provisional.

The oldest undisputed snake is *Lapparentophis defrennei*, from the boundary between early and late Cretaceous in Algeria (100 MaBP; Cuny et al. 1990). However, snakes are widespread later in the Cretaceous, including India (Rage and Prasad 1992), South America (Albino 1996), China (Gao and Hou 1996), and Africa (Rage and Wouters 1979). Several of these early fossils are assigned to extant families. Boidae is represented by *Madtsoia* (Madagascar, Africa, South America, and Australia; Rage 1987; Scanlon 1993), and aniliids are represented by *Coniophis* (North America, South America, and India). The earliest snake with a relatively complete skull is *Dinilysia patagonica* (Late Cretaceous of Argentina; Estes et al. 1970), which is thought to be a sister group to Alethinophidia (Rieppel 1988).

Tertiary fossil snakes include some exclusively extinct families (Russellopheidae, Nigeropheidae, and Anomalopheidae) whose relationships to other snakes are not clear. These all became extinct by the end of the Eocene (36 MaBP). Also in the Eocene, the earliest colubrid appears, an unnamed species from Thailand known from six fragmentary vertebrae (Rage et al. 1992). Other early colubrids appear in the early and middle Oligocene of France, Oman, and the United States (Rage 1988, Holman 1984). Viperids and elapids first appear in early Miocene deposits of Europe and North America and are common in deposits after that time. Snakes, especially caenophidians, are reasonably well represented in later Tertiary (Miocene–Recent) deposits of both Europe and North America (Szyndlar 1991a,b, Szyndlar and Bohme 1993, Holman 1995). Overviews of the snake fossil record are presented by Rage (1987), Albino (1996), and Szyndlar (1991a,b).

Systematics. Underwood (1967), Groombridge (1979a,b), McDowell (1986, 1987), Cadle (1988), Rieppel (1988), Heise et al. (1995).

Systematics and Phylogeny of Snakes

The phylogeny of snakes has been evaluated using both morphological and molecular data sets (Cadle 1988, Rieppel 1988, Kluge 1991, Cundall et al. 1993, Heise et al. 1995), and several major clades of extant snakes are supported by most of these analyses (Figure 4–24). These are a primary dichotomy between Scolecophidia and all other snakes (Alethinophidia); within Alethinophidia, a primary dichotomy between Anilioidea (Aniliidae + Uropeltidae) and Macrostomata (all other Alethinophidia); monophyly of Caenophidia (Acrochordidae + Colubroidea) and Colubroidea (Viperidae + Elapidae + Colubridae); and monophyly of Elapidae + Colubridae with respect to Viperidae. Major disagreements concern the relationships of the assemblage of relatively primitive snakes (often referred to as "Booidea" or "Henophidia") such as *Loxocemus*, *Xenopeltis*, Tropidophiidae, Bolyeriidae, and Boidae. Some evidence supports the monophyly of different subsets of these taxa (Groombridge 1979a, Rieppel 1988, Cundall et al. 1993), whereas other analyses show them as a nonmonophyletic assemblage between Anilioidea and Caenophidia (Kluge 1993). All of these alternative hypotheses are weakly supported at present. The monophyly of Colubridae, which includes about 70 percent of all extant species of snakes, has not been established.

The monophyly of Scolecophidia is supported by several characters of the eyes, skull structure, and soft anatomy (Rieppel 1988). The 1:1 relationship between ventral scale rows and vertebrae, characteristic of all other snakes, does not hold in Typhlopidae and Anomalepididae, in which the ratios vary from 1.5:1 to 2.3:1. These two families also share basic similarity in skull morphology and dentitional pattern (see below) that suggests their sister relationship relative to Leptotyphlopidae. Alethinophidia is supported by characters of the skull, nervous system, and axial muscles. Within that group, the Macrostomata are characterized by several features permitting increased gape (hence the name), including a supratemporal bone with a free posterior end, a long mandible, and elongation of the quadrate. In addition, macrostomatans have enlarged ventral scutes (reversed in *Acrochordus* and Hydrophiinae). Caenophidia is characterized by several osteological characters (loss of coronoid bones and details of skull structure) and the passage of the facial carotid artery dorsal to the mandibular branch of the trigeminal nerve (Groombridge 1984). The monophyly of Colubroidea is supported by characters of body segmentation (segmental arrangement of intercostal arteries, separation of spinalis and semispinalis muscles in the trunk), peculiarly shaped costal cartilages (Hardaway and Williams 1976, Persky et al. 1976), and several skull characters (Rieppel 1988).

Anomalepididae and Typhlopidae

All scolecophidians have blunt heads and short, blunt tails (often tipped with a spine), smooth scales, and lack enlarged ventral scales (Figure 4–25). The eyes are vestigial. Anomalepidids are similar to typhlopids in having toothed, movable maxillae but differ in lacking pelvic vestiges and in having the prefrontal bones extend posteriorly over the orbits (unique among squamates). The dentary rarely bears more than a single tooth. In typhlopids the premaxilla is toothless and firmly articulated with the snout (Figure 4–26). The maxillae bear several teeth and are attached to the skull via mobile articulations. The lower jaw is composed of an enormous compound bone, a large separate coronoid, a small splenial and angular, and a small toothless dentary.

Rhamphotyphlops is unusual among squamates in having a solid protrusible hemipenis rather than an eversible hollow structure (Robb 1966). Few species of scolecophidans are more than about 30 centimeters in total length. However, the largest species, *Typhlops schlegeli* of southern Africa, attains a length of nearly a meter. All typhlopids lay eggs, but egg retention is common, and freshly laid eggs may contain advanced embryos.

Scolecophidians are exceptions to the generalization that snakes consume relatively few relatively large prey items per feeding bout (i.e., most snake stomachs contain only one or a few prey items at a time). Typhlopid diets consist mainly of ant or termite pupae, larvae, eggs, and (less frequently) adults. However, the only known food of one New Guinea species is earthworms. An individual snake may consume an enormous number of individual prey items at once, and over 1,500 items have been found in the guts of individual snakes. These snakes frequently appear to gorge themselves inside ant or termite nests, for entire guts from esophagus to cloaca are packed with food.

Content and Distribution. Anomalepididae: 4 genera (*Liotyphlops*, *Anomalepis*, *Typhlophis*, *Helminthophis*), approximately 20 species. Southern Central America and northern South America (Figure 4–27). Typhlopidae: 3–6 genera (*Typhlops*, *Rhamphotyphlops*, *Acutotyphlops*, *Xenotyphlops*, *Rhinotyphlops*, *Cyclotyphlops*), approximately 200 species. Central and South America, West Indies, southern Africa and Eurasia, Australasia, and Australia (Figure 4–28).

Systematics. Laurent (1964), Myers (1967), McDowell (1974), Roux-Estève (1974, 1975), Dixon and Hendricks (1979), Dixon and Kofron (1983), Kofron (1988), Wallach (1993).

Figure 4–24 **Phylogenetic relationships among snake families.**

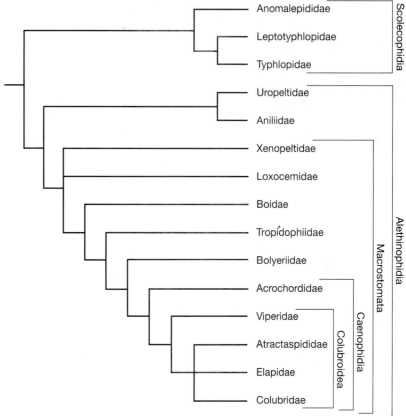

Leptotyphlopidae

The cranium and upper jaws (maxillae, palatines, and pterygoids) of leptotyphlopids are immobile and teeth are present only on the dentary bone (Figure 4–26). The lower jaw consists of an enormous, horizontally displaced quadrate, a tiny compound bone, and relatively larger dentary, angular, splenial, and coronoid.

Leptotyphlopids do not attain the large sizes of some typhlopids, the largest species being *Leptotyphlops macrolepis* (South America) and *L. occidentalis* (Africa), which reach lengths of about 30 centimeters. Most species are much smaller, some no more than about 10 centimeters. All species are oviparous. Females of *Leptotyphlops dulcis* are known to tend the eggs. Although they are generally burrowing snakes, *Leptotyphlops dulcis* has been found in dense, apparently stable populations in nests of predatory birds high in trees. The snakes are apparently brought alive to the nests as food for the young birds, escape, and simply subsist on the abundant insect fauna living in the nests (Gehlbach and Baldridge 1987).

Content and Distribution. 2 genera (*Leptotyphlops*, *Rhinoleptus*), approximately 80 species. Africa and the Middle East, northern South America to the southwestern United States (Figure 4–28).

Systematics. Klauber (1940), Hahn (1978).

Aniliidae

The single species in the family, *Anilius scytale*, is usually bright red with black bands or irregular markings. It reaches about a meter in length and is viviparous. Its eyes are very small and lie beneath a large head shield. *Anilius* is a burrower and is sometimes associated with soft soil along rain forest streams, but it is also surface-active and may be diurnal or nocturnal. *Anilius* feeds on elongate vertebrates, primarily caecilians and amphisbaenians.

Content and Distribution. 1 species, *Anilius scytale*. Amazon Basin and Guianan region of South America (Figure 4–28).

Systematics. McDowell (1975).

Figure 4–25   **Diversity of snakes.** (a) *Typhlops schlegelii* (Typhlopidae, Africa). (b) *Anilius scytale* (Aniliidae, South America). (c) *Cylindrophis rufus* (Uropeltidae, Indochina and Southeast Asia). (d) *Tropidophis taczanowskyi* (Tropidophiidae, South America). (e) *Eryx jaculus* (Erycinae, Boidae, Asia Minor to North Africa). (f) *Acrochordus javanicus* (Acrochordidae, Australia). *(Photographs by (a,b) John Cadle, (c,e,f) F. Harvey Pough, (d) Alan Savitzky.)*

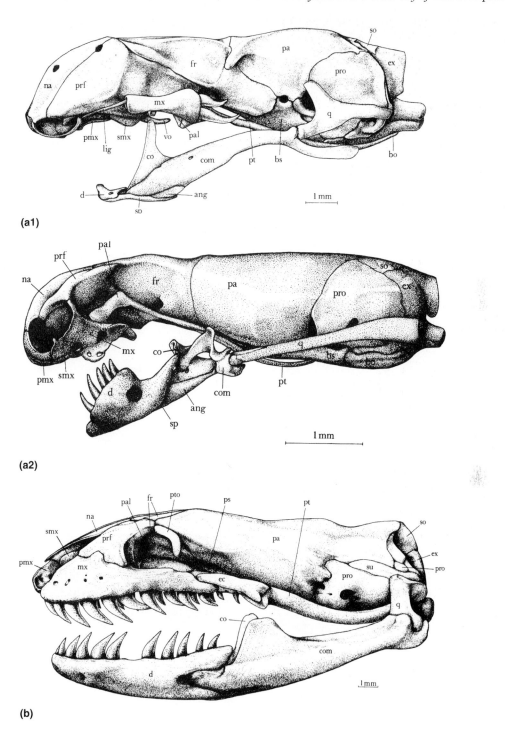

Figure 4–26 **Skulls of snakes showing structural characteristics of major groups.**
(a) Scolecophidia (Typhlopidae (*Typhlops*) and Leptotyphlopidae (*Leptotyphlops*)). (b) Cylin-
drophiine uropeltid (*Cylindrophis*). *(continued)*

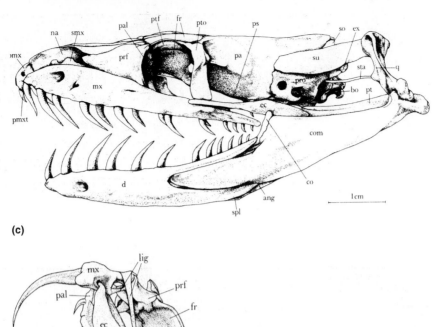

(c)

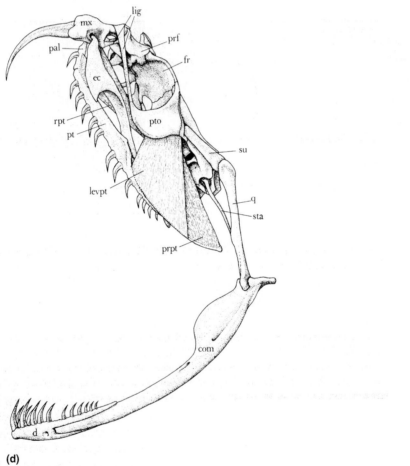

(d)

Figure 4–26   *(continued)* (c) Boidae (*Python*). (d) Viperidae (*Bitis*). *(continued)*

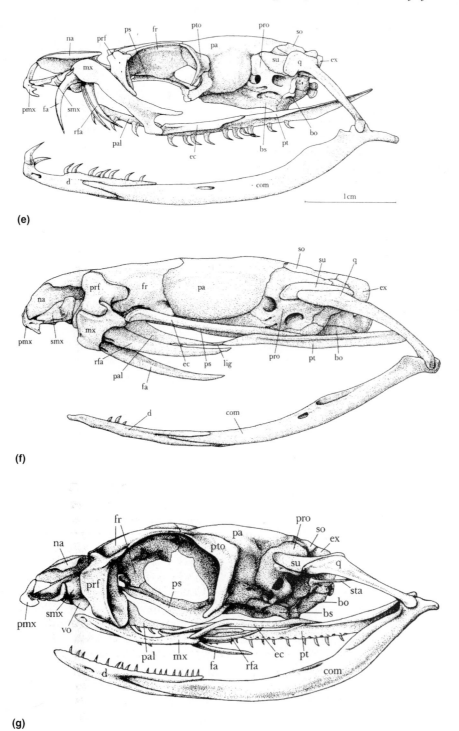

Figure 4–26 (continued) (e) Elapidae (*Dendroaspis*). (f) Atractaspididae (*Atractaspis*). (g) Colubridae (*Dispholidus*). *(Source: Parker and Grandison 1997.)*

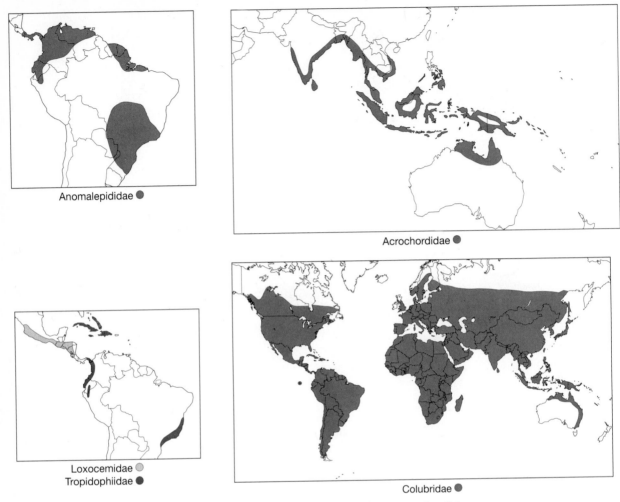

Figure 4–27 **Distribution of the snake families Anomalepididae, Acrochordidae, Loxocemidae, Tropidophiidae, and Colubridae.**

Uropeltidae

Uropeltidae includes *Cylindrophis*, a burrowing to terrestrial snake of southern Asia, and a highly specialized group of burrowers (uropeltines) restricted to India and Sri Lanka (Gans 1976). These two groups are sister taxa within the family. The heads of uropeltines are conical and slender, and often much narrower than their relatively thick trunks. The head is often provided with a distinct keel. The tails of uropeltines are blunt and in many species are capped with a single large scale with a rough surface. The tail cap collects a plug of dirt and protects the snake from behind when burrowing in a tunnel. Uropeltines have many specializations for burrowing, including anterior body musculature richly supplied with myoglobin, catalytic enzymes, and mitochondria (Gans et al. 1978). These biochemical specializations permit sustained activity of the anterior trunk muscles during burrowing. The vertebral column, body wall muscles, and viscera of uropeltines can move relative to the outer skin, allowing the snakes to use a portion of the body as a frictional point in the tunnel, while the head and anterior body move forward relatively friction-free (see Chapter 8).

The relationships of two species of *Anomochilus* (Indonesia) are unclear, but the genus has historically been associated with *Cylindrophis*. Cundall and Rossman (1993) and Cundall et al. (1993) analyzed the morphology and systematics of *Anomochilus* and hypothesized that it was the sister taxon to other Alethinophidia among extant snakes, and accordingly placed it within its own family, Anomochilidae.

Content and Distribution. 9 genera, about 45 species. 2 subfamilies: Cylindrophinae: *Cylindrophis* (7 species); Uropeltinae: 8 genera, about 45 species (*Uropeltis, Teretrurus, Rhinophis, Plectrurus, Platyplectrurus, Melanophidium, Pseudotyphlops, Brachyophidium*). All genera except *Cylindrophis* are restricted to western India and Sri Lanka. *Cylindrophis* has a disjunct distribution, with 1 species in Sri Lanka and about 5 others in southern Asia and the Indoaustralian Archipelago (Figure 4–28).

Systematics. Cadle et al. (1990).

Xenopeltidae and Loxocemidae

Xenopeltis is a burrowing, nocturnal rain forest snake reaching slightly over a meter in length. The dorsal scales are black or very dark brown and are highly iridescent. Unusually for a relatively primitive snake, the head scales consist of large plates similar to those of colubrids and the ventral scutes are only slightly reduced. Pelvic vestiges are absent. *Loxocemus* is a Central American snake with a somewhat pointed snout, and it is at least partially fossorial. It reaches about 1.3 meters in length and is known to prey on mammals and reptile eggs, including those of the sea turtle, *Lepidochelys kempi*. *Xenopeltis* and *Loxocemus* have sometimes been considered sister taxa, but most recent interpretations do not support that hypothesis (Figure 4–24).

Content and Distribution. Xenopeltidae: 1 species, *Xenopeltis unicolor*. India and southern China to Borneo and Celebes (Figure 4–28). Loxocemidae: 1 species, *Loxocemus bicolor*. Southern Mexico to Costa Rica (Figure 4–27).

Systematics. Haas (1955), Nelson and Meyer (1967).

Boidae

Boidae includes snakes commonly known as boas (Boinae), pythons (Pythoninae), and sand boas (Erycinae). This family includes the largest extant snakes, but many boids are quite small. Adults of *Exiliboa* are less than 50 centimeters in total length, and several other species (e.g., *Charina*) are nearly as small. Boids occur in rain forests, dry tropical forests, mountain cloud forests (*Exiliboa*), sandy and rocky deserts (Erycinae and some Pythoninae), and temperate coniferous forests (*Charina*). Terrestrial, arboreal,

aquatic, and semifossorial species are included, and body forms reflect these diverse habitats. As the size range and diversity of habitats occupied by boids suggest, diets within the group are extremely varied. Larger species consume a wide variety of vertebrate prey (mammals, reptiles), but known diets of some species are rather restricted. For example, the only known prey of *Exiliboa placata*, a cloud forest species of southern Mexico, is salamanders and frogs (Campbell and Camarillo 1992).

Specialized infrared-sensitive pits are variable in their presence, position, and number in boids. They are present in some boines (*Corallus*, some species of *Epicrates, Sanzinia*) and in many pythons. They are commonly present in either the upper or lower labial scales, but also occur on the rostral scales, and may occur between labial scales of some individuals of *Boa constrictor* (Kluge 1993a). Such pits are absent in erycines.

All Pythoninae are oviparous, whereas all Boinae and Erycinae are viviparous. Many pythons construct nests of leaves or lay their eggs in burrows. Females coil around the clutches and generate heat for incubation by muscular contractions. Up to about 100 eggs are laid by the larger pythons.

Eunectes, the anacondas, include two species of large boids that may occur in virtually any aquatic environment (rivers, ponds, flooded savannas). Individuals bask during the day on floating or overhanging vegetation or on the shore. *Eunectes murinus* is a contender for the longest snake in the world, as is *Python reticulatus* of Southeast Asia. Both species may reach 10 meters or more, and *P. molurus* (Asia) and *P. sebae* (Africa) are not far behind. However, *E. murinus* is by far the most massive of these. Diets of *Eunectes* include a wide variety of vertebrates, including fish, amphibians, turtles, caimans, birds, and moderately large mammals (pacas, capybaras, deer). They are ambush predators and await prey while submerged in water.

Like iguanian lizards, boines show an unusual distribution pattern in which most species are Neotropical, but three species are found in Madagascar, and several species of the genus *Candoia* are in the southwest Pacific (Fiji, New Guinea, and the Solomon Islands).

Content and Distribution. Approximately 20 genera, 63 species. 3 subfamilies: Boinae, 9 genera (*Boa, Candoia, Corallus, Epicrates, Eunectes, Sanzinia, Acrantophis, Exiliboa, Ungaliophis*), 26 species. Western North America through Central and South America, West Indies, southwest Pacific

Figure 4–28 **Distribution of the snake families Typhlopidae, Leptotyphlopidae, Xenopeltidae, Aniliidae, Uropeltidae, Bolyerridae, and Atractaspididae.**

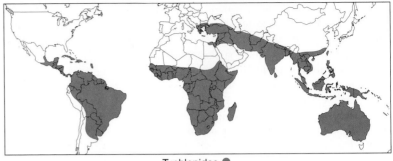

Typhlopidae ●

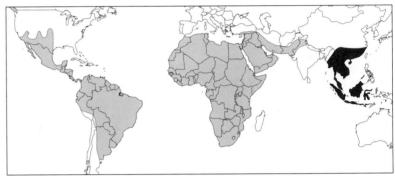

Leptotyphlopidae ●
Xenopeltidae ●

Aniliidae ●
Atractaspididae ○
Bolyeriidae ○
Uropeltidae ●

islands, including New Guinea and the Solomon Islands, Madagascar (Figure 4–29). Pythoninae, 5–8 genera (*Aspidites, Antaresia, Bothrochilus, Leiopython, Liasis, Apodora, Morelia, Python*), 24 species. Southern Africa, southern Asia from India, eastward throughout the Indoaustralian Archipelago and Australia (Figure 4–29). Erycinae, 5 genera (*Calabaria, Lichanura, Charina, Eryx, Gongylophis*), 13 species. Arid areas of northern Africa, southern Europe, and southwestern Asia (Figure 4–29).

Genera of boids recognized by various workers differ. Kluge (1991, 1993b), for example, synonymized *Sanzinia* and *Acrantophis* with *Boa*, and *Calabaria* and *Lichanura* with *Charina. Gongylophis* is variously recognized as distinct from *Eryx* (e.g., Tokar 1995). We include *Exiliboa* and *Ungaliophis* in Boinae following Zaher (1994), a position also supported by some molecular evidence (Dessauer et al. 1987). Other hypotheses place these genera with tropidophiids (Kluge 1993a). Erycinae is possibly a clade distinct from other Boidae (Kluge 1991).

Systematics. Underwood (1976), Rieppel (1977, 1978), McDowell (1979), Underwood and Stimson (1990), Kluge (1991, 1993a,b).

Tropidophiidae

Tropidophiids have a well-developed tracheal lung and the left lung is substantially reduced or absent. Both of these are derived characters. Vestiges of the pelvic girdle are present in all species except *Tropidophis semicinctus*. Species of West Indian *Tropidophis* are small (34 centimeters to slightly more than 1 meter in total length), nocturnal, terrestrial or arboreal snakes (Schwartz and Henderson 1991). They occur from xeric scrub habitats to wet rain forests. The diets of most species consist of frogs and lizards, and the largest species, *Tropidophis melanurus*, also eats birds and rodents. A peculiar defensive behavior has been noted for several West Indian *Tropidophis* species (e.g., Hecht et al. 1955, Iverson 1986). These snakes spontaneously hemorrhage from the mouth and eyes when disturbed, but the function of this behavior and its effect on potential predators have not been studied. Tropidophiids also use a more common snake defensive behavior, rolling into a tight ball with the head hidden. Very little is known of the natural history of the mainland species of tropidophiids. *Trachyboa boulengeri* is viviparous and is known to consume fish.

Content and Distribution. 2 genera (*Tropidophis, Trachyboa*), 18 species. West Indies (Greater Antilles and Bahama Islands) (*Tropidophis*); Panama and northwestern South America (*Trachyboa*); Ecuador, Peru, Brazil (*Tropi-* *dophis*) (Figure 4–27). The South American species are rare and their distributions are poorly known.

Systematics. Schwartz and Marsh (1960), McDowell (1975), Hedges and Garrido (1992), Zaher (1994).

Bolyeriidae

Bolyeriidae includes two highly unusual snakes. Vestiges of the pelvic girdle and hind limbs are entirely absent. The left lung of bolyeriids is reduced more than in boids, *Xenopeltis*, or *Loxocemus*. Like Caenophidia, bolyeriids have hypapophyses (ventral projections of the vertebral centrum) on the posterior trunk vertebrae (absent in all other snakes), and they retain a separate coronoid bone in the lower jaw, an ancestral character in snakes. Bolyeriids are unique among tetrapods in having divided maxillary bones, and *Casarea* can depress its snout. This condition is otherwise found only in sea snakes (hydrophiine Elapidae). On the basis of behavioral and anatomical observations, Cundall and Irish (1989) suggested that the divided maxillae aid in gripping skinks and geckos. The arrangement is functionally similar to the arched maxillae with enlarged teeth and diastemata seen in other skink-eating snakes, such as *Lycodon* and *Psammophis*. No examples of *Bolyeria* have been seen alive since 1975 (Bullock 1986). In the late 1970s the estimated population size of *Casarea* was approximately 75 individuals (Cundall and Irish 1989). *Casarea* is oviparous, but reproduction in *Bolyeria* is unknown.

Content and Distribution. 2 species, *Bolyeria multocarinata* and *Casarea dussumieri*. Round Island (in the Indian Ocean near Mauritius) (Figure 4–28).

Systematics. Anthony and Guibé (1952).

Acrochordidae

Species of *Acrochordus* are aquatic snakes with very small, strongly keeled scales that give the skin the texture of a coarse file or sandpaper (Figure 4–25). The skin is loose and baggy, and *Acrochordus* is nearly incapable of movement on land. The ventral scutes are only feebly enlarged, and a slightly compressed tail is present. The ventral skin of *Acrochordus granulatus* is raised into a low ridge that functions as a median fin in swimming. Unlike some other aquatic snakes (e.g., homalopsine colubrids and sea snakes), which have valvular nostrils, *Acrochordus* has a tissue flap inside the mouth that closes off the choanae. The metabolic

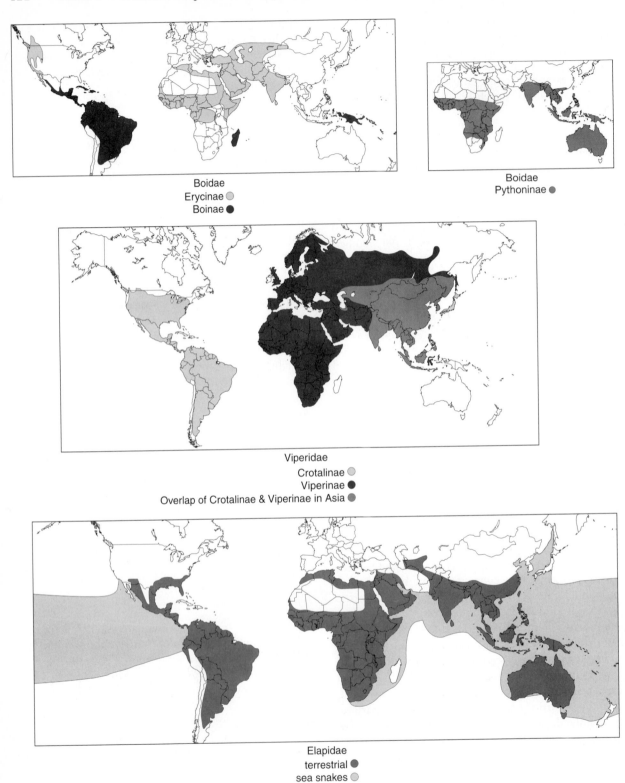

Boidae
Erycinae ◯
Boinae ●

Boidae
Pythoninae ●

Viperidae
Crotalinae ◯
Viperinae ●
Overlap of Crotalinae & Viperinae in Asia ●

Elapidae
terrestrial ●
sea snakes ◯

Figure 4–29   **Distribution of the snake families Boidae, Viperidae, and Elapidae.**

rates of acrochordids are lower than those of most other snakes, and they appear to reproduce and feed more infrequently than most other snakes (Shine 1986).

Acrochordus eats fish and crabs. All species are nocturnal and viviparous, producing up to 30 young in a litter. *Acrochordus arafurae*, up to about 2.5 meters in length, is primarily a freshwater species, although it enters estuaries and the open ocean. *Acrochordus granulatus* is primarily marine and estuarine and usually reaches no more than a meter in length. *Acrochordus javanicus* is a freshwater species reaching somewhat more than a meter in length. *Acrochordus arafurae* can reach very high population densities (100 snakes per hectare) in some areas of northern Australia (Shine 1986).

Content and Distribution. 1 genus (*Acrochordus*), 3 species. India to northern Australia and the Solomon Islands, primarily in coastal rivers, estuaries, and marine habitats (Figure 4–27).

Systematics. McDowell (1979), Groombridge (1979a,b).

Viperidae

Vipers are familiar snakes that may be terrestrial or arboreal and are found in habitats ranging from rain forests to deserts and high mountains (Figure 4–30). They may be oviparous or viviparous, and the females of some species remain with their eggs or young for some time (Greene 1992).

Several species in the genus *Causus* (Africa) are possibly a sister clade to other viperids (Groombridge 1984). They have large, symmetrical head plates rather than the fragmented plates or small scales of other vipers. They are nocturnal, oviparous snakes that feed primarily on toads (Bufonidae). Most other vipers generally feed on lizards as juveniles and, at least in larger species, on mammals as adults. Some species, such as *Agkistrodon piscivorus*, have extremely broad diets that even include carrion.

The most familiar of the Crotalinae are the rattlesnakes (*Crotalus*, *Sistrurus*). In these snakes the tip of the tail is modified into a rattle composed of a set of interlocking segments of keratin, the material that makes up the external portion of scales (Figure 4–31). Vibration of the tail causes portions of the segments to rub against one another, producing a sound.

The largest New World crotaline is the bushmaster, *Lachesis muta*, a species found in the lowland rain forests of southern Central America and northern South America. It reaches more than 3 meters in length. *Lachesis* is oviparous, whereas all other New World crotalines are viviparous.

The Southeast Asian viper *Azemiops feae* has traditionally been considered the most primitive extant viper (Liem et al. 1971) but recent molecular phylogenies show that it is closely related to, or phyletically a part of, the Crotalinae (Cadle 1992, Knight and Mindell 1993, Heise et al. 1995). *Azemiops* is a pitless viper, reaching about 70 centimeters in length, that feeds on mammals (Greene 1992).

Content and Distribution. 20–27 genera, approximately 215 species. 2 subfamilies: Crotalinae (pit vipers) (10–15 genera; e.g., *Crotalus*, *Agkistrodon*, *Bothrops*, *Lachesis*, *Porthidium*, *Trimeresurus*, *Tropidolaemus*, *Deinagkistrodon*, *Calloselasma*), about 150 species. North, Central and South America, eastern Asia from India to Japan, the Philippines, and islands of the Sunda Shelf. Viperinae (9–12 genera; e.g., *Atheris*, *Bitis*, *Causus*, *Echis*, *Vipera*, *Cerastes*), about 43 species. Africa (absent from Madagascar), Eurasia, and islands of the Sunda Shelf. *Azemiops feae* (southern China, Tibet, and northern Burma) is often placed in its own subfamily (Figure 4–29).

Systematics. Klauber (1972), Zhao and Zhao (1981), Groombridge (1984), Nilson and Andrén (1986), Campbell and Lamar (1989), Gloyd and Conant (1990), Thorpe et al. (1997).

Elapidae

Elapidae includes familiar terrestrial snakes such as cobras (*Naja*, *Ophiophagus*), mambas (*Dendroaspis*), New World coral snakes (*Micrurus*, *Micruroides*), kraits (*Bungarus*), a large radiation of Australasian endemics (e.g., *Notechis*, *Acanthophis*, *Demansia*, *Pseudechis*), and an extensive marine radiation (Hydrophiinae) (Figure 4–30). All elapids are venomous. They are characterized by proteroglyph dentition and a maxilla that is relatively longer than that of vipers, is relatively nonrotatable, and may bear teeth posterior to the fang.

Although all elapids are venomous, many species are very small and, except in exceptional circumstances, pose no danger to humans. However, because of their large size, active nature, and potent venoms, many elapids, including the mambas (*Dendroaspis*), some sea snakes, and several Australian elapids (*Notechis*, *Oxyuranus*, *Pseudechis*), are among the world's most dangerous snakes. *Ophiophagus hannah*, the king cobra, is the longest (but not the most massive) highly venomous snake in the world, attaining

Figure 4–30 **Diversity of snakes.** (a) *Bothriechis schlegelii* (Viperidae, Central and South America). (b) *Atractaspis microlepidota* (Atractaspididae, Africa). (c) *Bungarus multicinctus* (Elapidae, Southeast Asia). (d) *Naja naja* (Elapidae, India to the Philippines). (e) *Acalyptophis peronii* (Hydrophiinae, Elapidae, Australia). (f) *Pseudoxyrhopus tritaeniatus* (Colubridae, Madagascar). *(Photographs by (a,e) Harvey Pough, (b,c) Harry Greene, (d) David Northcott/Nature's Lens, (f) John Cadle.)*

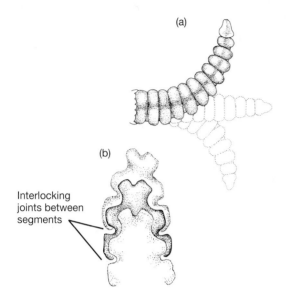

(a)

(b)

Interlocking
joints between
segments

Figure 4–31 **Structure of a rattlesnake rattle (*Crotalus*, Viperidae).** (a) External appearance. (b) Internal structures in vertical and horizontal sections showing the interlocking segments. *(Source: Klauber 1972.)*

more than 5 meters in total length. *Ophiophagus* and the African forest cobra, *Naja melanoleuca*, are among the few snakes that construct a nest for egg laying and provide parental care to the eggs. Both parents are involved in guarding the nests, which are constructed of leaves and decaying vegetation. Most terrestrial elapids are oviparous, but viviparity occurs in some African species (e.g., *Boulengerina*, *Hemachatus*) and in many Australian elapids (e.g., most species of *Pseudechis* are oviparous, but *P. porphyriacus* is viviparous). All sea snakes except *Laticauda* are viviparous. Elapids may be terrestrial (most species), arboreal (*Dendroaspis*, *Pseudohaje*), aquatic (*Boulengerina*), semi-fossorial (some *Micrurus*, *Elapsoidea*, *Aspidelaps*), or marine (Hydrophiinae).

Species of *Naja*, *Ophiophagus*, and *Aspidelaps* are known for their hood-spreading defensive displays, which is little more than the common neck-flattening behavior of many other elapids and colubrids, which are incapable of spreading extensive hoods. A hood is formed by spreading elongate ribs in the neck region. Several species of *Naja* (e.g., *N. nigricollis*, *N. mossambica*, and *N. sputatrix*) and the southern African *Hemachatus* are known as spitting cobras. In these

species the openings through which venom is ejected from the fangs point forward rather than downward as in other cobras (Bogert 1943, Wüster and Thorpe 1992b). *Hemachatus* feigns death if spitting fails to deter an attacker. Many species of coral snakes (about 65 species of the primarily Neotropical genera *Leptomicrurus* and *Micrurus*) are mimicked by nonvenomous to mildly venomous colubrids in genera such as *Pliocercus*, *Lampropeltis*, and *Erythrolamprus* (Greene and McDiarmid 1981).

Approximately 70 species of sea snakes (Hydrophiinae) have specializations for a marine existence. All species have a salt gland surrounding the tongue sheath to help maintain osmotic balance in sea water (see Chapter 5). Two putative clades of sea snakes are generally recognized, the *Laticauda* group, including only *Laticauda*, and the *Hydrophis* group, which contains all other genera. The two groups differ somewhat in the degree of specialization for marine life. *Laticauda* retains the wide ventral scales characteristic of most terrestrial elapids, and these have a one-to-one correspondence with the vertebrae. In *Laticauda* the tail fin is poorly developed and is not supported internally with elongated vertebral processes. Species of *Laticauda* live in inshore marine areas, spend considerable time on land, and come ashore to lay eggs in caves above the tide line in coral reefs.

Genera of the *Hydrophis* group are more fully adapted to a marine existence. All species are viviparous and give birth at sea. Ventral scales are much reduced, almost nonexistent in most species, and the body is often laterally compressed, so much so that some species are long and ribbonlike. Species of the *Hydrophis* group have lost the one-to-one association of ventral scales with vertebrae (Voris 1975), and their well-developed tail fins are internally supported by elongate, expanded neural spines and haemapophyses. The nostrils of snakes in the *Hydrophis* group are dorsally located and can be closed with valves. The mouth can be sealed by depression of the tip of the snout, which is facilitated by a loose articulation of the premaxilla to the rest of the skull. The differing degrees of marine adaptation shown by *Laticauda* in comparison to the *Hydrophis* group, along with subtle internal differences, primarily of the head glands and muscles, have led some scientists to propose that the *Hydrophis* group and *Laticauda* are each more closely related to distinct terrestrial elapid lineages than to each other. With the exception of *Pelamis*, sea snakes are primarily snakes of coastal marine and estuarine waters and coral reefs. Two species of sea snakes, *Hydrophis semperi* and *Laticauda crockeri*, are restricted to freshwater lakes in the Philippine Islands. They presumably entered the lakes when there was a connection to the sea.

Content and Distribution. Approximately 62 genera, 280 species. 2 subfamilies: Elapinae (terrestrial elapids, 220 species; representative genera: *Acanthophis, Austrelaps, Boulengerina, Bungarus, Dendroaspis, Elapsoidea, Hemachatus, Homoroselaps, Maticora, Micrurus, Naja, Notechis, Ophiophagus, Oxyuranus, Pseudechis, Pseudohaje, Vermicella*); Hydrophiinae (sea snakes, 60 species; representative genera: *Aipysurus, Emydocephalus, Enhydrina, Hydrophis, Laticauda, Pelamis*). Elapinae: southern North America, throughout Central and South America; Africa (absent from Madagascar), southern Asia, the Indoaustralian Archipelago to New Guinea, Australia, Tasmania, Solomon Islands, and Fiji. Hydrophiinae: the Indian Ocean and southwest Pacific Ocean; 1 species, *Pelamis platurus*, extending from Asia across the Pacific to the coasts of Central and South America (Figure 4–29).

Molecular and morphological data suggest that sea snakes (Hydrophiinae) are more closely related to Australasian terrestrial elapids than to other terrestrial elapids. Thus, Elapinae is assuredly paraphyletic. However, neither the monophyly of the Australian radiation nor of Hydrophiinae with respect to it is firmly established. Other subfamilial arrangements for Elapidae are recognized, but no compelling phylogenetic reasons for the alternatives yet exist.

Systematics. Elapinae: McDowell (1967, 1986), McCarthy (1985), Wüster and Thorpe (1989, 1992a), Szyndlar and Rage (1990), Slowinski (1994, 1995), Wüster et al. (1995). Hydrophiinae: Smith (1926), McDowell (1969, 1972), Voris (1977), Mengden (1983), McCarthy (1986), Slowinski (1989), Roze (1996), Rasmussen (1997).

Atractaspididae

Species of *Atractaspis* are small to moderate-sized African snakes (lengths to about 1 meter) with slender bodies (Figure 4–30). In external appearance they are similar to many terrestrial colubrids. Their most peculiar features are internal, particularly characters of the dentition, jaw apparatus, venom gland, and venom. The maxilla is extremely reduced and bears an enormous hollow fang. As in viperids, the maxilla of *Atractaspis* has a complex articulation with the pre-frontal bone and can be erected. However, unlike in viperids, in which the fangs are erected in a posterior-anterior direction, the peculiar maxilla-prefrontal articulation and the enormous length of the fangs relative to head size permit only lateral rotation and erection in *Atractaspis*. Thus, the erect fangs barely protrude from the lip margin, with the fang tip directed posteriorly, and *Atractaspis* does not need to open its mouth to erect its fangs. Envenomation is accomplished by a lateral and posterior stabbing motion of the head. Species of *Atractaspis* feed on small mammals and nestling birds. The ability to erect the fangs without extensive opening of the mouth probably facilitates subterranean foraging within tunnels.

The venom glands of *Atractaspis* are similar in their overall structure to those of elapids, but the venom contains unusual components. In particular, one class of toxins, the sarafotoxins, are unique to *Atractaspis*. These are low-molecular-weight compounds that are structurally related to a family of hormones (endothelins) that act on smooth muscles in blood vessels. As in some elapids, the venom glands of several species of *Atractaspis* are extremely elongate and extend into the dorsal neck region behind the head.

The relationships of *Atractaspis* have proved extremely difficult to establish. *Atractaspis* was originally considered a rather aberrant viperid, but Bourgeois (1965) hypothesized a relationship between *Atractaspis* and an assemblage of African colubrids known as aparallactines (e.g., *Aparallactus, Xenocalamus, Chilorhinophis, Macrelaps*). Some herpetologists (e.g., Underwood and Kochva 1993) continue to associate the genus with aparallactines, whereas others (e.g., Cadle 1994) dispute this hypothesis and even question the monophyly of Aparallactinae (Cadle 1983, 1994, McDowell 1986). Some molecular evidence and aspects of venom gland structure and venom suggest elapid affinities for *Atractaspis* (Cadle 1988, Heise et al. 1995), but no hypothesis of relationships is strongly supported at present.

Content and Distribution. 1 genus (*Atractaspis*), about 18 species. Africa and the Middle East (Figure 4–28). Some herpetologists include the African aparallactine colubrids within this family, but interpretations of the supportive data differ and the monophyly of aparallactines is questionable.

Systematics. Laurent (1950), McDowell (1968, 1987), Cadle (1983, 1994), Underwood and Kochva (1993).

Colubridae

Colubridae has no known unique shared derived characters and may be paraphyletic with respect to Elapidae and *Atractaspis* (Cadle 1994). This huge family includes approximately 70 percent of all species of snakes, or more than 1,700 species. Basically, colubrids are colubroids that are not elapids, vipers, or *Atractaspis*. Reproductive modes, life histories, and habitats exploited by colubrids encompass those seen in Alethinophidia as a whole.

Several species of colubrids are dangerously venomous to humans. These include the African colubrines *Dispholidus typus*, *Thelotornis* (four species), and several species of the Asian natricine *Rhabdophis*. *Dispholidus* and *Thelotornis* are very closely related to several other genera that are not known to be venomous (Cadle 1994). Two famous herpetologists, Karl P. Schmidt and Robert Mertens, died from bites of *Dispholidus* and *Thelotornis*, respectively (see Pope 1958).

The systematics of Colubridae is poorly resolved and we mention only six infrafamilial groups (sometimes given formal family or subfamily status). Some core of each of these groups is probably monophyletic, but their limits are not well established. A vast array of genera are of uncertain relationships within Colubridae (see Cadle 1994).

Content and Distribution. Approximately 320 genera, >1,700 species. Worldwide (Figure 4–27).

Systematics. Dowling et al. (1983, 1996), Cadle (1988, 1994), Lopez and Maxson (1995, 1996).

Natricines. Natricines are small to moderately large colubrids that may be terrestrial, aquatic, or semifossorial. All New World natricines are viviparous, whereas most Eurasian species are oviparous, and Africa has natricines with both modes. Natricines occur in North and Central America, Africa, Eurasia and the Indoaustralian Archipelago into northern Australia. Representative genera are *Nerodia*, *Thamnophis*, *Storeria*, *Virginia*, *Regina*, *Natrix*, *Rhabdophis*, *Sinonatrix*, *Tropidonophis*, and *Limnophis*.

Colubrines. This is the largest clade of colubrids, including more than half of the genera within the family. Many of these are familiar to North American students because they are prevalent (along with natricines) in the North American fauna. Colubrines are cosmopolitan except for most of Australia. Representative genera are *Lampropeltis*, *Elaphe*, *Tantilla*, *Opheodrys*, *Coluber*, *Dasypeltis*, *Coronella*, *Philothamnus*, *Boiga*, *Dendrelaphis*, and *Chrysopelea*.

Xenodontines. Several major clades are included within this group, but it has historically been recognized on the basis of shared ancestral characters. Two large, primarily Neotropical, clades, Dipsadinae and Xenodontinae, are monophyletic within the broader group. Five exclusively North American genera (*Heterodon*, *Diadophis*, *Contia*, *Farancia*, and *Carphophis*) are xenodontines, but their relationship to the Neotropical clades is unclear (Cadle 1984). The biogeography and radiation of xenodontines was summarized by Cadle and Greene (1993). Xenodontines are found in North, Central, and South America. Representative Neotropical genera include the Central American xenodontines or Dipsadinae (*Rhadinaea*, *Dipsas*, *Sibon*, *Coniophanes*, *Geophis*, *Ninia*, and *Pliocercus*) and the South American xenodontines, or Xenodontinae in the strict sense (*Clelia*, *Xenodon*, *Erythrolamprus*, *Alsophis*, *Philodryas*, and *Liophis*).

Homalopsines. Homalopsines include about 35 species of estuarine, marine, and freshwater snakes. Their nostrils contain valves and the glottis can be plugged into the choanae so that breathing can take place with only the nostrils above water. Most species appear to be viviparous. Homalopsines are rear-fanged and consume a variety of aquatic vertebrates and invertebrates, including hard-bodied crustaceans such as crabs (see Chapter 9). Homalopsines occur from Southeast Asia to northern Australia. Representative genera include *Homalopsis*, *Enhydris*, *Fordonia*, *Myron*, *Herpeton*, and *Cerberus* (Gyi 1970).

Pareatines. Pareatines are Asian ecological equivalents of New World snakes of the genera *Dipsas*, *Sibon*, and their relatives that feed on gastropods. They are found in Southeast Asia and islands of the Sunda Shelf. Representative genera are *Pareas* and *Haplopeltura*.

Aparallactines. We have already mentioned this group in connection with *Atractaspis*. The monophyly of the group is not well established (McDowell 1986, Underwood and Kochva 1993, Cadle 1994), but we mention the group because it figures prominently in discussions of the origin of venom delivery systems of snakes (Cadle 1983). These are relatively small snakes, many of them burrowers, and some with highly unusual diets (e.g., centipedes in *Aparallactus*). They occur in sub-Saharan Africa. Representative genera are *Aparallactus*, *Amblyodipsas*, *Xenocalamus*, *Chilorhinophis*, and *Macrelaps*.

Crocodilians (Crocodylia)

Crocodilians have heavily armored, elongate, lizard-like bodies with long snouts and powerful tails. The armor is formed by heavy plates of bone called osteo-derms that lie within the dorsal skin, underneath the heavy epidermal scales covering the body. Osteo-derms are also present ventrally in many species. The teeth of crocodilians are thecodont (set into sockets) rather than attached to the sides or top of the bones. The limbs of crocodilians are relatively short, and all feet are webbed. All species are aquatic to varying de-grees. The eye has a transparent membrane that is drawn across the eye underwater.

Crocodilians have several morphological features that allow breathing in an aquatic setting. The nostrils are dorsally located at the tip of the snout and are closed by valves during diving. A well-developed sec-ondary palate completely separates the buccal and res-piratory passages (Figure 4–32). Internally, the nasal passages (choanae or internal nares) open in the throat behind the secondary palate and can be closed off from the throat by fleshy folds on the back of the tongue and palate. These modifications allow crocodilians to breathe in water while holding prey in the mouth. Ad-ditional changes, such as a muscular partition separat-ing the pectoral and abdominal cavities (analogous to the diaphragm of mammals) and well-developed alve-oli in the lungs, are unique among reptiles and are also associated with respiratory efficiency. Crocodilians have a four-chambered heart, similar to the heart in their avian relatives, but mixing of oxygenated and de-oxygenated blood can occur via an aperture, the fora-men of Panizza, that connects the pulmonary and sys-temic circulations as they leave the heart (see Chapter 6). Crocodilians have exceptional abilities to adjust pe-ripheral and core (heart, brain) blood flow in response to the demands of diving or thermoregulation.

The dorsal and ventral scutes of crocodylids and gavialids (including *Tomistoma*) have conspicuous

pits that may have an osmosensory function (Jackson et al. 1996), although mechanosensory functions have also been suggested. Lingual salt glands are present on the tongues of crocodylids and alligatorids (see Chapter 5).

Crocodilians are excellent swimmers and can move surprisingly fast on land, where they generally use a walk in which the belly is held high off the ground and the limbs are placed underneath the body. However, they also use a belly crawl when plunging into water, and *Crocodylus johnsoni* is known to gallop. In water, crocodilians fold the limbs against the body and use lateral undulation of the body and tail to swim.

All crocodilians are oviparous and lay their eggs either in mound nests constructed from vegetation and soil (alligatorids, *Tomistoma*, and most croco-dylids) or directly in soil on beaches or other exposed areas (*Gavialis* and several species of *Crocodylus*). Tem-perature-dependent sex determination occurs in all species that have been examined. Parents guard the nest, and some species break open the nest to release the hatchlings, carry the babies to water in their mouths, and remain with the young for weeks or months.

The social behavior of crocodilians is more com-plex than that of other reptiles. Hearing acuity is good, and vocalizations are used in a variety of social contexts including territorial bellowing during the breeding season, aggressive warnings to intruders, and signals given by neonates within the nest to elicit nest-opening behavior by attending adults. Circum-stantial evidence suggests that young within the eggs use vocalizations to communicate with one another, perhaps as a means of synchronizing hatching. Vocal-izations are used after hatching to maintain group cohesiveness and to alert adults to potential threats. Sounds produced by crocodilians are of low frequency and surprisingly diverse repertoire, including roars, grunts, coughs, and purrs. Most crocodilian sounds are produced by the vocal cords, but sounds barely perceptible to humans are produced by rapid con-traction of the body wall musculature underwater. The senses of smell and eyesight are also well developed in crocodilians, and social signals often in-volve combinations of visual, olfactory, and acoustic cues. Body postures and headslapping at the water surface are often used as dominance advertise-ments.

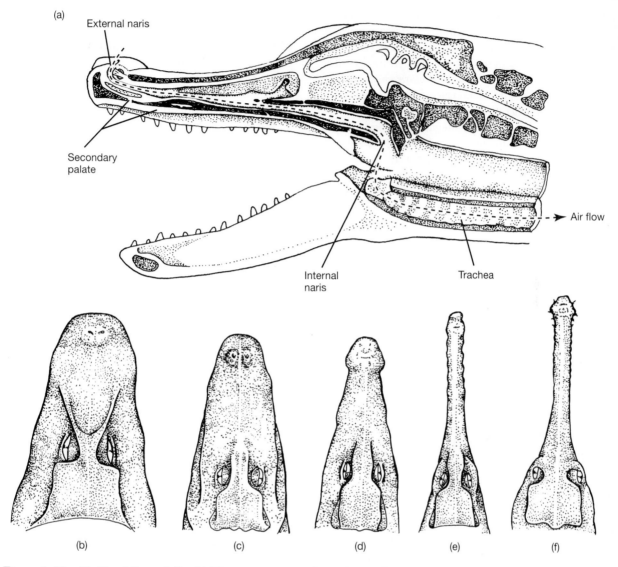

Figure 4–32 **Skulls of Crocodylia.** (a) The extensive secondary palate and positions of the external and internal nares, (b–f) structural diversity among crocodilians: (b) *Caiman latirostris.* (c) *Alligator mississippiensis,* (d) *Crocodylus porosus,* (e) *Tomistoma schlegii,* (f) *Gavialis gangeticus.* (*Source: (a) Modified from Goodrich 1930, (b–f) Modified from Bellairs 1970.*)

Crocodilians are efficient predators and eat a variety of prey, including fishes, turtles, birds, mammals, and shellfishes. Small crocodilians eat insects, frogs, and other small prey. Generally, predation is nocturnal, and most crocodilians either forage in water or drag their prey into water. Drowning is often used as a method of killing large prey. Crocodilians do not chew. Rather, they are food gulpers that must dis-member their prey into sizes suitable for swallowing (see Chapter 9).

Many species of crocodilians throughout the world have been exploited for their skins, which are made into leather, and for their meat. The leather trade has exploited species successively, beginning with *Alligator mississipiensis* in the 1700s and moving to other species as commercial hunting became unprofitable when populations were reduced to near extinction. Several species, including *Caiman niger* and *Crocodylus mindorensis*, have had their ranges vastly reduced as a result of hunting pressures. More recently some species, including *Alligator mississipiensis*, have recovered as a result of conservation efforts. Wild populations of crocodilians are protected by international agreements, although illegal poaching is still a major problem in many areas. Farming of several species of crocodilians for the leather trade has become a commercially successful venture in many countries (see Chapter 15).

Fossil Record

Crocodilians have an extensive fossil record spanning the last 215 million years (middle Triassic to Recent). As with turtles, their large size, aquatic habits, and extensive ossification have favored fossilization. Fossils of the modern families Alligatoridae and Crocodylidae appeared in the late Cretaceous, whereas the oldest gavialid fossil is from the Eocene. Like most of the extant crocodilian fauna, most fossil crocodilians lived in freshwater environments. However, diverse lineages radiated into more or less fully marine or terrestrial environments during their history. For example, metriorhynchid crocodilians (Jurassic to early Cretaceous) were efficient marine predators that had lost their dorsal armor. Their limbs were transformed into paddles and their tail bore a fishlike fin at the tip. Buffetaut (1985), Gasparini (1996), and Carroll (1988) have reviewed the fossil record.

Crocodilian evolution has been marked by two general trends that occurred in parallel in many different lineages. The first was the evolution of increased flexibility and strength of the spine, marked by a transition from amphicoelous vertebrae early in their evolution to procoelous vertebrae later. The second trend was the evolution of a complete secondary palate separating the buccal cavity from the respiratory passages, allowing breathing to continue when the mouth was open underwater.

Content and Distribution

Eight genera and 22 extant species of crocodilians are recognized. They are distributed throughout tropical and subtropical regions, but *Alligator* is found in warm temperate regions of the United States and China. Most species occur primarily in freshwater habitats, but several readily enter brackish or salt water. *Crocodylus porosus* of the Indopacific region and *C. acutus* of Central and South America and the Caribbean are frequently found in marine habitats.

Systematics. Medem and Marx (1955), Medem (1981, 1983), King and Burke (1989), Poe (1996).

Systematics and Phylogeny of Crocodilians

Three major lineages among extant crocodilians are universally recognized. These are sometimes considered subfamilies of a single family (Crocodylidae), but we use three separate families: Alligatoridae (alligators, caimans, and dwarf caimans), Crocodylidae (crocodiles and dwarf crocodiles), and Gavialidae (true and false gharials or gavials). A phylogeny of crocodilians based on combined molecular and morphological characters is presented in Figure 4–33. With one exception, there is little controversy over relationships among these groups from the standpoint of morphological and molecular data.

The major disagreement between morphological and molecular phylogenies concerns the relationships of the gharials, *Gavialis* and *Tomistoma*. Phylogenetic analysis of morphological characters (Frey et al. 1989, Norell 1989, Tarsitano et al. 1989) place *Tomistoma* firmly within the Crocodylidae, whereas a variety of molecular data sets (immunological studies of albumin and transferrin proteins, allozymes, and ribosomal DNA sequences and restriction mapping) equally firmly associate *Tomistoma* with *Gavialis* relative to other crocodilians (Densmore and White 1991, Hass et al. 1992, Gatesy and Amato 1992). Poe (1996) reanalyzed all previous data sets and found strong support for the monophyly of *Gavialis-Tomistoma* relative to other crocodilians. Nevertheless, he also concluded that there was significant conflict between the morphological data and other data sets concerning this aspect of crocodilian phylogeny. Relationships among the three extant crocodilian groups are not well resolved. Morphological data support a crocodylid-alligatorid clade relative to gavialids, whereas molecular data and combined analyses of all data suggest that alligatorids are a sister group to Crocodylidae-Gavialidae (Poe 1996). In any case, support for either of the alternative views is weak.

Alligatoridae

In alligatorids, the teeth of the lower jaw fit into pits in the upper jaw and cannot be seen when the mouth is closed. Alligatorids include some of the largest (*Caiman niger*, at more than 6 meters in total length) and the smallest (*Paleosuchus*, less than 1.7 meters in total length) extant crocodilians (Figure 4–34). With the exception of *Paleosuchus*, alligatorids are inhabitants of larger rivers, lakes, swamps, and lagoons. The two species of *Paleosuchus* characteristically inhabit small streams and pools of forested areas of the Amazon Basin. *Alligator mississipiensis* freely enters coastal marine waters. The term *alligator* is a corruption of the Spanish *El lagarto* (the lizard), which early Spanish explorers applied to these creatures. Caiman is the word commonly used throughout Latin America for any of these animals.

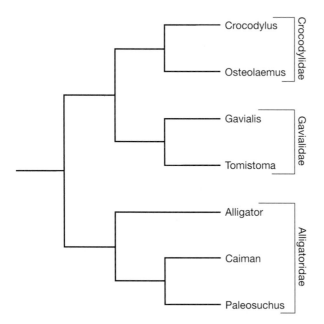

Figure 4–33 **Phylogeny of crocodilians based on a combined analysis of molecular and morphological data.** *(Source: Poe 1996.)*

Alligator mississipiensis of temperate areas of the United States is the only crocodilian inhabiting areas where freezing of surface waters may occur regularly during the winter. It was long thought that alligators sought refuge in underwater or underground dens during severe cold periods. However, radiotelemetry studies have shown that the alligators move into shallow water and position themselves with their nostrils exposed above the surface ice. They are incapable of lowering their metabolism sufficiently to depend entirely on anaerobic metabolism, and hence must breathe during these periods. They have even been found with their snouts frozen into surface ice.

The two species of dwarf caimans, *Paleosuchus*, are unlike other caimans in that they live in heavily forested areas of the Amazon Basin. *Paleosuchus palpebrosus* more often occurs in rivers and lakes than *P. trigonatus*, which prefers small streams (Magnusson et al. 1987). The may be found in the forest some distance from water. They are nocturnal and may remain undetected by local inhabitants where they occur. Little is known of the natural history of these caimans, but diets sometimes include substantial numbers of terrestrial vertebrates, compared to the diets of other sympatric crocodilians (Magnusson et al. 1987). Being deep-forest animals, *Paleosuchus* cannot rely on heat from the sun to warm eggs for hatching. *Paleosuchus trigonatus* lays eggs in mound nests usually constructed adjacent to termite mounds in the forest. The metabolic heat generated by the termites is the primary source of heat for egg incubation, in contrast to heat sources for other crocodilian nests (solar heat, decaying vegetation of the nest, and metabolic heat of the embryos themselves) (Magnusson et al. 1985).

Content and Distribution. 3 genera (*Alligator*, *Caiman*, *Paleosuchus*), 7 species. All genera except *Alligator* restricted to Central and South America. *Alligator* includes *A. mississipiensis* (coastal plain of the southeastern United States) and *A. sinensis* (lower Yangtze River of eastern China) (Figure 4–35). Poe (1996) synonymized *Melanosuchus* with *Caiman*.

Systematics. Medem (1958, 1963, 1981, 1983), Hass et al. (1992), Poe (1996).

Crocodylidae

In crocodylids, the fourth tooth in the lower jaw is accommodated in a notch in the upper jaw and is visible

(a)

(b)

(c)

(d)

Figure 4–34   **Diversity of crocodilians.** (a) *Caiman crocodilus* (Alligatoridae, South America). (b) *Paleosuchus palpebrosus* (Alligatoridae, South America). (c) *Crocodilus rhombifer* (Crocodylidae, Australia). (d) *Gavialis gangeticus* (Gavialidae, India). *(Photographs by (a, c) R. Wayne Van Devender; (b) John Cadle, (d) Michael Fogden/DRK Photo.)*

when the mouth is closed (Figure 4–34). The saltwater crocodile, *Crocodylus porosus*, grows to more than 7 meters and is probably the largest extant crocodilian. The other extreme of the size range is represented by the West African dwarf crocodile, *Osteolaemus tetraspis*, which reaches about 2 meters. Nothing is known of the natural history of *Osteolaemus*, but it is an inhabitant of small streams within rain forests, and its natural history may be similar to that of the small alligatorids *Paleosuchus* (see below).

Crocodylus porosus has the widest distribution of all crocodilians, from India to Australia and perhaps as far as Fiji in the southwest Pacific Ocean. It has been observed in open ocean far from land and has colonized small islands nearly 1,000 kilometers from the nearest land. This species and the Nile crocodile, *Crocodylus niloticus*, are nearly exclusively responsible for crocodiles' reputation for eating humans, and for these two species the reputation is well-deserved.

Content and Distribution. 2 genera, 13 species. *Crocodylus* (12 species) and *Osteolaemus tetraspis* (West Africa). Southern Mexico to northern South America (Orinoco River Basin); West Indies and the southern tip of Florida; coastal Madagascar and nearly all of Africa; southern Asia from Iran east throughout India, Southeast Asia, and the Indoaustralian Archipelago to New Guinea and northern Australia (Figure 4–35). Molecular data suggest the monophyly of *Crocodylus* relative to *Osteolaemus* (Hass et al. 1992).

Systematics. King and Burke (1989), Poe (1996).

Figure 4–35 **Distribution of the crocodilian families Alligatoridae, Gavialidae and Crocodylidae.**

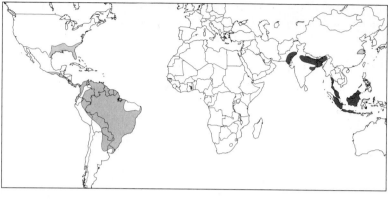

Alligatoridae ◯
Gavialidae ⬤

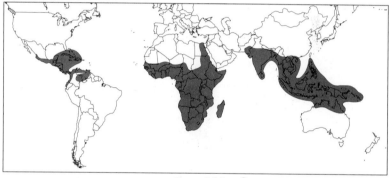

Crocodylidae ⬤

Gavialidae

Gavialis and *Tomistoma* reach lengths of about 6.5 meters and 4 meters, respectively. Both have very elongate, narrow snouts, and consume primarily fish (Figure 4–34). *Gavialis* is possibly the most aquatic of crocodilians, has relatively weak limbs, and lives in fast-flowing rivers. Nests of *Gavialis* consist of holes in sandbanks along rivers, whereas *Tomistoma* constructs a mound nest.

Content and Distribution. 2 genera, 2 species: *Gavialis gangeticus* (northern India, Pakistan, Nepal, Bangladesh, Bhutan, east to Burma); *Tomistoma schlegelii* (southern Thailand and Malaysia, Sumatra, Borneo, and Java) (Figure 4–35).

Systematics. King and Burke (1989), Poe (1996).

Summary

Extant reptiles (aside from birds) comprise more than 7,000 species, of which squamates (lizards, snakes, and amphisbaenians) make up approximately 95 percent. Most of the rest of the species diversity is made up of turtles (3.5 percent), followed by crocodilians and the two species of rhynchocephalians. Turtles have a unique body plan among vertebrates in which the pectoral and pelvic girdles are located within the rib cage. Turtles are primarily adapted for aquatic lifestyles, but the tortoises are primarily terrestrial. Rhynchocephalia is represented by two species restricted to New Zealand, and both are threatened with extinction.

Rhynchocephalia is the sister taxon to Squamata, and together they make up the Lepidosauria. Although we generally think of snakes and amphisbaenians as distinct taxonomic groups from lizards, both of these mono-phyletic clades are nested phylogenetically within lizards. About two thirds of squamates are lizards, and the most species-rich groups of lizards are skinks, geckos, and iguanids in the broad sense. Most snakes (nearly 70 percent) belong to the family Colubridae. As in amphibians, reproductive modes are diverse in reptiles and include viviparous and oviparous species.

How Do They Work?

5

Temperature and Water Relations

Animals are aqueous systems. In other words, 70 to 80 percent of the body mass of an animal is water and the chemical processes that allow the animal to function take place in a solution of inorganic and organic compounds in water. Biochemical reactions depend on having the proper concentrations of substrates, salts, and other cofactors in the cells; the spatial relationships of intracellular organelles are determined in part by the volume of water in the cell; and blood (which is about 90 percent water) carries chemical substrates, reaction products, and other biologically active compounds such as hormones from one part of the body to another. All of these components must be kept in balance, and that balance is shifted when the amount of water or salt in the body changes. Cells swell or shrink as the volume of water in them increases or decreases, and those changes alter the concentrations of dissolved substances within the cells as well as the configuration of intracellular structures. Blood becomes more viscous as its water content drops, flowing more sluggishly and requiring more effort from the heart to pump it. Thus, as we noted in Chapter 1, a departure from the normal water and salt content of an animal upsets the balance of biochemical and physiological processes and interferes with the integration of those processes.

The effect of body temperature is equally profound. The rates of many biochemical reactions approximately double when the temperature increases by 10°C and fall to half the original rate when the temperature decreases by 10°C. This phenomenon is known as the Q_{10} effect, and a doubling or halving of the rate of a reaction for a temperature change of 10°C corresponds to a Q_{10} effect of 2. Not all reactions have the same temperature sensitivity, and the Q_{10} values of biochemical reactions extend from 1 (i.e., no change in reaction rate as temperature changes) to values of 3 or more. Furthermore, the effects of temperature are not necessarily linear over the entire range of temperature an animal may encounter. That is, the same reaction may have different Q_{10} values at different temperatures.

An organism, viewed from the perspective of biochemistry, is a series of linked chemical reactions, and the product of each reaction is the substrate for the next reaction in the series. If each reaction has a different Q_{10} value, you can imagine what havoc a variable body temperature causes with the integration of cellular processes. To complicate the situation still further, these reactions take place in a cellular environment that also changes with temperature. As temperature increases, the viscosity of cytoplasm decreases, the lipid membranes of cells and cell organelles become more permeable, the rate at which electrical impulses travel along nerve axons increases, and the speed and force of muscular contractions increase. How could an animal function in the face of such chaos?

The solution, of course, is for an animal to regulate its body temperature and water content so as to minimize the disruptive effects of variation. Early studies of temperature regulation focused on reptiles and were carried out in the Soviet Union and in North America (Sergeyev 1939, Cowles and Bogert 1944). Our knowledge of temperature regulation by reptiles and amphibians has increased greatly since then (Brattstrom 1979, Avery 1982, Bartholomew 1982, Huey 1982, Hutchison and Dupré 1992). Studies have shown that amphibians use behavioral and physiological mechanisms to regulate water content in much the same way that reptiles use behavior and physiology to control body temperature (Brattstrom 1979, Feder 1982, Pough et al. 1983, Hutchison and Dupré 1992).

Pathways of Energy and Water Exchange

Evaporation is the phenomenon that links temperature and water regulation. Evaporation is the loss of water, of course, but it is also a way to lose heat. The rate at which an animal evaporates water has a major influence on its body temperature, and its body temperature influences its rate of evaporation.

The study of the way animals exchange heat and water with their environments is known as biophysical ecology, and reptiles and amphibians have been especially important in the development of this field because interaction with the physical environment is a conspicuous part of their lives. Understanding the routes by which an animal gains and loses heat and water is an enormous help in understanding the regulatory mechanisms used by amphibians and reptiles, and it is far from an abstract topic. If you have the opportunity to watch frogs or lizards in the field, or even in cages in the laboratory or at a zoo, you'll see them changing posture or color in ways that modify the rates at which water or energy is gained and lost. Even preserved specimens display features that you can relate to biophysical principles, such as the colors and textures of the skin and scales on different parts of their bodies.

Although water balance and temperature regulation are closely intertwined, we will discuss them separately for the sake of clarity.

Water Uptake and Loss

Regulating the amount of water in the body requires balancing gain and loss. In a steady state (i.e., no change in body water content), the total intake of water must equal the total water loss, and each side of the water balance equation has several components:

Gain: Liquid water + preformed water + metabolic water =
Loss: Evaporation + urine + feces + salt glands.

This is a general equation that fits terrestrial or aquatic amphibians or reptiles, but some of the details of water movement apply only to particular situations or certain kinds of animals.

Routes of Water Gain

All of the water an organism needs for its metabolic processes and to replenish water lost by evaporation or in the feces and urine enters through the mouth or across the skin, and some amphibians and reptiles have structures and behaviors that enhance collection of water. Three sources of water are available to animals: liquid water, preformed water, and metabolic water.

Liquid Water. Liquid water is what you normally think of as water—that is, enough water molecules collected in one place to form a pool, puddle, drop, or at least a film of water on the surface of a rock or leaf. Amphibians and reptiles make use of all those sources of water. Many reptiles drink from pools or puddles, just as birds and mammals do. More intriguing are the ways some reptiles have of obtaining liquid water in habitats where puddles rarely form. In deserts, water often disappears into the soil as quickly as it falls, and some reptiles catch water droplets before they reach the ground. In the Namib Desert of southern Africa, for example, the small lacertid lizard *Aporosaura anchiatae* and the viper *Bitis peringueyi* collect droplets of fog that form when moist air from the sea blows across the cold Benguella Current. The lizard drinks droplets of water that collect on vegetation and can consume nearly 15 percent of its body mass in 3 minutes. The snake uses its own body surface to collect water, flattening to present a large surface area. The skin of the viper, like that of most snakes, is hydrophobic (i.e., it repels water, much as a freshly waxed automobile does), and small droplets of water on the skin flow together to form large drops. The snake moves its head along the length of its body, swallowing the water it has collected (Figure 5–1). *Aporosaura anchiatae* and *Bitis peringueyi* occur only in the fog belt region of the Namib, suggesting that the collection of water from fog plays a critical role in their overall water balance.

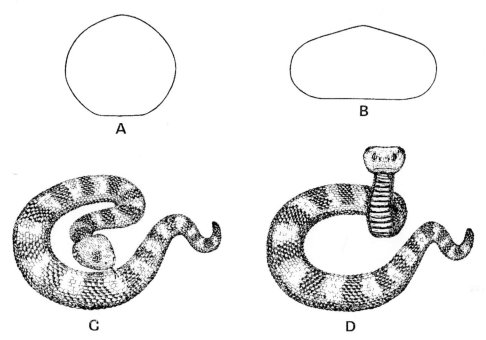

Figure 5–1 **The Namib Desert viper *B. peringueyi* collecting and swallowing fog droplets.** (a) The cross section of the body is normally round. To collect fog, the snake flattens its body (b), drinks the drops of water that form (c), and raises its head to assist the flow of water into the digestive tract (d). *(Source: Louw 1972.)*

Foggy deserts are limited to coastal locations, but dew forms at night in inland deserts. Some reptiles and amphibians probably collect dew as a source of water. When drops of water are sprayed on the body surface of the Australian lizard *Moloch horridus* or the Asian lizard *Phrynocephalus helioscopus*, the water moves through channels between the scales to the corners of the mouth. Soon after its dorsal surface has been wetted, the lizard begins to make swallowing movements.

Some turtles use their shells to channel rain into their mouths. The marginal scutes of some turtles curve upward to form a structure rather like a gutter that extends around the margin of the carapace. These turtles adopt a rain-collecting posture, extending their hind legs to raise the posterior end of the shell and stretching their forelegs parallel to their extended necks (Figure 5–2). Rain runs off the carapace into the gutters formed by the marginal scutes and flows to the front of the shell, where the forelegs channel it to the turtle's mouth. Other turtles, such as the desert tortoise (*Gopherus agassizii*), build water catchment basins at the bottoms of slopes. The tortoises move to the basins after a rain and drink enough water to increase their body mass, sometimes by as much as 17 percent.

Amphibians do not drink; instead they absorb water through their skins. In general, terrestrial species of amphibians have more permeable skins than do aquatic and arboreal species. Most terrestrial anurans have an area of skin called the pelvic patch that is relatively thin and is underlain by a dense network of capillaries. When a dehydrated frog is on a moist surface it splays its hind legs, presses the pelvic patch against the substrate, and absorbs water.

Spadefoot toads, *Scaphiopus* and *Spea*, make particularly dramatic use of the ability to absorb water through the skin. These anurans, which inhabit dry desert regions in the American Southwest, spend about 9 months of the year in burrows they construct by digging with the hind limbs. A western spadefoot

Figure 5–2   **Turtles in water-collecting postures.** Drinking stances of (a) *Psammobates tentorius* and (b) *Kinixys homeana. (continued)*

(a)

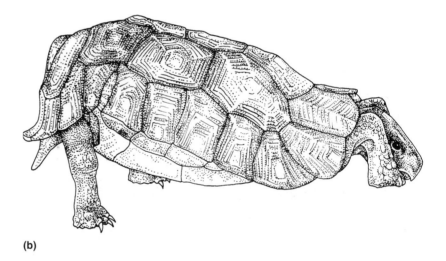

(b)

toad (*Spea hammondi*) is fully hydrated when it burrows into the ground in September or October. The bladder is filled with dilute urine (about 100 mmole·kg H_2O^{-1}), and the blood plasma is about 300 mmole·kg H_2O^{-1}. From September through March, the concentration of urea and the osmolalities of urine and plasma remain stable, even though urea is being produced by metabolism of protein. (Osmolality refers to the concentration of small molecules or ions in solution. Osmolality is measured in millimoles of solute per kilogram of water—mmole·kg H_2O^{-1}.) The lack of change in fluid concentrations during these 7 months indicates that the toads are absorbing water from the surrounding soil.

Between March and June, the situation changes and the concentration of urea rises in both the plasma and urine. As urea accumulates, the osmolalities of the plasma and urine increase in parallel, with the plasma remaining about 200 mmole·kg H_2O^{-1} more concentrated than the urine. By June, just before the summer rains begin, urea concentrations in urine can exceed 300 mmole·kg H_2O^{-1} and plasma osmolality can reach 500 to 600 mmole·kg H_2O^{-1}. These high internal osmolalities may allow the toads to continue to absorb water from the nearly dry soil. Throughout dormancy, the urine, which is always dilute compared to the plasma, provides a source of water to limit the increase in plasma osmolality. A similar accumulation of urea during dormancy has been demonstrated for the desert-dwelling tiger salamanders, *Ambystoma tigrinum*, and may be widespread among anurans and salamanders in arid regions.

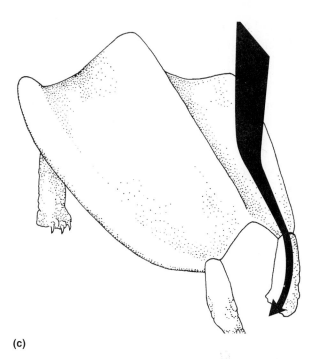

(c)

Figure 5–2   (continued) (c) The flow of water for *Kinixys*. The gutter formed by the marginal scutes of the shell and the role of the forelimbs in directing water to the mouth are shown. *(Source: Redrawn from Auffenberg 1963.)*

The amphibian hormone arginine vasotocin (AVT), which is closely related to mammalian antidiuretic hormone (ADH) and is produced by the posterior lobe of the pituitary (the neurohypophysis), plays a major role in water balance by stimulating osmotic uptake of water through the skin of semiaquatic and terrestrial amphibians (Boutilier et al. 1992). AVT also reduces water loss by reducing the rate at which urine is produced and increasing reabsorption of water from the bladder.

When they are in freshwater, amphibians face an osmotic influx of water (Boutilier et al. 1992). The body fluids of amphibians have an osmolality of around 250 mmole·kg H_2O^{-1} whereas the osmolality of freshwater is often as low as 1 or 2 mmole·kg H_2O^{-1}. That difference establishes a steep gradient for inward water flow, and the high permeability of amphibian skin means that there is little resistance to the inward movement of water. Excess water must be eliminated by producing a large volume of dilute urine. Concentrations of salts inside the body are higher than those in freshwater, so ions diffuse outward. These losses are counteracted by active uptake

of ions from the water via the skin (and gills, when they are present).

Salt water is a difficult environment for an amphibian, and only 13 species of salamanders and 61 species of anurans have been reported to inhabit (or at least to tolerate) brackish conditions. These species raise their internal osmolalities by accumulating inorganic and organic solutes in the body fluids. The African clawed frog, *Xenopus laevis*, maintains an internal osmolality higher than that of its surroundings, up to 500 mmole·kg H_2O^{-1} (equivalent to half-strength seawater). Accumulation of urea accounts for most of the adjustment. The green toad, *Bufo viridis*, ventures into brackish lakes in Israel, also maintaining an internal osmolality above that of its environment. Concentrations of sodium, chloride, and urea increase in the blood plasma, but in muscle, free amino acids account for most of the change in osmotic activity. Some populations of salamanders in the genus *Batrachoseps* live on the margins of beaches where ocean spray deposits salt. Both coastal and inland populations of these salamanders tolerate high environmental salt concentrations by increasing plasma concentrations of sodium and urea.

Movement of water through the skin is probably of minor significance for reptiles, because reptilian skin is relatively impermeable to water and ions (Lillywhite and Maderson 1982). Freshwater reptiles have a small osmotic influx of water and a correspondingly small loss of ions through the skin. Marine reptiles have similarly low outward movements of water and inward flows of ions. Active uptake of ions via the skin is not known to occur among reptiles, and it appears that the salt content of the prey, combined with the salt-conserving mechanisms of the kidney, is sufficient to achieve balance. Lipids in the skin form the primary barrier to water movement, and removal of lipids from shed skins by treatment with organic solvents increases permeability 15- to 30-fold.

Preformed Water.   Preformed water is a less visible part of the water budget of an organism than is liquid water, but it is vital for the survival of many reptiles. Most of the food animals eat contains large amounts of water: 70 to 80 percent of the body mass of vertebrates is water, and many insects are in the same range. Plants are more variable in their water content than are animals, and the water content of some plant tissues changes seasonally or even from day to night. In early May plants eaten by chuckwallas (*Sauromalus obesus*) contain 72 percent water, but by late May the water content has fallen to 51 percent. When their food contains less than 63 percent water, chuckwallas lose more water in their urine and feces than they gain from their food, and they stop eating when the water content of their food plants drops below that level.

The salt content of the diet is important, because if an animal consumes too much of a particular ion it must excrete the excess, and excretion of ions requires water and sometimes nitrogen as well. Other vertebrates are the best prey for a vertebrate to eat in terms of salt and water balance, because the water content and ionic composition of all vertebrates are quite similar. Consequently, when a snake eats a lizard its meal is about 70 percent water and contains sodium, potassium, chloride, and other ions in essentially the same concentrations as those of the snake. Insects usually have higher concentrations of potassium than do vertebrates, and a lizard that eats insects must have a way to eliminate excess potassium. Plants usually have higher concentrations of both potassium and bicarbonate than do vertebrates, and herbivorous vertebrates must excrete both those ions. Salt glands, described in the following section, provide a way to excrete excess ions with little loss of water.

Metabolic Water.   Metabolic water is the least obvious source of water in the diet of an animal, because it is not in the form of water molecules until hydrogen from chemicals in the food is combined with oxygen to form water. The water formed by cellular metabolism can be substantial: a gram of starch yields 0.556 gram of water, and a gram of fat produces 1.071 grams of water. The amount of water derived from the oxidation of protein depends on whether the end product of protein metabolism is urea or uric acid. Metabolism of protein to urea produces 0.396 gram of water per gram of protein. Uric acid contains fewer hydrogen atoms per nitrogen atom than does urea, so more hydrogen is converted to water, and the yield is 0.499 gram of water from 1 gram of protein.

Routes of Water Loss

An amphibian or reptile can use behavioral and physiological mechanisms to control the magnitude of its water loss and, to some extent, to control the routes by which water leaves the body. Control of water loss may limit the ability of an organism to be in a particular habitat at a particular time, or to engage in some kinds of activities.

Evaporation.   Evaporation is the major route of water loss for terrestrial amphibians (Shoemaker et al. 1992). Although most amphibians lose water rapidly by cutaneous evaporation, some arboreal frogs have rates of evaporative water loss that are one-half to one-third those of typical amphibians. Cutaneous resistance is measured as the time water takes to diffuse through the skin and is expressed as $sec \cdot cm^{-1}$ of skin thickness. On this scale, low numbers mean rapid diffusion (i.e., permeable skin), and high numbers indicate slow diffusion and impermeable skin. A comparison of the skin resistances of amphibians suggests that there are three groups with different skin properties (Figure 5–3). Terrestrial and aquatic amphibians have very small cutaneous resistances (0.05 to 0.07 $sec \cdot cm^{-1}$), and these are the species that evaporate water essentially as rapidly as a free water surface. A second group, mostly arboreal hylids from North and Central America but also including arboreal species of Asian rhacophorids and African hyperoliids, have cutaneous resistances around 2 $sec \cdot cm^{-1}$. That is about a 35–fold difference in skin resistance, and it produces substantially lower rates of evaporative water loss. For example, seven species of North American hylids lost water at rates of 10 to 16 mg $H_2O \cdot cm^{-2} \cdot sec^{-1}$, compared to rates of 25 to 31 mg $H_2O \cdot cm^{-2} \cdot sec^{-1}$ for 11 nonarboreal species under identical conditions. Australian tree frogs (*Litoria*) and African reed frogs (*Hyperolius*) have cutaneous resistances of 10 to 100 $sec \cdot cm^{-1}$ and a few frogs, such as *Chiromantis* (Rhacophoridae) and *Phyllomedusa* (Hylidae), have resistances as high as 300 $sec \cdot cm^{-1}$.

The mechanisms that produce these high cutaneous resistances are mostly unknown. Several observations suggest that dried mucus on the surface of the skin may contribute to the resistance of arboreal hylids such as *Hyla cinerea*. Cutaneous resistance increases at low humidity and is accompanied by bursts of evaporation that may correspond to the release of mucus from cutaneous glands. Mucus may also be involved in the high skin resistance of *Hyperolius viridiflavus*, and lipids may play a role for *Hyla kivuensis* and *Chiromantis rufescens*. Only in the case of *Phyllomedusa sauvagei* is the mechanism of skin resistance well understood. This large South American hylid frog lives

Cutaneous Resistance to Water Vapor

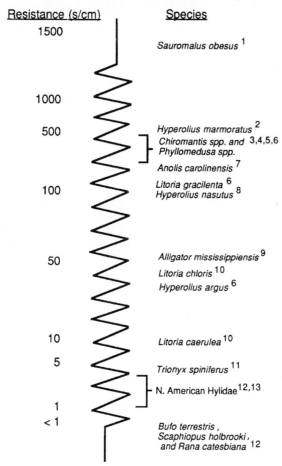

Figure 5–3 **Cutaneous resistance to evaporation for anurans and reptiles.** The scale is logarithmic. *(Source: Buttemer 1990.)*

species secrete watery mucus from cutaneous glands to increase evaporative cooling. Two additional methods of increasing evaporative cooling have been demonstrated for *Phyllomedusa:* (1) The epidermal layer of wax that waterproofs a resting frog melts at high temperatures, making the body surface permeable to water and allowing evaporative cooling to take place. (2) A frog opens its eyes as its brain temperature approaches a stressfully high level, and evaporation from the moist surface of the cornea cools the brain (Shoemaker and Sigurdson 1989).

Cocooning is another way that amphibians reduce evaporative water loss. Several species of anurans, including Australian frogs in the genera *Cyclorana* (Hylidae), *Limnodynastes,* and *Neobatrachus* (Myobatrachidae), the North American hylids *Pternohyla fodiens* and *Smilisca baudinii,* some South American leptodactylid frogs (*Ceratophrys* and *Lepidobatrachus*), and the African frogs *Leptopelis bocagei* (Hyperoliidae) and *Pyxicephalus adspersus* (Ranidae), bury themselves in soil and form a cocoon by shedding layers of the stratum corneum (the outer layer of the skin) (Figure 5–5). The salamander *Siren intermedia* also forms a cocoon, which probably has the same composition as the cocoons of anurans.

The cocoon of *Lepidobatrachus llanensis* accumulates at a rate of nearly one layer per day for the first 40 days after the frog buries, and each layer increases the resistance to water loss by 3 sec·cm^{-1}. Fully formed cocoons reduced the rate of evaporative water loss from 10 mg $H_2O·g^{-1}·h^{-1}$ to 0.9 mg $H_2O·g^{-1}·h^{-1}$. The total resistance of a cocooned *Lepidobatrachus llanensis* after 35 days of cocoon formation was 128 sec·cm^{-1} and resistances of 100 sec·cm^{-1} or higher are probably characteristic of cocooning amphibians in general.

Urine and Feces. Urine is a fluid that contains nitrogenous waste products plus ions (mostly sodium, potassium, chloride, and bicarbonate). It is produced in the kidney in a structure called the glomerulus, where blood pressure forces water and small molecules out through the capillary wall. This ultrafiltrate of the blood flows through the kidney tubules, where it can be modified by reabsorption of water and excretion of ions and small molecules. Ultimately the urine is gathered by the collecting duct, which empties either into the bladder or directly into the cloaca.

in seasonally arid regions and spends the day perched in bushes exposed to the wind. The skin of *P. sauvagei* contains an abundance of small glands that secrete a mixture of wax esters, oleic acid, and long-chain alcohols. When a frog assumes its resting posture, the glands secrete this mixture onto the skin, and the frog uses its forelimbs and hind limbs to spread the fluid over its entire body (Figure 5–4). The dried coating is shiny, and the rate of evaporative water loss from a resting *P. sauvagei* is nearly as low as that from a desert iguana. The waterproof coating is so effective that the frog may overheat on a sunny day. In that situation both *Phyllomedusa* and *Chiromantis* can bypass their waterproofing and use evaporative cooling. Both

(a)   **(b)**

Figure 5–4   **Wiping behavior of the tree frog *Phyllomedusa sauvagei*.** *(Courtesy of R. Ruibal.)*

Nitrogenous wastes are produced when amino groups are removed from proteins in the process of digestion and metabolism. Some of the nitrogen is used to synthesize new protein, and the rest is excreted. Ammonia, urea, and uric acid are the end products of nitrogen metabolism and are found in varying proportions in the urine of amphibians and reptiles.

Ammonia (NH_3) is the chemical produced by the removal of an amino group from a protein, and it is very soluble in water. Those are desirable attributes for a nitrogenous waste product because it doesn't require an energetically expensive biochemical system to synthesize it and a lot of nitrogen can be excreted in a small volume of urine. Ammonia has a serious drawback as a waste product, however: it is extremely toxic, and animals that use ammonia as an end product of nitrogen metabolism must excrete it rapidly. Aquatic species do this; ammonia leaves the body across the skin and gills soon after it is formed. Terrestrial species use a different method; they convert ammonia to nontoxic compounds.

A biochemical pathway called the urea cycle converts two molecules of ammonia to urea (CH_4ON_2), using two molecules of adenosine triphosphate (ATP) in the process. Urea is even more soluble in water than ammonia, and it is not toxic. Thus, animals can accumulate high concentrations of urea without damage.

An unusual species of frog from Southeast Asia illustrates an aspect of the complex life history of amphibians. The crab-eating frog *Rana cancrivora* lives in the mangrove zone at the mouths of rivers, where the daily tidal cycle exposes it to a changing external osmotic environment. The influx of seawater at high tide immerses the frogs in water with a high osmolality, whereas at low tide freshwater flow reduces the osmolality of the frogs' surroundings. The frogs are osmoconformers. That is, they maintain internal osmolalities that match the osmolality of the water around them and change their internal concentrations in parallel with the tides. The high internal osmolalities they require during high tide are achieved primarily by retaining urea in the body fluids. Concentrations of urea in the blood plasma reach 350 mmole·kg H_2O^{-1}. Sodium and chloride ions contribute additional osmotic activity, bringing the frogs into equilibrium with the water. With these adjustments, frogs can survive exposure to 80 percent seawater (approximately 800 mmole·kg H_2O^{-1}), although when given a choice, they prefer 50 percent seawater or less.

Tadpoles of the crab-eating frog live in the mangrove zone with the adults, and are confined to pools of water during low tide. Baking under the tropical sun, the pools lose water by evaporation, and the os-

(a)

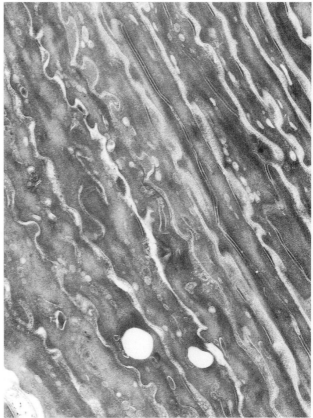

(b)

Figure 5–5 **A frog's cocoon.** (a) The empty cocoon of the South American frog *Lepidobatrachus llanensis*, and (b) a cross-section of the cocoon showing the layered structure. *(Courtesy of Rodolfo Ruibal.)*

molality can rise to about 750 mmole·kg H_2O^{-1}. The tadpoles are unable to synthesize urea because the urea cycle enzymes do not become active in tadpoles until shortly before metamorphosis. Instead, the tadpoles accumulate salt to raise their internal osmolality. Unlike the adult frogs, however, the tadpoles maintain internal osmolalities that are substantially lower than environmental levels. As a result of the differences in osmolality and ion concentration inside and outside the tadpoles, water moves outward and ions move inward across the skin. The tadpoles are believed to compensate for these passive movements of salt and water by drinking seawater to replace the water they lose, and by excreting ions via the gills to get rid of excess salt. The shift to the adult pattern of conforming to the external osmotic activity by adjusting the concentration of urea in the plasma occurs simultaneously with metamorphosis to adult body form.

Reptiles excrete mostly uric acid ($C_5H_4O_3N_4$). Two waterproof frogs, *Phyllomedusa sauvagei* and *Chiromantis xeramplina*, excrete nitrogen in the form of uric acid. Turtles and crocodilians shift from excreting urea when water is plentiful to uric acid when water is scarce.

Uric acid is a purine and is synthesized via several interlocking pathways. Unlike ammonia and urea, uric acid is very insoluble in water and is excreted as a semisolid. Urine in the reptilian kidney tubule is dilute, so the uric acid remains in solution. When the urine enters the bladder or the cloaca, water is reabsorbed, making the solution more concentrated, and some of the uric acid precipitates. This precipitation reduces the concentration of the uric acid, allowing more water to be reabsorbed, and that leads to additional precipitation of more uric acid. The end result of this process of reabsorption of water and precipitation of uric acid is a white or gray semisolid material containing uric acid and enough water to give it a pasty consistency. Thus, the insolubility of uric acid combined with the process of precipitation in the bladder or cloaca allows nitrogen to be excreted with relatively little water.

The semisolid urine produced by reptiles is a complex chemical mixture of uric acid, ions, and other compounds. The potassium salt of uric acid (potassium urate) is an important component of this mixture, and many reptiles excrete a substantial portion of their excess potassium as potassium urate. Indeed, excretion of potassium actually competes with protein synthesis for the nitrogen in a reptile's diet. Desert tortoises (*Gopherus agassizii*) eating food with a high potassium concentration were unable to retain nitrogen, even on high-nitrogen diets (Oftedal et al. 1994). Simply to excrete the potassium it eats, a tortoise needs 1.3 grams of nitrogen for each 1 gram of potassium in its food. If the tortoise is to synthesize protein and grow, the diet must contain additional nitrogen. Relatively few desert plants have nitrogen-potassium ratios higher than 1.3:1, and these desirable species include plants that tortoises have been reported to feed on preferentially in nature.

Urine and feces are excreted together through the cloaca (Latin for sewer; the Cloaca Maxima was the main sewer of ancient Rome). Thus, excretory products of reptiles emerge as a semisolid mixture of white or gray salts of uric acid and dark fecal material. That color combination will be familiar to anyone who has ever parked a car under a tree where birds roost.

Amphibians produce large volumes of dilute urine, and this urine is stored in a bladder that can be astronishingly large, especially among anurans. These animals use their urine as a store of water that can be reabsorbed if it is needed, and the size of the bladder is related to how terrestrial a species is and how arid a habitat it lives in. Aquatic anurans and salamanders such as *Xenopus*, *Triturus*, and *Necturus* have bladders that hold a volume of urine equivalent to 1 to 5 percent of the body mass of the frog (i.e., 1 to 5 grams of urine for a frog that weighs 100 grams), whereas terrestrial anurans have bladders that can hold a volume of urine equivalent to 20 to 60 percent of the frog's mass. Australian aborigines use the water-holding frog, *Cyclorana*, as a water source during the dry season. They dig the frogs from their burrows and squeeze them, emptying their bladders. An adult frog can produce half a glass of clear, tasteless urine. Of course, that's the death knell for the frog, which was counting on the urine in its bladder to survive until the next rainy season.

Salt Glands. Many reptiles have a step in the process of urine formation that appears paradoxical: they reabsorb potassium and sodium ions from urine after it has passed into the bladder or cloaca. Transporting ions across cell membranes against an electrochemical gradient requires ATP. Specific transport sites in the cell membrane use ATP to move ions from the urine into the cell and thence into the blood. What makes that process peculiar is that the ions were in the urine; why is the animal using energy to bring them back into the blood?

The solution of this paradox lies in the presence of an extrarenal (i.e., outside the kidney) route of salt excretion—a salt gland. Many reptiles have a specialized gland that can excrete sodium and potassium at high concentrations. These animals can save water by using the salt gland instead of the kidney to excrete potassium and sodium. The total concentrations of salt gland secretions can be as high as 2,000 mmole·kg H_2O^{-1}, and excreting salt via the salt gland conserves water (Table 5–1). Free-ranging desert iguanas excrete more salt via the salt gland than in the urine: 43 percent of the potassium, 49 percent of the sodium, and 93 percent of the chloride are excreted this way. In fact, without the salt gland a desert iguana could not eliminate the salt it ingests in its diet of plants.

The locations and embryonic origins of reptilian salt glands are diverse despite their similarity in function (Table 5–2). Sea turtles and the diamondback terrapin (*Malaclemys terrapin*) have a salt-secreting gland in the orbit of the eye. (Desert tortoises do not have salt glands, and they depend on the kidney to excrete salts. That is why the nitrogen-potassium ratio of their food plants is so critical.) Saltwater crocodiles (*Crocodylus porosus*), and probably other crocodilians as well, have glands on the surface of the tongue that secrete salt. Many lizards secrete salt from glands that empty into the nasal passages, and sea snakes have a salt-secreting gland on the tongue sheath. The diversity in the salt glands of extant reptiles and the patterns of presence and absence of salt glands among species indicate that salt glands were evolved, lost, and evolved anew several times during the phylogenetic history of the groups.

Reptilian salt glands contain cells specialized for salt transport (Figure 5–6). These secretory cells have large quantities of mitochondria and oxidative enzymes and an enormous surface area formed by slender evaginations, or outpouchings, of the cell surface.

Salt glands actively transport potassium and sodium (both cations) using ATP as the source of energy. Anions, primarily chloride and sometimes bicarbonate, move passively following the electrochemical

Table 5–1   **Secretory characteristics of some reptilian salt glands.**

Species	Concentrations (mmole•kg H_2O^{-1})			Maximum Rate of Secretion (μmole•100 g^{-1}•h^{-1})		
	Na	K	Cl	Na	K	Cl
Lizards						
Dipsosaurus dorsalis Terrestrial herbivore, free-ranging, estimate	180	1,700	1,000	1.9	17.5	10.8
Sauromalus obesus Terrestrial herbivore, captive, fed natural diet	150	1,102	827	0.7	12.5	8.6
Ctenosaura pectinata Terrestrial herbivore,						
captive, spontaneous	76	320	—	—	—	—
captive, NaCl-loaded	90	420	—	—	—	—
captive, KCl-loaded	67	537	—	—	—	—
Amblyrhynchus cristatus Marine herbivore, free-ranging	1,434	235	1256	108	17	100
Uma scoparia Terrestrial insectivore, free-ranging, estimate	639	734	465	4.0	4.7	3.0
Varanus semiremex Estuarine carnivore, captive, NaCl-loaded	686	57	745	33.9	2.7	—
Snakes **Hydrophiidae**						
Laticauda semifasciata Marine, basks on shore, captive, NaCl-loaded	686	57	745	72	3	72
Aipysurus laevis Marine, reef-dweller, captive, NaCl-loaded	798	28	791	165	6	157
Pelamis platurus Marine, pelagic, captive, NaCl-loaded	620	28	635	218	9.2	169
Acrochordidae						
Acrochordus granulatus Estuarine, captive, NaCl-loaded	483	15	492	—	—	—
Homalopsinae						
Cerberus rhynchops Esturarine, captive, NaCl-loaded	414	56	—	15	2.7	—
Turtles						
Caretta caretta Marine, carnivore, captive, NaCl-loaded, secretion stimulated with methacholine	732–878	18–31	810–992	—	—	—
Lepidochelys olivacea Marine, carnivorous, captive, natural	713	28.8	782	—	—	—
Malaclemys terrapin Estuarine, carnivore,						
freshwater, NaCl-loaded	288	29	—	—	—	—
saltwater, NaCl-loaded	682	—	—	—	—	—

(continued)

Table 5–1   **Secretory characteristics of some reptilian salt glands.** *(continued)*

Species	Concentrations (mmole•kg H_2O^{-1})			Maximum Rate of Secretion ($\mu mole•100\ g^{-1}•h^{-1}$)		
	Na	K	Cl	Na	K	Cl
Crocodilians						
Crocodylus porosus Estuarine to marine, freshly captured, secretion stimulated with metacholine	663	21	632	49	1.6	47
Birds						
Ring-billed gull Marine carnivore	808	43	779	1,160	61.9	—
Herring gull Marine carnivore	718	24	720	2,100	72	—

Based on Dunson (1976) and Minnich (1982).

Table 5–2   **Locations of reptilian salt glands.**

Lineages	Salt-Secreting Gland	Homologies
Turtles (cheloniids, dermochelyids, and *Malaclemys terrapin*)	Lacrymal gland	Lacrymal salt gland of birds
Lizards (agamids, iguanids, lacertids, scincids, teiids, varanids, xantusiids)	Nasal gland	None
Snakes (hydrophines and *Acrochordus granulatus*)	Posterior sublingual gland	None
(*Cerberus rhynchops*)	Premaxillary gland	None
Crocodilians (*Crocodylus porosus*)	Lingual glands	None

Source: Dunson (1976) and Minnich (1982).

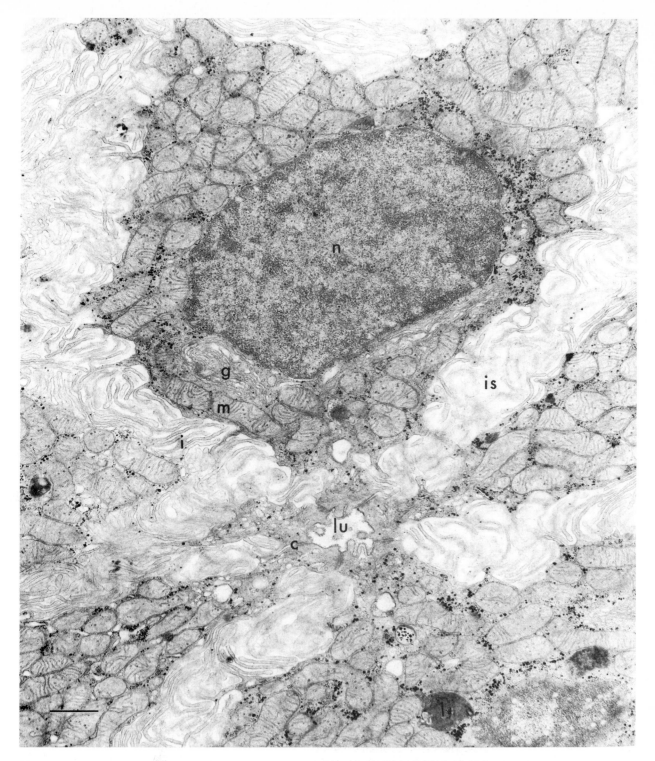

Figure 5–6 **Ultrastructure of a reptilian salt gland.** A cross-section of the sublingual salt gland from a yellow-bellied sea snake (*Pelamis platurus*) showing five cells around the central lumen. Note the high density of mitochondria and the complex interdigitations of cells. Scale = 0.5 μm. Key: c = junctional complex, g = golgi complex, i = interdigitations between cells, is = interstitial space, lu = lumen, m = mitochondria, n = nucleus. *(Photograph by M. K. Dunson, courtesy of William A. Dunson.)*

gradient. The proportions of sodium and potassium in gland secretions probably reflect their relative abundance in the diet of an animal, and the sodium-potassium ratio of the gland secretions of lizards can be altered by changing the ratio of the ions in the diet.

Heat Gain and Loss

The study of energy relations is based largely on steady-state models—that is, situations in which the organism is assumed to be at an equilibrium temperature and is neither warming nor cooling. The energy balance equation summarizes the routes of exchange between an animal and its environment. In its simplest form, the energy balance equation for a steady-state situation is

$$\text{Heat energy gained} = \text{Heat energy lost.} \quad \textbf{(1)}$$

The full energy balance equation includes a term for each of the major routes of heat gain or loss for an organism:

$$\text{Heat energy gained} = Q_{abs} + M \pm R \pm C \pm LE \pm G \quad \textbf{(2)}$$

All these values are rates, expressed in units of watts per square meter ($W \cdot m^{-2}$) of animal body surface. (A watt is 1 joule·sec^{-1}.) Temperatures are expressed as absolute temperature (°K), which is equal to the temperature in degrees Centigrade plus 273.16. The symbols have the following meanings:

Q_{abs} = radiation absorbed by the surface of the animal

M = metabolic heat production

R = infrared radiation received or emitted by the surface of the animal

C = heat gained or lost by convection to the fluid surrounding the animal (the fluid is air for a terrestrial animal and water for an aquatic species)

LE = heat gained by condensation or lost by evaporation

G = heat gained or lost by conduction through direct physical contact of the animal with the substrate it is resting on (soil, rock, tree trunk, etc.)

A detailed treatment of biophysical models is beyond the scope of this textbook; you can find a readable account of the assumptions and approximations that underlie the models in O'Connor and Spotila (1992), and more details in Tracy (1982) and Spotila et al. (1992). Here we will briefly explain each pathway of energy exchange and give examples of how amphibians and reptiles use them during temperature regulation. A summary of the pathways of energy exchange is presented in Figure 5–7.

Absorption of Solar Radiation, Q_{abs}. This term refers to the solar energy absorbed by the animal. The entire spectrum of solar radiation extends from very short wavelengths, such as x-rays, to very long wavelengths, such as radio waves. However, only a portion of the solar spectrum is relevant to the thermoregulatory mechanisms of amphibians and reptiles. The wavelengths we are concerned with extend from approximately 400 to 700 nanometers (visible light) and 700 to 1,500 nanometers (near infrared light). The total amount of light energy reaching Earth's surface is divided about equally between the visible and the infrared, and these are the wavelengths that amphibians and reptiles use for thermoregulation. Ultraviolet light—that is, wavelengths from about 200 to 400 nanometers—does not play a significant role in thermoregulation. Quanta of ultraviolet light have high energy, which is why they can damage living tissue, but the amount of ultraviolet light (i.e., the number of quanta) reaching Earth's surface is too small to change the temperature of an organism significantly.

Q_{abs}, the rate of solar energy absorption by an animal, is given by the equation

$$Q_{abs} = S \cdot A \cdot vf_s \cdot a, \quad \textbf{(3)}$$

where

S = intensity of solar radiation ($W \cdot m^{-2}$)

A = surface area of the animal in m^2

vf_s = view factor (i.e., the proportion of the animal's surface that is receiving solar radiation)

a = absorptivity to solar radiation (i.e., the proportion of solar energy striking the surface that is absorbed rather than being reflected)

These factors offer an animal substantial control over the amount of solar radiation it absorbs. The easiest component of the Q_{abs} term for an amphibian or reptile to control with behavioral adjustments is S,

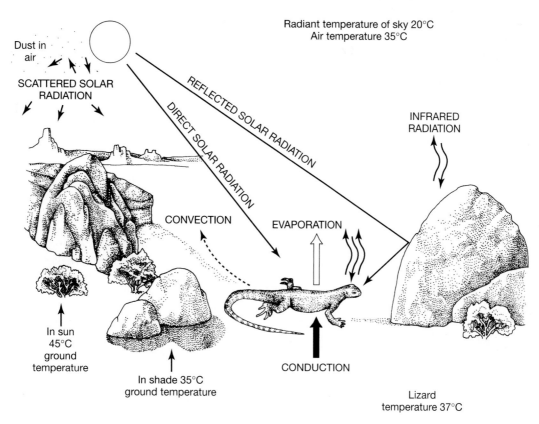

Figure 5–7   **Biophysical pathways of energy gain and loss.**

the intensity of solar radiation to which it is exposed, and the animal does it exactly the way you would—by moving between sun and shade.

Reptiles and amphibians can also change the amount of their body surface that they expose to solar radiation. For example, a lizard can change its surface area by spreading or compressing its rib cage, and it can change its view factor by changing its orientation to the sun. When a lizard is cool and trying to warm up, it adopts a positive orientation to the sun, spreading its ribs widely and orienting itself so the long axis of its body is perpendicular to the rays of the sun (Figure 5–8a). In this situation, the sun strikes the entire dorsal surface of the lizard's body and the lizard casts a large shadow on the ground. A lizard in this posture intercepts the maximum amount of solar radiation possible. At the opposite extreme, when the lizard is hot and trying to minimize the amount of

solar radiation it receives, it adopts a negative orientation, compressing its ribs against its body and facing directly into the sun (Figure 5–8b). In this position the sun strikes perpendicularly only on the lizard's head and shoulders, and the lizard casts a much smaller shadow. The positive and negative orientations can be compared by looking at the size of the lizard's shadow, which is proportional to the amount of sun intercepted by the lizard. If the area of the shadow when the lizard is holding its ribs in their normal relaxed position and the sun is overhead is defined as 1.00 unit (Figure 5–8c), the area of the shadow when the lizard is in positive orientation is 1.62 units, and the area in negative orientation is 0.28 unit. In other words, the lizard has changed the amount of solar radiation it intercepts by a factor of 5.8 (1.62 units/0.28 unit) simply by changing its body shape and orientation.

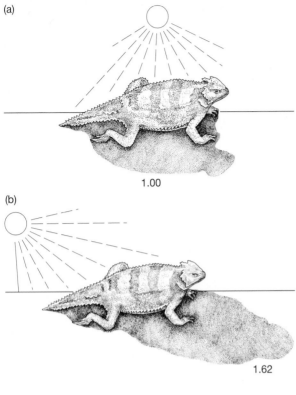

(a)

1.00

(b)

1.62

(c)

0.28

Figure 5–8 **The effect of posture on the amount of solar energy a lizard intercepts.** The diagrams show the area of the shadow cast by a horned lizard (*Phrynosoma cornutum*) in three postures. (a) Positive orientation: the ribs are spread to make the body nearly circular, and the long axis of the body is perpendicular to the sun's rays. (b) Negative orientation: ribs are compressed to make the body narrow, and the long axis of the body is parallel to the sun's rays. (c) Not oriented: the ribs are relaxed, the sun is overhead. (*Source: Modified from Heath 1965.*)

Many amphibians and reptiles can change color, and what you are seeing when you perceive a change in color is a change in absorptivity. A light-colored object appears light because it is reflecting much of the light energy that falls on it, and a dark object looks dark because it is absorbing light instead of reflecting it. Lightening and darkening is accomplished by movements of granules of melanin, a black pigment, in cells called melanophores that are found in the dermal layer of the skin. When the melanin is concentrated in the body of the melanophore, the reflective pigments in other cells are exposed. A lizard in this condition appears light-colored. When the melanin is dispersed toward the surface of the skin, the melanin granules cover the other pigments and absorb some of the light that strikes the lizard, making it appear darker. Light absorbed by the melanin is converted into heat, warming the lizard. Desert iguanas (*Dipsosaurus dorsalis*) can change color substantially (Figure 5–9). In their darkest phase they absorb 73.7 percent of the visible light that strikes them, compared to 57.5 percent when they are in their lightest phase. That difference in absorptivity changes the rate at which a desert iguana heats by $0.42°C \cdot min^{-1}$ (Norris 1967).

Metabolic Heat Production, M.   Metabolic processes are not 100 percent efficient: some chemical energy is lost as heat when ATP is synthesized and again when it is hydrolyzed. A few reptiles use metabolic heat production to accelerate heating and retard cooling, or even to maintain their body temperatures above ambient levels. Pure ectothermy and pure endothermy are the ends of a continuum of thermoregulatory patterns, and some species occupy intermediate positions. All of the reptiles that have a substantial endothermal component to their thermoregulation are large species, which means that their surface-to-mass ratios are relatively low. A small surface area for heat loss in relation to the mass of tissue that is producing heat is a first step toward retaining enough metabolic heat to raise the body temperature.

The largest of the species that have been shown to retain metabolic heat, and perhaps the largest living reptile, is the leatherback turtle (*Dermochelys coriacea*). Adults of this pelagic sea turtle can weigh more than 850 kilograms, and they range north and south into water as cold as 8 to 15°C. In these cold seas the body

Figure 5–9 **Color change by the desert iguana, *Dipsosaurus dorsalis*.** (a) Dark color adopted during heating. (b) Light color characterstic of a lizard within its activity temperature range. *(Photographs by (a) E. R. Degginger/Photo Researchers, Inc., (b) Tom Brakefield/Bruce Coleman.)*

(a)

(b)

temperatures of the turtles are 18°C or more above water temperature. That temperature difference results from metabolic heat produced by muscular activity as the turtles swim.

Several species of pythons use metabolic heat production to maintain body temperatures around 32°C while they are incubating eggs. This use of en-

dothermy is seen only in female pythons when they are preparing to lay eggs and brooding their eggs. The female python gathers her eggs into a heap and curls around them, enclosing them with her body (Figure 5–10). Heat is produced by spasmodic muscular contractions; the snake twitches visibly with each contraction. These twitches are produced by coordinated contraction of muscle fibers, and they are different from the uncoordinated muscle contractions that are responsible for mammalian shivering. The frequency of contractions and the rate of oxygen consumption by the snake are at a minimum around 31°C and increase as temperature falls (Figure 5–11). While she is brooding eggs, the snake's metabolic rate at an air temperature of 23°C is about 20 times the rate of a nonbrooding snake at the same temperature. The heat produced by the snake keeps the temperature of her eggs close to 30°C at air temperatures from 23 to 30°C.

Infrared (Thermal) Radiative Exchange, R. Heat exchange takes place continuously between an animal and its environment in the near infrared part of the electromagnetic spectrum (700 to 1,500 nanometers). Any object with a surface temperature above absolute zero (i.e., 0°K, or −273.16°C) radiates energy in the infrared. Thus, every moment of its life an organism radiates infrared energy from its surface to its surroundings and receives infrared (thermal) radiation from its surroundings. The net heat transfer is from the hotter surface to the colder surface. You can easily perceive this phenomenon when you stand beside a

(a)

(b)

Figure 5–10   **A ball python (*Python regius*) brooding its eggs.** Females of this small species of python coil around the eggs and display the same brooding behavior as do large species, but do not exhibit muscular contractions or increased metabolic rates. (a) A female incubating a clutch of eggs. The eggs are enclosed within the coils of the snake's body. (b) Hatchling pythons emerging from the eggs, still enclosed by the female's coils. (*Photographs by Mark A. Chappell.*)

wall that has been heated by the sun. The warmth you feel is infrared radiation from the wall.

The magnitude of heat flow depends on three variables:

1. The difference between the fourth powers of the absolute temperatures of the two surfaces—i.e., $T_s^4 - T_e^4$, where T_s is the surface temperature of the animal and T_e is the surface temperature of the object in the environment with which it is exchanging radiation, both measured in °K.

2. The area of the animal exposed to thermal radiation, i.e., $A \cdot vf_t$, where A = surface area and vf_t = the view factor for thermal radiation.

3. The infrared emissivity of the skin (represented by ϵ, the Greek letter epsilon) provides information about how readily a surface radiates and absorbs infrared radiation.

It is difficult to have an intuitive understanding of the thermal emissivity of the skin because few of the structural characteristics that determine emissivity are visible. Color does not provide a clue to how a surface behaves in the infrared, because color is the result of a surface's response to visible wavelengths. Thus, the statement that a black surface absorbs infrared radiation and a white surface reflects it is simply wrong,

despite its appearance in some biology textbooks. The best visible cue to the infrared absorptivity and emissivity of a surface is its texture. Matte surfaces generally absorb and emit infrared radiation well, whereas smooth, shiny surfaces are poor absorbers and emitters of infrared.

Many lizards, especially desert species like the fringe-footed sand lizards (*Uma*) and the closely related gridiron-tailed lizard (*Callisaurus*), which live in open desert habitats in North America, have scales with different infrared absorptivities on their dorsal and ventral surfaces. For several hours during the day, the surface temperature of the sand is 60°C or even higher. That's lethally hot for these lizards, which maintain body temperatures near 38°C during activity. The scales on the ventral surface of the lizards are smooth and shiny, and they reflect the infrared radiation that streams up from the sand, reducing the heat absorbed by the lizard. The scales on the dorsal surface of the lizard's body, in contrast to the ventral scales, have a matte surface that is a good absorber and emitter of infrared radiation. When the lizards are basking in the sun in the morning, the high infrared absorptivity of these scales increases the rate of heating. Later in the day, when avoiding overheating is the problem, a lizard perched on the shady side of a rock can use the high emissivity of its dorsal scales to exchange thermal radiation with the sky. A clear sky

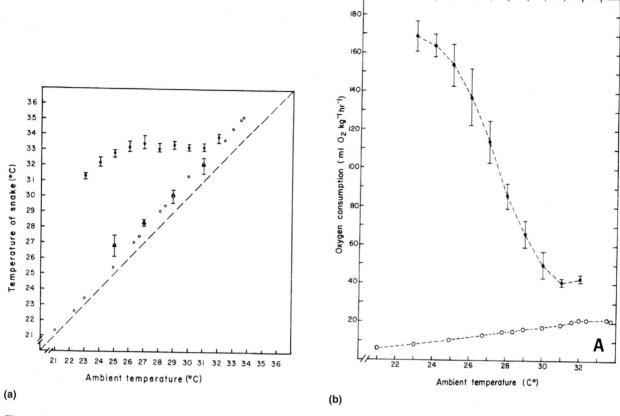

(a)

(b)

Figure 5–11 **Body temperature and oxygen consumption of a female Indian python,** *Python molurus,* **brooding eggs.** (a) Body temperature compared to air temperature. The diagonal line shows the line of equality of body temperature and air temperature. Values for a nonbrooding female (open circles) fall close to that line, whereas a brooding female (solid circles) maintains a body temperature substantially higher than air temperature. The female did not maintain an elevated body temperature before she laid her eggs (triangles). (b) Oxygen consumption of a nonbrooding female (open circles) and a brooding female (solid circles). The brooding female increases her rate of oxygen consumption as ambient temperature falls below 30°C. (*Source: Van Mierop and Barnard 1978.*)

behaves like an object with a surface temperature about 23°C, so the net movement of heat is from the lizard, which has a surface temperature about 38°C, to the sky.

Convective Exchange of Heat, C. Convective exchange occurs between an animal and the fluid medium surrounding it. For terrestrial animals the fluid is air, whereas for aquatic animals it is water. The principles of convective exchange are the same for either fluid, and for the sake of simplicity our discussion will consider a terrestrial animal in air.

Three variables are important in determining the magnitude of convective heat exchange between an animal and its environment:

1. The temperature difference between the animal and the air, i.e., $T_s - T_a$, where T_s = animal surface temperature and T_a = air temperature.

2. The surface area of the animal exposed to convection, which can be changed by altering posture or body shape.

3. The convective coefficient of the animal, which depends on (a) the velocity of air movement and

(b) the size of the animal, expressed as a measurement known as the characteristic dimension, which is roughly equivalent to the diameter of the animal in the direction parallel to air flow.

Animals can adjust convective heat exchange by shuttling back and forth between still and moving air. One way to do this is to climb a bush. Wind speed increases and air temperature decreases as you move away from the ground surface, and lizards can take advantage of that thermal profile by climbing into bushes during the hot part of the day. Air temperatures and wind speed change rapidly in the first half meter above the ground, and a lizard does not have to climb very far to be in cooler air and a stronger breeze.

The effect of body size on convective exchange results from the formation of a boundary layer of relatively still air close to the surface of an animal. Movement of heat in the boundary layer is slow, so a thick boundary layer means low convective heat exchange and a thin boundary layer means high exchange. The body size of the animal is important in this relationship because small animals have thin boundary layers and large animals have thick boundary layers.

The relation among body size, boundary layer thickness, and the magnitude of convective heat exchange has important implications for the ways animals behave and the microenvironments they can use. In general, convective heat exchange dominates the energy balance equation for very small animals and radiative exchange dominates for large animals. In other words, the body temperatures of small animals will depend mostly on air temperature, whereas body temperatures of large animals will depend on solar and thermal radiation. That difference is manifested in the things large and small animals can do.

Consider a small animal such as the side-blotched lizard (*Uta stansburiana*), which weighs about 2 grams, and a larger species such as the chuckwalla (*Sauromalus obesus*), which weighs 200 to 500 grams (Figure 5–12). Both species occur in rocky desert foothills of western North America and northern Mexico, often living on the same rocks. The behaviors and activities of the two lizards differ in many ways, and some of these differences are the result of differences in the ways large and small animals interact with their physical environments.

Convection is a major factor determining the body temperature of the side-blotched lizard. In the morning, when the sun is shining but the air is still cold, a side-blotched lizard must find a spot that is sheltered from the wind. It can do that by taking advantage of the thick boundary layer of a large object such as a boulder, because the lizard is small enough to submerge itself in the boundary layer of air around the rock. If the lizard moves away from the rock, into moving air, it will be surrounded only by its own thin boundary layer, and it will quickly cool to air temperature. In contrast, the body temperature of a chuckwalla is less sensitive to convection because its boundary layer is thicker, and a chuckwalla can move about more freely.

Later in the day, the close coupling of the side-blotched lizard's body temperature to air temperature can work to its advantage. When the ground and rocks are too hot to touch, a small lizard can climb into a bush and convection will cool it to air temperature. Even in full sun, the lizard loses heat by convection as rapidly as it gains heat from radiation. As a result, a side-blotched lizard can remain in the sun all day. A chuckwalla's temperature is more strongly influenced by solar radiation, and a chuckwalla might have to seek shade to avoid overheating in midday.

Evaporative Cooling, LE.   In addition to being an important part of water balance, evaporative cooling plays a role in the energy balance of an organism because a substantial quantity of heat is needed to change water from a liquid to a vapor. The amount of energy required varies with temperature, but $128 \, \text{J} \cdot \text{g}^{-1}$ of water is a reasonable average for biologically relevant temperatures.

The evaporation of water occurs in two stages: initially water or water vapor must cross the barrier imposed by the skin, and then water vapor must be moved away from the outer surface of the skin by convection. Either process can be the one that limits the rate of evaporation, depending on the permeability of the skin, the humidity of the air, and the rate of air movement.

Movement of water through the skin is the step that limits the rate of evaporation for reptiles. As we've noted, these animals have lipids in their skins that reduce permeability to water. As a result, evaporative water loss normally plays a small role in the energy balance of most reptiles. In many cases the heat lost by evaporation is approximately equal to the heat produced by metabolism.

Figure 5–12 **Body size and temper-ature regulation.** The small side-blotched lizard, (a) *Uta stansburiana*, lives on the same the rock outcrops as the chuckwalla, (b) *Sauromalus obesus*, but the difference in body size of the two lizards result in differences in their thermoregu-latory behavior. *(Photographs by (a) Lawrence E. Naylor/Photo Researchers, Inc., (b) Tom McHugh/Photo Researchers, Inc.)*

(a)

(b)

Some reptiles increase evaporative water loss as a thermoregulatory device. Many lizards, especially large species such as the New World iguanas (*Ctenosaura, Iguana*) and chuckwallas (*Sauromalus*) and the Old World mastigures (*Uromastyx*), pant when they are overheated. Monitor lizards (*Varanus*) have an analogous mechanism called gular fluttering in which the animal holds its mouth open and rapidly vi-brates the tissues of the throat, evaporating water and cooling the blood passing through those tissues.

Evaporative cooling is a major element of the en-ergy balance of most amphibians because their rates of cutaneous evaporation are high. The body temper-atures of amphibians are usually slightly lower than air temperature at night and during the day when they are in shade. Even when amphibians are in direct sun-light, evaporative cooling prevents most species from

reaching the high body temperatures that are charac-teristic of lizards. Bullfrogs, for example, may spend the entire day in the sun at the edge of a pond, evapo-rating water through the skin to prevent overheating. As long as a frog's ventral surface is in the pond, it can absorb water as fast as it evaporates.

Conduction, G. Conduction refers to the transfer of heat to or from the surface on which an animal is resting or the fluid medium it is in. For some reptiles and amphibians, especially nocturnal species, warm surfaces can be an important source of heat. An im-portant factor that determines the magnitude of con-ductive heat exchange is the surface area of the animal that is in contact with the substrate. In forest habitats you can watch foraging lizards move into a patch of sunlight and flop down on their bellies, maximizing conduction of heat from the warm ground by increasing the area of contact. In hot deserts you can often see lizards standing with their legs extended and their toes in the air. When a lizard stands this way, only a small part of its foot is actually in contact with the hot soil and conduction of heat is minimized.

Animals exchange heat with the fluid media that surround them as well as with the substrate. For animals that live on land, this route of exchange is relatively unimportant, because air has a low heat capacity and low heat conductivity. For aquatic animals, however, conduction is a major source of heat exchange, and it is difficult for most small aquatic animals to sustain a gradient between their internal temperature and water temperature. The body temperatures of most species of aquatic amphibians and reptiles are close to the temperature of the water surrounding them. When there are temperature gradients in the water, an aquatic species can control its body temperature by selecting a suitable position in the gradient, but small aquatic species that live in habitats without temperature gradients have little or no capacity to control their body temperatures.

Behavioral Aspects of Thermoregulation

Ectotherms regulate their body temperatures either at a specific set-point temperature or within the range between an upper and a lower set-point. The preoptic nucleus of the hypothalamus is the temperature-sensitive region of the brain, and experiments have shown that the temperature of this region controls thermoregulatory behavior. The hypothalamus receives blood from the internal carotid arteries, which carry blood that has just passed through the heart. Thus, a sensor in the hypothalamus can integrate information about the temperature of the body (via blood flow) and of the brain. When that temperature is below the low-temperature set-point, cold-sensitive neurons fire and stimulate behavioral and physiological adjustments that result in raising the body temperature—for example, increasing exposure to sunlight, darkening the skin, and adopting a positive orientation. When the hypothalamic temperature exceeds the upper set-point, heat-sensitive neurons fire and stimulate a set of behaviors that will reduce body temperature. In general these responses are the opposite of the previous ones—movement into shade, lightening the skin, adopting a negative orientation, and so on.

As a result of these responses to stimulation of the hypothalamic temperature sensor, the body temperature of the lizard varies up and down within a range that may be as narrow as 2 or 3°C during the portion of the day when the animal is thermoregulating. This is the set-point temperature range, and the most informative way to describe the thermoregulatory characteristics of a species is to say that it has a set-point temperature range of, for example, 35 to 38°C (Hertz et al. 1993) (Figure 5–13).

Set-point temperatures are not absolutely fixed. They change seasonally and differ between the sexes. Body temperatures are higher when animals are digesting food than when they are fasting, pregnant female reptiles often maintain different body temperatures than nonpregnant ones, and reptiles and amphibians with bacterial infections maintain higher than normal body temperatures—a behavioral fever that increases survival rate (Regal 1966, Kluger 1979, Firth and Turner 1982, Charland 1993, Andrews et al. 1997).

Cardiovascular Control of Heating and Cooling

The thermoregulatory processes we have discussed so far determine how much radiant energy reaches the body surface of an animal (movement between shade and sun, orientation with respect to sources of radiant energy, changing body contour), how much energy is absorbed and reflected (color change), and how much energy is gained or lost from the skin surface by convection, conduction, and evaporation (postural changes and shuttling between moving and still air). A second group of thermoregulatory mechanisms—changes in heart rate, in blood flow through the heart, and in the distribution of peripheral circulation—allow reptiles to accelerate heating and retard cooling (Stevenson 1985b).

Intracardiac Shunts and Peripheral Vasodilation

When a cold reptile (body temperature = 20°C) is placed in a warm environmental chamber (air temperature = 40°C) it warms up, and when it is moved back to a cold chamber (20°C) it cools off. That's hardly surprising, but it is surprising that the rate of heating

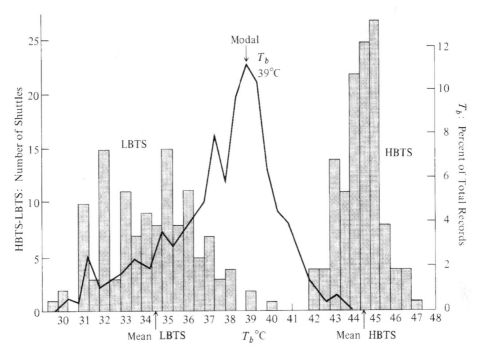

Figure 5–13 **Set-point temperatures of the desert iguana, *Dipsosaurus dorsalis*.** The histograms show the temperatures at which lizards in a laboratory temperature gradient changed their behavior to warm up (at the low body temperature set-points [LBTS], mean value 34.6°C) or cool off (at the high body temperature set-points [HBTS], mean value 44.6°C). The solid line shows the distribution of body temperatures of lizards in the gradient. Most measurements of body temperature fall between 34 and 43°C, with a modal value of 39°C. *(Source: Withers 1992.)*

can be twice as fast as the rate of cooling. For example, when an Australian bearded dragon (*Pogona barbata*) was moved from a chamber at 20°C to one at 40°C, it warmed to 40°C in approximately 38 minutes, but 50 minutes after it was moved back to the 20°C chamber it still had not cooled to 20°C (Figure 5–14). By accelerating heating and retarding cooling, a reptile is able to increase the time it spends within its set-point temperature range.

All reptiles can probably control their rates of heating and cooling to some extent. Large animals have more capacity for these adjustments than do small animals, but even some very small lizards such as the Carolina anole (*Anolis carolinensis*, 2–7 grams) and the six-lined racerunner (*Cnemidophorus sexlineatus*, 4–7 grams) can exert some control over their rates of heating and cooling. In contrast to reptiles, amphibians probably have little opportunity for this kind of control of heating and cooling rates because evaporative water loss dominates their energy exchange.

The difference in the rates of heating and cooling by reptiles is produced by three kinds of circulatory adjustments (Bartholomew 1982, Tracy 1982). First, most reptiles have a higher heart rate during heating than during cooling (Figure 5–15). The increased blood flow produced by the high heart rate carries heat from the warm body surface to the cooler body core. Second, a transfer of blood within the heart increases the proportion of blood going to the skin and body and decreases the proportion going to the lungs (a right-to-left intracardiac shunt), further increasing the amount of heat transported by the blood. Finally, blood vessels dilate in areas of the skin that are warm, and as a result the amount of blood flowing through the warm skin increases. This blood is warmed, and carries the heat to the core of the body. Parts of the body with high surface-to-mass ratios, especially the limbs, are crucial to accelerated heating. Both experiments and computer simulations have shown that the difference between heating and cooling rates is abol-

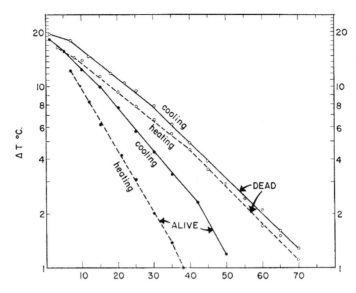

Figure 5–14 **Heating and cooling rates of the Australian bearded dragon,** ***Pogona barbata.*** The vertical axis (ΔT in °C) is a log scale showing the absolute value of the difference between air temperature (T_a) and the lizard's body temperature (T_b). At the start of a heating period, T_a in the chamber is 40°C and T_b is 20°C, so $\Delta T = |40°C - 20°| = 20°C$. ΔT decreases as T_a increases. When a warm lizard ($T_b = 40°C$) is put in a cold chamber, ΔT is $|20°C - 40°C| = 20°C$, and decreases as the lizards cools. A live lizard heats from 20 to 39°C in about 38 minutes, but requires more than 50 minutes to cool from 40°C to 21°C. Repeating the experiment with a dead lizard shows the role that circulation of blood plays in heat transport. The dead lizard heats and cools more slowly that the live lizard, requiring about 70 minutes to reduce ΔT to 1°C, and there is little difference in the rates of heating and cooling. *(Source: Bartholomew and Tucker 1963.)*

ished by shutting off blood flow to the limbs. Metabolic heat production may also help to speed heating and retard cooling for some large lizards such as the Australian blue-tongue skink, *Tiliqua scincoides*, and Gould's monitor lizard, *Varanus gouldi*, in laboratory experiments, but its importance in a natural context has not been established.

An elegant illustration of the flexibility of behavioral and cardiovascular thermoregulatory mechanisms is provided by the Galápagos marine iguana, *Amblyrhynchus cristatus*. Marine iguanas live at the edge of the sea on barren lava flows where there is no shade at midday. Male marine iguanas are territorial, and a successful male must remain in his territory to deter intruders.

Marine iguanas spend the night in sheltered locations, sometimes forming large piles as the animals climb on top of each other. In the morning, they are cold and use positive orientation to maximize the area of their body exposed to solar radiation. As the sun warms an iguana's back and limbs, increased blood flow to these regions carries the warmth to the core of the iguana's body. When an iguana reaches its setpoint temperature range, it adopts a negative orientation to minimize its heat load.

The Galápagos Islands lie on the equator, and by midday the black lava rock has become very hot. Female and juvenile iguanas move into shaded cracks and crevices, but moving to shade would take a male iguana off its territory and risk allowing an intruder to establish itself. Instead of seeking shade, a male iguana remains on its territory and uses circulatory adjustments and the cool trade winds to form a heat shunt that absorbs solar radiation on the dorsal surface of the body and dumps it through the ventral surface.

The iguana faces into the sun, with the forepart of the body held off the ground (Figure 5–16). The iguana's ventral surface is exposed to the cool wind blowing off the ocean, and a patch of lava under its

Figure 5–15 **Heart rate of the Australian bearded dragon, *Pogona barbata*, during heating and cooling.** The lizard has a higher heart rate during heating than during cooling at all body temperatures. *(Source: Bartholomew and Tucker 1963.)*

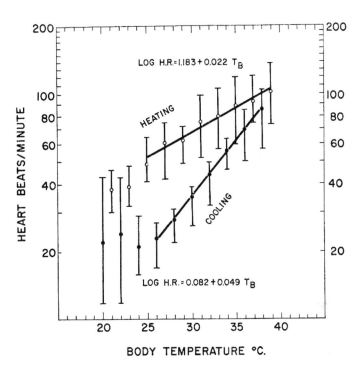

body is shaded. The warmth of the sun shining on the dorsal skin causes blood vessels in the skin to dilate, increasing blood flow to the dorsal surface and carrying heat away. As that warm blood flows through vessels on the ventral surface, it causes them to dilate. Warm ventral skin loses heat by convection to the cool breeze and by infrared radiation to the cool lava. By combining these physiological adjustments with appropriate behaviors, such as selecting a site where the breeze is strong, a male iguana can remain in its territory all day.

The Thermal Ecology of Amphibians and Reptiles

The multiple routes of energy exchange between an organism and its environment give amphibians and reptiles the potential to control their body temperatures with considerable precision, but they do not always do so. A combination of phylogenetic history, body size, and the ecological and behavioral costs and benefits of thermoregulation determines when, how, and with what precision an ectotherm controls its body temperature (Huey 1982, 1987, Stevenson 1985a,b; Dial and Grismer 1992).

You will have noticed that most of the examples of thermoregulation in the previous section were drawn from lizards, and especially from species of lizards that live in open, sunny habitats. That bias is real: lizards display more diversity of thermoregulatory behavior than do other reptiles or any amphibians, and sunny habitats offer excellent opportunities for ectotherms to manipulate energy exchange.

The costs of thermoregulation in open habitats are usually low because the environment provides an array of thermal conditions for an animal to exploit. Deserts, for example, receive intense solar radiation nearly every day, and lizards can bask. The sunny sides of rocks are hot, and they provide sources of infrared heat gain for a lizard, whereas their shady sides are cool and a lizard can lose heat via thermal radiation. The clear desert sky has a radiative temperature lower than the set-point temperature range of most diurnal lizards and is another heat sink for thermal radiation from a lizard. The widely spaced clumps of vegetation in deserts provide shade that allows a lizard to escape direct solar radiation, and climbing the plants gives a lizard access to cooler air and greater convection than it can find at ground level. Lizards that take advantage of these conditions to maintain high, stable body temperatures during activity are re-

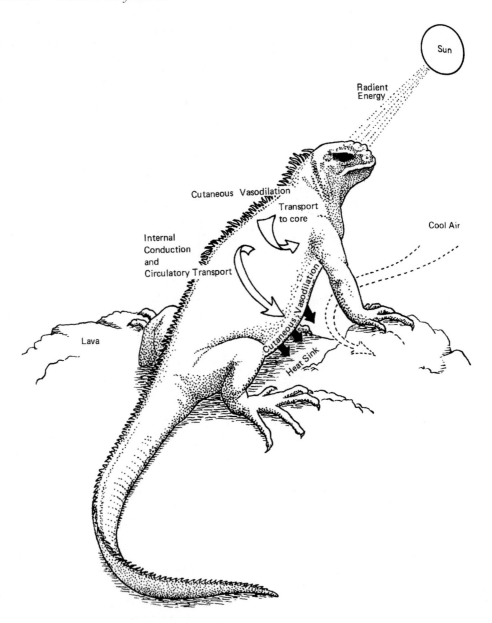

Figure 5–16   **A marine iguana, *Amblyrhynchus cristatus*, acting as regulated heat shunt.**
Solar energy absorbed at surfaces illuminated by the sun is transported through the body by
blood flow and lost from the ventral surface by convection and infrared radiation. *(Source: Modified from White 1973.)*

ferred to as heliothermic (*helio* = Gr. sun, *therm* = Gr.
heat).

Some habitats do not provide the heat sources and
sinks that deserts do, and the costs of thermoregulation in these habitats are high. In forests, for example,
sunlight penetrates the tree canopy only in patches.

Because the intensity of solar radiation beneath a forest canopy is low and relatively uniform, there is little
variation in temperature from place to place, and
lizards do not have the sources and sinks for thermal
radiation that occur in open habitats. A lizard may
have to leave its home range to find a patch of sun,

risking exposure to predators and aggression from owners of intervening territories in the process. Reptiles and amphibians in these situations generally have body temperatures close to air temperature and are called thermoconformers (Shine and Madsen 1996).

Nocturnal amphibians and reptiles cannot use solar radiation to control their body temperature when they are active, but they may thermoregulate during the day when they are in their retreat sites. The granite night lizard, *Xantusia henshawi*, lives in crevices in rocks that provide natural temperature gradients. During the day, the lizards maintain body temperatures averaging 27.2°C by moving to warm areas within the crevices.

Thermoconformity and heliothermy are the ends of a spectrum of thermoregulatory behaviors, and the position of a particular species on that continuum appears to be set partly by phylogeny and partly by the habitat it occupies (van Berkum et al. 1986). Some lineages of reptiles contain numerous species that are heliothermic, whereas others are composed of species that lie predominantly toward the thermoconformity end of the spectrum (see Avery 1982, for details). Thermoregulation is influenced by the presence of competitors and predators as well as by the availability and location of food and retreat sites, and variation in thermoregulatory characteristics can be found among the species within any family. In some cases a single species exhibits different thermoregulatory behaviors in different habitats, being heliothermic in sunny, open habitats and thermoconforming in shaded forest settings (Huey 1982). Amphibians face the additional complication of a powerful interaction between temperature regulation and water balance: the exposed sites and high rates of air movement that offer the best opportunities for radiative and convective heat exchange also lead to high rates of evaporative water loss. The sections that follow illustrate the effects of some of these variables, using particular species as examples.

Thermoregulation in Low-Cost Environments

The complex interactions of heliothermic thermoregulation are illustrated for a horned lizard, *Phrynosoma cornutum*, in Figure 5–17. Like most lizards, the horned lizard begins activity after sunrise, when it emerges from its nighttime shelter. In this case the lizard has spent the night buried in loose soil, and at first it pokes only its head out of the soil. Blood is pooled in the head by constricting the internal jugular veins, and this mechanism accelerates warming of the brain.

After an initial period of basking with only its head exposed, the lizard emerges fully from the soil. At this stage the lizard's body temperature is still low, and it maximizes its rate of heating by darkening, spreading its ribs to increase its surface area, and adopting a positive orientation to intercept as much solar radiation as possible.

The activities of a horned lizard diagrammed in Figure 5–17 are limited in space and time by its thermoregulatory requirements. For example, during the early morning and late afternoon the lizard must be in the sun, whereas during the middle of the day the environment is so hot that it must remain in the shade. On particularly hot days the lizard might have to climb into a bush to take advantage of lower air temperature and greater convection at that height. Finally, on extremely hot days the lizard may cease activity entirely and burrow into cool soil. This behavior protects it from high temperature but precludes feeding and the other activities the lizard would normally engage in.

Thermoregulation by terrestrial amphibians is often dominated by the cooling effect of evaporation, and they can use that fact to their advantage. For example, the canyon treefrog of Arizona, *Hyla arenicolor*, lives near streams in small canyons in desert foothills. The frogs spend the day resting on boulders several meters away from the stream, and a frog may be in full sun for several hours during the day (Figure 5–18). The rocks themselves become nearly too hot to touch, but the frogs remain cool because they are losing water by evaporation. In the course of a day a frog may evaporate an amount of water equal to its entire bladder reserve. A plausible explanation of this seemingly unfroglike behavior is based on the risk of predation during the day. Black-headed garter snakes (*Thamnophis cyrtopsis*) are predators of the frogs, and the snakes forage in the water and along the banks of the stream. The rocks beside the stream are too hot for the snakes to cross during the heat of the day, and a frog on a boulder is safe. Because of the cooling effect of evaporative water loss through its permeable skin, the frog can remain on the boulder even when the sun is shining directly on it.

Figure 5–17 **The interrelationships of thermoregulatory mechanisms of a horned lizard (*Phrynosoma cornutum*).** The diagram shows the range of body temperatures at which each response occurs. *(Source: Heath 1965.)*

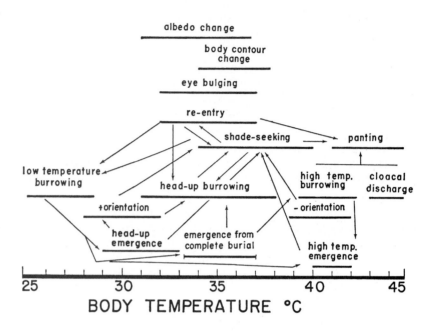

BODY TEMPERATURE °C

Thermoregulation in High-Cost Environments

Tropical lizards that live beneath the forest canopy where the costs of thermoregulation are high usually have body temperatures that are nearly the same as air temperature, whereas related species at the edge of the forest have lower costs of thermoregulation and control their body temperatures more precisely. The role of environmental conditions in shaping the thermoregulatory characteristics of a lizard is clear in a comparison of temperature regulation by the small polychrotid lizard *Anolis sagrei* on the islands of Jamaica and Abaco. Three species of anoles are common on Jamaica, and they occupy different habitats: *A. sagrei* occurs on tree trunks, fence posts, and buildings in open habitats such as pastures and around houses, *A. grahami* lives at the edge of forests, and *A. lineatopus* lives beneath the forest canopy. Thermoregulation is easiest for *A. sagrei*, which can move from sun to shade simply by moving around the circumference of a tree trunk, and most difficult for *A. lineatopus*, which may have to move several meters to find a patch of sunlight. The thermoregulatory patterns of the three species of lizards parallel these costs of thermoregulation (Figure 5–19). *Anolis sagrei* has body temperatures between 32 and 34°C, whereas *A. lineatopis* is substantially cooler (26 to 29°C), and not significantly different from air temperature. The costs of thermoregulation for *A. grahami* are intermediate between those of the other two species, and its activity temperature range is also intermediate (30 to 33°C).

Effectiveness of Thermoregulation

The thermoregulatory activities of amphibians and reptiles are often conspicuous, and it is easy to carry out experiments that show that a particular species runs fastest or catches prey most effectively at a particular temperature, but defining the effectiveness of thermoregulation has been difficult because several related but distinct questions are involved (Hertz et al. 1993).

Three elements are necessary to evaluate the thermoregulation of a particular species in a specific habitat. First, we must know the body temperatures (T_b) actually experienced by animals in the population we are studying during the part of the day when the animals are actively thermoregulating.

Figure 5–18 **Two canyon treefrogs (*Hyla arenicolor*) on a sunny rock in midday.** The frogs have adopted postures that minimizes the exposed surface area and have turned a chalky white color that reflects sunlight and closely matches the appearance of the rock. *(Photograph by R. Wayne Van Devender.)*

The second element is information about the body temperatures that nonregulating animals would have. An environment is a thermal mosaic, and animals placed randomly in different parts of the environment would have different temperatures even if they were not thermoregulating. The operative temperature (T_e) is defined as the temperature of an inanimate object of the same size, shape, and reflectivity as the animal being studied. T_e can be calculated using complex biophysical equations, or it can be measured using actual physical models (manikins) that duplicate the biophysical characteristics of the species being studied (Bakken 1992).

The third element needed to evaluate the effectiveness of thermoregulation is information about the set-point temperature range (T_{set}) of the species being studied. The set-point range represents the target body temperatures that an animal is trying to achieve, and can be measured in the laboratory (see Figure 5–13). The effectiveness of thermoregulation can be determined by comparing how well T_b matches T_{set}.

The data needed to calculate the effectiveness of thermoregulation for three populations of *Anolis* from Puerto Rico are shown in Figure 5–20. The upper graph of each pair shows the distribution of body temperatures (T_b) for active lizards. The lower graph shows the distribution of operative temperatures (T_e) in the same environments as the lizards. The vertical bar shows the set-point temperature range for lizards from that population; in each case the lower set-point temperature is about 2°C below the upper set-point.

Clearly the ranges of T_b and T_e in this example are broader than T_{set}, even for *A. gundlachi*, which has the narrowest range of T_b (Figure 5–20c). The deviations of T_b from T_{set} (called d_b) and T_e from T_{set} (called d_e) are calculated as the absolute value of each observation (T_b or T_e) minus the nearest limit of T_{set}. For example, in Figure 5–20a the lower limit of T_{set} is 28.1°C and the upper limit is 29.7°C. A T_b of 26.5°C would have a d_b value of 1.6°C (i.e., $d_b = |26.5 - 28.1| = 1.6$) and a Tb of 33.0°C would correspond to $d_b = 3.3°C$. When a T_b or T_e value falls within the set-point temperature range, the value of d_b or d_e is defined as zero.

The mean value of d_b ($\bar{d}_b$) is an expression of how closely the animals approach T_{set}. A low $\bar{d}_b$ means that animals in the population studied are often close to T_{set}, whereas a high value means that they often experience temperatures outside the set-point range. Similarly, the mean value of d_e ($\bar{d}_e$) shows how closely the operative temperatures in the environment approach the set-point temperature range. The effectiveness of thermoregulation, E, is defined as $E = 1 - (\bar{d}_b/\bar{d}_e)$ and usually has a value between 0 and 1. When animals are acting as thermoconformers (i.e., when they select microhabitats without regard for temperature), $\bar{d}_b$ will be similar to $\bar{d}_e$, and E will be close to 0. This situation is shown for *A. gundlachi* in Figure 5–20c. *Anolis gundlachi* lives beneath the forest canopy, where sunlight penetrates only in patches. It is a typical thermoconformer and rarely basks. In this example d_b was $0.6 \pm 0.1°C$ and d_e was $0.7 \pm 0.1°C$. Note that although *A. gundlachi* is a thermoconformer and has an E value of only 0.14, its mean body temperature (25.7°C) falls within its set-point temperature range (24.3 to 26.1°C).

Anolis cristatellus is a wide-ranging species that lives in open habitats on Puerto Rico from sea level to over 1,000 meters. It was studied at a high-altitude site (1,150 meters above sea level) in August and January. A comparison of Figures 5–20a and b shows that both T_b and T_e were higher in August than in January (Table 5–3). The average body temperature was about 1°C below the set-point range in August and nearly 5°C below it in January. A comparison of $\bar{d}_b$ to $\bar{d}_e$ shows that the lizards were thermoregulating in both months, and the effectiveness of thermoregulation was approximately the same in August and January (0.50 versus 0.46).

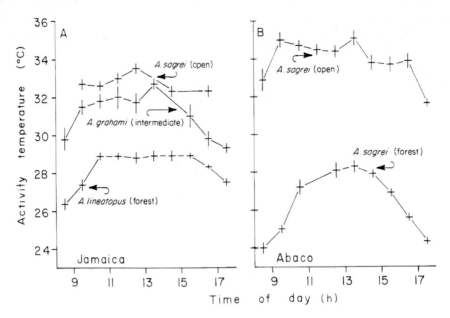

Figure 5–19   **Thermoregulatory strategies of *Anolis*.** (a) On the island of Jamaica *A. sagrei* is found in open habitats, *A. grahami* at the edge of forests, and *A. lineatopus* beneath the forest canopy. *Anolis sagrei* and *Anolis grahami* are heliothermic, whereas *A. lineatopus* is a thermoconformer. (b) On the island of Abaco *A. sagrei* occupies both open habitats and forests. It behaves as a helioterm in the open habitats and as a thermoconformer in forests. *(Source: Huey 1982.)*

These examples emphasize that we are focusing on the effectiveness of thermoregulation, not on whether the lizards are within their set-point temperature ranges. *Anolis gundlachi* is a thermoconformer, but it lives in a habitat that allows its average body temperature to fall within its set-point range without effective thermoregulation. *Anolis cristatellus* is a moderately effective thermoregulator, but it always has a mean body temperature below its set-point range. What it achieves by thermoregulation is the ability to live in a thermally unfavorable habitat ($\bar{d}_e$ values of 5.0 and 9.2 for *A. cristatellus*, compared to 0.7 for *A. gundlachi*). Thus, the thermoregulatory activities of *A. cristatellus* allow the species to live in habitats where it could not live without effective thermoregulation.

Table 5–3   **Temperatures of Puerto Rican *Anolis*.***

Species and Month (T_{set})	Mean T_b	Mean T_e	Mean d_b	Mean d_e	E
Anolis gundlachi (24.3–26.1)	25.7 ± 0.1	25.1 ± 0.1	0.6 ± 0.1	0.7 ± 0.1	.14
Anolis cristatellus (28.1–29.7) August	27.2 ± 0.3	23.4 ± 0.1	2.5 ± 0.2	5.0 ± 0.1	.50
January	23.5 ± 0.4	19.0 ± 0.1	5.0 ± 0.4	9.2 ± 0.1	.46

*Values are means ± standard error.
Source: Hertz et al. (1993).

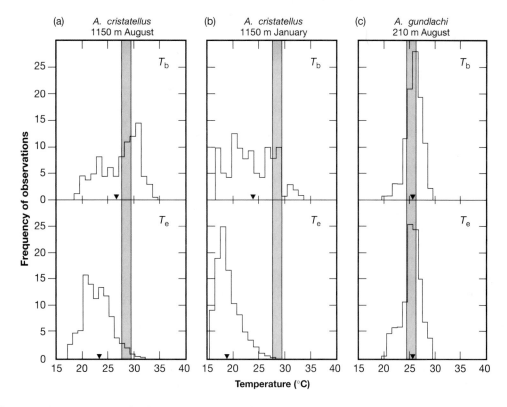

Figure 5–20 **Distributions of T$_b$ (body temperature) and T$_e$ (operative temperature) in *Anolis* populations.** (a) *A. cristatellus* at an altitude of 1,150 meters in summer (August). (b) The same population in winter (January). (c) *A. gundlachi* at 210 meters in summer. *A. cristatellus* lives in open habitats and is a heliotherm, whereas *A. gundlachi* lives beneath the forest canopy and is a thermoconformer. The upper graph of each pair shows the distribution of T$_b$ and the lower graph shows T$_e$. The vertical bar shows T$_{set}$ measured in the laboratory, and the arrowheads indicate mean values of T$_b$ and T$_e$. *(Source: Hertz et al. 1993)*

Freezing Resistance and Freezing Tolerance

Temperatures at or below freezing present a challenge for terrestrial ectotherms. Without the metabolic capacity to raise their body temperatures, ectotherms hibernate in cold seasons (Gregory 1982, Pinder et al. 1992). That usually means going deep into the ground, below the frost line, or into the mud on the bottom of a pond. Animals hibernating in these sites are protected from low temperatures, but they are too far from the surface to be active during periods of warm weather during the winter or to resume activity at the first sign of spring. Some species of amphibians and reptiles use a different strategy: for at least part of the cold season they select less protected retreat sites where they will be exposed to freezing temperatures. These species are able either to resist freezing or to tolerate freezing and thawing (Storey and Storey 1996).

Freezing Resistance.   Supercooling is the phenomenon by which a fluid such as water remains in its liquid state at a temperature below its freezing point. Water freezes when the molecules of water align themselves in the precise spatial relationship that produces an ice crystal. In the absence of an ice crystal to serve as a template, water can be cooled to several degrees below 0°C and still remain liquid. Freezing is a stochastic event (i.e., an event that occurs by chance and cannot be predicted)—it depends on two water molecules coming into the proper alignment, and then more water molecules adhering to them before they drift apart. This process is called nucleation. Thus, animals that depend on supercooling to avoid freezing are betting their lives (literally) that nucleation will not occur in their body tissues.

Yarrow's spiny lizard (*Sceloporus jarrovi*) is found in the mountains of southern Arizona, where it occurs at high altitudes. Although temperatures fall well below freezing in winter, sunlight is intense, and lizards can bask and reach activity temperatures during the day. These lizards spend the nights in relatively shallow rock crevices where the temperature sometimes falls below 0°C. When that happens, the animals supercool. Laboratory studies have shown that the lizards can be chilled to −3°C without freezing—usually, that is. Freezing of a supercooled solution is an unpredictable event, and every spring there are some dead lizards in the crevices. Those are the ones that lost their bets.

Newly hatched turtles sometimes spend the winter in the nest, not emerging until the following spring. Turtle nests are relatively shallow, and the soil around them can freeze solid. Newly hatched painted turtles (*Chrysemys picta*) overwinter in nests where temperatures remain below 0°C for several weeks and sometimes drop as low as −5 to −10°C. The hatchling turtles appear to depend on supercooling to survive. In a laboratory simulation, turtles in soil from the nest site survived exposure to temperatures between −8 and −9°C. Figure 5–21 shows temperature traces from four turtles. Thermocouples measured soil temperature and the temperature of each turtle. Freezing is an exothermic (heat-releasing) process. When supercooled water freezes, ice crystals spread rapidly through the solution and a measurable exotherm (release of heat) occurs. That phenomenon allowed the status of the soil and turtles to be followed. The turtles in jars with soil from the nest sites were placed in a chamber and the temperature was lowered in a series of steps. The soil froze at temperatures between −1.5 and −2.2°C, releasing heat as the soil exotherm. Freezing of the turtles in records B and D at temperatures of −2.6 and −2.8°C, respectively, is revealed by the animal exotherms. These turtles died, whereas the turtles in records A and C survived by remaining unfrozen in a supercooled state for 16 to 18 days.

Freezing Tolerance.   The formation of ice crystals damages intracellular structures, and most animals are killed by freezing. A few vertebrates, however, including about a dozen species of amphibians, are able to

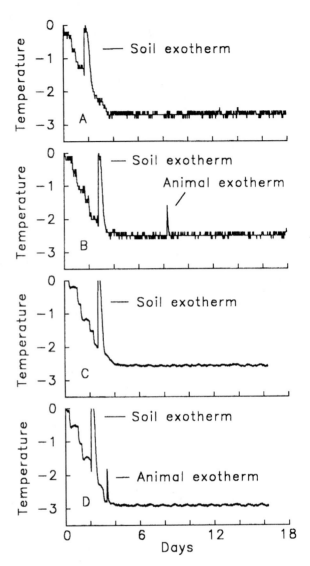

Figure 5–21   **Temperatures recorded at the surface of the carapace of hatchling painted turtles, *Chrysemys picta*.** The soil exotherms show that the soil in each container froze at a temperature near −2°C. Animal exotherms show that turtles B and D froze at temperatures of −2.5° and −2.8°C, respectively. Turtles A and C remained in a supercooled state for 16 to 18 days. (*Source: Packard and Packard 1995*).

tolerate repeated freezing and thawing (Pinder et al. 1992). These animals have a complex biochemical response to freezing that protects intracellular structures from damage. The cells face two problems as the water around them freezes: First, they are at risk of freezing themselves. Second, they risk dehydration as some of the water in the extracellular spaces freezes, raising the osmolality of the remaining extracellular water. The increased osmolality of the extracellular water will draw water from the cells unless it is counteracted by a mechanism that raises the intracellular osmolality.

The mechanism of freeze tolerance is illustrated by the wood frog, *Rana sylvatica*. As its body temperature falls, the frog becomes immobile, and actual freezing of some tissue (usually in the toes, because of their high surface-to-mass ratio) is the stimulus for a protective response. As soon as the first freezing occurs, glycogen in the frog's liver is converted to glucose that is rapidly transported throughout the body. The process is remarkably fast—within 10 or 15 minutes after its toes start to freeze the frog has dispersed glucose throughout its body. Two other species of frogs, *Pseudacris crucifer* and *P. triseriata*, use glucose as an antifreeze, whereas *Hyla versicolor* and and the salamander *Hynobius keyserlingi* use glycerol.

Accumulation of glucose or glycerol inside the cells increases the intracellular osmolality and lowers the freezing point of water in the cells. As water in the extracellular spaces turns to ice, the osmolality of the remaining extracellular fluid increases. The high concentration of glucose or glycerol inside the cells prevents water from being drawn into the extracellular space by osmosis and also prevents to formation of ice crystals inside the cells.

Complete freezing takes about 24 hours for an adult wood frog. A frozen frog is stiff and white, and more than half the water in its body has turned to ice. Respiratory movements, circulation, and heartbeat have stopped. The frog can remain in this state for a week or two. If freezing is extended for longer periods, the probability of survival decreases. Thus, freeze tolerance is probably a mechanism that allows animals to be active early in the spring or late in the fall, when cold and warm periods alternate. During sustained cold, even freeze-resistant animals probably must retreat below the frost line.

Water in the Lives of Amphibians and Reptiles

The availability of water in an animal's habitat and the behavioral and physiological mechanisms it uses to obtain water can play a profound role in shaping an animal's daily life. Any driver in a rural area has observed that amphibians are likely to be out on the roads on wet nights, and these nights are also when nocturnal species of frogs and salamanders are foraging most actively. The eastern redbacked salamander (*Plethodon cinereus*) emerges from the leaf litter on the forest floor on wet nights and searches for prey on the stems of plants. Plants appear to be good places to forage—salamanders on plants had significantly more prey items in their stomachs than salamanders foraging in the leaf litter—but it's only on wet nights that salamanders can climb plants to reach those prey items.

When you remember that most amphibians have skins that provide little barrier to evaporative water loss, it's not surprising that their water balance is often in a state of flux. Most amphibians appear to balance their water budgets on a time scale of hours to days, and they use a variety of behavioral and physiological mechanisms to do this. In contrast, reptiles have less permeable skins than do amphibians, and water flux is slower for most reptiles than it is for most amphibians. Some reptiles may balance their water budgets on time scales that range from days and weeks to months. A few reptiles appear to be out of water balance for such long periods that it may be more useful to think of them as having inherently variable salt and water concentrations rather than a set concentration from which they deviate.

Short-Term Water Balance

The costs and benefits of adjusting behavior to control water loss on a daily basis can be seen in the night-to-night activities of the Puerto Rican coquí (*Eleutherodactylus coqui*). This small, arboreal leptodactylid frog spends the day in a sheltered retreat in the understory vegetation of the forest. Soon after dusk, male coquís emerge and move to their calling stations, which are tree trunks or leaves only a meter or two from their retreat sites. Up to this point, the behavior of the frogs is nearly independent of the weather—male frogs move to their calling stations soon after dusk on both wet and dry nights.

The subsequent behavior of the frogs, however, does depend on the weather, and especially on the availability of water in the forest understory. During the rainy season in Puerto Rico, thunderclouds build through the morning and early afternoon, and by late afternoon and early evening it is often raining heavily. The rain penetrates the forest canopy and wets the frogs' calling sites in the understory vegetation. The rain usually stops in late afternoon, but the forest is still wet at dusk. On nights like this, a male coquí stands in the high alert posture with its legs extended, holding its trunk above the substrate (Figure 5–22d). A frog in this position is exposing nearly its entire surface area to the atmosphere (only the soles of the feet are in contact with the substrate). Furthermore, each time the frog calls the trunk and vocal sac inflate and deflate (Figure 5–22e), disrupting the boundary layer of air around the frog. This pumping action mixes the air that is in contact with the frog's skin (which is saturated with water vapor) with the ambient air (which is not saturated with water vapor). Thus, the combined effects of exposing the entire body surface to the atmosphere and disrupting the boundary layer by calling produce the highest rate of evaporation.

Some days rain does not fall, even in the rainy season, and the forest is dry in the evening when coquís emerge from their daytime retreats, or only a little rain has fallen and the leaf surfaces dry off before midnight. Under these conditions, male coquís do not call very often, and spend most of the evening in one or more of several postures that reduce the rate of evaporative water loss by exposing only part of the body surface to air. The water-conserving posture (Figure 5–22a) is the most extreme of these postures, and is adopted by many anurans when they are in dry situations. In the water-conserving posture a frog presses its trunk and chin against the substrate, draws its legs and feet into contact with the body, and closes its eyes. A frog in water-conserving posture exposes only half as much of its surface area to the air as it does in high alert posture, and its rate of evaporative water loss is correspondingly reduced.

Although a frog benefits from adopting the water-conserving posture by reducing its rate of evaporative water loss, it pays a price. Male frogs in the water-conserving posture are not vocalizing, and as a result they have no chance of attracting a mate. That may not be a serious cost, however, because on dry nights female coquís are not moving through the forest understory seeking mates. Like male coquís, females are resting in the water-conserving posture. However, male coquís also sacrifice the opportunity to capture prey when they are in the water-conserving position. Resting with their eyes closed, they do not respond when an insect passes close to them. Insects are at least as abundant in the forest understory on dry nights as they are on wet nights, but many coquís return to their daytime retreats with empty stomachs after a dry night, whereas at the end of a wet night nearly all frogs have prey in their stomachs.

Thus, behavioral control of the rate of evaporation via postural adjustments allows male coquís to remain at their calling stations on a dry night, but only at the cost of not vocalizing or feeding under those conditions. The water-conserving posture minimizes water loss, but does not stop it entirely. On dry nights male coquís lose an average of 8 percent of their initial body mass between dusk and dawn. That change in weight represents water lost by evaporation, but the tissue water content of the frogs does not change. The frogs reabsorb water from urine in their bladders to maintain the normal water content of their body tissues. Thus, they combine a behavioral response that minimizes evaporative water loss with a physiological response that preserves the normal intracellular water content to withstand the rigors of a dry night. The frogs return to moist retreat sites at dawn and rehydrate through their pelvic patches during the day. Thus, they are ready to resume activity with a fresh water reserve in the bladder the next evening. A daily cycle of water loss and rehydration of this sort is probably characteristic of many terrestrial anurans, and it may apply to some salamanders as well.

The fringe-footed lizard, *Uma notata*, also has a daily cycle of water balance. This lizard lives in wind-blown sand dunes in southern California and adjacent Mexico. The habitat is open and relatively barren because the shifting sands do not allow many plants to become established in the dunes. The lizards bury themselves in the sand when they are inactive, and

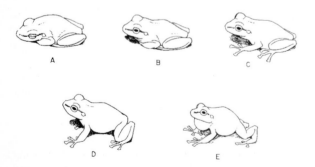

Figure 5–22 **Behavioral control of evaporative water loss.** Postures of coquís (*Eleutherodactylus coqui*) in order of increasing exposure of body surface to evaporation. (a) Water conserving. (b) Chin up. (c) Low alert. (d) High alert. (e) Calling. *(Source: Pough et al. 1983.)*

spend about 19 hours a day under the sand. This period of burial turns out to be a critical element in their daily water balance. When a lizard is moving on the surface of the dunes with a body temperature of about 39°C, it evaporates water at an average rate of 0.883 mg·g^{-1}·h^{-1}. Lizards buried in the sand at an average body temperature of 34.3°C have an average rate of evaporation of 0.358 mg·g^{-1}·h^{-1} which is less than half that of a lizard on the surface. Part of the reduction in evaporation is produced by the lower body temperature, but more important is the high relative humidity of air beneath the sand—85 percent, compared to 10 percent relative humidity or less for the air above the sand. Forty percent of a lizard's total daily water loss occurs during the average 4.7 hours it spends above ground, and adjusting the time spent aboveground is a behavioral mechanism that controls water balance.

Long-Term Water Balance

We have already mentioned some amphibians from deserts and brackish water environments that tolerate prolonged periods of elevated solute levels in the body fluids. Some reptiles also tolerate long periods of high concentrations (Bradshaw 1986). For example, the small Australian lizard *Amphibolurus ornatus* eats mainly ants, which have a high sodium content. These lizards do not have salt-secreting nasal glands, so the only way to excrete this sodium is in the urine, and that requires a substantial loss of water. Instead of excreting sodium and losing water during dry spells, the lizards allow the ion to accumulate to twice its normal concentration in the extracellular fluids. Only when it finally rains and the lizards are able to drink do they excrete the excess sodium and return their body fluids to normal levels.

The desert tortoise (*Gopherus agassizii*) is a large herbivorous reptile found in the Mojave and Sonoran deserts of North America. The variability of rainfall in these deserts produces year-to-year changes in the availability of food and water, and the tortoises are frequently out of water and salt balance for months at a time (Nagy and Medica 1986). A study of the water budget of tortoises at sites in California and Nevada revealed large and prolonged deviations from normal body fluid and solute concentrations (Peterson 1996). In May and June of 1989 about 65 percent of the tortoises' body mass was water, but a drought in the summer of 1989 greatly reduced the growth of the annual plants that form the mainstay of the tortoises' sum-

mer diet. During this time the tortoises lost as much as 40 percent of their initial body mass, and their average body water content dropped below 60 percent. Sodium, chloride, and urea concentrations in the blood plasma increased, and the average osmolality of the blood plasma rose from about 300 mmole·kg H$_2$O^{-1} to 400 mmole·kg H$_2$O^{-1}. Some animals had osmolalities above 500 mmole·kg H$_2$O^{-1}, which are among the highest values ever recorded for reptiles.

Urine in the bladder is a key element of the salt and water balance of desert tortoises. When rain falls, tortoises drink and then defecate, emptying the cloaca and bladder. The bladder is refilled with dilute urine and serves as a repository for potassium and nitrogenous wastes and as a source of water for the tortoise. The osmolality of the plasma is initially maintained at about 300 mmole·kg H$_2$O^{-1} by withdrawing water from the

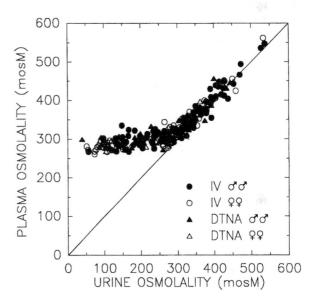

Figure 5–23 **Urine as a water reserve for the desert tortoise, *Gopherus agassizii*.** Withdrawal of water from the urine to stabilize plasma osmolality is illustrated by these measurements for individual tortoises. The diagonal line shows equal osmotic activities of plasma and urine. Urine osmolalities are initially as low as 50 mmole·kg H$_2$O^{-1}. Tortoises withdraw water from the bladder to maintain plasma osmolality near 300 mmole·kg H$_2$O^{-1} and the osmolality of urine rises from 50 to 300 mmole·kg H$_2$O^{-1}. As additional water is withdrawn from the bladder, the osmolalities of urine and plasma rise in parallel, ultimately reaching 550 mmole·kg H$_2$O^{-1}. The symbols identify male and female tortoises from two study sites in the Mojave Desert, Ivanpah Valley (IV) and the Desert Tortoise Natural Area (DTNA). *(Source: Peterson 1996)*

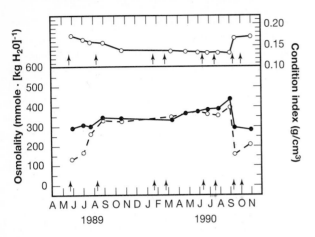

Figure 5–24   **The effects of rainfall and drinking on a male desert tortoise,** *Gopherus agassizii,* **in Ivanpah Valley.** Arrows indicate rainfall of 5 millimeters or greater. The upper graph shows an index of body condition (body mass divided by carapace length cubed). The lower graph shows the osmolalities of plasma (solid symbols) and urine (open symbols). Opportunities to drink are reflected by sudden drops in urine osmotic activity (e.g., in September 1990). Note that plasma osmolalities are more stable than urine osmolalities, and that water is withdrawn from the bladder after drinking. *(Source: Peterson 1996.)*

bladder. Once the urine in the bladder is as concentrated as the blood plasma, the osmolalities of the two body compartments increase in parallel (Figure 5–23).

The importance of summer showers in the water budget of desert tortoises is shown clearly by changes in plasma and urine osmotic activity before and after a rainstorm (Figure 5–24). The tortoise in this example drank once during a rainstorm in May 1989, and filled its bladder with dilute urine. During June, July, and August, the tortoise was able to maintain its plasma osmolality at 300 mmole·kg H_2O^{-1} by withdrawing water from the bladder. By September 1989 the os-

molality of the urine had risen to 300 mmole·kg H_2O^{-1}, and after that urine and plasma concentrations increased in parallel until the late summer of 1990. The tortoise obtained a single drink in September 1990, which allowed it to flush out the wastes it had been accumulating and return its plasma osmotic activity to 300 mmole·kg H_2O^{-1} and refill its bladder with dilute urine. Thus, the tortoise went 16 months between drinks, and it tolerated an elevated osmolality for an entire year.

Summary

The interactions of amphibians and reptiles with their physical environment are often a conspicuous part of their behavior. For example, when a lizard's body temperature falls below its lower set-point, the lizard moves into the sun, and that behavioral response is easy to see. Similarly, when a tree frog stops calling and adopts the water-conserving posture, the frog is visibly responding to the dryness of its environment.

Temperature- and water-regulating behaviors have a direct effect on how well an animal can carry out essential activities. An animal's running and striking speed and its success or failure in escaping from predators, capturing prey, or defending a territory are all mediated by its interactions with the physical environment. The intricate relations between amphibians and reptiles and their physical environments set the context within which behavioral and ecological phenomena occur. Whatever else a frog or lizard is doing—pursuing prey, vocalizing, or defending a territory—it is always interacting with the physical environment. Physiological ecology is the study of these sorts of interactions and their consequences in ecological and evolutionary time, and amphibians and reptiles have played a prominent role in the development of this field because their interactions with the physical environment are such a conspicuous part of their daily lives.

6

Energetics and Performance

One of the fascinating aspects of herpetology is how clearly physiological characteristics of amphibians and reptiles are reflected in their behavior and ecology. As we saw in the preceding chapter, regulation of body temperature and body water content is often critical in determining what amphibians and reptiles do and when and where they can be active. Metabolic characteristics have equally profound effects on the day-to-day activities of amphibians and reptiles.

Cellular metabolism depends on extracting oxygen from air or water and transporting it to body tissues. This chapter traces some of those relationships, emphasizing the close connections between structures and processes at the molecular, cellular, and organ system levels and the behaviors and ecological characteristics of amphibians and reptiles in the field.

We will start with the structures used for gas exchange and then consider the cardiovascular system, which transports oxygen and other substances throughout the body. Next we will consider how the production of adenosine triphosphate is shaped to meet the needs of different ways of life, the energetic costs and rewards of natural activities, and finally how an individual's performance may determine its Darwinian fitness.

Sites of Gas Exchange

Water and air are the respiratory media for aquatic and terrestrial animals, respectively. Birds and mammals use only air as a respiratory medium, but all amphibians and some reptiles can use both media, often simultaneously. The sites at which gas exchange takes place include the lungs (pulmonary gas exchange) and the skin surface, gills, pharynx, and cloaca (collectively called nonpulmonary routes of gas exchange).

Nonpulmonary Gas Exchange

We mentioned the role of amphibian skin in gas exchange when we were discussing evaporative water loss. Permeability to oxygen and carbon dioxide is inseparable from permeability to water in biological systems, and the skin of amphibians plays a major role in both gas exchange and water balance (Feder and Burggren 1985). The buccal region of the throat can be a site of gas exchange, and salamanders in the family Plethodontidae (which is characterized by the absence of lungs) carry out all of their gas exchange via the skin and buccal region. Most other amphibians have lungs as adults, although the lungs of some aquatic species of salamanders and frogs appear to be more important for adjusting buoyancy than for respiration. Some aquatic amphibians have folds of skin that increase the surface area (Figure 6–1). These folds are highly vascularized, and the capillaries run close to the surface of the skin. In a stream of moving water the current carries away carbon dioxide and brings water containing oxygen into contact with these folds, but in still water the animals must renew the layer of water in contact with the skin by moving. Aquatic amphibians such as *Cryptobranchus* and *Telmatobius* do this by swaying from side to side with a rippling of their skin folds that mixes the water around their bodies.

Figure 6–1 **Skin folds as gas exchange structures.** Some aquatic amphibians use folds of skin to increase the surface area available for gas exchange. (a) the Hellbender, *Cryptobranchus alleghaniensis*, and (b) the Lake Titicaca frog, *Telmatobius culeus*. (Photographs by (a) Alvin E. Staffan/Photo Researchers, Inc., (b) Tom McHugh/Photo Researchers, Inc.)

(a)

(b)

Species of amphibians that are primarily lung-breathers may have specialized sites for aquatic gas exchange that are important for particular activities. During the breeding season, adult males of the African hairy frog, *Trichobatrachus robustus*, grow filaments of skin from the posterior part of the trunk that, like gills, have an extensive blood supply and a large surface area (Figure 6–2). These filaments increase the surface area for gas exchange with water. Several functions have been suggested for these structures, which are seen only in males and only during the breeding season. Initially Noble (1925) proposed that the extra surface area for gas exchange provided by the filaments helped male frogs sustain high levels of activity associated with breeding. Subsequently it was proposed that the hairs increase oxygen uptake

Figure 6–2 **Supplementary gas exchange structures of amphibians that are associated with reproduction.** (a) Male hairy frog, *Trichobatrachus robustus*, showing the projections from the skin that develop while it is attending its eggs. *(continued)*

from water and allow males to remain with their eggs in underwater nests. Yet another possibility is that the hairs may actually release oxygen absorbed through the lungs into the water in the nest, thereby aerating the egg mass and promoting embryonic development (Zippel 1997).

The dramatic crests and tail fins developed by male European newts, *Triturus*, during the breeding season probably increase aquatic gas exchange as well as play a role in the courtship display. Male newts court females underwater, leading them through an extended series of activities (see Chapter 12), and the longer a male can remain underwater the better are his chances of mating. A male newt that stops to swim to the surface and breathe is unlikely to resume courtship successfully. The surface area of the crest and tail fin may increase oxygen uptake from water, thereby allowing a male to sustain courtship.

All aquatic larval amphibians have gills, and adults of some paedomorphic species of salamanders such as *Pseudobranchus*, *Necturus*, *Proteus*, and some *Ambystoma* retain gills as adults. Gills are effective for gas ex-

Figure 6–2   (continued) (b) A male common newt, *Triturus vulgaris*, in breeding condition. *(Source: (a) Boulenger 1902, (b) Holt Studios International/Photo Researchers, Inc.)*

(b)

change in water because water is dense and supports the individual gill filaments. The embryos of direct-developing and viviparous species of amphibians also respire with gills supported by fluid within the egg capsule (Figure 6–3).

Reptiles have dry skins with a layer of lipids that reduces water loss. The skins of reptiles are less permeable to water and gases than are the skins of amphibians, and cutaneous gas exchange is less important for most reptiles than it is for amphibians. Nonetheless, nonpulmonary respiration is significant for some aquatic reptiles. Sea snakes (Hydrophiinae) can obtain nearly all their oxygen via diffusion across the skin, and some freshwater turtles have specialized gas exchange regions in the pharynx or cloaca. Short, filamentous projections in the pharynx of *Apalone* (first reported by Louis Agassiz in 1857) appear to increase the surface area for gas exchange, and 30 percent of nonpulmonary gas exchange by *A. triunguis* occurs through the mouth, compared to 7.4 percent in the cloaca. An Australian cryptodiran turtle, *Rheodytes leukops*, has evaginations from the cloaca (bursae) that are lined with villi. These turtles, which rarely come to the surface to breathe air, swim with the cloaca open and pump water in and out of the bursae at rates of 15 to 80 cycles per minute. A South American cryptodiran turtle, *Podocnemis*, also has been reported to obtain nearly all of its oxygen via cloacal bursae.

Lungs

Excellent as gills are in water, they are useless in air. Without water to support them, gill filaments collapse on each other and the surface area available for gas ex-

change is drastically reduced. Air-breathing animals rely on lungs, which are sacs of air inside the body.

Buccal Pumping (Positive-Pressure Ventilation). Pumping movements of the buccal region of the mouth and throat are an ancestral character of tetrapods, and amphibians and reptiles use buccal movements to move both water and air. Amphibians use buccal pumping to force air into the lungs (Burggren 1989). This method of breathing is called positive-pressure ventilation, because air is forced into the lungs by raising the pressure in the buccal region higher than the pressure in the lungs. (In contrast, reptiles—including birds—and mammals use a negative-pressure system to ventilate the lungs.)

A cycle of lung ventilation begins by opening valves in the glottis and nostrils and lowering the floor of the buccal cavity. Elastic contraction of the body wall forces air out of the lungs, and this exhaled air passes above the freshly inspired air in the buccal cavity and flows out the nostrils. Next the nostrils are closed, while the glottis remains open and the buccal cavity contracts, forcing air from the floor of the buccal cavity into the lungs. The glottis then closes, the nostrils open, and the floor of the buccal cavity is depressed, drawing more air in through the nostrils (Figure 6–4).

Reptiles also use buccal pumping, but in this case its function is to inflate the lungs for defense rather than for respiration. Chuckwallas (*Sauromalus obesus*) inflate their lungs to wedge themselves into rock crevices. A chuckwalla can wedge itself so tightly into a crevice that a predator cannot pull it out.

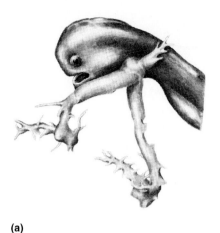

(a)

(b)

Figure 6–3 **Modified gills of embryos of direct-developing and viviparous amphibians.** (a) The plethodontid salamander *Pseudoeurycea nigromaculata*. (b) The caecilian *Typhlonectes compressicauda*. In some species other structures are used for embryonic gas exchange: (c) The expanded tail of the leptodactylid *Tomodactylus nitidus*. *(Source: Duellman and Trueb 1986.)*

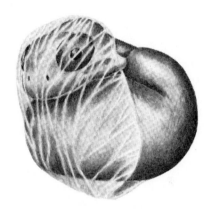

(c)

Aspiration (Negative-Pressure Ventilation). All reptiles use a negative-pressure pumping system to breathe. Inhalation is accomplished by expanding the thoracic cavity, thereby lowering the pressure in the lungs below the pressure of the external environment so that air flows passively into the lungs. Beyond that level of generalization, however, different lineages of reptiles have different ways of producing the negative pressure.

Respiration by tuatara, lizards, and snakes is triphasic: exhalation, inhalation, and relaxation. Active contraction of specific regions of two hypaxial trunk muscles, the transversalis and retrahentes costarum, compresses the rib cage, raising the pressure in the lungs and forcing air out through the nostrils. Air is inhaled by contracting a specific region of a different set of trunk muscles, the external and in-

ternal intercostals, to expand the rib cage. This expansion drops the pressure in the lungs below atmospheric pressure and draws air into the lungs via the nostrils. When these muscles relax, the elasticity of the rib cage causes it to contract slightly. Reptiles usually pause after the relaxation phase of ventilation, sometimes waiting for several minutes before they initiate the next respiratory cycle by exhaling (Shelton and Boutilier 1982). These pauses in ventilation (apneic periods) cause the oxygen concentration in the lungs to vary during a respiratory cycle. The oxygen concentration is highest immediately after fresh air has been drawn into the lungs and decreases during the apneic period.

Using trunk muscles to ventilate the lungs creates problems for lizards during locomotion because the muscles that change the volume of a lizard's thoracic

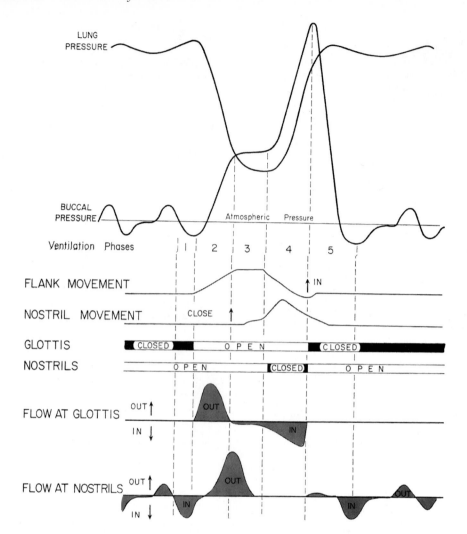

Figure 6–4   **Five phases of respiratory movements and air flow in a bullfrog (*Rana catesbeiana*).** 1. Muscular contraction increases the volume of the posterior portion of the buccal cavity, reducing buccal pressure. 2. Opening the glottis allows air to flow from the lungs to the buccal cavity, equalizing the pressure. 3. Lung pressure drops slightly below buccal pressure during the third phase. 4. The nostrils close and muscles in the floor of the mouth contract, raising buccal pressure and forcing air into the lungs. 5. The glottis closes and the nostrils open. Air flows into the buccal cavity as muscles in the floor of the mouth relax. *(Source: de Jongh and Gans 1969.)*

cavity during breathing are the same muscles that produce lateral bending and stabilize the trunk during locomotion. As a result, locomotion and lung ventilation are incompatible—a lizard cannot breathe and run simultaneously (Carrier 1986). Oxygen and carbon dioxide exchange via the lungs nearly ceases when a lizard walks or runs, and resumes when it stops (Figure 6–5). The stop-and-go locomotor pattern that is typical of many lizards may be related to this conflict between locomotion and respiration, with lizards stopping to breathe between bouts of movement.

The exchange of oxygen and carbon dioxide between air and blood takes place across the lung surface, and lungs of amphibians and reptiles range from simple sacs with few if any internal divisions to complex structures with interior walls and passages that direct air flow and increase the surface area. This morphological variation is related to physiological differences in the importance of lungs for gas exchange and in the metabolic requirements of different species. Amphibian lungs are generally simple, reflecting the large role the skin plays in gas exchange for these animals. In general the lungs of frogs are more complex than those of salamanders, and have a larger surface area for gas exchange. Sedentary lizards also have lungs with only a few internal divisions, whereas varanids, the most active lizards, have complex lungs with many chambers.

The right lung of snakes is the site of gas exchange; the left lung is extremely small (in boids) or entirely absent (other snakes). Two regions can be distinguished in the lungs of snakes—a vascular lung and an air sac (Figure 6–6). The vascular lung is in the anterior part of the body and is well supplied with blood vessels. Its walls are elaborated into chambers that provide a large surface area for gas exchange. The air sac is posterior to the vascular lung. It lacks partitions or other structures to increase the surface area, and it has a limited blood supply. Gas exchange takes place in the vascular lung, whereas the air sac appears to regulate air flow. Oxygen and carbon dioxide pressures in the vascular lung rise and fall with each breath, but they show little variation in the air sac. Both the air sac and the vascular lung are especially long in some marine snakes (Hydrophiinae) where the air sac extends almost to the cloaca and is used to adjust buoyancy.

Crocodilians use a different mechanism of lung ventilation. Although crocodilians have a well-developed rib cage, trunk muscles do not appear to play a major role in lung ventilation. Instead, the liver (which lies posterior to the lungs) acts as a plunger, compressing and expanding the lungs. During expiration, the liver is pulled anteriorly by several abdominal muscles, compressing the lungs and forcing air out. Inhalation is accomplished by a posterior movement of the liver that is produced by contraction of the diaphragmaticus muscles, which originate on the pelvis. Connective tissues attach the liver to the lungs, so that the lungs are expanded when the liver is pulled backward.

Turtles face unique problems in breathing: How can an animal change the volume of its thoracic cavity when it is enclosed in a rigid shell? The ribs of turtles are fused to the dermal bone forming the carapace, and no movement of the ribs is possible. Only the skin and muscle at the anterior and posterior openings of the shell provide the flexibility needed to change the volume of the lungs to draw air in and force it out.

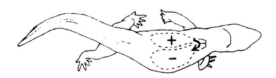 

Figure 6–5   **Axial bending during locomotion by a lizard.** Activity of the trunk muscles during locomotion interferes with lung ventilation. The axis of bending is along the spinal column, between the left and right lungs. As the lizard bends, the lung on the concave side is compressed and the lung on the convex side is expanded. Air pressure in the lung on the concave side increases (indicated by +), whereas pressure in the lung on the convex side decreases (–). As a result of these pressure changes, air may be pumped between the lungs, but little or no air moves in or out through the trachea. (*Source: Modified from Carrier 1987.*)

Figure 6–6 **The structure and gas content of a snake's lung.** The anterior (vascular) region of the lung is ventilated during breathing movements, whereas the posterior region (air sac) is not. Oxygen pressure rises during inspiration in the vascularized region and transition zone and falls during the period of breath-holding. Carbon dioxide pressure shows reciprocal changes, rising during breath holding and falling after each inspiration. The pressures of oxygen and carbon dioxide in the air sac show little change. (*Source: Redrawn from Withers 1992.*)

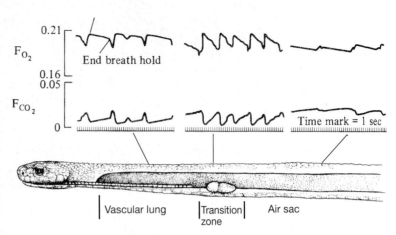

The dorsal surface of the lungs of a turtle is attached to the carapace, and the ventral surface of the lungs is attached to a sheet of connective tissue that is in turn attached to the viscera (Figure 6–7). Turtles exhale by forcing the viscera upward against the lungs and driving air out. Enlarging the visceral cavity allows the viscera to slump downward, and the connective tissue connecting the viscera to the lungs pulls the ventral surface of the lungs down, increasing the volume and drawing air inward.

Gas Exchange by Eggs

The gelatinous material that surrounds the eggs of many amphibians is a potential barrier to diffusion of oxygen from the water (Seymour and Bradford 1995). Many frogs deposit their eggs in relatively loose masses, and channels running among the eggs allow a convective flow of oxygen-rich water that replaces water from which oxygen has been removed by respiration of the embryos. Water flow increases as the eggs develop, because the gelatinous material becomes more fluid and the egg mass expands.

Many tropical frogs deposit their eggs in nests of foam that are created by the parents during the mating process. Skin secretions from the adults are whipped into a foam with water, and the eggs are deposited inside this blob of foam. The African frog *Chiromantis xeramplina* is an example of this reproductive mode (Seymour and Loveridge 1994). Pairs of *Chiromantis* deposit their eggs in foam nests in trees overhanging water. When the eggs hatch, the tadpoles swim about in the foam nest. Oxygen in the foam sustains the eggs until well after hatching, and as that initial store of oxygen is depleted by the tadpoles it is renewed by diffusion. As water evaporates from the nest, the volume of foam decreases and the tadpoles become increasingly crowded. The oxygen pressure within the foam nest declines, eventually falling to less than half of atmospheric oxygen pressure. A decreasing oxygen concentration may be the stimulus that causes tadpoles to drop out of the nest to complete development in the water below.

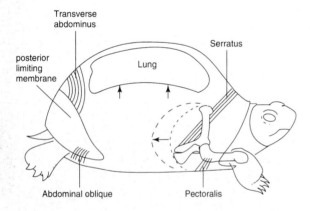

Figure 6–7 **Schematic view of the lungs and respiratory movements of a tortoise.** Expiration is produced by reducing the volume of the visceral cavity. Contraction of the transverse abdominis muscle draws the posterior limiting membrane anteriorly and dorsally, and contraction of the pectoralis muscle draws the pectoral girdle posteriorly. These movements force the viscera against the ventral surface of the lungs, driving air out. Inspiration is produced by enlarging the visceral cavity. Contraction of the abdominal oblique muscle pulls the posterior limiting membrane posteriorly and ventrally, and contraction of the serratus muscle moves the pectoral girdle anteriorly. (*Source: Pough et al. 1996.*)

Gas exchange of reptile eggs occurs by diffusion through the shell (Packard and Packard 1988). Crocodilians and many turtles have rigid shells formed by crystals of calcium minerals. Pores extend through the crystalline layer of the eggshell, allowing oxygen to diffuse in and carbon dioxide to diffuse out. Other turtles and many lepidosaurs have eggs with flexible, fibrous shells that lack discrete pores. Oxygen and carbon dioxide diffuse through gaps between fibers of these shells.

Patterns of Blood Flow

The circulatory system (which includes the heart, blood vessels, and the blood itself) carries oxygen from the sites of gas exchange to metabolically active tissues and brings carbon dioxide produced by metabolism to sites where it is released. For an animal with gills, blood passes from the heart through the gills to the tissues and back to the heart in a single loop. The heart receives venous (that is, oxygen-poor) blood from the vena cava (the major vein that returns blood to the heart) and pumps the blood through the gills, where it releases carbon dioxide and absorbs oxygen. The oxygen-rich blood that leaves the gills is bright red and contains 10 to 20 milliliters of oxygen per 100 milliliters of blood, depending on the species of animal and its sex, age, and physiological condition. This oxygen-rich blood flows through arteries and arterioles to the capillaries, which are the sites of oxygen exchange with the tissues. In the capillaries the blood releases oxygen, which diffuses outward through the walls of the capillaries, and absorbs the carbon dioxide produced by metabolism in the tissues. The oxygen-poor blood that leaves the capillaries and flows through venules into veins and ultimately into the vena cava and back to the heart is dark red and contains only about half as much oxygen as it did when it left the gills

The pattern of blood flow becomes more complicated in animals with lungs, because these animals have dual circuits that act in parallel. The circulatory system of a lung-breathing animal can be pictured as a figure 8 with the heart at the intersection of two loops, which represent the pulmonary (lungs) and systemic (head and body) circuits. Oxygen-poor blood in the pulmonary loop flows from the right side of the heart via the pulmonary artery to the lungs, where it is oxygenated, and then back via the pulmonary vein to the left side of the heart. Oxygen-rich blood in the systemic loop flows from the left side of the heart via the aortic arches to the body and back via the vena cava to the right side of the heart.

Pulmonary and Systemic Blood Circulation

Anatomical separation of the pulmonary and systemic circuits allows the pressures in the two systems to be different. The systemic circulation is immense, and tiny capillaries—so narrow that red blood cells must squeeze through them—run close to every living cell in the body. Forceful contraction by the heart is needed to drive blood through these capillaries, and the systemic system operates under high pressure.

The blood pressure in the pulmonary circuit is substantially lower than the pressure in the systemic circuit. The cells forming the walls of the pulmonary capillaries and the cells lining the gas exchange sites of the lungs are thin, minimizing the distance between the blood in the capillaries and the air in the lung. Because these tissues are so thin, high blood pressure in the pulmonary capillaries would force fluid into the lungs, where it would interfere with gas exchange. Thus, blood pressure in the pulmonary circuit is lower than pressure in the systemic circuit.

Multiple Sites of Gas Exchange. Amphibians are the only vertebrates in which the cardiovascular system carries oxygen-poor blood to the skin as well as to the lungs. Adult amphibians have paired pulmocutaneous arteries that divide into two branches. The pulmonary branch carries blood to the lungs, and the cutaneous branch carries blood to the skin, especially to the flanks and dorsal surface of the trunk. Cutaneous respiration accounts for 20 to 90 percent of the total oxygen uptake and 30 to 100 percent of the total carbon dioxide release for a variety of adult amphibians (Feder and Burggren 1985). Oxygen-rich blood returning from the cutaneous arteries enters the heart via the vena cava, where it is mixed with oxygen-poor blood returning from the systemic circuit.

Blood Flow in the Heart

Separation of the pulmonary and systemic circuits requires a mechanism to keep oxygen-rich and oxygen-poor blood from mixing as they pass through the heart. Birds and mammals have a septum that divides the ventricle into systemic (left) and pulmonary (right) chambers. The hearts of amphibians, and of all reptiles except crocodilians, lack a ventricular septum. Nonetheless, the hearts of amphibians and reptiles achieve excellent separation of oxygen-rich and oxygen-poor blood, and can even maintain different pressures in the systemic and pulmonary circuits.

Amphibians.   The hearts of anurans differ in structure from the hearts of salamanders and caecilians (Boutilier et al. 1992). The functional aspects of blood flow through the heart are best understood for anurans, and we will focus on them and note some differences found in urodeles and caecilians. Figure 6–8 shows the major features of blood flow through the heart of the African clawed frog, *Xenopus laevis*. Blood from the left and right atria enters the single ventricle, where sheets of tissue (**trabeculae**) extend from the walls of the ventricle into the central space. The trabeculae form compartments that are believed to limit mixing by creating separate channels for oxygen-rich and oxygen-poor blood. As the ventricle contracts, the trabeculae on the left side of the heart trap oxygen-rich blood and direct it into the conus arteriosus. A spiral valve runs longitudinally along the conus from the ventricle to the point at which the conus divides into the carotid, systemic, and pulmocutaneous arteries. Oxygen-rich blood enters the conus on one side of the spiral valve and flows into the carotid and systemic arches. Oxygen-poor blood is directed by trabeculae into the other side of the conus and flows into the pulmocutaneous arteries.

The hearts of urodeles and caecilians are structurally less complex than those of anurans. The atria of salamander and caecilian hearts are not always completely separated—the wall between the atria may be pierced by openings called fenestrae (Latin for windows). Nonetheless, salamanders and caecilians can maintain separation of oxygen-rich and oxygen-poor blood. Trabeculae in the ventricle are believed to channel oxygen-rich and oxygen-poor blood as described for anurans.

Plethodontid salamanders do not have lungs, and all gas exchange is carried out via the skin and buccal region of the throat. The hearts of plethodontid salamanders have no structures for separating oxygen-rich and oxygen-poor blood (Burggren 1988). The septum dividing the left and right atria is greatly reduced, and oxygen-rich blood returning from the skin mixes with oxygen-poor blood in the heart.

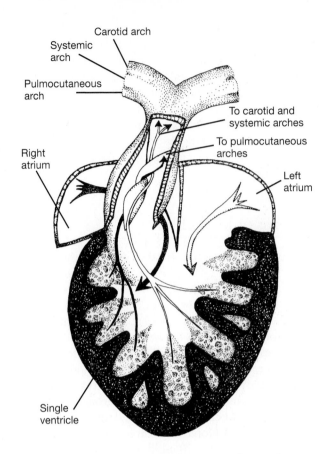

Figure 6–8   Blood flow in the anuran heart. Oxygen-rich blood (open arrows) entering the ventricle from the left atrium is largely separated from oxygen-poor blood (dark arrows) entering from the right atrium by the complex internal structure of the ventricle. When the ventricle contracts, the spiral valve in the conus arteriosus directs oxygen-poor blood primarily to the pulmocutaneous arch and oxygen-rich blood primarily to the carotid and systemic arches. *(Source: Redrawn from Shelton and Boutilier 1982.)*

Turtles and Squamates.   The hearts of turtles and squamates lack the conus and spiral valve found in the hearts of amphibians; instead, two systemic arches and a pulmonary artery open from the ventricle. The heart of a turtle or squamate is anatomically three-chambered, consisting of a left and right atrium and a single ventricle, but it is functionally five-chambered because contraction of the ventricle turns its single chamber into three compartments. Heart morphology has been described for only a few species of turtles and squamates, and variation exists even within that small sample.

Figure 6–9 shows a schematic view of the heart of a turtle or squamate. The left and right atria are completely separate, and three subcompartments can be

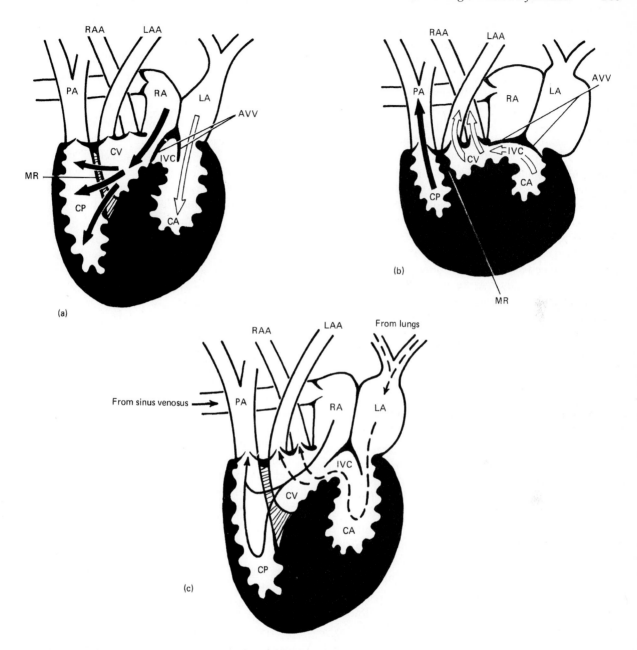

Figure 6–9 **Blood flow in the heart of a turtle or squamate.** (a) As the atria contract, oxygen-rich blood (open arrows) from the left atrium (LA) enters the cavum arteriosum (CA), while oxygen-poor blood (dark arrows) from the right atrium (RA) first enters the cavum venosum (CV) and then crosses the muscular ridge (MR) and enters the cavum pulmonale (CP). The atrioventricular valve (AVV) blocks the intraventricular canal (IVC) and prevents mixing of oxygen-rich and oxygen-poor blood. (b) As the ventricle contracts, the oxygen-poor blood in the cavum pulmonale is expelled through the pulmonary arteries; the atrioventricular valve closes and no longer obstructs the intraventricular canal, and the oxygen-rich blood in the cavum arteriosum is forced into the cavum venosum and expelled through the aortic arches. Contact between the wall of the ventricle and the muscular ridge prevents mixing of oxygen-rich and oxygen-poor blood. (c) A summary of the pattern of blood flow through the heart of a turtle. (*Source: Modified from Heisler et al. 1983.*)

distinguished in the ventricle. A muscular ridge in the core of the heart divides the ventricle into two spaces, the cavum pulmonale and the cavum venosum. The muscular ridge is not fused to the wall of the ventricle, and thus the cavum pulmonale and the cavum venosum are only partly separated. A third subcompartment of the ventricle, the cavum arteriosum, is located dorsal to the cavum pulmonale and cavum venosum. The cavum arteriosum communicates with the cavum venosum through an intraventricular canal.

The heart has two inflow routes (the right and left atria) and three outflows (the pulmonary artery and the left and right aortic arches). The right atrium receives oxygen-poor blood from the body via the sinus venosus and empties into the cavum venosum, and the left atrium receives oxygen-rich blood from the lungs and empties into the cavum arteriosum. The pulmonary artery opens from the cavum pulmonale, and both aortic arches open from the cavum venosum.

The mechanisms responsible for keeping oxygen-rich and oxygen-poor blood separate as they pass through the heart can be understood by tracing the movement of blood during a cardiac cycle.

1. When the atria contract, the atrioventricular valves open and allow blood to flow into the ventricle.

2. At this stage in the cardiac cycle the large median flaps of the valve between the right atrium and the cavum venosum are pressed against the opening of the intraventricular canal, sealing it off from the cavum venosum. As a result, the oxygen-rich blood from the left atrium is confined to the cavum arteriosum. Oxygen-poor blood from the right atrium fills the cavum venosum and then continues over the muscular ridge into the cavum pulmonale.

3. When the ventricle contracts, blood pressure inside the heart increases. Blood begins to flow into the pulmonary circuit before it flows into the systemic circuit because resistance is lower in the pulmonary circuit.

4. As oxygen-poor blood flows out of the cavum pulmonale into the pulmonary artery, the displacement of oxygen-poor blood from the cavum venosum across the muscular ridge into the cavum pulmonale continues. As the ventricle shortens during contraction, the muscular ridge comes into contact with the wall of the ventricle and closes off the passage for blood between the cavum venosum and cavum pulmonale.

5. Simultaneously, blood pressure inside the ventricle increases, and the flaps of the right atrioventricular valve are forced into the closed position, preventing backflow of blood from the cavum arteriosum and cavum venosum into the atrium.

6. When the atrioventricular valve closes, it no longer blocks the intraventricular canal. Oxygen-rich blood from the cavum arteriosum can now flow through the intraventricular canal and into the cavum venosum. At this stage in the heartbeat, the wall of the ventricle is pressed firmly against the muscular ridge, separating the oxygen-rich blood in the cavum venosum from the oxygen-poor blood in the cavum pulmonale.

7. As the pressure in the ventricle continues to rise, the oxygen-rich blood in the cavum venosum is ejected into the aortic arches. At this time blood pressure in the cavum venosum and the aortic arches is more than twice that in the cavum pulmonale and pulmonary artery.

The ability of turtles and squamates to keep oxygen-rich and oxygen-poor blood separate in the ventricle and to maintain different pressures in the pulmonary and systemic circuits is remarkable, but the ability to move blood between the two circuits within the heart (intracardiac shunting) is probably the most significant function of anatomy of the ventricle. Cardiac shunts can be descibed as right to left or left to right. A right-to-left shunt means that a portion of the oxygen-poor blood from the right atrium that would normally go to the pulmonary artery is expelled through the aortic arches into the systemic circulation. Thus, a right-to-left shunt increases the volume of blood going to the body and lowers its oxygen content. A left-to-right shunt means that oxygen-rich blood is diverted from the aortic arches into the pulmonary artery, increasing the volume and oxygen content of blood going to the lungs.

The direction and degree of shunting are controlled by two mechanisms, pressure differences between the pulmonary and systemic circuits and washout of blood remaining in the cavum venosum (Hicks et al. 1996). Valves on the pulmonary artery and aortic arches open when pressure in the ventricle exceeds pressure in the blood vessel. Thus, blood begins to flow first into the circuit with lower pressure, which is normally the pulmonary circuit. The pulmonary valve opens 40 to 100 milliseconds earlier than the valves on the aortic arches in the snakes *Tropidonotus natrix* and *Thamnophis* sp. and in the turtle *Trachemys scripta*. That difference in timing may promote the flow of oxygen-poor blood from the cavum venosum around the muscular ridge into the cavum pulmonale and out the pulmonary artery. If the delay between the opening of the pulmonary and aortic valves is long enough, oxygen-rich blood from the cavum arteriosum may be forced into the cavum pulmonale, creating a left-to-right shunt.

If the pressure difference beween the pulmonary and systemic circuits decreases, the time between opening of the pulmonary and aortic valves will also decrease. In this situation some oxygen-poor blood remains in the cavum venosum and is expelled into the systemic circulation, creating a right-to-left shunt.

Crocodilians. Crocodilians, too, are able to shunt blood between the systemic and pulmonary circuits, but the heart morphology of crocodilians and the mechanism of shunting differ from those seen in turtles and squamates. In the crocodilian heart, the right aortic arch opens from the left ventricle and receives oxygen-rich blood (Figure 6–10). The left aortic arch and the pulmonary artery both open from the right ventricle. Oxygen-poor blood enters the pulmonary artery, but it does not necessarily flow into the left aortic arch. The pattern of blood flow in the heart depends on what the animal is doing. When an alligator is at rest, blood pressure is approximately the same in the right and left ventricles. This situation creates a right-to-left shunt, and oxygen-poor blood flows from the right ventricle into the left aortic arch and then posteriorly to the viscera. Note that the blood supply to the head comes from the right aortic arch, so the brain receives oxygen-rich blood even when a right-to-left shunt is operating.

A different pattern of blood flow occurs when the alligator is diving and pressure in the left ventricle rises above that of the right ventricle. The left and right aortic arches are connected via the foramen of Panizza. When pressure in the right aortic arch exceeds that in the left, blood flows through this passage from the right aortic arch into the left. The increased pressure in the left aortic arch holds the ventricular valve closed, preventing entry of oxygen-poor blood from the right ventricle, and both aortic arches receive oxygen-rich blood.

Functions of Intracardiac Shunts

Why do the hearts of amphibians and reptiles have elaborate morphological provisions that allow blood to flow between the pulmonary and systemic circuits?

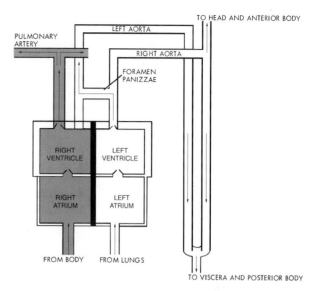

Figure 6–10 **Relationship of the heart and major vessels of a crocodilian.** The right aortic arch opens from the left ventricle and receives oxygen-rich blood, which flows to both the anterior and posterior parts of the body. The left aortic arch opens from the right ventricle and normally receives oxygen-poor blood that flows via the pulmonary artery to the lungs. When pressure in the right ventricle exceeds pressure in the left ventricle, however, the atrioventricular valve opens and oxygen-poor blood flows into the left aorta, which carries only to the posterior part of the body. When pressure in the left ventricle exceeds pressure in the right ventricle, the right atrioventricular valve is held shut, and oxygen-rich blood flows via the foramen of Panniza into the left aortic arch. *(Source: Kluge 1977.)*

Clearly it is not only that these animals have retained ancestral vertebrate characters, because the structures that control the degree of mixing of oxygen-rich and oxygen-poor blood vary among lineages. In other words, evolution has fine-tuned an ancestral condition to meet the needs of each group of animals.

Intracardiac shunts probably have at least three important functions. First, they may stabilize the oxygen content of the blood: both left-to-right and right-to-left shunts may be used to control the oxygen content of blood during the pauses in lung ventilation that are a normal part of respiratory patterns of amphibians and reptiles. Second, a right-to-left shunt is partly responsible for the increased blood flow to the systemic circuit that reptiles use to increase their rates of heating. Third, a right-to-left shunt directs blood away from the lungs when an animal is holding its breath.

Stabilization of Blood Oxygen Content

Adjusting the oxygen content of blood and the delivery of oxygen to the tissues may be the most general use of intracardiac blood shunts by amphibians and reptiles. Both the quantity of oxygen bound by hemoglobin and the readiness with which it is released from the blood to the tissues are affected by temperature and pH. Behavioral thermoregulation and the metabolic demands of high levels of activity result in substantial variation in body temperature and blood and tissue pH of amphibians and reptiles.

Furthermore, as we noted in the previous section, amphibians and reptiles normally breathe intermittently, alternating periods of lung ventilation with nonbreathing periods. This pattern of respiration leads to fluctuating oxygen concentrations in the lungs—high at the end of a period of ventilation and declining until the next series of breaths. A mathematical model of this system indicates that variable intracardiac shunting could be used to stabilize oxygen pressure in the blood while oxygen pressure in the lungs varies (Wood 1984, Hicks and Wood 1989).

Intracardiac Shunts and Thermoregulation

As we discussed in the previous chapter, many reptiles, including species weighing only a few grams, can heat faster than they cool, and a right-to-left intracardiac blood shunt that increases the volume of blood flowing through the systemic circuit is part of this control of heating and cooling rates.

Bypassing the Lungs During Breath-Holding.

Breath-holding occurs most conspicuously when an animal dives, and a right-to-left shunt is a common (but not universal) response to voluntary diving. Reptilian divers include not only aquatic species but also many terrestrial species that forage in or near water. Green iguanas (*Iguana iguana*), for example, escape predators by diving from branches overhanging water. An iguana can remain underwater for 30 minutes or more, long enough for a discouraged predator to seek more rewarding prey.

An animal does not have to be underwater to stop breathing, however, and other defensive behaviors used by amphibians involve breath-holding. When turtles retract into their shells they are unable to make the movements of the pectoral girdles that ventilate their lungs. Similarly, a lizard that has inflated its lungs to wedge itself into a rock crevice or to make itself appear large and threatening is not breathing. In all of these situations the animals are believed to reduce blood flow through the lungs by slowing the heart rate and initiating a right-to-left intracardiac blood shunt.

ATP Synthesis: Oxidative and Glycolytic Metabolism

The metabolic processes that support muscular activity take place in stages. The first seconds of muscle contraction are powered by energy stored in the cell as adenosine triphosphate (ATP) and phosphocreatine. The concentration of phosphocreatine in amphibian skeletal muscle is about six times that of ATP. Phosphocreatine is used to regenerate ATP as it is consumed by muscle contraction during the initial 20 or 30 seconds of exercise. The total energy stored in ATP and phosphocreatine in the limb muscles of anurans is sufficient for about 100 muscle contractions, which is enough to allow *Xenopus laevis* to swim about 1.4 meters in about 20 seconds. When phosphocreatine is exhausted, new ATP can be synthesized by oxidative (aerobic) and glycolytic (anaerobic) metabolic pathways.

Oxidative metabolism uses energy in substrate molecules more efficiently, but glycolysis starts working faster. Oxidative metabolism of 1 mole of glycogen to carbon dioxide and water yields 35 moles of ATP, whereas gylcolytic metabolism of the same quantity of glycogen to lactic acid produces only 3 moles of ATP. But accelerating oxidative metabolism in a muscle has a lag time of about 30 seconds while the circulatory system brings glucose from the liver and oxygen from the lungs. In contrast, glycolysis springs into action immediately because all of the components are present in muscle cells.

Thus, despite its energetic inefficiency, glycolysis is the most effective metabolic pathway for situations that demand a quick response, such as escaping from a predator. Glycolysis is not good in situations that require sustained ATP synthesis, because the glycogen stored in a cell is quickly used up. Activities that continue for long periods are supported by oxidative metabolism because the circulatory system can replace glucose and oxygen as they are used. Some muscles are specialized for oxidative metabolism and others for glycolytic metabolism. The muscles that male frogs use for vocalization, for example, are capable of very high rates of oxidative metabolism, whereas the limb muscles of the same frogs are entirely glycolytic.

Red and White Muscle

Muscle is composed of different types of fibers with different responses to stimulation (e.g., fast twitch, slow twitch, tonic) and different biochemical characteristics (e.g., oxidative, glycolytic, and oxidative–glycolytic). Oxidative fibers contain more myoglobin, a pigment that facilitates uptake of oxygen from blood, than do glycolytic fibers. Red muscle contains primarily oxidative fibers and white muscle is primarily glycolytic. (When you ask for dark meat or light meat from a chicken you are choosing oxidative or glycolytic muscle, respectively.)

Individual muscles and even parts of muscles differ in the proportions of different fibers. In the desert iguana (*Dipsosaurus dorsalis*), for example, the iliofibularis (a muscle in the thigh) is largely white, but has a central region of red muscle running its entire length. Other limb muscles, such as the gastrocnemius, have red fibers near the joints. The white regions are composed primarily of fast-twitch glycolytic fibers, whereas the red regions are fast-twitch oxidative–glycolytic fibers with a substantial proportion of tonic fibers. Oxidative fibers are capable of more sustained contraction than glycolytic or oxidative–glycolytic fibers. The discrete pockets of red muscle in muscles such as the gastrocnemius may strengthen and stabilize the joints. Red fibers that run parallel to the limb bones, as in the iliofibularis, may be limb stabilizers, or they may be directly involved in locomotion.

The total metabolic energy available to an animal is the sum of the ATP synthesized by both oxidative and glycolytic metabolic pathways. Oxidative metabolism is usually determined by measuring oxygen consumption, and glycolysis is measured as lactic acid production.

Comparisons of the total energy use by different species and of the relative importance of oxidative versus glycolytic metabolic pathways have revealed a variety of correlations that link phylogeny, ecology, and behavior. Certain evolutionary lineages of reptiles have high oxidative metabolic rates, for example, and other lineages have especially low rates. High versus low reliance on glycolytic metabolic pathways is associated with different methods of hunting for prey and is reflected in other aspects of an animal's life, including diet, social behavior, reproductive mode, and defenses against predators (see Chapter 13). Superior individual metabolic performance may be linked to high reproductive success. The following sections explore some of these relationships.

Metabolic Rates

Animals consume energy every moment of their lives. Even when a lizard is motionless on a rock, it is using muscular contractions to breathe and to pump blood, intracellular transport systems are consuming ATP as they move molecules across membranes, and a myriad of biochemical processes are consuming energy. The lizard's rate of energy consumption would increase if it moved its head to scan its surroundings, increase a bit more if it stood erect, still more if it ran a few centimeters, and so on. The metabolic rate of an animal is not a single value; rather, it is a continuum of values, and an animal's metabolic rate is affected by many variables simultaneously—what it is doing, its body temperature, whether it is digesting food, the time of day, and its emotional state, to list only a few. In an attempt to define repeatable conditions, physiologists have focused on three situations for measurement of oxidative metabolism: standard metabolism, resting metabolism, and maximum metabolism.

Standard Metabolism. The standard metabolic rate ($\dot{V}O_{2standard}$) is the minimum energy consumption an animal needs to remain alive. (The symbol VO indicates volume of oxygen, and the dot over the V is a convention that indicates a rate. Thus, $\dot{V}O_2$, which is read "V dot O 2," refers to oxygen consumption per unit time, most often $cm^3 \cdot min^{-1}$ or $cm^3 \cdot h^{-1}$. A subscript is often used to indicate the conditions under which the measurement was made.) Standard metabolic rates are, by definition, measured from postabsorptive animals (that is, without food in the gut) during the time of day when they are normally inactive (such as at night for an animal that is normally active during the day).

Resting Metabolism. Resting metabolic rates ($\dot{V}O_{2rest}$) are measured from postabsorptive animals that are motionless in the metabolic chambers, but the measurements are made during the time of day that the animals would usually be active. These animals are alert and may make small postural adjustments during the measurement period. Resting metabolic rates are usually about 10 percent higher than standard rates.

Both phylogeny and ecology are reflected in resting metabolism. Among squamates, for example, varanid and lacertid lizards have high metabolic rates and boid snakes have low rates. Surface-dwelling squamates have higher metabolic rates than fossorial (burrowing) species, and species of lizards that eat insects or other vertebrates have higher rates than do herbivorous species (Andrews and Pough 1985).

Maximum Metabolism.   Measurements of maximum metabolic rates are difficult to standardize. The goal is to induce an animal to reach a high level of activity in a way that allows comparison of different levels of an experimental variable, such as temperature. Natural forms of activity—walking or running—provide the most ecologically relevant measurements, and locomotion is the method most commonly used to elicit high levels of activity.

Figure 6–11 illustrates the general pattern of oxygen consumption in relation to speed of locomotion. $\dot{V}O_2$ increases as the speed of locomotion increases, and the

energetic cost of locomotion can be expressed mathematically as the slope of the increase in $\dot{V}O_2$ with speed.

At some speed (which depends on the species of animal being tested, its size, body temperature, and many other variables), the animal is consuming oxygen as rapidly as it can—that is, it has reached its maximum aerobic metabolic rate, $\dot{V}O_{2max}$. The speed that produces $\dot{V}O_{2max}$ is called the maximum aerobic speed. To go faster than the maximum aerobic speed the animal must use glycolytic (anaerobic) energy production to supply part of the total ATP.

Glycolytic Metabolism

Glycolysis provides as much as 80 percent of the ATP used for sprint locomotion, and ATP production by glycolysis is most rapid in the initial 30 seconds or so of activity. Glycogen stored in the muscles is used up during exercise, and depletion of glycogen is one of the factors that produces exhaustion following high levels of activity. Lactate, the ionic form of lactic acid, is the metabolic product of glycolysis, and whole-body lactate concentrations can increase as much as 20-fold during intense activity.

Most studies of glycolytic metabolism have been carried out in laboratory settings in which animals are forced to engage in activity, but a few studies have

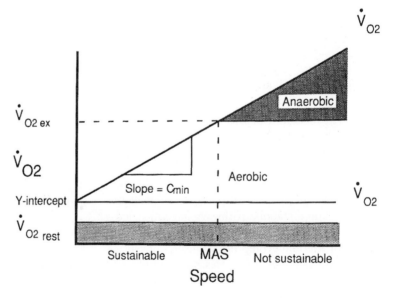

Figure 6–11 **Relation of the rate of oxygen consumption ($\dot{V}O_2$) to speed.** The cost of standing erect is illustrated by the difference between the value of $\dot{V}O_{2rest}$ and the *y*-intercept, which represents an animal standing motionless. $\dot{V}O_2$ increases with increasing speed, and the slope of the increase represents the minimal cost of locomotion (C_{min}). At the maximum aerobic speed (MAS), the rate of oxygen consumption reaches its maximum ($\dot{V}O_{2ex}$). This is the highest speed that can be sustained by oxidative (aerobic) metabolism. To run faster, an animal must use glycolysis (anaerobic metabolism). *(Source: Gatten et al. 1992.)*

shown that reptiles engaged in natural activities in the field use glycolysis as a source of ATP. Female green sea turtles (*Chelonia mydas*), for example, accumulate moderate levels of lactic acid when they come ashore to dig nests and deposit eggs.

Two studies have shown that free-ranging lizards use glycolysis when they defend territories and engage in other kinds of activity (Bennett et al. 1981, Pough and Andrews 1985b). For example, glycolysis fuels territorial defense by male lizards. Male Yarrow's spiny lizards (*Sceloporus jarrovi*) survey their territories from perches on large boulders, and attack intruding males (Figure 6–12). Lactate concentrations of male *S. jarrovi* were low when they emerged in the morning, and remained low as the lizards moved to their perch sites and engaged in feeding and thermoregulatory activities (Figure 6–13). That situation changed, however, when another male lizard, tethered to a pole, was moved into a male's territory. The resident male approached the intruder, gave threat displays, and ultimately attacked. The high level of activity during territorial defense was fueled partly by glycolysis, and the lactic acid concentrations of the territorial defenders rose to four times resting levels. Lactate concentrations were correlated with the intensity of the fight, which was measured as the number of bites delivered by the territorial male. The same study showed that another species of *Sceloporus*, the striped plateau lizard (*S. virgatus*), uses glycolysis during routine locomotion and for swallowing prey.

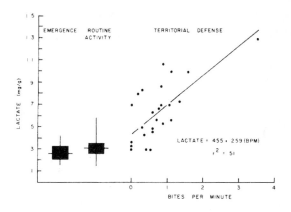

Figure 6–13 **Use of glycolysis by free-ranging lizards.** Male *Sceloporus jarrovi* emerge in the morning with low whole-body lactic acid concentrations, and routine activity does not increase lactate concentration. Territorial defense, however, requires glycolytic metabolism and can result in a 6-fold increase in lactate. The increase in lactate concentration is proportional to the intensity of a territorial defense measured as the rate at which the resident male bites the intruder. (*Source: Pough and Andrews 1985b.*)

Metabolism of Lactic Acid

Most species of amphibians and reptiles become exhausted when they are forced to sustain intense levels of activity for more than 2 or 3 minutes. An exhausted lizard or frog is nearly inert, and an animal's inability to right itself when it is turned on its back is a common criterion of exhaustion in laboratory studies. Whole-body lactic acid concentrations at exhaustion approach 2 mg lactate·g^{-1} body mass for some lizards, and concentrations exceeding 6 mg·g^{-1} have been reported for amphibians.

Maximum lactate concentrations measured in laboratory studies are higher than any reported for animals engaged in natural activities, and amphibians and reptiles may normally use their entire glycolytic capacity only in emergencies. Nonetheless, a submaximal use of glycolytic metabolism depletes muscle glycogen stores and reduces the capacity for additional activity. For example, the male *S. jarrovi* defending their territories in the previous example accumulated lactic acid concentrations averaging 1.3 mg·g^{-1}. Those lizards had used a substantial part of their muscle glycogen store and would have a reduced capacity to defend their territories against another intruder until their muscle glycogen was replenished.

Figure 6–12 **Measurement of glycolysis during territorial defense.** A resident male *Sceloporus jarrovi* (on the right) is attacking a tethered male lizard that has been moved into the resident's territory. (*Photograph by Harvey Pough.*)

Thus, the speed with which it can replenish glycogen may have a direct impact on an animal's behavior.

Much of the lactate produced by glycolysis is used to replace glycogen stores in the muscles themselves. The changes in glycogen and lactate during a period of intense exercise followed by 2 hours of recovery are shown by a study of desert iguanas (*Dipsosaurus dorsalis*) that compared red and white regions of the iliofibularis muscle of the hind limb (Gleeson and Dalessio 1990). Five minutes of running on a treadmill at speeds of 1.5 to 3 km·h^{-1} exhausted the lizards. Glycogen was depleted in red and, especially, in white muscle, and lactate concentrations increased correspondingly (Figure 6–14).

Following exercise, lactate concentrations declined and glycogen stores were replenished. The rate of resynthesis of glycogen in red muscle was five times faster than in white muscle. Two hours after exercise, the glycogen content of red muscle was higher than it had been at the start of exercise, but glycogen stores in white muscle were still depleted.

Reptiles and amphibians convert about 50 percent of the lactate back to glycogen in the muscles and oxidize less than 20 percent of it. This pattern is strikingly different from the metabolic fate of lactate in mammals, where as much as 90 percent of the lactate formed in muscles during activity is oxidized to carbon dioxide and water. Muscle glycogen stores of mammals are replenished by glucose from the liver (Gleeson 1996).

Total ATP Production and Activity

Understanding the costs and benefits of relying on glycolysis to sustain activity requires a perspective that integrates long-term energy efficiency and short-term energy needs. As we explained in Chapter 1, the low energy requirements of amphibians and reptiles are associated with their ectothermy. Because they do not depend on metabolic heat production to raise their body temperatures, amphibians and reptiles can have low resting rates of oxygen consumption and convert 40 to 80 percent of the energy in their food into new body tissue. In contrast, birds and mammals (endotherms) have high resting metabolic rates and use about 98 percent of the energy in their food for temperature regulation and activity. Thus, the low resting metabolic rates of amphibians and reptiles promote efficient use of energy.

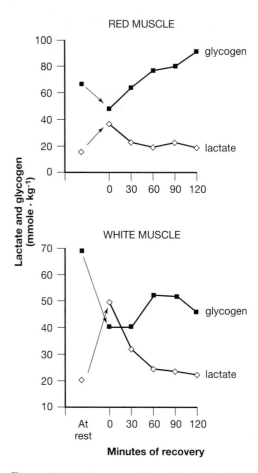

Figure 6–14   **Recovery from glycolysis.** In vivo changes in lactate and glycogen concentration in red and white muscle of *Dipsosaurus* following exercise. Five minutes of running on a treadmill produced a substantial drop in glycogen concentration and a corresponding increase in lactate compared to the levels for animals at rest. Lactate was reconverted to glycogen during the 2-hour recovery period. *(Source: Based on data from Gleeson and Dalessio 1990.)*

A different aspect of low resting metabolic rates is seen during activity. Amphibians and reptiles can increase their resting rates of oxidative metabolism about 10-fold during activity. Mammals also increase resting rates of oxygen consumption about 10-fold when they are active. Because the resting metabolic rates of mammals are 7 to 10 times higher than those of amphibians and reptiles (Figure 1–7), the maximum rates of oxidative metabolism by mammals during activity are also 7 to 10 times higher, and mammals synthesize ATP via oxidative pathways more rapidly than do reptiles or amphibians.

By combining glycolytic and oxidative pathways of ATP synthesis, amphibians and reptiles achieve a short-term total metabolic capacity equivalent to that of a mammal (Table 6–1). During 30 seconds of running, a desert iguana actually produced 32 percent more ATP than a kangaroo rat, but 76 percent of the lizard's ATP was derived from glycolytic metabolism, compared to only 30 percent for the mammal. The lizard depletes its muscle glycogen in about 5 minutes and must stop running, whereas the mammal can continue.

A lizard uses high levels of activity to sprint away from a predator, and power (total ATP production) for a few seconds is important. It doesn't matter that the lizard will be exhausted in 5 minutes, because long before that it will be either in its burrow or in the predator's stomach.

In general, reptiles and amphibians are intermittently active—they simply do not engage in behaviors that require sustained high levels of ATP synthesis. The few exceptions to this generalization, such as vocalization by male frogs, result from specialized physiological characteristics that are described in the following section.

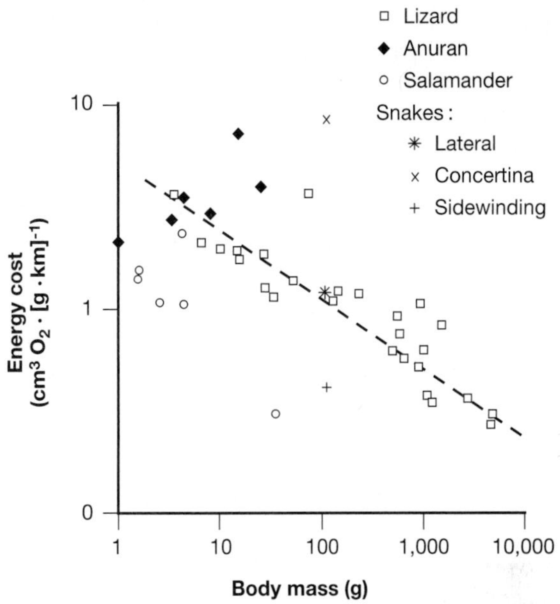

Energy Costs of Natural Activities

It has been possible to measure the amount of energy that an amphibian or reptile uses to carry out some natural behaviors. Locomotion is a prominent feature of the lives of amphibians and reptiles, and the costs of hopping and walking by anurans and of running and crawling by squamates have been measured. Measurements of the energy costs of consuming prey for lizards and of vocalization by anurans have been particularly fruitful because they have allowed tests of hypotheses about optimal foraging and sexual selection. Field metabolic rates integrate the costs of all

Figure 6–15 **Net cost of locomotion for amphibians and reptiles.** Several types of locomotion are shown: walking and running for lizards and salamanders, jumping and walking for anurans, and lateral undulation, concertina locomotion, and sidewinding for snakes. The dashed line shows the relationship for mammals. *(Source: Based on data in Walton et al. 1990.)*

the activities an animal engages in and provide an estimate of the daily and annual cost of living.

Locomotion

The locomotor modes of amphibians and reptiles are diverse (Chapter 8), and the net cost of transport (ml $O_2 \cdot [g \cdot km]^{-1}$) shows some variation among and within groups (Figure 6–15). Salamanders use substantially less energy walking than do anurans, either walking or

Table 6–1 **Estimated ATP production via oxidative and glycolytic pathways during 30 seconds of activity by a mammal (the kangaroo rat *Dipodomys merriami*) and a lizard (the desert iguana *Dipsosaurus dorsalis*).**

Species	ATP Production (mmol ATP•g^{-1})		
	Oxidative	**Glycolytic**	**Total**
Dipodomys	0.0098 (70%)	0.0043 (30%)	0.0141
Dipsosaurus	0.0044 (24%)	0.0142 (76%)	0.0186

Source: Based on Ruben and Battalia (1979).

hopping (Gatten et al. 1992). Two slow-moving lizards, the North American Gila monster (*Heloderma suspectum*) and the Australian shingle-back lizard (*Trachydosaurus rugosus*), have net costs of locomotion that are substantially below those of other lizards.

Most species of snakes can shift among several forms of locomotion (Chapter 8), and these different locomotor modes have different net costs of transport. Lateral undulation by snakes is as energetically efficient as locomotion with limbs by lizards, but no more so (Walton et al. 1990). Concertina locomotion by black racers (*Coluber constrictor*) was about seven times as expensive as lateral undulation by the same species. The net cost of transport for sidewinding rattlesnakes (*Crotalus cerastes*) was less than half that of lateral undulation by the racers (Secor et al. 1992).

Feeding

An animal invests both time and energy in obtaining food, and either time or energy could be an ecologically important cost. Theoretical models of optimal foraging make predictions about an animal's predatory behavior. Observations of lizards were important in the development of optimal foraging theory (Schoener 1969), and measurements of the energy costs of prey handling by lizards have been used to test some optimal-foraging theories.

The time and energy a lizard needs to subdue and swallow a prey item can can be measured by feeding the lizard in a metabolism chamber (Figure 6–16). Both the time and the energy used by a skink (*Chal-*

cides ocellatus) to crush and swallow a cricket increase as cricket size increases (Figure 6–17). More than 90 percent of the ATP used during feeding comes from oxidative pathways. Glycolysis makes an insignificant contribution to the total energy cost of lizards feeding on insects (Preest 1991), but it may be important for other combinations of predator and prey. For exam-

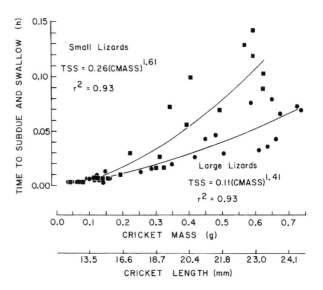

(a)

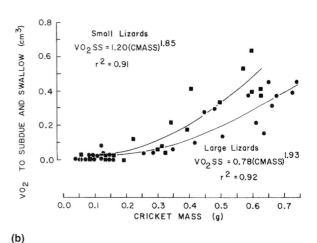

(b)

Figure 6–17 **Costs of prey handling.** (a) Time required and (b) energy used by skinks (*Chalcides ocellatus*) to consume crickets. Both time and energy increase with increasing prey size, and small lizards require more time and use more energy than do larger lizards. TSS, time to subdue and swallow. (*Source: Pough and Andrews 1985a.*)

Figure 6–16 **Measuring oxygen consumption during prey handling.** A *Chalcides ocellatus* eating a cricket in a metabolism chamber during studies of the energy cost of feeding. (*Photograph by F. Harvey Pough.*)

ple, when garter snakes (*Thamnophis elegans*) ate salamanders (*Plethodon jordani*), both participants accumulated substantial amounts of lactic acid as the snakes struggled to eat the salamanders and the salamanders tried to escape from the snakes (Feder and Arnold 1982).

The morphological characteristics of lizards and their prey affect the energy cost of feeding. A large-headed species of skink (*Eumeces inexpectatus*) was able to eat crickets faster than a small-headed species (*Chalcides ocellatus*). The *Eumeces* used less energy in feeding than did *Chalcides* and were able to consume larger prey relative to their own body size (Andrews et al. 1987). Insects with heavy exoskeletons, such as beetles, are harder for lizards to eat than are soft-bodied insects (Grimmond et al. 1994). *Chalcides* requires 50 percent more energy to crush and swallow a beetle than to eat a soft-bodied insect larva.

The net energy a lizard gains by eating a prey item can be calculated by subtracting the energy used during feeding from an estimate of the energy obtained by digesting the insect. In every case—even for a small skink eating a large beetle—these calculations show that the energy cost of prey handling is less than 1 percent of the energy the skink obtains by digesting the prey. Thus, the energy cost of feeding is trivial and probably is not a factor in determining what sizes of prey a lizard attacks.

Time spent feeding and the risk of predation during feeding might influence prey selection by lizards. The time needed to subdue and swallow an insect increases rapidly as the size of the insect increases (Figure 6–17). A lizard has only a certain amount of activity time every day, and minimizing the time spent feeding might be desirable. Lizards could minimize the time they devote to feeding by attacking prey of the size that allows them to meet their daily energy requirement in the shortest possible time. Predators that behave this way are called time minimizers.

The risk of predation that a lizard faces while it is eating could play a role in prey selection. Some lizards spend most of their time resting motionless on a perch. When a prey item comes into view the lizard makes a quick dash to capture and eat the prey, and then returns to its perch. Lizards that behave this way have cryptic colors and patterns, and they are hard to see while they are motionless. As soon as they move, however, they become conspicuous to their own predators. Perhaps minimizing the number of prey captures reduces a lizard's own risk of predation. If

this is true, a lizard should select large prey to meet its daily energy requirement with as few captures as possible. For example, the Neotropical forest lizard *Corytophanes cristatus* eats insects that are half its own body length (Figure 13–10), and less than one capture per day would satisfy the lizard's energy requirement. Animals with this feeding strategy are movement minimizers.

Vocalization by anurans

Vocalizing during the breeding season is the hardest work most male anurans do in their lives, and the acoustic energy in frog calls is substantial (Prestwich 1994). Sound energy is expressed as decibels (dB), and sound pressure levels of the calls of seven species of frogs, most with body masses less than 10 grams and some as small as 2 or 3 grams, averaged 91 dB. The vocalizations of these frogs contained as much energy as the calls of 17 species of songbirds (average, 89 dB) that weighed an average of 23 grams (Pough et al. 1992).

The rates of oxygen consumption of calling frogs are as much as 25 times resting rates and are substantially higher than the rates measured during even the most rapid locomotion (Figure 6–18). Furthermore, a male frog may call every few seconds for several hours during the night and return to the chorus night after

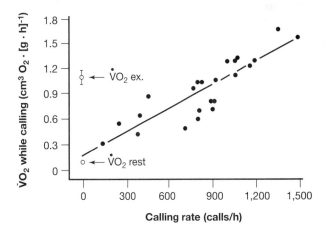

Figure 6–18 **Energy cost of vocalization.** Rates of oxygen consumption of frogs calling inside metabolic chambers as a function of calling rate. The metabolic rates of frogs calling at high rates were substantially greater than the highest rates that were produced by forced locomotion. *(Source: Taigen and Wells 1985.)*

night. Vocalization by anurans is the most dramatic exception to the generalization we made earlier that most amphibians and reptiles do not engage in activities that require sustained high levels of energy expenditure. The remarkable energy costs of vocalization by anurans are supported by a series of anatomical and biochemical specializations.

Characteristics of Trunk Muscles.   The high rates of oxygen consumption by calling frogs are associated with morphological and biochemical characteristics of the trunk muscles. These specializations are found only in males and only during the breeding season. The trunk muscles of males undergo anatomical and biochemical changes before the breeding season starts and regress to a resting state when the season is over. The most conspicuous change is hypertrophy of the trunk muscles responsible for vocalization. At the height of their development, just two trunk muscles—the internal and external oblique—account for 2 to 15 percent of the total body mass in males of nine species of frogs. The size of the trunk muscles relative to the size of the frog is roughly proportional to the calling effort (seconds of vocalization per night): males of species with high calling effort have larger muscles than males of species with lower efforts.

The trunk muscles of calling frogs are highly oxidative, with many mitochondria, high densities of capillaries, and high activities of two enzymes associated with oxidative metabolism, citrate synthase and β-hydroxyacyl co-A dehydrogenase (Ressel 1996).

Lipid is the primary metabolic substrate used by species of anurans with high rates of vocalization, whereas glycogen appears to be the primary substrate used by the trunk muscles of anurans with lower rates of calling. At the start of the breeding season, lipids account for as much as 45 percent of the volume of trunk muscles of male spring peepers (*Pseudacris crucifer*) and nearly that much in two species of Panamanian frogs (*Hyla microcephala*, Hylidae, and *Physalaemus pustulosus*, Leptodactylidae), all of which emit 3,000 to 6,000 notes per hour. In contrast, very little lipid is stored in the trunk muscles of the North American wood frog (*Rana sylvatica*), which emits only 500 to 600 notes per hour (Bevier 1995, Wells and Bevier 1997).

Energetics and Chorus Tenure.   The energy cost of vocalization by some anurans is so high that it might limit a male frog's capacity to call. Several species of anurans are known to lose weight and deplete glycogen or lipid reserves during a night of calling and across the length of a breeding season. Anuran vocalization is a social behavior, and it is influenced by the presence and behavior of other male frogs. For males of most species of anurans, the availability of receptive females at the breeding sites is what limits reproductive success. On a given night the number of males at a breeding site is usually much greater than the number of females, so only a few males are able to mate. A strongly skewed mating success is typical of male frogs—that is, most individuals never mate and a few individuals achieve multiple matings. Under these conditions sexual selection should be intense, and male frogs should behave in ways that maximize their attractiveness to females and, hence, their chances of mating.

Females of many species of frogs appear to make a choice among males, probably on the basis of one or more characteristics of their vocalizations. For example, female *Hyla microcephala* mate preferentially with males that call at high rates, and the calling rates of male *H. microcephala* can exceed 6,000 calls per hour. Male frogs in a breeding chorus often respond to the presence of a female by increasing the energy output of their vocalization—by calling faster, by giving longer calls, or by adding additional elements to the call. When one male increases its calling effort, nearby males often follow suit, and these interactions among males can produce substantial variation in calling effort in different parts of a chorus. An analysis of one chorus of *H. microcephala* found that the calling effort of individual males varied from 205 to 6,330 notes per hour, corresponding to a 300 percent variation in energy expenditure (Wells and Taigen 1989).

Chorus Tenure.   The number of nights a male anuran spends in a chorus is an important factor in determining whether it will get a mate (Figure 6–19). In every species of anuran that has been studied, the probability that a male will obtain a mate increases with the number of nights spent in the chorus. For example, in a 4-year study Christopher Murphy found that male barking tree frogs (*Hyla gratiosa*) average about one mating for every 5 nights in a chorus (Murphy 1994a,b). The breeding season of *H. gratiosa* lasts for several months, but most male barking tree frogs call for only 1 to 3 nights (Figure 6–20). These short chorus tenures are characteristic of anurans that call at high rates. The median chorus tenure for individual males among 20 species of anurans is only 20 percent of the breeding season.

Depletion of energy reserves was the most likely explanation of the short chorus tenures of males. In 2 successive years Murphy captured male frogs as they left the breeding pond, placed them in cages and fed them crickets, and then released them. Control males were captured and held in cages while the fed males ate their crickets, but the control males were not given crickets. Individuals that ate crickets returned to the chorus sooner than did controls (medians of 2.4 and 2.6 nights for fed males versus 4.0 and 5.5 nights for control males). The fed males also returned to the chorus more often than did the controls (medians of 4 and 5 nights for the fed males versus 1 and 2 nights for the controls).

Depletion of energy stores as a result of the high energy demands of vocalization may be a widespread phenomenon among male anurans. Individual variation in the vocal behavior of male anurans, such as adjusting calling patterns as chorus density changes or in response to the presence of females, allows them to use their limited energy stores effectively and may contribute to their success in attracting mates.

Annual Energy Budgets

The standard metabolism or the cost of locomotion for a species can be measured in the laboratory at a partic-

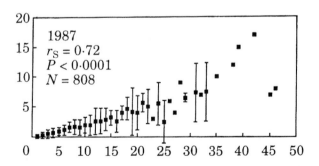

Figure 6–19   **Mating success of male tree frogs.** The average number of mates obtained by male barking treefrogs (*Hyla gratiosa*) increases with the number of nights a male spends in the chorus. *(Source: Murphy 1994a.)*

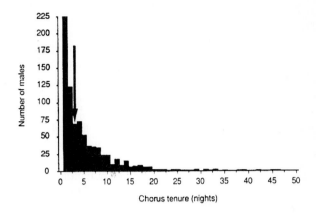

Figure 6–20   **Chorus tenure of male tree frogs.** Number of nights that male barking treefrogs (*Hyla gratiosa*) called in the breeding chorus. The median chorus tenure (indicated by the arrow) was only 3 nights. *(Source: Murphy 1994b.)*

ular temperature, but these measurements cannot readily be extrapolated to natural conditions. Free-ranging animals are exposed to variable temperatures, they alternate periods of activity and rest, and sometimes they are digesting food. Measuring the field metabolic rate—that is, the energy expenditure of a free-ranging animal—requires radioactive tracers, and doubly labeled water is the method used most often.

The term doubly labeled refers to water in which some hydrogen atoms (^1H) are replaced by deuterium (^2H) or tritium (^3H), and some oxygen atoms (^{16}O) are replaced by radioactive oxygen (^{18}O). A small amount of doubly labeled water is injected into an animal and the disappearance of labeled hydrogen and oxygen is followed as the animal is recaptured at intervals for several weeks or months. Labeled oxygen is lost in the form of carbon dioxide and water, whereas labeled hydrogen is lost only as water. Subtracting the rate of hydrogen loss from the rate of oxygen loss provides an estimate of carbon dioxide production, and that value can be converted to an estimate of oxygen consumption. (For a discussion of the assumptions and potential sources of error in studies with doubly labeled water, see Nagy [1992].)

Annual energy budgets of free-ranging animals can be constructed from measurements of field metabolic rates, and these studies reveal the impact of environmental conditions. Steven Beaupre studied mottled rock rattlesnakes (*Crotalus lepidus*) at different altitudes in Big Bend National Park, Texas (Beaupre 1995, 1996). He found that temperature has an enormous effect on activity patterns, food intake, and the partitioning of energy between maintence and production. Boquillas, the low-altitude site, is substantially warmer during the summer than the high-altitude site, Grapevine Hills. The body temperatures of free-ranging snakes were monitored by surgically implanting small, temperature-sensitive radio transmitters in their bodies. The average body temperature of snakes at Boquillas was 29.9°C, compared to an average of 28.8°C at Grapevine Hills. Small as it is, that 1.1°C difference in average body temperature between the sites is statistically significant and has important biological consequences.

The high temperatures at the Boquillas site affect the energy balance of the rattlesnakes in three ways. First, because body temperatures of snakes at Boquillas are high, their metabolic rates are high. As a result, approximately 31 percent of the annual energy expenditure of snakes at Boquillas is resting metabolism, compared to 19 percent for snakes at Grapevine Hills (Figure 6–21).

The second response to the high temperatures at Boquillas is behavioral: the snakes spend more time in relatively cool underground shelters and less time searching for prey on the surface at Boquillas than they do at Grapevine Hills. Snakes at Boquillas were on the surface in only 33 percent of censuses, whereas snakes at Grapevine Hills were on the surface 62 percent of the time.

The third effect of the temperature difference between the low- and high-altitude sites is ecological. Mottled rock rattlesnakes at Boquillas consumed less food than did snakes at Grapevine Hills, probably because the high temperatures at Boquillas limit the amount of time a snake can be on the surface searching for prey. Snakes from Boquillas had fed recently only 44 percent of the times they were examined, compared to 90 percent of examinations for snakes from Grapevine Hills.

Because snakes from Boquillas captured less prey and used more energy for maintenance than did snakes at Grapevine Hills, they grew more slowly than the Grapevine Hills snakes. The number of rattle segments is a measure of age, and a male mottled rock rattlesnake from Boquillas with 10 segments in its rattle weighed about 50 grams, whereas a male snake of the same age from Grapevine Hills weighed about 125 grams. Furthermore, snakes from Boquillas stopped growing when they reached sexual maturity, whereas snakes at Grapevine Hills continued to grow.

Environmental Variables and Performance

In biology, performance means how well an organism carries out an activity—how fast it runs, how successfully it captures prey, how rapidly it grows, and so on. Both the physical and biological environments of an animal can affect performance in many ways, and the responses of amphibians and reptiles to these environments can be seen at the level of individuals, populations, and communities (e.g., Congdon 1989, Dunham et al. 1989).

Temperature

As we noted in Chapter 1, temperature exerts a profound effect on virtually everything an amphibian or reptile does, and the rates of many physiological processes reach maxima within a species' activity temperature range. Ectotherms cannot always maintain body temperatures within the activity range,

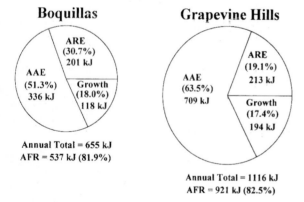

Figure 6–21 **Annual energy budget.** Estimated annual energy budgets for a 100-g adult male *Crotalus lepidus* from Boquillas and Grapevine Hills. Key: ARE = annual resting metabolism, AAR = annual activity metabolism, AFR = annual field metabolism. (AFR = ARE + AAR). The size of the circle represents the total annual energy budget for a snake at each site. (*Source: Beaupre 1996.*)

however, and at times they must function with body temperatures above or below the optima. What effects do nonoptimal temperatures have on performance?

Consequences of Thermoregulation

In general, capacity for performance initially increases as body temperature rises, reaches a maximum, and then declines (Figure 6–22). A teiid lizard, *Ameiva festiva*, studied in Costa Rica by Fredrica van Berkum provides an example of the effect of temperature on performance (van Berkum et al. 1986). An *A. festiva* basks in the sun, raising its body temperature to the upper set-point (average, 39.4°C), and then begins a bout of foraging during which it cools off as it moves between sun and shade (Figure 6–23). When its body temperature falls to the lower set-point (average, 34.5°C), the lizard stops foraging and basks again.

The sprint speed of Ameiva festiva is temperature sensitive, reaching a maximum of about 2.3 m·sec^{-1} at a body temperature of 37.5°C (Figure 6–24). The optimum temperature for sprinting is close to the mean body temperature of *Ameiva* in the field (35.9°C), and an *Ameiva* with a body temperature anywhere be-

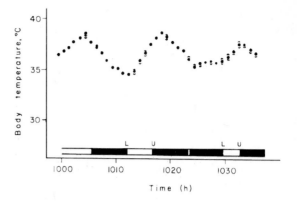

Figure 6–23 **Body temperature regulation by *Ameiva festiva*.** A lizard basks in the sun until its body temperature rises to the upper set-point and then forages in shade until body temperature falls to the lower set-point, when it resumes basking. (*Source: van Berkum et al. 1986.*)

tween the lower and upper set-points can achieve at least 90 percent of its maximum speed.

Speed may be most critical for *Ameiva festiva* when it is are escaping from a predator. During van Berkum's study an *Ameiva* was captured by a snake that overtook the running lizard in a 6-meter dash, and increasing the ability to escape from predators may be an important consequence of thermoregula-

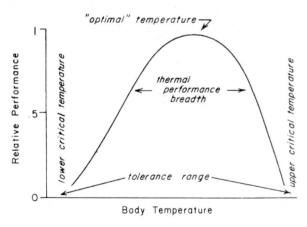

Figure 6–22 **Hypothetical performance curve of an ectotherm as a function of body temperature.** The vertical axis can represent any activity (e.g., sprint speed, rate of digestion). The optimum T$_b$ is the body temperature that maximizes that rate. Performance breadth describes the range of temperature over which the rate is an arbitrarily chosen percentage of the maximum rate (e. g., 80 percent). The tolerance zone is the range of temperatures outside of which performance falls to zero. (*Source: Huey and Stevenson 1979.*)

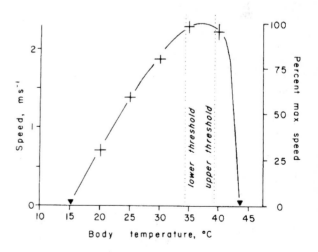

Figure 6–24 **Sprint speed of *Ameiva festiva* as a function of body temperature.** The lower and upper thresholds indicate the range of body temperatures within which lizards can run at 90 percent or more of their maximum speed. (*Source: van Berkum et al. 1986.*)

tion for many reptiles and amphibians. The sprint speed of Galápagos land iguanas (*Conolophus pallidus*) increases as body temperature rises from 15°C to 32°C, and is stable at higher temperatures. Galápagos hawks prey on juvenile iguanas (the adults are too large for the hawks to attack), and the ability of iguanas to escape is temperature dependent. Iguanas escape from only 33 percent of the attacks made when the body temperatures of the animals are below 32°C, whereas they escape from 81 percent of the attacks made when body temperatures are 32°C or above (Christian and Tracy 1981).

Synergism of Temperature and Water

Loss of body water can affect the performance of amphibians and reptiles. Amphibians, because of their permeable skins, are probably affected by dehydration more often than are reptiles. Many terrestrial amphibians undergo substantial changes in body water content on a daily basis, and these changes in hydration state affect locomotor performance. For example, dehydration reduces the jumping speed and endurance of leopard frogs (*Rana pipiens*). Fully hydrated frogs can travel nearly 35 meters and sustain activity for 2.5 minutes, whereas frogs that have lost 30 percent of their body water travel only 12 meters and become exhausted in just over 1 minute (Figure 6–25).

Water balance and body temperature interact to determine the locomotor capacity of American toads (*Bufo americanus*). The optimal temperature for loco-motion by American toads at 100 and 90 percent of full hydration is 30°C. At 80 percent of full hydration it falls to 25°C, and at 70 percent of full hydration it falls again, to 20°C (Figure 6–26). This interaction between temperature and hydration state adds an important dimension to studies of performance. It means that there is no one optimal body temperature for performance—the body temperature that produces most rapid locomotion depends on hydration state. Rather than regulating only body temperature or only water content, ectotherms are probably making the best compromises possible among a series of ecological and physiological forces.

Nest Environment

The temperature and moisture that reptile eggs experience in their nest can have long-term effects on the hatchlings that emerge. For many reptiles the sex of the adult is determined by the temperature of the embryo (temperature-dependent sex determination is discussed in Chapter 7). In addition, the size of hatchlings is affected by nest temperature and by the availability of water to the eggs during embryonic development.

The embryos of egg-laying reptiles metabolize yolk, using some of it as energy for metabolism and growth and converting the rest to fat. Temperature and moisture in the nest interact to determine how much yolk is metabolized and how the energy is divided between maintenance and growth (Packard and

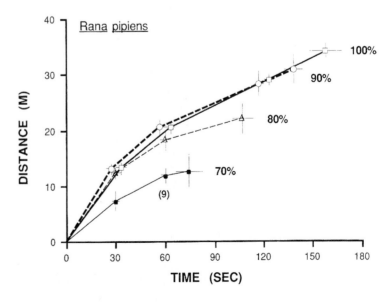

Figure 6–25 **Locomotor performance of *Rana pipiens* as a function of hydration state.** Frogs were tested at at 70, 80, 90, and 100 percent of standard mass. Horizontal lines show mean distance jumped and vertical lines show the 95 percent confidence intervals for the mean distances. The final point on each curve represents the mean distance and time to exhaustion, and the lengths of the horizontal lines at those points represent the 95 percent confidence intervals for time to exhaustion. (*Source: Moore and Gatten 1989.*)

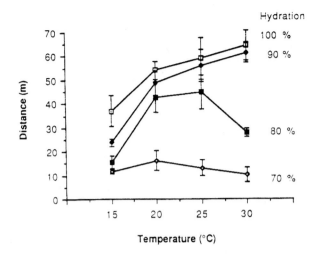

Figure 6–26 **Combined effects of temperature and hydration.** Simultaneous variation in body temperature and hydration state of locomotor performance of *Bufo americanus*. (a) Distance moved in 10 minutes of forced locomotion as a function of body temperature at four hydration states. Symbols represent mean values ± 1 standard error. *(Source: Preest and Pough 1989.)*

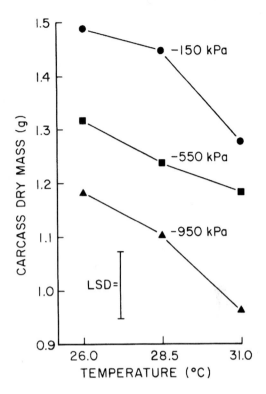

Figure 6–27 **Effect of incubation conditions on hatchling size.** Mean body masses of hatchling snapping turtles, *Chelydra serpentina*, incubated at different combinations of temperature and water potential. The least significant difference (LSD) is shown for alpha = 0.05; differences in means ≥ the LSD are statistically significant. The largest hatchlings are produced by eggs incubated in wet substrate (−150 kPa) at low temperature (26°C), and the smallest hatchlings emerge from eggs incubated in dry substrate (−950 kPa) at high temperature (31°C). Intermediate conditions produce hatchlings of intermediate sizes. *(Source: Packard and Packard 1988.)*

Packard 1988). Figure 6–27 shows the combined effects of these two variables on the sizes of hatchling snapping turtles (*Chelydra serpentina*). Water potential is expressed as negative kilopascals (−kPa); more negative numbers indicate drier substrate. At each temperature the wettest substrate (−150 kPa) produced the largest hatchlings and the driest substrate (−950 kPa) produced the smallest. Within each level of substrate moisture, the lowest incubation temperature (26°C) produced larger hatchlings than the highest temperature (31°C).

To test the effect of moisture during embryonic development on the performance capacity of hatchling turtles, snapping turtle eggs were taken from natural nests and incubated in the lab in wet (−150 kPa) or dry (−850 kPa) substrates at 29°C (Miller et al. 1987, Miller 1993). The wet substrate produced larger hatchlings—an average mass of 8.00 grams and an average carapace length of 27.0 millimeters for the wet substrate versus 6.69 grams and 24.6 millimeters for hatchlings from the dry substrate. This difference persisted after the turtles began to eat and grow (Figure 6–28). Furthermore, hatchling snapping turtles from wet substrates are able to run and swim faster than hatchlings from dry nests. By 50 days after hatching, turtles from wet substrates swim 23 percent faster than turtles from dry substrates (Figure 6–29).

Amphibian eggs in terrestrial nests are affected by the amount of water available in the nest. Eggs of the Puerto Rican coquí (*Eleutherodactylus coqui*) produce larger hatchlings in wet nests than dry nests. Coquís lay their eggs in fallen leaves and palm fronds. The gelatinous material in which the eggs are enclosed has a high water content when the clutch is laid, but it loses water by evaporation during the 3-week period of embryonic development. Male coquís, like adults of other species of anurans with terrestrial eggs, attend their eggs during development, and male coquís apparently add water to the egg mass, either by transfer-

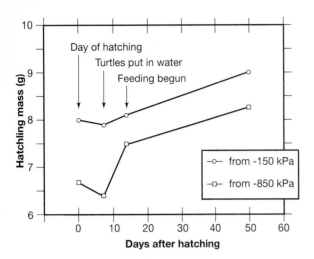

Figure 6–28 **Long-term effects of incubation conditions.** Mass of hatchling snapping turtles (*Chelydra serpentina*) hatchling from eggs incubated in wet (–150 kPa) and dry (–850 kPa) substrates. The size difference persists after hatching. *(Source: Miller 1993.)*

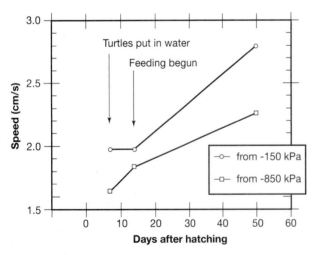

Figure 6–29 **Effect of incubation conditions on locomotor performance.** Swimming speeds of snapping turtles, *Chelydra serpentina*, hatched from wet and dry substrates. The turtles from the wet substrate swim faster than those from the dry substrate, and the difference is maintained after hatching. *(Source: Miller 1993.)*

ing water by omsmotic flow through the male's skin or by urinating on the eggs (Taigen et al. 1984). A male coquí should be able to ensure that his eggs produce large hatchlings by assiduously keeping the egg jelly moist.

Performance and Fitness

Variation in physiological capacity among individuals may determine how well they are able to perform certain activities. If inherited aspects of performance determine an individual's chances of survival or reproducing, physiological capacity might play a role in determining the evolutionary fitness of one individual relative to another. The evolutionary consequences of individual variation in performance capacity have become a major focus of studies of the physiological ecology of amphibians and reptiles (Bennett 1987). Actual measurements of individual fitness are usually impossible to make in wild populations, so mating success, growth rate, survival, speed, endurance, and a variety of other short-term phenomena have been used to estimate fitness.

Two conceptual models have been used to study the relationships between individual variation in performance and fitness. The first method, the gene-to-performance approach, begins by identifying genetic variation and then looking for effects of that variation on performance and fitness. The second method, performance-outward studies, works in the opposite direction. It starts with an observation of individual variation in performance and looks for both the physiological and genetic basis of the variation and for its fitness consequences (Arnold 1987).

Gene-to-Performance Studies

We do not yet have any gene-to-performance studies of amphibians or reptiles. The value of this approach is illustrated by elegant studies of sulfur butterflies (*Colias*) that have demonstrated mechanistic links among specific variants of the enzyme phosphoglucoisomerase, the ability to fly at certain temperatures, and mating success (Watt et al. 1986). Male sulfur butterflies fly in search of females, and males that carry the thermally favored form of the enzyme are able to fly more than males with other forms of the enzyme. Furthermore, the thermally favored males are disproportionately successful in siring broods of young. At Gunnison, Colorado, for example, thermally favored males made up 40.5 percent of the population, but these males sired 58.0 percent of the broods produced by females at this site.

A gene-to-performance study for amphibians might begin with identification of variation at the genetic locus responsible for the enzymes citrate synthase and β-hydroxyacyl co-A dehydrogenase. These enzymes are involved in the metabolism of glycogen and lipid, respectively, and high activities of these enzymes correlate with high calling rates for male frogs. Some species of frogs with long breeding seasons call at low body temperatures in early spring and at much higher temperatures later in the season. Spring peepers (*P. crucifer*), for example, call at body temperatures from 4 to 23°C (Wells and Bevier 1997). If different forms of the enzymes have different temperature sensitivities, it would be possible to analyze calling and mating success to see if males with the thermally favored forms of the enzymes were disproportionately successful in attracting females.

Performance-Outward Studies

The performance-outward approach has played a prominent role in studies of amphibians and reptiles, and examples are cited in Bennett (1987), Pough (1989), Bennett and Huey (1990), Garland and Carter (1994), and Garland and Losos (1994). No one study has yet made all the links between heritable variation in performance and variation in fitness, but many of the intermediate steps have been defined.

Repeatable variation in performance among individuals has been demonstrated for a variety of activities. Locomotor capacity has frequently been studied because it is relatively easy to measure and can be plausibly related to the ability of an animal to capture prey or to escape its own predators. The repeatability of sprint speed or endurance is usually high when subsequent measurements are made within a few days of the first measurement. Repeatability of locomotor performance is high for squamate reptiles, and lower—but still statistically significant—for amphibians (Austin and Shaffer 1992). Repeatability usually declines when the period between the initial measurement and subsequent measurements is lengthened, but canyon lizards (*Sceloporus merriami*) from two populations showed repeatable individual variation in sprint speed after a year's interval (Huey et al. 1990).

We know that variation in locomotor performance can be produced by environmental variables, such as the effect of moisture in the nest on snapping turtle hatchlings. Environmentally induced variation is not heritable—the offspring of a snapping turtle that was speedy because it developed in a wet nest will not necessarily be good sprinters. Does the individual variation in sprint speed seen in *Sceloporus* have a genetic basis? Probably it does, at least in part. That conclusion is supported by the observation that hatchling western fence lizards (*Sceloporus occidentalis*) from 13 different clutches of eggs incubated under the same conditions had mean sprint speeds ranging from 57.4 to 115.8 cm·sec^{-1}, and hatchlings from the same clutch were more similar to each other in sprint speed than they were to hatchlings from different clutches (van Berkum and Tsuji 1987).

In some cases individual variation in performance has been correlated with variation in morphological and physiological characters. The endurance of 17 black iguanas (*Ctenosaura similis*) walking at 1 km·h^{-1} ranged from 3.7 to 35.8 minutes, and the lizards traveled total distances of 25 to 103 meters before they were exhausted (Garland 1984). Body size has an effect on both of these measurements. Larger lizards were able to walk longer and farther than smaller ones, but lizards that were very similar in size differed by sixfold in endurance and by more than twofold in distance traveled (Figure 6–30). This individual variation in locomotor performance is correlated with individual variation in several physiological and biochemical characteristics. A statistical model that incorporates individual variation in thigh muscle mass, heart mass, maximum rate of oxygen consumption, and citrate synthase activity in the liver accounted for 89 percent of the individual variation in how long locomotion could be sustained.

Interpretation of the evolutionary significance of these correlations between physiological characteristics and performance remains speculative because it is hard to demonstrate a cause-and-effect relationship between individual variation in performance and individual variation in fitness. For example, individual variation in the sprint speed of canyon lizards did not correlate with survivorship over a 5-year period (Bennett and Huey 1990), and variation in aerobic capacity of male toads (*Bufo americanus*) was not correlated

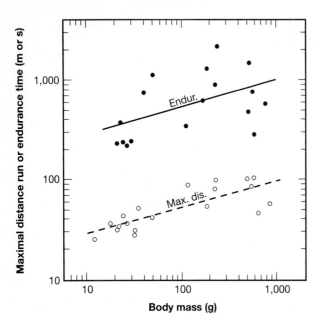

Figure 6–30   **Individual variation in locomotor performance.** Ctenosaurs (*Ctenosaura similis*) varied nearly 10-fold in the distance they could walk at 1 km·h⁻¹ (called endurance, and shown by closed circles) and 4-fold in the maximum distance, they could run on a circular track (called maximum distance and shown by open circles). Body size accounted for 24 percent of the variation in endurance and for 64 percent of the variation in maximum distance. Differences among individuals in the physiological variables described in the text accounted for an additional 65 percent of the variation in endurance and 34 percent of the variation in maximum distance. *(Source: Garland 1984.)*

with variation in the distance a toad moved in the chorus or the number of attempts it made to clasp females (Wells and Taigen 1984).

Trade-offs

Performance-outward studies assume that high rates in behavioral or physiological measures of performance are advantageous to individual animals. Probably that assumption is often correct, but it is not always true. The environment in which animals function is not constant, and year-to-year differences may reverse selective gradients. For example, individuals of a small Australian lizard, *Amphibolurus ornatus*, show extreme individual variation in growth rate (Bradshaw 1986). About 50 percent of the individuals in the population at Baker's Hill, Western Australia, grow rapidly and reach sexual maturity in 9 months,

40 percent of the lizards grow more slowly and reach maturity in their second year, and 10 percent do not mature until they are 37 months old. The fast-growing individuals are potentially able to produce offspring in their first year of life, whereas the slower-growing individuals do not reproduce until they are 2 or 3 years old.

If early reproduction increases fitness, fast-growing lizards should dominate the population, but fast- and slow-growing individuals are about equally represented. Differential mortality accounts for the balance between lizards with fast and slow growth rates. Fast-growing lizards are particularly sensitive to hot, dry summers, apparently because they have difficulty regulating salt and water balance. Rapid growth may be associated with enhanced secretion of growth hormone and prolactin by the pituitary gland, and high rates of synthesis of these hormones may reduce the ability of the pituitary to secrete the hormones required for salt and water regulation, such as adrenocorticotropic hormone and arginine vasotosin. If that is the case, fast-growing lizards have made an evolutionary trade-off that is advantageous in wet years and deleterious during droughts.

Slow-growing lizards survive hot, dry summers, but they have a higher mortality than fast-growing lizards on frosty winter nights. To survive cold nights, lizards must seek refuge in shelters. The number of shelters is limited, and fast-growing lizards, because they are larger, displace the smaller slow-growing lizards from shelters.

Thus, it is plausible that individual variation in physiological or behavioral performance can be a cause of individual variation in fitness, but it is hard to demonstrate cause-and-effect relationships because so many nongenetic factors can affect performance. To complicate analyses even more, synergistic effects may appear when two or more variables act simultaneously, and the selective regime may change from year to year. Performance characteristics that increase fitness in one year may be deleterious the following year. We may discover that relationships of performance to fitness are hard to generalize beyond the situation that was studied.

Summary

The ectothermy of amphibians and reptiles has consequences that can be seen in many facets of their daily lives. Temperature has a pervasive effect on everything an animal does. In general, the rates of biochemical, physiological, and behavioral processes initially increase as body temperature rises, reach a plateau of relative insensity to temperature, and decline at higher body temperatures. Thermoregulatory mechanisms allow an animal to remain within the range of temperatures at which most processes are at or near their maximum rates.

The benefits of thermoregulation are as diverse as increased ability to escape from predators and increased growth rates, but animals cannot always find ideal thermal environments. Environments that are too cold or too hot can limit growth rates by limiting annual food intake and by diverting energy from growth to maintenance.

Amphibians and reptiles cannot synthesize ATP via oxidative metabolic pathways (aerobic metabolism) fast enough to support high levels of muscular activity. Instead, most amphibians and reptiles rely on glycolytic metabolic pathways (anaerobic metabolism) that convert glycogen in the cells to lactic acid. Glycolysis is an effective way to synthesize ATP for brief bursts of activity, and the total ATP available to a lizard in the first 30 seconds of activity is slightly greater than that available to a mammal. Thus, a lizard can sprint to escape from a predator or fight to defend a territory.

Glycolysis cannot sustain high rates of ATP synthesis indefinitely. Five minutes of intense activity depletes glycogen in the muscle cells and leaves an animal nearly inert. Short periods of activity use only part of the glycogen store and do not exhaust an animal. Thus, the activity of most amphibians and reptiles is episodic—bursts of activity alternate with periods of rest.

A few amphibians and reptiles have biochemical and physiological specializations that do permit high levels of activity. Vocalization by male anurans is the most dramatic example of sustained high oxidative metabolic rates. The trunk muscles used for vocalization increase in size during the breeding season. Lipids are the primary metabolic substrates for species of anurans that call at high rates, and depletion of energy reserves can limit the number of days an individual male anuran spends in the breeding chorus.

Time in the breeding chorus is the best predictor of number of matings for a male anuran, and physiological characteristics that allow a male frog to remain in the chorus for many days should increase his Darwinian fitness. The link between performance capacity and fitness is currently a subject of intense study, and work with amphibians and reptiles has demonstrated statisitically significant associations between genetic and physiological characteristics and behavioral measures of perfomance, such as the ability to run fast or far. The final step in the process, demonstrating that speed or endurance is positively correlated with a measure of fitness such as reproductive success, growth rate, or survival, has so far been elusive.

CHAPTER

7

Reproduction and Life History

Many amphibians and reptiles are largely solitary in their habits and seldom encounter, let alone interact with, other members of their species. Others are highly social and interact intensively with other members of their species on a daily basis, even to the extent of recognizing their neighbors as individuals. Regardless of the extent of day-to-day social interaction, the annual cycles of amphibians and reptiles typically involve one specific social interaction between males and females, and that is mating. Mating can be viewed as the pivotal event of reproduction. Physiological, morphological, and social events that lead up to mating include the maturation of gonads, the development of secondary sexual structures, and courtship. Surprisingly, even asexual species engage in mating behaviors. Although mating is often associated with high levels of gonadal activity, the act of mating for some species occurs at a time when the levels of sex steroids are low and the gonads are undeveloped. In this situation, mating is not followed immediately by fertilization, and sperm must be stored for later use. The events that follow mating include fertilization (or the formation of an asexually generated egg) and embryonic development. Postmating events may also include parental care of the eggs, and perhaps care of the offspring as well. Parental care may also include viviparity. The annual cycle and the life cycle make up the life history of a species, and important life history characteristics are directly or indirectly related to reproduction. For example, the total volume or mass of a clutch is a component of reproductive effort, and the amount of reproductive effort is negatively related to adult survival.

This chapter takes up the interrelated themes of reproduction and life history. For many aspects of reproduction and life history, amphibians and reptiles exhibit very similar patterns. For others, their biology is so different that common patterns are not at all evident. This latter disparity comes about because amphibians are nonamniotes and reptiles are amniotes. This fundamental difference in the eggs of amphibians and reptiles has important consequences for their reproduction and life histories.

Sexual and Asexual Reproduction

Most amphibians and reptiles reproduce sexually, and virtually all individuals are either female or male. (This condition stands in contrast to hermaphroditic species, in which a single individual may be both female and male, either simultaneously or sequentially.) Each sex produces a distinctive gamete, a cell that contains a haploid chromosome complement. Fusion of the female gamete, the ovum or egg, and the male gamete, the sperm, forms the zygote, a cell with the full diploid complement of chromosomes. Because of the recombination of genes during meiosis, offspring are genetically variable and unlike their parents.

Despite the seeming advantage of sexual reproduction, asexual reproduction occurs sporadically among fishes, amphibians, and reptiles. Amphibians and reptiles that exhibit asexual reproduction share three features: they usually originate from hybridization of two species, their populations are usually all

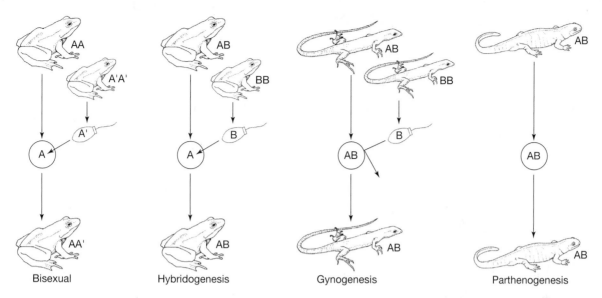

Bisexual   Hybridogenesis   Gynogenesis   Parthenogenesis

Figure 7–1   **Sexual and asexual reproduction.** Each capital letter represents a haploid complement of chromosomes. In hybridogenesis, females transmit only the maternal genome (A) into gametes. In gynogenesis, sperm is required to initiate development, but not for fertilization. *(Source: Modified from Zug 1993.)*

female, and reproduction is via clonal inheritance (Dawley and Bogart 1989).

Asexual reproduction by reptiles is called parthenogenesis. Parthenogenetic females produce diploid eggs from unreduced (2n) gametes, and all resultant individuals are genetically identical and intermediate in characteristics between the two parental species (Figure 7–1). Parthenogenesis has been reported in about 30 species of squamates, mostly gekkonids, lacertids, and teiids. The best-studied examples are teiids in the genus *Cnemidophorus* of the southwestern United States and Mexico, in which nearly a third of the more than 45 species are parthenogenetic (Wright 1993). Some parthenogenetic species are diploid, and genetic analyses allow identification of the bisexual species that hybridized to form the unisexual species. Matings between a diploid unisexual and males of one of the parental bisexual species, or a third species, have produced triploid lineages as well. Parthenogenetic lineages of *Cnemidophorus* are of recent origin, on the order of hundreds of years. This matches the theoretical prediction that, because of the potential accumulation of deleterious mutations, asexual species will have shorter lifetimes than sexual species.

The absence of fertilization does not mean that sexual behavior does not occur. Female *Cnemidophorus* engage in pseudocopulation, a behavior in which one individual plays the role of the male and the other the female. Courtship and copulatory behavior of parthenogenetic *Cnemidophorus* are almost identical to those of closely related bisexual species (Figure 7–2).

Figure 7–2   **Pseudocopulation by captive *Cnemidophorus uniparens*.** Both individuals are females. *(Photograph by David Crews.)*

The benefit of this behavior has been demonstrated in the laboratory. Females that engage in courtship and pseudocopulation produce more eggs than females held in isolation. The continuation of sexual behavior by asexual species is consistent with the observation that courtship behavior of bisexual species of vertebrates enhances gonadal activity (Crews and Moore 1993).

The amphibian parallel to parthenogenesis is hybridogenesis. In this process, the genome of the female parent is passed unchanged from one generation to the next. Hybrid females mate with males of one of the parental species and produce individuals with an intermediate phenotype. However, the paternal genome is discarded at the time of gametogenesis, and the gametes produced by the hybrid female contain only the maternal genome. This means that hybridogenetic genotypes are reconstituted in each generation and the maternal genome is inherited clonally. The males that mate with either hybrid females or nonhybrid females of another species in a hybridogenetic complex do not pass their genes on to succeeding generations.

Hybridogenetic reproduction by frogs of the *Rana esculenta* complex has been studied extensively. *Rana esculenta*, the common edible frog of Europe, was first described by Linnaeus in the eighteenth century, but it was not until the 1960s that Leszek Berger (1977) discovered that populations of this frog in Poland are actually hybrids derived from crosses between the pool frog (*Rana lessonae*) and the marsh frog (*Rana ridibunda*). In contrast to parthenogenetic lizards, both male and female hybrid frogs are produced. Populations of hybrids can occur with both parental species, but this situation is rare because the ecology of the parental species differs. More commonly hybrids are found in association with one parental species, with which they interbreed. Experimental crosses demonstrated that matings between *R. lessonae* and *R. esculenta* always produce an *esculenta* phenotype, whereas matings between *R. ridibunda* and *R. esculenta* always produce the *ridibunda* phenotype.

The interpretation of these experimental crosses is that the *lessonae* genome is discarded before meiosis when *R. esculenta* females form eggs. The *ridibunda* genome that remains is then duplicated. During normal meiosis, haploid eggs are formed that, when combined with *R. lessonae* sperm, reconstitute the *R. esculenta* genome. If females mate with *R. ridibunda* males, the fertilized eggs end up with two *ridibunda* genomes

and a *ridibunda* genotype and phenotype. This system means that the *ridibunda* genome from *R. esculenta* females is passed from generation to generation unchanged (clonal inheritance). In a few populations where *R. esculenta* is associated with *R. ridibunda*, it is the *lessonae* genome that is retained and the *ridibunda* genome that is renewed in each generation. To further complicate the picture, some *R. esculenta* females produce both haploid and diploid eggs, and subsequent matings with one of the parental species yield some triploid offspring with a double dose of one genome.

A similar system occurs in populations of salamanders in the *Ambystoma jeffersonianum* complex in North America. Matings between Jefferson's salamander (*A. jeffersonianum*) and the blue-spotted salamander (*A. laterale*) produce viable hybrids. Indeed, many populations consist largely of hybrids associated with a few individuals of one parental species. Normally only female hybrids are formed, and they depend on males of the parental species for reproduction. Early studies suggested that these salamanders reproduced by gynogenesis, a process in which females produce diploid eggs that are stimulated to develop by contact with sperm but the genome of the sperm is not incorporated into the egg. More recent studies have shown that the reproductive system of at least some species is more like the hybridogenetic systems of water frogs, although some details differ, and this interpretation is controversial (Dawley and Bogart 1989).

Reproductive Cycles

Successful reproduction by an amphibian or reptile requires the coordination of many different internal processes and external events. Obviously, males and females must be physiologically and behaviorally ready for mating at the same time. Environmental conditions must also be suitable for the successful development of embryos and neonates. A complex network of neural and hormonal mechanisms integrates internal processes and environmental conditions (Figure 7–3). Reviews of reproductive cycles of amphibians and reptiles can be found in F. L. Moore (1987), Whittier and Tokarz (1992), M. C. Moore and Lindzey (1992), and Houck and Woodley (1995).

External factors act as cues for the induction and release of reproductive hormones and for alterations in the responsiveness of target organs to the presence

Figure 7–3 **Control of reproduction.** Some of the reciprocal interactions of external and internal factors that influence reproduction by amphibians and reptiles. The hypothalamic–pituitary–gonadal axis is shown in the center of the diagram. Environmental stimuli include abiotic factors (light, heat, moisture), the physical and biological environments (space, habitat, food), and the social environment (behavior, population density, position of an individual in the social structure of the population). These stimuli are transmitted as nerve impulses and are integrated largely in the hypothalamus. The hypothalamus produces gonadotropin-releasing hormone (GnRH), which causes the pituitary to release follicle-stimulating hormone (FSH) and luteinizing hormone (LH). These substances in turn stimulate the gonads to produce eggs and sperm and to release sex steroids (estrogen and testosterone). The sex steroids affect the development of secondary sexual characters and reproductive behavior *(Source: Houck and Woodley 1995.)*

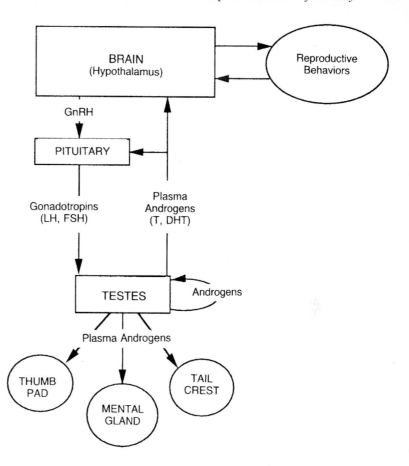

of these hormones. Because of their predictable seasonal changes, temperature and photoperiod play important roles in the timing of reproductive cycles. Rising temperatures and increasing day length, for example, stimulate gonadal activity, especially among temperate zone species. Rainfall acts as the most proximate stimulus for the breeding behavior of amphibians in both temperate and tropical zones and for oviposition by some squamates.

Amphibians and reptiles exhibit three general types of reproductive cycles: associated, dissociated, and continuous (Figure 7–4). Associated cycles are the most common and are characteristic of species living in predictable environments with moderately long activity seasons. In these species, the maturation and shedding of gametes and the secretion of sex steroid hormones are associated with mating and fertilization. Spermatogenesis and oogenesis are initiated more or

less simultaneously, and high levels of circulating sex hormones are associated with mating and fertilization. Most amphibians and reptiles exhibit this pattern. The green anole *Anolis carolinensis* of the southeastern United States provides a good example of an associated reproductive cycle (Crews 1980). Growth of the gonads occurs in the spring, mating and egg production occur over the summer, and gonads of both sexes regress almost simultaneously in the fall.

Dissociated cycles are characteristic of species living in habitats in which the breeding season is brief and the best time for mating may not be the best time for gonadal activity or for the production of young. Red-sided garter snakes (*Thamnophis sirtalis parietalis*) in Canada provide a well-documented example of a dissociated reproductive cycle (Crews and Garstka 1982). The snakes emerge from hibernation in May and return to their dens in September. Mating occurs

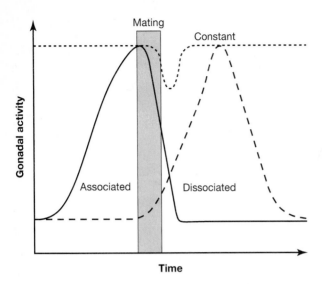

Figure 7–4   **Three reproductive patterns of vertebrates.** In species with associated cycles, gonadal activity increases shortly before mating. In species with dissociated cycles, gonadal activity is low when mating occurs and increases at other times of the year. In species with constant reproduction, gonadal activity is maintained at or near maximum levels most of the year *(Source: Whittier and Crews 1987.)*

when the snakes emerge in spring. At this time, the testes and ovaries are small and the level of sex steroids in the blood is low. After mating, the snakes leave the den site and move to foraging areas. During the next 3 months, testosterone levels of males are high and sperm are produced in the testes. These sperm are stored in the ductus deferens, to be used for mating the following May. Gonadal maturation of females also occurs during the summer, and the eggs are fertilized by stored sperm. For red-sided garter snakes, a dissociated mating cycle allows individuals to mate at the time when males and females are together at denning sites and to use the rest of their brief activity period for feeding, gonadal maturation, and gestation.

Dissociated reproductive cycles require that sperm be stored until fertilization. Among snakes, where mating often precedes fertilization, sperm are stored in the ductus deferens of males and in crypts in the anteriormost part of the oviduct (infundibulum) of females (Saint Girons 1985). Fertilization occurs when ova enter the oviduct and pass through the infundibulum. Female lizards store sperm at some distance from the spot where fertilization occurs. For example, *Anolis carolinensis* stores sperm in tubules in the lower part of the oviduct (Conner and Crews 1980). Because the eggshell has formed by the time the egg reaches these tubules, sperm must move out of the tubules and up the oviduct to reach unfertilized ova.

Continuous reproductive cycles are exhibited by some anurans and squamates that live in aseasonal tropical habitats. Gonadal activity is high nearly year-round or high during a lengthy period, typically the wet season. For example, male salamanders in Central America have sperm in the ductus deferens and in the testes in all months of the year. Presumably, courtship and mating occur year-round as well. Despite the continuous reproductive activity of males, however, the reproductive cycle of female salamanders at some localities is seasonal. For example, at one site in Guatemala, females of most species laid eggs in November or December and brooded their eggs until May or June, when hatching occurred (Houck 1977).

For species that produce more than one clutch per reproductive season, each clutch represents a distinct cycle of vitellogenesis, mating, ovulation, and egg laying. For example, female *Anolis carolinensis* produce single eggs at intervals of 7 to 14 days during the summer. The reproductive behavior of the female is correlated with the stage of development of the next egg that will be produced. A female *A. carolinensis* becomes receptive to courtship by a male when a developing egg follicle in the ovary reaches a diameter of between 3.5 and 6 millimeters, and she remains receptive for about a week while the egg grows to a diameter of 8 millimeters. At this point the egg is released from the ovary (ovulated), and the female stops responding to courting males. She remains unreceptive to males for several days, during which time she deposits the egg, and she becomes receptive again when the next egg follicle reaches a diameter of about 3.5 millimeters (Crews 1980). All members of the genus *Anolis* have a clutch of a single egg, and all geckos have a clutch of either one or two eggs. The small clutch sizes of these arboreal lizards may reflect a limit to the load-bearing capacity of females (Andrews and Rand 1974). Anoles and geckos have adhesive toe pads that enable them to climb on smooth surfaces, but that may limit the amount of extra load that a female can carry and still climb effectively.

In contrast to species that reproduce at least on an annual basis, females of many species of snakes exhibit reproductive cycles of 2 years or more (Saint Girons

1985). Such extended cycles are related to the ability of female to accumulate energy for reproduction. Stored lipid is used to synthesize yolk, and reproduction is inhibited when the lipid stores are low or depleted. For example, some species of rattlesnakes (*Crotalus*) reproduce every second or third year in the colder parts of their geographic ranges and more frequently in warmer areas where the activity season is longer and a female is able to replenish her lipid stores faster.

Gametes, Fertilization, and Development

Amphibians and reptiles have fundamentally different eggs (Figure 7–5), and this difference affects their ecology and reproductive biology. Amphibians produce jelly-covered eggs. These eggs are typically laid in water or in moist sites on land. The only embryonic membrane in an amphibian egg is the yolk sac. In contrast, reptiles produce amniotic eggs. An amniotic egg has a leathery or calcareous shell that makes the egg resistant to water loss. Some reptile eggs take up water during development (most squamates and some turtles), and others develop normally with only the water initially in the egg (crocodilians and some turtles). The shell provides mechanical protection to the embryo while allowing movement of respiratory gases and water vapor. The amniotic egg has three unique membranes that grow outward from the embryo. The amnion is the innermost membrane and surrounds the embryo. The chorion is the outermost, and eventually expands to cover the entire inside of the egg. The allantois forms as an outgrowth from the rear of the gut. It expands to lie beneath the chorion and fuses to it. This combined chorioallantoic membrane provides a large surface for gas exchange through the overlying shell, and the allantois itself serves as a storage site for uric acid or urea, the waste products formed from the metabolic breakdown of protein.

Despite the differences in the structure of the egg itself, the basic processes of oogenesis and spermatogenesis are typical of vertebrates in general. Ova are provided with yolk , a nutrient substance consisting of lipoproteins, phosphorylated proteins, and glycogen. Yolk is synthesized in the liver (vitellogenesis) and transported by the circulatory system to the maturing ovum. Ova may be characterized by the presence of little, moderate, or large quantities of yolk (Blackburn 1992). A species in which the embryonic energy source consists largely of yolk is said to be lecithotrophic. In oviparous (egg-laying) species, yolk is the only source of nutrition available to the developing embryo. In some viviparous (live-bearing) species, the female may replace yolk almost completely by nutrients transferred to the embryo during its development. This process may involve the transport of nutrients across a placenta (placentotrophy),

Figure 7–5 **Eggs of amphibians and reptiles.** Note the jelly layers surrounding the amphibian egg and the tough shell of the reptilian egg. In addition to the yolk sac, the reptilian embryos have three additional extraembryonic membranes.

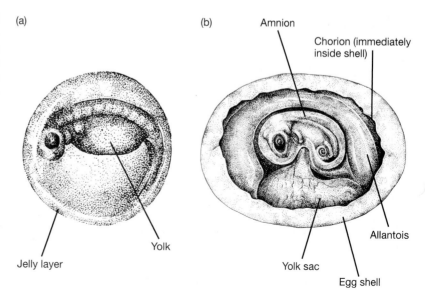

(a)

(b)

Amnion

Chorion (immediately inside shell)

Allantois

Jelly layer

Yolk

Yolk sac

Egg shell

and other mechanisms may be involved, as discussed in the following section on viviparity.

As in most vertebrates, ova are released at the surface of the ovary and enter one of the paired oviducts by way of its anterior opening, the ostium. The oviduct contains glands that, in amphibians, produce several layers of jelly-like proteins that surround each egg and, in reptiles, produce albumin and the protein fibers and calcium compounds that form the shell. In amphibians, a region of the oviduct (ovisac) is enlarged for storage of the ova prior to their release (oviposition). Similarly, in reptiles a wide region of the oviduct (uterus) accommodates the eggs of oviparous species prior to oviposition and supports gestation in viviparous species.

External fertilization is the ancestral condition among amphibians. Internal fertilization has evolved independently in the caecilians, some salamanders, and a few species of frogs. Each of these groups, however, has a different mechanism by which the sperm are introduced to the reproductive tract of the female. Male caecilians possess a copulatory organ, the phallodeum, which is formed from the cloacal wall. Salamanders other than cryptobranchids, hynobiids, and sirenids have internal fertilization but no intromittent organs. Instead, males produce spermatophores, packets of spermatozoa atop a proteinaceous base, which they attach to the substrate. The female picks up a spermatophore with the lips of her cloaca, usually as a part of courtship, and stores the spermatozoa in a spermotheca in the roof of the cloaca. Among anurans, only the tailed frog, *Ascaphus*, possesses an intromittent organ. In other anurans, such as the Puerto Rican frog *Eleutherodactylus coqui*, internal fertilization is accomplished by cloacal apposition (Townsend et al. 1981).

All reptiles have internal fertilization because eggs must be fertilized before the shell forms. Male turtles and crocodilians have a penis that lies in the floor of the cloaca. In contrast, a penis apparently was absent from the lineage leading to extant lepidosauromorphs. *Sphenodon*, the sister group of squamates, lacks an intromittent organ. Internal fertilization is accomplished by cloacal apposition. Squamates have evolved unique intromittent organs, the paired hemipenes that develop as paired evaginations from the rear wall of the cloaca. When engorged with blood, a hemipenis is everted, turning inside out and exposing a surface ornamented with folds, papillae, or calcified spines (Dowling and Savage 1960). Only one

hemipenis is used during copulation. Male *Anolis carolinensis* tend to use alternate hemipenes in sequential matings (Crews 1980). Because the two reproductive tracts of males are independent, alternation of the hemipenes may ensure an adequate number of sperm for each mating. Less explicable, however, is the observation that the male uses his right hemipenis if the ovum ready to be fertilized is on the right size of the female, and vice versa. Which sex controls this association is unknown.

Most of the process of differentiation, the commitment of cells to become specific tissues, and a considerable increase in the size of the embryo, takes place during embryogenesis. Embryonic development is typically initiated at fertilization and continues until hatching or birth. Sometimes a considerable amount of development occurs while the eggs are still in the oviduct. Most squamates lay eggs when 25 to 50 percent of their development has been completed, but turtles and crocodilians lay eggs at very early stages of development (Shine 1983a, DeMarco 1993).

Hatching or birth is an important event during an individual's ontogeny, but development continues beyond that point. For amphibians, a period of larval development is interposed between embryogenesis and metamorphosis to the adult form. Because amphibian eggs are relatively small compared with those of reptiles, larvae form a bridge between the embryo and the adult form. For reptiles, the neonate is a fully formed but miniature edition of the adult at hatching or birth.

Sex Determination

In humans as in other mammals, the sex of an individual is established at the moment of conception by the particular complement of genes received from its parents. Such a system is known as genotypic sex determination, or GSD. In mammals, the male is heterogametic, that is, the sex with different sex chromosomes, an X and a Y. In birds, on the other hand, the female is the heterogametic sex. In this case, by convention, the female is designated ZW and the male ZZ. The sex chromosomes in such systems often, but not always, differ from each other in gross size, shape, or staining properties. Although XY and ZW are the most common forms of heterogamety, they are not the only ones. Some species exhibit male heterogamety in which the chromosome complement is XO (the male lacks one sex chromosome) or XXY, whereas some exhibit ZZW or ZWW systems of female heterogamety. Both male and female heterogamety occur in amphibians and reptiles (Table 7–1). Female heterogamety is the ancestral condition for frogs and salamanders, and male heterogamety has evolved at least seven times, with only one subsequent reversal from male to female heterogamety (Hillis and Green 1990). The species of lizards examined thus far exhibit XY, XXY, or ZW systems in roughly equal numbers. Most snakes exhibit a ZW system of heterogamety, as do varanid lizards (Janzen and Paukstis 1991).

From our mammalian perspective, GSD seems not only normal but necessary. This perspective is misleading because the sex of some lizards, most turtles, all crocodilians, and sphenodontians is determined by the temperature of the nest, not by genotype. At some temperatures only males are produced and at others only females. The range of temperatures that produce both sexes is small, typically about 1°C. Within this range, the temperature that produces 50 percent of each sex is called the pivotal temperature (Figure 7–6).

Three major patterns of temperature-dependent sex determination (TSD) have been reported (Bull 1983). Most turtles exhibit pattern Ia, in which males are produced at cooler temperatures and females at warmer ones (Figure 7–7). In pattern Ib the opposite relationship obtains: females are produced at cooler temperatures and males at warmer ones. This pattern is found in a number of species of lizards. Pattern II takes a more complex form, with females produced at the coolest temperatures, males at intermediate temperatures, and females again at the warmest ones. This pattern has been found in some species of turtles and crocodiles as well as in alligators and several species of lizards previously thought to exhibit pattern

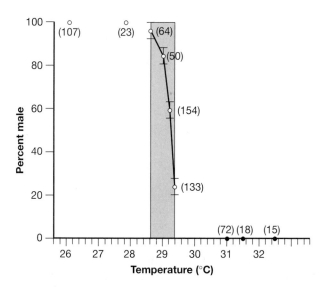

Figure 7–6 **Relationship between incubation temperature and sex ratio in the red-eared slider turtle (*Trachemys scripta*).** Values are the mean sex ratios obtained over a 4-year period. The total number of individuals is shown in parentheses and the standard error is indicated by the vertical bar. Gray area indicates the range of temperature that produces both sexes *(Source: Crews et al. 1994.)*

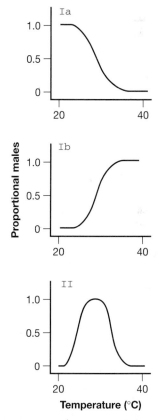

Figure 7–7 **Three patterns of response to incubation temperature in reptiles.** (a) Pattern Ia: males develop at low temperatures and females at high temperatures. (b) Pattern Ib: males develop at high temperatures and females at low temperatures. (c) Pattern II: males are produced at intermediate temperatures and females are produced at the coolest and the warmest temperatures *(Source: Modified from Bull 1983.)*

Table 7–1 **Distribution of genetic sex determination (GSD) and temperature-dependent sex determination (TSD) in amphibians and reptiles. Patterns are summarized for each family in which the condition is known. Data indicate only the occurrence of each pattern, not their relative proportions or distribution among lineages within a family. Only those families for which GSD or TSD have been studied are listed.**

Taxon	GSD			TSD
	Male Heterogamety	Female Heterogamety	Heteromorphy Absent	
Amphibia				
Urodela				
Ambystomatidae		ZW		
Plethodontidae	XY	ZW		
Proteidae	XY			
Salamandridae	XY	ZW		
Sirenidae		AW		
Anura				
Bombinatoridae	XY			
Bufonidae		ZW		
Discoglossidae		ZW		
Hylidae	XY			
Leiopelmatidae		ZW, OW		
Leptodactylidae	XY			
Pelodytidae	XY			
Pipidae		ZW		
Ranidae	XY	ZW		
Reptilia				
Chelonia				
Chelidae	XY		Yes	
Pelomedusidae				Yes
Bataguridae	XY	ZW		Yes
Carettochelyidae				Yes
Cheloniidae				Yes
Chelydridae				Yes
Dermatemydidae				Yes
Dermochelyidae				Yes
Emydidae			Yes	Yes
Kinosternidae				Yes
Staurotypidae	XY		Yes	
Testudinidae				Yes
Trionychidae			Yes	Yes
Crocodylia				
Alligatoridae				Yes
Crocodylidae				Yes
Gavialidae				Yes

(continued)

Table 7–1   *(continued)*

| Taxon | GSD | | | TSD |
	Male Heterogamety	Female Heterogamety	Heteromorphy Absent	
Sphenodontia				Yes
Squamata				
Iguanidae	XY, XXY		Yes	Yes
Agamidae				Yes
Gekkonidae	XY, XXY	ZW, ZZW	Yes	Yes
Pygopodidae	XY, XXY			
Lacertidae		ZW, ZZW	Yes	Yes
Teiidae	XY		Yes	
Scincidae	XY, XXY		Yes	
Amphisbaenia		ZW		
Varanidae		ZW		
Boidae		ZW		
Colubridae		ZW	Yes	
Elapidae		ZW, ZZW, ZWW	Yes	
Viperidae		ZW		

Source: Cree et al. 1995, Hillis and Green 1990, Janzen and Paukstis 1991, Lang and Andrews 1994, and Viets et al. 1994.

Ib (Lang and Andrews 1994, Viets et al. 1994). It remains possible that the assignment of other species to pattern Ib reflects experimental designs that did not fully test for sex determination at the highest viable temperatures. In a few pattern II species, such as *Crocodylus johnstoni* (Crocodylidae) and *Macroclemys temminckii* (Chelydridae), at least some males are produced at all temperatures (Ewert et al. 1994).

The processes of sexual differentiation of amphibians and reptiles with GSD are not well studied compared to those of mammals, but we assume that the basic processes are similar. In mammals, the gonads mature as ovaries unless a gene product encoded on the Y chromosome acts to cause their differentiation as testes. This process of gonadal differentiation is known as primary sexual differentiation. It is followed later by the development of the secondary sexual characteristics, such as copulatory organs and display structures. A gene located on the mammalian Y chromosome and known as Sex-Determining Region Y is implicated as the testis-determining factor in the mammalian genome. Its expression is limited to the gonad-forming tissues of males at the time of differentiation of the testes (Wachtel and Tiersch 1994). Subsequent changes in the sexual phenotypes result from a cascade of hormone-mediated interactions between the gonads, endocrine glands, and the diverse target tissues.

For species with TSD, temperature sensitivity begins prior to differentiation of the gonads and extends to a time when their differentiation into testes and ovaries is evident. This period corresponds to the middle one third to one half of embryonic development (Wibbels et al. 1994). What is the mechanism by which temperature determines sex? A current model proposes that the activity of genes encoding enzymes that convert steroids from one form to another is temperature dependent (Crews et al. 1994). In this model, the steroid precursor of the gonad-differentiating hormones is testosterone (presumably deposited in the egg yolk by the mother). At temperatures that produce females, the enzyme aromatase is induced and converts testosterone to estradiol (an estrogen). Estradiol binds to estrogen receptors on the undifferentiated gonads, triggering their differentiation into ovaries, which in turn synthesize more estrogens. Estrogen production causes a positive feedback that induces additional aromatase production. The result is a cascade of events leading to differentiation of the remaining structures involved in female reproduction. At temperatures that produce males the enzyme 5α-reductase is induced and converts testosterone to dihydrotestosterone. Dihydrotestosterone binds to androgen receptors on the undifferentiated gonads and triggers the differentiation of testes, which in turn produce additional androgens. As in the female, these events feed back to increase reductase induction and result in a cascade of events leading to the differentiation of the structures involved in male reproduction. Experimental studies support the main components of this model, especially with regard to the effects of the enzyme aromatase (Wibbels et al. 1994).

The evolutionary origins of TSD are not clear. TSD may be the ancestral condition for reptiles, although the evidence is far from conclusive (Janzen and Paukstis 1991). If TSD is ancestral, its retention in many lineages would not be surprising. Several benefits of TSD have been proposed. TSD would be favored by natural selection if the fitness of either sex responded differently to incubation temperature—that is, if males and females were each more fit as a result of development at different temperatures (Charnov and Bull 1977). A recent study of the snapping turtle, *Chelydra serpentina*, tends to support this hypothesis. Fredric Janzen (1995) incubated eggs at three temperatures, 26°C (male-producing), 30°C (female-producing), and 28°C (male- and female-producing). He determined the sex of hatchlings by examination of their gonads, weighed and measured them, and tested swimming and running performance. After these initial observations, he released the hatchlings in an experimental pond. As predicted by the Charnov-Bull hypothesis, males incubated at 26°C and females incubated at 30°C had higher survivorship after 7 months than did either sex from the 28°C treatment (Figure 7–8).

Whatever the evolutionary basis for TSD, its ecological consequences are considerable. Field studies confirm that the sex ratio of hatchlings of species with TSD vary with nest temperature, which depends on nest location. Nests of the American alligator (*Alligator mississippiensis*) constructed on levees are warm and produce mainly males, whereas those constructed in marshes are cooler and produce mainly females. The sex ratio over a 4-year period was five females to one male (Ferguson and Joanen 1982). Species with TSD have the potential to produce sex ratios that are greatly skewed toward one sex or the other, whereas species with GSD produce sex ratios of approximately 1:1. For example, 94 percent of the hatchlings pro-

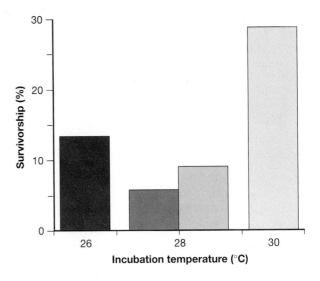

Figure 7–8   **Survivorship of hatchling snapping turtles *Chelydra serpentina* as a function of incubation temperature.** Only males (solid bars) were produced at 26°C and only females (open bars) were produced at 30°C. Both sexes were produced at 28°C. After incubation, hatchlings were released into an outdoor pond and the number of survivors was determined after 7 months. The survival of males incubated at 26°C was higher than the survival of males incubated at 28°C, and the survival of females incubated at 30°C was higher than the survival of females incubated at 28°C. This result indicates a sex-specific advantage of temperature-dependent sex determination. *(Source: Janzen 1995.)*

duced over 3 years at one of the major nesting beaches of the loggerhead turtle (*Caretta caretta*) were female (Mrosovsky 1994). Sand temperatures at nest depth remained above the pivotal temperature for most of the three nesting seasons, and few males were produced. While some populations of sea turtles have more balanced sex ratios, the apparently fixed pivotal temperatures of turtles (approximately 29°C) suggest that global warming could have negative consequences. An analysis of the effect of global climate change on the painted turtle (*Chrysemys picta*), a freshwater turtle widespread in North America, suggests that a temperature increase of less than 2°C would drastically alter the sex ratio and an increase of 4°C would virtually eliminate production of males (Janzen 1994). Such temperature increases are within the range of forecasts for the next 100 years, too short a period to allow long-lived species such as turtles to evolve shifts in their thermal biology.

Life Cycles and Reproductive Modes

Amphibians and reptiles have very different life cycles. The most common life cycle of amphibians involves three distinct stages: egg, larva, and adult. Jelly-covered eggs in an aquatic habitat hatch into larvae that subsequently metamorphose to the adult form. Amphibians also exhibit several other life cycles. Some amphibians do not have an independent larval stage. Usually the three-stage life cycle is simplified by loss of the larval or adult stage. For example, embryonic and larval development may be completed within the egg (direct development), either in a nest or within the body of one of the parents. Paedomorphic species have lost the adult body form and reproduce as aquatic larvae. In contrast to these simplified life cycles, newts in the North American genus *Notophthalmus* have added a fourth stage to their life cycle. Newt larvae transform into a nonreproductive terrestrial eft stage that persists as long as 14 years before transforming into the aquatic adult.

Metamorphosis is more dramatic for anurans than it is for salamanders or caecilians. Tadpoles look completely different from frogs. Tadpoles develop their hind limbs first, with the front limbs erupting just before metamorphosis. Species that live in poorly oxygenated water, such as warm tropical ponds, develop lungs early and can breathe air; those that live in well-oxygenated water, such as cool streams, often do not develop lungs until just before metamorphosis (Burggren and Just 1992). At metamorphosis, tadpoles resorb their tails. Individuals of most species move onto land at this time as miniature adults. A few species remain in the water throughout life but go through the same type of metamorphosis. Among the changes that occur at metamorphosis, the transition from a suspension feeding tadpole to a carnivorous adult requires major modifications of the morphology and physiology of the digestive tract (Hourdry et al. 1996).

In contrast to anurans, larval salamanders and caecilians look like small adults, although changes occur in internal organs and in the skin during development. Salamanders develop limbs early in the larval period and often develop lungs early as well. Because larval salamanders are carnivorous, like adults, metamorphosis does not involve major changes in the digestive system, although some modifications of the head and tongue occur (Reilly 1994).

Larvae of most species of anurans develop in aquatic sites that are unpredictable in terms of both duration and the quantity and quality of available food. Tadpoles that develop in unpredictable habitats generally have extremely variable developmental times, and they can metamorphose over a wide range of body sizes. Species that develop in relatively stable, permanent sites generally have a narrower range of developmental times and body sizes at metamorphosis. The changes that occur during metamorphosis are controlled by hormones, although biotic and abiotic environmental factors can strongly influence the timing of metamorphosis. Examples of such factors include temperature, food availability, and the density of larvae.

The term reproductive mode refers to the site of egg deposition and the type of larval development found in a particular species or lineage. Reptiles have only two reproductive modes—egg deposition (oviparity) and birth of fully formed young (viviparity). In contrast, a diversity of reproductive modes have been described for amphibians. For example, Salthe (1969) divided the reproductive modes of salamanders into three categories: pond breeding (mode I), stream breeding (mode II), and nonaquatic eggs (mode III) (Table 7–2). Stream-breeding salamanders lay larger eggs than do pond-breeding species because oxygen concentrations are higher in streams than in ponds (Bradford 1990). In addition, the food available to stream-dwelling larvae often comes in relatively large packages, such as aquatic insects, that can be eaten only by relatively large larvae, whereas the food of pond-dwelling larvae is zooplankton and other small prey (Nussbaum 1985a). Salamanders that lay eggs on land produce even larger eggs than do stream dwellers, perhaps because a large egg has a small surface-to-volume ratio and thus loses water slowly.

Some salamanders, including the four-toed salamander (*Hemidactylium scutatum*) and several species of dusky salamanders (*Desmognathus*), have terrestrial eggs that hatch into larvae that make their way to water. The eggs of *Desmognathus aeneus* hatch into larvae that remain in the moist nest site and develop to metamorphosis without feeding. Terrestrial reproduction has gone a step further in *Desmognathus wrighti* and all members of the tribes Plethodontini and Bolitoglossini. Eggs of these species undergo direct development and hatch into miniature adults with no intermediate larval stage. The embryos develop external gills similar to those of aquatic larvae, and these gills

Table 7–2   **Modes of egg deposition and development in salamanders. Modes of reproduction from Salthe (1969) are given in roman numerals.**

Egg Deposition Site	Larval Development	Selected Examples
Aquatic Eggs		
Still water (I)	Feeding in ponds	Sirenids, *Amphiuma*; some hynobiids, ambystomatids, salamandrids, plethodontids
Flowing water (II)	Feeding in streams	*Cryptobranchus, Necturus, Dicamptodon, Rhyacotriton;* some hynobiids, ambystomatids, salamandrids, plethodontids
Nonaquatic Eggs		
Terrestrial nest	Feeding in ponds	*Ambystoma opacum, Ambystoma cingulatum*
Terrestrial nest	Larvae move to water	*Hemidactylium,* some *Desmognathus*
Terrestrial nest	Nonfeeding in nest	*Desmognathus aeneus*
Terrestrial nest	Direct development	*Desmognathus wrighti, Plethodon, Ensatina,* some bolitoglossines
Arboreal nest	Direct development	*Aneides lugubris,* some bolitoglossines
Retained in oviducts	Give birth to larvae, or terrestrial young	*Salamandra, Mertensiella*

may enhance gas exchange inside the egg. Plethodontids evolved direct development in North American before invading Central and South America, and their success in moist tropical habitats, especially montane regions that lack ponds, may be due to this reproductive mode (D. B. Wake 1987).

The reproductive modes of frogs are more diverse than those of any other group of amphibians (Table 7–3). Oviposition in ponds and swamps may be the ancestral mode of reproduction for anurans, and is it widespread in large families such as ranids, bufonids, and hylids. Even among pond-breeding anurans, there is considerable variation in the way eggs are deposited. Differences in the structure of the egg mass are related to the availability of oxygen; the jelly layers of the egg form a barrier to the diffusion of respiratory gases (Seymour and Bradford 1995). Species that breed in cold water that is relatively well oxygenated often lay eggs in compact clumps surrounded by a thick jelly coat, and even form large communal egg masses (Figure 7–9a). Species that lay eggs in warm or poorly oxygenated water tend to lay tiny eggs that are attached individually to plants, distributed in long strings, or deposited in thin films on the water's surface, maximizing exposure of individual eggs to the air (Figure 7–9b). Some frogs, especially in the families Leptodactylidae and Myobatrachidae, lay eggs in foam nests floating on the water's surface (Figure 7–10). The eggs are surrounded by air bubbles that not only provide some oxygen for the eggs, but also keep the eggs from sinking into poorly oxygenated deeper water (Seymour and Roberts 1991). As in salamanders, frogs that lay eggs in flowing water, such as the tailed frog (*Ascaphus truei*), have relatively large eggs compared to those of pond-breeding species (Brown 1989).

The shift from depositing eggs in ponds or streams to breeding in small pools or on land is a common theme in the reproductive biology of anurans. A reduction in predation on eggs and larvae is thought to be the benefit of this behavior. Small pools do not have the predators found in large permanent bodies of water. Many tropical frogs lay eggs in small bodies of water, sometimes with volumes no greater than a teacup. Certain Neotropical tree frogs, such as *Hyla boans* and *Hyla rosenbergi*, lay eggs in basins constructed by the male at the edge of a stream. Other

Table 7–3   **Modes of egg deposition and development of anurans. Only a few examples are given for each mode. Not all members of genus necessarily exhibit a given reproductive mode, and some genera are represented in more than one mode.**

Egg Deposition Site	Tadpole Development	Selected Examples
Aquatic Eggs		
Still water	Feeding in ponds	*Rana* (Ranidae) *Bufo* (Bufonidae)
Flowing water	Feeding in streams	*Ascaphus* (Ascaphidae) *Atelopus* (Bufonidae)
Streamside basin	Feeding in streams	*Hyla boans* (Hylidae)
Tree holes, bromeliads	Feeding in water	*Anotheca* (Hylidae) *Mertensophryne* (Bufonidae)
Water-filled depressions	Feeding in water	*Eupsophus* (Leptodactylidae) *Leiopelma* (Leiopelmatidae)
Tree holes, leaf axils	Nonfeeding in water	*Anodonthyla* (Microhylidae) *Syncope* (Microhylidae)
Ground or water	Develop in stomach	*Rheobatrachus* (Myobatrachidae)
Foam nest in ponds	Feeding in ponds	*Physalaemus* (Leptodactylidae)
Foam nest in pools	Feeding in streams	*Megistolotis* (Myobatrachidae)
Dorsum of female	Feeding in ponds	*Pipa carvalhoi* (Pipidae)
Dorsum of female	Direct development	*Pipa pipa* (Pipidae)
Nonaquatic Eggs		
Terrestrial nest	Feeding in ponds	*Pseudophryne* (Myobactrachidae)
Ground, rock, nest, depressions	Tadpoles move to water	*Hemisus* (Hemisotidae) *Centrolene* (Centrolenidae) *Leptopelis* (Hyperoliidae)
Terrestrial nest	Carried to water	*Dendrobates* (Dendrobatidae)
Terrestrial nest	Nonfeeding in nest	*Leiopelma* (Leiopelmatidae)
Terrestrial nest	On dorsum, in vocal sac, in pouches	*Sooglossus* (Sooglossidae) *Rhinoderma* (Rhinodeermatidae) *Assa* (Myobatrachidae)
Terrestrial nest	Direct development	*Eleutherodactylus* (Leptodactylidae)
Leaves over water	Feeding in water	*Hyla, Agalychnis* (Hylidae) *Cochranella* (Centrolenidae)
Walls of tree holes	Feeding in tree hole	*Acanthixalus* (Hyperoliidae) *Chirixalus* (Rhacophoridae)
Arboreal nest	Direct development	*Platymantis* (Ranidae) *Eleutherodactylus* (Leptodactylidae)
Foam in burrow	Feeding in ponds	*Heleioporus* (Myobatrachidae)
Foam in burrow	Nonfeeding in burrow	*Adenomera* (Leptodactylidae)
Foam in trees	Feeding in ponds	*Chiromantis* (Rhacophoridae)
On male's legs	Feeding in ponds	*Alytes* (Discoglossidae)

(continued)

Table 7–3   (continued)

Egg Deposition Site	Tadpole Development	Selected Examples
Female dorsal pouch	Feeding in ponds	*Gastrotheca* (Hylidae)
Female dorsum or pouch	In bromeliads or bamboo stems	*Flectonotus, Fritziana* (Hylidae)
Female dorsum or pouch	Direct development	*Hemiphractus, Stefania* (Hylidae)
Eggs retained in oviducts	Nutrition provided by yolk	*Eleutherodactylus jasperi* (Leptodactylidae)
Eggs retained in oviducts	Nutrition provided by secretions	*Nectophrynoides occidentalis* (Bufonidae)

Source: Duellman and Trueb 1986.

(a)

(b)

Figure 7–9   **Egg masses of two North American ranid frogs.** (a) Communal egg mass of the wood frog *Rana sylvatica* and (b) egg mass of the green frog *Rana clamitans.* *(Photographs by Kentwood Wells.)*

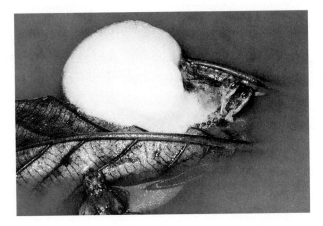

Figure 7–10   **Foam nest being constructed by male and female Túngara frogs (*Physalaemus pustulosus*) in Panama.** *(Photograph by Kentwood Wells.)*

anurans lay eggs in small water-filled depressions on the ground. These include such species as the golden toad of Costa Rica (*Bufo periglenes*) and small leptodactylid frogs in the genus *Eupsophus*.

Frogs in several families lay eggs in the water that collects in the leaf axils of bromeliads and other plants or in tree holes. These sites are often poorly supplied with oxygen and contain relatively little food. Tadpoles in these sites usually develop lungs very early and become obligate air-breathers (Lannoo et al. 1987). The food problem is circumvented in several ways. Some frogs produce tadpoles that do not have to eat because their energy needs are provided by a relatively large yolk. For example, the tadpoles of *Bufo periglenes* can metamorphose successfully without food, although tadpoles that were fed grew larger than tadpoles that were not fed (Crump 1989). Other frogs feed their tadpoles (see Parental Care).

Many frogs lay their eggs entirely out of water. Some terrestrial breeders retain an aquatic tadpole stage. Some tree frogs, for example, lay eggs on vegetation overhanging ponds or streams (Figure 7–11). As the eggs hatch, the tadpoles drop into the water and complete a life cycle similar to that of typical pond- or stream-breeding frogs. Groups with this type of life history include centrolenids, phyllomedusine hylids such as *Agalychnis*, and other hylids such as *Hyla ebraccata*. The rhacophorid tree frog *Chiromantis* lays its eggs in foam nests overhanging temporary ponds. These nests provide protection from desiccation in the hot tropical sun (Seymour and Loveridge 1994).

Many frogs have eliminated the aquatic larval stage altogether. Tadpoles of a few species, including the dendrobatid *Colostethus stepheni*, remain in a terrestrial nest without feeding until metamorphosis (Juncá et al. 1994). Eggs of other species undergo direct development and hatch into small froglets. In contrast to salamanders, in which direct development has evolved only in the family Plethodontidae, direct development has evolved independently in many different lineages of frogs. The best-known examples of direct development are leptodactylid frogs in the genus *Eleutherodactylus*, which is the single largest genus of vertebrates, with over 600 described species. Much of the ecological success of this group may be attributable to their mode of reproduction, which allows them to occupy terrestrial and arboreal habitats not available to frogs with aquatic larvae. Parental care of eggs is common in this genus, although not universal (Townsend 1996). In the Old World, the closest ecological equivalents of *Eleutherodactylus* are microhylid frogs from lowland and montane forests of New Guinea and northern Australia. All are thought to have direct development, but their biology is poorly known. One of the most bizarre of all frogs is the turtle frog, *Myobatrachus gouldii*, which lives almost entirely underground in very dry Australian deserts. The turtle frog lays direct-developing eggs buried more than a meter under the surface in an environment that seems most inhospitable to amphibians.

Parental Care

Many species of amphibians and reptiles provide some degree of care to their eggs, and sometimes to

Figure 7–11   **Egg mass of the red-eyed tree frog *Agalychnis callidryas* in Panama.** When the eggs hatch, the tadpoles drop into the pond below. *(Photograph by Kentwood Wells.)*

their offspring as well. The evolution and persistence of parental care depends on the balance between the benefits it brings (usually increased survival of eggs or young) and its costs (which include the investment of time and energy and the increased risk of predation for the attending parent).

Parental Care by Amphibians

Parental care is particularly common among amphibians that breed on land. Because the terrestrial environment is not hospitable to the permeable, jelly-covered eggs of amphibians or their larvae, parental behaviors may be particularly beneficial to the survival of eggs and tadpoles of terrestrial breeders. Parental care is exhibited by caecilians, salamanders, and frogs (Crump 1996). Parental care by caecilians and salamanders consists only of egg attendance. That is, one of the parents remains with the eggs during part or all of their development. The benefits of egg attendance include deterrence of parasites and small predators and maintenance of suitable moisture conditions. Limited field observations suggest that females of all oviparous caecilians attend eggs, whereas only about 20 percent of salamander species attend their eggs after oviposition. In contrast to caecilians and salamanders, the forms of parental care by frogs are exceedingly diverse, although parental care is exhibited by only about 10 percent of anuran species.

Egg attendance is the most common form of parental care by frogs (Figure 7–12). The benefits of egg attendance are best known for the Puerto Rican coquí (Townsend et al. 1984). Female coquís lay eggs in a sheltered retreat within a male's territory, and males attend the eggs assiduously. To determine the benefits of egg attendance, Daniel Townsend and his colleagues removed attending males from some clutches to determine the fates of attended and nonattended eggs. The results of their experimental manipulations were clear (Table 7–4). Unattended clutches failed more frequently than attended clutches. The main causes of mortality of unattended clutches were being eaten by other male coquís and drying out. Male coquís are important predators on the eggs of other coquís, and attending males aggressively defend their retreats from intrusion by other males. Attending males also protect their clutches from desiccation.

Similarly, males of species such as *Hyalinobatrachium valerioi* (Centrolenidae) attend eggs that are at-

Figure 7–12   **A male Puerto Rican frog, *Eleutherodactylus coqui*, attending its clutch of eggs.** *(Photograph by F. Harvey Pough.)*

tached to leaves or stems hanging over streams, and the tadpoles drop into the water when they hatch. Midwife toads (*Alytes*, Discoglossidae) do more than simply attend their eggs at the site of fertilization; males carry strings of eggs representing the clutches of one or more females entwined about their hind legs until the eggs are ready to hatch. At this time, the male enters the water and the tadpoles swim off to complete their development.

For other frogs that breed terrestrially, parental duties do not end once the eggs hatch. The newly hatched tadpoles of dendrobatids wiggle onto the back of one of the parents, which transports them from the nest to water. Depending on the species,

Table 7–4   **The effect of paternal care on the hatching success of eggs of the Puerto Rican frog, *Eleutherodactylus coqui*. The attending male was removed from some (experimental) clutches and not from other (control) clutches. Data from 1980 and 1982 are combined. For the experimental and control clutches, the percentage of that group with particular fates follows the number of clutches in parentheses. Significantly more clutches failed in the experimental than the control groups.**

			No. of Clutches That Failed as a Result of:		
Treatment	Total No. of Clutches	No. of Clutches That Hatched	Desiccation	Cannibalism by Other Males	Predation by Invertebrates
Experimental	104	24 (23%)	43 (41%)	33 (32%)	4 (4%)
Control	175	135 (77%)	2 (1%)	25 (14%)	13 (7%)

Source: Modified from Townsend et al. 1984.

these bodies of water range from streams to small pools in tree holes or in the axils of bromeliads. Parental care does not end even then for female *Dendrobates pumilio* and several closely related species and the hylid *Osteopilus brunneus*. These species place their tadpoles in the axils of bromeliads and females visit their tadpoles on a regular basis and feed them with eggs that the female deposits in the pool of water (Brust 1993, Thompson 1996). Tadpoles of the Asian rhacophorid *Chirixalus eiffingeri* are also fed unfertilized eggs by the female (Kam et al. 1996).

Egg attendance is extended to tadpole attendance by some aquatic frogs. Male African bullfrogs (*Pyxicephalus adspersus*) remain with their tadpoles until metamorphosis and defend them against potential predators. They also construct channels to free the tadpoles from entrapment in small pools (Kok et al. 1989). In the New World, females of several species in the genus *Leptodactylus*, including *L. ocellatus* and *L. insularum*, also remain with their tadpoles. All of these frogs are large, and an adult can act effectively against some predators and some physical hazards.

Some amphibians complete embryonic and larval development within the body of one of the parents but outside of the reproductive tract. External fertilization may explain why frogs are so apt to retain eggs in nonoviductal sites. Both males and females are present at oviposition, and both have the opportunity to sequester the eggs. Sequester they do, seemingly in all possible places (Figure 7–13). Female marsupial frogs carry their eggs on their backs. In the genus *Hemiphractus*, the eggs are attached to the female's back with mucus and are carried until the young hatch as froglets. In the genus *Gastrotheca*, the eggs are enclosed in a dorsal pouch. Either fully developed froglets or well-developed larvae are released to complete their development in a body of water (Duellman et al. 1988). This mode of egg transport is paralleled by the aquatic Surinam toad *Pipa pipa*. Eggs are embedded in the back of the female by overgrowth of the epithelium. Some species of *Pipa* release young as larvae and others produce fully developed froglets. Perhaps the most unusual form of parental care is exhibited by Australian frogs in the genus *Rheobatrachus*, in which the female swallows her eggs and broods them in her stomach (Tyler 1983). Froglets emerge from the female's mouth following metamorphosis. They secrete prostaglandin E_2, which inhibits the production of gastric acids by the mother.

Parental Care by Reptiles

Parental care is distributed unevenly among the major reptilian clades (Shine 1988, Gans 1996). Turtles rarely exhibit parental care, whereas parental care may be universal among crocodilians. The most plausible explanation for the lack of parental care among turtles and its universality among crocodilians is based on the effectiveness of defense against predators. On the one hand, it is difficult to see how a turtle could effectively defend its eggs or offspring from predators. For example, when nesting *Manouria emys*, a tortoise from Southeast Asia, were confronted with human and simulated animal predators, the females either positioned themselves on top of the nest or pushed the intruder away from the nests. They did not bite, and the nest-guarding behavior lasted only 2 to 3 days. Parental care by turtles may thus have costs (reduced survival of adults) and no benefits. On the other hand, adult crocodilians are formidable animals and can easily defend their eggs and young.

Parental care by crocodilians is extensive, and although females are usually the caregivers, males are often involved (Lang 1989). Females typically remain at the nest site after oviposition and defend the nest against potential predators. At the end of the incubation period, the young vocalize and attract the female or both parents to the nest. The female, and sometimes the male as well, opens the nest and helps the young to escape from the eggshell, and may also carry the young to water (Figure 7–14). Young crocodilians remain together for periods as long as a year or more and are guarded by one or both parents during this time. Distress calls by the young instigate defensive behaviors by adults.

About 100 species of squamates exhibit some form of parental care (Shine 1988). Parental care by squamates includes defense of the nest site, egg attendance, and egg brooding by muscular thermogenesis. Each of these behaviors is characteristic of particular lineages. For example, attending the eggs and keeping them warm is characteristic of some pythons. Squamates seldom extend parental care to hatchlings, although females of viviparous squamates are commonly reported to help the young emerge from fetal membranes, and other behaviors may have some characteristics of parental care. For example, a female and her young may remain together until the young disperse from the nest site. During this period, the

Figure 7–13 **Frogs that carry eggs on or in their bodies but outside the female reproductive tract.** (a) Female marsupial frog, *Gastrotheca ovifera*, releasing young from pouch on her back. (b) Male Darwin's frog, *Rhinoderma darwinii*, with young just released from his vocal sacs. (c) Female Surinam toad, *Pipa pipa*, carrying eggs embedded in thickened skin on her back. *(Photographs by (a, b) Michael and Patricia Fogden, (c) John D. Groves.)*

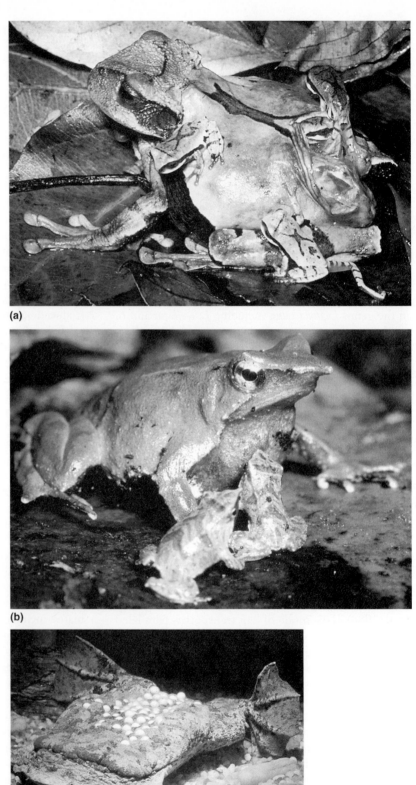

(a)

(b)

(c)

Figure 7–14 **Parental care by the mugger crocodile,** *Crocodylus palustris.* The male parent is carrying a hatchling from the nest to the water, which is about 9 meters away. *(Photograph by Jeffery Lang.)*

presence of the female may keep predators away from the young, although the association between young and female may be largely fortuitous.

The most common type of parental care by squamates is egg attendance. This behavior is characteristic of pythons, all members of the scincid genus *Eumeces*, and many species in the families Colubridae, Elapidae, and Viperidae. Snakes that exhibit egg attendance tend to be large or venomous, or both. Egg attendance has been reported for about half of the oviparous genera of the Viperidae and Elapidae and for some large and venomous species of colubrids. These snakes can defend eggs from vertebrate predators. In contrast, the major benefit of egg attendance for small lizards may be to protect eggs from other small vertebrates and invertebrates.

Who Cares?

Parental care is most likely to evolve in the sex that is most closely associated with the eggs (Gross and Shine 1981). When fertilization is external, either the female or the male may be the caregiving parent because both are present when the eggs are laid. In this case, the relative costs and benefits of parental care appear to determine who cares. For example, male frogs tend to be the caregivers when eggs are laid within the male's territory and the male can continue to attract and mate with additional females while guarding eggs. On the other hand, female frogs cannot immediately produce another clutch of eggs, whereas a male can fertilize many clutches in quick succession. Female frogs tend to be the caregivers when the calling site of the male and the oviposition site are in different places.

The great majority of amphibians and reptiles do not exhibit parental care at all. This observation suggests that parental care has costs to the parents that may outweigh the enhanced survival of offspring. Reduced reproductive output is one cost of parental care. Species that exhibit parental care usually produce fewer eggs per clutch than related species that do not have parental care. Females typically do not eat when they are guarding nests or eggs, and females that remain with their clutch produce fewer clutches overall than do noncaring females. Pythons, for example, do not feed when they are brooding eggs, and also expend energy for heat production. By not feeding, females may reduce the size of future clutches or may extend the time they need to accumulate the energy to reproduce again. The potential benefits of parental care should be greatest for species whose reproductive output is already limited for other reasons. If one clutch per season is the norm, caring for that clutch might not reduce reproductive output appreciably.

Reduced survival of the parent is another potential cost of parental care. Remaining with the eggs could put a parent at risk of death. Most amphibians and reptiles are small and have ineffective defenses against vertebrate predators. Parental care by these species would increase the risk of their own death and would not save their eggs or young. On the other hand, species that attend eggs in a concealed nest may actually have less exposure to predators than they would when normally active. In this situation, parental care could increase survival of adults.

Viviparity

Viviparity, the retention of embryos in the oviducts until development is complete, is popularly considered unique to mammals, but this is not the case. Many sharks and bony fishes are viviparous. Among amphibians and reptiles, many caecilians and squamates exhibit viviparity as well as a few frogs and salamanders (Shine 1985, M. H. Wake 1993). Internal fertilization may explain how viviparity can be common in the caecilians and squamates and extremely uncommon in the anurans, but why is viviparity uncommon in salamanders and nonexistent in turtles and crocodilians?

Comparative studies indicate that viviparity has evolved through gradual increases in the length of time that eggs are retained in the oviduct (Guillette 1987). Viviparous taxa exhibit specializations for embryonic development in the uterine environment. One set of specializations enhances gas exchange between maternal and fetal tissues. During the evolution of viviparity, the thickness of the jelly layers of the egg (amphibians) or the shell membranes (squamates) is reduced. The total length of blood vessels in the oviduct and in the respiratory structures of embryos is increased. Because the ancestral patterns of development of amphibians and reptiles are different, the embryonic respiratory structures of these taxa are not homologous. For viviparous amphibians, previously existing larval structures are modified and used for respiration. Larvae of viviparous species have greatly elaborated and highly vascularized gills or tails. These respiratory structures are pressed against the walls of the oviduct and may enhance gas exchange.

In contrast, embryonic respiration by viviparous squamates involves the embryonic membranes (Figure 7–15). Highly vascularized embryonic and maternal membranes are closely associated and are sites for the exchange of water, gases, and other nutrients (Blackburn 1993). The chorioallantoic placenta of squamates is convergent with the mammalian placenta, but various types of placentas formed from the yolk sac are unique features found only among squamates (Stewart 1993).

Other specializations provide the hormonal support for extended gestation in the uterus (Guillette 1987). The secretory activity of the corpus luteum (a structure formed from the walls of the ovarian follicle after ovulation in all vertebrates) is usually limited to the preovulatory period in oviparous species. Corpora lutea secrete progesterone, a hormone that inhibits the contractions of the uterine wall that expel eggs from the oviduct. The corpora lutea remain active after ovulation in viviparous species, thus maintaining gestation. The corpora lutea of amphibians that brood embryos outside of the female's reproductive tract also persist during at least part of gestation and presumably secrete progesterone.

Species that are viviparous exhibit all grades of embryonic nutrition, from total dependence on the original stores of yolk in the egg (lecithotrophy) to almost total dependence on nutrients transferred from the mother. Most viviparous amphibians and reptiles are lecithotrophic, but for the few in which substantial nutrient transfer from the mother occurs, the process differs between amphibians and reptiles. For viviparous amphibians, larvae ingest nutritive secretions of the oviduct through their mouths (matrotrophy). In contrast, for squamates, nutrients are transferred through the placenta (placentotrophy). Embryos of viviparous squamates may receive some or nearly all nutritive support from the female. For example, a recently ovulated egg of the South American skink *Mabuya heathi* is only about 1 millimeter in diameter, and placental transport accounts for more than 99 percent of the dry mass of the embryo just before birth (Blackburn et al. 1984). Earlier works identified a mode of reproduction known as ovoviviparity, in which yolked eggs were held by the female until live birth occurred, in contrast to viviparity, then defined as the retention of embryos with little or no yolk. Recent studies have revealed the artificial nature of that dichotomy and the presence of many intermediate conditions (Blackburn 1992). For this reason, the mode of parity (oviparity or viviparity) is best distinguished from the mode of embryonic nutrition (lecithotrophy or placentotrophy).

Viviparity is restricted to a few lineages of amphibians and reptiles. The comparatively small number of origins of viviparity suggests that the costs of viviparity may outweigh the benefit for most taxa. Viviparity, like parental care, reduces total reproductive output because it usually limits reproduction to one clutch per season. Furthermore, clutch size is usually lower in viviparous species than in oviparous species because space within the female may be limited. Additionally, pregnant females may be burdened by their clutches to the extent that they cannot move as rapidly as nonpregnant individuals and thus are more vulnerable to predators (Shine 1980a). In apparent compensation for reduced agility, female *Lacerta vivipara* remain closer to refuges when pregnant than when nonpregnant (Bauwens and Thoen 1981). While reducing the risk of death, this behavior means that pregnant females may have reduced opportunities for thermoregulation or for feeding, and these may be additional costs of viviparity.

Figure 7–15 **Placentation in lizards.**
(a) Diagram of a viviparous squamate embryo illustrating the position of the embryo with regard to the uterus (UT), shell membrane (SM), fetal membranes, and two types of placenta. Apposition of the chorioallantoic membrane (CA) to the inner lining of the uterus constitutes the chorioallantoic placenta. An omphaloplacenta is formed when mesodermal tissue invades the yolk sac and isolates part of it from the rest of the yolk. The inner margin of the isolated yolk mass (IYM) is vascularized and the outer margin is made of embryonic ectoderm and endoderm (BI OMP). (b) Section of the chorioallantoic placenta of the skink *Chalcides chalcides* . This highly vascularized area lies dorsal to the fetus (between CA and UT in (a) and consists of interdigitated chorioallantoic (CA) and uterine (U) tissue. Note the absence of a shell membrane. IA, inner allantoic membrane; M, mesometrium. *(Source: (a) Stewart 1993, (b) Blackburn 1993.)*

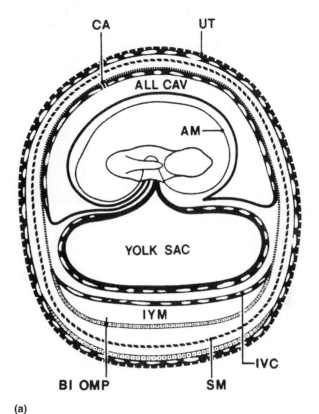

(a)

(b)

Viviparous Amphibians

Approximately 75 percent of caecilian species are viviparous and matrotrophic (M. Wake 1993). Larvae of *Dermophis mexicanus* are retained in the oviduct of the female for a total gestation period of 11 to 12 months. The larvae use the yolk stores of the eggs for the first 3 months of development. When the yolk stores are exhausted, the larvae feed on a lipid-rich material that is secreted by a greatly proliferated and vascularized oviductal epithelium. Larvae have an unusual fetal dentition (epidermal structures on the lips, rather than true teeth) that they use to graze on the epithelium (Figure 7–16). Some researchers speculate that grazing by larvae on this oviductal epithelium may stimulate its secretory activity.

In contrast to caecilians, only four species of salamanders and six species of anurans are viviparous. Viviparity in salamanders is confined to the fire and Caucasus salamanders (*Salamandra* and *Mertensiella*) in Europe and the Middle East. Embryos of the viviparous *Salamandra salamandra* are nourished entirely from yolk, whereas embryos of *Salamandra atra* feed on oviductal secretions after the yolk is exhausted. One of the viviparous anurans is the Puerto Rican frog *Eleutherodactylus jasperi*. Eggs of *E. jasperi* undergo direct development, like those of other members of the genus, but development occurs inside the mother (M. Wake 1978). The African tree toads *Nectophrynoides* exhibit the entire sequence in the evolution of viviparity. One species is oviparous with external fertilization and aquatic tadpoles, another has direct development, and four are viviparous with internal fertilization. Of the viviparous species, embryos of two species depend only on yolk for food and embryos of the other two feed on oviductal secretions after depleting the yolk.

The ecological circumstances that favor the evolution of viviparity in amphibians are difficult to evaluate. The limited number of viviparous salamanders and anurans makes any sort of statistical correlation between environmental factors and viviparity impossible. The observation that viviparous anurans are tropical may reflect only the fact that most anurans are tropical, and that both parental care and viviparity are associated with terrestrial breeding, which is largely a tropical phenomenon.

Viviparous Reptiles

The only extant viviparous reptiles are squamates. Despite vigorous debate (Packard et al. 1977, Black-

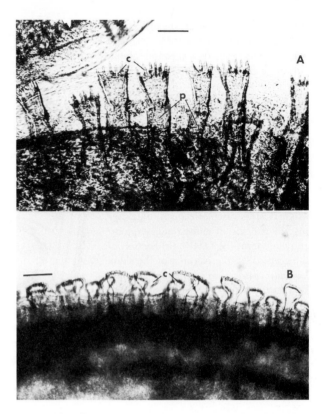

Figure 7–16   **Fetal dentition of two species of caecilian larvae.** Note differences in crown shapes between the species. (*Photographs by Marvalee Wake.*)

burn and Evans 1986, Anderson et al. 1987), there is no generally accepted explanation for the absence of viviparity among turtles, crocodilians, and birds. Viviparity is associated only with some squamate lineages. Some families are entirely oviparous (Teiidae, Varanidae, Pygopodidae, Helodermatidae, Dibamidae, Leptotyphlopidae) and others are entirely viviparous (Anniellidae, Xenosauridae, Xantusiidae, Aniliidae, Uropeltidae, Acrochordidae, Tropidophidae). Among the 15 families with both oviparous and viviparous species, viviparity has had at least 100 independent origins. Of these, at least 34 occurred in the Scincidae and 14 in the Colubridae (Shine 1985).

The richness in individual case histories has facilitated evaluation of the ecological context in which viviparity evolves (Packard et al. 1977, Tinkle and Gibbons 1977, Shine 1985). Viviparity is associated with cold climates (high elevations and latitudes). Recent origins of viviparity support the hypothesis that viviparity evolves in cold climates. The best-documented examples of recent origins of viviparity involve geographic variation in reproductive mode of the European viviparous lizard *Lacerta vivipara* (Huelin et al. 1993), the Australian skink *Lerista bougainvillii* (Qualls and Shine 1995), and the Mexican bunchgrass lizard *Sceloporus bicanthalis* and its close relative, *Sceloporus aeneus* (Mink and Sites 1996, Guillette 1982). In each of these cases, the viviparous population or species is found in colder climates than its oviparous relative. For example, the viviparous *S. bicanthalis* is found at altitudes of 3,000 meters or higher in the transvolcanic axis of Mexico, whereas its close relative, the oviparous *Sceloporus aeneus*, is found at lower elevations. The altitudinal ranges of these two species do not overlap, and viviparity is presumed to have evolved at the high elevations where *S. bicanthalis* now occurs (Shine and Bull 1979).

Viviparity may be beneficial in cold climates because embryos retained within a thermoregulating female experience warmer temperatures and develop more rapidly than embryos developing in nests in the ground. (Shine 1983b). The benefit of enhanced development in cold climates would accrue to the intermediate stages of egg retention as well as to fully viviparous species. For example, oviparous females that extend the length of egg retention might enhance the survival of their offspring. Even small increases in the rate of embryonic development would result in earlier hatching, and earlier hatching would give neonates extra time for growth and fat storage before winter. In more extreme climates, the faster developmental rates of embryos in utero enable viviparous species to live in areas too cold for successful reproduction by oviparous species.

Life History Variation

The life history of an organism is the composite of traits that describe its life cycle. Quantitative traits include size at hatching or birth, clutch size (number of eggs) and frequency of reproduction, age and size at maturity, age- and size-specific mortality rates, and longevity. Qualitative traits, such as mode of reproduction (e.g., oviparity versus viviparity, sexual versus asexual reproduction, aquatic versus terrestrial breeding), are life history traits as well. Particular combinations of traits often occur together, suggesting that they evolve in a correlated manner.

Covariation among life history traits of lizards has received particular attention (Dunham et al. 1988).

Covariation in life history traits for this group can be represented by a multiply branching array, with three major groupings of species and a number of subgroups (Figure 7–17). One group includes species with a single clutch per year and delayed maturity (more than 1 year after hatching). Viviparous species form a distinct subgroup within this group. A second group includes species with multiple clutches per year and early maturity (within a year of hatching). A third group includes species with multiple clutches and delayed maturity. Subgroups of species with multiple clutches are defined by clutch size and body size—big-bodied species have bigger clutches than small-bodied species. The simplistic idea that life histories can be described by a simple dichotomy between large-bodied species with one set of traits and small-bodied species with another set clearly is not correct.

We will focus on two well-studied questions about life history traits: How much energy should an individual allocate to reproduction, and should that energy be put into many small packages or a few large ones? The answers to these questions involve trade-offs among traits. Trade-offs occur when traits are linked so that changes in one trait affect changes in the other. The mechanism of a trade-off is particularly easy to visualize for an individual organism. For example, a female *Batrachoseps attenuatus* salamander that loses its tail during an encounter with a predator can usually grow a new one, but there is a physiological cost. While the tail is regenerating, reproduction may be curtailed because energy allocated to the growth of the tail is not available for producing eggs.

Variation in Life History Traits: Reproductive Effort

A central question about life histories concerns how much of an individual's available resources should be allocated to reproduction. Why don't individuals always maximize their reproductive effort (amount of current reproduction)? This question is answered by the observation that reproduction decreases survival or future reproduction, or both (Schwarzkopf 1994). The mechanistic bases for these trade-offs are clear. High allocation of energy to reproduction can entail activities that attract predators or make individuals more vulnerable to predators. Courtship makes both males and females conspicuous, and pregnant females run relatively slowly. Moreover, a high allocation of

Figure 7–17 **Graphical analysis of lizard life histories.** Species are linked by their similarities in five life history traits: snout–vent length at maturity, number of eggs per clutch, age at maturity, oviparity or viviparity, and clutch frequency. Major groups and some subgroups are labeled, and some representative North American iguanids are indicated. Dashed lines separate single from multiple clutched species. Viviparous species are those in the lower right sector. *(Source: Modified from Dunham et al. 1988.)*

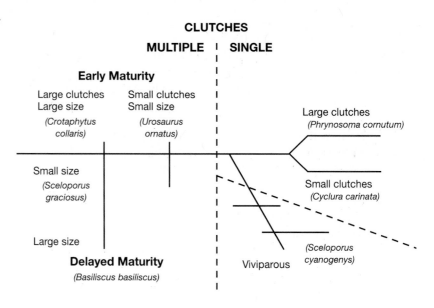

energy to reproduction in 1 year might reduce the amount of energy available for future growth or reproduction. Observations of the tree lizard *Urosaurus ornatus* in New Mexico document both survival and energetic costs of reproduction. Females of this species lay two clutches during the summer. When the number of eggs in their first clutch was reduced by half by surgical reduction in the number of follicles, females had higher survival and also grew faster and had larger subsequent clutches than did females with normal-sized first clutches (Landwer 1994).

A trade-off between reproduction and survival is also evident in comparisons among species (Figure 7–18). Because small species generally have shorter life spans than large species, and thus a smaller chance of reproduction in the future, they should allocate relatively more energy to reproduction per season than do large species and take greater risks associated with reproduction. This prediction is supported by observations of lizards. Small-bodied species often have multiple clutches per season and produce a relatively greater mass of eggs per season than do large-bodied species.

The mass of the clutch divided by the mass of the gravid female is often used as an index of the female's reproductive effort. This index is called relative clutch mass (RCM). The use of this index is justified because the energy in a clutch is proportional to its mass. (A better index of reproductive effort would include not only costs associated with current reproduction but also the effect of current reproduction on future reproduction.) Studies of RCM have provided novel in-

sights into the costs of reproduction. One of these insights is that the cost of reproduction is related to foraging mode and escape behaviors (Vitt and Price 1982). Lizards such as whiptails (*Cnemidophorus*) that forage actively have lower RCMs than species that sit and wait passively for prey to come within range (0.15 versus 0.26, respectively). Within sit-and-wait foragers, species that use rapid flight for escape have

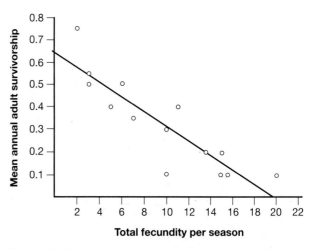

Figure 7–18 **Reproductive effort and survival.** The negative relationship between annual adult survival (the probability of surviving from one reproductive season to the next) and annual fecundity (number of eggs per clutch multiplied by the number of clutches per year) for 14 species of lizards. *(Source: Tinkle 1969.)*

lower RCMs than species that use crypsis (concealing colors and patterns) to escape predators (RCMs of 0.19 versus 0.26, respectively). RCM thus represents a trade-off between the benefits of a high investment in reproduction and the costs of that investment with respect to foraging and escape (see Chapter 13).

In general, the volume of the clutch is correlated with the size of the female, or, more directly, with the space available in the female's abdominal cavity (Kaplan and Salthe 1979, Shine 1992). For example, actively foraging species have low RCMs and slim bodies, and sit-and-wait lizards have high RCM and chunky bodies. Snakes have higher RCMs than lizards of the same mass because snakes have larger abdomens relative to their mass. Are reproductive females always so full of eggs that their abdomens can hold no more? If the answer to this question is yes, then RCM could limit reproductive effort. On the other hand, if RCM is below the maximum, then some other factor must limit reproductive effort (Shine 1992). Intraspecific comparisons suggest that space in the abdomen does not always limit RCM. Sand lizards, *Lacerta agilis*, in Sweden had larger clutches in some years than others, suggesting that the energy available for reproduction varied from year to year and that low energy availability limited clutch sizes in some years to less than that allowed by the space available for eggs (Olsson and Shine 1997).

Variation in Life History Traits: Expenditure per Progeny

A question that follows directly from the consideration of reproductive effort is, How much of the total allocation to a clutch should a parent allocate to a single offspring? A female can produce a clutch consisting of one or a few large eggs or many small eggs. For example, tiny Cuban frogs in the genus *Eleutherodactylus* produce one large egg at a time, whereas North American bullfrogs (*Rana catesbeiana*) produce enormous numbers of eggs; one female had a clutch of almost 50,000 eggs. The size of the egg can affect the offspring's probability of survival, and the optimal combination of egg and clutch size for the parent is one that maximizes the total number of surviving offspring.

Parental care and viviparity are associated with relatively large eggs and a small number of eggs per clutch for both amphibians and reptiles. The increased investment per offspring (large eggs, parental care, or both) must increase the survival of eggs enough to compensate for the reduction in the numbers of offspring produced.

The idea that egg size is optimized is supported by the relatively narrow range of egg sizes produced by females of most species. For example, variance in food availability is associated with success of different-size eggs for the California newt, *Taricha torosa* (Kaplan 1985). When food is availabe in excess, larvae that hatch from relatively large eggs metamorphose sooner and at larger sizes than those that hatch from relatively small eggs. When food is limited, however, larvae that hatch from small eggs metamorphose considerably earlier than larvae from large eggs. Thus, large eggs should be favored over small eggs when food is abundant because large eggs hatch sooner and produce larger metamorphs. When food is limited, however, the slow development of larvae from large eggs exposes them to predators for longer periods and puts them at risk of dying if their pond dries out before they have metamorphosed. Thus, small eggs would be favored over large ones when food is scarce.

Egg size cannot always be optimized. Several studies suggest that the eggs of some species are smaller than the optimal size because of a structural constraint imposed by the size of the female's pelvis. For example, eggs of painted turtles *Chrysemys picta* from Michigan and chicken turtles *Deirochelys reticularia* from South Carolina are proportional in size to the width of the female's pelvic opening—the larger females lay larger eggs (Congdon and Gibbons 1987). This observation suggests that females of these small-bodied turtles (mean plastron lengths of 130 and 160 millimeters, respectively) would lay even larger eggs if females attained larger body sizes.

Experimental studies of the side-blotched lizard *Uta stansburiana* provide direct evidence of the negative consequences of producing eggs that are too large (Sinervo and Licht 1991). Side-blotched lizards normally produce clutches of two to nine eggs, and the larger the clutch, the smaller the eggs. When all but one developing follicle was removed surgically, extra yolk was deposited in that follicle and females ovulated one abnormally large egg. Thirty-six percent of these eggs burst while being laid (the embryo died) or became bound in the oviduct (the embryo and female both died). The reason why side-blotched lizards do not normally lay clutches of one large egg is clear.

In contrast to side-blotched lizards, the entire lizard genus *Anolis* is characterized by a clutch size of one egg. An *Anolis* of the same body size as *Uta* produces an egg that is only slightly smaller than the one-egg clutch of *Uta*. This observation suggests that *Anolis* produces the largest egg that can safely pass through a female's pelvis.

The size of the egg typically increases with the size of the species, but the relationship is not directly proportional. Big species have smaller offspring relative to their adult body size than do small species. For example, hatchlings of small species of lacertid lizards have snout–vent lengths that average 44 percent of the mean length of adult females, whereas hatchlings of large species average only 29 percent of the mean length of adult females (Bauwens and Díaz-Uriarte 1997). These authors suggested that hatchlings must attain some minimal size to be viable, but the survival benefits of increases in size above this minimum do not outweigh the costs of reduced fecundity. Because clutch mass of lacertids increases in proportion to body mass and large species produce hatchlings that are relatively small, large species produce considerably more offspring than do small species.

Heterochrony

The development of an individual starts at fertilization and continues until it reaches its final adult size. The timing of genetic and epigenetic (environmentally controlled) events largely determines the path of development (i.e., the developmental trajectory). The time when a developmental event occurs is central to an individual's eventual phenotype. A difference in developmental timing between an ancestral organism and its descendant is known as heterochrony (*heteros*-different, *chronos*-time), and this phenomenon is a particularly active focus of study among evolutionary biologists and one to which the study of amphibians has made substantial contributions. Many biologists believe that the study of heterochrony affords a unifying explanation for diverse evolutionary phenomena (reviewed by Reilly et al. 1997).

Any developmental process can be characterized by three variables related to timing: when the event begins (onset time), when it ends (offset time), and the rate at which the process occurs (Figure 7–19).

Heterochrony results from changes in one or more of these variables. The simplest heterochronic changes involve shifts in only one variable. For example, if a descendant organism exhibits a decrease in the rate of a particular developmental event (such as, the ossification of a bone) and retains the same onset and offset times, the slower rate of ossification will produce a less completely ossified bone in the descendant. Thus, the descendant will resemble the immature form of the ancestor (paedomorphosis). A similar result is achieved when the rate is not changed but the time during which it takes place is shortened by delaying the onset time or advancing the offset time. Alternatively, if one or more of the three variables are shifted so as to extend development, the descendant form develops beyond that seen in the ancestor. This condition is known as peramorphosis.

Salamanders in the genus *Ambystoma* present some of the best-known examples of heterochrony. *Ambystoma tigrinum* and *A. mexicanum* are closely related species (Figure 7–20). *Ambystoma tigrinum* typically passes through an aquatic larval stage and metamorphoses to a terrestrial adult before the gonads mature. *Ambystoma mexicanum*, in contrast, fails to metamorphose and the gonads mature while the body remains in the larval form. Undergoing metamorphosis is the ancestral condition for *Ambystoma*, and *A. tigrinum* retains the ancestral condition of metamorphosis. The retention of larval morphology in sexually mature *A. mexicanum* is the derived condition. *Ambystoma mexicanum* is paedomorphic, resembling the juvenile form of its *A. tigrinum*-like ancestor.

By what heterochronic process has *A. mexicanum* achieved its paedomorphic condition? To answer this question we must compare the timing of the development of body form of the two species (Figure 7–21). *Ambystoma tigrinum* retains its larval form until its abrupt metamorphosis at about 6 months of age. During metamorphosis its shape changes rapidly until it attains the terrestrial form. Growth in size continues, but its shape changes little after metamorphosis, when the salamander reaches its offset time. *Ambystoma mexicanum*, in contrast, fails to change shape once it has achieved the larval form; its offset time has been shifted to a much earlier age. Both species become sexually mature at approximately the same chronological age, but *A. mexicanum* stopped changing shape without passing through metamorphosis.

Because the developmental mechanisms that produce paedomorphosis are so general, distantly related paedomorphic salamanders have features in common, as seen among *A. mexicanum* (Ambystomatidae), *Nec-*

Figure 7–19 **Six heterochronic processes.** Changes in timing of developmental events can be identified by comparing ontogenetic trajectories of ancestral (a) versus descendant (d) ontogenies. Ontogenetic trajectories are defined by the rate of shape development (k) from the age of onset of growth (α) to the age when the offset shape is attained (β). Arrows on the shape (vertical) axis indicate patterns of truncated (top) or extended (bottom) development. Shape is defined as one or a combination of dimensions of the trait of interest. *(Source: Adapted from Reilly et al. 1997.)*

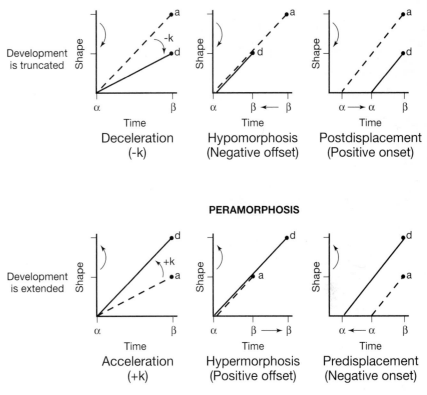

turus (Proteidae), and *Cryptobranchus* and *Andrias* (Cryptobranchidae). Differences in the degree of paedomorphosis, however, may lead to different shapes among paedomorphic species (Alberch et al. 1979). For example, the highly paedomorphic *Necturus* (Proteidae) retains large external gills, and it has a pointed snout because the maxillary bone does not form. The two maxillae are among the last bones to appear during the development of the nonpaedomorphic salamander skull, and development of *Necturus* is truncated before the maxillae appear. In contrast, *Cryptobranchus* has a rounded snout that reflects the curve of the underlying maxillary bones because *Cryptobranchus* has a less strongly truncated developmental sequence than *Necturus*.

The evolution of paedomorphosis may be favored when the aquatic habitat of larvae is more stable or more productive than the terrestrial habitat of adults (Whiteman 1994). These conditions might favor reproduction in the aquatic larval form. For example, paedomorphosis is commonly seen in populations of *Ambystoma* living in permanent ponds in dry terres-

trial environments (Rose and Armentrout 1976). Paedomorphic salamanders have also evolved repeatedly in the aquatic environments of caves and groundwater systems where terrestrial food sources are limited. However, transforming and paedomorphic populations of *Ambystoma* sometimes occur in close proximity, and more complex selective forces may be at work in some species.

Heterochronic processes also occur within species. For example, some populations of *Ambystoma talpoideum* show extreme phenotypic plasticity in their development (Semlitsch 1985). In temporary ponds all individuals exhibit the ancestral life history in which larvae hatch in the spring and metamorphose in response to drying of the pond in the summer. They reproduce the following winter and may live up to 7 years. Individuals that hatch in semipermanent ponds, however, frequently delay metamorphosis and breed in their first year as larvae and later metamorphose into terrestrial adults.

The evolution of webbed feet by arboreal species of *Bolitoglossa* (Plethodontidae) is another example of

(a)

(b)

(c)

Figure 7–20   **Paedomorphosis.** The paedomorphic ax-olotl, *Ambystoma mexicanum* and its nonpaedomorphic rela-tive, *A. tigrinum.* (a) *A. mexicanum,* (b) larval and (c) adult *A. tigrinum.* (*Photographs by R. Wayne Van Devender.*)

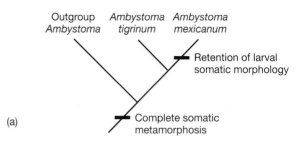

(a)

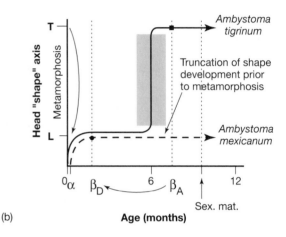

(b)

Figure 7–21   **Paedomorphosis revealed by head mor-phology of the axolotl *Ambystoma mexicanum* relative to *A. tigrinum.** The ontogenetic trajectory of the axolotl (dotted line) ends at the larval somatic morphology (L), compared to the ancestral ontogeny (solid line), which con-tinues through metamorphosis (gray box) to the trans-formed head shape (T). Note that the axolotl has an earlier offset time (β_D) than its ancestor (β_A), and thus it retains the larval head shape even as a reproductive adult. Sex. mat., onset of sexual maturity. (*Source: Reilly et al. 1997.*)

paedomorphosis (Alberch 1981, Alberch and Alberch 1981). Large species with webbed feet use suction when climbing smooth surfaces, but smaller species, such as *B. occidentalis*, can be supported simply by ad-hesion between their webbed feet and the substrate, owing to their small body mass (see Chapter 8). The extensive webbing on the feet of *B. occidentalis* is a re-sult of truncation of foot development before the toes grow to their ancestral length. (The toes of amphib-ians form by growing out from the paddle-like em-bryonic foot, unlike the toes of amniotes, which form by death of the cells that lie between adjacent digits.) *Bolitoglossa occidentalis* exhibits other signs of truncated development, including reduction or absence of cra-nial bones that form late in the development of re-lated species. Thus, the webbed feet of *B. occidentalis* did not arise as an adaptation to arboreality, but rather as a consequence of early offset of shape change.

Peramorphosis has received much less attention than paedomorphosis. In theory, it should be no less common. Once developmental sequences of more species are available and have been placed in a phylogenetic context, it is likely that many structures will be found to have evolved through peramorphic processes. Elaborate display structures such as the hypertrophied crests and dewlaps of certain lizards might be candidates for such a study.

Studies of heterochrony in reptiles have lagged behind those in amphibians, although it has been suggested that several groups of reptiles may exhibit paedomorphosis. The reduced ossification observed in the shell and skull of *Dermochelys* (Dermochelyidae) and the reduced shell of the Trionychidae may reflect paedomorphic trends. Such interpretations, however, await confirmation by more detailed developmental studies.

Summary

The vast majority of amphibians and reptiles reproduce sexually. The few asexual species are the result of interspecific hybridization and reproduce by parthenogenesis or hybridogenesis. Parthenogenetic species are all-female and produce unreduced (diploid) gametes. Hybridogenetic species include males and females. Sperm are required for fertilization, but the paternal genome is discarded when gametes are formed and only the maternal genome is transmitted clonally to the next generation.

Sexual behavior and mating is part of the annual cycle of the great majority of amphibians and reptiles. Even some asexual species have a pseudocopulatory behavior. Species with associated reproductive cycles mate when they have high levels of sex steroids and mature gonads. Species with dissociated reproductive cycles mate at a time when hormonal activity is low and gonads are undeveloped. Dissociated reproductive cycles are characterized by the storage of sperm by one or both sexes. Some amphibians and reptiles in the aseasonal tropics may reproduce more or less continuously throughout the year.

Female amphibians and reptiles produce relatively large ova that provide all (oviparous species) or at least some (viviparous species) of the nutrition needed for embryonic development. The eggs of amphibians and reptiles are vastly different, and those differences affect broad aspects of the ecology and reproduction of members of these two groups. The amphibian ovum is covered with jelly. Because the jelly-covered eggs are vulnerable to drying out, they are laid in water or in moist places on land. Amphibian eggs are typically small, and growth during the larval stage provides a bridge to the adult form. In contrast, reptilian eggs are covered by a desiccation-resistant leathery or calcareous shell and are laid on land. Eggs may take up water (most squamates and some turtles) or develop normally with only the water initially in the egg (crocodilians and some turtles). Reptile eggs are relatively large, and newly hatched individuals are small editions of adults.

Sex is determined genetically for amphibians. In contrast, the sex of many reptiles is determined by the temperature at which eggs are incubated (all crocodilians, most turtles, and a few lizards). Individuals of both sexes are produced over a very narrow temperature range (as narrow as 1°C), and only one sex is produced above or below the pivotal temperature. Temperature-dependent sex determination means that the sex ratio in a clutch of eggs varies as a function of the temperature of the nest site. Sex ratios at hatching in wild populations of turtles and crocodilians deviate considerably from the 1:1 ratio usually observed for species in which sex is determined by the genotype of the individual.

The most commonly observed life cycles of amphibians include aquatic egg and larval stages and an aquatic or terrestrial postmetamorphic stage. The larval stage may be eliminated or another stage may be added to the life cycle. An important evolutionary trend for amphibians is a shift from depositing eggs in ponds or streams to breeding on land. This shift often involves parental care, in which the attending parent protects eggs from desiccation or from predators. Most commonly one of the parents stays with the eggs until they hatch. Other forms of parental care include carrying tadpoles to water, guarding tadpoles, providing tadpoles with food by depositing eggs for them to eat, and sequestering eggs and tadpoles within external pouches or inside vocal sacs. A few amphibians are viviparous, and offspring are retained within the oviduct until embryonic and larval development is complete.

The life cycle of reptiles includes an egg and the adult form. Parental care is widespread. Crocodilians exhibit the most elaborate parental care, which extends from guarding the nest to guarding the offspring for several years after hatching. Parental care by squamates consists largely of nest guarding. A shift from oviparity to viviparity is an important evolutionary trend among squamates. Viviparity has evolved in at least 100 different lineages and includes about 20 percent of squamate species.

Amphibians and reptiles exhibit a wide diversity of life history traits such as the age at sexual maturity, the number of clutches produced per year, the size and number of eggs in each clutch, and length of life. Successful offspring can be produced by various combinations of life history traits. Nonetheless, some life history traits repeatedly occur together, and these patterns of covariation form the basis for groupings of species with similar life histories. For example, species that have high annual reproductive effort, as measured both by their relative energy allocation to reproduction and by the intensity of behaviors related to reproduction, have lower adult survival than species that reproduce more modestly. Some life history characteristics of squamates, such as mode of reproduction and the mass of the clutch relative to the body mass of the female, are related to foraging mode. Species that have sedentary foraging modes have larger relative clutch masses than species that forage actively, and are more likely to be viviparous. These patterns of covariation probably reflect the increased risk of predation to gravid or pregnant females.

8

Body Support and Locomotion

The earliest tetrapods were large, robust descendants of fleshy-finned fishes that came ashore in tropical habitats during the Devonian Period, some 370 million years ago. Their stocky limbs presumably moved in a cadence augmented by fishlike undulations of the vertebral column to provide maximum stride length and propulsive force. Although many living species of amphibians and reptiles have retained that ancestral quadrupedal locomotor pattern, both groups have evolved a remarkable variety of locomotor modes and a corresponding diversity of limb and body morphologies. Such locomotor diversity might be expected from the taxonomic and geographic diversity of amphibians and reptiles, and the relaxation of limits on small body size and elongate shape permitted by ectothermy has facilitated this diversification. Living species of amphibians and reptiles include terrestrial forms with quadrupedal, bipedal, or limbless locomotion. Many species are largely or entirely fossorial (burrowing), whereas others are specialized for climbing. Many are aquatic, including larval and adult amphibians and many reptiles. Some aquatic species use undulation of the body as their primary means of propulsion, whereas others use their limbs to generate thrust. Finally, a small number of species take to the air, at least briefly, and have morphological adaptations to generate lift and to break their fall.

This chapter explores the diversity of locomotor modes among amphibians and reptiles from a functional perspective, emphasizing the mechanics of locomotion and examining the underlying similarities among diverse taxa that have converged on similar locomotor patterns. First, however, we will consider some physical principles that govern support and locomotion in animals.

Body Support, Thrust, and Gait

An understanding of animal locomotion requires a modest appreciation for the fundamental laws of physics, specifically Newtonian mechanics. Newton's first law states that a body at rest will remain at rest and one that is in motion will remain so unless acted on by an external force. The second law states that force (F) equals mass (m) times acceleration (a), or $F = m \cdot a$. Finally, the third law states that for every action there must be an equal but opposite reaction. Together, the first and third laws mean that any change in the state of motion of an animal must be the result of an external force acting on it. Furthermore, motion results not directly from the actions of an animal's muscles, but from the reactive force the environment exerts on the animal. During locomotion, the limb of a terrestrial animal pushes on the ground at an angle, eliciting an equal and opposite reaction force on the foot. That force can be resolved into a vertical component, which resists the force of gravity, and a forward propulsive component, which generates thrust in the direction of motion (Figure 8–1).

The problem of support is a very real one for animals, and it strongly affects their locomotor morphology and behavior. Any object within Earth's gravitational field accelerates toward the center of Earth at 980 cm·s^{-2}. The mass of an object times the gravitational acceleration is its weight. Gravity acts on an organism as if all the organism's weight were concen-

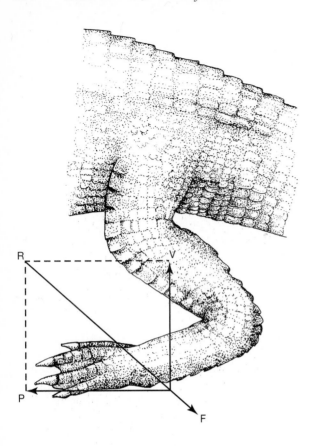

Figure 8-1   **Diagram of the forces acting on the limb of a tetrapod.** Muscular action generates a force (F) by the limb against the substrate. That force elicits a reaction force (R), which is equal in magnitude and opposite in direction. R can be resolved into a vertical component (V), which provides support against the force of gravity, and a forward propulsive component (P). Note that this example is simplified to two dimensions. A more realistic example would include a third component acting in a lateral direction.

trated at a specific point, the center of gravity. The location of the center of gravity depends on the shape of the animal. An animal's center of gravity must remain above its base of support, which is the area enclosed by the points of contact between its feet and the ground. Most amphibians and reptiles have a broad, quadrangular base of support when they stand on all fours (Figure 8–2a), but as they lift their feet during locomotion, the base of support shrinks. When one leg is lifted, for example, the base changes to a triangle. If the animal's center of gravity now lies outside its base of support, it begins to fall. (This is, in fact, the way turtles walk.)

Quadrupeds moving at slow to moderate speeds, such as many salamanders, lizards, and crocodilians, must maintain their center of gravity over their shifting base of support as they move their limbs. The pattern in which their limbs are lifted and placed during locomotion, the footfall pattern, reflects that requirement. Most tetrapods employ a lateral footfall pattern in which movement of the forefoot on each side follows movement of the hind foot on the same side: RH, RF, LH, LF (Figure 8–2b). Such a sequence typically produces the largest triangles of support. (A diagonal sequence, in which the forefoot follows the hind foot of the opposite side, results in narrower triangles; Figure 8–2c.) The base of support is further enhanced by lateral flexion of the vertebral column and by the sprawling limb position of most amphibians and reptiles, in which the upper segment of the limb (humerus or femur) extends horizontally outward from the trunk at right angles to the lower segment (the radius and ulna of the forelimb or tibia and fibula of the hind limb). Keeping the body low to the ground also enhances stability by lowering the center of gravity, making it less likely to lean beyond the base of support.

A gait is a specific pattern and timing of limb movements, and gait involves more than footfall pattern. In a walk each foot is on the ground for at least half of a gait cycle (one complete sequence of footfalls); in a run the feet are in contact for less than half a sequence. The percentage of time a specific foot is in contact with the ground is known as its duty factor. For any given limb the cycle of its movement is known as a stride. Stride can be divided into a propulsive phase, when forward thrust is being generated, and a recovery phase, in which the limb is repositioned before regaining contact with the substrate. Although the stability provided by the lateral footfall pattern is important in a slow gait, animals moving rapidly often use less stable gaits. Salamanders and lizards may advance to a trot, in which diagonal limbs move in synchrony. In a trot the body may be supported only by two diagonal limbs for much of the gait cycle. For example, the body of *Dicamptodon tenebrosus*, a large terrestrial salamander, is supported by two limbs for 92 percent of the time when it is trotting, versus only 41 percent of the time during a lateral sequence walk (Ashley-Ross 1994). When less stable gaits are employed, the center of gravity may fall briefly before the feet reestablish contact beneath the body, or the body may be propelled forward and upward at each step.

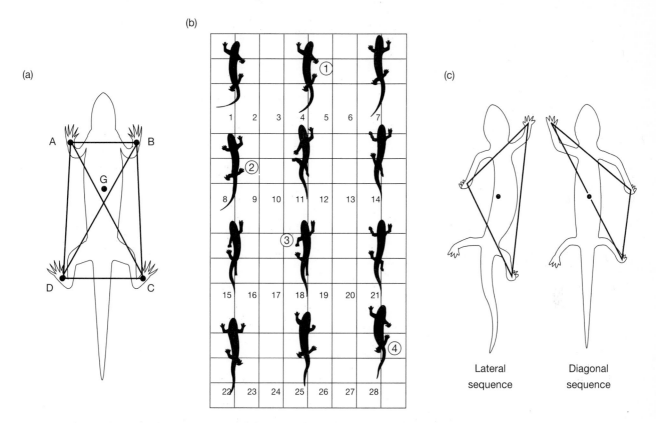

Figure 8-2   **The relationship between center of gravity, base of support, and gait.** (a) A tetrapod standing on four feet, providing a broad rectangular base of support (ABCD). The center of gravity (G) lies well within that base. If the right hind foot is lifted, the result would be a triangular base of support (ABD), which would also underlie the center of gravity and thus would be stable. If the right forefoot were lifted, however, the center of gravity would lie outside the resulting triangle of support (ACD) and the body would fall toward the right. (b) Lateral sequence walk of *Triturus cristatus*, drawn from a cinematographic record. Subsequent images are displaced to the right. The circled numbers indicate the lifting of a specific limb at the beginning of its recovery phase. In a lateral sequence the forelimb follows movement of the hind limb on the same side (RH, RF, LH, LF). Interval between frames is 0.08 second. (c) The relationship between the center of gravity (solid circle) and the triangular base of support in a lateral sequence walk compared with a diagonal sequence walk. (*Source: (a, b) Modified from Gray 1968, (c) Hildebrand 1985.*)

Lever Systems

Consideration of musculoskeletal systems as simple levers provides useful insights into their function. A lever is a simple machine that consists of a rigid bar that pivots around a fulcrum (Figure 8–3). Force applied at one end of the lever is reflected in force generated at the other end. Bones can be considered levers to which muscles can apply force. How that muscular force is modified depends on the configura-

tion of the lever. Every lever consists of two arms, the in-lever (or lever arm, L_i) and the out-lever (or moment arm, L_o). The in-lever lies between the point where force is applied (the in-force, F_i) and the fulcrum. The out-lever lies between the fulcrum and the point where force is produced (the out-force, F_o). Levers are classified by the arrangement of the in-force, out-force, and fulcrum. The simplest case is a first-class lever, in which the in-force and out-force are applied on opposite sides of the fulcrum, so the

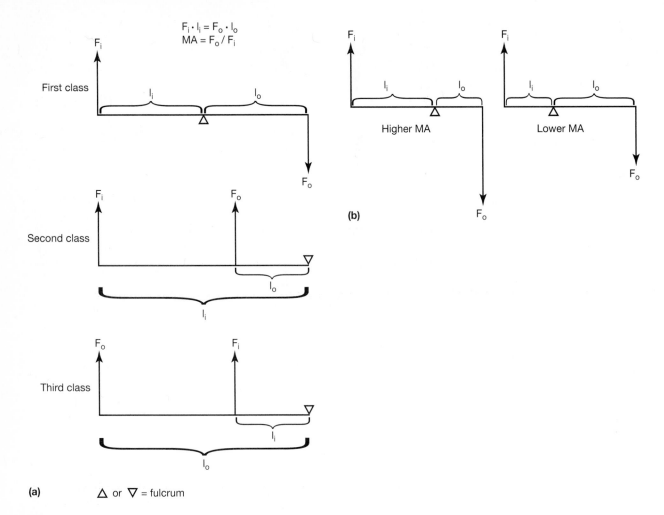

Figure 8-3   **Lever mechanics and their applicability to musculoskeletal systems.** (a) The three classes of levers, which are distinguished by the linear order of the in-force (F_i), out-force (F_o), and fulcrum (which in musculoskeletal systems is usually, but not always, a joint.) In a first-class lever the in-lever (L_i) and out-lever (L_o) are separated by the fulcrum and the in-force and out-force act in opposite directions. In second- and third-class levers the in-lever and out-lever overlap in space, and the in-force and out-force act in the same direction. However, the mathematical relationship between lengths of lever arms and forces is the same for all three classes. The length of the force vectors reflects their relative magnitude, based on the Law of the Lever ($F_i \cdot L_i = F_o \cdot L_o$). Second class levers are uncommon in animals, but first- and third-class levers often result from antagonistic pairs of muscles with opposing actions around a joint. (b) The relationship between relative length of the lever arms and out-force as a consequence of the Law of the Lever. Note that, for a given in-force, the longer the in-lever relative to the out-lever, the greater the force produced. The ratio of the out-force to the in-force is known as the mechanical advantage ($MA = F_o/F_i$) and is a measure of the relative output of the lever. (c) Antagonistic first-class and third-class levers acting on the foot; the ankle joint is the fulcrum. (Skeleton based on that of *Alligator mississippiensis*.) (d) Antagonistic first- and third-class levers acting on the lower jaw; the quadratomandibular joint is the fulcrum. (Skeleton based on that of *Varanus griseus*.) In reality, the effective in-force transmitted to the in-lever by each of the muscles shown would be somewhat less than the muscle's total force, due to the angle at which the muscle attaches. *(Source: (c) Modified from Gatesy 1991; (d) modified from Savitzky 1980.)*

in-lever and out-lever are physically distinct. (A see-saw is a first-class lever.) In a second-class lever the fulcrum lies at one end of the rigid bar and the in-force is applied at the other; the out-force is generated between the two (e.g., a nutcracker). In a third-class lever the fulcrum also lies at one end, but the out-force is generated at the other, with the in-force being applied between the two (e.g., a pair of forceps). In both second- and third-class levers, the in-lever and out-lever physically overlap each other for some distance.

There are three ways in which to generate more force from a musculoskeletal system. The most obvious way is to increase the in-force—that is, to put more muscular force in. Less obvious, however, is that out-force can be increased by increasing the length of the in-lever or decreasing the length of the out-lever. Another way of stating this relationship is in terms of the mechanical advantage (MA), defined as F_o/F_i, the amount of force generated by the lever for the force applied to it. From the equation in Figure 8–3 it is clear that for any given lever $MA = L_i/L_o$. Although this formula is easiest to visualize with a first-class lever, it acts precisely the same way for second- and third-class levers. Second-class levers are rather uncommon in animals, but first- and third-class levers are common and often reflect the actions of antagonistic pairs of muscles that produce opposing motions around the same articulation.

Terrestrial Locomotion with Limbs

Locomotion by means of sprawling limbs is characteristic of most salamanders, lizards, and to a lesser degree crocodilians when they move on land. The sprawling posture not only promotes stability by providing a broad base of support, it also works in tandem with lateral undulations of the body that early tetrapods inherited from their osteichthyan ancestors. Although both lateral undulation of the vertebral column and sprawling limb position are ancestral features, the two attributes work together as a sophisticated system. Most amphibians and reptiles do not remain standing for long periods while motionless. Instead they allow their bodies to rest on the substrate, minimizing the energy required for postural support. Some groups have evolved a more nearly vertical limb position (erect posture), often associated

with the demands of supporting great weight (dinosaurs), achieving a narrow body profile (chameleons), or moving rapidly (crocodilians).

A sprawling limb may be moved in any of three separate ways, and together those movements contribute to the total propulsive force of the limb (Figure 8–4). First, the upper segment can be swung anteroposteriorly (from front to back) relative to the limb girdle. Second, the girdle itself can be thrust forward as the body axis bends laterally. Finally, the upper element can be rotated about its long axis. When that happens the lower segment, which lies at a right angle to the upper, swings forward or backward like the spoke of a wheel. James Edwards (1977) filmed 48 species of terrestrial salamanders and quantified the contribution that each mode of limb motion made to propulsion. Across various speeds and gaits, retraction of the limbs relative to the girdles produced 56 to 62 percent of the propulsion, rotation of the girdles due to axial flexion produced 10 to 18 percent, and rotation of the upper limbs around their long axes produced 26 to 28 percent. As speed increased the effect of axial flexion became greater and the contribution from the other two sources decreased.

Lizards

Lizards have important specializations of the vertebrae, pectoral girdle, and hind limb that enhance the function of their sprawling limbs. The vertebrae of lepidosauromorphs allow substantial lateral bending, which promotes longer stride length. The pectoral girdle was studied in the savanna monitor, *Varanus exanthematicus*, by Farish Jenkins and George Goslow (1983), who combined detailed dissections of the muscles and ligaments with electromyography and cineradiography. Electromyography involves monitoring the actions of muscles using fine electrodes, usually synchronized with motion pictures or videography. An alternative is cineradiography, in which an x-ray movie is recorded. Jenkins and Goslow trained lizards to walk on a treadmill, which kept them walking in place during filming. Electromyography revealed that seemingly antagonistic muscles, those with opposite actions around a joint, sometimes contracted simultaneously. Although such muscles might be expected to contract sequentially to generate opposing motions around a joint, their simultaneous action in a

Figure 8-4 **Diagram showing movement in a sprawling limb, such as that of a salamander or lizard.** (a) Axial flexion, resulting in movement of the limb girdle. (b) Retraction of the limb relative to the girdle. (c) Rotation of the upper element of the limb. Straightening of the limb itself also contributes to propulsion.

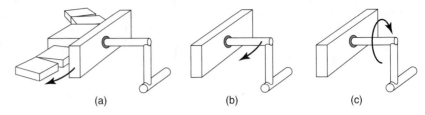

(a)   (b)   (c)

sprawling limb probably stabilizes the elbow or shoulder. Vertebral flexion moves the shoulder region through an arc of about 40 to 60 degrees from side to side (Figure 8–5a). That motion is coupled with retraction of the humerus through an arc of 40 to 55 degrees and rotation of the humerus 30 to 40 degrees around its long axis.

An additional increase in stride length results from an unusual joint between the pectoral girdle and sternum in lepidosaurs (Figure 8–5b). The humerus articulates with the pectoral girdle at a socket formed by the scapula and the coracoid. Behind the paired coracoids, in the ventral midline, lies a shield-shaped sternum. Together, the coracoid and sternum form a

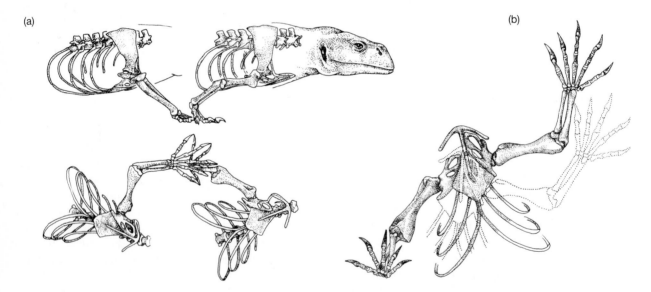

(a)   (b)

Figure 8-5 **The function of the pectoral girdle of the savanna monitor lizard, *Varanus exanthematicus*, during locomotion.** (a) Lateral (top) and ventral (bottom) views of the pectoral girdle and forelimb at the beginning (left) and end (right) of the propulsive phase. The T-shaped medial bone in ventral view is the interclavicle, the broad element behind it is the sternum, and the coracoids lie against either side of the sternum and articulate with the limb. Note the posterior sliding of the coracoid relative to the sternum by the end of the propulsive phase. (b) Ventral view of the pectoral girdle and sternum, with the right forelimb (left side of figure) at the end of the propulsive phase and the left forelimb (right side of figure) at the beginning of that phase. Note the posterior position of the coracoid at the end of the propulsive phase. The dotted lines indicate the position of the left forelimb if there were no movement at the coracosternal joint, indicating the contribution of that joint to the length of the stride. (*Source: Jenkins and Goslow 1983.*)

tongue-and-groove joint that allows the coracoid to slide posteriorly relative to the sternum. Anterior movement is prevented by the clavicles and interclavicle, which are firmly attached to the sternum. The cineradiographs demonstrated that while the forefoot of one side is in contact with the substrate during its propulsive phase, the ipsilateral (same side) coracoid slides posteriorly for a distance equal to about 40 percent of the length of its resting contact with the sternum. The result is that the contralateral (opposite side) limb is propelled forward for that same distance, significantly increasing the stride length.

The hind limbs of lizards also have specializations that enhance the efficiency of sprawling locomotion, including an unusual condition of the tarsal bones and a characteristic shape of the fifth metatarsal. The hind limbs of lizards are longer and usually more robust than the forelimbs and provide most of the propulsive force. However, the ability of a hind limb to maximize its propulsive force in the direction of movement is compromised by the lateral orientation of the femur and its swing through a horizontal arc relative to the body as the limb is retracted (Rewcastle 1980). The knee joint, which is a hinge, is oriented transversely when the femur faces forward at the start of the propulsive phase, and in that position extension of the knee would generate force primarily in a forward direction. However, as the femur is retracted, the knee joint changes orientation. With the femur at a right angle to the body, the knee is parallel to the body axis, and extension of the joint should generate force mainly toward the midline of the body, rather than in the direction of movement.

The hind limbs of lizards function effectively in spite of this geometric paradox, but the mechanisms are not yet fully understood. Recent functional studies have come to different conclusions than earlier ones based solely on comparative morphology, and the differences may be due not only to technique but also to differences between the taxa studied. Morphological studies based primarily on large genera such as *Iguana* and *Varanus* suggested that asymmetry in the knee joint and complex articular surfaces in the ankle combine to produce thrust primarily in an anterior direction, with little lateral component of force (Brinkman 1980a,b, Rewcastle 1980). Certainly the ankle joint of lizards is very different from that of mammals, in which the tibia and fibula rotate against the proximal tarsal bones, the astragalus and calcaneum. The astragalus and calcaneum of most lizards are fused into a single bone, the astragalocalcaneum, which is rigidly sutured to the tibia and fibula. The astragalocalcaneum rotates against the more distal tarsals, forming an intratarsal joint with complex surfaces that limit the movement of the foot as the ankle straightens. Earlier workers argued that those joint surfaces ensured anteriorly directed propulsive forces. Functional studies on a much smaller species, however, suggest that substantial forces are in fact generated in a medial direction. Using high-speed videography and electromyography, Stephen Reilly (1995) found that the hind limbs of *Sceloporus clarki* remain in a strongly sprawling position throughout the gait cycle and apparently produce substantial lateral forces, eliciting a reaction force directed toward the midline of the lizard's body. Those forces from the right and left hind limbs counter each other, leaving a net forward propulsive force. As with the forelimbs, vertebral flexion makes an important contribution to propulsion by shifting the limb girdle, while the large caudofemoralis muscle, which runs from the base of the tail to the femur, provides much of the force for retraction of the limb relative to the girdle (Reilly and DeLancey 1997).

The lizard hind foot is unusual because the first four metatarsal bones are tightly bound together. The fifth toe looks as though it had been added as an afterthought, often lying splayed behind at an awkward angle to the others (Figure 8–6a), and the fifth metatarsal bone is distinctly L-shaped, or hooked (Figure 8–6b). An additional ventral inflection of that bone forms a pair of plantar tubercles. Brinkman (1980b) studied the role of the fifth metatarsal during walking in the green iguana, *Iguana iguana*. Late in the propulsive phase of the stride the peroneus brevis muscle apparently contracts. This muscle inserts on the outer corner of the fifth metatarsal, powering a first-class lever that lifts the foot onto the first four metatarsals. Once the foot is elevated, final straightening of the ankle involves the powerful gastrocnemius muscle, which inserts on the plantar tubercles of the fifth metatarsal, forming another first-class lever with a different orientation. The function of the hooked fifth metatarsal as a first-class lever powering extension of the ankle has a close analogue in mammals, in which the gastrocnemius muscle inserts by the strong Achilles tendon on the calcaneal tuber, the bony projection that forms the heel in humans. Contraction of the gastrocnemius lifts up on the heel, pressing the ball of the foot against the substrate and eliciting a reaction force for propulsion.

Figure 8-6   **The hind foot of a lizard.**
(a) Photograph of the hind foot of the
iguanid *Ctenosaura quinquecarinata* showing the first four metatarsals bound together and the divergent fifth digit. (b)
Skeleton of the hind foot of *Iguana
iguana*, showing the hooked fifth metatarsal (V), in dorsal view (left) and posterior view (right). The peroneus brevis
muscle attaches to the outer rear corner
of the bone (large arrow), and the gastrocnemius muscle inserts on the plantar
tubercles on its ventral surface (small arrows). *(Source: (a) Photograph by R. Wayne
Van Devender. (b) Modified from Brinkman
1980b.)*

(a)

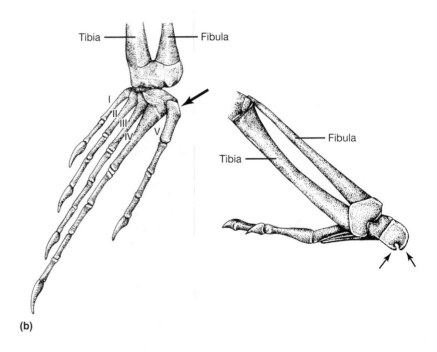

(b)

The complex mechanics that govern the sprawling limbs of lizards clearly are capable of generating substantial propulsive forces, and some lizards can achieve considerable speeds. The Komodo monitor (*Varanus komodoensis*), which reaches a total length of about 1.7 meters, can briefly attain speeds of up to 18.5 kilometers per hour and can sustain a speed of about 14 kilometers per hour for over 0.5 kilometer, fast enough to ambush small deer and wild boar as they move along trails in the Indonesian forest.

Some species of lizards can run on their hind legs alone. Such species are considered dynamic bipeds, capable of supporting themselves on two legs only while running. Many belong to iguanian genera, such as *Basiliscus* (Corytophanidae), *Crotaphytus* (Crotaphytidae), and *Chlamydosaurus* (Agamidae). The Neotropical *Basiliscus* employs bipedal locomotion when avoiding predators, but does so by running across the surface of the water (Figure 8–7). These riparian lizards have fringed scales on the posterior edges of their toes (Luke 1986) that provide resistance as they strike the surface of the water. Air rushes in behind the foot to form a pocket in the surface within which the foot can be withdrawn on the recovery stroke without encountering resistance from the water (Glasheen and McMahon 1996). Its locomotor habits have earned *Basiliscus* the local name of *lagartija Jesu Cristo* (Jesus Christ lizard).

Bipedal locomotion requires no major structural changes. Bipedal species tend to have relatively long hind limbs, short forelimbs, a short presacral vertebral column (resulting in a posterior shift in their center of gravity), and a robust articulation between the pelvis and vertebral column. Many also have a relatively long tail, which serves as a counterweight to balance the unsupported head and body during bipedal locomotion. The fibers of the muscles are concentrated close to the body, with long tendons that extend down the leg. A longer hind limb swings through a greater arc and so moves farther in a given period of time. Likewise, a shorter muscle contracts more quickly. Both features increase locomotor speed but do so at the expense of force because the longer limb typically has a proportionately longer out-lever. The concentration of muscle mass proximally also reduces the energy needed to swing the leg during both the propulsive and recovery phases. Another common feature of bipeds is an elongation of the plantar tubercles of the fifth metatarsal, increasing the in-lever of the gastrocnemius for more powerful thrust.

Locomotion on loose sand presents a special problem because the substrate yields to a propulsive force, rather than producing a strong reaction force. Sand-dwelling (psammophilous) lizards in many lineages have a fringe of scales on the trailing edge of each of the toes (Luke 1986, Bauer and Russell 1991).

Figure 8-7 **Basiliscus** **running bipedally across the surface of the water.** *(Photograph by Stephen Dalton/NHPA.)*

Fringed toes are characteristic of many lizards, including *Uma* (Iguanidae), *Phrynocephalus* (Agamidae), *Ptenopus* (Gekkonidae), *Scincus* (Scincidae), and *Acanthodactylus* (Lacertidae).

Crocodilians

Although crocodilians appear at first to have sprawling limbs like those of lizards, the situation is somewhat different. Compared with lepidosauromorphs, archosauromorphs have vertebrae that apparently support stronger axial muscles. Although the limbs sprawl to the side when a crocodilian rests, on land crocodilians typically use a high walk, in which the limbs are placed in a semierect position, midway between the sprawling and erect postures. In that position the movement of the limb is more nearly parasagittal (parallel to the long axis of the body) than in a lizard. Crocodilians also may employ a sprawling gait or use a belly slide when entering water from a smooth bank. The tarsus is complex, with one joint between the astragalus (which is firmly articulated with the tibia and fibula) and the calcaneus and another between the proximal and distal tarsals (Brinkman 1980b). Although the intratarsal joint apparently functions during sprawling locomotion, the high walk involves mainly the astragalocalcaneal articulation. The calcaneum bears a posterolateral tuber, which functions with the gastrocnemius in a first-class lever analogous to that of mammals (see Figure 8–3c).

Functional studies of crocodilian locomotion in the laboratory have been limited to relatively young individuals (Brinkman 1980b, Gatesy 1991). Indeed, large crocodilians can reach a length of 7 meters and a body mass greater than 1,000 kilograms. Small individuals, however, are capable of rapid terrestrial locomotion, and a bounding gait has been described for small *Crocodylus porosus*, the saltwater crocodile, and *C. johnstoni*, the freshwater crocodile (Webb and Gans 1982). In a bound, both hind feet contact the ground almost simultaneously, as do both front feet. Dorsoventral flexion of the vertebral column, an unusual motion for reptiles, extends the stride of bounding crocodilians. In *C. johnstoni* each bound carries the animal about 1.3 times its body length, and the crocodiles attain speeds of 10 to 15 km·h⁻¹ (Webb and Gans 1982). That species has also been reported to gallop, a gait in which the members of each pair of feet land slightly out of step, primarily as they transition from a rapid high walk to a bound.

Turtles

Turtles face unusual problems when walking on land. Their vertebrae and ribs are fused to the shell. That means that turtles cannot bend the vertebral column to shift their center of gravity. Likewise, their limb girdles are enclosed within the shell, and the humerus and femur can move only within the limits of the openings between the carapace and plastron. Furthermore, the relatively slow locomotion of most terrestrial turtles is inherently unstable unless the center of gravity can be maintained over the base of support. Turtles exhibit some obvious concessions to their locomotor problems. Their limbs are in contact with the substrate for extended periods, with duty factors of about 80 percent, and their short, broad bodies and lateral footfall pattern provide wide triangles of support (Figure 8–8). However, a turtle typically begins to lift its hind foot before the contralateral forefoot has been placed on the substrate, leaving the center of gravity to fall forward before the triangle of support is established. As a result the body pitches and rolls through angles of about 10 degrees from the horizontal. Jayes and Alexander (1980) filmed several species of turtles as they walked across a force platform, a device that measures the magnitude and direction of the force exerted by a foot against the ground. The greatest component of force proved to be vertical, supporting the mass of the body. The turtle's forward velocity varied considerably and the body pitched and rolled, with the forefeet catching the turtle as it fell forward. Jayes and Alexander suggested that the slowly contracting muscles of turtles may not be capable of adjusting rapidly enough to the shifting weight to achieve greater stability.

Jumping

Jumping (saltation) is characteristic of the anuran amphibians. In jumping, propulsion is generated by the two hind limbs acting simultaneously rather than alternately. Saltation benefits from a short and relatively rigid vertebral column and greatly elongated hind limbs (Figure 8–9). Not only does a frog's hind limb have a long femur, lower leg (with a fused tibiofibula), and hind foot, but two additional functional segments are added to the limb. The astragalus and calcaneum of the ankle are elongate and articulate with the more distal tarsals, and the pelvic girdle itself

Figure 8-8 **Terrestrial locomotion by the painted turtle, *Chrysemys picta*.** Circles represent the center of gravity, and lines are drawn between limbs that are in contact with the substrate. Note the broad triangles of support and periods when the body is supported by only two diagonal limbs. *(Source: Walker 1971a.)*

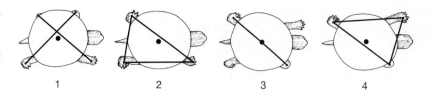

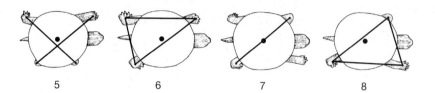

is movably articulated with the vertebral column via the diapophyses (the enlarged transverse processes) of the sacral vertebrae. The girdle is shaped like an elongate U, with very long ilia (the pelvic elements that articulate with the sacral vertebra), so that the hip joints lie far posterior to the articulation of the pelvis with the vertebral column. The urostyle (a rod composed of fused caudal vertebrae) lies between the ilia, and muscles passing between the ilia and the urostyle contribute to the jump. The overall result is a pair of very long, multiply jointed, and powerfully muscled hind limbs that launch the frog off the substrate.

The height and distance of the frog's jump depend on two variables, the takeoff angle and takeoff velocity. Jump height is maximized when the takeoff angle is 90 degrees from the horizontal (that is, the frog jumps straight up), and jump distance is greatest when the takeoff angle is 45 degrees. Takeoff velocity is more complex and depends on two independent variables: the acceleration generated by the action of the animal's muscles and the distance over which that acceleration acts. Both the animal's body mass and the downward acceleration due to gravity act to decrease the height and length of a jump. (From Newton's second law, $F = m \cdot a$, so $a = F/m$. That is, acceleration is inversely proportional to mass.)

This analysis suggests that jumpers should maximize muscular force and the distance through which that force acts, while minimizing body mass. The distance through which force is applied to the body can be increased by lengthening the limbs, thereby prolonging contact with the substrate while muscular force is applied (Figure 8–9). Hind limb length roughly correlates with relative jumping ability among taxa, although differences in relative muscle mass probably also contribute to differences in performance.

A detailed examination of jumping by *Rana temporaria* showed that the strongest jumps involved a takeoff velocity of about $1.8 \text{ m} \cdot \text{s}^{-1}$ at a takeoff angle of about 55 degrees, and resulted in a jump distance of about 0.4 meters on level ground (Calow and Alexander 1973). Although the contributions of individual muscles change in a complex fashion as the hind limbs straighten, the primary propulsive muscles include a series that retract the femur (semimembranosus, semitendinosus, and gracilis major), a pair that extend the knee (gluteus magnus and cruralis), and one that extends the joint between the tibiofibula and the astragalus and calcaneum (plantaris longus). Together these muscles compose 60 percent of the muscle mass of the hind limbs. A recent electromyographic study of both jumping and swimming by *Rana pipiens*, the leopard frog (Kamel et al. 1996), revealed that the hind limb extensor muscles continue to contract at a low level long after the limb is fully extended. Their action may maintain the hind limbs in a streamlined position.

A series of important papers by Sharon Emerson (1982, 1983, 1985a) has contributed to an understanding of the role of the limb girdles in anuran lo-

Figure 8-9 **A red-eyed tree frog (*Agalychnis callidryas*) illustrates the characteristics that maximize jump distance.** The very long hind limbs and feet extend the distance through which force can be applied, and the takeoff angle is approximately 45 degrees. (*Photograph by Stephen Dalton/Photo Researchers Inc.*)

comotion. Several features of the pelvic girdle and sacrum tend to vary in concert and constitute a functional unit (Figure 3–12). In strong jumpers, such as *Rana*, the ilia are relatively short and the sacral diapophyses tend to be round in cross section, allowing the ilia to rotate around them, straightening the trunk. Muscles that run from pronounced ilial crests and the dorsal surface of the trunk vertebrae to the urostyle extend the vertebral column during the early takeoff, straightening the trunk. This action aligns the trunk with the pelvis and aims the body about 45 degrees from the horizontal.

In contrast to strong jumpers, hopping and walking anurans such as *Bufo* tend to have wider iliosacral joints that permit the ilia to swing laterally, a motion that contributes to the lateral gait when such species walk. A third sacral condition consists of broadly expanded sacral diapophyses enveloped by a wide ligament that permits the ilia to slide anteroposteriorly and, to a lesser degree, laterally. This condition is found in a range of specialized taxa, including some branch-walking species such as *Agalychnis* (Hylidae), fossorial (burrowing) species such as *Scaphiopus* and *Spea* (Pelobatidae), and the specialized swimmers of the family Pipidae. In all of these taxa, pelvic sliding contributes to thrust.

Frogs land on their forelimbs, and the shock of landing is borne as compressive forces acting through the limbs on the pectoral girdle. Although two basic forms of pectoral girdle architecture have long been recognized among frogs (Figure 3–13), most attention centered on the phylogenetic significance of the two types. In the arciferal (arch-bearing) condition that is ancestral for anurans, the epicoracoid cartilages remain separate elements that form an arch between the medial ends of the coracoid bones and the clavicles, overlapping midventrally. In contrast, in the derived firmisternal condition the epicoracoids are fused into a midventral partition between the paired coracoid bones. Both girdle types occur among frogs with a range of jumping abilities and locomotor behaviors. Emerson (1983) studied movements in the girdle during landing using cineradiography and recorded landing forces using a force plate. In an arciferal girdle the two epicoracoid cartilages slide past one another during landing. The coracoids experience compressive stress, whereas the epicoracoid cartilages experience tension. In the firmisternal girdle the compressive loading of the coracoids is transmitted to the fused, medial epicoracoid cartilage. The alternative morphologies appear to be functionally equivalent in their role as shock absorbers.

Terrestrial Limbless Locomotion

The reduction or loss of limbs has been a recurrent evolutionary theme among amphibians and reptiles. Three major lineages—caecilians (except one extinct species), amphisbaenians (with one exception), and snakes—are functionally limbless from the standpoint of locomotion. In addition, many other families of squamates (such as the Pygopodidae, Scincidae, and Anguidae) include species that have partial or complete reduction of limbs, often with numerous intermediate conditions represented among living forms. Although most salamanders with reduced limbs are aquatic, many terrestrial species do not employ their limbs when harassed, but rather undulate their bodies in a manner similar to limbless reptiles (Edwards 1985). Limb reduction occurs in several genera in the plethodontid tribe Bolitoglossini, including *Lineatriton* and *Batrachoseps*.

Limb reduction is associated with elongation of the body in a complex and still incompletely understood fashion (Raynaud 1985). Limb reduction and loss follow different morphological patterns in different squamate taxa. For example, among the amphisbaenians the hind limbs appear to have been lost before the forelimbs, although vestiges of the former may remain. The only amphisbaenian retaining external limbs, *Bipes*, retains short but functional forelimbs. In contrast, most other squamates with reduced limbs have larger hind limb vestiges than forelimb ones, and often only hind limb vestiges remain, as in the Pygopodidae and some snakes. In some snakes the vestigial limbs, while useless in locomotion, function in courtship or male combat.

Snakes

Terrestrial limbless locomotion has been most extensively studied in snakes, in which six modes of locomotion are recognized (Cundall 1987). These six modes grade into one another, and several may be employed simultaneously at different points along the animal's body. The six locomotor types can be divided into those in which there are no static points of contact with the substrate (lateral undulation and slide-pushing) and those in which there are (rectilinear, concertina, sidewinding, and saltation). The anatomical basis for the locomotor system of snakes is a series of several hundred vertebrae and extraordinarily complex, multisegmental muscle chains comprised of elongated and interconnecting segmental muscles and tendons (Gasc 1981; Figure 8–10). The spinalis muscle, for example, can span as many as 45 vertebrae. Morphological, functional, and physiological studies have greatly clarified the mechanisms underlying locomotion in snakes, as well as its energetic costs.

Lateral undulation is the most widely used locomotor mode. Horizontal waves travel down alternate sides of the body axis and generate force at fixed points in the animal's physical environment (Figure 8–11). Those points may be surface irregularities—rocks, sticks, and the like—and are usually known by the French term *points d'appuis* (meaning points of application of force; also called pivot points). The body pushes posterolaterally on each point, generating a force that has both lateral and posterior components. The lateral components cancel each other, leaving a net rearward component pushing against the substrate, which elicits a forward reactive force for propulsion. A similar pattern is used by animals traveling through branches, but a vertical component to the force is also generated at each point.

Bruce Jayne (1988a,b) accomplished the daunting task of studying snake locomotion electromyographi-

cally and showed that contraction of the three primary lateral muscle masses (the spinalis-semispinalis, longissimus dorsi, and iliocostalis) on the same side is virtually synchronous at any point along the body. Surprisingly, the tail of a snake seems not to contribute greatly to the generation of propulsive force. A comparison of garter snakes, *Thamnophis sirtalis* (Colubridae), with both natural and shortened tails revealed little difference in locomotor performance (Jayne and Bennett 1989).

Lateral undulation differs in several important ways from limbed locomotion. First, there are no fixed points on the body at which propulsive force is generated. Rather, the body moves continuously past fixed points in the environment, against which more posterior body segments come to push. Second, there is no recovery phase analogous to the lifting and protraction of a limb, nor is there an obvious vertical component of force applied to provide support against gravity. Some workers have suggested that lateral undulation might be more efficient energetically than quadrupedal locomotion. That idea was dispelled by Walton et al. (1990), who demonstrated that lateral undulation by *Coluber constrictor* (Colubridae) is equal in energetic cost to quadrupedal locomotion of a lizard of similar mass. The inefficiencies of lateral undulation may include the loss of energy to the lateral components of axial motion, loss through friction between the extensive ventral surface and the substrate, and the likelihood that some muscular energy still must be expended to support the body through postural adjustments of the axial muscles.

Slide-pushing, employed on low-friction substrates, resembles lateral undulation in that it relies on alternating waves of body motion. It differs, however, in not using fixed points in the physical environment to generate forward reaction forces. Instead, body waves are propagated so rapidly that they generate sliding friction sufficient to produce a large enough reaction force to propel the snake forward (Figure 8–12a). The appearance is of a snake flailing about on a smooth surface, but gradual forward progress is made.

Concertina locomotion may also be used on low-friction surfaces. It involves repeatedly establishing a stable platform with one portion of the body while another portion is moved (Figure 8–12b). For example, the anterior region of the body may remain stationary while the posterior end is drawn up behind it in a series of tight curves. Next, with the posterior

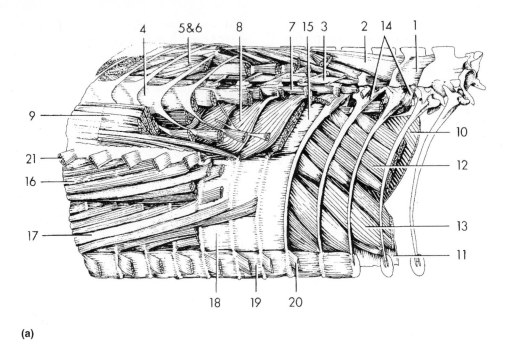

(a)

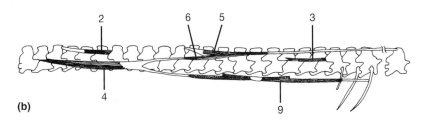

(b)

Figure 8-10   **The complex axial muscles of snakes.** (a) Slightly oblique right lateral view (anterior is to the right) and dissection of the axial muscles of the viperid *Cerastes vipera*. Note the overlapping segmental units of each muscle mass. (b) Right lateral view showing single segmental units of the major axial muscles of the colubrid *Nerodia fasciata*. Note that several of the muscles are linked through their tendons into chains that span many vertebrae. Key: 1 = interneuralis, 2 = multifidis, 3 = interartciularis superior, 4 = longissimus dorsi, 5 = spinalis, 6 = semispinalis, 7 = interarticularis inferior, 8 = levator costa, 9 = iliocostalis, 10 = transversus dorsalis, 11 = transversus ventralis, 12 = obliquus internus dorsalis, 13 = obliquus internus ventralis, 14 = tuberculocostalis, 15 = intercostalis quadrangularis, 16 = supracostalis lateralis superior, 17 = supracostalis lateralis inferior, 18 = intercostalis externus, 19 = intercostalis ventralis, 20 = costocutaneus inferior, 21 = costocutaneus superior. *(Source: (a) Cundall 1987. (b) Modified from Jayne 1985, 1988a.)*

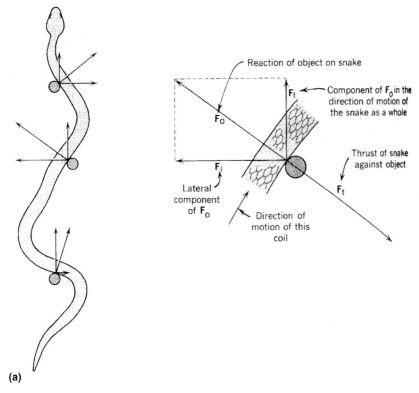

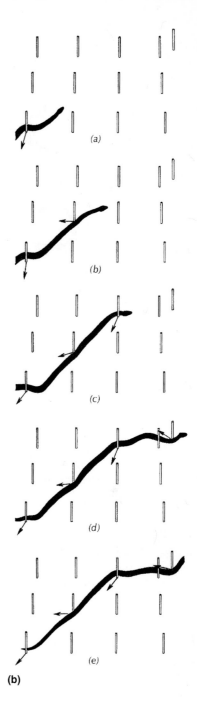

Figure 8-11 **Undulatory locomotion of snakes.** (a) A snake in contact with three *points d'appuis*, showing the magnitude and direction of forces acting against the body. Different points elicit different magnitudes and directions of force. The right and left laterally directed components of the reaction forces cancel each other, while the forward components are additive, contributing to forward movement. (b) Tracings of a rat snake (*Elaphe*) passing through a field of vertical pegs, viewed from a dorsolateral angle. Interval between tracings is 0.17 second. The arrows show the direction of the forces applied by the snake to the pegs. Note that the snake continues to use the same pegs as it moves, thus passing successive points on its body past fixed *points d'appuis*. (Sources:(a) Hildebrand 1995; (b) Gans 1974.)

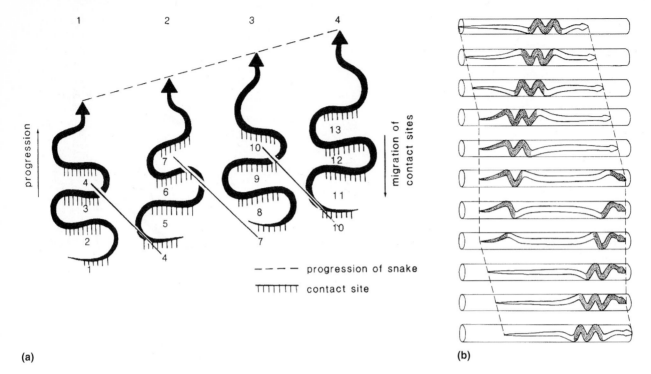

Figure 8-12 **Specialized locomotor modes of snakes.** (a) Slide-pushing, in which waves pass rapidly down the body, eliciting enough sliding friction to propel the body forward slowly. Four successive positions are shown (1–4), displaced to the right for illustration. The head points in the direction of travel. Note that the zones of contact formed by waves of the body (numbered 1-13) travel backward much more quickly than the snake progresses. (b) Concertina locomotion by a snake moving through a narrow tube. Note the use of tight bends of the body to provide regions static contact (shaded) against the sides of the tunnel. *(continued)*

end stationary, the anterior region is extended forward. The sequence is then repeated. This mode is effective on the ground, where static friction alone is used to provide resistance to rearward slippage, and in restricted areas such as burrows, where the body can be braced against the walls. Arboreal snakes, such as some boids and tree vipers, use their prehensile tail to provide anchorage while extending the body. Concertina locomotion is relatively slow. The banded water snake, *Nerodia fasciata*, can travel 1.88 total lengths per second using lateral undulation, but only 0.05 total lengths per second using concertina locomotion. It also is energetically costly, requiring seven times more energy than lateral undulation in the racer, *Coluber constrictor* (Walton et al. 1990).

Sidewinding is associated with low-friction or shifting substrates, generally sand dunes or mud. Most snakes appear to be capable of sidewinding, and some, such as *Cerberus*, an Australasian colubrid that inhabits mud flats, use that mode when frictional resistance is insufficient for lateral undulation. The most specialized sidewinders include the sidewinder rattlesnake, *Crotalus cerastes*, of the American Southwest, and several sand-dwelling Asian and African viperines. Watching a sidewinder can be a bewildering experience because loops of the body appear to be thrown in all directions. In fact, sidewinding is a highly ordered process in which virtually all forces directed against the substrate act vertically, avoiding the slipping that would result if the body contacted the ground at an angle. Sections of the body are alternately lifted, moved forward, and then set down (Figure 8–12c), producing a series of separate, parallel tracks, each oriented at an angle to the direction of travel. The snake itself usually is in static contact with the ground at two points. *Crotalus cerastes* can proba-

Figure 8-12 (continued) (c) Sidewinding by the sidewinder, *Crotalus cerastes*, showing five successive positions. All forces against the substrate are vertical, and sections of the body are lifted and moved laterally between tracks. The head faces approximately in the direction of motion, and the sidewinder travels at an angle to the unconnected tracks it leaves in the sand. *(continued)*

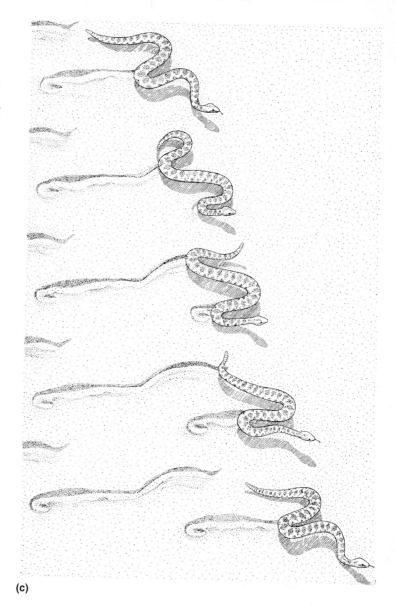

(c)

bly attain forward velocities of 2.0 total lengths per second.

Saltation is an extreme form of motion that has been reported in *Bitis caudalis*, a short, heavy-bodied viper from southern Africa. It occurs when a rapid straightening of the body from anterior to posterior lifts the entire body off the substrate, and is limited to very small individuals.

Unlike the preceding modes of ophidian locomotion, rectilinear locomotion does not rely on alternating contraction of the lateral muscle masses of the trunk. Instead, muscles on both sides of the body act synchronously, sequentially contracting and relaxing to draw the body forward in a more or less straight line. Rectilinear locomotion relies primarily on the two series of costocutaneous muscles, which run from the ribs to the ventral skin (Figure 8–12d). One pair, the costocutaneous superior, pulls the skin forward relative to the ribs, after which the ventral scales anchor the body to the substrate. The other pair, the costocutaneous inferior, then pulls the ribs—and with them the vertebral column, axial muscles, and viscera

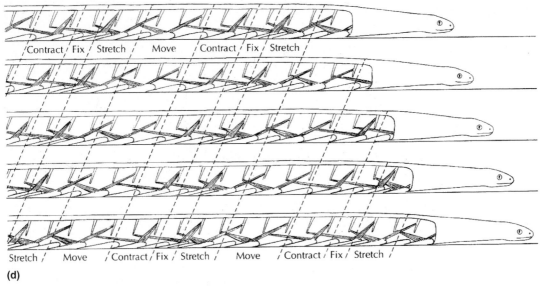

(d)

Figure 8-12   (continued) (d) Rectilinear locomotion by *Boa constrictor.* The body wall is shown as transparent to reveal the ribs, costocutaneus superior muscles (angling up and forward), and costocutaneus inferior muscles (angling up and back). Only every tenth rib, muscle, and ventral scale are shown. Waves of bilateral contraction of the costocutaneus muscles pass posteriorly, alternately stretching, fixing, contracting, and moving the skin. The costocutaneus superior pulls the skin anteriorly, after which the costocutaneus inferior pulls on the ribs, drawing the body forward. *(Sources: (a) Gans 1984; (b–d) Gans 1974.)*

—forward relative to the stationary ventral skin. Generally several waves of such symmetrical contractions pass down the body simultaneously, establishing several points of static contact with the substrate simultaneously and presenting an appearance that the ventrolateral skin is crawling on its own while the dorsal skin moves at a nearly even rate. Although most snakes can employ rectilinear locomotion, it is most commonly used by heavy-bodied snakes such as large boids and vipers (Edwards 1985).

Aquatic Locomotion

Aquatic locomotion differs in two important respects from that on land, and both reflect the physical properties of water. Water is both dense and viscous, features that simultaneously enhance and detract from locomotor efficiency. Water provides greater support against the force of gravity than does air, and its density makes it possible to elicit substantial reaction forces for propulsion. However, it is also heavy to move and difficult to push through, requiring considerable power for locomotion. The physics of locomo-

tion in water are extremely complex, varying with speed and temperature. We will take a general approach to the problem here, recognizing that greater complexity characterizes real situations.

Swimming vertebrates can conveniently be divided into undulatory and oscillatory propulsors. Undulatory propulsors generate waves of movement along their bodies, whereas oscillatory propulsors employ limbs or fins that move back and forth. Both amphibians and reptiles include species that fall into both categories.

All animals that swim encounter resistance due to the viscosity of water. That resistance is known as drag, and there are several sources of drag. The most important form of drag for amphibians and reptiles is surface drag, which results from the sliding of layers of water past the body. In fact, a thin layer of water known as the boundary layer effectively adheres to the body, while beyond that point we can think of the water being pulled along at a velocity that decreases with increasing distance from the moving body. When layers of moving water slide smoothly past each other they are said to exhibit laminar flow; when they mix roughly with each other they are said to exhibit turbulence. Surface drag operates around the entire surface of the animal, but it can be modified by altering the surface texture to improve retention of the boundary layer and thereby reduce turbulence or by changing the profile of the limb or body as it passes through the water. A larger surface will, of course, generate more surface drag.

Lift is another component of hydrodynamic force that can be generated when certain shapes pass through water. A thin structure passing at an angle through the water will tend to separate the fluid so that water passes more quickly around one side than the other (Figure 8–13). The effect can be enhanced by curving the structure so that it is convex on one side. The result is a pressure differential that moves the structure toward the convex surface. The component of that motion that acts at right angles to the surface drag is defined as lift. (The word comes from aerodynamics, in which the same principles apply.) Thus, the direction of lift is defined in reference to the direction of drag, and the latter is determined by the direction of motion. An object that generates lift is a hydrofoil (or airfoil). Its angle relative to the direction of motion is the angle of attack, and its degree of curvature is known as its camber. A hydrofoil traveling vertically through the water generates lift with a forward rather than an upward component. These properties are important because, although all motion in a fluid generates drag, not all motion generates lift. Nonetheless, an object lacking the characteristics of a hydrofoil can generate thrust by imparting a rearward acceleration to the water, thereby eliciting a forward reactive force.

Aquatic locomotion by lateral undulation is characteristic of many aquatic amphibians and reptiles. Undulatory swimming (Figure 8–14) differs from undulation on land in two related ways. First, every point on the body that pushes back against the water generates some propulsive force as well as some force lost to lateral displacement of the water. Second, the entire surface of the animal generates drag. Aquatic salamanders, including larvae, propel themselves by axial undulation. Some species, such as *Necturus* (Proteidae), have a laterally compressed tail that displaces a large water mass with each stroke. Elongate aquatic salamanders, such as *Amphiuma* (Amphiumidae) and *Siren* (Sirenidae), often have a laterally compressed cross section in the rear of the body, as do typhlonectid caecilians. Salamander larvae that occupy stillwater habitats have relatively deep caudal fins that often

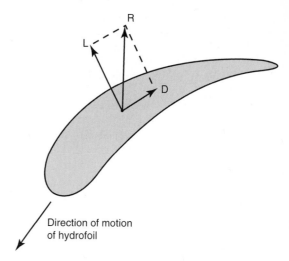

Direction of motion
of hydrofoil

Figure 8-13 **Generation of lift by a cambered hydrofoil passing downward through water.** The hydrodynamic forces are shown as if they were acting on the center of the hydrofoil. The passage of the hydrofoil elicits a hydrodynamic reaction force (R), which can resolved into a force acting in the direction opposite to the motion at that point, which is drag (D), and a force at right angles to D, which is lift (L). Lift is inclined forward, indicating that there is a forward propulsive component to its action. The magnitude and direction of R, and therefore its production and direction of lift, depend on the speed of the hydrofoil relative to the water (including any water current), the angle at which its surfaces meet the water, and the curvature of its surfaces (camber). A long, thin structure, such as the tail of a larval amphibian, can be modeled as an elongate hydrofoil, with each point on its surface generating a reaction force. In that case this figure would constitute a dorsal view, and lift would have both a lateral component (balanced by opposing curves of the body) and a forward propulsive component. Note that the same principles apply to the generation of lift by an airfoil moving through air.

extend anteriorly along the dorsum. These fins probably are thin and flexible enough to generate substantial lift. Conversely, larvae inhabiting rapidly flowing water have low caudal fins that reduce the likelihood of displacement by the current.

Similar differences in the height of the caudal fin are observed in anuran larvae from still and flowing water. Unlike salamander larvae, however, the tails of tadpoles are not supported by vertebrae, but only by a specialized caudal notochord. Despite their ungainly appearance, with a globose body and soft tail, tadpoles can turn as effectively and swim as quickly as fishes of

Figure 8-14 **Undulatory swimming by a banded water snake,** *Nerodia fasciata.* Lines of grid are 5 centimeters apart. Successive tracings from a movie have been displaced to the right by one square of the grid. Note increase in amplitude (lateral extent) of waves of the body posteriorly. Compare this figure, in which smooth waves move continuously along the length of the body, with Figure 8-13, in which the body forms irregular curves that remain fixed relative to the environment. *(Source: Jayne 1985.)*

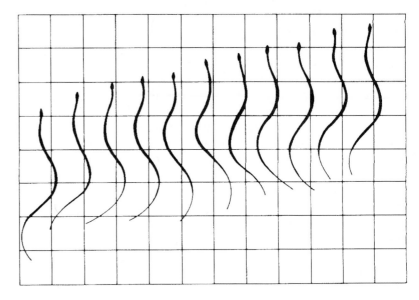

equivalent size, at least for short distances (Wassersug 1989). The unusual tails of tadpoles, with their reduced skeletal complexity, may have evolved for rapid degeneration during metamorphosis, thereby reducing the time spent in that vulnerable transition period. Use of the tail alone in controlling swimming allows the limbs to proceed farther toward the development of their adult form before metamorphosis begins. The forelimbs develop beneath a fold of skin, which prevents them from creating high drag while they grow.

Aquatic locomotion by snakes resembles terrestrial locomotion, but there are important differences. When swimming, snakes produce regular axial waves that increase in amplitude (lateral extent) posteriorly, whereas terrestrial locomotion involves the application of force at fixed points and a tendency for waves to dampen posteriorly. Many aquatic snakes have modifications that increase their lateral surface area. *Acrochordus* (Acrochordidae) has loose skin that hangs down when the animal swims to form a compressed ventral keel. Sea snakes (marine elapids such as *Laticauda* and other Hydrophiinae) have strongly compressed tails that probably generate substantial lift. Hydrophiines also have lungs that extend from 84 to 100 percent of the trunk length, providing buoyancy and preventing the rear of the body from sinking. Other reptiles scull using undulatory movements of their laterally compressed tails, including crocodilians

and *Amblyrhynchus,* the marine iguana of the Galápagos Islands.

Oscillatory propulsion involves use of the paired appendages rather than the axial skeleton, and may involve drag-based or lift-based mechanisms. Oscillatory propulsion characterizes two lineages incapable of lateral flexion of the vertebral column, frogs and turtles. Anurans usually swim by synchronous extension of their hind limbs, similar to the motions used during jumping, and aquatic species typically have extensive webbing of the hind feet. When the legs are extended with the toes spread, the webbing induces very high drag, forcing the frog forward. During the recovery phase of the oscillatory cycle the toes are drawn together and the entire limb is flexed, reducing surface drag during protraction. The unusual sacral articulation of the fully aquatic pipid frogs permits an extraordinary degree of anteroposterior (front to back) sliding of the ilia relative to the sacral diapophyses. *Xenopus laevis* can extend its body 18 percent beyond its contracted length, and the extra extension contributes to propulsion in the same way that the sliding seat of a rowing shell does (Figure 8–15).

Most aquatic turtles swim by drag-based propulsion, using their webbed feet to generate forward thrust. Typically the forelimb on one side and the hind limb on the other side move synchronously, a pattern that minimizes yaw. The most specialized freshwater species, especially the Trionychidae and Carettochelyidae, have extensively webbed digits and streamlined shells. The greatest morphological specialization is seen in the marine turtles of the families Cheloniidae and Dermochelyidae, in which the forelimbs are modified into elongate, winglike paddles, with the forearm and forefoot fully bound together into a blade borne on the short humerus. Rapid movement in open water involves synchronous movement of the forelimbs, combining retraction with a sweeping downstroke (Figure 8–16). Adjustments in the angle of the forelimbs during protraction and retraction apparently generate a forward component of lift on both the downstroke and the upstroke (Davenport et al. 1984). Birds also are capable of generating lift during both phases of the wing cycle, and marine turtles can be thought of as flying through the water. The much smaller hind limbs are modified into rounded paddles that are used as rudders and elevators for steering. Females emerging onto beaches to nest use one of two propulsive patterns. *Dermochelys* and the cheloniid *Chelonia* use synchronous retraction of the forelimbs, an action analogous to doing the butterfly stroke on land, whereas other cheloniids employ alternating retraction of the forelimbs. The hind limbs are also used as scoops to excavate the nest cavity.

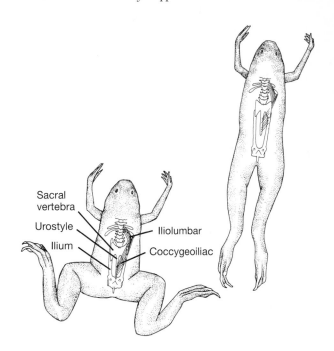

Figure 8-15   **Drawing of swimming *Xenopus laevis*.** The relationship of the pelvic girdle to the sacrum is shown during contraction (left) and extension (right) of the body. The extended length of the body is about 110 percent of the contracted length. *(Source: Modified from Videler and Jorna 1985.)*

Burrowing

Fossorial, or burrowing, amphibians and reptiles have evolved specializations for digging in substrates of varying consistency. Although the legs may be used in scratch-digging, fossoriality is especially common among limbless species, in which the slender body form is well-suited to penetration of soil or sand.

Frogs belonging to several families dig, and about 95 percent of those use their hind limbs for excavation. An enlarged metatarsal tubercle forms a sharp digging blade on the underside of the hind foot of many burrowing anurans. These frogs use a modification of the hind limb motion employed in jumping, in which the legs are extended alternately and motion is largely limited to powerful extension of the lower leg with the foot turned upward to expose the tubercle (Emerson 1976). The legs are relatively short, a condition that increases the in-lever of the digging muscles but sacrifices jumping ability. The pectoral limb

braces the frog against forward displacement during digging. *Hemisus* (Hemisotidae) is one of the few frogs that dig headfirst. It first flexes its head, bringing the snout in contact with the ground to serve as a brace while the forelimbs alternately scoop soil out and back. Major musculoskeletal modifications are associated with both flexion of the head and retraction of the forelimbs, including a threefold increase in the size of the retractor muscles. *Hemisus* also has an enlarged metatarsal tubercle and sometimes uses the hind limbs to initiate digging before switching to forelimb excavation. *Myobatrachus* (Myobatrachidae) uses its forelimbs to dig in search of termites and its large hind limbs to consolidate the soil.

Many fossorial amphibians and reptiles are limbless, and fossoriality often has been invoked as a primary selective force favoring the evolution of limb reduction leading to the origin of caecilians, amphisbaenians, and snakes. Although the early history of those groups remains unresolved, limblessness

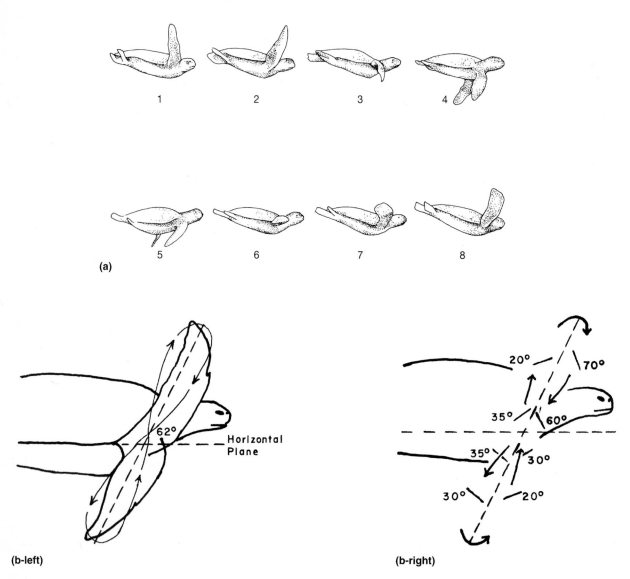

Figure 8-16   **Swimming by cheloniid sea turtles.** (a) A young green turtle, *Chelonia mydas*, swimming at moderate speed (traced from a video recording; the interval between frames is about 2 seconds). (b) Movements of the forelimb of an adult *Chelonia mydas* during swimming, drawn as if the body remained stationary, showing the path followed by the tip of the forelimb (left) and the angle of attack of the forelimb (right). The changing angle of the limb ensures a posterior component to the propulsive force during both the downstroke and the upstroke, generating lift with a forward component continuously during the limb cycle. *(Source: (a) Davenport et al. 1984, (b) Walker 1971b.)*

certainly is a frequent correlate of fossorial habits. Snakes, for example, include a substantial number of highly fossorial taxa. All of the Scolecophidia are primarily fossorial (or at least are frequent occupants of the nests of social insects), and their skulls are specialized, and distribute compressive forces around their surface (Cundall and Rossman 1993). The Uropeltidae are also highly derived burrowers, and their sharply pointed skulls are reinforced medially against compressive forces encountered during the penetration of soil. The Uropeltidae also have evolved a remarkable division of labor between the anterior and posterior ends of the body (Gans et al. 1978). The powerful anterior axial muscles force the head into the soil, and then draw the anterior end of the vertebral column into a series of concertina flexions (Figure 8-17). The skin is only loosely connected to the axial muscles, so vertebral flexion causes the integumentary envelope to expand, both widening the tunnel and providing a static platform for further penetration. The posterior end of the body, which has weak axial muscles, serves mainly as a passive vessel for the viscera. The muscles at the anterior end of the body have high oxidative metabolic capacity, whereas those at the posterior end are primarily glycolytic. This condition has been likened to the design of a freight train, in which a propulsive front end pulls a passive load of cargo behind it. Many other lineages of snakes, including the boids, colubrids, and elapids, have fossorial species. Adaptations to fossoriality include smooth skin, a reduced number of scales, and secondary contacts between bones of the normally highly flexible skull (Savitzky 1983).

Amphisbaenians have a variety of locomotor modes associated with a fossorial existence. Different lineages have cranial modifications that correlate with their burrowing behavior. *Bipes* (Bipedidae), the only amphisbaenian with limbs, has a blunt snout and uses its short front legs for scratch-digging (Figure 8-18), as well as for moving the front end of the body during surface activity. Limbless amphisbaenians use their heads to break the soil. Some amphisbaenids, such as *Ancylocranium* of Africa, have a vertically keeled skull and burrow with lateral movements of the head. Others, such as *Monopeltis*, also from Africa, have a dorsally compressed skull and compact the soil against the top of the burrow as they dig. The Trogonophidae, with shortened skulls, often have a heavy scale on the snout and employ a rotary motion when bur-

rowing. Amphisbaenians may employ undulatory, concertina, or rectilinear locomotion, and *Agamodon compressus*, an East African trogonophiid, uses an internal concertina similar to that of uropeltid snakes. Fossorial caecilians also employ internal concertina locomotion during burrowing. The axial musculature is separated from the skin and lateral musculature of the body wall. Thus, as in uropeltid snakes, concertina flexion of the vertebral column increases the diameter of the body, generating movements that enhance the already wormlike appearance of these animals. Some caecilians use preexisting tunnels in preference to forming new ones.

Unconsolidated sand presents unique challenges to locomotion. Loose sand collapses behind as an animal passes through it and generates friction around the entire body surface, much as happens in water. Many squamates engage in sand-diving (entering the sand intermittently to avoid predation) or sand-swimming (spending extended periods of time moving below the surface). Sand-diving occurs in species belonging to six families of lizards (Arnold 1995), many of which also have fringed toes, such as *Uma* (Phrynosomatidae), the fringe-toed lizards, and *Phrynocephalus* (Agamidae), the toad-headed lizards. Many also have a countersunk lower jaw, and some, such as *Uma*, have specialized labial scales that form a cutting edge to facilitate penetration of the sand. A great diversity of diving behaviors and propulsive mechanisms, both axial and appendicular, are employed by different sand-diving taxa.

Some skinks, such as the sandfish *Scincus*, are sand-swimmers, using strong axial flexion to generate propulsive thrust to travel beneath the sand. The inner ear of *Scincus* is specialized for sensitivity to seismic vibrations, such as those made when an insect walks on the sand surface (Hetherington 1989). The lizards localize an insect, travel toward it below the sand, and emerge at the precise point for a successful strike. While on the surface of the sand individuals push their heads below the surface, apparently listening for prey. Some snakes also are sand-swimmers, notably the colubrids *Chionactis*, the shovel-nosed snakes, and *Chilomeniscus*, the sand snakes, which resemble sand-swimming skinks in having smooth scales and countersunk lower jaws. Ventilation of the lungs poses a unique problem under sand. In most squamates exhalation involves compression of the ribs, which reduces the diameter of the body. In sand-

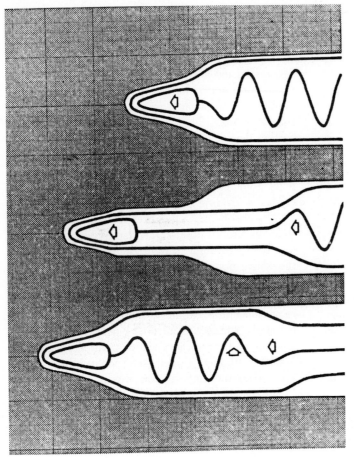

(a)

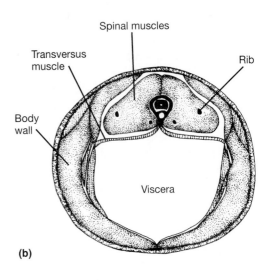

(b)

Figure 8-17   **Internal concertina locomotion of burrowing amphibians and reptiles.** (a) Diagram showing the role of the axial skeleton in burrowing by a uropeltid snake. The skin is loosely attached to the underlying musculature of the ribs and vertebrae. Axial flexion of the anterior vertebral column causes the diameter of the body to increase, providing a static point of contact against the burrow wall. The anterior vertebral column then straightens, pushing the pointed skull forward to lengthen the tunnel. (b) Diagrammatic cross section through the caecilian *Hypogeophis* showing the separation between the axial or spinal muscles and muscles associated with the body wall, which are attached mainly by the thin transversus muscle. The ribs are short and lie embedded among the axial muscles. The separation between those muscles and the body wall permits internal concertina motion of the vertebral column and expansion of the body diameter analogous to that of uropeltid snakes. *(Source: (a) Gans 1976; (b) Modified from Gaymer 1971.)*

swimmers, however, sand would immediately fill the surrounding space, preventing subsequent inhalation. Instead, some sand-divers, such as *Uma*, and sand-swimmers, such as *Scincus* and *Chionactis*, raise and lower their ventral surface, leaving a space free from falling sand for subsequent ventilatory movements.

Climbing

Many amphibians and reptiles are scansorial (climbing), and although most of them are arboreal (that is, they climb trees), others are saxicolous (rock-dwelling). The most common modifications are for either grasping or adhesion, using the feet, the tail, or

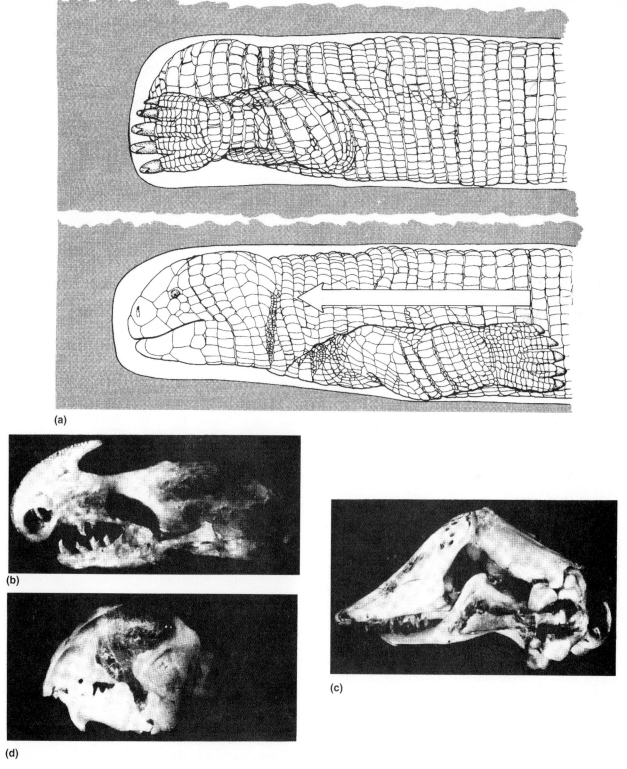

Figure 8-18 **Features associated with fossoriality in amphisbaenians.** (a) Scratch-digging by *Bipes*, the only amphisbaenian with limbs. (b–d) Skulls of amphisbaenians. (b) *Ancylocranium ionidesi*, a species that digs with lateral movements of the head; (c) *Monopeltis jugularis*, which uses vertical movements of the head; (d) *Agamodon compressum*, which burrows using rotational movements of the head. *(Source: Gans 1974.)*

both. In some cases there are modifications of body shape.

Adaptations for grasping are found in both amphibians and reptiles, although they are more common in reptiles. Prehensile tails occur in some large arboreal plethodontid salamanders of the genus *Bolitoglossa* and in a variety of arboreal squamates, including *Corucia* and *Prasinohaema* (Scincidae) and *Phelsuma* (Gekkonidae). The most extreme morphology is seen among chameleons, in which the tail can be tightly curled and straightened in the vertical plane. Many arboreal snakes grasp with their posterior body and tail, including *Corallus* (Figure 8-19), the tree boas. Some, such as the extraordinarily slender *Imantodes* (Colubridae) of the Neotropics, can move by gap-bridging. They are thin and light enough to remain perched on one branch while they cantilever up to half their body into open space until the head reaches another branch. Snakes that travel by gap-bridging often exhibit lateral compression of the body and enlarged vertebral scales, which may strengthen the body against bending in the dorsoventral plane. Vine snakes represent a specialized case of arboreality and have evolved in several colubrid lineages. *Oxybelis* (Central and South America), *Uromacer* (West Indies), *Ahaetulla* (Southeast Asia), and *Theletornis* (Africa) all include species with a slender, cryptically colored body and very narrow head. Some also show behavioral similarities, such as slow, waving body movements, which probably enhance their similarity to surrounding branches swaying in the wind.

Prehensile feet occur in the Chamaeleonidae, in which the digits are bound together with muscle and skin into groups of two or three (Figure 8-20). Each forefoot and hind foot has two groups of digits that oppose each other when grasping, a condition known as zygodactyly (*zygos* = yoke, *daktylos* = finger), analogous to the condition observed in parrots. Chameleons move in a very slow, deliberate fashion, often rocking fore and aft. Lateral undulation is greatly reduced in chameleons, accounting for only 6 percent of the length of the step for chameleons, versus 24 percent for *Agama*, a member of the related family Agamidae (Peterson 1984). Reduced lateral flexion allows chameleons to keep their center of gravity over their narrow base of support. On a stable perch they are even able to sustain brief periods when two legs are suspended, before a new triangle of support is established. Chameleons move their limbs in a lateral footfall pattern, but in contrast to the sprawling posture of most lizards, chameleons have vertically oriented limbs and highly mobile pectoral girdles, both of which increase step length and the force that can be transmitted parallel to the branch. Parasagittal movement of the girdle relative to the vertebral column accounts for 40 percent of the step length in chameleons, whereas most lizards have no girdle movement except sliding of the coracoids. These features give *Chamaeleo* a step length about 1.5 times that of a slowly moving *Agama*. The mobile girdle also helps chameleons reach far forward to grasp distant perches, often aided by support from the prehensile tail, thereby negotiating their complex, three-dimensional habitat. Prehensile feet with opposable digits have evolved among frogs as well. Species of the branch-walking hylid genus *Phyllomedusa* also move slowly, and their limbs follow a lateral footfall pattern different from the synchronous saltatory movements of most frogs.

Claws are used by some arboreal and saxicolous lizards to gain interlocking support from the substrate. Of greater interest, however, are specializations for adhesion found in the feet of certain climbing salamanders, frogs, and lizards. Pere Alberch and his colleagues (Alberch 1981, Alberch and Alberch 1981) have shown that there are two different adaptations to arboreality among bolitoglossine plethodontid salamanders, both of which are characterized by extensively webbed digits. In one group the salamanders are small, and simple capillary adhesion between a wet substrate and the smooth undersurface of the feet is sufficient to support the salamander. Capillary adhesion relies on surface tension in a lubricating fluid, and the total force produced is proportional to the area of contact. Although that mechanism cannot support a large mass, species such as *Bolitoglossa occidentalis* apparently have evolved through the process of hypomorphosis, which both reduces their body size and leaves them with fully webbed feet by truncating development of the digits (see Chapter 7). Larger arboreal species such as *B. dofleini* have too great a mass to be supported by capillary adhesion. Those species generate suction by lifting the center of the foot, which has a reinforced periphery that remains in contact with the substrate.

Specialized toe pads occur in arboreal frogs of many families, including the Hylidae, Rhacophoridae, and Hyperoliidae. Beneath the tip of each digit is an expanded toe pad (Figure 8-21) which, under high magnification, appears to consist of sharply polygonal

tiles separated by deep crevices. The tiles are epithelial cells, which in such frogs are highly columnar and have very flat exposed surfaces. The pads promote capillary adhesion, relying on a thin layer of liquid between the cells and the substrate. Mucous glands open occasionally between the cells and their product may contribute to adhesion, but water alone is sufficient to generate the necessary adhesive force. The role of capillary adhesion is supported by several observations. First, a dead frog can generate roughly the same adhesive force as a live one. Also, when detergent is applied to the substrate, adhesion is lost.

Although the expanded digital pads generally provide sufficient surface area for support, some large species, such as the hylids *Osteopilus septentrionalis*, the Cuban tree frog, and *Litoria caerulea*, White's tree frog, also have patches of similar tissue on their subarticular tubercles, which underlie the toe joints. Attachment is broken by peeling the toe pads from the substrate in a forward direction. That process is a passive one and results from the toes being lifted from base to tip as the foot is raised (Hanna and Barnes 1991). The same peeling action would result from gravity pulling on a frog facing head down, and in fact tree frogs are unable to stick in that position if their toes also point straight down. Even when resting at an angle tree frogs usually keep their toes pointing up for maximal adhesion. Members of some families, such as the Hylidae and Centrolenidae, have intercalary cartilages, small elements located between the last two phalanges of each toe. The purpose of those structures is unknown, but they may influence the forces acting on the toe pad. Toe pads similar to those of arboreal frogs occur in *Aneides*, a genus of climbing plethodontine salamanders.

A very different mechanism of support involves a dry adhesion system (scansors), in which the bonding strength results from molecular attraction between closely associated surfaces due to the sharing of electrons. Because those attractive forces are effective only at extremely small distances (less than 5×10^{-7} millimeters), animals that rely on dry adhesion must achieve extraordinarily close contact with the substrate. Remarkably, such a mechanism has arisen independently in three lineages of lizards—the Gekkonidae, *Anolis* (Polychrotidae), and *Prasinohaema*

Figure 8-19 **Body form of arboreal snakes.** *Corallus caninus*, the emerald tree boa of South America. The prehensile tail is used to maintain a grip on the branch while the snake constricts its prey. (*Photograph by Jany Sauvanet/ Photo Researchers, Inc.*)

Figure 8-20 **Chameleon limbs and locomotion.** (a) Zygodactylous foot of Jackson's chameleon, *Chamaeleo jacksonii.* (b) Diagram of successive body and limb positions in *Agama* (left) and *Chamaeleo* (right), showing reduced axial flexion and less sprawling limb positions in *Chamaeleo.* (c) Positions of the right forelimb near the beginning (top) and end (bottom) of the step cycle in *Chamaeleo.* Note the posterior shift in the position of the shoulder joint, reflecting parasagittal movement of the pectoral girdle, as well as the strongly vertical limb position and long stride. *(Source: (a) Photograph by Kevin Schafer; (b) Modified from Peterson 1984; (c) Source: Peterson 1984.)*

(a)

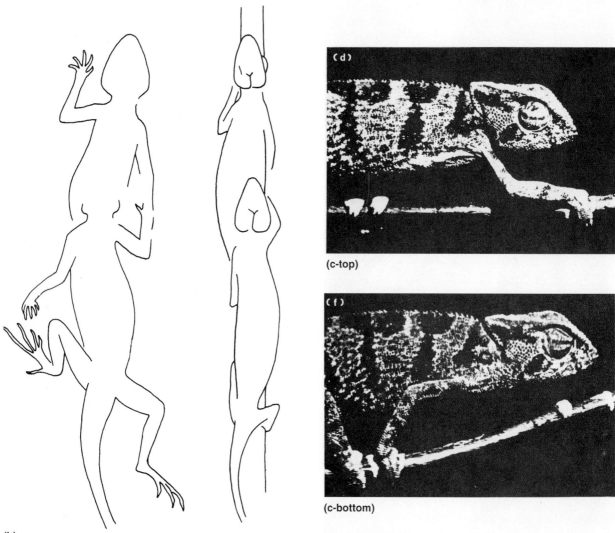

(b)

(c-top)

(c-bottom)

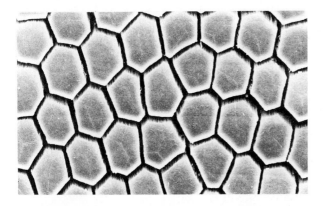

Figure 8-21   **Toe pad of *Hyla cinerea*.** Note the polygonal epithelial cells. (*Photograph by David Green.*)

virens (Scincidae). Scansors consist of a series of plate-like subdigital lamellae (Figure 8-22), each of which is covered with microscopic setae. Each seta bears at least one spatulate ending, and the endings make contact with the substrate. The details of setal morphology differ between lineages. The setae of geckos and *Anolis* are probably derived from the minute spines that cover most of the scales of the body in those taxa (Williams and Peterson 1982). Because adhesion relies on the nature of the integumentary materials, even a gecko that has recently died has feet that feel sticky. The system is further enhanced in some geckos by the presence of scansors beneath the tail.

Anthony Russell, Aaron Bauer, and their colleagues have studied the adhesive organs of geckos in great detail and have shown that the system is under complex tendinous and vascular control (Russell 1975, 1979a, 1981, 1986, Russell and Bauer 1990, Lauff et al. 1993). Scansors consist of a series of transversely expanded scales, or lamellae, that bear the adhesive setae. The scansors lie beneath highly flexible digits and are under the additional control of vascular sinuses. The size and shape of the scansors, the complex system of tendons that control them, and the bony skeleton of the digits vary among different gekkonid lineages, evidence that scansors have arisen and been elaborated multiple times within the Gekkonidae. In the tokay, *Gekko gecko*, a large Asian species and the best-studied form, a pair of lateral tendons send branches to each lamella. The tendons are assisted in their control of the scansors by a complex hydraulic system that includes a network of blood vessels deep within each scansor and a central venous sinus that may serve as a fluid reservoir. To-

gether those vascular adaptations probably facilitate close contact between the scansors and the substrate, and perhaps control their gradual release and re-attachment. The epidermis of the scansors is tightly bound to deeper layers of the skin and to the lateral tendons, coupling them to the skeletal and muscular system of the limb. Sense organs on the dorsal scales of the digits probably aid in control of the scansors by monitoring contact between adjacent scales and between scales and the substrate. In all, this remarkable and complex structural system permits geckos with scansors to climb up glass windows and across ceilings with remarkable agility, sometimes clinging by a single foot.

Aerial Locomotion

A few amphibians and reptiles have gone beyond simple arboreality and have taken locomotion in the vertical dimension to an extreme by evolving specializations for movement through the air. In addition to breaking their fall, such animals can travel a varying distance horizontally as they drop. Animals that fall at an angle greater than 45 degrees are considered to parachute, whereas those that fall at an angle less than 45 degrees are said to glide. At the extreme is the ability to remain in the air under the animal's own power, in which case flight has been achieved. Parachuting and gliding are sometimes considered passive flight, as distinguished from powered flight.

Several lineages of extant amphibians and reptiles include species that can parachute and a few that can glide. Some arboreal frogs, including *Hyla miliaria* (Hylidae) of Mexico, *Agalychnis spurrelli* (Hylidae) of Central America, and several species of the Southeast Asian genus *Rhacophorus* (Rhacophoridae), have large feet with extensive webbing between their fingers and toes, and sometimes fringes of skin on their limbs (Figure 8-23). By jumping from a tree, spreading their digits, and holding their limbs in a bent position at their sides, some can travel considerable distances at angles as low as 18 degrees (Emerson and Koehl 1990). Advanced gliding species of *Rhacophorus* exhibit considerable maneuverability while gliding and can execute banking turns of about 180 degrees. During heavy rainfall, *Agalychnis saltator* in Costa Rica uses parachuting to descend rapidly from trees to join large breeding aggregations (Roberts 1994). In Puerto Rico *Eleutherodactylus coqui* (Leptodactylidae), the

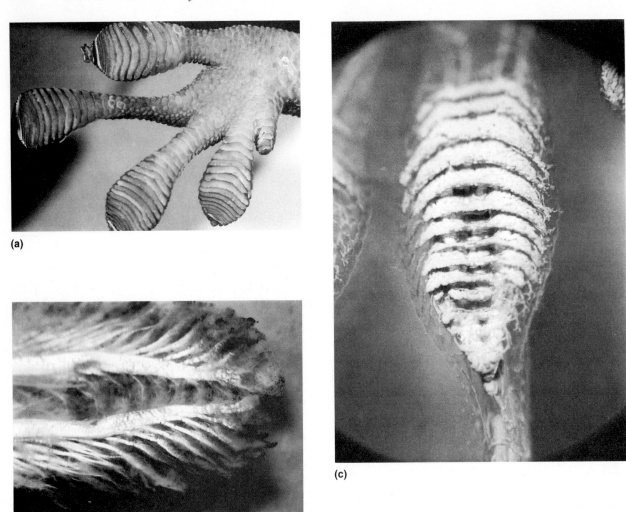

(a)

(b)

(c)

Figure 8-22   **The scansors of geckos.** (a) Ventral surface of the foot, showing the expanded scansors near the tips of the toes. (b) The relationship of the lateral digital tendons to the scansors. The phalanges of the toe extend down the center. (c) Ventral view of the foot, after injection with fine latex and subsequent clearing of soft tissues. Note the transverse network of blood vessels that lie within each scansor and presumably help control the pressure of the scansor against the substrate. *(Electron and light micrographs courtesy of Anthony Russell.)*

Figure 8-23 **Gliding amphibians and reptiles.** (a) *Rhacophorus pardalus,* the so-called flying frog, a rhacophorid of Southeast Asia. (b) *Draco volans,* the East Indian flying dragon, using its patagia in a territorial display. *(Photographs by (a) Michael Fogden, (b) Ken Preston-Mafham/Premaphotos.)*

(a)

(b)

coquí, uses parachuting to descend in the morning from nocturnal foraging sites in trees to diurnal retreats on the forest floor (Stewart 1985). The species has no special morphological features associated with aerial locomotion, but as the animals fall they spread their limbs in a manner similar to that of other parachuting frogs and slow their rate of descent. The frogs are so abundant and the behavior so common that, for a brief period at dawn, frogs can be heard raining from the trees.

Among lizards, strongly webbed digits are found in the genus *Ptychozoon,* a gecko from Southeast Asia, which also has broad flaps of skin on either side of the body that spread when the animal falls and small flaps of skin behind the head and along the tail. The webbing and skin flaps may have evolved from smaller cutaneous appendages associated with crypsis.

Ptychozoon can attain glide angles as shallow as 40 degrees from the horizontal, although its total glide path is usually steeper. The flight begins with a steep, rapid descent, and then becomes shallower after the lateral flaps open and drag slows the fall (Russell 1979b). If the flaps of *Ptychozoon* are immobilized to prevent them from opening, the rate of descent increases and the horizontal distance traveled decreases.

The most highly derived aerial species belong to the arboreal agamid genus *Draco,* known as flying dragons (Figure 8-23b). They have large patagia, flaps of skin supported by five to seven elongate ribs. As the lizard jumps, the iliocostalis muscle actively pulls the first two ribs forward. The remaining ribs are attached to the first two by ligaments, so the action spreads the entire airfoil. The flight can be divided into three phases. First, the lizard enters a rapid fall at an angle as steep as 80 degrees, during which time the long tail is raised behind the body. When enough kinetic energy has been accumulated the tail is lowered and the front of the patagium is raised to increase the angle of attack. The lizard enters the horizontal phase, during which it can glide at an angle as shallow as 15 degrees. Just before landing the tail is again raised and the animal sweeps sharply up before landing gently, usually on a tree trunk. Lizards jumping from 10 meters' elevation have glided as far as 60 meters, and even then they may land less than 2 meters below their launch height. Gliding by squamates is not limited to limbed species. Like *Draco*, the Asian colubrid snake *Chrysopelea* glides by spreading its ribs as it falls, broadening the body and achieving a deeply concave ventral surface, which may generate lift. The angle of descent can be as shallow as 30 degrees.

Summary

Amphibians and reptiles occupy a wide range of habitats and exhibit great diversity of form. It is not surprising, therefore, to find an equally diverse array of locomotor modes and associated morphologies. In some instances different features appear to represent obvious alternative solutions to a shared functional problem. For example, the sprawling posture of terrestrial reptiles has led to the origin of both the hooked fifth metatarsal of lizards and the calcaneal tuber of crocodilians as mechanisms for increasing thrust. Conversely, diverse taxa often have similar morphologies that reflect the strong influence of physical principles on the evolution of locomotor structures. Thus, undulatory propulsion in the aquatic environment has led to the elaboration of laterally flattened tails in salamanders, crocodilians, and sea snakes.

Some seemingly unrelated locomotor features prove to be united by common constraints, as in the consequences of a rigid trunk skeleton in turtles and frogs. Both lineages have evolved alternative morphological and behavioral mechanisms. On land, turtles rely on longer duty factors of the limbs, whereas frogs abandon quadrupedal locomotion altogether and use saltation (jumping).

Some habitats place particularly stringent demands on the locomotor system, evoking similar morphological responses from different lineages. For example, independent lineages of sand-swimmers have many similar features, as do lineages of gliders. Finally, we see the pervasive effects of both size and body shape. Gliding works for certain frogs and lizards in part because of their relatively small sizes. Numerous elongate and functionally limbless species of both amphibians and reptiles have fossorial habits. The evolutionary versatility of amphibians and reptiles is reflected in their responses to the challenge of moving through diverse environments.

9

Feeding

Amphibians and reptiles exhibit an extraordinary diversity of feeding modes, a diversity that mirrors their range of habitats and locomotor patterns. Included are mechanisms for feeding on a wide variety of food types in water or in air. Most species are carnivorous (including many that eat arthropods), and prey capture is enhanced by the use of projectile tongues in several groups. In others prey is engulfed by mobile cranial bones organized into complex linkages and controlled by a remarkable array of muscles. Venom delivery has evolved independently in several lineages of snakes and one of lizards, and the specializations associated with venom delivery can be spectacular. Finally, there are herbivores among both amphibians and reptiles, and processing plant matter often involves anatomical specializations.

This chapter takes a functional perspective, grouping organisms that exhibit similar feeding modes. Because this organization leads to consideration of independently evolved lineages, it emphasizes the convergence of different evolutionary lines on similar solutions to functional problems.

Suction and Suspension Feeding

The dynamics of feeding are different for animals that feed in water and those that feed on land. Food often is supported above the substrate by the density of water. On the positive side, food can be manipulated in water without expending effort to support it, and swallowing does not require saliva for lubrication.

However, movement of a predator may create pressure waves that push the prey away. This situation is like bobbing for apples, and suction often is the most effective way to capture food. Among amphibians and reptiles, suction feeding generally is employed by aquatic salamanders (including larvae), aquatic frogs, tadpoles, and some turtles. Food is usually sucked into the mouth with the surrounding water, and overcoming the inertia and viscosity of water requires considerable force.

Suction feeding relies on generating negative pressure within the buccal cavity, typically by expanding its volume. As simple as this process sounds, it presents several challenges. First, the negative pressure must be great enough to draw the prey into the mouth, and it must be achieved rapidly to capture elusive prey. Second, the opening of the mouth usually is small to permit directional suction. Aquatic feeding is greatly facilitated by a unidirectional flow of water through the buccopharyngeal cavity (Lauder and Shaffer 1993). Early in the history of the chordate lineage, when water was moved by cilia rather than by muscular pumping, unidirectionality was achieved by providing an outlet for fluid in the form of pharyngeal slits. Those openings are retained in fishes in the form of the branchial (or gill) slits, which allow water that is taken in for feeding and respiration to exit through the sides of the pharyngeal cavity. This unidirectional system is retained as an ancestral, but nevertheless highly functional, component of the feeding system of amphibian larvae and some paedomorphic adults (Figure 9–1). In contrast, fully metamorphosed

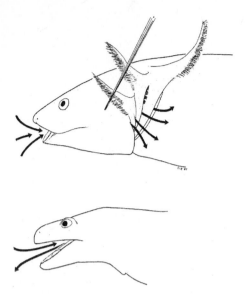

Figure 9–1   **Path of water flow during aquatic feeding by the tiger salamander, *Ambystoma tigrinum*.** (a) Larva, with unidirectional flow. (b) Adult, with bidirectional flow. *(Source: Lauder and Shaffer 1985.)*

amphibians and all reptiles lack pharyngeal slits, and the sides of the pharynx provide no route for fluids to exit. As a result of this morphology, suction feeding by transformed amphibians and by reptiles involves bidirectional flow—i.e., tidal movement of water in and out of the mouth.

Salamanders

Recent studies have revealed some details of the mechanics of suction feeding by larval and paedomorphic salamanders, as well as on the origins of the amphibian condition from that of fishes. Larval and strongly paedomorphic salamanders (which will be referred to as gilled salamanders) share with fishes a series of functional stages in the feeding sequence. The most dramatic is the expansive phase, during which the jaws open, the hyoid apparatus (which forms the floor of the mouth) drops, and the buccopharyngeal cavity expands (Lauder 1985), (Figure 9–2). Expansion proceeds from front to back, and results in a rapid flow of water into and through the buccopharyngeal cavity. A robust hyoid apparatus is a key to this system, contributing to expansion by movement in the vertical

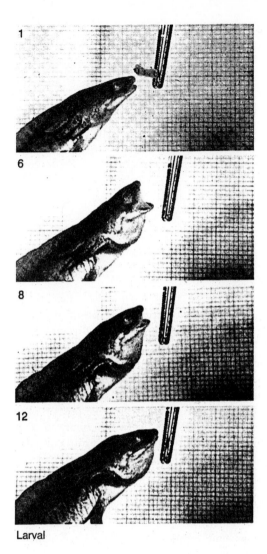

Larval

Figure 9–2   **Suction feeding by a larval tiger salamander, *Ambystoma tigrinum*.** Film frames from a high-speed movie record of a salamander offered a piece of earthworm from forceps. Frame numbers appear at upper left. Sequential frames are 5 milliseconds apart. Note the rapid depression of the hyobranchium during the expansion phase (frame 6). *(Source: Shaffer and Lauder 1988.)*

and horizontal planes. Early in the feeding sequence the depressor mandibulae muscle, which runs from the rear of the braincase to the retroarticular process at the posterior tip of the mandible (lower jaw bone), contracts and causes the mandible to drop (Figure 9–3). Simultaneously the hyoid apparatus is retracted by ventral muscles and the skull is lifted by dorsal ones.

Depression of the hyoid arch greatly expands the buccopharyngeal cavity and is discernible to the keen observer as a rapid drop in the floor of the mouth. As the floor drops, water rushes in. The expansive phase of the feeding sequence lasts only about 25 milliseconds. At that time water will rush into the buccal cavity through any available opening. Muscles draw the branchial arches in and cause the interdigitating gill rakers to occlude the branchial slits, however, directing all inflowing water through the mouth (Lauder and Shaffer 1985).

In the compressive phase, the mandibles are rapidly elevated by the large adductor mandibulae muscles, and the hyoid is elevated and protracted by several muscles. Toward the end of the compressive phase the branchial abductor muscles expand the branchial arches and water is ejected through the opened branchial slits as the buccal cavity is compressed by the rising floor of the mouth (Lauder 1985, Lauder and Shaffer 1985). The compressive phase lasts only about 35 milliseconds. The floor of the mouth is analogous to a piston that reduces buccal pressure by dropping and increases it by rising. The mouth serves as an intake valve that opens during expansion and closes during compression. The pharyngeal slits constitute the outflow valve, closing during expansion and opening during compression. Suturing the pharyngeal slits closed changes the flow system from a unidirectional to a bidirectional one and reduces the effectiveness of prey capture (Lauder and Reilly 1988).

Fleshy modifications of the lips, known as labial lobes, occur in a number of aquatic lineages, such as the paedomorphic North American salamanders *Amphiuma* (Amphiumidae), *Necturus* (Proteidae), and *Siren* (Sirenidae). Labial lobes prevent water and prey from escaping from the sides of the mouth during suction feeding (Figure 9–4). In species of salamandrids that return to water to breed, labial lobes appear each year with the resumption of aquatic habits (Özeti and Wake 1969).

Suction feeding has been studied and compared among members of the Ambystomatidae, Amphiumidae, Cryptobranchidae, Dicamptodontidae, Proteidae, and Sirenidae (Erdman and Cundall 1984, Reilly and Lauder 1992, Elwood and Cundall 1994). The large North American paedomorph *Cryptobranchus* and the related genus *Andrias*, found in Japan and China, are capable of asymmetrical suction feeding

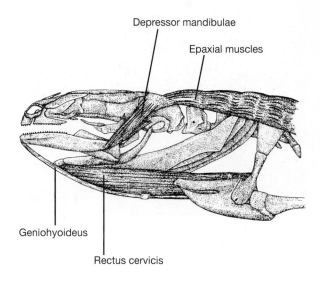

(a)

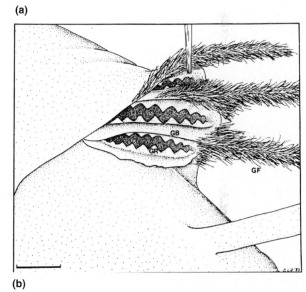

(b)

Figure 9–3 **The mechanics of suction feeding.** Skull, hyobranchial skeleton, and pectoral girdle of a larval tiger salamander and the major muscles involved in the expansion phase. (a) Contraction of the muscles shown opens the mouth and lowers the hyobranchium (stippled), expanding the oral cavity. DM = depressor mandibulae, EP = epaxial muscles, GH = geniohyoideus, RC = rectus cervicis. (b) Gill rakers of *Ambystoma mexicanum*, a paedomorphic salamander, in ventrolateral view, with the gills pulled forward. Note the interlocking gill rakers, GR. GB = gill bar, GF = gill filaments. *(Source: (a) Redrawn from Lauder and Shaffer, 1988. (b) Modified from Lauder and Shaffer, 1985.)*

Figure 9–4   **Head of *Amphiuma means* showing labial lobe.** *(Photograph by R. Wayne Van Devender.)*

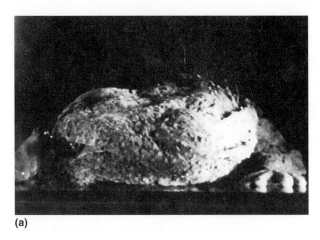

(a)

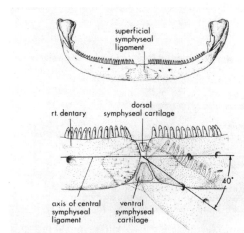

(b, c)

Figure 9–5   **Asymmetrical suction feeding by a hellbender salamander, *Cryptobranchus alleganiensis*.** (a) Unilateral depression of one side of the mandible, (b) Anterior view of lower jaw, (c) Enlargement of (b), showing location of symphyseal cartilages that permit independent movement between sides of the lower jaw. *(Source: (a) Photograph by D. Cundall. (b, c), Cundall et al. 1987.)*

(Figure 9–5). That feat is accomplished by a flexible mandibular symphysis that contains a pair of symphyseal cartilages. The larger ventral one is composed of elastic cartilage, and the two mandibular rami are joined by a complex series of ligaments that surround the symphyseal cartilages. *Cryptobranchus* typically depresses only one side of the mandible, dropping the jaw 10 to 40 degrees from the resting position (Figure 9–5).

Anurans

Adults of one group of frogs, the Pipidae, use suction feeding. All pipids remain fully aquatic as adults and are characterized by broad elements of the hyobranchial skeleton, especially the ceratohyals and ceratobranchials. Films of the small pipid *Hymenochirus boettgeri* suggest that the hyobranchium is fully depressed prior to mandibular depression, so that maximum suction is achieved as soon as the mouth is opened (Sokol 1969). Pipids are fully metamorphosed and have no external gill slits. Water is ejected from the mouth after prey capture. Larger species of pipids in the genera *Xenopus* and *Pipa* augment buccal suction by stuffing their prey into their mouths with their forelimbs, but *Hymenochirus* does not.

Tadpoles use suction feeding, but the mechanisms and anatomical structures that tadpoles use to produce a flow of water are entirely different from those of adults.

Turtles

Some aquatic turtles use a suction feeding mechanism that generates a negative buccopharyngeal pressure by rapid depression of the base of the hyobranchium. However, no pharyngeal slits remain in adult reptiles, so water cannot flow out through the rear of the buccopharyngeal cavity. Suction feeding by the snapping turtle (*Chelydra serpentina*), a bottom-dwelling freshwater turtle that is common in much of North America, involves the hyoid apparatus in association with several other structures (Lauder 1985). Elevation of the hyoid and constriction of the esophagus reduce the buccopharyngeal volume immediately before expansion begins. In that way the subsequent change in volume is rendered even more extreme. When a snapping turtle detects a prey item it darts its head forward, drops the lower jaw, and expands the buccal cavity to suck in the prey. As the mouth closes and the neck is retracted, the esophagus begins to expand, maintaining anteroposterior fluid flow while the mouth closes and prevents escape of the prey. Lacking pharyngeal slits, the turtle must expel the water it took in during the strike through its mouth. Extant turtles lack teeth and instead have sheaths known as rhamphothecae (*rhamphos* = beak, *theca* = sheath) composed of the horny protein keratin. These sheaths cover the margins of the upper and lower jaws and have sharp cutting edges. Many prey items, especially larger ones, are drawn in by suction and then forcefully bitten. At times they also are torn with the forelimbs. An aquatic turtle often will be observed alternately sucking in and then ejecting a food item, processing it with the jaws between fluid excursions.

Tadpoles

Most tadpoles do not employ momentary bursts of suction but are suspension feeders that pass water continuously across a filter designed to trap suspended particles (Sanderson and Wassersug 1993). Some tadpoles feed on phytoplankton that are uniformly distributed in midwater, whereas others scrape algae from the substrate with keratinized mouthparts to place it in suspension prior to ingestion. Tadpoles draw water into their mouths by creating negative buccal pressure. The feeding mechanism of tadpoles is a derived modification of the ancestral suction feeding mechanism seen in the larvae of other amphibians. Particles in the filter chamber may be captured by passing them through a fine strainer or by trapping them in mucus.

The fundamental mechanism of the suction pump of tadpoles is unlike that of any other amphibian and is based on a morphology radically different from that of the adult anuran (Figure 9–6). The cartilaginous skeletal support for the mouthparts of a typical tadpole includes a greatly elongated palatoquadrate cartilage that is strongly inclined forward from the rear of the skull (Gradwell 1972a,b). Its tip contacts a very short

Meckel's cartilage, the precursor of the mandible, and that cartilage contacts the inferior rostral cartilage, which bears the cornified lower beak. The superior rostral cartilage, with its upper beak, articulates with the front of the braincase and with the lower rostral cartilage. The hyobranchial skeleton includes a series of branchial arches fused to form a basket and a robust pair of transversely oriented ceratohyals. Each ceratohyal articulates with the corresponding palatoquadrate, which bears a tall orbital process from which large muscles arise. Those muscles insert onto the lateral tip of the ceratohyal and, by lifting that tip, drop the floor of the mouth and increase the buccal volume (Wassersug and Hoff 1979). A transverse ventral muscle raises the floor of the mouth and decreases buccal volume. The alternating contraction of those two muscles controls the rise and fall of the buccal floor. Their action first draws water in through the mouth and external nares and then, with the mouth and internal narial valves closed, forces the water back into the pharyngeal cavity, above the gill arches. A large flap of soft tissue on the floor of the mouth, the ventral velum, allows water to pass back to the gills and prevents backflow from the gills.

The branchial arches lie between two chambers, the pharyngeal cavity and the atrial (or opercular) cavity. (That cavity is formed by a fold of skin, the operculum, that grows over the gills early in larval life.) Water exits from the atrial chamber by one or two openings, the spiracles. A second muscular pump modifies the pressure in those two cavities, maintaining unidirectional flow.

The buccal cavity has a complex array of papillae that may direct large particles toward the esophagus while allowing smaller particles to pass toward the branchial filters (Wassersug 1980). Filtration occurs in the pharyngeal cavity by sieving through the branchial filters or, probably more often, by mucus entrapment. The branchial filters consist of branching ridges on the dorsal surface of the branchial arches (Figure 9–7), and the sieve that they produce has a pore size of about 5 micrometers. However, some tadpoles can capture particles as small as 0.126 micrometers, which are caught on strands of mucus in the pharynx. The most extensive branchial food traps are found on the undersurface of the ventral velum, where ridges capped with secretory cells form a complex array. As more and more particles are trapped, strands of mucus and food break loose and are carried back to the esophagus.

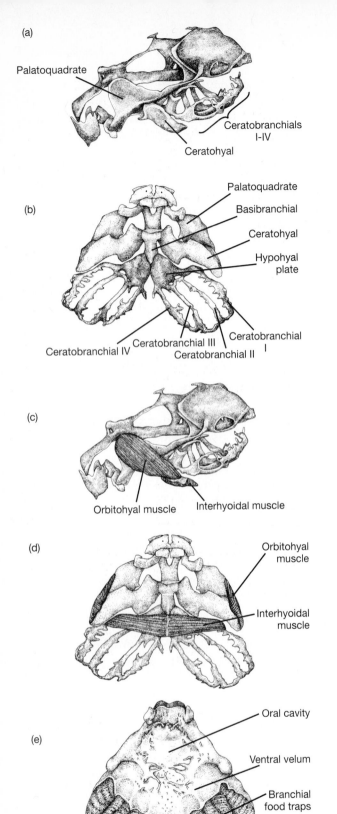

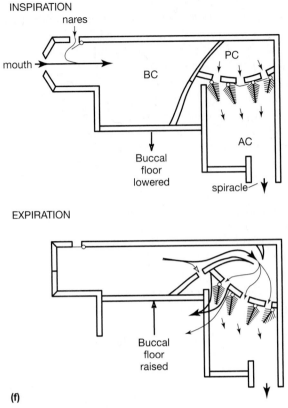

(f)

Figure 9–6 **Feeding morphology of tadpoles**. (a) Lateral and (b) ventral views of the chondrocranium (larval skull) of *Rana temporaria*; note the elongate palatoquadrate cartilage and the transverse ceratohyals. (c) Lateral and (d) ventral views of the major muscles involved in pumping water into the mouth and across the gills; the orbitohyal muscles raise the lateral tips of the ceratohyals, depressing the floor of the mouth and drawing in water; the interhyoid muscle lowers the lateral tip of the ceratohyal, raising the floor of the mouth and forcing water back across the gills. (e) Floor of the mouth and branchial baskets of *Hyla femoralis*, showing the ventral velum, which separates the oral cavity and the pharyngeal chamber; the branchial food traps lie on the underside of the velum. (f) Diagrammatic lateral view of a typical tadpole showing the relationships among the oral (or buccal) cavity (BC), pharyngeal chamber (PC), atrial (or opercular) cavity (AC), and spiracle. *(Source: (a–d) Based on figures in de Jongh 1968. (e) Wassersug 1980. (f) Modified from Gradwell 1971.)*

The body forms of tadpoles show repeated correlates between morphology and natural history (Altig and Johnston 1989). Body shape and mouth parts are related to the habitat (Figure 9–8), and mouth form is also related to diet (Figure 9–9). Most tadpoles have several rows of small keratinized structures known as labial teeth on the oral disk that surrounds the mouth. The number and configuration of the rows vary, and greater complexity is often associated with habitat specialization, such as living in fast-moving water. The teeth may be highly elaborate, as in the stream-dwelling ranid *Amolops larutensis*, in which each of the densely packed teeth bears 16 cusps per tooth, providing 1,800 cusps per millimeter of tooth row. It is not known whether the teeth are used for feeding or for grasping rocks in torrential streams. Tadpoles of the pelobatid *Megophrys* have upward-facing mouths with a broad oral funnel and feed on particles that float on the water surface (neuston). Carnivorous tadpoles often have beaks with sharp, pointed sheaths. Many tadpoles are arboreal, living in tree cavities or the axils of epiphytic plants, and these species eat mosquito larvae, frog eggs, or each other.

Macrophagous larvae, those that feed on large food items (such as carnivores), have a long lever arm of the ceratohyals (which offers a high mechanical advantage) and pump a large buccal volume with short movements of the broad floor of the mouth (Wassersug and Hoff 1979). *Lepidobatrachus* (Leptodactylidae), for example, is a specialized carnivore that often preys on tadpoles of roughly its own body size. In contrast, microphagous tadpoles such as midwater suspension feeders have low mechanical advantage and a small buccal floor. They pump a large buccal volume by moving the floor through a considerable vertical distance.

Tadpoles that live in torrential streams often show extraordinary modifications of their mouthparts. *Amolops* (Ranidae) and *Atelopus* (Bufonidae; Figure 9–10) have a suction device on the abdominal surface (they are said to be gastromyzophorous; *gastro* = stomach, *myzo* = suck, *phoreus* = bearer). Tadpoles of the tailed frog, *Ascaphus*, in contrast, have a greatly expanded oral disk (Figure 9–11) that applies suction to the rocky substrate of fast-flowing streams in the Pacific Northwest. A key element of the suction system of *Ascaphus* is an additional flap valve located just inside the mouth. After the tadpole applies its mouth to a rock, it begins buccal pumping, drawing water from the cavity formed by the large oral disk. As the pres-

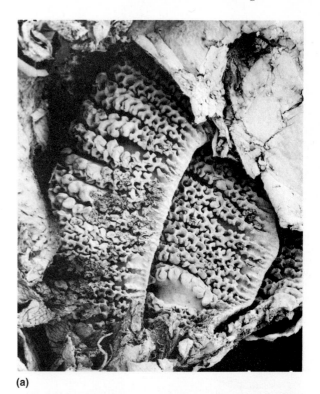

(a)

(b)

Figure 9–7 **Feeding structures of *Odontophryne robusta*.** Scanning electron micrographs of (a) the branchial filters and (b) the food traps on the ventral surface of the velum. Note the secretory ridges on the ventral velum, which generate the mucus that traps suspended particles. (*Source: Wassersug and Pyburn 1987; photographs provided by Richard Wassersug.*)

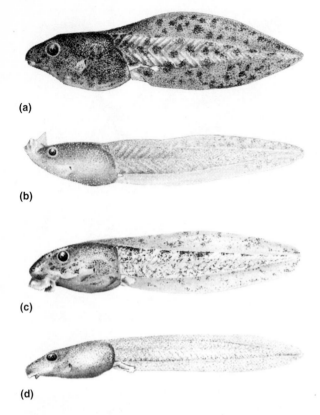

(a)

(b)

(c)

(d)

Figure 9–8   **Diversity of tadpole body forms.** (a) *Rana palmipes*, a generalized species founds in slow waters (note deep tail fin). (b) *Megophrys montana*, which feeds on neuston in slow waters (note dorsally facing oral funnel). (c) *Hyla rivularis*, a stream-dwelling tadpole with enlarged mouthparts. (d) *Hyla bromeliacia*, which lives in the axils of bromeliads. *(Source: Duellman and Trueb 1986.)*

sure within the disk drops, ambient pressure forces the tadpole's head against the rock. The oral valve flaps shut when the pressure in the disk falls below that of the buccal cavity, sealing the disk and allowing the tadpole to continue to respire through its nostrils. Even a dead tadpole can adhere for up to 30 minutes if it is pressed firmly against a smooth substrate. A tadpole presumably rasps food from the rock with its beak while it is attached. *Ascaphus* tadpoles remain attached most of the time, normally swimming only short distances before reattaching.

Tadpoles of *Otophryne robusta*, a microhylid from northern Amazonia, have large keratinized toothlike projections on the upper and lower jaws and an asymmetrical spiracle that is drawn out into an enormous tube (Figure 9–12). Their fierce appearance suggests a predatory diet, but gut contents revealed only small items such as bacteria and algae. The clue to their feeding habits is in their behavior. The tadpoles burrow in the sandy substrates of streams, leaving their long spiracle projecting into the water column. Water flowing past the opening of the spiracle apparently creates negative pressure, drawing water from the interstitial spaces in the sand into the mouth, across the filters, and out through the spiracle, while the large projections keep sand grains from entering the mouth.

Terrestrial Feeding Mechanisms

A broad range of mechanisms has evolved for feeding in the terrestrial environment. Furthermore, many terrestrial feeding methods can be used in water with little anatomical modification—e.g., crocodilians. Therefore, those groups also are discussed in this section.

We will recognize several broad classes of terrestrial feeding mechanisms based on the presence or absence of two important mechanical attributes, projectile mechanisms and cranial kinesis. Several lineages of amphibians and reptiles have evolved mechanisms of tongue projection, ranging from slight protrusion of a fleshy tongue to the rapid firing of a lengthy projectile. Other groups, most notably the squamate reptiles, have skulls that permit movement among the constituent bones (other than the simple hinging of the lower jaw against the cranium). Such movement is known as cranial kinesis and is widespread among vertebrates. Finally, some amphibians and reptiles exhibit neither projectile tongues nor cranial kinesis. We will examine this range of diversity, again taking a functional approach rather than a phylogenetic one.

Akinetic, Nonprojectile Feeding

The simplest tetrapod feeding system, although not necessarily the ancestral one, consists of a rigid skull and a hinged lower jaw. Even with such a simple mechanism, the possibilities for elaboration are legion, often reflecting differences in prey characteristics, feeding environment, or both. Many lineages with akinetic skulls have evolved tongue projection, and they are discussed below. Here we consider groups that rely primarily on akinetic, nonprojectile feeding mechanisms. Although such systems have

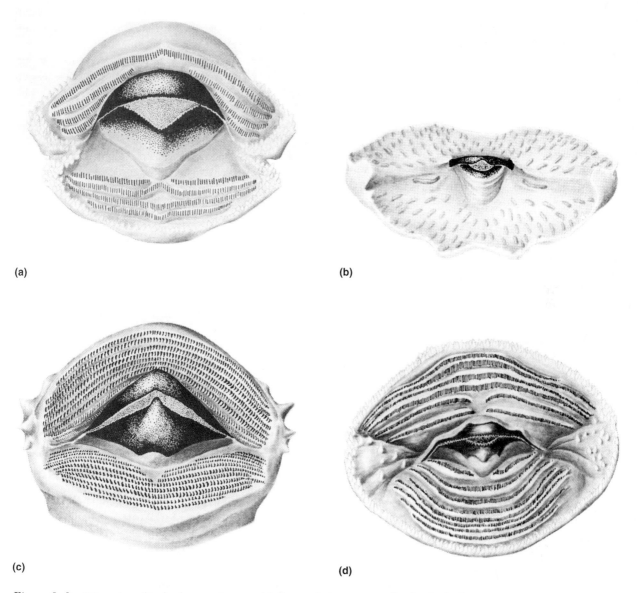

(a)

(b)

(c)

(d)

Figure 9–9   **Diversity of tadpole mouthparts.** (a) *Rana palmipes*, a generalized tadpole that grazes on epiphytic algae. (b) *Megophrys montana*, which feeds at the surface (note the wide oral funnel). (c) *Ceratophrys cornuta*, a carnivorous tadpole (note the large, pointed beak). (d) *Hyla lindae*, a stream-dwelling tadpole (note the large oral disk for attaching to rocks). *(Source: Duellman and Trueb 1986.)*

evolved in other groups, they are most typical of cae-cilians, amphisbaenians, turtles, and crocodilians, and we will concentrate on those groups here.

Caecilians.   Caecilians are notable for their rigid skulls, a feature that reflects the mechanical demands of burrowing through soil. Most researchers today

believe that the cranial rigidity of derived caecilians is the result of a secondary expansion of certain cranial bones, especially the squamosal, resulting in a fully roofed, or stegokrotaphic, skull. Cranial kinesis is seen among the Rhinotrematidae, and even some derived caecilians exhibit movement of the quadrate bone relative to the rest of the skull.

(a)

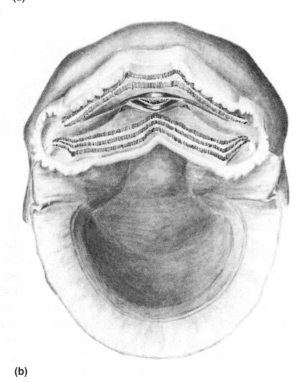

(b)

Figure 9–10 **The gastromyzophorous stream-dwelling tadpole of the tropical bufonid *Atelopus ig-nescens.*** (a) Lateral view of tadpole. (b) Ventral view of mouthparts and the enlarged suction disk on the ventral surface of the abdomen *(Source: Duellman and Trueb 1986.)*

Adult caecilians have relatively large, sharp, curved teeth. The upper ones are arranged in both marginal and palatal rows, between which the teeth of the lower jaw fit when the mouth is fully closed. The unique tentacle of caecilians may be used to locate prey, although its function has not been demonstrated conclusively.

The general gymnophionan feeding mechanism has been described, and a functional analysis has been conducted for *Dermophis mexicanus* (Bemis et al. 1983, Nussbaum 1983). The major jaw-closing muscle in most vertebrates with relatively simple skulls is the adductor (or levator) mandibulae, which lies anterior to the mandibular articulation. Thus, jaw closure operates as a third class lever. All else being equal, a thicker muscle provides greater force. Many vertebrates have bulging adductor muscles that are associated with rapid and/or powerful jaw closure (the pit bull terrier is an example). However, the adductor mandibulae of *Dermophis* is very small (Figure 9–13). Its size is apparently limited by the overlying skull roof and the narrow head needed to facilitate burrowing. However, caecilians have an unusual condition of the mandible, a very long retroarticular process, so named because it extends behind the joint that connects the mandible and the skull. The depressor mandibulae, a muscle that opens the jaw, inserts onto the dorsal surface of that process. In addition, an enormously enlarged interhyoideus muscle, which extends from its ancestral position below the throat and continues far back along the sides of the neck, pulls down on the retroarticular process. That arrangement forms a first-class lever that raises the mandible, a function normally served by the adductor mandibulae alone. Thus, hypertrophy of the retroarticular process and the interhyoideus muscle results in relocation of the major muscles used to close the jaws from the side of the skull, where space is at a premium, to the sides and floor of the neck.

Turtles.   Turtles have some of the most heavily built skulls among reptiles. As with caecilians, however, that condition appears to represent a secondary loss of cranial kinesis early in the history of the chelonian lineage. Apart from aquatic suction, the feeding mechanics of turtles are fairly simple. As in caecilians, the bulk of the adductor mandibulae muscles lie in a space between the braincase and the complete dermal skull roof (referred to in reptiles as the anapsid condition). The head size of turtles is not limited by the demands of burrowing, and the adductor muscles are relatively massive and, in combination with the rather short mandibles (short out-levers), are capable of generating substantial bite force.

Two limitations on the dimensions of the head in turtles do exist. First, the middle ear is greatly expanded in extant turtles, occupying some of the space that might otherwise be filled by adductor muscles. Second, the mechanism for retraction of the head into the shell limits head size. In most turtles the opening of a temporal notch, a posterior emargination of the

(a)

(b)

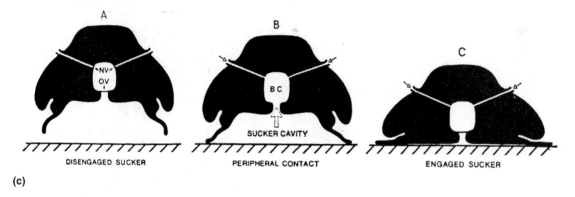

(c)

Figure 9–11   **The stream-dwelling tadpole of *Ascaphus truei*.** Tadpoles of this frog have greatly enlarged mouthparts for attaching to the rocky substrate. (a) Tadpole in its normal position on a rock, and (b) a ventral view of a tadpole adhering to glass. (c) Diagrammatic cross sections. *(Source: (a, b) Photographs by R. Wayne Van Devender, (c) Gradwell 1971.)*

skull roof, is coupled with a displacement of the adductor musculature toward the rear of the skull (Figure 9–14). That shift in muscle position creates a mechanical problem. A muscle transmits its force most effectively when it inserts perpendicular to a bone. How do turtles pull vertically on the lower jaw when the adductor muscles run horizontally forward from the rear of the skull?

Turtles have solved that problem twice, in an ingenious fashion. Extant turtles have a trochlear process (*trochlea* = pulley) in the skull, which acts as a mechanical pulley to change the direction of the muscle with minimal loss of force. Each of the two lineages of turtles has its own position for the trochlear process (Figure 9–15). In pleurodires, the pulley is

formed by the pterygoid bone of the palate, whereas in cryptodires it is formed by the quadrate and/or the prootic bones, in front of the inner ear. In both lineages the adductor muscle turns to insert at a much steeper angle on the mandible, providing maximal force transmission. The difference indicates that each group evolved a pulley system independently, presumably in response to the demands of cervical retraction (Gaffney 1975).

In comparison to the tongues of their suction-feeding aquatic counterparts, the tongues of terrestrial turtles have smaller hyobranchial skeletons and larger and fleshier tongue pads (Bramble and Wake 1985). The jaws of turtles may have broad crushing surfaces, as in *Graptemys* and *Malaclemys* (Emydidae),

Figure 9–12 **Tadpole of the South American microhylid frog *Otophryne robusta*.** (a) Scanning electron micrograph of the lower jaw, anterior view, showing the enlarged toothlike projections that apparently prevent sand from entering the mouth. (b) Lateral view of the tadpole showing the long tubelike spiracle. (c) The hypothesized resting position of the tadpole, buried in sand with the spiracle exposed in the water column. Bernoulli forces presumably draw water from between the sand grains into the mouth. *(Source: Wassersug and Pyburn 1987; photographs provided by Richard Wassersug.)*

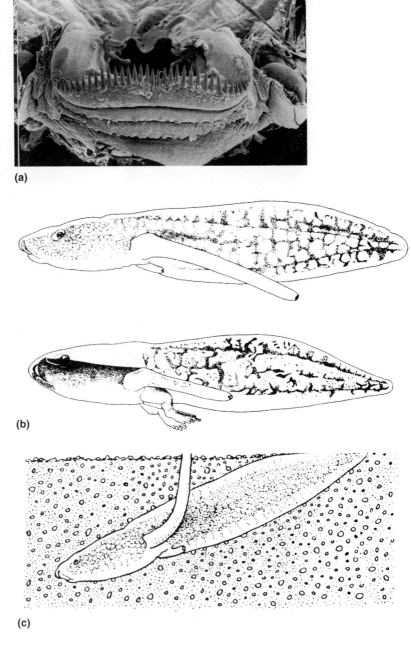

(a)

(b)

(c)

which feed primarily on mollusks, and in *Caretta* and *Lepidochelys* (Cheloniidae), which eat mollusks and crustaceans. (Animals that feed on hard-bodied prey are said to be durophagic; *durus* = hard, *phage* = to eat.) Alternatively, the jaws may be narrow and the rhamphothecae sharp, as is typical of some species of softshell turtles (*Apalone*: Trionychidae).

Many terrestrial tortoises feed on vegetation and exhibit adaptations for slicing through and transporting plant matter. An extreme example is seen in the gopher tortoises (*Gopherus*: Testudinidae) that occupy semiarid to arid environments in North America and feed on tough, xeric-adapted plants. Part of the adductor muscle maintains a strong dorsal pull on the mandible while a medial division of the same muscle draws back on the lower jaw (Bramble 1974b). The result is a slicing or milling motion in which the lower jaw slips rearward against the upper while maintaining pressure on the fibrous food between the ridged surfaces of the jaws. The degree of rearward movement roughly correlates with the toughness of the vegetation in the diet of the four species of *Gopherus*.

Crocodilians. Like turtles, crocodilians have akinetic skulls, but unlike turtles, crocodilians open their mouths by lifting their heads, not by lowering their jaws. A short but stout depressor mandibulae muscle extends from the rear of the skull to the large retroarticular process on the lower jaw. Contraction of this muscle, assisted by the dorsal neck muscles, elevates the skull and opens the mouth. These jaw-opening muscles have very little mechanical advantage, and that is why it is possible to keep a crocodilian's mouth closed by holding its snout.

Biting down is a different story. A complex but relatively small adductor mandibulae muscle is augmented by two massive pterygoideus muscles that arise from the roof of the mouth to insert on the mandible via a complex system of tendons shared with the external adductors. These muscles can produce a crushing force when a crocodile bites.

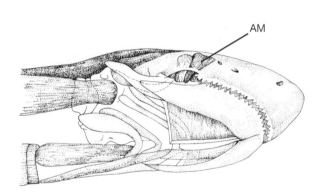

Figure 9–13 **Feeding mechanism of the caecilian *Dermophis mexicanus*.** Note the small adductor mandibulae muscles (AM), the size of which is limited by the skull roof. The unusually large interhyoideus muscle (IH), which inserts on ventral surface of the retroarticular process (RP) of the mandible, closes the mouth. The depressor mandibulae muscle (DM) inserts on the dorsal surface of the mandible and opens the mouth. *(Source: Modified from Bemis et al. 1983.)*

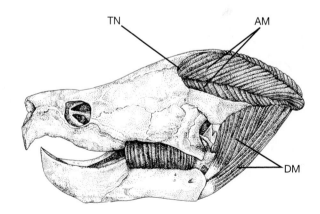

Figure 9–14 **Skull and major jaw muscles of the alligator snapping turtle, *Macroclemys temminckii*.** The temporal notch (TN) at the rear of the skull roof accommodates the large adductor mandibulae (AM) muscle that closes the jaws. The depressor mandibulae (DM) muscle opens the jaws. *(Source: Modified from Schumacher 1973.)*

Figure 9–15 **Trochlear processes of turtles.** (a) *Emydura* species, a pleurodire, showing trochlear process formed by pterygoid bone; (b) *Chelydra serpentina*, a cryptodire, showing trochlear process on the front of the bones housing the inner ear. The main tendon of the adductor muscle has been lifted slightly to reveal the underlying bones. Note that the trochlear process in each turtle forms a pulley that redirects the muscle so that, although the fibers originate on the rear of the skull, the tendon pulls in a nearly vertical direction, the most favorable direction for transmitting maximum force. *(Source: Gaffney and Meylan 1988.)*

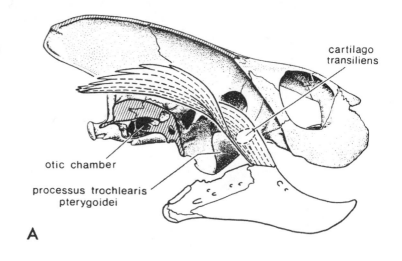

A

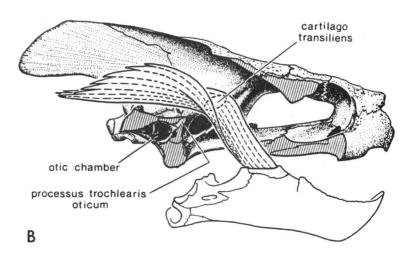

B

The relatively elongate snout of crocodilians and the insertion of adductor muscles close to the mandibular articulation produce rapid closure of the jaws. Species that feed predominantly on fishes, such as the African slender-snouted crocodile (*Crocodylus cataphractus*) and the gharial (*Gavialis gangeticus*), have long and narrow jaws that close rapidly. In contrast, species with more varied diets, such as the American alligator (*Alligator mississippiensis*), have broad and short snouts that generate great force at the expense of speed. Alligators are notorious for feeding on large turtles, a feat that requires considerable bite force.

Adult crocodilians may employ a behavior known as rotational feeding when feeding underwater on large prey items. They bite the prey and spin rapidly around their own longitudinal axis, tearing pieces from the carcass. Smaller prey may be crushed by the thick tongue, which presses food up toward the roof of the mouth. Crocodilians may also wait to eat large prey until decomposition makes it easier to tear apart. *Crocodylus moreletii* in Mexico regularly allows large prey to rot for a day or more before feeding on the bloated carcasses.

Extant crocodilians have a long secondary palate, a bony partition that separates the nasal passage from the buccal cavity. The rear of the palate has a soft transverse flap of tissue that contacts a transverse fold at the rear of the tongue. This arrangement seals the larynx and esophagus and allows an animal to breathe through its protuberant nostrils while floating at the surface of the water.

Digestion of large and bony prey items is facilitated by a highly muscular posterior region of the stomach, the pyloric gizzard. Stones are commonly ingested by crocodilians (the behavior is known as lithophagy; *lithos* = stone), and presumably help to grind hard food items.

Sphenodontians. *Sphenodon*, the sole surviving genus of sphenodontians, eats a variety of vertebrate and invertebrate prey. The skull of *Sphenodon* is structured more lightly than the skulls of turtles or crocodilians, with two large temporal fenestrae defining a pair of slender temporal arches. These arches appear to brace the skull so that kinesis does not occur in adult tuatara. The acrodont dentition of *Sphenodon* includes a pair of large incisorlike teeth at the front of the upper jaw, a pair of caninelike teeth near the front of the lower jaw, and a row of smaller marginal teeth extending posteriorly along the maxilla and dentary bones. Like many reptiles, *Sphenodon* also has teeth on the palate, in this case on the palatine bone. The palatal teeth form a row parallel to the maxillary teeth, leaving a groove between the two rows. The mandibular teeth fit into this groove when the mouth is closed. Small prey, such as insects, are captured by protruding the sticky tongue. Larger prey are impaled on the enlarged anterior teeth. After a prey item has been transferred to the rear teeth, a combination of vertical crushing movements and anteroposterior shearing movements of the lower jaw mash and slice the prey into a soft mass prior to swallowing (Gorniak et al. 1982).

Amphisbaenians. Amphisbaenians, like caecilians, are fossorial animals and have the reinforced cranial structure associated with burrowing. Amphisbaenians resemble caecilians in being secondarily akinetic. Unlike in caecilians, however, the skull of amphisbaenians is reinforced medial to the adductor muscles rather than lateral to them. Thus there is space for large jaw muscles, especially posteriorly. In some species the mandible bears a dorsal extension to which the adductor muscles from the rear of the skull attach at a favorable angle. Thus, both amphisbaenians and turtles have displaced the adductor muscles posteriorly, and both have evolved a solution to retaining a favorable angle of muscle insertion. However, in amphisbaenians that solution involves elevating the insertion point on the mandible, whereas in turtles it involves altering the path of the muscle.

Amphisbaenians feed on a variety of prey, including both invertebrates and vertebrates. Their acrodont and generally conical teeth are strong and sharp. As in caecilians, the teeth of the lower jaw lie between marginal and palatal upper rows when the mouth is closed. Amphisbaenians have a median tooth in the upper jaw, a feature seen in no other adult tetrapod. The median tooth lies between the anteriormost teeth of the two mandibular rami when the jaw closes, providing an interlocking grip. Unlike most other squamates, amphisbaenians often break off pieces of their prey, either by twisting or by repeatedly shearing off chunks (Gans 1974). One species, *Amphisbaena ridleyi*, has enlarged, blunt rear teeth that it presumably uses to crush terrestrial snails.

Projectile Feeding Mechanisms

Several distantly related lineages of amphibians and reptiles, including frogs, several lineages of salamanders, and the lizard family Chamaeleonidae, have independently evolved predatory mechanisms involving projection of the tongue. The mechanisms used to project the tongue vary among lineages. Recent studies have contributed greatly to our understanding of these mechanisms and shed light on the relationship between visual perception and tongue projection. In this section we will discuss the groups generally characterized by tongue projection systems, and also a few of the exceptions among members of those groups.

Salamanders. Terrestrial salamanders typically feed by slapping their prey with a large, moist, fleshy tongue. Prey capture by *Ambystoma tigrinum* (Ambystomatidae), for example, begins by flexing the entire head downward to within a few millimeters of the prey. With the tip of the mandible in contact with the substrate and further stabilized by ventral throat muscles, contraction of the depressor mandibulae muscles raises the skull rather than lowering the jaw, opening the mouth about 60 degrees. The hyobranchial skeleton is broad and roughly V-shaped, with the tongue supported on the anteriorly directed point of the V. The paired subarcualis rectus I muscles (Figure 9–16) run on each side from the ventral surface of the broad ceratohyal cartilages to the rear of the first epibranchials, which form the tips of the arms of the V. Those muscles perform most of the work of tongue protrusion, drawing the hypobranchium forward and pushing the tongue pad out of the mouth. As the tongue is protruded, complex interactions between the muscles, cartilages, and fluid-filled sinuses of the tongue alter its shape, producing a pronounced rim and a central trough rich in mucus glands (Figure 9–17). As the tongue contacts the prey a few millimeters beyond the jaw, the front of the rim is depressed and the prey adheres to the sticky tongue, which is then retracted by longitudinal muscles, especially the rectus cervicis (Larsen and Guthrie 1975).

Advanced tongue projection systems have arisen in several lineages of salamanders, including two genera of salamandrids and several of plethodontids. David Wake (1982) has argued convincingly that the evolution of a projectile tongue is facilitated by a shift in the mode of respiration, which in turn reflects the environment in which those salamander lineages evolved. The salamandrids *Chioglossa* and *Salamandrina* are lungless, like all of the Plethodontidae. That condition probably arose in response to the problems that lungs, which impart positive buoyancy, present to animals dwelling in rapidly flowing streams. Indeed, *Chioglossa* and *Salamandrina* now occur in such habitats, and the plethodontids are believed to have arisen in a similar environment in the present Appalachian Mountains. In the absence of lungs, the buccal floor is no longer involved in pumping air, and the hyobranchial apparatus is freed from its role in respiration. Absence of lungs is critical to the evolution of tongue projection, because the hyobranchial cartilages must be slender for tongue projection to occur, but slender hyobranchials are incompatible with pumping air into the lungs. The ability to modify the hyobranchium for tongue projection thus was facilitated by the evolution of lunglessness.

Although differing in detail and degree, the fundamental mechanism of tongue projection by plethodontids is clearly derived from that of other salamanders. The most extreme tongue projection systems in salamanders, as well as the best studied, occur in the family Plethodontidae (Figure 9–18). At least three major lineages have evolved tongue projection, and several functionally and anatomically distinct variations on the theme of tongue projection have been recognized (Lombard and Wake 1986). Tongue projection occurs when the specialized subarcualis rectus I muscles contract around the arms of a Y-shaped hy-

obranchial skeleton (Figure 9–19). In species with advanced tongue projection the tongue pad is at the anterior end of the median basibranchial cartilage and is not attached to the floor of the mouth.

From the rear of the basibranchial cartilage, the paired ceratobranchial cartilages I and II extend back and articulate with highly elongate, gently tapering, paired epibranchial cartilages. The subarcualis rectus I in these species is tightly wound into a spiral around each epibranchial cartilage. When it contracts, the muscle squeezes on the epibranchial, which slides forward. The two pairs of ceratobranchials rotate against the basibranchial as the arms of the Y slip beneath the throat, propelling the basibranchial and the tongue tip out of the mouth. Thus, it is the collapsing arms of the Y that propel the tongue, and the evolution of advanced tongue projection is accompanied by an elongation of the epibranchial cartilages.

The most extreme ability to project the tongue is observed in the tribe Bolitoglossini, and that ability seems to have arisen as a consequence of a seemingly unrelated adaptation. In plethodontid genera that have aquatic larvae (e.g., *Eurycea* and *Pseudotriton*), the

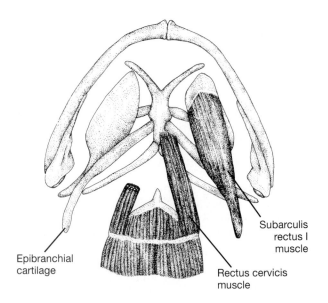

Figure 9–16 **Tongue protrusion by the tiger salamander *Ambystoma tigrinum*.** Ventral view of the hyobranchial skeleton and major muscles (shown only on one side) involved in tongue protrusion and retraction. Note the short epibranchial cartilages and straight fibers of the subarcualis rectus I muscle, which draw the tongue forward. The rectus cervicis muscle retracts the tongue. (*Source: Modified from Larsen and Guthrie 1975.*)

Figure 9–17 **Prey capture by the tiger salamander, *Ambystoma tigrinum*. The** tongue is protruded only a short distance. Note the trough formed by the rim of the protruded tongue. *(Source: Larsen and Guthrie 1975.)*

larvae use the hyobranchium for suction feeding and respiration. These movements are produced by depression of the buccal floor, which largely involves ceratobranchial I. That cartilage remains larger than ceratobranchial II even after metamorphosis. There-

fore, when the adult tongue is projected in such species, the arms of the Y fold until the larger first ceratobranchial binds against the second (Figure 9–20). Bolitoglossines, however, exhibit direct development. There are no free-living larvae, and young hatch as miniature replicas of the adults. Apparently freed of the constraint of larval suction feeding, bolitoglossines have greatly reduced the width of the first ceratobranchials, with the result that the arms of the hyobranchium can collapse still more, achieving a longer distance of projection before the basibranchials bind. Some bolitoglossine salamanders can project the tongue nearly half their body length.

Once the tongue has been projected and the sticky tongue pad has made contact with the prey, it must be retracted. That action involves contraction of a straplike derivative of the rectus abdominis and rectus cervicis muscles that extends from the pectoral girdle to the inside of the tongue tip. Ordinarily, one thinks of a sensory feedback system in which the tongue is fired, makes contact with the prey, signals that it has

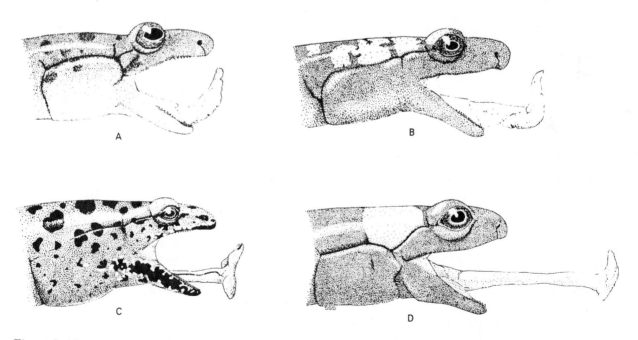

Figure 9–18 **Comparison of tongue protrusion and projection by plethodontid salamanders.** (a) *Desmognathus monticola*, a desmognathine with an attached tongue and minimal protrusion. (b) *Ensatina eschscholtzii*, a plethodontine with an attached tongue but a modest ability to protrude the tongue. (c) *Pseudotriton ruber*, a hemidactyliine with a free tongue and considerable projection. (d) *Pseudoeurycea belli*, a bolitoglossine with a free tongue and extreme projection. The tongues are drawn during projection but not at maximal distance. *(Source: Lombard and Wake 1977; drawing by S. S. Sweet.)*

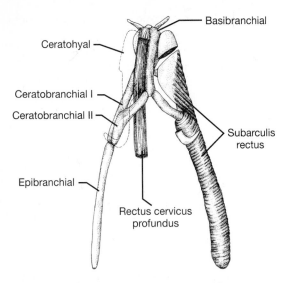

Figure 9–19 **Tongue protrusion by the two-lined salamander *Eurycea bislineata*.** Dorsal view of the hyobranchial skeleton and major muscles (shown only on one side) involved in tongue projection and retraction by a hemidactyliine plethodontid. Note the relatively long epibranchial cartilages and the spiral fibers of the subarcualis rectus I muscle, which contracts to project the tongue. The rectus cervicis muscle retracts the tongue. The glandular tongue pad would lie around the radial, lingual, and anterior ends of the basibranchial cartilages. *(Source: Modified from Lombard and Wake 1976.)*

made contact, and is retracted. However, high-speed cinematography of feeding by *Bolitoglossa* demonstrated that the entire feeding sequence took only 4 to 6 milliseconds, less time than would be required for sensory information from the tongue to reach the brain and a signal from the brain to return to the retractor muscle. Instead, the subarcualis rectus and retractor muscles fire simultaneously. Because the retractor lies slack within the throat, its contraction has no immediate effect on the rapidly moving tongue. Only as the tongue reaches its full extension does the retractor muscle pull taut, at which point the tongue begins its return.

Apart from the kinematic aspects of launching the tongue toward the prey, there are other consequences of tongue projection in plethodontids. For example, the nerves that run to the tissues of the tongue must extend greatly as projection occurs. The nerves serv-

ing the projectile tongue are tightly coiled and often looped like a telephone cord, presumably to allow them to stretch during prey capture. Interestingly, several distinct coiling patterns are observed, which, together with other features of the lingual apparatus, suggest that advanced tongue projection arose largely independently in at least four and possibly as many as six lineages of plethodontids.

Another consequence of advanced tongue projection is the requirement for special visual processing. A range-finding system is needed to determine prey distance, and plethodontids apparently use stereoscopic vision to judge distance. Stereoscopic vision involves transmission of visual information from each eye to both sides of the brain, which estimates distance based on the slight differences in the visual image from the vantage point of either eye. Bolitoglossine salamanders have such stereoscopic vision, which involves neural organization roughly comparable to the mammalian condition (Wiggers et al. 1995).

Anurans. Tongue projection in frogs proceeds by an entirely different mechanism. In most, the tongue is flipped from the mouth by a curious mechanism resembling a catapult (Gans and Gorniak 1982a,b). The body of the tongue is supported by the genioglossus muscles, which attach at the front of the mandible (Figure 9–21). As projection begins, the genioglossus contracts and stiffens, forming a stiff bar. Simultaneously a short, thick, transverse muscle at the front of the lower jaw, the submentalis, also contracts, bulging upward. As it bulges, the submentalis pushes on the now rigid tongue, propelling it upward, forward, and over the front of the mandible. The soft, sticky posterior tissues of the tongue are flung forward and downward, landing on the prey and adhering to it. The tongue is drawn back into the mouth by the hyoglossus muscle, whose straight fibers extend from below the hyoglossal skeleton to the interior of the tongue.

Anurans feed on a wide variety of prey, largely arthropods. Typical anuran prey capture is rapid, but not as fast as in plethodontid salamanders with projectile tongues. In a typical capture by *Bufo marinus* (Bufonidae), the toad approaches to within 1.5 head lengths of the prey and tongue projection takes only about 37 milliseconds. An entire tongue projection cycle takes about 143 milliseconds. The most fascinating aspect of anuran tongue projection is reliance on the temporary stiffening of the tongue musculature to provide rigidity to a structure that lacks inter-

Figure 9–20 **Two modes of tongue projection by plethodontid salamanders.** Hyobranchia (seen from below at nearly full projection) of (a) *Eurycea*, a hemidactyliine plethodontid and (b) *Pseudoeurycea*, a bolitoglossine. Note the relatively large first ceratobranchials in *Eurycea*, which cause the collapsing hyobranchium to bind while the epibranchials are still divergent. In contrast, the small first ceratobranchials in *Pseudoeurycea* allow the epibranchials to collape fully until they are parallel, allowing greater projection. BB = basibranchial, CBII = second ceratobranchial, L = ligament. (*Source: Modified from Lombard and Wake 1977.*)

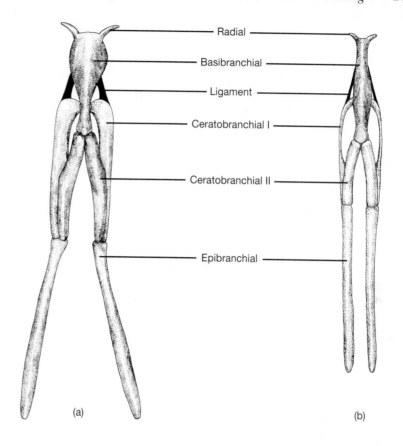

Radial

Basibranchial

Ligament

Ceratobranchial I

Ceratobranchial II

Epibranchial

(a)

(b)

nal skeletal support. Kiisa Nishikawa and David Cannatella (1991) studied tongue projection in *Ascaphus*, the sister taxon of other extant frogs, and demonstrated that the same basic mechanism was present in the common ancestor of the clade. Although the tongue of *Ascaphus* cannot be projected very far, apparently because it is tightly bound to the floor of the mouth, the frog compensates behaviorally by lunging forward toward the prey.

Not all of the many species of terrestrial frogs feed by true projection, and the system has been modified to reflect certain specialized predatory modes. The bizarre Mexican burrowing toad, *Rhinophrynus* (Rhinophrynidae), feeds on very small prey, ants and termites, that it apparently captures by tongue protrusion rather than projection. The base of the tongue is attached to a highly mobile hyobranchial apparatus, unlike the condition in other frogs. Apparently *Rhinophrynus* feeds by protruding just the tip of its glandular tongue, sliding it straight out of the arched mouth while the lips remain closed laterally. Presumably *Rhinophrynus* digs into a colony of termites or ants, pushes its narrow snout into a tunnel, and uses its tongue tip to snare its tiny prey (Trueb and Gans 1983).

Another atypical anuran feeding behavior occurs in *Tornierella* (Hyperoliidae), a genus of frogs from the Ethiopian highlands of Africa, in which adults feed primarily on snails and semislugs (snails with shells too small to withdraw into), which are swallowed whole. As in other durophagic species, *Tornierella* exhibits a variety of adaptations to its unusual diet, including stout, reinforced jaws and strong adductor muscles that may be used to grasp terrestrial snails and pull them from the substrate to which they adhere.

Chameleons. Tongue projection has arisen also among lizards. Members of the Chamaeleonidae, the Old World chameleons, can project their tongues at speeds up to 5.8 m·s^{-1} to twice their snout–vent length (Figure 9–22).

The long tongue of chameleons bears a sticky pad at its tip and a powerful circular muscle known as the

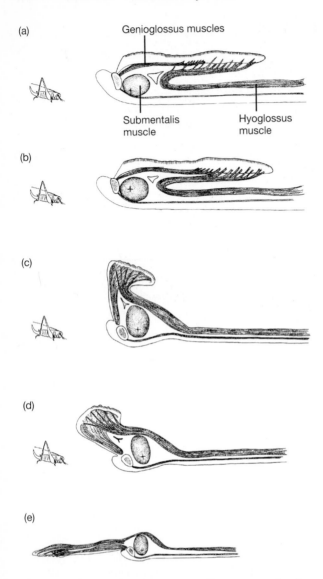

(a)

Genioglossus muscles

Submentalis muscle   Hyoglossus muscle

(b)

(c)

(d)

(e)

Figure 9–21   **Successive stages of tongue projection by the marine toad, *Bufo marinus*.** The stiffened genioglossus muscle pivots over the submentalis as the tongue flips forward. The tongue is retracted by the hyoglossus muscle. (*Source: Gans and Gorniak 1982a.*)

accelerator muscle (Figure 9–23). The long posterior region of the tongue contains a longitudinal muscle, the hyoglossus. At rest the accelerator muscle is wrapped around the tapered end of the processus entoglossus, a long extension of the hyobranchial skeleton, and the hyoglossus muscle lies loosely pleated behind it. When a prey item is sighted, the lower jaw is

depressed and the hyoid apparatus is drawn up and forward, slowly at first and then rapidly in a burst of muscular activity (Wainwright et al. 1991, Wainwright and Bennett 1992a,b). At the same time the accelerator muscle contracts, squeezing itself off the processus entoglossus and propelling the tongue forward. As the tongue leaves the mouth, the hyoglossus is stretched behind it. The prey adheres to the sticky tongue tip, and the hyoglossus contracts to return the tongue and prey to the mouth.

Although the tongue continues forward briefly after it contacts the prey, filmed sequences indicate that distance is measured rather precisely. Apparently the timing of contraction of the hyoglossus muscles determines how far the tongue travels. Because the tongue of a chameleon can be projected so far from the head, the hyoglossus muscles must be capable of an extraordinary degree of contraction. Most vertebrate muscles can contract to about 70 percent of their extended length, and their individual sarcomeres can be contracted to a limit of about 40 percent of their extended length. However, the hyoglossus of chameleons can contract to about 16 percent of its extended length, a feat that is attributable to a modification of the cellular structure of the sarcomeres. The Z-bands, which ordinarily fully separate adjacent sarcomeres of striated muscles, are perforated, allowing the contractile filaments to slide beyond the confines of individual sarcomeres. That condition results in supercontraction, a phenomenon otherwise known only in barnacles and certain insects.

The evolution of tongue projection in chameleons has been accompanied by the evolution of specialized distance perception. The eyes of chameleons are unique among those of vertebrates in their degree of mobility and independence (Figure 9–24). Located on movable turrets, the right and left eyes of chameleons can scan the environment independently. When a prey item is observed, both eyes are trained on the object. Although it would seem reasonable to expect that binocular vision is used for range finding, experiments showed that chameleons need only a single eye to determine prey distance and that distance is measured by accommodation (focusing). The horns of *Chamaeleo jacksoni* are ideal for mounting spectacles, and Lindesay Harkness (1977) fitted lizards with glasses that produced an image of a prey item either closer to or farther from the head than the prey itself. In either situation the chameleon projected its tongue to the distance of the optical image, overshooting the actual prey in the case of a distant image and under-

Figure 9–22 **The common chameleon, *Chamaeleo chamaeleon*, capturing prey by tongue projection.** Note the straight region at the base of the tongue, which results from the processus entoglossus, a part of the hyoglossal skeleton that lies inside the tongue. *(Photograph by Stephen Dalton/NHPA)*

shooting when receiving a near image. Furthermore, the lizard's performance was the same whether it used both eyes or only one, ruling out binocularity as a cue. The lens in a chameleon's eye focuses very rapidly and, uniquely among vertebrates, enlarges the visual image, acting like a telephoto lens to magnify a small region of the visual world (Ott and Schaeffel 1995). The relatively large retina and high density of cones produce an exceptionally large image on the retina, rendering accommodation and range finding more precise.

How did the extraordinary tongue projection system of chameleons arise? The projectile system seems to be an extreme modification of the use of the tongue in capturing prey, which in turn is the ancestral condition of squamates, being shared with *Sphenodon* and the other iguanian lizards, including the sister family to chameleons, the Agamidae (Schwenk and Throckmorton 1989). The agamid *Phrynocephalus helioscopus* occasionally attempts to capture prey with its tongue at a greater distance than usual, and in so doing its tongue tip extends well beyond the lower jaw and assumes a shape not unlike that of a chameleon's tongue just prior to projection. Indeed, many of the morphological precursors of the chameleon system are apparent in the Agamidae, including an elongate processus entoglossus, a circular muscle that generates pressure against that process, and a thick, sticky tongue pad. Apparently the in-

creased length and parallel sides of the processus entoglossus were the critical elements that resulted in the qualitatively different mode of prey capture that characterizes chameleons (Wainwright and Bennett 1992b).

Cranial Kinesis

Cranial kinesis is widespread among vertebrates, and its absence often reflects secondary loss. Many lineages of fishes exhibit cranial kinesis, and mammals are unique among the classes of tetrapods in having no members that exhibit any degree of kinesis. It is hard for us, as mammals, to appreciate how unusual our rigid (akinetic) skull is. Although only a few extant amphibians exhibit cranial kinesis, there is evidence that some lineages have secondarily lost that feature (caecilians, for example). Among extant reptiles, cranial kinesis is widespread among squamates, especially snakes.

Lizards

Lizards are a diverse group, and their feeding habits are correspondingly diverse. The vast majority of lizards prey on arthropods, including insects. Some, such as varanids, eat vertebrates, and a few are herbiv-

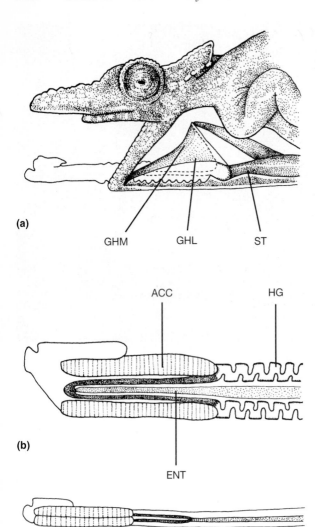

(a)

GHM   GHL   ST

ACC   HG

(b)

ENT

(c)

Figure 9–23 **Mechanism of tongue projection by a chameleon.** (a) The hyobranchium is first slowly and then rapidly protruded from the mouth by action of the sternothyroid (ST) and geniohyoideus lateralis (GHL) and geniohyoideus medialis (GHM) muscles. (b) Contraction of the accelerator muscle (ACC) causes it to decrease in diameter but lengthen, until it reaches the tapered tip of the processus entoglossus (ENT), at which point it is projected from the mouth. (c) As the tongue is projected the hyoglossus muscle (HG) is stretched and eventually hyperextended; it will later serve as the retractor for the tongue. *(Source: Modified from Wainwright et al. 1991, Wainwright and Bennett 1992a.)*

orous. Although lizards are phylogenetically diapsids, their skulls exhibit a modified diapsid condition. Specifically, lizards have lost the lower temporal arch, which immobilized the ventral end of the quadrate bone. As a result, the quadrate is free to move around its dorsal articulation with the squamosal bone. This condition, known as streptostyly, affects feeding because the lower jaw articulates with the ventral end of the quadrate. Streptostyly may have several mechanical functions.

An obvious result of streptostyly would be an increase in the gape of the jaws if the quadrate were swung posteriorly when the mouth opens. However, cineradiography shows that the quadrate actually swings forward as the mouth opens, which should reduce the gape. What, then, does streptostyly accomplish? Studies by Kathleen Smith (1982) suggest that streptostyly may increase biting force. The jaws are a third-class lever system, and their mechanical advantage would be increased if the pterygoideus (the large muscle that pulls the lower end of the quadrate forward) rotated the mandible-plus-quadrate around the joint between the quadrate and the rest of the skull. In other words, streptostyly may increase biting force by adding an extra segment that lengthens the in-lever of the mandible.

Lizards have additional forms of cranial kinesis beyond streptostyly. Two additional points of potential flexibility occur in the skull (Figure 9–25). One is between the frontal and parietal bones (mesokinetic articulation) and the other is between the parietal and supraoccipital bones (metakinetic articulation). Thus, the skull is said to be amphikinetic (Bramble and Wake 1985). The significance of mesokinesis and metakinesis during natural feeding has been debated and experimental studies have been contradictory (Frazzetta 1983, Smith and Hylander 1985). If flexibility does exist at either of those articulations, the result would be an increase in gape by elevation of the snout as the mouth is opened. A mesokinetic joint would also allow the snout to flex downward during jaw closing. This action has been demonstrated for the specialized pygopodid lizard *Lialis* (Patchell and Shine 1986a), and may occur in other small lizards that have relatively light cranial bones and flexible sutures.

Although most lizards exhibit at least potentially kinetic skulls, secondary reduction of kinesis occurs in some herbivorous species. Both mesokinesis and metakinesis have been lost by chameleons, although

Figure 9–24   **Eyes of chameleons.** (a) Eyes of *Chameleo vulgaris*, facing forward and down. (b) *Chamaeleo jacksoni* wearing spectacles to test binocular accommodation. *(Source: (a) Photograph by Dwight R. Kuhn/Bruce Coleman; (b) Photograph courtesy of L. Harkness.)*

streptostyly remains, at least to a limited degree. Several lizards eat hard-shelled prey, especially the South American caiman lizard (*Dracaena*), Australian blue-tongue skinks (*Tiliqua*), and some species of monitor lizards (*Varanus*). These species have large, blunt teeth at the rear of the jaws. *Dracaena* seems to be the most specialized of these durophagus lizards, preying on mollusks that it crushes in its jaws (Figure 9–26). *Dracaena* differs from the related teiid *Tupinambis* in having very large molar-shaped teeth, a stronger palatal skeleton, hypertrophied adductor muscles, and increased mechanical advantage at the rear of the tooth row. *Varanus olivaceus*, an unusual monitor lizard from the Philippines, eats both fruits and animals, especially land snails. Crushing dentition develops late in life, and it is adults that eat snails (Auffenberg 1988).

Most lizards use the tongue to move prey to the rear of the mouth in preparation for swallowing, but two lineages of such lizards (Scincomorpha and An-guimorpha) have independently evolved deeply forked tongues that are specialized for chemoreception (Schwenk 1994). These tongues are not suited to moving food in the mouth, and lizards such as tegus (*Tupinambis*) and monitors employ an alternative mechanism, inertial feeding (Figure 9–27), to transport large prey. The food is held aloft in the jaws, the head is drawn back, and then the prey is momentarily released while the head is rapidly thrust forward and the grip reestablished. The prey may be relatively stationary while the jaws are open (static inertial feeding), or a rearward motion may be imparted to it just before the jaws are opened (kinetic inertial feeding). In either case, the position of the prey is maintained by its own inertia while the jaws of the predator are open, hence the name of the process.

Snakes

The skull of snakes is highly modified and has the highest degree of cranial kinesis observed among tetrapods. Most snakes swallow relatively large prey by a system known as unilateral feeding, in which the

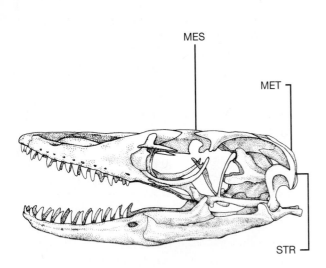

Figure 9–25 **Skull of the savanna monitor lizard,** ***Varanus exanthematicus.*** This lizard has three potentially kinetic joints: mesokinetic (MES), metakinetic (MET), and streptostylic (STR). *(Source: Modified from Smith 1982.)*

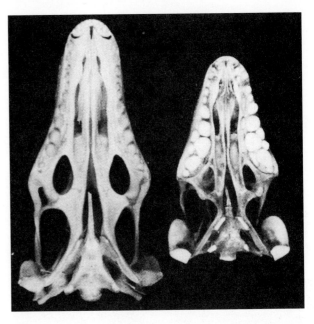

Figure 9–26 **The South American snail-eating teid lizard,** ***Dracaena guianensis.*** Ventral view of the skulls of the teid lizards *Tupinambis*, left, and *Dracaena*, right; note the large, molariform (molar-shaped) rear teeth of *Dracaena*. *(Source: Dalrymple 1979.)*

right and left bones of both the upper and lower jaws move alternately.

Our understanding of the mechanical consequences of the transition from lizards to snakes has been advanced by a number of important discussions in recent years, beginning with Carl Gans's (1961) suggestion that the evolution of the ophidian feeding system is related to increasing the size of the gape. The design of snakes presents a dilemma: the substantial body mass of a snake must be sustained by food that passes through a relatively small mouth. Gans noted that the gape of snakes is enlarged by the evolutionary loss of the mandibular symphysis, the bony articulation between the paired mandibular rami (Figure 9–28). The combination of independent fore-and-aft movement of the two sides of the lower jaw and a loosely articulated streptostylic quadrate that allows the jaws to be moved sideways permits large items to pass through the mouth. In most snakes an additional bone that links the quadrate to the braincase also is mobile, further increasing the gape.

Prey size is not a single variable. Harry Greene (1983) examined the diets and prey-handling behavior of a variety of snakes, including the extant descendants of the earliest lineages. He suggested that in

order to understand the evolution of ophidian feeding, prey size must be considered in terms of both its mass and its linear dimensions. Large prey may have a high weight ratio (the mass of the prey divided by the mass of the predator) without having a large ingestion ratio (diameter of the prey divided by the diameter of the head of the predator). The descendants of early snakes, the extant Aniliidae and Cylindrophiinae, feed on elongate prey, such as eels and caecilians, that have a high weight ratio but a low ingestion ratio (Figure 9–29). The greatest functional challenge to early snakes would have been dealing with the thrashing of heavy, elongate prey. Greene also argued that feeding on high weight ratio prey, regardless of its shape, favors the evolution of specializations for immobilizing the prey. Most primitive snakes immobilize their prey by constriction, a behavior in which coils of the snake are tightened around the prey and induce circulatory failure by compressing the thorax and thereby preventing the chambers of the heart from filling (Hardy 1994). Among the early ophidian lineages, constric-

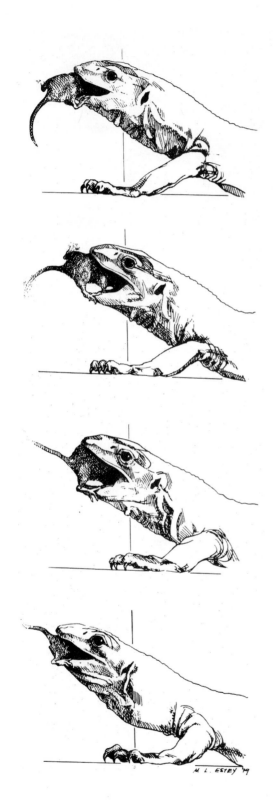

tion is characteristic of at least the Cylindrophiinae, Xenopeltidae, and Loxocemidae.

Although constriction seems to have arisen to handle elongate prey (low ingestion ratio), it proved well-suited to bulky prey (high ingestion ratio) such as mammals. Constriction requires tight bending of the vertebral column, and that means short vertebrae and muscle units that link nearby vertebrae. This anatomical arrangement appears to limit the speed of locomotion. Boas and pythons are relatively heavy-bodied, slow-moving snakes that often take very large prey.

Another lineage of snakes, exemplified by the Colubridae and related families, appears to have evolved more rapid locomotion, which requires long vertebrae and muscle units that span many vertebrae. These anatomical features create the broad body curves needed for rapid locomotion but are less effective at producing tight coils. Thus, these snakes could move fast but could not readily constrict prey. The appearance of Duvernoy's gland, a cephalic gland that immobilizes prey and is homologous to the venom gland of front-fanged snakes, occurred simultaneously with the evolution of rapid locomotion by snakes (Figure 9–30). Thus, constriction and venom delivery appear to be alternative strategies for prey immobilization that, in a broad sense, reflect differing solutions to balancing the functional demands of locomotion and feeding (Savitzky 1980). Some colubrid snakes such as the ratsnakes (*Elaphe*) and kingsnakes (*Lampropeltis*) that have secondarily evolved constriction have reduced Duvernoy's glands.

The unilateral feeding system allows snakes to swallow prey items too large to be transported by the tongue or by inertial feeding. Because there is no bony mandibular symphysis or, in most snakes, a rigid articulation at the front of the upper jaws, the right and left sides of the upper and lower jaws operate independently during swallowing, driven by the palatal muscles and the streptostylic quadrate (Haas 1973). The braincase of snakes is heavily ossified. The

Figure 9–27 **Inertial feeding by the savanna monitor lizard, *Varanus exanthematicus*.** Note position of the lizard's head and the prey item relative to the line, which represents a fixed point. First the head and prey are drawn back as the jaws are opened. As the jaws reach maximum gape, the head is quickly thrust forward as the prey continues to move back. *(Source: Smith 1982.)*

(a)

(b)

Figure 9–28   **The hognose snake, *Heterodon platyrhinos*, swallowing prey.** (a) Note the independent tips of the mandibles. (b) Note protraction of the palatomaxillary arch on one side. *(Photographs by R. Wayne Van Devender.)*

frontal and parietal bones surround the brain laterally (Figure 9–31) as well as dorsally, articulating rigidly with the floor of the skull and obliterating the ancestral mesokinetic articulation. The metakinetic articulation also is lost, and the result is a rigid braincase from which the powerful palatal muscles can pull on the bones of the upper jaw, driving the unilateral feeding sequence. A new point of flexion, the prokinetic articulation, arises between the bones of the snout (the premaxilla, nasals, septomaxillae, and vomers) and the frontal bones of the braincase, and some snakes also have flexible connections *among* the bones of the snout, a condition known as rhinokinesis (Cundall and Shardo 1995). Those new articulations allow the snout to move as the bones of the upper jaw draw prey into the mouth.

In most snakes the bones of the upper jaw are joined into a Y-shaped palatomaxillary arch in which the maxilla and the palatine and pterygoid bones bear teeth. The maxillae, together with the adductor muscles, are involved primarily in prey capture, whereas the palatal bones are used primarily in prey transport (Cundall and Gans 1979, Cundall 1983). The jaw muscles of snakes are complex, with a series of mandibular adductors that extend from the braincase to the mandible and three palatal muscles that drive the palatomaxillary arch forward and back. The quadrate is linked to the rear of the palatomaxillary arch by a ligament, mechanically linking the upper

and lower jaw on each side. An additional joint within the mandible provides further flexibility.

During transport the toothed palatal bones on one side of the head are lifted and protracted to gain a more forward grip on the prey. The final action on that side is to pull forward on the braincase as the palatal teeth are embedded in prey, effectively drawing the snake's head over the prey item. Those actions are then repeated by the bones of the opposite side (Figure 9–32). That sequence repeats itself until the prey is sufficiently far down the esophagus to be moved by waves of contraction of the axial muscles. The snake moves over the prey rather than drawing it in, clearly an effective strategy to avoid having to overcome the inertia and friction of relatively heavy prey lying on the ground.

Snakes have evolved an impressive diversity of prey capture mechanisms. The maxilla appears to play a major role in prey capture, and maxillary shape and dentition vary widely among snakes. That diversity of form often reflects very narrow prey preferences, with broad generalists in the minority among snakes. Various groups have evolved as specialists on toads, eels, fish eggs, and a wide variety of other prey.

The most divergent ophidian feeding mechanisms characterize the three scolecophidian families, the members of which prey on ants and termites. Unlike most snakes, they have short jaws, few teeth, and a very small gape (Figure 9–33a). They appear to com-

Figure 9–29 **Prey ratios.** A colubrid snake from Madagascar, *Dromicodryas bernieri*, feeding on a gecko (*Paroedura bastardi*), an elongate prey item (high weight ratio, low ingestion ratio). (b) Western diamondback rattlesnake (*Crotalus atrox*)snake, feeding on a wood rat (*Neotoma albigula*), a wide prey item (high weight ratio, high ingestion ratio). *(Photographs by (a) K. G. Preston-Mafham/Premaphotos, (b) Cancalosi/Tom Stack & Associates.)*

(a)

(b)

pensate for the small size of their prey by rapidly consuming enormous numbers of those social insects, which the snakes locate by following the pheromone trails laid down by the prey. The guts of four species of the Australian typhlopid genus *Rhamphotyphlops* contained primarily ant larvae and pupae, with pupae accounting for as much as 93 percent of the prey items of one species. Smaller numbers of eggs and a few adult ants also were taken, along with a very few termites and mites. Many of the snakes contained over 20 prey items, and one *R. nigrescens* contained an astonishing 1,431 items (Shine and Webb 1990).

Most snakes swallow their prey whole without crushing or chewing it, but a few genera of colubrids eat hard-bodied prey (Savitzky 1983). The rarity of durophagic snakes can be appreciated by comparing the incidence of insectivory among lizards and snakes. Although most small lizards are insectivorous, only a few small colubrids prey on insects or other small arthropods. The only major radiation of arthropod-eating colubrids is the Sonoratini (Figure 9–33b), centered in the xeric regions of southwestern North America, and its members exhibit only subtle differences in tooth and skull morphology that seem to reflect the toughness of their diet (Savitzky 1983).

Several lineages of snakes have spatulate teeth that are attached to the jaws by a connective tissue hinge rather than being firmly fused to the bone (Figure 9–34). These teeth fold down as a prey item passes

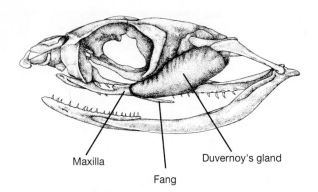

Figure 9–30 **The African boomslang, *Dispholidus typus*.** Note the Duvernoy's gland and the enlarged, grooved teeth at the rear of the maxilla. *(Source: Modified from Parker and Grandison 1977.)*

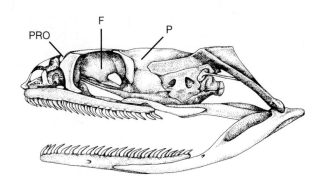

Figure 9–31 **Lateral view of the skull of *Nerodia rhombifer*, a colubrid snake.** The fully enclosed braincase results from downgrowth of the frontal (F) and parietal (P) bones. Note the location of the prokinetic articulation (PRO). *(Source: Modified from Cundall and Gans 1979.)*

into the mouth and lock in a vertical position if the prey tries to escape. Hinged teeth are found in three genera of colubrid snakes (*Liophidium*, *Scaphiodontophis*, and *Sibynophis*) that feed almost exclusively on skinks. In the genus *Xenopeltis* (Xenopeltidae) young individuals have hinged teeth and feed largely on skinks. Adult *Xenopeltis* eat mammals, and their teeth are pointed and firmly fixed to the jaw bones (Savitzky 1981).

Two species of *Regina* (Colubridae) that feed on hard-shelled crayfish have hinged teeth, whereas two other species of *Regina* that prey only on freshly molted crayfish retain the ancestral condition of pointed, firmly fixed teeth. An alternative method of dealing with crustaceans is seen in *Fordonia*, an estuarine colubrid from Indoaustralia. *Fordonia* eats crabs and mud-burrowing shrimp, which it crushes with its short, blunt teeth. *Fordonia* also has thick stomach muscles that may compact the shells of its prey, and a large salt-excreting gland.

Although many snakes will eat bird eggs, a few genera of colubrids specialize on that food source. The African egg-eating snake *Dasypeltis* can swallow hard-shelled eggs up to four times the diameter of its head (Figure 9–35), although they are usually not so large. The skin of the neck is extraordinarily elastic, and the teeth are reduced to a few small vestiges, being replaced with ridges of soft mucosal tissue. The

egg is engulfed intact, and the shell is not cracked until the egg has moved a short distance down the throat, where the esophagus lies below specially modified ventral processes of the vertebrae (hypapophyses: Figure 9–36). Rounded hypapophyses appear to crush the egg shell. Behind them are a series of long, anteriorly directed hypapophyses that actually protrude into the esophagus. These projections apparently slit the distended embryonic membranes of the egg, freeing its liquid contents and preventing the shell from passing farther back into the gut. The liquid contents are swallowed, and the egg shell is compressed and regurgitated.

Several species of colubrid snakes are specialized for feeding on squamate eggs, which differ from bird eggs in having flexible shells. These snakes typically have a pair of flattened, saberlike posterior maxillary teeth that apparently slice the leathery eggshells. Among the genera known to feed in that manner are the scarlet snake, *Cemophora*, and the leaf-nosed snakes, *Phyllorhynchus*, of North America, as well as the kukri snakes, *Oligodon*, of Asia, which receive their common name from the fancied resemblance of their rear teeth to a Gurkha knife. Other genera, such as *Umbrivaga* of South America, may feed in a similar manner. Some species of the Australian elapid genus *Simoselaps* also feed on lizard eggs. In this genus the maxillary teeth are modified for venom delivery, and

Figure 9–32 **Movements of cranial bones of a colubrid snake during unilateral feeding**. (a) Midway through opening of right side of jaws. (b) Protraction of right side. (c–d) During closing of right side. Cross designates a fixed point on surface of prey. br = braincase, md = mandible, mx = maxilla, p = palatine, pt = pterygoid, q = quadrate, sn = snout. *(Source: Cundall and Gans 1979.)*

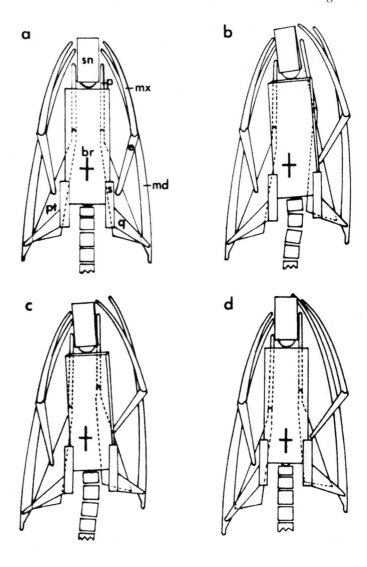

it is the rear mandibular teeth that are enlarged for slitting eggshells (Shine 1991).

Several colubrid groups feed on mollusks, both shelled and unshelled. Some genera, such as *Contia*, the sharp-tailed snake of the Pacific Northwest, and the Mexican genus *Chersodromus*, feed exclusively on slugs and have very long, narrow teeth. Two independent lineages, the Dipsadini of the Neotropics (Figure 9–37) and the Pareatinae of Southeast Asia, specialize on mollusks, including snails. They extract the body from the shell with long excursions of the mandibles, which are laterally compressed and bear a comblike array of teeth (Savitzky 1983).

Envenomation

Although venom delivery is not nearly as widespread among reptiles as most people imagine, complex and potent venoms have evolved several times, often accompanied by extraordinary mechanisms for their hypodermic injection. The resulting adaptations are among the most elegant examples of morphological and physiological correspondence to functional demands, and their multiple origin provides important insights into the selective advantages and evolutionary steps leading to a highly derived venom delivery apparatus. To understand the evolution of such systems it

(a)   (b)

Figure 9–33   **Dietary specialists.** (a) the Texas worm snake, *Leptotyphlops dulcis* (b) *Tantilla coronata* feeding on a large centipede. *(Photographs by R. Wayne Van Devender.)*

is important to recognize that venom delivery is fundamentally a strategy for prey capture and that its use in defense is only a secondary function, although sometimes an important one.

All extant venomous reptiles are varanoid squamates, and all but one are snakes. The only venomous lizards, the Gila monster (*Heloderma suspectum*) and the beaded lizard (*H. horridum*), belong to the family Helodermatidae. Gila monsters live in deserts of the southwestern United States and northwestern Mexico, and beaded lizards occupy thorn forest from southern Mexico to Guatemala. Helodermatids have a large venom gland along the lateral surface of each mandible (Figure 9–38) that secretes its product through one duct (in *H. horridum*) or three to five ducts (in *H. suspectum*) to the labial side of the mandibular tooth row. The venom gland of helodermatids appears to be homologous to the mandibular gland, or gland of Gabe, which is found in all varanid and lanthanotid lizards. The large mandibular teeth associated with the venom gland ducts of helodermatids bear deep grooves on their anterior and posterior surfaces. Venom is conducted into the bite wounds as the lizard tenaciously holds its prey, which consists of a variety of small animals, including nestling birds (Pregill et al. 1986). Helodermatid venom is about as toxic as cobra venom (Mebs 1978). A Gila monster bite causes a drop in blood pressure and difficulty breathing.

The kinetic cranial morphology of snakes has provided a rich substrate for the evolution of diverse venom delivery strategies. However, the venomous snakes in the families Elapidae and Viperidae represent only about 20 percent of extant snakes. Furthermore, many members of venomous lineages are so small that their bites are trivial for humans, whereas some generally nonvenomous lineages include species with venoms toxic to their natural prey and, in some cases, to animals as large as humans.

Venom delivery seems to have evolved incrementally, and varying degrees of specialization occur among snakes. For the purposes of discussion it is useful to recognize rear-fanged and front-fanged conditions, although even those categories are simplifications. Of the two, humans usually regard the front-fanged lineages as venomous, and that condition has arisen at least twice, and probably four times, in the history of snakes.

Duvernoy's gland is characteristic of most colubrid snakes, and often produces venom that immobilizes prey. Duvernoy's gland arises embryonically from the rear of the maxillary dental lamina, together with the posterior pair of maxillary teeth. In other words, the gland is developmentally linked to that pair of teeth, and these structures form a functional complex. Snakes with enlarged, grooved posterior maxillary teeth are described as opisthoglyphous (*opisthen* = behind, *glyphis* = knife), or rear-fanged (although that term is also frequently applied to taxa with enlarged teeth without grooves). The widespread occurrence of Duvernoy's glands and enlarged, grooved teeth among colubrid snakes suggests that the system

Figure 9–34   **Hinged teeth of skink-eating colubrid snakes.** (a) Scanning electron micrograph of maxillary teeth of *Liophidium rhodogaster*. Tooth on left is in erect position (note fibers of hinge at base of tooth); tooth on right was held in depressed position during preparation. (b) Cross section of tooth of *Scaphiodontophis venustissimus* showing fibers of hinge (arrow) connecting tooth and maxillary bone. *(Source: (a) Electron micrograph by Electron Microscopy Laboratory, National Museum of Natural History, provided by Alan Savitzky. (b) Photomicrograph by Alan Savitzky.)*

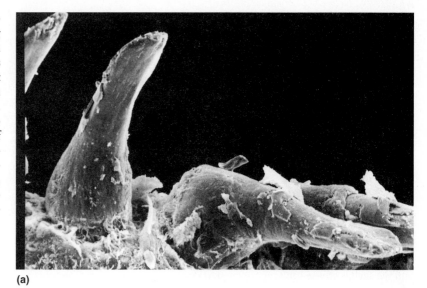

(a)

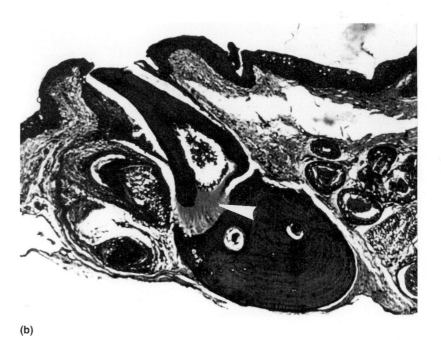

(b)

evolved early in the history of that group. Most instances of its absence can best be interpreted as secondary loss, sometimes in association with the secondary evolution of constriction.

Most rear-fanged snakes deliver relatively small volumes of venom relatively slowly. For that reason, the venoms of most such species still have not been well studied. The underlying jaw musculature may express venom from Duvernoy's gland (Jansen and Foehring 1983), but the venom probably flows through the groove in the fang and into the wound by capillary flow (Kardong and Lavín-Murcio 1993).

Figure 9–35   **The African egg-eating snake, *Dasypeltis scabra*, swallowing a bird egg.** The egg is swallowed whole, and the shell is cracked in the throat. The contents are swallowed, and the shell is crushed and regurgitated. *(Photographs by Michael and Patricia Fogden.)*

Only about 54 percent of venom delivered by *Boiga irregularis*, a rear-fanged colubrid from Asia, enters the wound, while the rest of the venom remains on the prey's skin. Venom delivery by vipers and elapids is more effective—89 percent for the pit viper *Crotalus viridis* and 92 to 97 percent for several species of Australian elapids (Hayes et al. 1992, 1993).

The secretions of Duvernoy's glands immobilize prey and aid digestion. North American lyre snakes (*Trimorphodon*) prey at night on lizards that sleep in rock crevices and defend themselves by inflating their

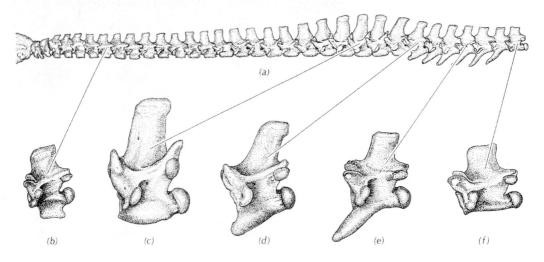

Figure 9–36   **Anterior vertebral column of the African egg-eating snake.** Anterior is to the left, and the rear of the skull is shown. Note vertebrae with thickened hypapophyses (ventral processes) used for crushing eggshells and those with long, anteriorly-directed hypapophyses that slit the egg membranes. *(Source: Gans 1974.)*

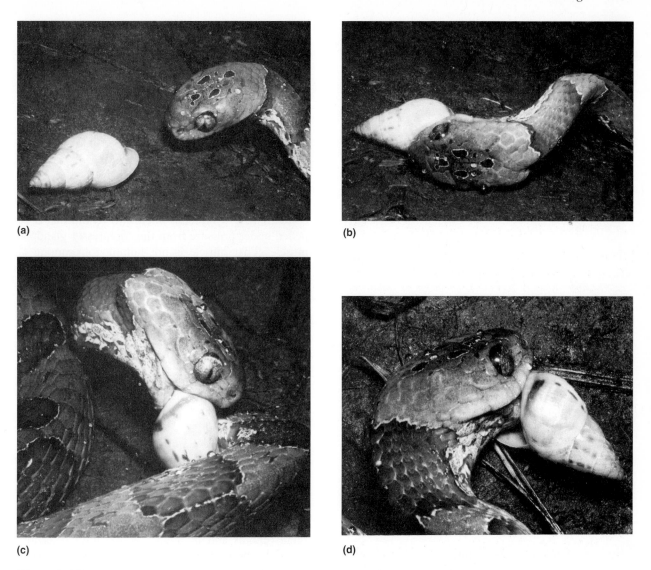

(a)

(b)

(c)

(d)

Figure 9–37   **A snail-eating snake, *Dipsas indica*, feeding on a snail.** (a) Waiting for the snail to reemerge from its shell after being tongue-flicked by the snake. (b) Grasping the snail's body with the teeth on the mandible. (c) Final phase of coiling around prey while still pushing mandible into shell. (d) Extracting snail, with about one third of the mandible into the shell while the mandibular units work the snail out of its shell. *(Source: Sazima 1989.)*

lungs to wedge their bodies in place. *Trimorphodon* bites a lizard and holds on, sometimes for hours, until the slow-acting toxins in its venom immobilize the prey, which is eventually withdrawn from the crevice and swallowed (Greene 1989). Digestion may be the primary function of the venom of the Puerto Rican colubrid *Alsophis portoricensis*. Lizards that had been envenomated before being swallowed were in a more advanced stage of digestion after 6 hours than lizards

that were swallowed without being envenomated (Rodriguez-Robles and Thomas 1992).

Several rear-fanged snakes are capable of inflicting significant bites on humans. The best known of these is the boomslang (*Dispholidus typus*), an arboreal African snake with a well-developed Duvernoy's gland. In 1957 a bite from a young boomslang claimed the life of Karl P. Schmidt, a prominent herpetologist. At that time the bite of the species was

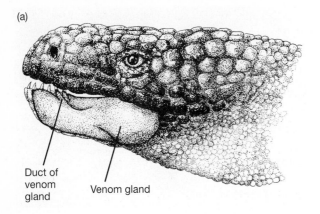

(a)

Duct of
venom
gland

Venom gland

(b)

Figure 9–38 **Venom gland and venom-conducting teeth of the Gila monster, *Heloderma suspectum.*** (a) Location of the venom gland. (b) Medial view of mandible, showing grooved teeth that conduct the venom. *(Source: (a) Based on Kochva 1978a.)*

poorly understood, and Schmidt was unconcerned that the snake had bitten him. The story of his bite (Pope 1958), based largely on his own notes, makes riveting reading. The clinical symptoms of *Dispholidus* envenomation involve extensive internal hemorrhage resulting from interference with normal blood clotting, which leads to loss of blood pressure and circulatory collapse. Species of *Rhabdophis*, common Southeast Asian colubrids, also produce life-threatening symptoms in humans. The clinical manifestations of bites from *Rhabdophis* also involve virtually complete suppression of the clotting mechanism.

Duvernoy's gland has given rise to more derived venom delivery systems in several groups of snakes. At least two, and probably four, lineages have evolved front-fanged venom delivery systems in which the an-

teriormost maxillary teeth conduct the venom (Figure 9–39). The demonstration that those glands, like Duvernoy's glands, arise embryonically from the dental lamina of the fangs suggests that fangs represent rear maxillary teeth whose position has been shifted forward by reduction of the anterior part of the maxilla (McDowell 1968). Furthermore, the fangs of front-fanged snakes have a tube for conduction of the venom that clearly evolved by closing off the ancestral groove.

The venom glands of all front-fanged snakes are relatively large and have a lumen lined with secretory epithelium. The lumen is filled with venom that is expelled under pressure from the compressor glandulae muscle. That muscle is not homologous among all groups of front-fanged snakes, and this lack of homology is one of several clues to the multiple origin of the front-fanged condition.

The front-fanged venomous snakes include two large and diverse families, Elapidae (cobras, coral snakes, and sea snakes) and Viperidae (true vipers and pit vipers), plus two African genera, *Atractaspis* (mole vipers) and *Homoroselaps* (dwarf garter snakes), whose systematic position is in dispute. Most of what we know of the biology of highly venomous snakes pertains to the Elapidae and Viperidae, both of which are species-rich and ecologically diverse families. Therefore, although broad patterns in their biology can be described, care must be taken not to read too much into such generalizations.

Differences in the mechanics of venom delivery by elapids and vipers reflect differences in the nature of their venoms and in their prey-handling behaviors. Most elapids have relatively short fangs in comparison to the size of their heads, and the fangs remain in a vertical position when the mouth is closed. Elapid venoms are rich in short-chain polypeptides that interfere with neuromuscular transmission and immobilize prey relatively rapidly as it is held in the jaws. Viperids, in contrast, have relatively long fangs that are the only teeth on the highly mobile maxillary bones (Figure 9–40). The fangs lie against the roof of the mouth at rest, and protraction of the palatomaxillary arch causes the fangs to rotate through an angle of about 120 degrees (Cundall 1983). The venom, which generally contains enzymes in addition to toxins, is injected deep into the prey, which is then released. The snake trails the prey and swallows it after it has become immobilized (Chiszar et al. 1992).

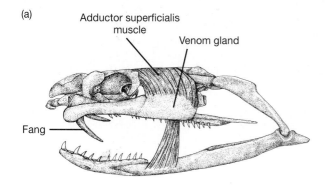

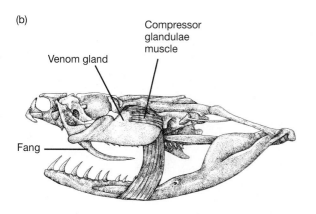

Figure 9–39   **Venom apparatus of elapid and viperid snakes.** (a) the common cobra, *Naja naja*, and (b) the water moccasin, *Agkistrodon piscivorus*, lateral view. Note short fang (F) and small venom gland (VG) of the elapid. The adductor superficialis muscle (AS) compresses the venom gland of elapids, whereas the compressor glandulae muscle (GC) performs that function in viperids.

Although the description of elapid fangs as fixed and viperid fangs as movable is generally true, exceptions occur. Among elapids, for example, a modest degree of rotation of the maxillary bone occurs in some cobras, which spray venom into the face of a predator as a defensive mechanism. Maxillary rotation by elapids is especially pronounced in *Acanthophis*, the death adders of Australia, which exhibit convergence on viperid foraging ecology, among other traits (Shine 1980b). In addition, the mambas (*Dendroaspis*) have relatively long fangs that are depressed during prey capture. Mambas are arboreal, and their long fangs may reflect the need for arboreal snakes to hold on to prey.

Most elapids eat elongate (low ingestion ratio) prey, such as snakes, elongate lizards, or eels. Viperids, in contrast, generally take prey that is both high weight ratio and high ingestion ratio. Viperid snakes have a suite of features associated with swallowing and digestion of high ingestion ratio prey (Pough and Groves 1983). Viperids have a wider gape than most other snakes, vipers' bodies are stouter, and vipers have many scale rows and thus very elastic skin. These features help vipers swallow large prey and accommodate it in their stout bodies. Furthermore, rectilinear locomotion, which is characteristic of vipers, is not disrupted by a bulge caused by prey in the gut.

Once swallowed, prey must be digested before the bacterial flora of the prey's gut causes putrefaction. It's a race between the snake's digestive enzymes, working from outside the prey inward, and the bacteria already present in the prey's gut. The long fangs of vipers deliver venom deep into the prey, and enzymes in the venom break down cell membranes and speed degradation (Thomas and Pough 1979). Snakes can adjust the amount of venom they deliver in a bite, and the prairie rattlesnake (*Crotalus viridis*) has been shown to inject more venom into large prey than small (Hayes et al. 1995).

Snake venoms consist of a complex blend of ingredients with diverse effects (Table 9–1). Smaller proteins include a variety of toxins, many of which attack neuromuscular junctions. Some interfere with lung ventilation and others with the contraction of heart muscle. The venom of *Echis carinatus*, the saw-scaled viper, directly inhibits clotting. In contrast, the venom of the Asian crotaline *Calloselasma rhodostoma* has a compound, known as ancrod, that mimics thrombin and promotes the formation of minute clots. As a result, the concentration of fibrinogen in the blood drops and larger clots cannot be produced. Ancrod therefore is used as an anticoagulant in treating human cardiovascular disease (Mebs 1978, Russell 1980).

Many venoms may immobilize prey simply by causing massive and nonspecific physiological damage, but some aspects of venom chemistry may be under relatively tight selective control. For example, populations of *Calloselasma rhodostoma* from different regions have different proportions of amphibians, reptiles, and mammals in their diets. Variation in venom composition among populations correlates with diet, but not with geographic proximity or phy-

Figure 9–40 **Fang erection by the African puff adder, *Bitis arietans*.** (a) Fang in retracted position. (b) Fang in erect position, involving roation of both the maxillary and prefrontal bones. Note action of the palatal muscles, which drive rotation of the maxillary bone by their action on the palatal bones. c = gland compressor muscle, com = compound bone, d = dentary, dm = depressor mandibulae muscle, ec = ectopterygoid, fa = fang, fr = frontal, levpt = levator pterygoidei muscle, lig = ligaments, mx = maxilla, na = nasal, pa = parietal, pal = palatine, prf = prefrontal, prpt = protractor pterygoidei muscle, pt = pterygoid, pto = postorbital, q = quadrate, rpt = retractor pterygoidei muscle, sta = stapes, su = supratemporal, vgl = venom gland. *(Source: Parker and Grandison 1977; drawings by B. Groombridge.)*

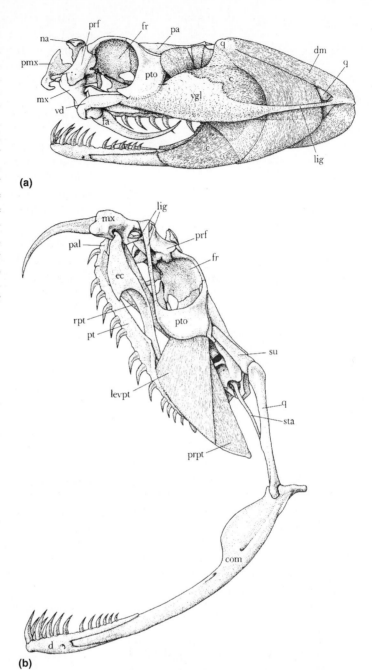

logenetic relatedness of snake populations (Daltry et al. 1996).

Because of the complexity of snake venoms, treatment of venomous snakebite can be a difficult medical problem. Most clinically significant bites are received from elapids or viperids, but a number of rear-fanged snakes have been reported to produce clinical mani-festations in humans. Gans (1978b) listed 41 species of colubrids, belonging to 32 genera, for which human symptoms have been reported. Although the bites of many of those species are inconsequential, some produce considerable local swelling, and bites from a few rear-fanged species have caused human deaths (Minton 1990).

Table 9-1   **Some components of snake venom.**

Compound	Occurrence	Effect
Proteolytic enzymes	All venomous squamates, especially vipers	Digest tissue proteins and peptides
Hyaluronidase	All venomous squamates	Reduces viscosity of connective tissue, increases permeability, hastens the spread of other consitutents of venom through the tissues
L-amino acid oxidase	All venomous squamates	Attacks a wide variety of substrates, causes great tissue destruction
Basic polypeptides	Elapids	Blocks neuromuscular transmission
Cholinesterase	High in elapids, low in viperids	Unknown. It is not responsible for the nerve-muscle-blocking effects of elapid venom
Phospholipase	All venomous squamates	Attacks cell membranes
Phosphatases	All venomous squamates	Attacks high-energy phosphate compounds such as ATP

Source: Russell 1980.

In the case of human bites by front-fanged species, the clinical outcome is complicated by variation in the volume of venom delivered in a defensive strike. Snakes can meter their venom and often do not expend much on defense. It has been estimated than anywhere from 10 to 73 percent of defensive snake bites against humans involve no clinically significant envenomation. The clinical symptoms of snakebite reflect the primary role of venom as a feeding adaptation. The enzyme-rich crotaline venoms often result in extensive tissue destruction, whereas elapid venoms paralyze respiratory muscles. The treatment of snakebite is a complex subject, and discussion and additional references can be found in Russell (1980) and Dart and Gomez (1996).

Venomous snakebite is a relatively trivial medical problem in the United States, with about 8,000 bites by venomous snakes and a dozen deaths. In tropical regions, where large rural populations lacking protective footwear coexist with a diverse reptilian fauna, venomous snakebite can be a significant public health concern. Estimates are difficult to come by, but as many as one million venomous bites and 30,000 to 40,000 deaths may occur worldwide each year.

Pit Organs

Pit vipers have a pair of sensory structures known as pit organs on the sides of the head, between the nostril and the eye, and analogous pit organs, forming a series of paired structures along the lips, are present in many boids and pythonids (Figure 9–41). Pit organs are exquisitely sensitive infrared receptors and convey spatial information about the thermal environment. The image is carried by the trigeminal nerves to the optic tectum of the brain, where it is superimposed on the visual image, presumably giving the snake a view of its world in a combination of light and heat (Hartline et al. 1978, Molenaar 1992).

Herbivory

Plant material has a lower energy content than animal tissue, plant cells are encased in a cell wall that vertebrates cannot digest, and many plants contain toxic compounds. Thus, it is no surprise that herbivory is a difficult feeding mode, and it has evolved in relatively few lineages of extant amphibians and reptiles. Some

Figure 9–41 **Infrared receptive pit organs.** (a) *Bothriechis schlegelii,* a crotaline viperid. All pitvipers have a single pair of facial pit organs located between the eye and nostril. (b) The green tree python, *Morelia viridis.* Many boas and pythons have series of pit organs on the labial scales. *(Photographs by (a) Michael & Patricia Fogden, (b) David Northcott/ DRK Photo.)*

(a)

(b)

turtles and lizards are omnivorous, consuming a substantial amount of vegetation in addition to animal prey. Examples include the North American box turtles (*Terrapene*), at least some of the large Australian skinks (*Tiliqua*), some xantusiid lizards (e.g., *Xantusia riversiana*), and the Angolan plated lizard (*Angolosaurus skoogi*), an opportunistic seed-eater from the Namib Desert of Africa. Still other groups, although restricted in their diet to plant matter, nevertheless show relatively little specialization of their feeding apparatus. Among those are certain species of turtles, such as *Pseudemys rubriventris* (Emydidae), the red-bellied turtle.

Fruit-eating animals have few problems processing food because fruits are characterized by large, energy-rich cells that are easily ruptured. The omnivorous *Varanus olivaceus* swallows large fruits whole. Leaf-eaters, in contrast, must release nutrients from within the cells of leaves. Many vertebrates have acquired symbiotic intestinal microorganisms that produce enzymes capable of breaking down cellulose into simpler compounds that can be absorbed by the host. Gut symbionts permit the green sea turtle (*Chelonia mydas*) to obtain 15 percent of its daily caloric intake from the fermentation of cellulose and hemicellulose. Each generation of animals must obtain those symbionts by ingesting feces of other individuals (coprophagy) or by eating soil contaminated by feces (see Chapter 13).

Among lizards, the most species-rich lineage of herbivores is the iguanid subfamily Iguaninae, a group of relatively large lizards distributed from the southwestern United States through the Neotropics, including the West Indies and the Galápagos Islands. A genus with just two species occurs on Fiji. Some iguanines are terrestrial, and others are arboreal leaf-eaters. One, the marine iguana (*Amblyrhynchus cristatus*) of the Galápagos Islands, forages underwater, scraping marine algae from submerged lava. Herbivorous lizards bite off pieces of plants but do not chew them. Iguanines have one to 11 valves at the anterior end of the greatly enlarged colon where fermentative cellulose digestion takes place (Iverson 1982). The valves may be circular or semilunar (Figure 9–42). The number of valves is species-specific and strongly correlated with body size. Although the exact function of the valves is uncertain, they clearly increase surface area in the hindgut and probably slow the passage of digesta giving more time for fermentation.

Although most species of herbivorous lizards are iguanines, fermentative herbivory has evolved independently in two other families of lizards. *Corucia zebrata*, the prehensile-tailed skink from the Solomon Islands near New Guinea, is an arboreal leaf-eater. Two genera of agamids, *Uromastyx* of northern Africa and southern Asia and *Hydrosaurus* of southeastern Asia, also are herbivorous. Like iguanines, all three are relatively large lizards and all have partitions in the colon. The association between herbivory and large body size in three families has prompted varied interpretations. Large lizards are not agile enough to make a living by eating small insects. Herbivory may have facilitated the evolution of large body size by iguanines and *Corucia*, and the low mass-specific metabolic rate that accompanies large body size makes a low-energy diet of plants feasible (Pough 1973, Van Devender 1982).

Summary

We find among amphibians and reptiles some of the most extraordinary feeding mechanisms known in vertebrates, from the projectile tongues of plethodontid salamanders and chameleons to the advanced venom delivery systems of elapid and viperid snakes. The diversity of feeding mechanisms of ectothermal tetrapods can be organized into categories based on where they feed and how they procure their food. Feeding mechanisms reflect not only a species' diet and habitat, however, but also its evolutionary history. Thus, for example, suction feeding by aquatic salamanders is very different from that seen in turtles. Salamanders often retain gill slits that permit a unidirectional flow of water through the pharynx, whereas turtles must create a bidirectional flow during the feeding cycle. Similarly, the mechanisms for tongue projection by salamanders, frogs, and chameleons are based on vastly different systems. Indeed, the differences between the salamander and anuran systems are greater than those between salamanders and chameleons.

Dietary specialization may involve any phase of the feeding sequence, from prey capture through oral transport and swallowing, to digestion and the processing of wastes. Sometimes features that seem unrelated to feeding prove to have an influence on jaw morphology. The evolution of a shell in turtles led to the retraction of the head, which in turn led to the evolution of different pul-

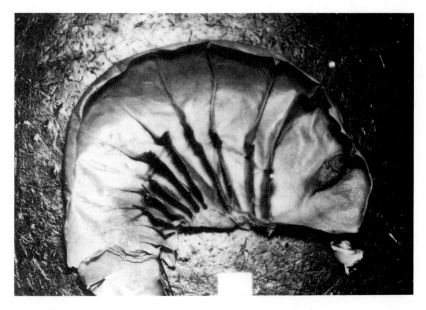

Figure 9–42 **The colon of a Caribbean land iguana, *Cyclura cornuta*.** This ventral view of a frontal section shows the partitions associated with a diet of leaves. *(Source: Iverson 1980; photograph courtesy of John Iverson.)*

ley systems for the jaw adductor musculature in the two lineages of extant turtles. Similarly, feeding specializations may have consequences for aspects of an animal's biology that are far removed functionally from their diets.

To be understood, feeding morphology must be studied in detail, with a variety of analytical tools brought to bear. Often the critical elements of a feeding mechanism reside in the details, and such studies remain to be done for most amphibians and reptiles. Once we have achieved an understanding of feeding mechanisms, we can better appreciate the evolutionary responses of diverse lineages to the challenge of getting enough to eat.

What Do They Do?

10

Movements and Orientation

When Alice fell down the rabbit hole in Lewis Carroll's famous story, she found herself in an unfamiliar world, shrunk to the size of mice and caterpillars. Flowers and mushrooms became the size of trees, while logs and fallen leaves seemed as large as buildings. This imaginary change in spatial scale, a popular theme in literature and movies from *Gulliver's Travels* to *Honey, I Shrunk the Kids,* is a useful exercise to help understand what it would be like to live as an amphibian or reptile. Most of these animals are small, less than 20 grams in body mass, and, being small, they don't move around very much. A red-backed salamander (*Plethodon cinereus*), for example, may spend its entire life in an area smaller than a tabletop in a college biology laboratory, and even a relatively active lizard or snake may range over an area no larger than a typical classroom or lecture hall. A 45-minute walk is sufficient for a person to traverse an average college campus, but even the most mobile amphibian or reptile may not cover that much ground in a period of weeks, months, or even years. The few exceptions, such as sea turtles that move over hundreds of kilometers of open ocean, are much larger than most amphibians and reptiles. This chapter describes patterns of movement in amphibians and reptiles, including the use of limited home ranges for normal daily activities, the defense of home ranges as territories, longer-distance movements such as seasonal migrations to breeding or hibernation sites, movements of juveniles from natal areas, homing behavior, and sensory mechanisms used by these animals for spatial orientation.

Ecological Consequences of Movement

Amphibians and reptiles move from one place to another for many reasons, but ultimately all movement is related to the acquisition of resources, including food and water, mates, basking or hibernation sites, nesting sites, shelter, and anything else required for survival and reproduction. Any movement is likely to entail some costs, including energy expenditure and exposure to unfavorable environmental conditions or predators. Most animals move only when absolutely necessary, but species differ dramatically in how often and how far they move. Patterns of movement affect, and are affected by, almost every other aspect of an animal's biology, including its water and temperature relations, foraging ecology and energetics, mating system, responses to predators, and interactions with other species.

The dynamics of populations are closely tied to patterns of movement. Mortality can be particularly high when individuals move from one habitat to another, as when tiny, newly metamorphosed toads move from ponds to terrestrial habitats or hatchling sea turtles make a mad dash for the ocean. Species that have eliminated this movement phase also may have lower juvenile mortality. This in turn can influence the evolution of life history traits, such as the number and size of eggs that are produced. Movement patterns affect the dynamics of metapopulations, in which some habitat patches are populated by individuals that immigrate from other habitat patches, not by recruitment of individuals that were born there

309

(Sjögren 1991, Sinsch and Seidel 1995). Movement also affects the genetic structure of populations. Strong natal philopatry (a tendency to return to a birthplace) in juveniles and adults can produce local genetic differentiation (Berven and Grudzien 1990, Smith and Scribner 1990). Recent work using DNA sequencing has shown that many local breeding populations of sea turtles are genetically distinct from one another, even though individuals from different nesting sites commonly share the same feeding grounds (Bowen and Karl 1997).

Patterns of movement in amphibian and reptile populations also have major conservation implications. Roads, housing developments, agricultural land, and other man-made barriers can limit movement between populations, increase the mortality of migrating animals (Fahrig et al. 1995), increase levels of inbreeding (Reh and Seitz 1990), and lead to the extinction of small, isolated populations (Laan and Verboom 1990). Consequently, maintaining dispersal corridors of suitable habitat and providing amphibians and reptiles with tunnels under roads and other means of traversing habitat barriers have become major issues in herpetological conservation (Langton 1989, Ruby et al. 1994).

Methods for Studying Movements

Two essentials for studying the movements of amphibians and reptiles are the ability to identify particular individuals and the ability to relocate them. This means that animals either must be marked or equipped with a device to track their movements. The method most commonly used to study movements of small amphibians and reptiles is to mark the animals, recapture them, and mark their positions on a map. A variety of methods have been used to mark amphibians, including a coded system of toe clips, colored or numbered waistbands and tags, tattooing, heat or freeze branding, marking with fluorescent powders, and recording of natural color patterns. Many of these techniques have been used with reptiles as well, especially lizards. Turtles usually are marked with small notches on the shell or with paint. Snakes are marked either by selectively removing scales or by branding. Unfortunately, marks sometimes disappear if toes regenerate, if the skin is shed, or if marks simply wear off. A more serious problem is the difficulty of relocating marked animals in hidden retreats. Animals usually are recaptured infrequently, and recapture data alone do not provide any information on the actual movements of the animals.

Several methods have been developed to follow the movements of animals more continuously. In the 1940s, Lucille Stickel (1950) began a classic study of box turtles (*Terrepene carolina*) in Maryland. She equipped her turtles with spools of thread that unwound as the turtles moved, revealing the precise path taken by each animal. This technique was later adapted by Dole (1965) to study movements of leopard frogs (*Rana pipiens*), and it has since been used for studies of other anurans and turtles. Some investigators have used tiny implanted passive integrated transponders (PITs) containing computer chips that produce unique identification numbers when activated with a special reading device. These devices allow an animal to be identified without handling, but animals cannot be located from very far away.

The best method for studying movements of large species is to equip animals with small radio transmitters, each of which broadcasts on a unique frequency. This allows frequent relocation of animals even in hidden retreats. Radio tracking has been used to study the movements of snakes, lizards, crocodilians, turtles, and large amphibians. Radio tracking usually provides a more complete picture of movement patterns than do mark–recapture studies (Weatherhead and Hoysak 1989). Sea turtles present a special problem because they move over very long distances. Their movements can be studied with transmitters that are monitored by satellites that transmit positional information to investigators on the ground.

Many amphibians and reptiles confine most of their activities to a limited home range. Home range size usually is estimated by marking all capture locations on a map and then drawing the smallest convex polygon that encompasses all of the capture points (Figure 10–1).

Variation in numbers of recaptures can have a major impact on calculations of home range sizes, with more accurate estimates usually being obtained with large numbers of recaptures. Consequently, some investigators have included a correction factor in their calculations of home range size that adjusts for numbers of recaptures. Other investigators have used more sophisticated computer programs to map the use of space by individuals. Unfortunately, there is no easy way to compare the results of studies that use different methods to estimate home range size.

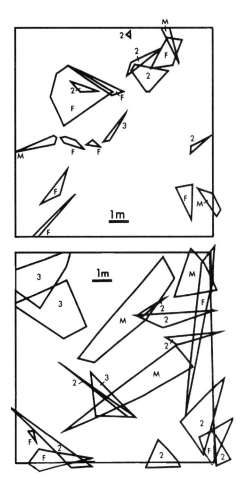

Figure 10–1 **Home ranges of salamanders (*Plethodon teyahalee*).** This map shows the minimum convex polygon method of home range estimation. M and F indicate home ranges of males and females. Numbers indicate ranges of 2- and 3-year-old juveniles. This type of map can be used to estimate both home range size and the amount of overlap between adjacent home ranges. *(Source: Nishikawa 1990.)*

Types of Movement

Terms such as home range, migration, and dispersal have been used differently for different kinds of animals. Dingle (1996) classified types of movement based on behavioral aspects of movement rather than its ecological consequences. Movements directly related to the acquisition of resources such as food, mates, basking sites, or retreat sites and that tend to keep an animal within a home range are termed station keeping. Foraging is one type of station keeping,

as is commuting, a pattern of movement back and forth from a fixed location to resource patches. Territorial defense of all or part of the home range also is considered station keeping. Movement outside of a home area for the purpose of exploring new habitats or resource patches is termed ranging behavior. This type of movement includes not only occasional forays outside of an established home range by adults (for example, when males are searching for mates) but also the movements of juveniles from the areas where they were born. The latter have traditionally been termed dispersal movements in the herpetological literature. Unfortunately, the term dispersal has been used to describe both the behavior of individual animals and population-level processes. For example, the departure of juvenile anurans from a breeding pond when they move into terrestrial habitats has been called juvenile dispersal, but many of these individuals eventually return to the same pond as adults to breed. Only those that move permanently to new ponds can be considered dispersers (Breden 1987). The rest range over an area, perhaps exploring it for new breeding sites, but do not permanently shift their locations. Dingle (1996) has argued that only those movements that tend to increase the space between individuals should be considered dispersal. Sometimes animals leave the area where they were born but do not increase their distance from other individuals. Juvenile lizards, for example, are sometimes attracted to areas where other lizards are already living, using their presence as an indicator of habitat quality (Stamps 1994).

The term dispersal is often used for any sort of one-way movement between habitat patches. Dispersal has been used this way to distinguish one-way movements from migration, which often is considered bidirectional movement. In contrast, Dingle (1996) defined migration as a specialized kind of movement that is not directly responsive to resources. Five key features of migration, which may not be evident in all migrating animals, are movement is persistent and of greater duration than during station keeping or ranging; the movement takes place along relatively straight-line paths; responses to resources that normally terminate movement, such as food, are temporarily suppressed; the animals engage in activity patterns related to departure and arrival; and animals often allocate energy reserves specifically to support migration. Many migratory animals exhibit bidirectional movement, but not all migration is bidirectional.

Local Movements and Home Range

Most amphibians and reptiles confine their routine activities to a limited home range, although the sizes of home ranges vary enormously (Table 10–1) and are affected by both body size and ecology. Some species exhibit remarkable fidelity to familiar areas, returning year after year to the same home ranges. For example, European fire salamanders (*Salamandra salamandra*) remained in the same home ranges for up to 7 years, and European common toads (*Bufo bufo*) were recaptured in the same summer home ranges for 9 years in a row (Heusser 1968, Joly 1968). Even more remarkable are the results of Stickel's study of home ranges in box turtles (*Terrepene carolina*), which she continued from 1944 through 1981 (Stickel 1989). Most of the turtles that were recaptured over periods of many years (up to 32 years, in one case) showed little or no change in either the location or size of their home ranges.

Not all animals confine their activities to stable home ranges. Some reptiles, especially snakes, shift from one center of activity to another and show little fidelity to one site. Other reptiles are nomadic, including some sea turtles and pelagic sea snakes that drift passively with winds and currents. How can we make sense of the variation in movement patterns and home range sizes of amphibians and reptiles? Why are some species closely tied to one area for most of their lives, while others move over large areas and seldom return to the same spot? Why do some species have large home ranges, while others have relatively small ones? What accounts for variation in home range size and stability among individuals of the same species? Some of these questions have been addressed for particular taxonomic groups, including lizards (Stamps 1977, 1994), snakes (Gregory et al. 1987), turtles (Gibbons et al. 1990, Schubauer et al. 1990), and amphibians (Mathis et al. 1995). What we need is a broader perspective that examines the use of space in all amphibians and reptiles in relation to the resources that they use.

Resource Dispersion and the Use of Space

Because all movement is ultimately related to the acquisition of resources, the way in which an animal moves through its habitat is closely related to the spatial and temporal distribution of resources, especially food. We can think of a habitat as being composed of a series of resource patches. As an animal moves

Table 10-1 **Approximate home range sizes of selected groups of reptiles and amphibians in square meters (10,000 square meters = 1 hectare). Most estimates were calculated by the minimum convex polygon method.**

Group	No. of Species Included	Range (m²)	Median (m²)	Source
Turtles (freshwater)	8	500–210,000	4,000	Schubauer et al. 1990
Snake (Colubridae)	17	9–210,000	6,500	Macartney et al. 1988
Snakes (Viperidae)	6	800–60,000	4,900	Macartney et al. 1988
Lizards* (males)	15	10–20,000	730	Turner et al. 1969
Lizards (females)	10	15–1,000	450	Turner et al. 1969
Anurans	21	1–1,900	40	K. Wells, unpublished
Caudates	13	0.1–87	4	K. Wells, unpublished

*Data for lizards include only small insectivorous species.

Behavior of Iguana During Rain Storms

Coleman J. Goin

While in Paramaribo, Surinam, during July 1966 I made some observations of the behavior of an iguana (*Iguana iguana*) that seem to be worth recording.

In the yard of our house was a pine tree (*Pinus carribea*), approximately eight inches in diameter at the base and thirty feet tall. On July 15, it began to rain and blow about 2:30 in the afternoon. As soon as the first drops started to fall, a large iguana came to the tree, climbed until it had reached one of the uppermost branches, crawled out to the end which was hardly three-fourths an inch thick, wrapped its fore limbs around the branch, braced its hind feet against tufts of needles and "held on for dear life" while the rain pelted and the tree swayed in the wind. After the storm abated about 5:00 o'clock PM the lizard settled down on the same limb and so far as we know did not move until it began to rouse a little after 10:00 the next morning. It finally made its descent shortly after noon.

After the above observations were made we started to keep more precise records and found that on afternoons in which we had even a light shower the iguana spent the night in the pine, climbing to its perch before sundown, but if there was no shower it spent the night elsewhere. For a period of about a week during which we had no rain the big iguana was not seen although a small one did feed in broad-leaved trees in the yard from time to time.

On July 27, the day before our departure, another heavy storm came up at 3:00 PM and once again the lizard came running across the lawn, climbed to one of the uppermost branches, and hung on while the rain pelted and the tree swayed as before. It stayed in the tree after the storm was over. Since we had to leave at 8:00 AM on the 28th we were not able to note when it finally descended.

These observations are not fully in agreement with those of Swanson (1950) in Panama, where he saw no iguanas in trees during the periods of cool, wet weather, although he added they seem to prefer more open trees when it is raining. It would be interesting to know whether the behavior noted above is typical or whether it simply represents the idiosyncratic behavior of this individual.

My work in Surinam was supported by a grant from the National Science Foundation GB-3644.

LITERATURE CITED

SWANSON, PAUL.   1950. The Iguana *Iguana iguana iguana* (L). Herpetologica, vol. 6, pp. 187-193.

Department of Zoology, University of Florida, Gainesville, Florida.

Quart. Jour. Florida Acad. Sci. 30(2) 1967 (1968)

through the habitat, it passes from one patch to another, making use of resources available in each patch. Resources can be divided into two categories, those that are depleted by the animal, and those that are not depleted. Food clearly is a depletable resource, whereas retreat sites such as burrows, rocks, or fallen logs are not. Once an animal has made use of a retreat site and then moves on, that resource is available for use by other individuals, or the same animal may return to the site again and again. In contrast, once an animal has eaten the food in a habitat patch, it cannot be used by that animal or another individual until those resources are replenished. The animal must make a decision about whether to remain in a given patch or move to a new patch based on the relative abundance of resources in the available patches, the distance between patches, the risks of moving from one patch to another, and a variety of other factors. Animals that live in restricted home ranges or territories revisit the same patches repeatedly, and the rate at which they do so depends in large part on how rapidly resources in those patches are replenished.

Animals are expected to occupy limited home ranges when resources are sufficiently abundant that they are depleted very slowly or when the rate at which resources are renewed is sufficiently high to allow the animal to return to the same patches again and again. To use a nonherpetological example, nectar-feeding animals such as hummingbirds and bumblebees frequently visit the same patches of flowers and even the same individual flowers every day for many days, because the flowers continually produce new nectar to replace that depleted by foragers. In contrast, if resources are renewed very slowly after being depleted, then animals will not be able to revisit patches frequently. For example, an animal that feeds on small fruiting bushes can deplete all of the fruit on a given bush in one visit. In that case, the animal must move to another bush to feed, and it may be a year or more before a bush produces fruit again. These types of resources often are patchily distributed in the habitat, and an animal must range over an area large enough to encompass all the patches needed to sustain it year-round. Hence, animals feeding on patchy resources with low renewal rates tend to have relatively large home ranges. Shifting from one activity center to another, or adopting a nomadic lifestyle, is characteristic of animals that feed on extremely patchy resources with very low renewal rates, such as swarming termites or local concentrations of pelagic

marine organisms. In short, the spatial strategies of animals form a continuum from strong site fidelity with very limited movement between resource patches to no site fidelity with extensive movement between patches (Waser and Wiley 1979).

Spatial Strategies of Amphibians and Reptiles

How well do these generalizations apply to amphibians and reptiles? In general, most of the spacing systems and patterns of movement that have been described for these animals can be predicted from the abundance, patchiness, and renewal rate of their food resources (Table 10–2). Unfortunately, quantitative data on the spatial and temporal distribution of resources used by amphibians and reptiles are scarce, so the examples given in Table 10–2 are based on a qualitative assessment of resource distribution patterns.

Animals that feed mostly on leaves (folivores) usually have abundant food readily at hand, although the quality of their food may vary seasonally. These animals are expected to occupy small, stable home ranges and exhibit relatively little movement. Reptiles in this category include the green iguana (*Iguana iguana*), marine iguana (*Amblyrhynchus cristatus*), and green turtle (*Chelonia mydas*). Outside of the breeding season, green iguanas live in small home ranges, show little aggression toward one another, and spend most of their time resting in one place (Dugan 1982). They make occasional forays out of their normal home ranges to feed on local concentrations of fruit, with many individuals sometimes feeding peacefully in the same tree. Marine iguanas exhibit many behavioral and ecological similarities to green iguanas but feed on algae in the intertidal zone or underwater. They move from terrestrial basking sites to foraging sites, and often return repeatedly to the same foraging area. Movements at foraging sites are minimal, and relatively little time is devoted to foraging each day (Wikelski and Trillmich 1994). Up to 90 percent of the diet of green turtles is composed of sea grasses, an abundant and uniformly distributed resource in shallow lagoons. These turtles have small activity centers and low rates of movement while foraging (Mendonça 1983).

Terrestrial herbivorous and omnivorous reptiles resemble folivores in having abundant food resources, but they often feed on a wider range of food types.

Table 10-2 **Feeding niches, patterns of resource distribution, and predicted spacing systems and movement patterns for selected amphibians and reptiles. When more than one spacing system is listed, the most common is given first.**

| Feeding Niche | Resources | | | Spacing | Movement | | Examples |
	Abundance	Patchiness	Renewal Rate		Within Patch	Between Patches	
Arboreal folivore	High	Low	Moderate	Home range	Low	Low	Green iguana
Marine folivore	High	Low	Moderate	Home range	Low	Low	Marine iguana, green turtle
Herbivore	High	Moderate	Moderate	Home range, temporary aggregations	Low	Moderate	Iguanid lizards, tortoises
Frugivore	High	Moderate	Moderate	Home range, temporary aggregations	Low	Low	Gray's monitor lizard
Terrestrial omnivore	High	Moderate	Moderate	Home range	Low	Moderate	Box and wood turtles, some skinks
Aquatic omnivore (rich habitats)	High	Low	High	Home range	Low	Low	Turtles in bogs & marshes
Aquatic omnivore (poor habitats)	Moderate	Moderate	Moderate	Home range, shifting activity center	Moderate	Variable	Turtles in ponds, lakes, and rivers
Sponge predator	Variable	Moderate	Low	Shifting activity center	Moderate	High	Hawksbill turtle
Jellyfish predator	Variable	High	Low	Nomadism	Low	High	Leatherback turtle
Active predator of vertebrates	Variable	Moderate	Low	Home range, shifting activity center	Moderate	Variable	Many snakes
Ambush predator of vertebrates	Variable	Variable	Low	Shifting activity center, home range	Low	Variable	Viperid snakes
Predator of large vertebrates	Variable	High	Moderate	Home range	Moderate	High	Komodo dragon (large)
Scavenger	Low	High	Low	Shifting activity center, temporary aggregations	Low	High	Komodo dragon
Sit-and-wait insectivore	Moderate	Moderate	High	Home range, territoriality	Moderate	Low	Iguanian lizards, some amphibians
Widely foraging insectivore	Moderate	High	High	Home range	Moderate	High	Scleroglossan insectivores
Earthworm specialist	Variable	Moderate	High	Home range, shifting activity center	Moderate	Moderate	Worm and garter snakes
Ant and termite specialist	Variable	High	Variable	Shifting activity center	Low	Variable	Horned lizards

True herbivores include the large ground iguanas of the Caribbean (*Cyclura*) and the Galápagos (*Conolophus*), a variety of smaller iguanids, such as *Dipsosaurus* and *Sauromalus*, several agamids (*Uromastyx* and *Hydrosaurus*), at least one skink (*Corucia zebrata*), and most tortoises. Terrestrial omnivores include box turtles (*Terrapene*), wood turtles (*Clemmys insculpta*), Australian sleepy lizards (*Tiliqua rugosa*), and the largely frugivorous Gray's monitor lizard, or butaan, of the Philippines (*Varanus olivaceus*). The spatial strategies of these reptiles are similar to those of folivores, with individuals occupying small home ranges, sometimes for many years, exhibiting low rates of movement between food patches, and generally remaining inactive for long periods of time (Satrawaha and Bull 1981, Auffenberg 1982, 1988, Werner 1982, Stickel 1989, Diemer 1992, Kaufmann 1995). Sometimes these species form temporary aggregations at bonanza resources, such as fallen fruit around trees, with little aggression among individuals. Home range size tends to increase as the patchiness of food increases. For example, tortoises that live in deserts have considerably larger home ranges than do those living in more productive habitats.

Aquatic turtles also are largely omnivorous, and their food usually is relatively abundant. Suitable habitats, however, often are aggregated into large patches within bodies of water, because turtles tend to center their activities in the shallow, vegetated zones of ponds, lakes, and rivers, especially in areas near logs and other basking sites. The predicted spatial system is one of overlapping home ranges with sharing of clumped resources such as basking logs. Turtles that live in highly productive environments, such as bogs, marshes, and shallow ponds, typically have home ranges of less than 2 hectares. Examples include bog, spotted, and Pacific pond turtles (genus *Clemmys*), Blanding's turtles (*Emydoidea blandingi*), and several kinosternid turtles (Ross and Anderson 1990, Schubauer et al. 1990, Rowe and Moll 1991, Lovich et al. 1992). Larger turtles that live in less productive ponds, lakes, and rivers usually have home ranges exceeding 2 hectares, and sometimes exceeding 10 hectares, as predicted from their larger body size and more patchily distributed resources. Examples include snapping turtles (*Chelydra, Macroclemys*), map turtles (*Graptemys*), sliders (*Trachemys*), and cooters and red-bellied turtles (*Pseudemys*) (Schubauer et al. 1990).

At the opposite end of the spectrum are species that use less abundant, patchily distributed resources

with low renewal rates. An extreme example is the marine leatherback turtle (*Dermochelys coriacea*), which feeds mostly on jellyfish and other pelagic invertebrates associated with drift lines and fronts in the ocean. Because prey patches are created by movements of currents, they are not likely to be rapidly renewed once depleted. Leatherbacks are nomadic (Figure 10–2), moving over thousands of kilometers of open ocean in a pattern reminiscent of some whales that feed on similar prey. Similar nomadism has been reported in other sea turtles feeding in the open ocean or on benthic invertebrates (Beavers and Cassano 1996).

Vertebrate prey animals tend to have a patchy distribution and low to moderate renewal rates. Hence the predators of vertebrates are expected to occupy relatively large home ranges, or to use a series of

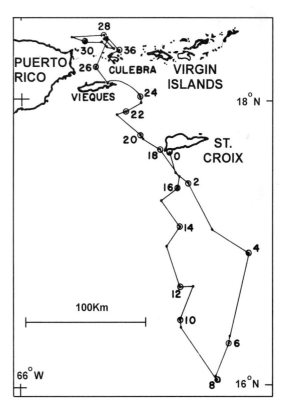

Figure 10–2 **Movements of a female leatherback turtle (*Dermochelys coriacea*).** Numbers indicate successive locations of the turtle in the Caribbean as determined from satellite tracking over an 18-day period. The turtle went ashore three times to nest, on St. Croix, Vieques Island, and Culebra Island. The total distance it moved was at least 512 kilometers. (*Source: Keinath and Musick 1993.*)

shifting activity centers with only weak attachment to a fixed home range. Many snakes that actively search for vertebrate prey have relatively large home ranges, including Australian blacksnakes (*Pseudechis*), rat snakes (*Elaphe*), and coachwhip snakes (*Masticophis*) (Shine 1987, Weatherhead and Hoysak 1989, Secor 1995). Black racers (*Coluber constrictor*) in South Carolina prey heavily on vertebrates and have larger home ranges than those in Kansas or Utah, where the snakes are more insectivorous (Plummer and Congdon 1994). Some widely foraging snakes, such as natricine water snakes, tend to use a limited foraging area for some period of time before shifting to a new area (Figure 10–3; Tiebout and Cary 1987).

Viperid snakes also prey on small vertebrates, but they are predominantly ambush predators that remain immobile for long periods of time while waiting to capture animals that pass by. Movements within habitat patches are infrequent, and individuals of some species remain in the same place for weeks or months at a time. Most of the movements these snakes make are between habitat patches as they shift from one activity center to another in search of good ambush sites (King and Duvall 1990, Secor 1995). Presumably they use chemical cues to monitor prey activity and to choose suitable ambush sites. Duvall et al. (1990) found that prairie rattlesnakes (*Crotalus viridis*) in Wyoming stopped searching and remained near stations that were provided with either caged rodent prey or prey odors.

Komodo dragons (*Varanus komodoensis*) are mainly predators of vertebrates. Individuals vary greatly in size and display a variety of foraging tactics and movement strategies. Only the largest lizards establish residence in permanent home ranges centered on burrows and basking sites. Radiating out from this core area is a large foraging range that overlaps those of other individuals. The lizards patrol familiar trails and lie in wait for prey along heavily used paths (Figure 10–4). Carrion is obtained from an even larger area, and lizards often travel long distances to carcasses, resulting in the formation of temporary aggregations. Smaller Komodo dragons are mostly transient, moving from place to place and feeding opportunistically on a variety of prey (Auffenberg 1981).

Reptile and bird eggs, nestling birds, and baby mammals are particularly patchy types of prey with low renewal rates because they tend to be aggregated at discrete nest sites. Once a nest site is depleted, a predator may have to move some distance to find a new one. Both the Mexican beaded lizard (*Heloderma horridum*) and the Gila monster (*H. suspectum*) specialize on this type of diet. These lizards have very large home ranges, averaging about 22 hectares for *H. horridum* (Beck and Lowe 1991) and up to 66 hectares for *H. suspectum* (Beck 1990). The latter is comparable to the home range size of a widely foraging snake, such as *Masticophis* (Secor 1995), despite the low metabolic rate and rather sluggish demeanor of these lizards. The difference is that Gila monsters and

Figure 10–3 **Movements of grass snakes (*Natrix natrix*).** Radio-tagged male (open circles) and female (solid circles) grass snakes were followed over a 4 month period on a plot in southern Sweden. Wavy lines indicate ponds. Sizes of circles indicate the number of days a snake stayed in each location. Note the tendency for the snakes to move between several major centers of activity. H, hibernation site. A, oviposition site used by the female. *(Source: Madsen 1984.)*

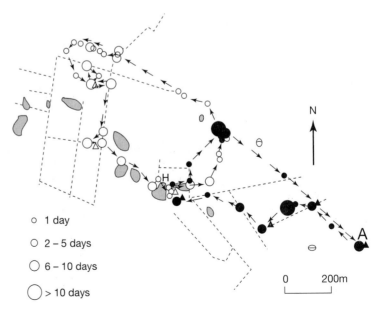

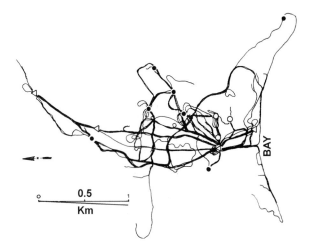

Figure 10–4 **Patterns of movement of a resident adult Komodo dragon (*Varanus komodoensis*).** Lines show paths taken by the animal over a 2-year period. Solid circles are shelters, open circles are water pools, triangles are carcasses of large animals placed at baiting stations. The thickness of the lines shows the frequency of use of each trail. Note the repeated use of familiar trails by the animal. *(Source: Auffenberg 1981.)*

beaded lizards feed on large food packages and remain relatively inactive between meals, moving on less than a third of all days during the activity season. In contrast, coachwhip snakes feed on smaller packages of food and move on more than two thirds of all days. Consequently, they make more intensive use of their large home ranges than do Gila monsters.

Most small species of amphibians and reptiles are insectivorous or specialize on other invertebrate prey such as earthworms. Such resources tend to be more abundant, less patchily distributed, and more rapidly renewed than vertebrate prey. Hence, these species are expected to occupy relatively small home ranges, a pattern seen in salamanders, frogs, lizards, and insectivorous and worm-eating snakes (Macartney et al. 1988, Mathis et al. 1995). Among insectivorous lizards, those that adopt a sit-and-wait foraging strategy and feed on moving prey tend to have smaller home ranges than widely foraging lizards that feed on hidden prey (See Chapter 13). Sit-and-wait foragers sometime chase down prey within habitat patches but seldom move between patches, whereas widely foraging lizards move frequently between patches, but spend little time in each one (Anderson 1993).

Most iguanian lizards fall into the sit-and-wait category, whereas many scleroglossans, especially tei-

ids, lacertids, scincids, and varanids, are widely foraging species. Nevertheless, even within these clades, there s some variation in foraging mode that is correlated with patterns of movement. For example, some lacertids have evolved a sit-and-wait foraging mode and tend to move less and have smaller home ranges than widely foraging relatives. Similarly, some cordylid lizards, a scleroglossan clade, are sit-and-wait predators that resemble some species of *Sceloporus* or *Agama* in their ecology in that they have relatively small home ranges (Burrage 1974). Some horned lizards (*Phrynosoma*) and Australian thorny devils (*Moloch horridus*) specialize on ants, which tend to be patchily distributed. They exhibit elements of both foraging strategies, moving periodically between patches and showing only loose attachment to a home range, but moving very little within patches and sometimes remaining in one area for long periods of time (Munger 1984, Withers and Dickman 1995).

Territoriality

Some amphibians and reptiles aggressively defend all or part of their home ranges as territories. There is a large literature on this subject, with good published reviews for lizards (Stamps 1977, 1983, 1994) and amphibians (Mathis et al. 1995). Patterns of home range defense can be predicted from the abundance, patchiness, and renewal rate of resources. Territoriality is part of a spectrum of spacing strategies from exclusive home range defense to complete home range overlap (Waser and Wiley 1979). The principal cost of sharing a home range with other individuals is that resources are depleted more rapidly, limiting the ability of a resident to return repeatedly to the same area. Defending a home range is worthwhile only if the increased availability of resources compensates for the costs of defense. Territoriality is favored when resources are moderately abundant, have an even or moderately patchy distribution, and have a high renewal rate. Very high or very low resource abundance, high patchiness, and low resource renewal rates tend to select against territoriality.

At very high levels of resource abundance, the slight increase in resources available to the defending animal probably will not be significant enough to compensate for the costs of defense. This probably accounts for the lack of home range defense in herbivorous and omnivorous reptiles such as iguanas and turtles, even in species that have relatively small home ranges. When resources are concentrated in large patches, such as fallen fruit around a tree, the costs of defense are likely to be too high to make territoriality economically feasible, because so many animals are attracted to the resource patch. Animals in such aggregations are likely to feed peacefully with little interaction, as do some iguanas, or fight over individual food items, as has been observed for Komodo dragons feeding on carrion (Auffenberg 1981).

When resources are patchily distributed or have low renewal rates, home ranges tend to be large and home range overlap extensive, with little evidence of home range defense. One reason for this pattern is that large home ranges are harder to monitor and defend than small home ranges. Visibility also may be limited in microhabitat patches where food is located (Stamps 1977). Perhaps more important is the fact that a patchy distribution of resources itself favors home range overlap rather than home range defense. This is because the home range must be large enough to meet the needs of the resident throughout the year, but for much of the year, only a small portion of the home range can provide sufficient resources for one animal. Hence, the cost of sharing the home range is minimal compared to the cost of defending a large area (MacDonald and Carr 1989). Consequently, species that feed on patchily distributed resources, such as small vertebrates or hidden insect prey, usually do not defend home ranges as territories. For example, home range defense is rare among widely foraging lizards and is completely unknown for snakes, which are derived from widely foraging lizards.

Indeed, home range defense is common in only one group of reptiles, insectivorous lizards that employ a sit-and-wait foraging mode. These include not only most iguanian lizards, but also some lacertids, teiids, scincids, and cordylids, which probably evolved a sit-and-wait foraging mode secondarily from widely foraging ancestors (Stamps 1977). The cordylids, which are restricted to sub-Saharan Africa, are particularly interesting because the most derived species have evolved a sit-and-wait foraging mode, aggressive home range defense, and pronounced sexual dimorphism in body size and coloration similar to that of ecologically convergent phrynosomatids and agamids. Home range defense also is common in geckos, especially diurnal species, but these lizards apparently retain the ancestral form of territorial behavior for lizards in general (Figure 10–5).

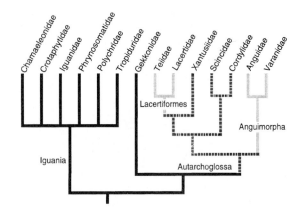

Figure 10–5 **Phylogeny of 14 families of lizards based on morphological characters, with spacing patterns mapped on the cladogram.** Black shows families that exhibit home range defense, the presumed ancestral condition. Dashed lines indicate site defense. Dark gray indicates lack of home range or site defense. More recent research has shown that some cordylids exhibit home range defense, and some varanids show limited site defense. *(Source: Martins 1994.)*

Several lines of evidence suggest that mates rather than food constitute the principal limited resource for most territorial lizards (see Chapter 12). In many species, males are territorial, but females are not. Even when both sexes are territorial, males often have larger territories that overlap those of several females (Stamps 1983). In some species, the territory sizes of males remain relatively constant throughout the year but the intensity of territory defense diminishes outside of the breeding season. Furthermore, experimental studies in which food was added to male territories often have failed to induce males to reduce the size of their territories (Stamps 1994). Nevertheless, the abundance, relatively even distribution, and high renewal rate of insect prey probably make home range defense possible for these lizards, because this pattern of resource distribution results in females having relatively small home ranges that can be encompassed within the territories of males.

Territorial lizards vary in the degree to which home ranges are defended as exclusive domains. In addition to having territories that overlap the home ranges of females, many male lizards tolerate subordinate males or juveniles on their territories, even during the breeding season (Stamps 1977). Presumably these invaders are tolerated because they cannot compete effectively with resident males for mates, and their impact on food resources in the territory is minimal. In some habitats, however, suitable territory sites can be severely limited, and intense competition over territories is expected. For example, Judy Stamps and her colleagues have made detailed studies of territorial behavior of juvenile anoles (*Anolis aeneus*) (Stamps 1994). Juveniles leave the woodlands where they are born and move into clearings not inhabited by adults. In years when many juveniles are produced, some cannot find suitable territories and become floaters. Floaters are at a disadvantage because they occupy home ranges that overlap those of other lizards. As a result, floaters have lower growth rates and lower survival than do residents. Floaters quickly move into vacated territories if residents are removed. When many juveniles establish territories in an area, the territories often overlap extensively, reducing food availability and growth rates and decreasing the survival of residents compared to lizards occupying small, exclusive territories (Figure 10–6). For many species of lizards, an uneven distribution of suitable habitat patches leads to variation in the quality of territories inhabited by adults, which in turn can affect their growth, survival, and reproduction.

Some amphibians and reptiles do not defend the entire home range as a territory but do defend specific sites within home ranges. Normally these are nondepletable resources such as basking perches, burrows, or retreat sites. Site defense is common in some groups of widely foraging lizards that do not defend home ranges (Martins 1994; see Figure 10–5). It also is common among plethodontid salamanders, which sometimes defend retreat sites under rocks or logs (Mathis et al. 1995). These sites provide refuges during dry weather and access to foraging areas nearby. *Plethodon cinereus* has been studied for many years by Robert Jaeger and his students. These salamanders prefer large logs, which are relatively scarce. Ownership of territories is advertised with pheromones deposited in fecal pellets, and territories are aggressively defended (see Chapters 11 and 12). The presence of a resident inhibits other salamanders from occupying a site, but if a resident is removed, new individuals rapidly move in. Some terrestrial salamanders that are ecologically similar to *P. cinereus*, such as *P. vehiculum*, show few signs of territorial behavior, perhaps because they live in wetter habitats where moist retreat sites are not limited. Some aquatic salamanders defend retreat sites that are used to hide from predators and to ambush prey. Site defense is less common among frogs than salamanders, although some species defend rock crevices, burrows, or other retreats in dry weather, while others defend suitable feeding perches. Territoriality among frogs is frequently related to defense of resources used for reproduction (see Chapter 12).

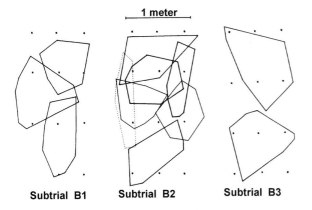

1 meter

Subtrial B1   **Subtrial B2**   **Subtrial B3**

Figure 10–6 **Territories established by juvenile *Anolis aeneus* after being introduced into a habitat patch.** In subtrials B1 and B3, 6 juveniles were added to the patch, whereas in B2, 12 were added. Note the greater overlap of territories in B2 with the larger number of competitors. Not all of the introduced animals established territories. The dotted line shows an area inhabited by a nonterritorial floater. *(Source: Stamps 1990.)*

Migration

Amphibians and reptiles sometimes leave their normal home ranges to move to different habitats some distance away. Often these migrations are made on an annual basis to and from breeding areas, nesting sites, or hibernation dens, and these are discussed in more detail below. Some migrations, however, are responses to changes in habitat caused by seasonal droughts, floods, or other weather events. For example, some turtles can move several kilometers from drying ponds to more favorable habitats, and aquatic snakes do so as well. Other aquatic reptiles, including crocodiles (*Crocodylus johnstoni*), caimans (*Caiman crocodilus*), file snakes (*Acrochordus arafurae*), and Australian long-necked turtles (*Chelodina longicollis*) make regular migrations between seasonally flooded swamps and more permanent bodies of water (Shine and Lambeck 1985, Ouboter and Nanhoe 1988, Graham et al. 1996). Madsen and Shine (1996) observed regular seasonal migrations of Australian water pythons (*Liasis fuscus*) in response to habitat shifts by their rodent prey. During the dry season rats were common in soil crevices in the drying floor of a large floodplain, and the snakes moved onto the floodplain to hunt. In the wet season the area was flooded, forcing the rats out and causing the snakes to follow them to higher ground. Some snakes moved as much as 12 kilometers between wet and dry season foraging areas.

Breeding Migrations of Amphibians

Sychronized migrations of amphibians to aquatic breeding sites usually are triggered by rainfall, with temperature also being important for temperate-zone species that migrate in early spring. Breeding migrations usually occur at night, but once a migration has started, movement often continues during the day as well. Most amphibians do not move great distances to breeding sites. The longest movement reported for any amphibian is for a mixed population of European water frogs, *Rana lessonae* and *R. esculenta* (a hybridogenetic species derived from hybridization between *R. lessonae* and *R. ridibunda*), breeding in a lake along the border of Hungary and Austria. Some frogs moved as far as 15 kilometers from hibernation sites in peat fens to breeding sites along the lakeshore, but most individuals moved only a few hundred meters (Tunner 1992). Migration distances for other anurans seldom exceed 1,500 meters, and most species of salamanders move less than 500 meters (Sinsch 1990).

Amphibians typically exhibit strong fidelity to breeding sites as adults, returning year after year to the same pond or stream segment. A study of newts (*Taricha rivularis*) in California conducted for many years by Victor Twitty and his colleagues revealed that individuals continued to return to their home stream segments up to 11 years after being marked, and they were never recaptured in adjacent streams (Twitty 1966). Berven and Grudzien (1990) marked thousands of wood frogs (*Rana sylvatica*) in ponds in the mountains of Virginia over a period of 7 years. Adults always were recaptured in their original breeding ponds, even though other suitable breeding ponds were available 250 to 1,000 meters away. Some populations of European common toads (*Bufo bufo*) exhibit such strong site fidelity that individuals continued to migrate to breeding areas several years after the ponds

had been drained or turned into parking lots . On the other hand, some species of toads that use more emphemeral ponds for breeding sometimes change breeding pools. Males in a population of natterjack toads (*Bufo calamita*) breeding in a network of temporary pools used as many as four different pools in a single breeding season, and some females used sites up to 3 kilometers apart in different years (Sinsch and Seidel 1995).

Nesting Migrations of Reptiles

Males and females of most species of reptiles do not migrate to common areas to mate, as do many amphibians (see Chapter 12), but there are a few exceptions. Male vipers (*Vipera berus*) in Finland moved rapidly from basking sites to a common mating area several hundred meters away. Male sea turtles mate in shallow waters near nesting beaches, having made the same migrations from distant feeding grounds as did the females (Miller 1997). Most reproductive migrations, however, are performed by females moving to specialized nesting sites. All aquatic reptiles that use terrestrial oviposition sites undergo migrations regularly, as do some terrestrial turtles, lizards, and snakes.

By far the most spectacular migrations are those of sea turtles (Figure 10–7), which often travel several thousand kilometers from feeding grounds to nesting beaches (Meylan 1982). Early work by Archie Carr starting in the mid-1950s demonstrated that sea turtles can move over enormous expanses of open ocean and yet return with great precision to the same nesting beaches year after year. Carr's work inspired a generation of later sea turtle biologists, and the general patterns of migration that he described for turtles in the Caribbean and southern Atlantic have been largely confirmed. One of the most famous colonies of sea turtles in the world is the population of green turtles (*Chelonia mydas*) nesting at Tortuguero, Costa Rica, where as many as 50,000 green turtles once were observed coming ashore in a single year. Tens of thousands of turtles have been tagged at this site since the 1950s by Carr and his successors, and individuals have been recovered on feeding grounds as far away as Panama, Colombia, Venezuela, and the Yucatan Peninsula in Mexico (Carr et al. 1978).

Another well-studied green turtle colony nests on Ascension Island, part of the Mid-Atlantic Ridge more than 2,000 kilometers off the coast of Brazil.

Figure 10–7 **Female olive ridley turtles, *Lepidochelys olivacea*, laying eggs on Nancite Beach, Costa Rica.** *(Photograph by David Hughes/Bruce Coleman Limited.)*

Turtles tagged on this island migrate to feeding grounds along the Brazilian coast (Mortimer and Carr 1987). Here they mix with turtles that nest along the coast of Surinam and French Guiana. These breeding populations are maintained as genetically distinct because of a strong tendency for females to return to their natal beaches to nest (Figure 10–8). Biologists have long puzzled over the origins of the isolated nesting colony on Ascension Island. Carr and Coleman (1974) proposed a radical hypothesis to explain the evolution of such long-distance migration. They suggested that seafloor spreading over the past 40 to 70 million years gradually separated Brazil from the Mid-Atlantic Ridge, requiring an increasingly long migration by the turtles. If this were true, then one would expect the colony on Ascension Island to exhibit considerable genetic divergence from other Atlantic green turtle colonies. Recent studies using mitochnodrial DNA have shown that the amount of genetic divergence is far too small to support the hypothesis of an ancient origin of the Ascension Island colony. Instead, it appears that this colony originated much more recently, on the order of a few tens of thousands of years ago (Bowen et al. 1989). How the turtles first happened to locate the tiny island and begin using it for nesting remains a mystery.

Green turtles in other parts of the world, and other species of sea turtles, also move long distances between feeding and nesting grounds (Meylan 1982). For example, green turtles from various colonies in the Pacific and Indian oceans regularly migrate more than 1,000 kilometers, while olive ridley turtles (*Lepidochelys olivacea*) tagged along the coast of Surinam have been found as far as 1,900 kilometers away. Loggerhead turtles (*Caretta caretta*) tagged in South Africa have turned up as far away as Zanzibar, nearly 3,000 kilometers away, and one female loggerhead tagged in Queensland, Australia, was recaptured nearly 1,800 kilometers away, in New Guinea. The most impressive travels are those of the highly nomadic leatherback turtle (*Dermochelys coriacea*). Individuals tagged in Surinam and French Guiana were recaptured more than 5,000 kilometers away in Mexico, Texas, South Carolina, and New Jersey. One individual swam all the way across the Atlantic to the west coast of Africa. Migrating sea turtles can cover these distances surprisingly quickly. Minimum travel speeds based on the time between sightings for several species range from about 20 to 80 kilometers per day. This value agrees well with estimates for radio-tracked green turtles of 2 to 3 kilometers per hour, or about 48 to 72 kilometers per day (Luschi et al. 1996).

The nesting migrations of other reptiles are less impressive than those of sea turtles. Freshwater turtles typically move to nesting sites over distances ranging from about 50 meters for small species, such as mud turtles (*Kinosternon*), to more than 5 kilometers for snapping turtles (*Chelydra serpentina*) (Gibbons et al. 1990). Females of some species exhibit strong fidelity to general nesting areas and also tend

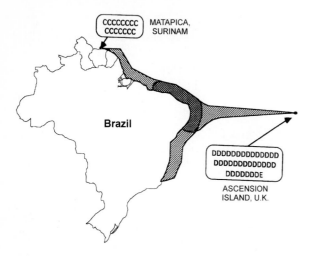

Figure 10–8   **Feeding ranges of green turtles (*Chelonia mydas*) nesting on Ascension Island and the coast of Surinam.** Cross-hatched area shows region of overlap between the two populations. Letters (C, D, E) indicate the mitochondrial DNA haplotypes of turtles found on the two nesting beaches. Females in the two populations do not share any haplotypes, indicating a strong tendency for females to return to their natal beaches to breed. (*Source: Bowen and Karl 1997.*)

to return to the same aquatic habitat each year after nesting. Migration distances for terrestrial turtles have not been well documented, but it seems likely that most species find suitable nesting sites within or near their normal home ranges. Female alligators and caimans sometimes increase the size of their home ranges during the breeding season, but nesting sites usually are located within the home range or only a few hundred meters away. Nile crocodiles (*Crocodylus niloticus*) nesting on islands in Lake Rudolf, on the other hand, often swim considerable distances to nesting beaches (Chelazzi 1992).

Most squamates probably do not move long distances to deposit eggs but find suitable sites within their normal home ranges. A few species of snakes have been reported to make directional movements of up to a few hundred meters from home ranges to specialized oviposition sites, such as manure piles and hollow trees. Nesting migrations also have been reported for several large iguanid lizards, especially arboreal species or those that live in habitats where nesting sites are scarce. Green iguanas (*Iguana iguana*) living in the forests of Barro Colorado Island in Panama nest on a small sandy island just offshore, but

most move less than a kilometer from their home ranges (Bock and Rand 1985). Ground iguanas (*Cyclura stejnegeri*) on Mona Island, Puerto Rico, live in a habitat dominated by limestone, with little soil suitable for nests. Females often migrate up to 6.5 kilometers to nesting areas (Wiewandt 1982). Galápagos land iguanas (*Conolophus subcristatus*) inhabiting the rocky terrain of Fernandina Island have a similar problem and sometimes move more than 15 kilometers to find nesting sites. Many females climbed to the rim of a 1,400-meter volcanic crater and then descended another 900 meters to the crater floor to nest. Werner (1983) estimated that the energetic cost of migration by these lizards equaled nearly half of a female's total reproductive effort. In all of these species, the shortage of nesting sites results in aggressive competition among females and territorial defense of nests.

Migrations to Overwintering Sites

Many amphibians and reptiles leave their home ranges in the fall and move to overwintering sites. For some anurans, these migrations are part of an extended round-trip between breeding ponds and summer home ranges. For example, European common toads (*Bufo bufo*) usually occupy summer home ranges that are 500 to 1,500 meters from breeding ponds. Some individuals initiate return movements toward breeding ponds in the fall and spend the winter in sites closer to the ponds (Sinsch 1988). This means that the toads have shorter distances to travel when breeding starts in the spring. Other amphibians make regular annual trips to specific overwintering areas, often turning up in the same sites year after year. Manitoba toads (*Bufo hemiophrys*) were found hibernating in the same prairie mounds for up to 6 years (Kelleher and Tester 1969), and individually marked fire salamanders (*Salamandra salamandra*) have turned up at hibernation sites in caves for up to 20 years. The most dramatic migrations to overwintering sites are those of snakes that use communal hibernation dens. These snakes often make relatively rapid, directed movements from foraging areas to hibernation dens and exhibit strong fidelity to specific den sites. Although they do not move in groups, snakes sometimes have been observed using similar paths to communal dens. Juveniles generally are unfamiliar with the locations of den sites but can follow the pheromone trails of adults to find them (Ford and Burghardt 1993).

Movements of Juveniles

Juvenile amphibians and reptiles often leave the areas in which they were born and move into new habitats. Such movements are necessary for many amphibians that lay eggs in water, because aquatic breeding sites do not provide suitable habitats for metamorphosed juveniles or adults. The same is true for aquatic and marine turtles emerging from terrestrial nests, but movement occurs in the opposite direction, from land to water. In both cases, some juveniles may later return to natal ponds or beaches to breed, but others move to new breeding areas. For amphibians and reptiles that are permanently terrestrial, movement of juveniles out of natal areas may enable them to find patches of habitat that are not already occupied by adults, or to avoid cannibalism by adults. Movement of juveniles from natal areas has important consequences for the genetics of populations. Limited movement can result in close inbreeding and reduced heterozygosity in the population, whereas movement from natal areas promotes outbreeding and increases genetic variation in populations.

Movements of Juvenile Amphibians

Almost nothing is known about the movements of juveniles of amphibians that breed on land, such as plethodontid salamanders, so we will focus on species that breed in water and live much of their lives on land. Many researchers have assumed that amphibians tend to return to their natal ponds to breed, but several studies have shown that a small but significant proportion of a population sometimes moves to new areas. Berven and Grudzien (1990) marked over 5,000 juvenile wood frogs (*Rana sylvatica*) and recaptured 18 percent of them in ponds other than their natal ponds as breeding adults. The dispersal rate of juvenile Fowler's toads (*Bufo fowleri*) was even higher (27 percent), but in that case, the breeding ponds were much closer together (Breden 1987). Occasionally, marked amphibians have been recaptured 2 to 5 kilometers from their natal areas, suggesting that long-distance dispersal sometimes occurs, but at frequencies that are too low to be detected in most mark–recapture studies.

Colonization of new habitats sometimes occurs very quickly. Investigators have placed shallow pans of water on the floor of tropical forests and found them rapidly occupied by frogs such as *Physalaemus pustulo-*sus and even the largely aquatic species *Pipa arrabali*. Other investigators have constructed larger artificial ponds in suitable habitats, and these were soon colonized by a variety of amphibians (Laan and Verboom 1990). Some amphibians specialize in invading newly opened habitats. The natterjack toad of Europe (*Bufo calamita*) is a pioneering species with tadpoles that do better in newly formed, nutrient-poor ponds than in established ponds with abundant predators and competitors (Banks and Beebee 1987). The Neotropical toad *Bufo marinus* also is a pioneering species, a trait that allowed it to exand its range rapidly after being introduced into northern Australia in the 1930s. Juveniles tend to show up in new localities before any breeding activity has been observed, suggesting that they are responsible for much of this dispersal (Freeland and Martin 1985).

Movements of Juvenile Reptiles

Many small lizards and snakes lay eggs within their home ranges. Hatchlings emerge from nests and enter the home range of their mothers or nearby individuals. Some remain near where they were born, whereas others move to new areas. The costs of dispersal include increased energy expenditure, exposure to predation, and the risk of being forced to occupy habitats of marginal quality because the best sites are already taken (Clobert et al. 1994). The costs of remaining in place include reduced growth and survival because of competition for scarce resources. Juvenile lizards sometimes are subject to cannibalism by adults, making it advantageous for them to move to habitats not frequented by adults (Auffenberg 1981, Stamps 1994, Castilla and van Damme 1996). Patterns of juvenile movement are likely to be affected by a variety of factors, including rates of adult mortality. If the population of adults turns over rapidly, then habitat may become available relatively quickly, whereas if adults are long-lived, patches of habitat may be occupied for years or even decades. Resource abundance and patchiness will influence the movements of juveniles as well. Juveniles of species that feed on abundant or rapidly renewing resources should be less likely to move long distances than those that feed on patchy, slowly renewing resources.

A lizard that exhibits very limited juvenile movement is *Anolis limifrons*, studied in Panama by Andrews and Rand (1983). This species lives in relatively homogeneous forested habitats and feeds on insects, an abundant and rapidly renewing resource. Both adults and juveniles are quite sedentary and occupy small home ranges. Furthermore, once a juvenile establishes a home range, it seldom moves to a new one as an adult. More than two thirds of individuals captured as both juveniles and adults had juvenile and adult home ranges less than 8 meters apart. This species has a high adult mortality rate, with most of the population dying within 1 year, so juveniles do not have to wait long for a home range site to become available. The behavior of this species contrasts with that of *Anolis aeneus* studied by Stamps (1994) in a more patchy habitat in Grenada. Hatchlings emerge from eggs laid in woodlands, but instead of remaining there, they move to clearings, where they establish juvenile territories that are occupied for 2 to 6 months. Once they reach a size that allows them to compete for adult territories and avoid being eaten by larger lizards, they return to the woodlands and establish new territories. Movement by young juveniles is not permanent dispersal, because they eventually return to the habitat where they were born, although not necessarily to exactly the same spot.

The effect of population density and a variety of other factors on juvenile dispersal has been studied in some detail for the European lizard *Larcerta vivipara* (Clobert et al. 1994). These lizards have home ranges about 20 meters in diameter. Dispersers were defined as individuals that moved more than 30 meters, whereas those moving less than 20 meters were considered to be philopatric. Juveniles from a good-quality habitat with a high density of lizards exhibited greater dispersal tendencies than those from an adjacent, poor-quality habitat with a lower population density. Clobert et al. (1994) attributed this pattern to a higher level of competition with adults in the good-quality habitat, where adults were expected to live longer, and to a greater competitive ability of individuals moving from high-quality habitats. Competitive ability can be a major determinant of eventual success in settling in a new habitat, because established residents usually have an advantage in competition for territory sites with new immigrants. Individuals that have poor competitive abilities may gain little from attempting to disperse into new habitats, while suffering all of the energetic and predation costs of moving.

Patterns of dispersal by reptiles from specialized nesting habitats are quite different from those of the relatively sedentary lizards described above. First, juveniles often are forced to move much longer distances to reach suitable habitats. Second, many juveniles emerge from a limited area, especially in species that form large nesting aggregations. If all of these animals remained near the nesting area, competition for limited food resources could be intense. Nesting aggregations of reptiles tend to attract predators that prey on hatchlings emerging from their nests. This means that rapid escape to more protected habitats is essential for juvenile survival. Reptiles with this type of dispersal pattern exhibit several distinctive behavioral traits, including a tendency toward sociality among the hatchlings and very rapid and direct movement to refuge habitats.

One example of this type of social dispersal is the behavior of green iguanas (*Iguana iguana*) emerging from nests on a small island next to Barro Colorado Island in Panama. Hatchlings emerge in groups and wait at nest entrances, observing the behavior of other individuals and scanning for predators before moving across open ground to protective vegetation (Figure 10–9). They usually move in groups, gathering in patches of reeds before moving across open water to the shore of the main island. Juveniles remain together for several weeks and often follow each other's movements as they gradually spread out in bushes and other vegetation on the island (Burghardt 1977). Baby crocodilians also tend to emerge from nests in groups and remain relatively social for up to several years. Group dispersal from the nest usually is aided by one or both parents, which help to dig the babies out of the nest and then transport them to a protected nursery area, where they are guarded for variable lengths of time.

Figure 10–9 **Hatchling green iguanas (*Iguana iguana*) emerging in a group from a nest in Panama.** (*Source: Burghardt 1977.*)

Perhaps the most dramatic example of group dispersal of nestlings is the behavior of baby sea turtles. They exhibit some social behavior even before emerging from nests, working together to dig their way out of the nest cavity. Once they reach the surface, usually early in the evening, they make a mad dash for the water. Often they are attacked by seabirds and other predators, so moving in groups of up to 120 individuals provides some protection. Dispersing turtles exhibit frenzied activity, moving as rapidly as possible across the beach and continuing very rapid swimming movements along a direct seaward course once they hit the water. They may maintain this course for many hours or even days (Lohmann et al. 1997).

For decades, the whereabouts of these young turtles once they left their natal beaches was a mystery. Archie Carr referred to this period of the life history as the lost year, although lost years would be a more accurate description. Recent evidence for several species indicates that juvenile sea turtles spend their early life floating passively in currents near the surface and feeding on pelagic food that accumulates in drift lines. Often they are found floating in mats of *Sargassum* and appear to actively seek out floating seaweed. This pelagic phase of the life history can last for many years (Figure 10–10). The young turtles are carried hundreds or even thousands of kilometers from their starting points and are distributed by current systems over thousands of square kilometers of open ocean. Eventually older juveniles settle into a more sedentary, bottom-dwelling lifestyle in shallow-water feeding areas, although olive ridley (*Lepidochelys olivacea*) and leatherback turtles (*Dermochelys coriacea*) retain a pelagic lifestyle for most of their lives (Musick and Limpus, 1997). Only after a very long time (30 to 50 years for some species) do the adult males and females leave these feeding areas and begin their long-distance migrations to mating and nesting areas (Miller 1997).

Homing Behavior

When amphibians and reptiles move between breeding sites and terrestrial home ranges, or make occasional movements out of their normal home ranges, they need some mechanism that allows them to find their way home again (Chelazzi 1992, Sinsch 1992). Homing also may be important for juveniles that initially leave their natal areas, search surrounding areas

for suitable habitats, and then return home if good-quality sites are not found (Phillips 1987). The ability to return after displacement has been well established for most of the major clades of amphibians and reptiles, but the distance over which the animals are able to home varies considerably. Homing performance can be affected by the age and sex of the animal, its reproductive condition, season of the year, and weather conditions.

Homing by Amphibians

Plethodontid salamanders generally home successfully when displaced up to 30 meters, but are much less successful when displaced more than 60 meters. In contrast, the red-bellied newt (*Taricha rivularis*), studied in California by Victor Twitty and his colleagues, exhibits the most impressive homing ability of any amphibian. Twitty displaced hundreds of newts about 2 kilometers upstream or downstream from where they were initially captured. Some individuals remained at the release site the first year, but after 5 years, about two thirds of the newts were recaptured, 90 percent of them in their home stream segments (Figure 10–11). Newts were later displaced over much longer distances to adjacent stream valleys, and many managed to return to their home streams from as far away as 8 kilometers, having moved overland across dry wooded ridges (Twitty 1966). This species clearly can return home even when displaced to unfamiliar terrain.

The homing performance of anurans depends in part on how displacements are carried out. When anurans are moved from breeding ponds to land, they usually return quickly from distances of several hundred meters, and they sometimes show homeward orientation at distances up to several kilometers (Sinsch 1990). In contrast, displacement from one terrestrial location to another often results in relatively poor homing over long distances, probably because the release site provides acceptable habitat so the animals are not motivated to make a long trek home. In an early study, Bogert (1947) displaced southern toads (*Bufo terrestris*) from their summer home ranges to locations 90 to 1,600 meters away. The toads successfully homed from all distances, but the percentage of animals recaptured was low at the longest distances (only about 20 percent returned home from a 1,600-meter displacement). Most males of a Puerto Rican tree frog, *Eleutherodactylus coqui*, that were displaced

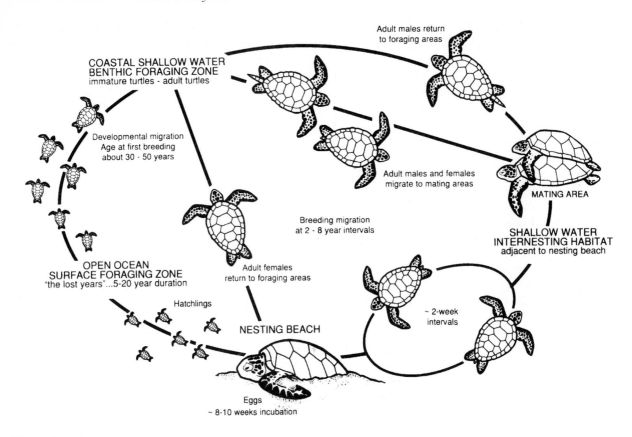

Figure 10–10   **Life cycle of a sea turtle such as *Chelonia mydas*.** The turtles move between distinct habitats at different stages of life. *(Source: Miller 1997.)*

up to 40 meters from their calling perches returned home, but far fewer returned from 100-meter displacements (Gonser and Woolbright 1995). Crump (1986) displaced males of *Atelopus varius* from their territories to locations on the forest floor about 10 meters away; 86 percent homed successfully.

Homing by Reptiles

Freshwater and terrestrial turtles vary in their homing performance, from species such as spotted turtles (*Clemmys guttata*) that do poorly when displaced more than a few hundred meters, to species that can return home from displacements of several kilometers (Chelazzi 1992). Aquatic turtles displaced along watercourses tend to home from greater distances than do turtles or tortoises displaced over land, perhaps because they often have larger home ranges, or because they can move faster and with less energy expenditure

in water. For example, some map turtles (*Graptemys pulchra*) returned home after displacements of up to 24 kilometers along a river, although the total number of individuals returning home was low. Displacements of several kilometers probably put most turtles outside of their usual home ranges. One must keep in mind, however, that turtles can live an exceedingly long time, so individuals have time to become familiar with areas outside their normal home ranges. Turtles frequently make long-distance movements away from their home ranges. Some giant tortoises (*Geochelone gigantea*) on Aldabra Atoll in the Indian Ocean make regular seasonal movements of several kilometers between inland areas and the coast (Swingland and Lessells 1979). Aquatic turtles sometimes move many kilometers in search of mates or nesting sites, or to escape drying ponds.

Evidence for long-distance homing by sea turtles comes mainly from studies in which turtles tagged at nesting beaches were later recaptured on their feeding

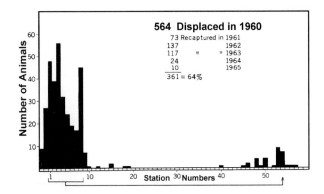

Figure 10–11   **Recaptures of newts (*Taricha rivularis*).** When adult newts were displaced approximately one mile downstream along a stream in California a few individuals remained at the release site, but two thirds of the displaced animals returned to their home stream segment within five years. *(Source: Twitty 1966.)*

grounds (Meylan 1982). Multiple recaptures of the same individuals at both nesting and feeding areas are very rare, in part because tags usually are recovered on feeding grounds from turtles that have been killed by fishermen. One of the few examples of multiple recaptures is a single female green turtle (*Chelonia mydas*) marked on a nesting beach in the Galápagos Islands and recaptured 3 months later on the coast of Ecuador, more than 1,000 kilometers to the east. The same turtle was recaptured 4 years later back in the Galápagos (Green 1984). Only a few attempts have been made to investigate homing in sea turtles by displacing them, and most such experiments have involved relatively short-distance displacements. Luschi et al. (1996) displaced a single female green turtle twice from her nesting beach off the coast of Malaysia, once to a site about 10 kilometers north across open water, and again more than 280 kilometers south along the coastline. In both instances the female quickly returned to the nesting beach along a relatively straight path.

Most crocodilians are relatively sedentary for much of the year. Females of some species move overland to nesting sites, and both juveniles and adults of several species make extensive overland migrations in response to seasonal flooding or drought. A study of *Caiman crocodilus* in Surinam revealed that most individuals eventually returned to the precise spot where they were originally captured, having traveled up to 3 kilometers away (Ouboter and Nanhoe 1988).

Gorzula (1978) displaced individuals of the same species more than 2 kilometers and reported that more than 80 percent returned home, with most having traveled over land. The homing of adult saltwater crocodiles (*Crocodylus porosus*) along a river in northern Australia was even more impressive, with some individuals returning from displacements 30 kilometers upstream from their original capture sites.

The most detailed study of homing by crocodilians was done by Gordon Rodda on juvenile American alligators (*Alligator mississippiensis*). Individuals displaced from 1 to 10 home range diameters (up to 7 kilometers) either returned home rapidly or showed strong homeward orientation. Subsequent experiments revealed that juvenile alligators maintained a homeward orientation even when displaced much farther (12 to 34 kilometers), with the accuracy of orientation depending on the method used to displace them (see Orientation). Radio tracking studies of juveniles released up to 16 kilometers from their home ranges also demonstrated an ability to home from unfamiliar areas, although homing performance decreased at distances greater than 5 kilometers (Rodda 1985).

Relatively little detailed work has been done on homing by lizards, which is surprising in light of the enormous literature on home ranges and territoriality in these animals. Most studies have focused on North American phrynosomatids and European lacertids (Chelazzi 1992). *Sceloporus, Uta*, and *Phrynosoma* can home from displacements of less than 300 meters from their normal home ranges, but homing from greater distances has not been tested. Most of these studies were done in relatively open, desert habitats. The animals were released outside of their home ranges, but not necessarily out of view of familiar landscape features. Some lizards, however, do appear to be capable of returning home even when familiar landmarks are not visible. For example, Barbara Ellis-Quinn and Carol Simon (1989) displaced both adult and juvenile *Sceloporus jarrovi* up to 200 meters in a riparian habitat at the bottom of a canyon. Adults homed successfully from all displacement distances, but juveniles performed poorly if moved more than 50 meters. This result suggests that juveniles had not become strongly attached to home ranges where they originally were captured and therefore were not motivated to return home, that they had not yet traveled very far from where they were born and were unfamiliar with more distant areas, or that they had not yet developed the ability to detect environmental cues needed for distant orientation.

Limited evidence from homing studies with adult lizards suggests that the strength of attachment to a home range can affect homing performance. For example, homing performance was similar in males and females of *Sceloporous jarrovi*, both of which are territorial (Ellis-Quinn and Simon 1989). In contrast, females of *S. orcutti*, which are not territorial, exhibited poorer homing performance than did males, which are territorial (Weintraub 1970). Homing performance was relatively poor in *Phrynosoma douglassi*, a species that exhibits little evidence of territoriality and only weak attachment to a home range. Individuals of *Sceloporus graciosus*, a territorial species living in the same habitat, exhibited much better homing performance (Guyer 1991).

Studies of homing by snakes have produced inconsistent results (Gregory et al. 1987, Chelazzi 1992), but this inconsistency may have more to do with the design of the studies than real interspecific differences in homing ability. Several early studies revealed little evidence of homing behavior. Most of these studies were based on displacements of only a few individuals from summer home ranges to similar habitats. Displaced individuals tended to remain near the site where they were released, perhaps because the areas provided sufficient food. Garter snakes (*Thamnophis ordinoides*) that feed mainly on earthworms, a rather uniformly distributed resource, showed little evidence of homeward orientation (Lawson 1994). In contrast, snakes usually exhibit good homing ability when displaced from communal den sites (Brown and Parker 1976), or from rich food sources such as a fish hatchery or the shore of a lake (Lawson 1994). These observations suggest that snakes are more motivated to return home when removed from especially valuable resource patches, but they do not necessarily indicate that other species lack homing ability.

Mechanisms of Orientation

Directed movements require, at a minimum, that an animal know which way it must go to reach its goal. Amphibians and reptiles use several methods of orientation, ranging from familiarity with local landmarks to an ability to sense Earth's magnetic field.

Orientation to Local Environmental Cues

Amphibians and reptiles use several kinds of orientation cues to find their way around their home ranges, to locate breeding ponds, nesting sites, or overwintering sites, or to find their way home after long migrations (Chelazzi 1992, Sinsch 1992). The behavior of animals within their home ranges often reveals a detailed familiarity with the local environment and suggests that movements are guided by local landmarks. For example, Dole (1965) reported a tendency for leopard frogs (*Rana pipiens*) to return repeatedly to exactly the same spot, even after relatively long excursions. Repeated use of familiar foraging perches, display sites, retreats, burrows, and other features within a home range has been observed in many species of amphibians and reptiles. Ciofi and Chelazzi (1994) reported that European rat snakes (*Coluber viridiflavus*) occupied shelter sites from which they made short foraging excursions. Some individuals made single-day excursions and returned to the same shelter each night, whereas others made longer excursions and occupied secondary shelters before eventually returning to the main shelter (Figure 10–12). Tortoises appear to be familiar with a number of burrows used as retreat sites within their home ranges but return most frequently to a single primary burrow (Diemer 1992).

The types of cues that amphibians and reptiles use for local orientation depend on the sensory capabilities of the animals. Plethodontid salamanders probably use a combination of visual and chemical cues to identify familiar foraging areas or retreat sites, with chemical cues probably being the most important. The European fire salamander (*Salamandra salamandra*) appears to find its way home after displacement by following a circuitous route connecting landmarks such as rocks and logs rather than by returning along a direct straight-line path (Figure 10–13). Whether these landmarks are identified by visual or chemical cues is not known, but in the laboratory, this species learned to orient to retreat boxes that were associated with unique patterns of stripes and squares (Plasa 1979).

Tree frogs moving through a complex three-dimensional environment of leaves and branches probably orient almost exclusively by visual cues. Orientation toward aquatic breeding sites, on the other hand, probably involves several sensory cues, including visual recognition of local landmarks, identification of odors emanating from ponds, and responses to sounds

Figure 10–12 **Patterns of movement of a European yellow-green racer (*Coluber viridiflavus*).** Activity is concentrated around a central shelter site (triangle) and secondary shelters (circles). Both short excursions from the shelter (box at left) and complex looping excursions were made by this snake. *(Source: Ciofi and Chelazzi 1994.)*

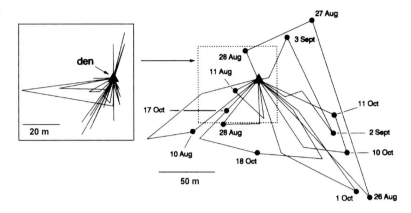

produced by calling conspecific males. Experimental studies in which these cues were manipulated or sensory inputs were impaired suggest that the absence of one type of cue can reduce orientation ability, but that no single cue is absolutely essential for the animals to locate breeding sites (Sinsch 1990).

Tortoises may use olfactory cues to orient toward home ranges after displacements of a few hundred meters (Chelazzi and Delfino 1986), and it seems likely that both turtles and crocodilians can use olfactory cues to locate bodies of water at some distance. For several decades, sea turtle biologists have specu-

Figure 10–13 **Movement and homing of European fire salamanders (*Salamandra salamandra*).** Dashed line at right shows movements of an individual on a single night. Solid line on left shows return path of an individual displaced from its home shelter (straight arrow). Hatched areas indicate logs and trees. Note the circuitous path taken by the displaced animal to return home as it moved between familiar landmarks. *(Source: Plasa 1979.)*

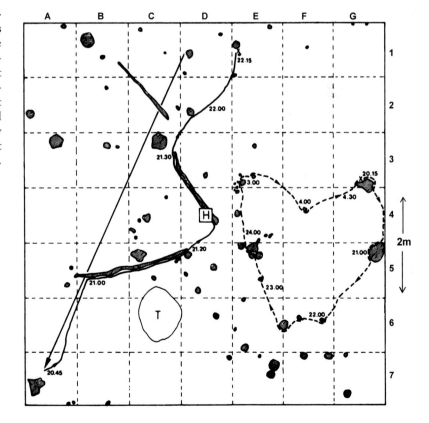

lated that hatchling turtles might imprint on chemical cues associated with natal beaches and use these cues to identify beaches when they return as adults. Some have even suggested that sea turtles might use chemical cues for long distance homing. Although the results of some experimental studies are suggestive, they do not definitively support either hypothesis (Lohmann et al. 1997). Most squamates probably rely on visual cues for orientation at moderate distances, but chemical cues probably are important for identifying retreat sites, burrows, or communal hibernations dens, especially for scleroglossan lizards and snakes. European vipers (*Vipera aspis*) and granite spiny lizards (*Sceloporus orcutti*) displaced within sight of their home areas exhibited better homing performance than those displaced out of sight of home areas. Displaced *S. orcutti* behaved in a manner similar to that observed for *Salamandra*, returning home by moving between conspicuous rock outcrops (Figure 10–14).

Animals can use local environmental cues for orientation even if they are not familiar with an area, provided the cues are reliable indicators of suitable habitats. For example, Omland (1996) found that redspotted newts (*Notophthalmus viridescens*) tend to orient downhill (geotaxis) when placed in unfamiliar surroundings, presumably because moving downhill would generally lead them to water. Adult newts collected during spring migrations and placed in an apparatus that allowed the slope angle to be varied exhibited a strong downhill orientation. Red efts, the terrestrial juvenile stage of this species, did not exhibit downhill orientation under the same conditions and showed no tendency to move toward ponds. Omland also reported that juveniles of two anuran species emigrating from breeding ponds exhibited uphill orientation in similar test situations, which suggests that they use reverse geotaxis to orient toward terrestrial habitats.

The sea-finding behavior of hatchling sea turtles is another example of the use of local cues (Lohmann et al. 1997). Because hatchlings must move rapidly to water to avoid predation and desiccation, they need a reliable cue to find the ocean. They seem to rely mainly on visual cues. Hatchlings with their eyes covered and those released in areas where visual cues are obscured do poorly in orientation tests. The turtles tend to orient toward the brightest area on the horizon. Brightness provides a good indicator of seaward direction, because reflection of sunlight, moonlight,

or starlight off the water makes the sky over the ocean somewhat brighter than over land. Hatchlings exhibit a preference for blue or purple wavelengths over red, orange, or yellow light, and the loggerhead (*Caretta caretta*) shows a strong avoidance of yellow light. A possible advantage of this response is to avoid orienting toward a rising or setting sun, which produces light rich in yellow. Hatchlings appear to pay most attention to the brightness of light closest to the horizon and are not much affected by other sources of light, such as the sun, moon, or stars.

Hatchlings also tend to avoid dark, elevated silhouettes, a tendency that usually would direct them away from dunes or trees and toward the flatter ocean shore. Sometimes brightness cues can override such shape cues. For example, artificial lights from coastal cities and roads often are much brighter than the light reflected off the ocean and can cause hatchling turtles to move inland instead of seaward. Construction of light barriers or even the placement of rows of trees along urban beaches may significantly enhance seaward orientation and the survival of hatchling turtles (Salmon et al. 1995). Once baby sea turtles reach the water, they apparently use the direction of wave movement to guide them as they move directly out to sea from the shoreline (Figure 10–15).

Compass Orientation

Many amphibians and reptiles can orient in a homeward direction even when placed in situations where local cues are not detectable (Chelazzi 1992, Sinsch 1992). This requires, at a minimum, a well-developed compass sense. In some cases, a compass sense may enable an animal to orient relative to familiar features in the environment even when those features are not visible. For example, frogs that live in vegetation along the shores of ponds often escape from predators by jumping directly into the water. When frogs are collected along a shoreline with a particular compass orientation and placed in a featureless arena, they usually orient as if they were moving perpendicular to their home shore, instead of moving in a homeward direction (Figure 10–16). This so-called Y-axis orientation has been a useful tool to investigate the sensory basis of orientation because animals can be trained to an artificial shoreline in a tank and then tested for orientation in an arena after environmental cues or sensory imputs have been manipulated (Adler 1976).

Figure 10–14 **Homing by *Sceloporus orcutti.*** Solid lines show displacments of individuals. Dashed lines show return paths. Irregular outlines are rocky outcrops. Most individuals took circuitous paths home, moving between rocky outcrops. *(Source: Weintraub 1970.)*

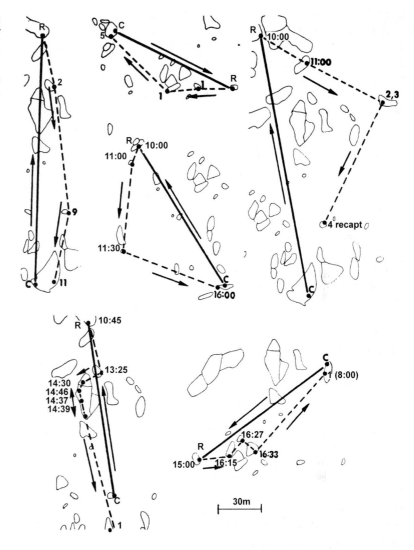

Amphibians also exhibit a compass sense during migrations to and from breeding ponds. In several studies, the movements of animals have been monitored by placing drift fences around ponds. Migrating animals fall into traps distributed at intervals around the fence. This makes it possible to determine the compass headings of animals as they enter the pond and again as they leave. In many cases, individuals enter and exit at approximately the same spot. Migrating amphibians also tend to move in and out of ponds along relatively straight paths, and they usually maintain their original compass headings when transferred to test arenas. Natural migration along straight paths could be due to compass orientation or to an integration of compass orientation with information about local landmarks as animals retrace familiar paths (Sinsch 1990). The ability to maintain the same directional heading when placed in an arena some distance away can be explained only by the presence of a compass sense.

Studies conducted by Denzel Ferguson, Kraig Adler, Douglas Taylor, and others during the 1960s and 1970s demonstrated that amphibians can orient using celestial cues, especially the position of the sun. In most of these experiments, animals were trained to a particular shoreline direction and then tested for Y-axis orientation in an arena that afforded a view of the sky but not the horizon. Nearly all species tested under sunny conditions exhibited strong orientation to the trained direction, but most showed random ori-

Figure 10–15 **Modes of orientation in baby loggerhead turtles (*Caretta caretta*).** In step 1, turtles on a beach on the east coast of Florida use visual cues to locate the ocean. In step 2, they use the direction of waves to orient in a seaward direction to swim offshore. In step 3, hatchlings are hypothesized to transfer seaward orientation to a magnetic compass. *(Source: Lohmann et al. 1997.)*

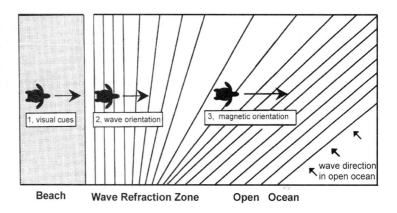

entation when tested under an overcast sky. A few species can orient under moonlight or starlight, but most are disoriented under an overcast night sky or on very dark nights. Tests with migrating animals placed in arenas produced similar results, with most species orienting in the direction of migration under a clear sky, but not always under an overcast sky . Similar results have been obtained in experiments with turtles, crocodilians, lizards, and snakes (Chelazzi 1992).

Because the position of the sun in the sky is not fixed but changes both daily and seasonally, an animal must have a built-in time sense to use the sun's position for orientation. If an animal moves on a fixed compass course its correct angle of movement relative to the sun will depend on the time of day, so it must know the local time to orient correctly (Figure 10–17). An animal in the Northern Hemisphere moving south would head to the right of the sun in the morning and to the left of the sun in the afternoon. If the photoperiod to which the animal is exposed is experimentally delayed by 6 hours, the animal is expected to behave as if the afternoon sun were in the morning position and shift its movement 90 degrees in a clockwise direction. A 6-hour advance in the photoperiod should produce a 90-degree shift in counterclockwise direction. Experimental studies of several species of amphibians (Adler 1976) and reptiles (Un-

Figure 10–16 **Y-axis orientation by aquatic and terrestrial vertebrates.** Terrestrial toads (left) tend to orient toward shore in a natural pond, whereas tadpoles orient toward the water. When placed in a featureless arena some distance from the pond, they are expected to orient in a direction corresponding to their normal preferences (circle). A metamorphosed salamander trained in a tank with an artificial shore (right) is expected to orient in the same direction in the arena. *(Source: Adler 1970.)*

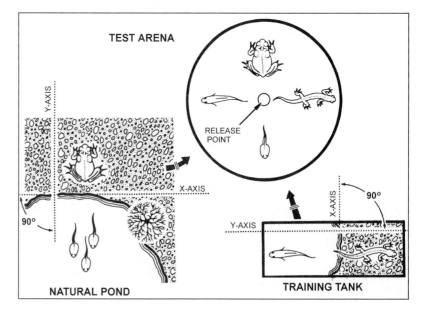

Figure 10–17 **Effect of clock shifting on compass orientation in a newt in a north-temperate location.** The diagram at top shows 6-hour advances or delays in photoperiod compared to the normal control situation. Control animals are expected to take the correct compass heading when tested in an arena, regardless of the time of day, because their clock is in synchrony with the natural photoperiod. Phase-advanced animals are expected to move in a direction 90 degress counterclockwise to the correct direction, whereas phase-delayed animals are expected to move 90 degrees clockwise to the correct direction. *(Source: Adler 1976.)*

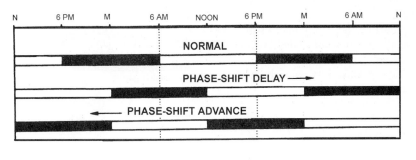

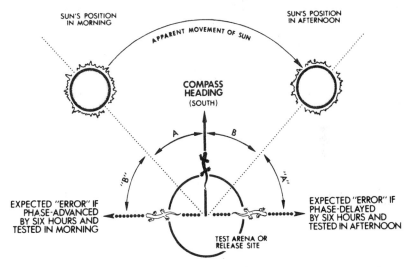

derwood 1992, Lawson 1994) demonstrated exactly the expected shift in orientation with a change in photoperiod.

Exposure to the sun is necessary to maintain an animal's internal clock, and animals held for several days in total darkness become disoriented. The cues that maintain the clock are detected by the pineal complex in the dorsal region of the brain. Frogs have a third eye, or pineal end organ, located on the top of the skull that connects to the pineal body (Figure 10–18). Salamanders lack this organ, but the pineal body is sensitive to light transmitted through the skull. The animals can no longer entrain to a photoperiod when these organs are covered by opaque plastic or removed (Adler 1976). All reptiles except crocodilians also have pineal organs that regulate a variety of circadian and circannual rhythms, entrainment to a photoperiod, and regulation of an internal clock (Underwood 1992). Some lizards have a parietal eye, which serves some of the same functions as the pineal end organ of frogs. The parietal eye of lizards,

however, is much more similar to a true eye in that it has a well-developed retina, lens, and cornea.

Early studies of celestial orientation in amphibians and reptiles suggested that the position of the sun was detected visually. However, some experiments showed that animals that had been experimentally blinded still exhibited directional orientation, and several studies revealed that blind animals returned home successfully. Some investigators took this as evidence that olfactory cues played an important role in homeward orientation. Subsequent work demonstrated that amphibians are capable of detecting polarized light through extraoptic photoreceptors and can use this information to determine the location of the sun. Polarized light is a product of atmospheric scattering of sunlight, which causes light waves to vibrate primarily in a plane perpendicular to the direction of wave propagation. At any given time of day, the direction of vibration, called the e-vector of polarized light, is directly related to the position of the sun (Figure 10–19). An ability to detect the direction of the e-vec-

Figure 10–18 **Frontal organ and pineal region.** Diagram of a sagittal section of the brain of a ranid frog, showing the frontal organ (also called the pineal end organ), an extraoptic photoreceptor, and its relationship to the pineal region of the brain. *(Source: Adler 1976.)*

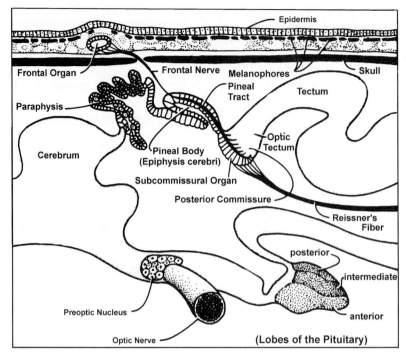

tor would give an animal a compass sense (Adler 1976).

The first demonstration of orientation by polarized light was a study of *Ambystoma tigrinum* (Figure 10–20). Polarized light is detected by the pineal body, not the eyes, which explains the ability of blinded animals to orient under polarized light. The use of polarized light as a directional cue is especially advantageous for animals that do not have a direct view of the sun. This would often be true for animals in heavily vegetated habitats. It would also be true at twilight, when the sun is low on the horizon, but the polariza-

Figure 10–19 **Diagram of training tank used to train salamanders under polarized light.** (A) Light box with polarizing filter. The newts are trained to an artificial shoreline with the e-vector of polarized light (parallel lines) perpendicular to the long axis of the tank and parallel to the shoreline. In a test arena (C), the newts orient in the same direction relative to the e-vector of polarized light, which would place them in a shoreward direction in the training tank. *(Source: Taylor and Adler 1973.)*

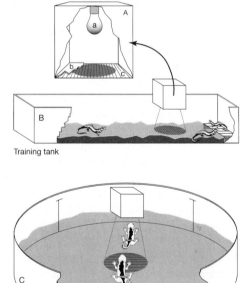

Figure 10–20 **Orientation responses of tiger salamanders (*Ambystoma tigrinum*) under polarized light.** Each dot represents the final orientation of an animal in a test arena. Animals were trained to a shoreline perpendicular to the e-vector of polarized light. Shaded areas on the periphery of the circles show the expected direction of orientation in an arena if polarized light can be detected. Both blinded animals and those with eyes intact could orient in the proper direction if the top of the head was covered with clear plastic, but not if the top of the head was covered with opaque plastic, which blocked light reception by the pineal body. This experiment demonstrates the role of the pineal body in the perception of polarized light. (*Source: Adler 1976.*)

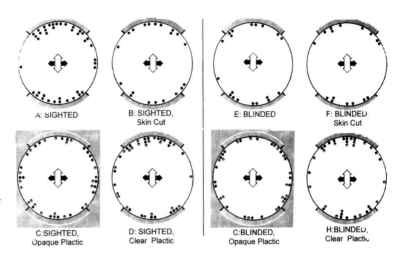

A: SIGHTED   B: SIGHTED, Skin Cut   E: BLINDED   F: BLINDED, Skin Cut

C: SIGHTED, Opaque Plactic   D: SIGHTED, Clear Plactic   C: BLINDED, Opaque Plactic   H: BLINDED, Clear Plactic

tion of sunlight is greatest directly overhead. Polarization is obscured by complete overcast, which explains why many amphibians cannot orient under a completely cloudy sky (Adler 1976). One species of desert lizard, *Uma notata*, can orient to polarized light, but whether other reptiles can do so is not known.

Magnetic Orientation and Navigation

One of the great difficulties in studying the sensory basis of orientation in amphibians and reptiles is that new sensory capabilities are constantly being identified. For example, Victor Twitty (1966) spend a considerable amount of time trying to demonstrate that newts (*Taricha rivularis*) can smell their way home, based on the observation that blinded newts exhibited homing ability. His work, however, preceded the discovery that newts can detect polarized light, a sensory system he did not even consider. More recently, John Phillips (1987) demonstrated that red-spotted newts (*Notophthalmus viridescens*) can use Earth's magnetic field for orientation, and it seems likely that other amphibians can do so as well (Sinsch, 1992).

Newts collected during migration to ponds exhibited true homeward orientation, even when they were tested indoors at sites up to 30 kilometers from their home ponds after having been displaced in containers that deprived them of all visual and olfactory cues (Phillips et al. 1995). This observation shows that these newts are capable of true navigation, the ability to find their way home from an unfamiliar location. To do this, animals must have not only a compass sense, but also a map sense. That is, they must have information about where they are relative to where they want to go. Humans would have great difficulty doing this. A person placed on an oceanic island could readily determine north from south or east from west, but determining how to return home would be impossible without information on the latitude and longitude of the island. The only sources of this type of map information for animals would be the stars and Earth's magnetic field. The structure of the visual systems of amphibians and reptiles makes it unlikely that they can resolve the spatial relationships of stars, and magnetic orientation is currently the only other viable explanation. The precise sensory system used to detect magnetic information is still being investigated, but recent work suggests that newts have two separate magnetic receptors, one of which is sensitive to a change in wavelengths of ambient light and therefore probably is related to the visual system. Possibly inputs from both types of magnetic receptors are necessary for navigation (Phillips and Borland 1994).

Sensitivity to Earth's magnetic field has been demonstrated for alligators (Rodda 1984) and sea turtles (Lohmann et al. 1997). Baby sea turtles are carried hundreds or even thousands of kilometers from natal beaches by winds and currents, yet manage to find their way home to breed, sometimes 30 to 50 years later. This observation suggests that sea turtles develop a magnetic map as they leave their natal beaches that can be extrapolated to unfamiliar areas. Hatchlings of both loggerhead and leatherback turtles are sensitive to Earth's magnetic field. When trained to swim in a particular compass direction toward a light source, they maintain the same compass orientation when tested in the dark. When the magnetic field is reversed, the turtles reverse their orientation. It seems likely that baby turtles set their initial compass direction as they leave the beach and swim out to sea, and then maintain that direction through magnetic orientation as they swim farther offshore.

Other experiments have shown that young turtles in the open ocean are sensitive to changes in the inclination angle of the magnetic field (this is a measure of the relationship of the magnetic field to Earth's gravitational field), and this information may be used to approximate the latitude of a turtle's position. Sea turtles may also detect local changes in the intensity of the magnetic field. Magnetic sensitivity plus the detection of changes in inclination angle would provide enough information for a magnetic map sense. If such a map sense does exist, it is not yet clear whether it provides the turtles with a very precise ability to locate specific nesting beaches or simply guides them to the general region of the natal beach, at which point chemical or visual cues would allow the turtles to find their way home.

Summary

Patterns of movement by amphibians and reptiles affect, and are affected by, nearly all other aspects of their biology. Animals move to acquire resources, including food, water, mates, shelter, and nesting sites, but generally avoid moving unless absolutely necessary. Many amphibians and reptiles remain immobile for long periods of time and often occupy relatively small home ranges. Variation in home range size is related to the spatial and temporal distribution of resources, especially food. When food is abundant or is renewed at a high rate, animals tend to occupy relatively small, stable home ranges. This type of spacing is characteristic of herbivorous and omnivorous lizards and turtles as well as insectivorous amphibians and squamates. A patchy distribution of food, coupled with a low renewal rate, results in animals occupying larger and less stable home ranges. This pattern is characteristic of widely foraging lizards and many snakes. Extreme patchiness of resources and very low renewal rates often result in animals occupying a series of shifting activity centers or adopting a nomadic lifestyle, a pattern that is characteristic of some snakes and sea turtles.

Some amphibians and reptiles defend all or part of their home ranges as territories. Territorial defense is particularly common among insectivorous lizards with a sit-and-wait foraging style, which use a relatively evenly distributed, rapidly renewing food resource. As home range size increases, territory defense becomes less feasible because the costs of defense can exceed the benefits of exclusive control of resources, and a system of overlapping home ranges is more likely to evolve. This behavior is characteristic of most widely foraging reptiles. Territoriality also is rare among species that use very abundant resources, such as leaves, even when home range size is small, because defense of resources provides little benefit to residents.

When amphibians and reptiles move out of their normal home ranges, they usually do so to locate suitable breeding sites, nesting sites, or overwintering sites. These movements range from short excursions of a few dozen meters by some amphibians to vast migrations over thousands of kilometers by sea turtles. Many juveniles also make relatively long movements, especially when it is advantageous to leave the habitat in which they were born and move to other habitats. The most dramatic examples are the movements of hatchling sea turtles from nesting beaches to the open ocean, where they are carried over thousands of kilometers by currents. The juveniles of other amphibians and reptiles generally move no more than a few hundred meters from their birthplace. Most species exhibit well-developed homing ability, and even sea turtles reliably return to their natal beaches to breed, sometimes decades after leaving them.

Amphibians and reptiles use a variety of senses for orientation. Local environmental cues probably are most important for short movements within a home range and for movements to and from nearby breeding or hibernation sites, with vision and chemical senses playing the largest role. Longer movements and migrations may require the use of a compass sense, and many amphibians and reptiles can determine direction from celestial cues, especially the position of the sun. Some can do this visually, and others use extraoptic photoreceptors to detect polarized light, which provides reliable directional cues even when the sun is not directly visible. Some amphibians and reptiles appear to be capable of true navigation, which enables them to find their way home from an unfamiliar location, sometimes over long distances. The only likely cue that could be used for such navigation is Earth's magnetic field, and an ability to detect changes in the magnetic field has been demonstrated for several amphibians and for alligators and sea turtles. The precise mechanism of magnetic orientation has yet to be determined.

C H A P T E R

11

Communication

Anyone who has passed by a swamp or pond on a summer evening probably is familiar with the sounds produced by frogs and toads. Upon encountering frog choruses in southern Africa, the nineteenth-century English explorer and naturalist William Burchell wrote, "No sooner does the delightful element moisten the earth, and replenish the hollows, than every pool becomes a concert-room, in which frogs of all sizes, old and young, seem contending with each other for a musical prize. Some in deep tones perform their croaking bass, while the young ones, or some of a different species, lead in higher notes of a whistling kind. Tenors and trebles, countertenors, sopranos, and altos, may be distinguished in this singular orchestra. . . . The noise produced, particularly in the evenings, is truly astonishing, and nearly stunning" (Burchell 1822, p. 352). Others have been less sympathetic. During the seventeenth century, the French nobility reportedly found their sleep so disturbed by the nighttime calling of frogs that peasants were sent into the swamps to keep the animals quiet.

The sounds produced by frogs and toads are not, of course, designed either to please or irritate humans but are used by males to communicate with members of their own species. Animals can communicate with members of other species as well (for example, when a poisonous frog displays aposematic coloration, or a rattlesnake sounds a warning to predators; see Chapter 13). We will be concerned only with intraspecific communication in this chapter. Communication has been defined in many different ways by different investigators, but central to any definition is the idea that communication involves interactions between at least two individuals, a signaler and a receiver. In fact, there may be more than one signaler and more than one receiver, or individuals may alternate between roles as signaler and receiver.

Some animal behaviorists define communication in terms of the way in which a signal alters the behavior of another individual. A receiver may respond to a signal by approaching the signaler, retreating from the signaler, attacking, initiating courtship, or in many other ways. Other researchers consider communication an exchange of information between two individuals (Halliday 1983). The use that a receiver makes of information conveyed by a signal depends on the context in which the signal is given and the nature of the receiver. Information about the signaler is termed the signal message and is a function of the physical properties of the signal and the way in which it is delivered. A signal may have several different meanings that can be discerned only by studying the reactions of other animals to the signal (Smith 1977). For example, the advertisement call of the North American bullfrog (*Rana catesbeiana*) sends messages about the species identity, sex, spatial location, size, reproductive state, territorial status, and individual identity of the caller. The call may have very different meanings to receivers, depending on their sex and reproductive state. To sexually mature males it is a keep-out signal, an advertisement of territory ownership, whereas to gravid females it is an attractive signal that indicates the location of a potential mate (Wiewandt 1969; Davis 1987).

Herpetologists have used many different terms to describe the signals and displays of amphibians and reptiles, but most can be classified into a few simple categories, depending on the context in which they are produced. Perhaps the most widespread are advertisement signals, usually given by males to advertise ownership of territories or to attract mates. Also common are courtship signals, which are used during close-range sexual interactions between males and females. Most species also have distinctive aggressive signals that are used during contests between individuals, and some have submissive signals as well. Less common are contact signals, used by animals to keep in touch with other individuals in a social group, and alarm signals, given when a predator threatens.

Modes of Communication

Virtually any sensory modality can be used for communication, but the suitability of different modes depends on the sensory capabilities of animals and characteristics of the environments in which they live. Acoustic signals are ideally suited for communication over long distances because sound propagates rapidly in air or water. Usually the source of an acoustic signal is relatively easy to locate, and the signals can be perceived at night and when physical barriers separate the signaler and receiver. Messages conveyed in sound can be altered through changes in intensity, frequency (pitch), or timing of various signal components, so animals can use a rich repertoire of acoustic signals to communicate subtle changes in behavior. Visual signals are effective at short to moderate distances and are easily located, but they also are easily obstructed by physical barriers and are highly dependent on light levels and the wavelengths of light present in the environment. Visual signals can be modified to convey complex messages by combining display elements such as color, movement, and changes in posture (Fleischman 1992). Chemical signals (pheromones) are equally effective at night or during the day, but such signals broadcast into air or water are transmitted very inefficiently unless the medium is moving. Chemical signals often are not very directional, so the source of the signal can be difficult to locate. For the most part, chemical signals appear to convey relatively simple messages and are less easily modified or elaborated than acoustic or visual signals (Alberts 1992). Tactile signals, produced by touching another animal or through surface waves and vibrations, are used mainly for short-range communication.

Constraints on Signal Production

The production of communication signals by amphibians and reptiles is subject to a number of biological and physical constraints. Most of these animals are relatively small, and small size can limit the distance over which an animal can communicate. For example, the distance at which a visual signal can be detected is proportional to the size of the signaler and the amplitude of the display, so a small lizard performing a head-bob display will be visible to other animals at shorter distances than a large lizard performing a similar display (Fleischman 1992). Small animals generally produce relatively high-frequency sounds, and these tend to be more readily absorbed by the ground or vegetation during transmission than low-frequency sounds (Gerhardt 1994). Huge animals such as elephants and whales can communicate over distances of many kilometers using low-frequency sound, but even the largest reptiles, such as crocodilians, probably do not communicate over distances of more than a few kilometers. The signals of smaller species probably cannot be detected more than a few hundred meters away.

Changes in temperature affect most physiological processes of amphibians and reptiles (see Chapter 5), including the production of communication signals. Temperature usually has a major effect on the rate of signal production, as do the temporal features of signals that are determined by active muscle contraction (Figure 11–1). For example, features such as calling rate, call duration, and the number of sound pulses in calls are highly temperature dependent. These temperature-dependent changes in call structure can have a major effect on the energetic cost of calling, which increases substantially at warmer temperatures. Energetic costs probably represent the most important constraint on the rate at which signals can be produced and the length of time an animal can continue signaling. In frogs, for example, conversion of metabolic energy to sound energy is relatively inefficient (Prestwich 1994), and males that call at very high rates can incur substantial energetic costs (see Chapter 6).

The other major cost of signaling is exposure to predation. Communication signals, by their very nature, must make animals conspicuous to other individuals. Unfortunately for the signalers, some predators make use of these signals to home in on prey. For example, a Neotropical bat, *Trachops cirrhosus*, hunts for frogs by homing in on their calls. The frogs may

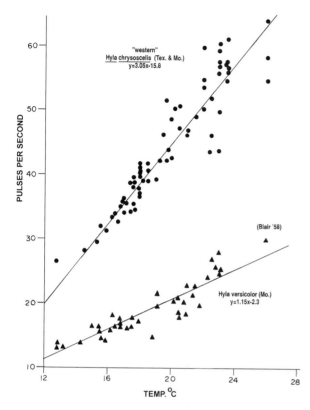

Figure 11–1 **The effect of temperature on pulse rate of the calls of two species of gray tree frogs, *Hyla versicolor* and *Hyla chrysoscelis*.** At the same temperature, females of each species can readily distinguish conspecific from heterospecific males by differences in pulse rate alone. *(Source: Gerhardt 1978.)*

as white, yellow, or pink, whereas those living in more open, higher-light environments have red, blue, or purple dewlaps (Fleischmann 1992).

Communication and Noise

Once a signal is produced, it must be transmitted to a receiver. The process of transmission is subject to interference by environmental noise that makes it hard for a receiver to detect or recognize the signal. We commonly associate noise with sound, but any communication channel can be affected. For frogs and other animals that use sound to communicate, sources of background noise include wind, rain, waterfalls, and the sounds of other animals. These types of noise often fluctuate in intensity and do not have a predictable temporal pattern. One way to overcome such noise is to produce a signal that contrasts with the unpredictable variations of background noise. The best such

alter the rate or timing of their calls to reduce the chances of being eaten, even though these changes make it more difficult for females to find the males (see Chapter 13). Predation probably has been an important selective force in the evolution of the display behavior of lizards. Many lizards have inconspicuous colors that blend in with their surroundings. Some of these lizards have brightly colored display structures that are kept hidden except during courtship or aggressive displays. For example, small Neotropical lizards in the genus *Anolis* usually are brown or green, but they reveal brightly colored throat fans (dewlaps) during sexual or aggressive displays (Figure 11–2). Often these display structures have colors that contrast with the prevailing background to make them especially visible to receivers. *Anolis* lizards that live in forests dominated by low-light conditions often have dewlaps that emphasize highly reflective colors such

Figure 11–2 **Dewlap display by a male *Anolis aquaticus*.** *(Photograph by K. G. Preston-Mafham.)*

signals are stereotyped calls (that is, calls that exhibit little variation) that are repeated frequently. For example, most frog calls are stereotyped, and call repetition rates tend to be highest in the noisiest environments, such as large choruses (Narins and Zelick 1988).

One effective way of reducing acoustic interference in large choruses is for males to call during periods when other frogs are quiet. Males of the small Central American tree frog *Hyla ebraccata* call in temporary ponds and often are surrounded by dense choruses of another small frog, *Hyla microcephala*. The latter species calls in distinct bouts, separated by silent periods. *Hyla ebraccata* males insert their calls into silent periods between the calling bouts of *H. microcephala*. Experiments in which calls were broadcast to females in an indoor arena showed that males that call during silent periods are more likely to attract mates because females have difficulty hearing calls that are overlapped by the calls of the other species (Schwartz and Wells 1983).

Another way to avoid acoustic interference is to use relatively noise-free channels for communication. For example, the Puerto Rican white-lipped frog (*Leptodactylus albilabris*) calls from burrows. In addition to emitting an airborne signal, males calling underground produce ground-borne vibrations that can be detected by other frogs (Figure 11–3). Indeed, the organ in the white-lipped frog's ear that is used to detect seismic signals is one of the most sensitive vibration detectors

in any vertebrate and is tuned to detect the specific frequencies produced by calling males (Narins 1990). The advantage of calling in burrows is that very few other animals use the seismic communication channel, and the frog is protected from predators as well.

Animals that use visual signals to communicate often have to make themselves conspicuous against a visually complex background. One way to do this is to move, and most visual displays involve some sort of movement, such as jumping, changes in posture, waving of the legs or tail, movement of the head, and expansion of special display structures such as crests and dewlaps. Some elegant work by Leo Fleischman has shown how the displays of one lizard, *Anolis auratus*, are adapted to overcoming visual noise in the environment. This species is found in grassy habitats in Panama and must make itself conspicuous against a background of windblown vegetation. In laboratory experiments, Fleishman found that the lizard's visual system is most sensitive to movements that combine high velocity and high acceleration (jerky movements), features that are characteristic of their head-bobbing displays, but not the movements of windblown vegetation. Furthermore, the displays emphasize movement frequencies to which the visual system is most sensitive, enabling the lizards to detect the displays even under windy conditions. Fleischman also found that head movements used in long-distance displays were more pronounced and more jerky than those used at close range, suggesting that these elements are used to attract the attention of other individuals (Fleischman 1992).

Communication in the Major Clades of Amphibians and Reptiles

The major clades of amphibians and reptiles have undergone millions of years of independent evolution, and it is not surprising that they have diverged considerably in their modes of communication. Chemical communication has assumed a dominant role in some clades, such as caecilians, salamanders, turtles, and some groups of squamates, but in others, such as anurans and crocodilians, the importance of chemical communication has diminished.

All groups of amphibians and reptiles use some visual signals. Acoustic signals are largely restricted to frogs, crocodilians, some tortoises, and geckos and seem to be associated either with nocturnal habits (most frogs and geckos) or with signaling over relatively long distances (frogs and crocodilians).

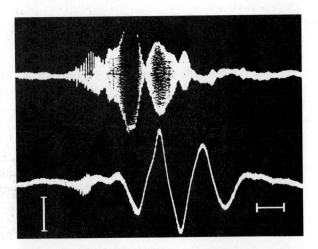

Figure 11–3   Airborne and ground-borne sound. Oscillograph showing a single chirp vocalization of *Leptodactylus albilabris* recorded with a microphone (top) and a geophone (bottom) located 1 meter from the calling male. The oscillograph shows the change in amplitude of the signal over time. Horizontal time bar = 10 milliseconds. (*Source: Narins 1990.*)

Communication by Salamanders

Chemical, visual, and tactile signals dominate the communication behavior of salamanders. These animals are severely near-sighted (Roth et al. 1992), and the distance over which visual signals can be detected is limited. Most communication occurs at close range, especially during courtship and aggressive interactions. The communication behavior of most salamander families has not been studied in detail. Ambystomatids have been studied mainly during courtship and mating; some of their behavior is discussed in Chapter 12 and is reviewed by Arnold (1977) and Mathis et al. (1995). The behavior of plethodontids and salamandrids has been studied in most detail, and we will focus our discussion on those families.

Plethodontid salamanders are active mainly at night and make only limited use of visual signals. Red-backed salamanders (*Plethodon cinereus*) signal aggression and submission by elevating or flattening the body (Mathis et al. 1995). Chemical signals are much more important. All salamanders have two types of chemical-sensing organs in their noses, each with neurons projecting to different parts of the brain. The main olfactory epithelium is stimulated by small, volatile molecules (airborne odors), whereas the vomeronasal epithelium is stimulated by larger, nonvolatile molecules. Plethodontids frequently tap their snouts on the substrate. This behavior facilitates uptake of nonvolatile chemicals by the nasolabial grooves, which run from the upper lip to the corners of the external nares (Figure 11–4). Liquids move through the nasolabial grooves by capillary action and are directed primarily to the vomeronasal organs, not to the main olfactory epithelium (Dawley and Bass 1989). The vomeronasal organ of plethodontids is thickest in the region adjacent to the nasolabial grooves. In at least one species, *Plethodon cinereus*, the organ also is much larger in males than in females (Dawley 1992).

Plethodontid salamanders use their chemical senses to detect prey, for courtship and other social interactions, and to identify home sites. The chemical structure of pheromones used by plethodontids for communication has not been determined for any species, but they are derived mainly from secretions of specialized glands in the head and cloacal region, and possibly from the skin (Houck and Sever 1994). Some species defend territories around retreat sites and mark their territories with pheromones deposited

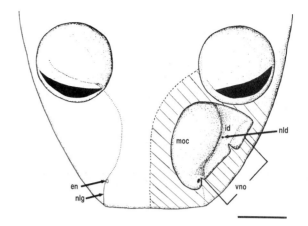

Figure 11–4   **Diagram of the nasal organ of the red backed salamander (*Plethodon cinereus*).** The dashed lines shows where the skin and skull were removed to reveal the nasal organ. Abbreviations: en, external nares; ld, lateral diverticulum containing the vomeronasal organ; moc, main olfactory chamber; nld, nasolacrimal duct; nlg, nasolabial groove; vno, vomeronasal organ. Horizontal bar, 1 mm. *(Source: Dawley and Bass 1989.)*

on fecal pellets by glands in the cloacal region (Mathis et al. 1995). Red-backed salamanders not only can distinguish their own territories from those of other individuals, they can also distinguish the odors of different individuals of their own species. Territory owners tend to be more aggressive toward strangers than toward familiar neighbors, presumably because strangers are more likely to try to take over the territory (Jaeger 1986). Females deposit fecal pellets most quickly when exposed to substrates marked with their own odors, whereas males tend to mark most vigorously when exposed to odors of females. Females apparently can use chemical cues in fecal pellets of males to assess the quality of food the males are eating. Termites are more profitable prey than ants because they are more digestible, and females seem to associate more readily with males that have termites in their territories, perhaps because termites provide better resources for their young (see Chapter 12).

Plethodontid salamanders also are capable of detecting and responding to airborne odors. Ellen Dawley (1984b) used several large species of *Plethodon* to test sex and species recognition using chemical cues. The animals were placed in a glass Y-tube apparatus (Figure 11–5) and given a choice of odors from airstreams emanating from live salamanders of differ-

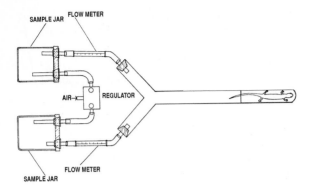

Figure 11–5   **Y-tube apparatus used for olfactory tests.** Each sample jar is used to hold salamanders of different species or sexes. The test animal is placed in the apparatus facing away from the odor sources to force it to turn around before making a choice of odor streams, making it less likely the animal will run rapidly down the tube to try to escape. *(Source: Dawley 1984b.)*

ent sexes or species. The experiments showed that the salamanders can distinguish the odors of males from those of females. Furthermore, males and females of four species of black and white salamanders once considered to be a single species (*Plethodon glutinosus, P. aureolus, P. kentucki,* and *P. teyahalee*) all preferred odors of their own species to those of other species living in the same area. The same was true when some sympatric populations of *P. jordani* and *P. teyahalee* were tested against each other. Genetic studies have shown that the interactions between these two species are complex, with some populations exhibiting extensive interspecific hybridization, whereas others appear to be completely isolated. Behavioral discrimination was strongest in those populations that do not hybridize in the field (Dawley 1987). Similar work by Paul Verrell (1989a) showed that two sympatric species in another subfamily, *Desmognathus imitator* and *D. ochrophaeus,* also can distinguish their own species from other species by means of chemical cues.

Plethodontid courtship begins with a male approaching a female and touching her with his snout, probably to determine her species and sex. Usually there is a preliminary exchange of tactile signals before the female enters into a tail-straddling walk, a behavior unique to plethodontids (Figure 11–6). The female positions her chin over the base of the male's tail. Some male plethodontids have special glands in the skin at the base of the tail that probably produce a

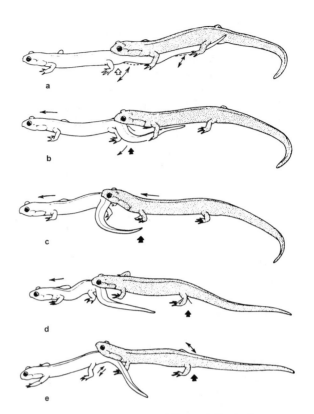

Figure 11–6   **The tail-straddling walk during courtship of a plethodontid salamander,** *Plethodon jordani.* The stippled animal is the female. After depositing a spermatophore (black arrow), the male moves forward, leading the female over the spermatophore until it is positioned beneath her cloaca. The male moves his tail to the side and extends his hind legs, pushing the base of his tail into the chin of the female. This probably helps to maintain the female in position over the spermatophore. *(Source: Arnold 1976.)*

pheromone that is transmitted to the female by way of her nasolabial grooves. The nature of these pheromones and their precise function have not been determined (Houck and Sever 1994). The tail-straddling walk also positions the female so she can use her cloaca to pick up a spermatophore deposited by the male (Arnold 1977).

Often females are not immediately responsive but require a period of courtship by the male before picking up a spermatophore. During this persuasion phase of courtship, the male transfers courtship pheromones to the female. Courtship glands used by male plethodontids are located on the chin and are called mental glands (Figure 11–7). In large species of *Plethodon,*

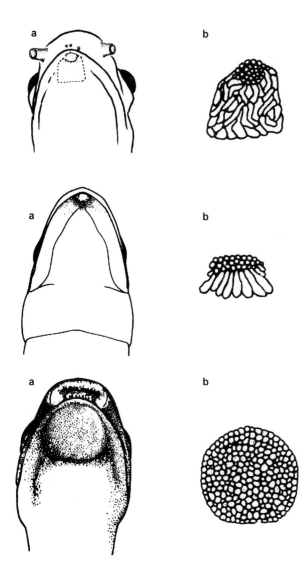

these glands are large padlike structures. The male delivers pheromones during the tail-straddling walk by turning around and slapping his chin on the female's snout (Figure 11–8). In most other plethodontids, the glands are smaller. Males develop enlarged premaxillary teeth during the breeding season that are either raked across the female's skin or used to puncture the skin when the male presses his chin against the back of the female and then flips his body away from the female ("snapping"). These actions cause glandular secretions to be applied to the wounds, delivering pheromones directly into the female's bloodstream. Two very small species, *Desmognathus wrighti* and *D. aeneus*, carry this process to an extreme. The male actually bites the female, sometimes holding on for several hours at a time to inject pheromones into her bloodstream.

A phylogenetic analysis of the evolution of mental glands and the mode of pheromone delivery in plethodontids showed that different types of mental glands and different modes of pheromone delivery apparently have evolved independently a number of times within the family (Figure 11–9). Mode of pheromone delivery roughly correlates with the type of mental gland. Only the largest species of *Plethodon* with large pad-shaped glands deliver the secretions by slapping the female's snout. All other use some form

Figure 11–7 **Mental courtship glands of plethodontid salamanders.** In each set of drawings, (a) shows the gross structure of the gland, while (b) shows the details of gland structure. Top: (a) Mental gland of *Eurycea bislineata*. The dashed line shows the portion of the gland below the skin; the small circular area is the exposed secretory portion of the gland. (b) Diagram of gland, with horizontally oriented tubules leading to secretory pores. Center: (a) Mental gland of *Desmognathus*, a small protrusion at the tip of the chin. (b) Diagram of gland showing vertical tubules and secretory pores. Bottom: (a) Padlike mental gland of *Bolitoglossa*. (b) Cross-section of gland, showing verticle tubules. *(Source: Houck and Sever 1994.)*

Figure 11–8 **Male *Plethodon jordani* slapping his mental gland on the snout of the female (stippled) during courtship.** This behavior is characteristic of salamanders with padlike glands shown in Figure 11–7. *(Source: Arnold 1976.)*

of delivery with enlarged teeth. The two very small species of *Desmognathus* in which males restrain females by biting are not closely related to each other, indicating that this type of pheromone delivery probably evolved twice. Some plethodontids have secondarily lost mental glands and the courtship behavior associated with pheromone delivery, but the reasons for this loss are unknown (Houck and Sever 1994).

Although it has long been assumed that courtship pheromones increase female receptivity, this effect has been demonstrated conclusively for only one species, *Desmognathus ochrophaeus*. Houck and Reagan (1990) treated females either with an extract of mental gland secretions or a control saline solution before allowing them to be courted by a male. Males successfully courted both experimental and control females, but females engaged in a tail-straddling walk and picked up spermatophores more quickly when treated with the pheromone. Presumably the advantage to a male in increasing female receptivity is that it reduces courtship time, thereby reducing the chances of interference by another male (see Chapter 12) or the length of time a courting pair is vulnerable to predators.

Most salamandrids are aquatic during the breeding season. Territorial behavior is rare in this family, although some newts defend small temporary territories during the mating season (see Chapter 12). Courtship interactions are complex and involve a combination of visual, chemical, and tactile cues. The male and female of some species have considerable physical contact (Figure 11–10). A male *Salamandra* clasps the female from below by looping his front legs over hers. Once the female signals her receptivity, the male deposits a spermatophore and then moves the rear of his body to the side, lowering the female over

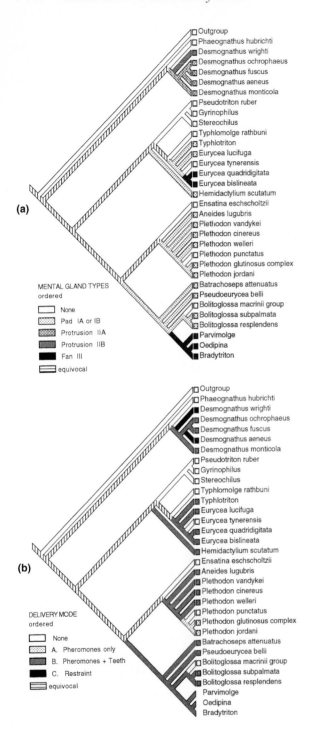

Figure 11–9 The evolution of mental glands and mode of pheromone delivery among plethodontid salamanders. The cladograms show the phylogenetic relationships of the salamanders derived from independent morphological and molecular characters. Not all species are shown. (a) Distribution of types of mental glands, as shown in Figure 11–7, in relation to phylogeny. The cluster of species at the top are the desmognathines, in which the protrusion-type glands predominate. All of the other species are in the subfamily Plethodontinae. The padlike gland may have been ancestral for this subfamily, but the protrusion type has evolved in some species of *Eurycea*, and some species have secondarily lost their mental glands and the associated courtship behaviors. (b) Distribution of modes of pheromone delivery in relation to phylogeny. Most plethodontids deliver pheromones by raking or biting the female with enlarged teeth and depositing glandular secretions into the wounds. *(Source: Houck and Sever 1994.)*

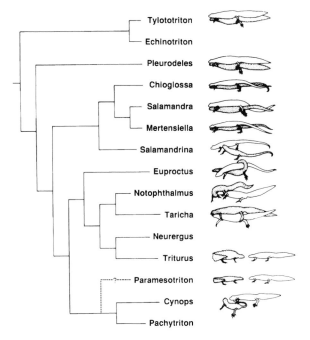

Tylototriton	
Echinotriton	
Pleurodeles	
Chioglossa	
Salamandra	
Mertensiella	
Salamandrina	
Euproctus	
Notophthalmus	
Taricha	
Neurergus	
Triturus	
Paramesotriton	
Cynops	
Pachytriton	

Figure 11–10 **Phylogenetic relationships of the genera of salamandrids, derived from morphological characters, with types of amplexus and courtship behavior shown.** The male is stippled in each drawing. The male does not clasp the female during courtship in Old World newts (*Triturus, Paramesotriton, Cynops*). *(Source: Halliday 1990.)*

the spermatophore. Male North American newts (*Notophthalmus* and *Taricha*) clasp the female from above and hold on only during the initial persuasion stages of courtship, releasing the female to deposit a spermatophore. The male of two genera of Old World newts (*Triturus* and *Paramesotriton*) does not clasp the female at all (Halliday 1990).

The aquatic newts are the only salamandrids in which the functions of signals have been studied experimentally. Generally courtship begins with a male approaching a female, attracted either by the movements of the female or by chemical signals. Experiments using an aquatic Y-tube apparatus have shown that male newts are attracted to chemicals emanating from females, even in the absence of visual cues (Dawley 1984a). These chemical cues can be used by males to identify both the species and the sex of potential mates (Cogalniceanu 1994).

Once the male and female have come together, there usually is a prolonged period of courtship in

which the male stimulates the female with chemical, visual, and tactile signals. Male newts from western North America (*Taricha*) court females by rubbing a chin gland against the female's snout, whereas a male of eastern North American newts (*Notophthalmus*) clasps a female around her neck with his hind feet while rubbing her with glands on the sides of his head (see Figure 12.7 in Chapter 12). Males of Old World newts (*Triturus*) position themselves in front of a female and waft pheromones produced in the cloacal region toward the female with movements of the tail. The importance of chemical signals has been demonstrated for *Triturus* in experiments in which the olfactory bulbs of females were cut or the nares blocked. Females that received these treatments showed little or no response to male courtship (Belvedere et al. 1988). The most elaborate courtship occurs in some of the largest species of newts, which have mating systems in which many males display in the same area and highly sexually dimorphic display structures (see Chapter 12).

Females of most species of *Triturus* signal receptivity after initial courtship by approaching the male, and the male responds by retreating from the female while continuing to display and direct pheromones toward her (Figure 11–11). The retreat display apparently allows the male to assess the female's level of receptivity, as shown by her willingness to follow him, and it may also enable a male to lead a female away from rival males that might interfere with courtship (see Chapter 12). Eventually the male turns around and moves away from the female. If the female follows, the male will stop, deposit a spermatophore, and then move forward just enough to allow the female to move over the spermatophore. The male then stops the female's movements with a braking action of his tail. Both visual and chemical signals probably are important in inducing females to follow males in the spermatophore-deposition phase of courtship. Females of some species have been observed to sniff at the male's cloaca while following. A male may also wiggle the tip of his tail, displaying a brightly colored spot that attracts the female (Halliday 1990).

Why is courtship by salamanders so elaborate? Complex courtship probably is not needed for species recognition, which is accomplished rapidly by means of chemical cues. Stevan Arnold (1977) and Tim Halliday (1990) have suggested that extended courtship is necessary because the difficulties associated with transferring spermatophores to the female require

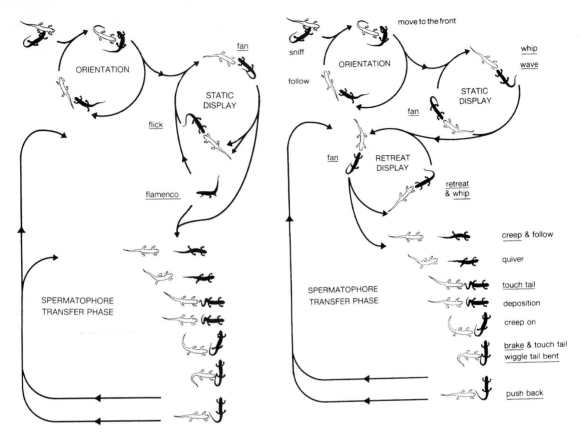

Figure 11–11   **Courtship behavior of two species of European newts.** The male is shown in black. (left) Courtship of *Triturus boscai*. The male orients toward the female and then performs a series of static displays involving flicking or fanning his tail toward the female. Males sometimes perform another dancelike visual display called flamenco. Eventually the male moves away from the female, deposits a spermatophore, and moves foward, stopping the female with his tail when she is positioned over the spermatophore. The male then turns and may push the female back to place her cloaca over the spermatophore. (right) The more complex courtship of *Triturus vulgaris*. Many elements are similar to courtship of *T. boscai*, but in this species the static display is followed by a retreat display in which the male fans or whips his tail while backing away from the female. *(Source: Arntzen and Sparreboom 1989.)*

close coordination between the male and female. Male salamanders appear to have a limited supply of spermatophores, so anything the male can do to increase the reliability of spermatophore transfer will be advantageous. The prolonged period of courtship persuasion seen among plethodontids and salamandrids represents an adaptation to improve the chances of spermatophore transfer by first ensuring that the female is ready to receive them. For example, female smooth newts (*Triturus vulgaris*) are most likely to pick up a spermatophore if they are courted vigorously by males. Male red-spotted newts (*Notophthal-*

mus viridescens) skip the preliminary stages of courtship if females are very receptive and use more elaborate courtship if they are not.

Communication by Anurans

In contrast to salamanders, most frogs and toads make little or no use of chemical signals for communication. Indeed, no experimental studies have demonstrated the use of chemical signals by anurans. Frogs also make only limited use of visual signals. Females of a South American dendrobatid frog, *Colostethus trinitatis*, pulsate their bright yellow throats when challenging intruders in their territories. Males of this species turn pitch black when calling and jump up and down on prominent display perches to make themselves conspicuous (Wells 1980). Males of several ranid frogs from Borneo in the genus *Staurois* have white or light blue webbing on their hind feet, which contrasts with the green of the surrounding forest. A displaying male periodically extends one of his back legs and spreads the webbing, producing a conspicuous flash of light color. These frogs live along streams with noisy rapids and waterfalls, where vocal communication may be difficult, and the visual display probably attracts the attention of females or other males (Harding 1982). Some frogs signal aggression by changes in posture or movements of their limbs. For example, territorial males of the South American frog *Atelopus varius* threaten other males by raising their front feet and waving them in a circle, usually as a prelude to pouncing on an opponent (Crump 1988).

Calling and Call Production.

Vocalizations are by far the most important communication signals for most frogs. Sound production is closely linked to respiration. Frogs force air into their lungs with positive pressure exerted by throat muscles. When muscles in the trunk region are contracted, air is forced out of the lungs and into the buccal cavity. Only males of most anuran species produce calls, and the trunk muscles of males are much larger than those of females. The muscles are richly supplied with mitochondria, the organelles where aerobic respiration takes place, and with capillaries that supply oxygen to the mitochondria. These muscles also contain energy stores in the form of lipids and carbohydrate (glycogen), which support the high energetic cost of calling (see Chapter 6).

The airstream produced by contraction of the trunk muscles moves through the larynx (Figure 11–12), causing the vocal cords and associated cartilages to vibrate at a characteristic frequency. This vibration determines the frequency characteristics (pitch) of the call. Most frogs have vocal sacs that couple the buccal cavity to the air (Figure 11–13). Frogs with large vocal sacs have much louder calls than those that have small vocal sacs or that lack vocal sacs altogether. Pipid frogs, which are strictly aquatic, lack both vocal cords and vocal sacs and have a unique

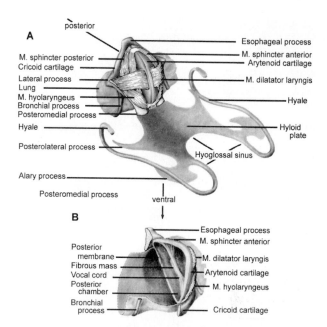

Figure 11–12 **Structures associated with vocalization by anurans.** (a) Diagram of the layrngeal apparatus of an anuran, showing the hyoid apparatus and the cartilages and muscles of the larynx. The dilator muscles open the cartilages of the larynx to let through an airstream from the lungs. The sphincter and hyolaryngeus muscles close the larynx. (b) Cross section through the larynx, showing the position of the vocal cords that vibrate to produce sound. (*Source: Duellman and Trueb 1986.*)

method of producing sound. All species in this family have clicklike calls that are produced when the two sides of the larynx, which is modified into a cartilaginous box, are suddenly pulled apart by contractions of laryngeal muscles (Yager 1992b).

Frog calls come in many different forms: clicklike calls with a broad frequency spectrum, tonelike calls with a narrow frequency spectrum, whistles that change in frequency, and a variety of squawks and trills composed of a series of distinct sound pulses (Figure 11–14). For frogs other than pipids, the dominant frequency of the call (i.e., the frequency of highest sound energy) is determined mainly by the mass of the vocal cords and the size of the buccal cavity. Large frogs usually have low-frequency calls. Tonelike calls are produced when the larynx is opened gradually, passing an airstream over the vocal cords and causing them to vibrate at a constant frequency. In contrast, clicklike calls are produced by very rapid opening of the larynx, sending a burst of air across the vocal cords. Whistled calls are produced by changes in the

(a)

(b)

(c)

(d)

Figure 11–13   **Different kinds of vocal sacs in frogs.** (a) Median external vocal sac of *Hyla ebraccata*. (b) Large median external vocal sac of *Physalaemus pustulosus*. (c) Internal vocal sac of *Rana clamitans*. (d) Paired lateral vocal sacs of *Rana virgatipes*. *(Photographs (a–c) by Kentwood Wells, (d) by Mac Given.)*

tension of the vocal cords through contraction of laryngeal muscles. Many frogs have calls made up of a series of distinct sound pulses (Figure 11–14). Relatively low pulse rates usually are produced when the laryngeal muscles actively open and close the larynx, allowing bursts of air to pass over the vocal cords. Calls with very high pulse rates are produced when the airstream passing through the larynx causes the edges of the laryngeal cartilages to vibrate.

Advertisement Calls and Chorusing Behavior.
Most species of anurans have several distinct types of calls that are used in different behavioral contexts (Gerhardt 1994). The calls given most frequently by male frogs during the breeding season are termed advertisement calls because these signals often serve the dual function of attracting mates and advertising a male's ownership of a territory or calling perch to other males. The role of these calls in attracting females has been known for many years, but only recently has it become clear how much information these calls convey to other males in a chorus. Males of some species use the intensity of their neighbors' calls to assess the distance between males in a chorus (Wilczynski and Brenowitz 1988). Other species use differences in the pitch of calls to assess the body sizes of potential competitors, and this in turn can affect

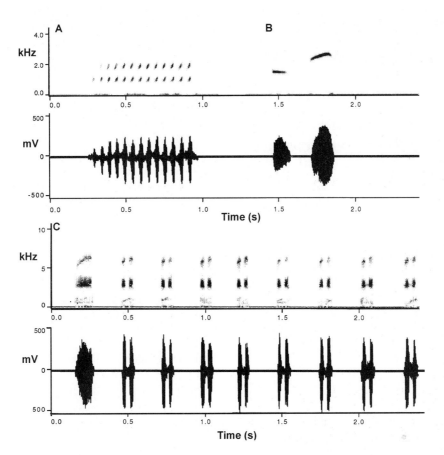

Figure 11–14 **Examples of calls of anurans.** Sonagrams on top of each part show changes in frequency (kHz) over time. Oscillographs on bottom show changes in amplitude (mV) over time. (a) Pulsed advertisement call of *Hyla versicolor*. (b) Two-note advertisement call of *Eleutherodactylus coqui*, with frequency-modulated second note. (c) Multinote advertisement call of *Hyla microcephala*, composed of a buzzlike introductory note and eight secondary click notes.

the outcome of fights (Wagner 1989). Male bullfrogs (*Rana catesbeiana*) can even use subtle differences in call structure to identify other males individually (Davis 1987).

The advertisement calls of most frogs and toads consist of a simple whistle, trill, or other type of note repeated many times in succession. However, some frogs have more complex calls in which several different types of notes are combined. For example, a small Panamanian tree frog, *Hyla ebraccata*, has an advertisement call composed of a buzz-like introductory note followed by one or more secondary click notes (see Figure 11–17). Males calling from relatively isolated positions give mostly single-note calls, but they add

secondary notes to calls when calling in dense choruses. Females are more likely to be attracted to more complex calls (Wells 1988). The multinote calls of *Hyla microcephala* are even more impressive, sometimes containing up to 30 notes (Figure 11–14). Males of other species, such as *Hyla versicolor*, increase the duration of call notes as chorus density increases. In general, longer or more complex calls require more energy to produce or make males more conspicuous to predators, so it makes sense for males to give these calls only when competition for females is intense.

Chorusing frogs often adjust the timing of their calls to reduce acoustic interference with calls of their neighbors (Figure 11–15). Male *Hyla microcephala* fur-

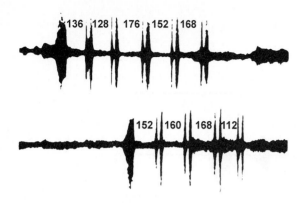

Figure 11–15 **Call note alternation by a Panamanian tree frog, *Hyla microcephala*.** The figure shows two traces from an oscilloscope with the multinote calls of two neighboring males in a chorus. The numbers give the duration of gaps between notes in milliseconds. The two males alternate notes within calls so that their calls do not overlap. Males further reduce the chances of acoustic interference by increasing the duration of the interval between notes when interacting with other males in the chorus. *(Source: Schwartz and Wells 1985.)*

ther reduce the chances of notes overlapping by increasing the gaps between call notes, responding in only a few thousandths of a second to the notes of another male (Schwartz and Wells 1985). Peter Narins (1992) has found that males of the Puerto Rican coquí (*Eleutherodactylus coqui*) are remarkably precise in their ability to avoid call overlap with neighboring males. They can insert their own calls into silent periods only three-quarters of a second long, even when the occurrence of these periods is completely unpredictable. Males respond only to a few close neighbors, with the rest of the chorus simply contributing to the general background noise. Clearly a frog chorus is not the disorganized cacophony it first appears to be.

The adaptive significance of call alternation was clarified in experiments by Joshua Schwartz (1987) with *Hyla versicolor*, *Hyla ebraccata*, and *Pseudacris crucifer*. He tested three hypotheses: First, call alternation makes a male's calls easier to locate in a chorus. Second, call alternation allows males to communicate more effectively with other males in the chorus. Third, call alternation reduces interference that can obscure key features of calls used by females in species recognition.

Schwartz did not find any evidence to support the first hypothesis. Males of all three species gave more aggressive calls when presented with playbacks of calls that did not overlap their own calls. This suggested that males have some difficulty hearing other males when they are calling themselves. Call notes of the two *Hyla* have a pulsed structure, and the distinctive pattern of pulse timing is disrupted if calls are overlapped. When females were presented with a choice of overlapping or nonoverlapping calls from two sets of loudspeakers, they showed a strong preference for nonoverlapped calls. In contrast, female spring peepers (*Pseudacris crucifer*) were equally likely to approach overlapped or nonoverlapped calls. The calls of this species are simple tones that lack an internal temporal structure, so overlapping of calls does not interfere with call recognition by females. Hence, the second hypothesis was supported for all three species, and the third hypothesis was supported for the two species of *Hyla* but not for *Pseudacris*.

Advertisement Calls and Species Recognition. Herpetologists have long recognized that each species of frog has a call distinct from that of sympatric species. In fact, many new species of frogs have been discovered because they have calls that differ from calls of other species that are very similar in appearance. For example, the relationships of North American leopard frogs were not fully understood until differences in their calls were discovered. Other species that look very similar, including the Puerto Rican frogs *Eleutherodactylus coqui* and *E. portoricensis* and the North American gray tree frogs *Hyla versicolor* and *H. chrysoscelis*, were first recognized as distinct species because of differences in their calls.

Dozens of experimental studies have now demonstrated that females recognize the calls of their own species. In choice tests, females invariably prefer the calls of their own species to the calls of a different species. The calls of two species often differ in several ways, such as dominant frequency, duration, pulse repetition rate, or pulse shape. One cannot simply assume that all of these differences are important, because some features of calls may be irrelevant to females. Only experiments in which different features of calls are varied independently can show definitively which ones are crucial for species recognition.

Both temporal and frequency information in calls can be important for species recognition (Gerhardt

1994; Gerhardt and Schwartz 1995). Differences in pulse repetition rate often are sufficient for females of species with pulsed calls to distinguish calls of their own species from those of other species. Other features of calls, such as dominant frequency, call repetition rate, and loudness, can be varied considerably and females will still choose calls with the correct pulse rate. However, this does not necessarily mean that females ignore other call features. For example, the calls of two North American tree frogs, *Hyla versicolor* and *H. chrysoscelis*, differ not only in pulse rate but also in pulse shape, which is a function of the way in which the larynx is opened during sound production. Females of *H. versicolor* show strong preferences for calls with the correct pulse rate, but if pulse rate is held constant, they prefer calls with the pulse shape of their own species to those with pulses similar to *H. chrysoscelis* calls.

In general, frogs are most responsive to calls with peak frequencies similar to those of their own species' calls. This was first demonstrated in pioneering work by Robert Capranica (1965), who investigated the acoustic properties of bullfrog (*Rana catesbeiana*) calls required to elicit vocal responses from males. Bullfrog advertisement calls have two frequency peaks, one at about 200 Hz (Hertz = cycles per second) and another at around 1,400 Hz. Capranica found that electronically synthesized calls with both frequency bands present were much more effective in eliciting calls from males than were calls from which either the high- or low-frequency peak was absent. Subsequent work by Carl Gerhardt (1988) on the green tree frog (*Hyla cinerea*) showed that females preferred calls with one peak of energy at about 900 Hertz and another between 2,700 and 3,300 Hertz, which mirrors the structure of their own species' calls. When calls were presented at low intensity, however, females did not discriminate between calls with both spectral peaks present and those with high frequencies absent. High frequencies would tend to fade out faster than low frequencies in a natural chorus, so the low-intensity playback simulated the calling of a male some distance away. Low frequencies appear to be sufficient to attract females to a distant chorus but are not sufficient for species recognition at close range.

Neural Basis of Call Recognition.   The highly selective way in which frogs respond to their own species calls is related to features of the auditory sys-

tem. Robert Capranica's work on bullfrogs showed that the neurons that transmit information from the ear to the brain are most sensitive to the frequencies of the bullfrog's call. Subsequent work has shown this to be a general pattern in many anuran species. Frogs have two sensory organs in the middle ear that are tuned to different frequency ranges. The amphibian papilla detects relatively low-frequency sounds, whereas the basilar papilla is sensitive to higher frequencies. Large frogs, such as bullfrogs, often have calls that stimulate both sensory organs, and this accounts for the greater response elicited by calls that include both high- and low-frequency sounds. Smaller frogs, such as spring peepers (*Pseudacris crucifer*), have higher-pitched calls that stimulate only the basilar papilla. These organs respond to a broad range of frequencies, but they respond best to sounds in a relatively narrow frequency range that usually corresponds to frequencies of the advertisement call (Figure 11–16). The Puerto Rican coquí (*Eleutherodactylus coqui*) has a two-note call—*co-quí*. Female coquís have auditory fibers tuned to the frequency of the second note of the call (*quí*), but males lack a peak of sensitivity in this range. Behavioral experiments showed that the *quí* note is essential for attracting females (Narins and Capranica 1978). Neurons in the central auditory pathway in a frog's brain also are selectively tuned to frequencies found in the advertisement call, which further enhances the ability of frogs to recognize their own species' calls (Hall 1994).

For many species of frogs, the frequency sensitivity of the ear determines the range of frequencies the frog can hear best. Species recognition is based mainly on temporal features of calls, such as pulse repetition rate. The middle ear organs and the nerve fibers that carry information to the brain are not sensitive to particular pulse rates. Instead, auditory nerve fibers fire in synchrony with each burst of sound in the call. Temporal information is processed in parts of the brain where there are neurons programmed to detect certain pulse rates and filter out others (Hall 1994). For example, the temporal processing centers in the brains of *Hyla versicolor* respond more strongly to pulse repetition rates characteristic of this species' calls than to the much faster pulse rates of *Hyla chrysoscelis* calls.

Aggressive Calls and the Defense of Calling Sites.
Male frogs often defend territories or calling perches

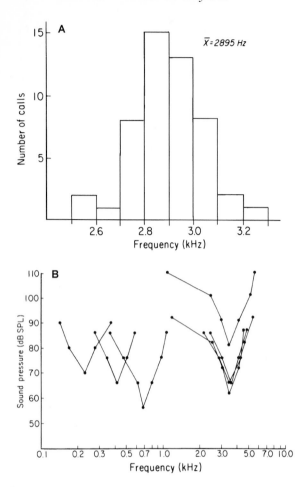

Figure 11–16 **Call frequency and tuning of the ear in the spring peeper (*Pseudacris crucifer*).** (a) Distribution of dominant frequencies in a sample of calls of males from a population in Ithaca, New York. The average dominant frequency for the population is 2,895 Hertz. (b) Tuning curves of the auditory nerve of a male from the same population. Each point indicates the sound pressure level (SPL) of sound required to elicit a firing response from a nerve fiber. Each curve is for a different nerve fiber. The best frequency to which a fiber is tuned is at the lowest point on the curve. The group of nerve fibers on the right are those that innervate the basilar papilla, the organ that detects high frequency sounds. In this frog, the best frequency of this organ is about 3,500 Hertz, slightly above the average frequency of calls in the population. *(Source: Wilczynski et al. 1984.)*

against other males (see Chapter 12), and they often have specialized aggressive calls. These calls usually differ from advertisement calls in temporal features such as duration or pulse repetition rate rather than in frequency, but there is no particular structure common to all aggressive calls. The aggressive call of the spring peeper (*Pseudacris crucifer*) is a long trill, which is readily distinguished from the short peep of the advertisement call (Figure 11–17). Aggressive calls of the Panamanian tree frogs *Hyla ebraccata* and *H. microcephala* are similar to advertisement calls, but pulse repetition rates of the introductory notes are much faster. Males of the Puerto Rican coquí have aggressive calls composed of an introductory *co* note and a series of secondary *quí* notes (*co-quí-quí-quí-quí*). Aggressive calls often form a graded signaling system in which changes in the duration or complexity of the calls are related to changes in aggressive behavior. For example, in both *Hyla ebraccata* and *H. microcephala*, males greatly increase the length of the introductory notes of aggressive calls as the distance between opponents decreases (Wells 1988). In spring peepers, the length of the trill increases as males move closer together (Schwartz 1989). Male *Eleutherodactylus coqui* give short aggressive calls when defending calling perches and longer calls when defending retreat sites or nests with eggs (Stewart and Rand 1991).

Other Types of Calls.   Some frogs have courtship calls that are given in close-range interactions with females (Wells 1988). Sometimes these are simply modified versions of the advertisement call. Male *Hyla versicolor* give very long calls when females are nearby. Male *Hyla ebraccata* and *H. microcephala* give a rapid series of single-note calls. These longer or more repetitious calls make it easier for females to locate a calling male in a noisy chorus. Males of other species give courtship calls that are quite distinct from the advertisement call. Females of a few species, such as the carpenter frog (*Rana virgatipes*), respond to males by giving soft calls of their own (Given 1993). Other types of anuran calls include release calls given by males or unreceptive females when they are clasped by males and defensive calls given by frogs confronted by predators. The latter probably startle predators, but there is no evidence that they function in intraspecific communication.

Communication by Turtles

Most information about the communication behavior of turtles comes from studies of courtship, and to a lesser extent, studies of aggressive interactions. Courtship by aquatic turtles usually is initiated by the male, probably attracted by the movements of a female. Generally the male pursues the female and

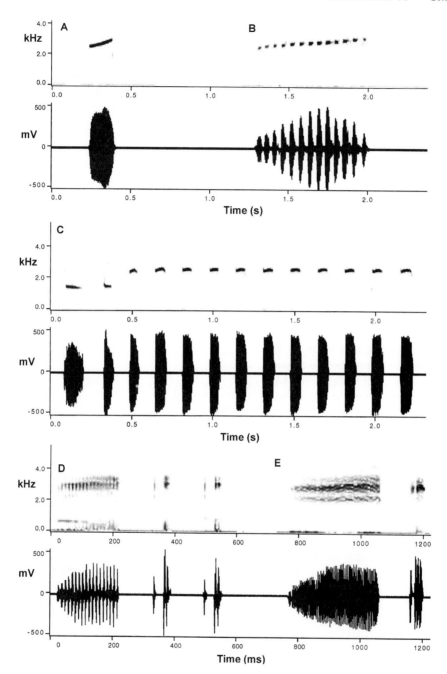

Figure 11–17 **Examples of advertisement and aggressive calls of anurans.** Sonagrams and oscillograms as in Figure 11–14. (a) Advertisement call of *Pseudacris crucifer*, a frequency-modulated peep. (b) Aggressive call of *P. crucifer*, an amplitude-modulated trill. (c) Multinote aggressive call of *Eleutherodactylus coqui* (compare with advertisement call in Figure 11–14b). (d) Advertisement call of *Hyla ebraccata*, with long introductory note and secondary click notes. (e) Aggressive call of *H. ebraccata*, with a longer duration and higher pulse rate of the introductory note.

sniffs her cloacal region, presumably using chemical cues to identify the species and sex of prospective mates. Male green sea turtles (*Chelonia mydas*) respond differently to breeding and nonbreeding females, suggesting the presence of some chemical indicator of reproductive condition (Crowell Comuzzie and Owens 1990). Often a male's biggest problem is getting the female to stay still, because it is difficult for a male turtle to mount the shell of a moving female. A common technique for immobilizing the female is to bite at her shell, head, and legs, evidently an ancestral form of courtship behavior that has been observed in many species of turtles (Jackson and Davis 1972b; Carpenter and Ferguson 1977). Unreceptive females usually reject courting males simply by swimming away, but females may resort to aggression if males are particularly persistent.

Courtship by some turtles, such as kinosternids, consists of little more than the male chasing and biting the female, but other turtles use more elaborate tactile signals. In general, turtles that swim in the water perform more courtship actions before mounting the female than do turtles that walk on the bottom; the latter often mount the female before performing most courtship actions (Bels and Crama 1994). Bottom-walkers include Blanding's turtle (*Emydoidea blandingi*), which performs several courtship actions while mounted on the female, including rubbing the chin on the snout of the female, swaying the head back and forth over the female's head while mounted on her back, and expelling water from the mouth and nostrils over the snout of the female (Baker and Gillingham 1983). Swimming turtles include some chelid turtles, in which males and females engage in mutual head-bobbing and rub together barbels on their chins (Murphy and Lamoreaux 1978). Other sorts of head rubbing or mutual head-bobbing by a male and female have been observed in a variety of other turtle species.

Some of the most elaborate courtship behavior by aquatic turtles occurs in emydid turtles in the genera *Chrysemys*, *Trachemys*, *Pseudemys*, and *Graptemys*. Males swim toward females and face them head-on, displaying species-specific patterns of colored stripes and blotches on their front legs and necks. Males of most species also have elongated claws on the front feet that are used to stroke or tap the female's head during courtship (Figure 11–18). This behavior has been termed titillation and appears to increase the receptivity of a female. A male painted turtle (*Chrysemys*) or slider (*Trachemys*) approaches a female

Figure 11–18 **Courtship of aquatic turtles.** Top: Courtship of the red-eared slider (*Trachemys scripta elegans*), with the male (the smaller individual) approaching the female from the front. Bottom: Courtship of the Florida cooter (*Pseudemys floridana*), with the male approaching the female from above. (*Source: Oliver 1955.*)

from the front, extends his front feet, and tap his claws on the sides and top of the female's head for a variable length of time until the female is ready to mate (Jackson and Davis 1972b). Similar behavior has been described for map turtles (*Graptemys*), but a male of some species in this genus stimulates the female by rubbing or vibrating his head against the side of the female's head. Male cooters and red-bellied turtles (*Pseudemys*) approach females from above, extend their front feet over the female's head, and rapidly move their elongated claws on or in front of the female's head (Jackson and Davis 1972a).

Tortoises (family Testudinidae) have more complex communication than other turtles, with a rich repertoire of visual, tactile, chemical, and acoustic signals (Auffenberg 1977; Ruby and Niblick 1994). As in other turtles, courtship is initiated by the male, attracted by either the movement or scent of a female.

There is no clear evidence that male tortoises produce advertisement signals to attract females, but males of at least one species call in choruses in rainy weather, and males sometimes vocalize while following females. Males generally investigate any tortoise-shaped object by sniffing. Most tortoises apparently use cloacal odors to determine the species, sex, and reproductive condition of other individuals. Males and females of the North American genus *Gopherus* have subdentary glands on their chins. Males have larger glands than females, and dominant males have larger glands than subordinates. The glands become enlarged during the breeding season, especially in males, and secretory activity increases. Tortoises sometimes rub glandular secretions onto enlarged scales on the front feet and extend the feet to allow the courting partner to sniff them (Figure 11–19). They also appear to broadcast chemical signals by elevating the chin and moving the head up and down. Gland secretions contain more than a dozen protein compounds of different molecular weights, as well as several kinds of lipids and volatile free fatty acids. Individual variation in composition of these secretions may allow individual recognition, which in turn would facilitate the establishment of a stable dominance hierarchy in a local population (Alberts et al. 1994).

A male tortoise usually responds to the sight of another tortoise by moving his head up and down or side to side. Other males often respond with reciprocal head movements, eventually leading to fighting, whereas females usually do not. These head movements are similar to olfactory movements used by tortoises to sample airborne odors, and these movements may have evolved into communication signals. Walter Auffenberg (1965) hypothesized that the ancestral mode of sex and species recognition by tortoises was based entirely on olfactory cues, which also are used to locate food. He proposed that some species secondarily evolved visual signals to identify males. In a study of sex and species recognition by two closely related South American species, *Geochelone denticulata* and *G. carbonaria*, Auffenberg found that horizontal head movements by males are slightly different in the two species. *G. denticulata* moves the head to the side and then back to the midline in one smooth movement, whereas *G. carbonaria* moves the head sideways and back in a jerky motion. Females are initially detected visually, but sex and species identity are confirmed by chemical cues. When empty tortoise shells were smeared with cloacal odor from females of the appropriate species, males attempted to mate with

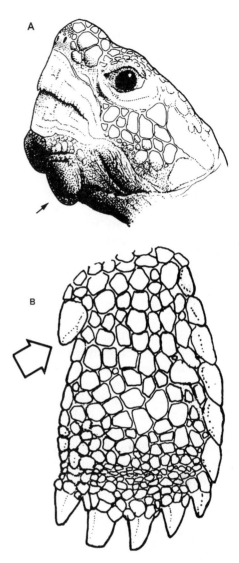

Figure 11–19 **Structures associated with courtship by tortoises.** (a) Subdentary glands of an adult male tortoise (*Gopherus berlandieri*) which enlarge during the breeding season. (b) Enlarged claw on front foot of a male tortoise (*Gopherus polyphemus*) used to rake the subdentary glands and present pheromones to females. *(Source: Auffenberg 1977.)*

them. Auffenberg even observed one male trying to mate with a head of lettuce that a female had just crawled over.

Once a male has identified a sexually active female, he may follow her for many days or proceed immediately to courtship. Courtship by tortoises is not subtle; a male attempts to immobilize a female by repeatedly biting her on the shell, head, and legs, and by ramming her with an enlarged bony process at the front of his plastron (epiplastral ramming) (Figure 11–20). This behavior may go on for hours until the female signals her receptivity by lifting the rear of her shell and exposing her cloaca. A female can reject a male's advances by simply dropping the rear of her shell to the ground. After a variable period of biting and ramming the female, the male will attempt to mount. He may then deliver an additional series of tactile signals in which he repeatedly slams an enlarged area at the rear of the plastron against the shell of the female (xiphiplastral ramming) (Auffenberg 1977).

Male tortoises often are very aggressive toward one another, especially when they are competing for females. Aggressive behavior includes epiplastral ramming like that used in courtship, hooking the epiplastral extension under the shell and attempting to overturn an opponent, and biting. Communication signals used in aggressive encounters include the head-bobbing displays described earlier, open-mouthed threat displays, and dominance displays in which males elevate their heads as high as possible above the ground. In some cases, these displays are accompanied by vocalizations (Auffenberg 1977). Aggressive encounters among giant tortoises of the Galápagos Islands are generally won by the male that can raise his head to the greatest height. Some populations of these tortoises have a saddle-shaped carapace with a notch at the front of the shell that allows the neck to be more fully extended than is the case for tortoises with dome-shaped shells (Schafer and Krekorian 1983). The saddle-shaped carapace probably did not evolve to facilitate display, however. These tortoises feed by browsing on tall plants, whereas the ones with dome-shaped shells live in a wetter environment and graze on grasses and other low vegetation.

Communication by Crocodilians

Two features of crocodilian communication set these animals apart from other reptiles. Crocodilians are the most vocal of reptiles, employing a wide variety of vocalizations for both long-range and close-range communication. They also are almost unique in the extent to which they combine acoustic, visual, and tactile signals into complex displays. Crocodilian communication signals can be divided into four general categories, based on the context in which the signals are given: advertisement displays, given by males or females, that announce ownership of a territory or attract mates; aggressive displays, usually given during encounters between males; courtship displays, given during interactions between males and females; and signals exchanged between parents and their offspring.

The American alligator (*Alligator mississippiensis*), an inhabitant of swamps, marshes, lakes, and ponds in the southeastern United States, is the best-studied crocodilian (Vliet 1989). The most conspicuous display by alligators is a loud, repetitive roar that can be

Figure 11–20 **Courtship by a tortoise (*Gopherus polyphemus*).** (a) The male walks in a circle and bobs his head toward the female. (b) The male bites the female on the shell and legs to immobilize her. (c) The male engages in vigorous biting until the female moves backward and stretches her hind legs. (d) The male mounts the female and copulates with her. *(Source: Auffenberg 1966.)*

heard nearly 200 meters away. This display attracted the attention of early explorers in the Southeast, like William Bartram (1791), who said that the roaring of alligators caused the air and water to shake and the earth to tremble. Perhaps he exaggerated, but the bellowing display does in fact include both airborne sound and waterborne vibrations. It is produced by the contraction of trunk muscles that force air through the vocal cords, producing a low-frequency sound. When males give this display, the audible bellow is preceded by subaudible vibrations (i.e., vibrations of a frequency too low to be heard by humans) that cause water droplets to dance on the alligator's back while propagating underwater waves for some distance (Figure 11–21). Females also bellow but do not produce the subaudible vibrations. Bellowing advertises an animal's ownership of a territory and attracts potential mates. Male alligators often engage in a bellowing chorus lasting for an hour or more on mornings during the breeding season.

A less frequent advertisement display is head-slapping (Figure 11–21), which is performed mostly by males. This display actually has a number of separate components that may or may not be included in every display. These include clapping the jaws together, slapping the chin on the surface of the water, subaudible vibrations, a low-pitched growl, an inflated posture in which the back and tail are held above the water, and vigorous undulations of the tail. Male alligators often respond aggressively to either bellowing or head-slapping by other males. Aggressive behavior includes chasing, lunging, and biting at the opponent, as well as a variety of visual displays. A male signals aggression by lifting his head out of the water with his tail arched above the water surface. Often this display is followed by subaudible vibrations, bellowing, and clapping the jaws together. Submission is signaled by lifting the snout out of the water at an oblique angle or by simply submerging and swimming away. Victorious males sometimes assume an inflated posture with the body high in the water.

Female alligators are attracted by these displays. A female alligator initiates courtship by placing her snout on the male's head or snout. A period of mutual head and snout rubbing follows, during which the male and female both emit a series of distinctive coughlike vocalizations. Males sometimes also produce subaudible vibrations during courtship. If the female is receptive, the male and female circle one another in the water, rubbing their heads together, moving over the back of the partner, blowing bubbles through the mouth, and blowing a stream of water into the air from the nostrils. If a female is not receptive, she often responds with a series of bellow–growl vocalizations and submerges or swims away from the male.

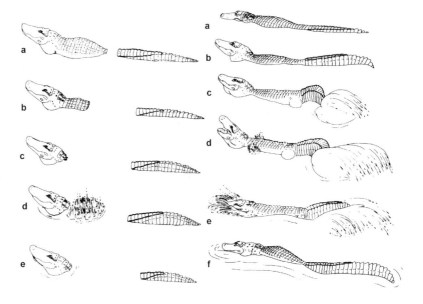

Figure 11–21 **Visual and acoustic displays of the American alligator.** Left: Bellowing display of male. (a) Inhalation. (b) Lowering. (c) Head-oblique, tail arched posture. (d) Subaudible vibrations. (e) Audible bellow. This display lasts about 5 seconds. Right: Head-slap display of male. (a) Elevated posture. (b) Head-oblique, tail-arched posture. (c) Tail wag begins. (d) Head slap with subaudible vibrations. (e) Jawclap. (f) Inflated posture. This display lasts about 35 seconds. *(Source: Vliet 1989.)*

Many of the displays described for American alligators have been observed in other crocodilians, but the structure, timing, and sequence of display components vary among species. Alligators tend to be more vocal than crocodiles, probably because they live in more enclosed, marshy habitats that cause them to be out of visual contact with other alligators (Garrick and Lang 1977). Crocodiles often live along open riverbanks or lakeshores, where close-range visual displays are more useful. Nile crocodiles (*Crocodylus niloticus*) can occur at very high densities and constantly encounter other individuals. This situation requires frequent use of submissive displays, such as snout-lifting by males approached by a larger individual or by females during the initial stages of courtship. Snout-lifting also is performed by male crocodiles during courtship, but it has never been seen in this context among alligators. In general, crocodiles tend to have more sex-specific displays (e.g., females apparently do not bellow or head slap), whereas in alligators the two sexes perform many of the same displays. This difference may be related to pronounced territorial behavior of female alligators.

Crocodilians are almost unique among reptiles in the degree to which communication signals are used in interactions between juveniles or between juveniles and adults. Baby crocodilians begin exchanging vocal signals while still inside the egg, responding to the calls of siblings in the nest. This communication may help to coordinate synchronous hatching by members of the same brood, as it does with birds. The calls attract the attention of the female, which usually remains near the nest and guards it during incubation. The female digs the hatchlings out of the nest and often transports them to a nursery pool. The young then remain with the mother for several weeks in crocodiles and up to 2 years in the American alligator. The juveniles often live together in a relatively cohesive group, responding to each other's movements with grunting vocalizations that may serve as contact calls. Juvenile crocodilians also produce distress calls that grade into the contact grunts. Experimental studies with several species have shown that distress calls elicit protective responses from parents (Lang 1989).

A poorly understood aspect of crocodilian communication is the extent to which they make use of chemical signals (Mason 1992). All crocodilians have glands under the chin that produce a musky odor, and there are scent glands in the cloacal region as well. Musky odors have been detected during bellowing displays of alligators, and an oily substance thought to be produced by the cloacal glands was observed during head-slapping displays and also was accompanied by a musky odor. In addition, there has been some speculation that mutual chin- and head-rubbing during courtship of many species is related in part to detection of chemical signals. Experiments using both behavioral assays (rate of olfactory throat pumping) and recordings from the olfactory bulbs have shown that crocodilians can detect and respond to gland secretions, but their precise function as communication signals has not been established.

Communication by Lepidosaurs

The only nonsquamate lepidosaurs are tuatara (*Sphenodon*), which are unusual in being active mainly at night. Their behavior is poorly known, but they have some visual displays similar to those of iguanian lizards, including a limited amount of color change, head-bobbing and head-shaking displays, inflation of the body and throat region, erection of dorsal crests on the head and back, and open-mouthed threat displays (Figure 11–22). Mating behavior is similar to that of lizards as well (Gillingham et al. 1995). The nocturnal habits of tuatara suggest that they also might employ chemical signals, but this has not been established experimentally. These animals have vomeronasal organs, but these are not directly connected to the oral cavity as they are in squamates, but instead open into the nasal cavity. This morphology probably represents the ancestral condition for lepidosaurs. The tongue lacks the specializations for chemosensory function seen in squamates, and tuatara do not engage in tongue-flicking behavior to sample chemicals in the environment (Schwenk 1986).

Most squamates have well-developed visual systems, and they have three distinct chemosensory systems: the olfactory system, taste buds (lost in some lineages), and the vomeronasal system (Schwenk 1995). The latter opens directly into the oral cavity and receives chemical stimuli by way of the tongue (Figure 11–23). The tongue picks up chemicals by licking the substrate or by rapid tongue-flicking in the air. Squamates generally use both visual and chemical signals for communication (geckos also use acoustic signals), but the relative importance of each mode of communication varies in different lineages and is correlated with the sensory modality used to locate food

Iguanians also use the tongue to capture prey and transport it into the mouth, which is the ancestral condition for all lepidosaurs (see Chapter 9). Use of the tongue in feeding constrains its evolution as a specialized chemosensory organ. Scleroglossans capture prey with their jaws. This evolutionary innovation freed the tongue from its ancestral role in feeding and set the stage for the evolution of enhanced chemosensory functions (Schwenk 1993). The specialization of the tongue as a chemosensory organ reaches its peak in the varanid lizards and snakes, both of which have deeply forked tongues that are used exclusively for chemoreception (Figure 11–24). Scleroglossans generally rely more heavily on chemical signals for communication than do iguanians, but they use some visual signals as well. The communication behavior of these clades is discussed in more detail in the following sections.

Visual Communication by Iguanians. Visual communication is the best-studied mode of communication by iguanian lizards (Stamps 1977). Males of many species are territorial, at least during the breeding season (see Chapters 10 and 12). A territorial lizard typically spends much of its time surveying its territory from a conspicuous perch, or patrolling the territory by moving from perch to perch. Periodically, the lizard performs a series of displays that vary from simple vertical movements of the head with the rest of the body stationary to elaborate two-legged or four-legged push-ups. The displays often are performed spontaneously rather than being directed at a particular receiver (Martins 1993). These have been called assertion displays in the lizard behavior literature, but

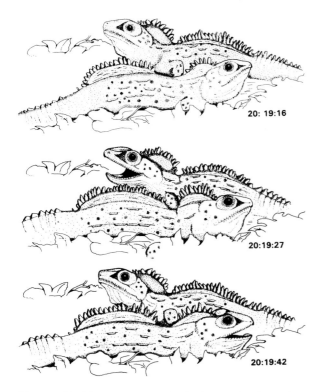

Figure 11–22 **Open-mouthed threat display of tuatara.** *(Source: Gillingham et al. 1995.)*

and with feeding behavior. Visual location of food probably is the ancestral condition for squamates, although both olfaction and taste may play a role in the identification of suitable food items. The ancestral condition is retained by iguanian lizards, and these animals rely mainly on visual signals for communication, although they make some use of chemical signals as well.

Figure 11–23 **Diagram of the chemosensory systems of squamates, shown for a generalized lizard.** *(Source: Schwenk 1995. Drawing by George Schwenk.)*

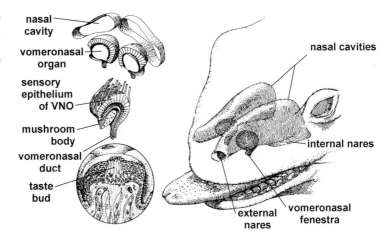

Figure 11–24 **Variation in the form of the squamate tongue.** Tongues are shown in dorsal view with the front of each tongue at the top. (a) *Xantusia* (Xantusiidae, Scincomorpha), (b) *Abronia* (Anguidae, Anguimorpha), (c) *Podarcis* (Laceridae, Scincomorpha), (d) *Coleonyx* (Gekkonidae, Gekkota), (e) *Varanus* (Varanidae, Anguimorpha), (f) *Gonocephalus* (Agamidae, Iguania), (g) *Crotaphytus* (Crotaphytidae, Iguania), (h) *Cnemidophorus* (Teiidae, Scincomorpha), (i) *Cordylus* (Cordylidae, Scincomorpha), (j) *Dasia* (Scincidae, Scincomorpha). *(Source: Schwenk 1995. Drawing by M. J. Spring.)*

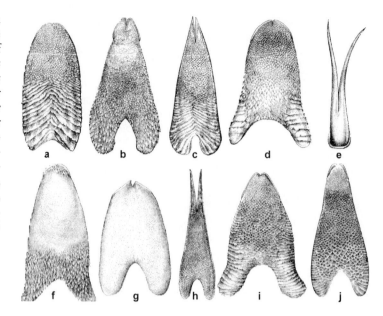

they serve the same function as advertisement displays in other animals. The displays usually include a stereotyped component known as the signature display or signature bob that encodes the species identity of the performer in the temporal pattern of the display (Figure 11–25). Species living in the same region have distinct head-bobbing displays, and there is little doubt that these displays are important in maintaining reproductive isolation among sympatric species. Other components of the display may exhibit individual or gender differences.

The advertisement displays of *Anolis* lizards have been studied in more detail than those of most other lizards (Jenssen 1977). These largely arboreal lizards are common throughout the Neotropics. They generally are cryptically colored, but they have dewlaps that can be extended to reveal bright patches of color ranging from purple and blue to red, orange, yellow, or white. Some species have elaborately colored dewlaps with spots of one color superimposed on a background of another color (Figure 11–25). Species living in the same geographic region usually have different colors on their dewlaps, except in cases where two species are very different in size, shape, or body color. It has long been assumed that these color differences are important for species recognition, and this has been demonstrated experimentally for a pair of sympatric species in the Dominican Republic, *Anolis marcanoi* and *A. cybotes*. Males of these species look very similar, but *A. marcanoi* has a red dewlap,

whereas that of *A. cybotes* is yellow or white. Jonathan Losos (1985) painted the dewlaps of each species to resemble the other species. The altered males were readily attacked by males of the other species, indicating that color is important for species recognition. In subsequent experiments, projected video recordings of displays of both species elicited displays from *A. marcanoi* males, but they synchronized their display behavior more closely with stimuli from their own species (Macedonia et al. 1994). Even stronger evidence of species recognition was obtained in similar experiments with the Jamaican species *A. grahami*. Males of this species gave more displays to playbacks of displays of their own species than to displays by other species (Macedonia and Stamps 1994).

While the advertisement or assertion displays of territorial male lizards elicit aggressive responses from other males or cause them to retreat, these same displays are attractive to females. This phenomenon was demonstrated by Thomas Jenssen (1970), who presented caged *Anolis nebulosus* females with films of displaying males projected onto tiny screens at the ends of a cage. Females were given a choice of normal displays or displays altered by playing the film backward or by adding and subtracting head bobs and dewlap extensions. The females showed a strong tendency to move toward the screen showing the normal display. Once a female lizard approaches a displaying male, he usually responds with a series of courtship displays. Courtship displays of some lizards are very

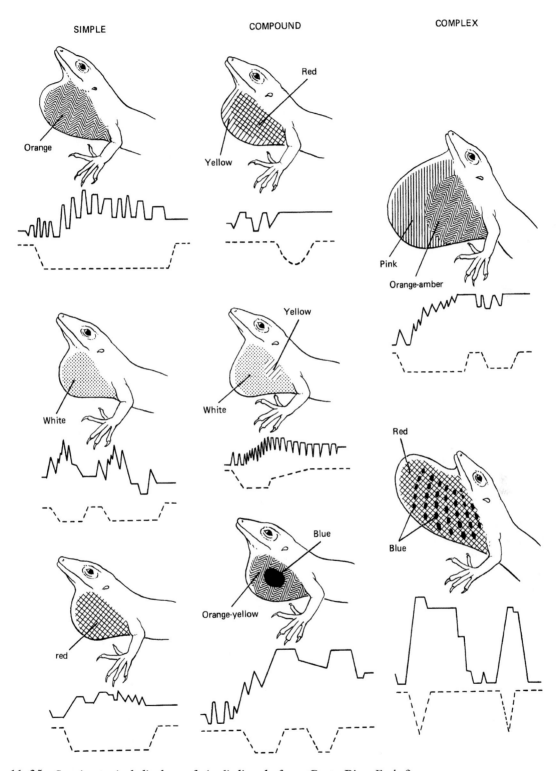

Figure 11–25 **Species-typical displays of *Anolis* lizards from Costa Rica.** Each figure shows the color of the dewlap and the pattern of head-bobbing and dewlap extension. The solid lines below each lizard show about 10 seconds of head-bobbing, with the height of the line indicating the extent to which the head is raised or lowered. The dashed lines show dewlap extension, with the dewlap being extended when the line is below the horizontal. *(Source: Echelle et al. 1971.)*

similar to those used in territorial advertisement, but the courtship display of many species has a distinctly different tempo of head-bobbing and dewlap extension. For example, male lizards in several families approach females while giving a rapid series of shallow head bobs (a display variously described as jiggling, shudder-bobbing, or courtship nodding) that appears to convey a nonaggressive message to females (Carpenter and Ferguson 1977).

These courtship displays serve the same function as the pheromonal courtship of salamanders, namely, to increase the receptivity of the female. This has been investigated in considerable detail by David Crews and his colleagues, working with the common North American anole, *Anolis carolinensis* (Crews 1980). A male *Anolis* advances toward a female while rapidly bobbing his head and periodically extending his bright red dewlap. Extended periods of courtship accelerate development of the ovaries by stimulating production of gonadotropic hormones in the pituitary. Crews demonstrated this by collecting female anoles during the winter, when they are not reproductively active, and exposing them to courting males in cages. Rates of gonadal development were much faster in females exposed to courtship than in females housed alone. Other experiments showed that movement of the dewlap during courtship was important for inducing receptivity in females, but the color of the dewlap was not. While courtship by males increased female receptivity, exposure of females to groups of males that fought among themselves actually inhibited gonadal development, even under temperature and photoperiod conditions that normally stimulate reproduction. When females were exposed first to male–male aggression and then to courtship, they switched from complete inhibition of gonadal development to rapid ovarian growth. These experiments show not only that courtship can induce physiological and behavioral changes in females, but also that males are better off if they establish territories prior to the courtship season.

In addition to using visual displays to advertise territories and court females, male lizards display to one another in aggressive interactions. The form of these displays is quite similar in many families in both the iguanian and scleroglossan clades, suggesting that evolution of aggressive displays has been relatively conservative (Carpenter and Ferguson 1977; Carpenter 1978). In general, the aggressive displays of lizards, as of most animals, increase the apparent size of an individual or exhibit weapons used in actual fighting. Common displays include raising the body high on all four legs, arching the back, inflating the trunk region, laterally compressing the trunk, inflating the head and throat or engorging these regions with blood, extending a dewlap or gular flap, lowering the head, and opening the mouth to display the teeth and an engorged tongue. Lizards in some families also move their tails during aggressive displays, and iguanians use the same sort of head-bobbing and push-up displays seen during territorial advertisement. A particularly spectacular display is given by male Australian frillneck lizards (*Chlamydosaurus kingii*), which erect a huge frill around the head while gaping the mouth. There is even an acoustic component to the display, because males sometime lash their tails against the bark of a tree, producing a rasping sound (Shine 1990).

Some iguanian lizards make more elaborate use of color in territorial advertisement and aggressive interactions than the simple display of colored dewlaps seen in *Anolis* lizards (Cooper and Greenberg 1992). Males often develop bright colors during the breeding season, and in some species, differences in color pattern reflect differences in dominance status. For example, male tree lizards (*Urosaurus ornatus*) from the southwestern United States have color patterns that vary both within and among populations. In most populations, males have an orange or yellow area on the throat with a central green or blue spot that varies in size. Throat colors of males change as they mature, a sequence that is under hormonal control. Both males and females develop orange throats first, and young males have only the orange color. In older males, the central spot progresses from green to blue or blue-green. Color is not strictly correlated with size, so some relatively large males can have green or even orange throats. Experimental manipulation of throat color demonstrated that males with more blue on their throats dominate those with less blue (Thompson and Moore 1991). Throat color is not, however, the only determinant of the outcome of fights. Larger males consistently dominate smaller individuals, even when throat color is not as blue, and males that were dominant in previous fights were more likely to win contests as well (G. C. Carpenter 1995).

In addition to throat color, males in some populations of tree lizards turn very dark or even black during the breeding season. Darkening can occur in a matter of minutes and is used by territorial males to advertise their dominance status. In a comparative study of two populations in New Mexico, Geoffrey

Carpenter (1995) found that males in a high-density population were more aggressive and exhibited a greater degree of dorsal darkening than males in a lower-density population. The males from the high-density population also had less polymorphic throat coloration, because nearly all mature males had blue throats. When throat color was similar, darker males won contests with lighter males. The degree of dorsal darkening is not fixed in these populations. When males from the low-density population were placed in enclosures at high density, they gradually assumed a darker dorsal coloration (Zucker 1994a). The dark dorsal coloration is a more effective long-distance territorial signal than the blue throat coloration because it provides more contrast with the background colors in the environment. Presumably the dark color also makes males more conspicuous to predators. Naida Zucker (1994b) has shown experimentally that dominant territorial males darken only when females are available, perhaps because the potential increase in mating success outweighs the risks.

The use of rapid color change for communication is best developed in agamids and chamaeleontids. Male rainbow lizards (*Agama agama*) have distinctive courtship, aggressive, and submissive colors, and they can change color very rapidly. Dominant males typically have bright orange heads that contrast with the dark, blue-black body coloration, whereas subordinate males assume a duller color, with less contrast between the head and body colors. Subordinates take on brighter colors as they move farther away from dominant males (Madsen and Loman 1987). Chameleons are well known for their ability to change color rapidly, but only a few studies have examined the role of color change in social interactions. Male chameleons from Madagascar change from a cryptic color pattern to much brighter colors when courting females, and they have other distinctive color patterns that are used during aggressive interactions with other males. Males of one Madagascar species, *Chamaeleo pardalis*, change rapidly from their normal green color to a pattern of bright red, orange, and yellow markings when confronting other males (Martin and Wolfe 1992).

Chemical Communication by Iguanians. The chemical signals of iguanian lizards have received much less attention than their visual signals. Many species have femoral glands located on the undersides of their back legs. These glands generally are better developed in males than in females and become more active during the breeding season, secreting a nonvolatile combination of lipids and proteins. The lizards smear these secretions on rocks and other substrates in their territories, where they are actively investigated by other lizards with tongue-flicking or licks. The femoral gland secretions of at least one species, *Dipsosaurus dorsalis*, absorb ultraviolet radiation, making them visible to lizards that are sensitive to these wavelengths. These scent marks almost certainly serve as territorial markers analogous to those of many mammals (Alberts 1989). The secretions themselves may provide a considerable amount of information about the territory owner. Alberts et al. (1993) found sufficient variation in the protein composition of femoral gland secretions from different individual green iguanas (*Iguana iguana*) to provide information about sex, individual identity, and kinship relationships, assuming that the lizards can detect these differences. In green iguanas, these chemical signals may play a role in the establishment of dominance relationships among males during the breeding season, when they defend small mating territories in close proximity to one another (see Chapter 12).

Chemical Communication by Scleroglossan Lizards. Males of some scleroglossan lizards, including some skinks and cordylids, develop bright colors during the breeding season (Cooper and Greenberg 1992), and all scleroglossans use some visual displays in aggressive and courtship interactions. Nevertheless, chemical communication seems to predominate in most scleroglossan families. Scleroglossans have well-developed olfactory systems, but most chemical signals are detected by the vomeronasal system. Chemical stimuli are sampled by touching the substrate or other lizards with the tongue. Sampling of airborne or substrate chemicals by tongue-flicking is common among teiids, lacertids, helodermatids, varanids, and snakes (Mason 1992). Rates of tongue-flicking or tongue-touching are frequently used as indices of the ability of lizards and snakes to make chemical discriminations. It is assumed that squamates convey nonvolatile chemicals to the vomeronasal organs when they lick a substrate or touch it with their tongue. Other possible modes of chemical detection have largely been ignored. Morphological work by Kurt Schwenk (1985) showed that many lizards have taste buds on the tongue or the lining of the mouth. Nearly all families examined had some taste buds, with iguanians, scincids, and cordylids all having concentrations of taste buds near the tip of the tongue. Schwenk has argued that many studies that have used tongue touching or licking as measures of chemosensory response do not distinguish between vomeronasal functioning and detection of chemical cues directly by the taste buds.

The use of chemical signals by skinks has been studied in some detail (Cooper and Vitt 1986). Male North American skinks of the genus *Eumeces* are aggressive toward other males of their species during the mating season, but they do not defend territories. Males determine the sex and species of other lizards by touching them with their tongues, and males of other species generally are ignored. When male *Eumeces fasciatus* were smeared with cloacal material from male *E. inexpectatus*, they were attacked by *E. inexpectatus* males. Despite the similar appearance of some sympatric skinks, males seldom try to court females of the wrong species, but will do so if the females are smeared with cloacal material from the male's species. Chemical information sometimes can even override other types of signals. For example, the heads of some *Eumeces* males turn bright orange during the mating season, and this color serves as an aggressive advertisement to other males. Females with heads painted orange were initially attacked by males, but were courted once the males had touched them with their tongues. Male skinks can follow chemical trails of females produced by a specialized cloacal gland (the urodeal gland) that enlarges during the breeding season.

Male gekkonids, lacertids, teiids, helodermatids, and varanids also lick or touch females with their tongues during courtship. Male varanids direct their attention mostly to the sides of the head, the region where the back legs join the body, and sometimes the female's back. Male geckos also lick a female's skin, and experimental studies of the leopard gecko (*Eublepharis macularius*) have shown that males use skin pheromones to distinguish males from females (Mason 1992). Some scleroglossan lizards mark substrates with chemical signals, a behavior also seen in iguanian lizards. For example, Komodo monitors (*Varanus komodoensis*) deposit feces in conspicuous places along regularly used trails and especially at basking sites, where they are investigated intensively by other monitors. Although this species does not defend its home range as an exclusive territory, these chemical markers may enable individuals to avoid areas currently being used by other lizards (Auffenberg 1981).

Chemical communication almost certainly is the principal mode of communication for amphisbaenians, but their behavior in the wild is almost completely unknown. There has been only one experimental study dealing with chemical communication. Cooper et al. (1994) used tongue-flicking behavior as an assay of possible pheromone discrimination by males of *Blanus cinereus*, an amphisbaenian from Spain. These animals have pores near the cloaca that might produce pheromonal secretions. Males gave more tongue flicks to extracts of cloacal pores of females than of males, and they often bit cotton swabs that contained male extracts. They did not appear to discriminate between swabs rubbed over the skin of males or females.

Acoustic Communication by Geckos. Most lizards can detect airborne sounds, but only the nocturnal geckos (family Gekkonidae, subfamily Gekkoninae) regularly use acoustic signals for communication. Geckos are unique among lizards in having vocal cords and therefore are the only lizards that can produce sounds more complex than simple hisses and gasps (Marcellini 1977). Males of many species of geckos are territorial and vocalize to advertise territory ownership and to attract females. These vocalizations have been termed multichirp calls in the gecko behavior literature, and are functionally equivalent to the advertisement calls of frogs. They generally consist of a repeated series of short notes with a relatively broad frequency spectrum. Although call structure is similar in many geckos, calls of sympatric species appear to be sufficiently different to be species-specific mate recognition signals. Most reports indicate that these calls are directed toward particular males or females, but in some cases they are given spontaneously as well. Male barking geckos from southern Africa (*Ptenopus*) emerge from burrows in the desert at dusk and call in choruses reminiscent of frog choruses (Haacke 1969). Males of some species of geckos may answer one another, but organized chorusing has not been reported. Male geckos tend to move away from speakers playing multichirp calls, but playbacks elicit no response from females. The reproductive state of the females tested is not known, so this experiment does not necessarily show that the calls are not attractive to females.

Some geckos give distinct churr calls during aggressive interaction between males. These often consist of a series of notes that are longer than those of the multichirp calls, with the calls being broken into a series of pulses. Churr calls are used both in intraspecific aggressive encounters and to threaten humans or other potential predators. Both males and females of some species also produce single-chirp calls that resemble individual notes of the advertisement call; the function of these is unclear, but they appear to be used in a defensive or aggressive context (Marcellini 1977). An especially complex vocal repertoire is found in the large Asian tokay gecko, *Gekko gecko*. This species is unusual in having a multipart advertisement call that begins with a rattling sound that is followed by a series of two-syllable chirps, the *to-kay* sound that gives the lizard its common name (Brillet and Paillette 1991). Nothing is known about the functional significance of these call components, but it is possible that this species resembles some frogs in using different parts of the advertisement call to convey aggressive messages to males and attractive messages to females.

Communication by Snakes. Snakes rely heavily on chemosensory information to locate prey and to communicate with members of their own species. Their highly forked tongues function exclusively as chemosensory organs, and the vomeronasal organ is the most important chemosensory system. Snakes employ a variety of tactile signals, but make only limited use of visual signals (Carpenter 1977). They lack external ear openings, and acoustic signals are completely absent.

Male snakes do not produce signals to attract mates. Some species, such as the red-sided garter snake (*Thamnophis sirtalis parietalis*), form large mating aggregations as they emerge from communal hibernation dens, but males of most species search for widely dispersed females (see Chapter 12). More than 50 years ago, G. Kingsley Noble (1937) showed that male garter snakes could follow chemical trails produced by pheromones in the female's skin. These pheromones are derived from skin lipids that probably evolved originally to retard cutaneous water loss, and secondarily acquired a communication function (Mason 1992). Pheromones produced by female red-sided garter snakes are composed of nonvolatile chemicals called methyl ketones, which also are com-

mon in pheromones produced by mammals and insects. They differ in structure from the skin pheromones produced by males, which contain the compound squalene. Some males have been found to lack squalene and mimic the pheromones of females. These so-called she-males are sometimes courted by other males in mating aggregations. This chemical mimicry may enable males to avoid being pushed aside by other males and increase their chances of mating with a female (see Chapter 12 for additional discussion of the mating system of this snake).

Skin pheromones produced by females appear to be sufficiently long-lived to allow males to follow pheromone trails for some distance. Males of some populations tested in the laboratory discriminated between the trails of their own species and those of other species living in the same area. Closely related species apparently have similar pheromones, and when males are not given their own species as a choice they sometimes follow the trails of related species. In addition to garter snakes, chemical trail-following has been reported for other colubrids, leptotyphlopids, typhlopids, boids, and viperids. Trail pheromones are detected by the vomeronasal system. If the tongue is removed or the functioning of the vomeronasal organs experimentally impaired, male snakes do not trail or court receptive females (Halpern 1992). The ability to follow trails probably is enhanced by the deeply forked tongue, which allows the snake to sample two points simultaneously. This two-point sample should enable a male to determine when he has reached the edge of a trail (Schwenk 1994).

Once a male snake has located a female, a period of courtship follows. This usually begins with the male exploring the female's body with tongue flicks, particularly on the dorsal surface. The male then engages in a period of tactile stimulation, the details of which vary among families. In most species, the male moves forward along the back of the female, rubbing his chin against her body and sometimes making a series of jerking movements with his head (Figure 11–26). In some colubrid snakes, the male also lies on top of the female and sends waves of muscle contractions forward or backward along his body. Other types of body contractions, sometimes described as twitching or spasmodic contractions, have been observed for colubrids, elapids, and viperids. Some male colubrids also bite the female gently, usually when the female is coiled up and not in the proper position to

Figure 11–26   **Courtship behavior of the garter snake,** ***Thamnophis sirtalis.*** The male moves along the top of the female, rubbing his chin against her skin while using his tail to search for the female's cloaca. *(Source: Whitfield 1979.)*

be courted by the male. Male boids have hard spurs that are remnants of the pelvic girdle. The spurs are used to stroke the female during courtship. The final stage of courtship begins with the male using his tail to search for the female's cloacal opening before inserting one of his hemipenes for copulation (Gillingham 1987).

Chemical communication does not end once copulation has been completed. Males of some snakes produce a hard copulatory plug composed of proteinaceous and lipid compounds that is inserted into the female's cloaca after she is inseminated. Copulatory plugs have been studied mainly with garter snakes. The plug may remain in the female's cloaca for up to 4 days and inhibits mating by other males. Initially it was thought that the plug simply served as a mechanical block that prevented other males from copulating with a recently mated female, but studies of *Thamnophis radix* demonstrated that the plug contains a pheromone that inhibits courtship by males. When mated females had the plug removed, they again became attractive to males (Ross and Crews 1977).

Chemical signals are important in aggressive interactions between males. Aggression in snakes almost always occurs in the presence of females; male snakes do not defend territories (see Chapter 12). Males respond to the movements of other snakes by approaching them, but apparently recognize males of their own species by chemical cues. Blockage of the vomeronasal system of the European adder, *Vipera berus*, prevents males from fighting with one another (Andrén 1982). Actual combat is rather similar in

most snakes. Biting is rare, with most fights consisting of males coiling around each other and attempting to pin the opponent to the ground. Each snake apparently tries to maintain a position on top of his opponent, with the individual doing so the longest usually winning the fight. The form of these combat rituals varies among families. Most colubrids fight while lying flat on the ground with only the heads of the males slightly elevated. Elapids also entwine their bodies, but usually have their heads or front portions of their bodies elevated. Viperids usually fight with the front half of the body in a vertical position, sometimes twisting their necks around one another, but with the rest of the body only occasionally entwined (see Figure 12–8). Boids fight by entwining their bodies and scratching each other with their pelvic spurs, and each male also tries to keep its head above that of its opponent (Gillingham 1987). Snakes that elevate their bodies during fights sometimes engage in mutual head-swaying behavior, a visual display that might facilitate assessment of an opponent's size or strength (Carpenter 1977).

Summary

Amphibians and reptiles use a variety of signals to communicate with members of their own species. The production of signals is constrained by the small body size of most amphibians and reptiles, which limits the distance over which a signal can be transmitted, by their ectothermal physiology, and by the costs of signal production, especially energetic costs and exposure to predators. The transmission of signals is subject to interference by noise, and amphibians and reptiles have developed behavioral strategies to overcome the problems of acoustic or visual noise in the environment.

Salamanders and newts use a variety of chemical signals to mark territories, identify the species and sex of potential mates, and stimulate partners during courtship interactions. Some plethodontids mark terrestrial territories with pheromones produced in cloacal glands. Courtship pheromones are produced mainly by glands on the chin. The mode of pheromone delivery varies among different clades of plethodontids, and the mode of pheromone delivery and the structure of courtship glands have evolved in tandem. Salamandrids also use courtship glands. Aquatic newts have especially elaborate courtship rituals that combine visual, chemical, and tactile signals into complex displays.

In contrast to salamanders, anurans seldom use chemical signals, and visual signals are not very important. Acoustic signals are produced by most anurans, which often have a rich repertoire of advertisement, aggressive, and courtship calls. Advertisement calls attract females and advertise a male's ownership of a calling site. Many frogs have graded aggressive calls in which the temporal structure of the call changes as the intensity of an aggressive encounter increases.

The various clades of reptiles have undergone millions of years of independent evolution and have diverged in their use of communication signals. Turtles rely mainly on chemical and visual signals, but some tortoises vocalize during courtship and aggressive interactions. Turtles probably identify the species and sex of other individuals by means of chemical cues, and chemical signals may be used by some male tortoises to increase the receptivity of females. Aquatic turtles often have courtship displays that combine visual signals, especially displays of color, with tactile stimulation of the female. Male tortoises often are very aggressive toward one another and use visual displays in fighting.

Crocodilians are the most vocal of living reptiles (excluding birds), and they often have a complex repertoire of displays that combine vocalizations, acoustic signals produced by slapping the water, subaudible vibrations, and postural changes. American alligators make more use of long-distance acoustic signals than do some species of crocodiles that live in open habitats in proximity to one another. Juveniles produce acoustic signals that attract a parent to dig them out of the nest, contact signals that are used to communicate with other juveniles in nursery groups, and alarm signals that elicit defensive reactions from parents.

The two major clades of squamates, iguanians and scleroglossans, have diverged to some extent in their communication signals. Iguanians rely heavily on visual displays, including the use of color and color change, for communication, but they also have chemical signals that are used for marking territories and for courtship. Most scleroglossans have tongues specialized for delivery of chemicals to the vomeronasal organ and make greater use of chemical signals than do iguanians, although some groups, such as cordylid lizards, have converged with iguanians in the use of visual signals. The use of chemical communication is most fully developed in snakes, which make very limited use of visual signals.

12

Mating Systems
and Sexual Selection

In 1871, Charles Darwin published his book *The Descent of Man and Selection in Relation to Sex*. The book set off a storm of controversy because of its detailed discussion of human evolution. However, more than half the book was devoted to the study of animal behavior, and in particular Darwin's theory of sexual selection. Although this part of the book was much less controversial at the time, it has proved to be one of Darwin's most enduring contributions to biology. Indeed, the continuing influence of his theory of sexual selection on modern biology is second only to his theory of natural selection presented in his more famous book, *On the Origin of Species*, published in 1859.

Darwin formulated his theory of sexual selection to explain the evolution of certain types of traits in animals that could not be accounted for by the action of natural selection. Darwin had long recognized that males and females of many animals do not look exactly alike. Often males are larger, more brightly colored, or equipped with special weapons, such as horns and antlers. He argued that such traits could not have evolved through the action of natural selection, which generally would favor traits that increase the survivorship of individuals. Indeed, many of these traits might actually decrease individual survivorship—for example, brightly colored males might be more vulnerable to predators, or males might injure or kill one another with their dangerous weapons. He concluded that such traits, which he called secondary sex characters, had evolved because they increase a male's chances of mating with females, thereby increasing his reproduc-

tive success even at a possible cost of reduced survivorship.

Darwin recognized two components of sexual selection that could lead to the evolution of secondary sex characters and other sexually dimorphic traits, such as differences in body size. The first was competition among males for access to females. This idea was relatively uncontroversial, because nineteenth-century naturalists were well aware that males of many animals fight vigorously during the breeding season for control of females. Such fighting clearly could lead to selection for effective weapons. However, Darwin suggested that other, more subtle forms of competition were important as well, such as displays directed at other males. The second component of sexual selection was female choice. Darwin argued that females of many animal species can compare the displays or other traits of males and choose to mate with those that are judged to be the most attractive or strongest individuals. This idea was more controversial at the time, in part because Darwin attributed to animals a sense of beauty equivalent to the aesthetic sense of humans. Even some of Darwin's strongest supporters parted company with him on this issue, and the study of female choice languished for nearly a century before eventually becoming a major focus of modern studies of sexual selection.

Darwin's book provided an encyclopedic survey of sexual dimorphism and secondary sex characters of all sorts of animals. Although he focused heavily on birds and mammals, he included some herpetological examples. The elaborate horns of male chameleons, the

368

colorful tail fins of male newts, and even the voices of male frogs were attributed to the action of sexual selection (Figure 12–1). However, for nearly a century the study of sexual selection was not a major focus of herpetological research. One of the few early herpetologists who did understand the importance of sexual selection was G. Kingsley Noble, who devoted an entire chapter of his book, *The Biology of the Amphibia* (1931), to secondary sex characters. He also conducted the first detailed experimental study of the functions of sexually dimorphic colors of lizards (Noble and Bradley 1933).

Over the next 40 years, a number of excellent field studies focused on the social behavior and mating sys-

tems of lizards and frogs, but other groups of amphibians and reptiles were largely neglected. The modern era of research on mating systems and sexual selection began in the late 1970s with the appearance of several major papers on the social behavior of salamanders (Arnold 1976, Halliday 1977), frogs (Wells 1977a,b, 1978, Howard 1978a), and lizards (Stamps 1977). More recently, there have been additional reviews dealing with mating systems and sexual selection in amphibians (Verrell 1989, Halliday and Tejedo 1995, Sullivan et al. 1995), lizards (Stamps 1983, Tokarz 1995), crocodilians (Lang 1989), and snakes (Duvall et al. 1993). Little work has been done on turtles.

Figure 12–1 **Examples of sexually dimorphic structures of amphibians and reptiles.** These illustrations are from Charles Darwin's *The Descent of Man and Selection in Relation to Sex* (1871). (a) *Triturus cristatus*. (b) *Chamaeleo owenii*. (c) *Chamaeleo bifurca*. Males are shown at the top of each figure. *(Source: Darwin 1871.)*

This chapter focuses on the mating systems of amphibians and reptiles and the relationship of these mating systems to the process of sexual selection. Rather than discussing each taxonomic group separately, the chapter is organized around several major conceptual themes. The first section reviews modern thinking about sexual selection in the context of mating system organization. This is followed by a review of the major types of mating systems found in amphibians and reptiles. The third section discusses the major determinants of variation in male mating success in different types of mating systems, including the importance of female choice. The final section deals with the evolution of sexual dimorphism and the influence of sexual selection on particular morphological, physiological, and behavioral traits.

The Relationship of Mating Systems to Sexual Selection

Sexual selection is a form of directional selection that acts on genetically variable phenotypic traits that affect the reproductive success of individuals of one sex. In other words, if increased fighting ability, brighter coloration, or louder calls tend to increase a male's chances of acquiring a mate and contributing offspring to the next generation, then these traits can be considered sexually selected traits (Andersson 1994). Sexual selection, therefore, is like natural selection in being an evolutionary process that results in changes in gene frequencies in populations. However, different components of selection often work in opposite directions because of the costs imposed on an animal by sexually selected traits. For example, the production of loud calls by male frogs may be favored by sexual selection if this enhances their attractiveness to females. However, the calls may also attract predators, which reduces a male's chances of surviving for many nights, or the high energetic cost of calling may deplete a male's energy reserves and further reduce his chances of survival (Ryan 1988).

Males are usually subject to more intense sexual selection than females because individual variation in mating success is greater for males than for females. This difference derives in part from differences in the amount of energy that males and females invest in the production of gametes (sperm and eggs). Females typically produce eggs that are provisioned with yolk reserves to support growth and development, and energetic investment may be even greater if embryos are retained in the female's body and provided with nutrition, or if the female engages in extensive parental care after eggs are laid. Consequently, the reproductive success of females is limited mainly by energy intake, and it may take some time for a female to produce a new clutch of eggs once she has laid her first batch. If one male can fertilize all of her eggs, then a female's reproductive success is not likely to be strongly affected by the number of times she mates. Hence, selection for acquisition of multiple mates will be relatively weak in females.

Males, on the other hand, produce sperm that have only enough energy reserves to make their way to an egg to be fertilized. Hence, the cost of sperm production generally is lower than the cost of egg production, even though males produce millions of sperm. Males usually have sufficient sperm supplies to fertilize the eggs of many females, and their reproductive success is largely dependent on the number of mates they acquire. Consequently, traits that enhance a male's ability to acquire many mates are likely to be strongly affected by sexual selection (Andersson 1994, Arnold and Duvall 1994). This is particularly true when the male contributes little or nothing to parental care, which is the case for most amphibians and reptiles. Under such circumstances, males tend to be polygynous—that is, they attempt to acquire several mates in each breeding season, although usually only a small proportion of the males in a population succeeds in doing so. Other mating systems, such as polyandry (a female mates with several males) and polygamy (both males and females acquire multiple mates) are rare among amphibians and reptiles. Also rare is monogamy (both males and females have only one mate). In fact, only one reptile, the sleepy lizard of Australia (*Tiliqua rugosa*), has been shown to have a truly monogamous mating system (Bull 1994); its unusual behavior is discussed in more detail later in the chapter.

Mating Systems of Amphibians and Reptiles

Most amphibians and reptiles have polygynous mating systems, but the precise form of these mating systems varies. The behavioral tactics used by males to acquire mates depend on the temporal and spatial distribution of females (Wells 1977a, Duvall et al. 1993, Mathis et al. 1995, Sullivan et al. 1995). Females can be aggregated or dispersed in space, and they may be available to males for a long time or only very briefly. When the breeding period is very short and large numbers of females are found together in one place, the usual mating system is an explosive mating aggregation characterized by furious competition among males for possession of females. At the opposite extreme, females may be widely dispersed but available over a longer breeding season. In such cases males have few options other than searching for receptive females over large areas. When females are moderately aggregated in space and available over a relatively long period, a variety of mating systems are possible. One option is for males to search for individual females and actively guard them from other males until the females are ready to mate. Another is for males to use signals to attract mates. Sometimes signaling males are widely dispersed, but many species aggregate into choruses or gather at traditional display areas (leks) to attract females. Finally, if males can monopolize resources needed by females for successful reproduction, such as good egg-laying sites or areas with abundant food, then the predominant mating system is for males to defend territories containing those resources.

This classification of mating systems is somewhat arbitrary, and certain categories tend to grade into one another. For example, mate searching and mate guarding differ mainly in the likelihood that more than one male will encounter a female simultaneously. If they do, then males usually will defend the female. Mate guarding can occur in other types of mating systems as well. For example, toads that form explosive mating aggregations guard females by holding them in amplexus and fighting off other males that try to displace them. Sometimes different males in a population use different tactics to acquire mates. For example, some males might defend resource-based territories, while other males search for females in areas where they are likely to be concentrated. We organize mating systems into discrete categories simply because it provides a convenient framework for understanding common features of mating systems that may have evolved in response to similar selective pressures.

Explosive Mating Aggregations

Explosive mating aggregations are common for anurans that breed in temporary ponds in early spring,

such as North American wood frogs (*Rana sylvatica*), European common frogs (*Rana temporaria*), American toads (*Bufo americanus*), and European common toads (*Bufo bufo*) (Davies and Halliday 1979, Howard 1980, 1988, Ryser 1989). This type of mating system also is found among desert-dwelling anurans that use temporary rain pools for breeding, such as spadefoot toads (*Scaphiopus* and *Spea*), and tropical species that breed only after heavy rains, such as the Neotropical tree frog *Agalychnis saltator* (Sullivan 1989, Roberts 1994). The selective pressure favoring a very short breeding period often is the short life of the breeding pond, but other factors, such as the danger of cannibalism by larger members of the same species or predation by other species, can select for synchronous breeding as well (Petranka and Thomas 1995).

Sometimes hundreds of individuals arrive at a pond on a single night. Breeding periods range from one night in some species to nearly 2 weeks in others. Males generally arrive slightly earlier than females and usually greatly outnumber them (Figure 12–2). Often there is a mad scramble among males searching for females. Males move around the pond while calling, grabbing any other individuals they encounter. Males are quickly released, but females are tightly clasped and defended against other males attempting to displace the clasping male. This can result in mating balls in which a half-dozen or more males struggle for control of a single female (Figure 12–3), sometimes with fatal consequences for the female. In lower-density choruses, males tend to reduce the amount of time devoted to searching and spend more time calling from stationary positions and waiting for females to come to them (Woolbright et al. 1990).

Some ambystomatid salamanders that breed in winter or early spring also form explosive mating aggregations (Verrell 1989, Sullivan et al. 1995). In contrast to anurans, these salamanders have internal fertilization, so that mating may be separated from egg laying by several days. Breeding migrations of male and female salamanders usually are triggered by warm rains, with hundreds of individuals sometimes moving into ponds on a single night. As in explosive-breeding frogs, males usually outnumber females in breeding ponds. The behavior of the spotted salamander (*Ambystoma maculatum*), studied by Stevan Arnold (1976), is typical of explosive-breeding salamanders. Males mill about on the bottom of the pond, searching for females. When a female is encountered, a male deposits a number of mushroom-shaped spermato-

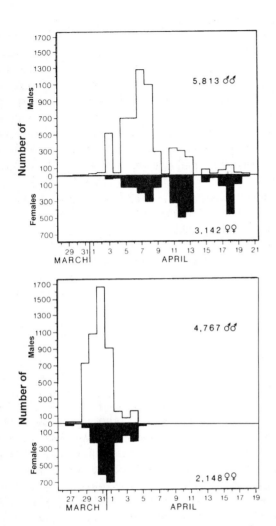

Figure 12–3 **Mating ball of male European common toads (*Bufo bufo*) competing for possession of a female.** *(Photograph by Tim Halliday.)*

Figure 12–2 **Arrival of male and female wood frogs (*Rana sylvatica*) at breeding ponds.** Data for 1980 (top) and 1981 (bottom). Males arrived earlier and in greater numbers than females in both years, but the breeding period was more explosive in 1981. *(Source: Howard and Kluge 1985.)*

phores scattered around the female (Figure 12–4). Fertilization occurs when the female picks up the cap of the spermatophore, which contains the sperm mass, with the lips of her cloaca. Sometimes several males court a female simultaneously, resulting in a large number of spermatophores being clustered in a small area. Some hynobiid salamanders form explosive mating aggregations, but these salamanders have external fertilization, so rather than scrambling to court females, groups of males form mating balls around egg sacs immediately after they are deposited

by a female and compete to release sperm onto the eggs (Hasumi 1994).

The best example of explosive mating aggregations among reptiles comes from studies of red-sided garter snakes (*Thamnophis sirtalis parietalis*) in Manitoba. This species has been used by David Crews and his colleagues as a model system to investigate the behavior and endocrinology of reproduction by snakes. These snakes gather by the thousands in communal hibernation dens during the winter. When spring arrives, males emerge before females and wait around the entrances of communal dens to intercept females. Females produce a pheromone in the skin that is attractive to males. Each female is mobbed by dozens of males, each frantically attempting to get close enough to the female to insert his hemipenis into her cloaca. The result is a writhing ball of snakes, with many males jostling one another, but with little overt aggression (Figure 12–5). Eventually one male succeeds in mating with the female. After transferring sperm, the male leaves a waxy mating plug in the female's cloaca that is impregnated with a pheromone that renders the female unattractive to other males. They quickly lose interest in her and move on to search for other females (Crews and Garstka 1982). Similar mating balls have been observed in other natricine snakes, such as the European grass snake (*Natrix natrix*) (Luiselli 1996).

All of these mating systems share certain characteristics that have important implications for sexual selection. First, males usually outnumber females, sometimes by as much as 10 to 1. There are several

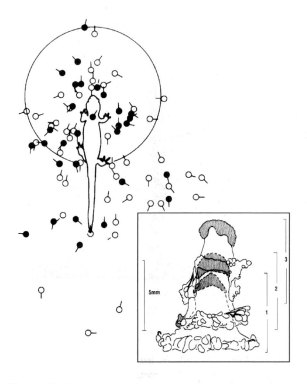

Figure 12–4 **Pattern of spermatophore deposition by the spotted salamander (*Ambystoma maculatum*).** Males scatter large numbers of spermatophores around a female on the bottom of the pond. Solid circles indicate single spermatophores. Open circles indicate spermatophores deposited on other spermatophores (see inset), a form of sexual interference. Males are oriented randomly with respect to the female during spermatophore deposition, as shown by short lines on circles. The large circle designates a diameter of one body length around the female's snout. *(Source: Arnold 1976.)*

Figure 12–5 **Mating aggregation of red-sided garter snakes (*Thamnophis sirtalis parietalis*) after emerging from hibernation in the spring.** *(Source: Photograph by Bianca Lavies/National Geographic Society.)*

possible reasons for these highly skewed sex ratios, including greater annual mortality of females, delayed sexual maturity in females, and a tendency for males to breed every year, whereas females sometimes do not. In addition, individual females usually mate only once and are present for only a short time at the breeding site. The result is intense competition among males, which often are very indiscriminate in trying to mate with anything that resembles a female, including other males, dead individuals, and females that have already mated (Wells 1977a, Garstka and Crews 1985). Because many females are mating simultaneously, most males that mate get only one mate in a season, and many do not mate at all (Figure 12–6). This type of mating system also provides little opportunity for females to compare the available

males and choose the best-quality mate; in general, male–male competition is expected to be a more important determinant of variation in male mating success than female choice (Sullivan et al. 1995).

Mate Searching

For animals that are essentially solitary and widely dispersed in their habitats, males and females may have some difficulty finding one another. Many aquatic turtles, for example, live in large overlapping home ranges (see Chapter 10), and males probably find mates by searching for them over large areas. Mate searching also is a common mating system for snakes, which tend to be solitary (Gregory et al. 1987). A well-studied example is a population of prairie rattlesnakes (*Crotalus viridis viridis*) studied in Wyoming (Duvall et al. 1992). Males and females emerge from hibernation dens in the spring and immediately begin moving over relatively long distances in search of rodents, their principal prey. This behavior results in individual males and females being widely dispersed over the available habitat by the time the mating season begins in midsummer. For a period of about 6 to 8 weeks, males move around more than females, searching for mates. Data from individuals equipped with radio transmitters show that most males do not succeed in finding even a single female, whereas most females do mate and presumably produce young. Because the location of females is unpredictable, models of searching behavior suggested that

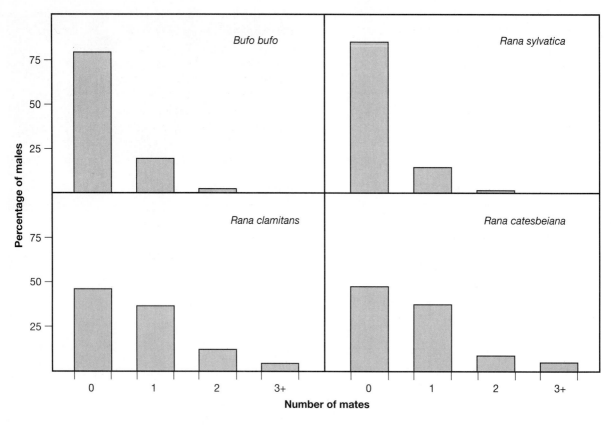

Figure 12–6 **Mating success of male anurans.** Distribution of number of matings per male of four pond-breeding anurans with explosive (top) and prolonged (bottom) breeding seasons. *(Sources: Davies and Halliday 1979, Howard 1980, Wells 1977b, Howard 1978a.)*

the best strategy for males would be to move in a relatively straight line until a female is encountered, and this appears to be what prairie rattlesnakes do.

The likelihood of male rattlesnakes finding females once they are nearby is enhanced by their ability to follow chemical trails produced by pheromones in the female's skin. Males of many species of snakes have this ability (Mason 1992; see Chapter 11). Once a male finds a female, he may remain with her for several days until she is ready to mate. Although this mating strategy apparently has a relatively low payoff for males, it is a viable strategy for relatively long-lived animals that potentially have many breeding seasons in which to locate mates. It also may be the only option for an animal that lacks long-distance signals that could be used to attract mates.

Mate searching occurs in a population of diamond pythons (*Morelia spilota spilota*) studied by Slip and Shine (1988) in southern Australia. Females do not

aggregate, as red-sided garter snakes do, but males engage in a similar kind of scramble competition. Males are attracted to females, apparently by a pheromone, and several males sometimes remain with a single female, competing for mating opportunities but without much violent fighting. Both males and females tend to mate with multiple partners, and individual females often are involved in several different mating aggregations. In a more northern subspecies, *Morelia spilota variegata* (the carpet python), the mating system is quite different. Males fight over females but do not engage in a prolonged search for mates and do not form large aggregations, nor do they spend very much time with each female. This difference in behavior may be due to the higher density of snakes in northern populations and the greater availability of potential mates (Shine and Fitzgerald 1995).

Males of lizards that forage over wide areas for food, such as teiids, lacertids, varanids, and scincids, usually search for females rather than defend territories (Stamps 1977, 1983). Skinks of the genus *Eumeces* have this type of mating system, and males follow pheromone trails just as snakes do (Cooper and Vitt 1986). Male white-throated monitors (*Varanus albigularis*) in the Namibian Desert sometimes move more than 6 kilometers in a single day while looking for females, and their home ranges may be up to 18 square kilometers in area. Receptive females are reported to make themselves conspicuous by climbing into trees, where they release an airborne pheromone that attracts males (Phillips 1995a,b). Male Komodo dragons (*Varanus komodoensis*) often take advantage of temporary aggregations of females around food sources such as dead deer to engage in courtship. Because all of the males in a given area appear to be familiar with one another and organized into an established dominance hierarchy, the dominant male usually does most of the courting (Auffenberg 1981).

The Australian sleepy lizard, *Tiliqua rugosa* (Scincidae) has an unusual type of mate searching. These lizards are large and relatively sluggish, with a life span exceeding 12 years. Both males and females are nonterritorial, living in home ranges that overlap with those of other sleepy lizards. During the mating season, males seek out females and remain with them for long periods of time, following them while they search for food and occupying the same retreat sites. What is unusual about this lizard is that males and females repeatedly pair with the same mates in successive years, even though their home ranges overlap those of other potential mates. Although males are occasionally seen fighting with one another, aggressive defense of mates is rare, and indeed, even when males leave their mates for some period of time, other males show little interest in them. When mates separate temporarily, they appear to find each other by following chemical trails on the ground, as well as airborne odors. They also search in known locations where their mates are likely to be found. This is the only clear example of long-term monogamy for any amphibian or reptile, but the precise advantage of this unusual mating system is not yet understood (Bull 1994).

Some salamanders find their mates by searching. One example is the red-spotted newt, *Notophthalmus viridescens*. Newts often occur at high densities in ponds, and mating assemblages resemble explosive mating aggregations even though the mating season can last for several months. Males move about on the bottom of the pond, approaching any newt seen moving nearby (Massey 1988). Usually males seize females and hold them in amplexus, with the male's rear legs wrapped around the female's neck (Figure 12–7). This posture allows the male to stimulate the female with courtship glands on the side of his head, which he rubs against the female's snout. It also is a form of mate-guarding behavior, because clasping the female makes it more difficult for other males to interfere with courtship. Males are more likely to clasp females in amplexus when there are many other males nearby. Mate searching similar to that seen in red-spotted newts also occurs in European smooth newts (*Triturus vulgaris*) (Verrell 1989), but males of this species and all other species in this genus court the female without clasping her in amplexus (see Chapter 11).

Mate Guarding

Once a male has located a female after a prolonged search, he often will remain with the female and defend her against other males, especially if additional mates are scarce. The amplexus behavior of explosive-breeding anurans and red-spotted newts, which enhances the ability of a male to monopolize a female, is one form of mate guarding. The prolonged copulation of some snakes, which can last for hours or even days, also can be considered a form of mate guarding because it prevents mating by other males (Olsson et al. 1996). Mate guarding takes a more aggressive form in other species, including violent fighting for possession of individual females. Male European adders (*Vipera berus*) often accompany females for several

Figure 12–7 **Courtship behavior of the red-spotted newt (*Notophthalmus viridescens*).** The male has seized the female in amplexus and is stimulating her with pheromones produced by glands on the side of the head. *(Source: Halliday 1990.)*

days until they are ready to mate. Males respond very aggressively toward other males and may engage in prolonged wrestling bouts in which the males coil around one another, each attempting to force the opponent to the ground (Figure 12–8); they never bite one another, however. On rare occasions, three or even four males may fight one another simultaneously, but most fights involve only two males (Madsen et al. 1993). Fights between males that differ considerably in body size generally are short, with the larger male always winning. Fights between more evenly matched opponents are much longer, sometimes lasting for hours, but again, the larger male almost always wins. Similar fighting among males for possession of females has been reported for many other species of snakes, especially viperids, elapids, colubrids, and pythonids (Gillingham 1987, Shine 1994).

Mate guarding can occur after copulation as a way to prevent other males from fertilizing some of a female's eggs. For example, male European sand lizards (*Lacerta agilis*) are not territorial, but they do fight over females during the breeding season. Copulation lasts only a few minutes, but males stay with females for up to 3 days before searching for new mates. This

extended mate guarding requires a substantial energetic investment by a male and reduces the time he has to find other mates. However, the payoff for such guarding can be large, because mating by a second male anytime within 24 hours of a first mating can result in multiple paternity of egg clutches (Olsson et al. 1994). Females are promiscuous and will mate with any males that court them. Large males acquire more mates and guard females longer than do small males. Prolonged mate guarding does not always ensure exclusive paternity of a female's offspring, but it may be a worthwhile strategy for males to pursue while replenishing their sperm reserves (Olsson et al. 1996).

Aggressive contests for possession of females have been reported in some turtles. For example, Kaufmann (1992) observed numerous aggressive encounters between male wood turtles (*Clemmys insculpta*), with larger individual almost always winning fights. Most of the fights occurred when a male attempting to court a female was challenged by another male. Males of some species of tortoises have elongated, forked projections on the front of the plastron (the lower part of the shell). These projections are used to ram opponents in fights and to flip them over on their backs (Figure 12–9). Tortoises usually inhabit overlapping home ranges, and such contests take place almost exclusively during the breeding season in the presence of receptive females (Auffenberg 1977, Ruby and Niblick 1994). Male desert tortoises (*Gopherus agassizii*) move around more than females during the breeding season, and a male often visits a number of different females. Males sometimes remain for several days in the same burrow with a female. Such cohabitation is rare outside the breeding season and suggests that males are actively guarding females from other males while engaging in periodic courtship behavior to bring them into sexual receptivity (Niblick et al. 1994).

Several types of mate guarding behavior have been reported for salamanders, in addition to the clasping of females by male red-spotted newts (Verrell 1989, Halliday and Tejedo 1995). Male tiger salamanders (*Ambystoma tigrinum*) court females individually and, if approached by other males, sometimes physically push a female away from competitors. Males of other ambystomatids and some salamandrids (*Pleurodeles, Taricha*) clasp females during early stages of courtship and sometimes carry the female away from competing males. Wrestling bouts between intruding and clasping males also have been observed (Arnold 1977). Males of *Rhyacotriton* and many species of plethodon-

Figure 12–8   **Wrestling contest between two male European adders (*Vipera berus*).** The fight is a test of strength, with each male trying to press his opponent to the ground. *(Source: Smith 1969.)*

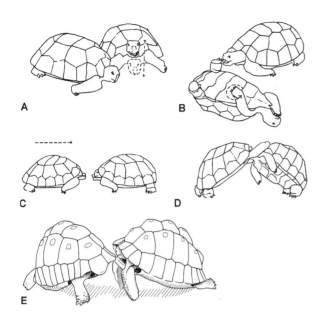

Figure 12–10 **Male and female *Atelopus varius* in amplexus.** Males often become emaciated after long periods of amplexus. *(Photograph by Martha L. Crump.)*

Figure 12–9 **Aggressive behavior of male tortoises fighting over females.** (a) Aggressive head-bobbing display. (b) One male bites the foot of another male. (c) Frontal ramming with the anterior projection of the plastron. (d) Two males elevated above the ground after ramming into each other. (e) The male on the right is hooking the other male with projections of the plastron, attempting to flip his opponent over. *(Sources: (a–d), Ruby and Niblick 1994; (e), Auffenberg 1977.)*

tid salamanders are very aggressive toward other males, chasing and biting any individuals that intrude on their courtship attempts. Males of some of these species have enlarged teeth that are used in aggressive encounters between males and during courtship of females.

An unusual form of mate guarding occurs in Neotropical anurans of the genus *Atelopus* (Bufonidae). These toads are terrestrial for most of the year but move briefly to streams to breed during the wet season. Males defend territories against other males throughout the year, but allow females into their territories. When a male encounters a female, he usually tries to clasp her in amplexus, even weeks or months before the breeding season (Figure 12–10) Females resist these attempts, but sometimes males succeed in clasping females and then remain in amplexus for several weeks, or for several months in some species (Crump 1988). Clearly this is disadvantageous for the females, because they must carry males around on their backs for long periods of time,

so it is not surprising that they try to fight off males during the nonbreeding season. Such prolonged mate guarding has costs for the males as well, because they have trouble feeding while in amplexus and may become emaciated. Presumably males have adopted this unusual mating strategy because females are relatively scarce and do not aggregate in large numbers near breeding sites. Males therefore grab females whenever they are encountered, rather than waiting until the females are ready to mate and risking not finding a mate at all.

Choruses and Leks

The predominant mating system of frogs and toads is for males to gather at aquatic breeding sites and form choruses to attract females (Sullivan et al. 1995). Usually males space themselves throughout the available habitat and defend their calling sites with aggressive calls and physical attacks on intruders (Halliday and Tejedo 1995). Depending on the species, males call from the water or from elevated perches such as rocks or tree branches. In most species, these calling sites are not immediately adjacent to suitable oviposition sites, so females approach calling males, enter into amplexus, and then carry the males on their backs to egg-laying sites. Once a male has mated with one female, he usually returns to the chorus and advertises for additional mates. Competition among males in such choruses often takes the form of vocal interactions, with individuals attempting to outsignal their neighbors by increasing the rate, intensity, or complexity of their calls. Direct aggressive competition among males for possession of females is rare, and females have an opportunity to move through the chorus and compare males before selecting a mate.

This type of mating system resembles the leks of some birds and mammals. The term lek is a Scandinavian word used to describe a gathering of males on a traditional display ground to attract females. Typically males defend very small, closely packed territories that do not contain any resources that can be used by females, and often do not contain sufficient resources even to support the male for long periods of time. Females are free to move about the lek and choose their mates, although males often try to keep females from leaving their territories (Höglund and Alatalo 1995). Birds such as ruffs and black grouse, and mammals such as the Uganda kob and hammer-headed bats, often form leks in open, highly visible areas, in places with especially favorable display sites, or in locations where the home ranges of individual females or groups of females overlap. These sites are not necessarily near nesting areas used by females. Male frogs, on the other hand, aggregate at ponds, marshes, or streams because females must lay eggs at those sites. Hence, despite their similarity, frog choruses and bird leks may have quite different evolutionary origins.

Lek mating systems are not limited to animals that use acoustic signals to attract mates. In several species of European newts (*Triturus*), males gather on leks in limited parts of ponds and defend small territories against other males with displays and violent fighting. Females passing through the area are greeted with elaborate visual displays, including handstands that accentuate the broad tail fins and colorful markings of the males (Figure 12–11). Because these display structures and colors are important for both male–male competition and attracting females, newts with lek mating systems, including *Triturus cristatus*, *T. vittatus*, and *T. marmoratus*, are the most sexually dimorphic newts and, indeed, among the most colorful of all salamanders (Hedlund and Robertson 1989). This is similar to patterns of sexual dimorphism seen in birds, with lek species being the most colorful and equipped with the most elaborate display structures (Höglund and Alatalo 1995).

The mating systems of large, herbivorous iguanid lizards are quite variable, ranging from mate searching to leks to resource defense (Stamps 1983). Most herbivorous lizards feed primarily on leaves, which provide an abundant but relatively low-energy food

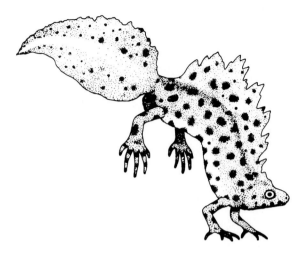

Figure 12–11 **Courtship display of the great crested newt (*Triturus cristatus*).** This species has a lek mating system. The display emphasizes the size of the male's body, tail, and crest, which are important cues for females choosing mates. *(Source: Halliday 1977.)*

source. A mating system based on territorial defense of food resources is unlikely for two reasons. First, with an abundant food supply, variation in the quality of male territories is likely to be small and therefore of little significance for females in choosing mates. Second, defense of a large territory is incompatible with the low-energy lifestyles of these lizards. Indeed, males of most large herbivorous lizards are territorial only during the breeding season, and they often defend very small territories aggregated into leks. Male green iguanas (*Iguana iguana*) often establish breeding territories in dead trees, which provide conspicuous display sites but no food (Rodda 1992). Males spend most of their time giving head-bob and dewlap displays, defending the boundaries of their territories against other males, and courting females, and they can become quite emaciated by the end of the breeding season. Females aggregate in the territories of the largest males, where they eventually mate. Smaller males cannot defend territories but remain at the periphery of territories defended by larger males and attempt to copulate with females as they pass by.

The marine iguana (*Amblyrhynchus cristatus*) of the Galápagos Islands also has a lek mating system. Originally the mating system of this species was described as a resource-defense system. Males defend territories in areas used by groups of females as basking sites when they emerge from the water after feeding on algae. Characteristics of male territories, including the presence of suitable basking rocks and sheltered sites used by females to escape from midday heat, were thought to be correlated with male mating success (Trillmich 1983). More recent work has shown that male territories tend to be clustered even when population densities are relatively low, and the location of territories does not appear to be related to physical characteristics of the habitat (Wikelski et al. 1996). Males on clustered lek territories obtained more mates than peripheral males that defended individual territories (Figure 12–12). Females preferred to mate with the largest territorial males, but smaller nonterritorial males attempted to sneak copulations when territorial males were absent.

Male Galápagos land iguanas (*Conolophus subcristatus*) establish breeding territories in areas near burrows constructed by the males that are used by both males and females as nighttime retreats. Females are attracted to territories with burrows, and most courtship activity occurs near burrow entrances (Werner 1982). The burrows are not used for nesting, however, and so do not represent resources used for reproduction by females. Hence, the mating system of this species is more similar to a lek than a resource-based mating system.

Some crocodilians also have leklike mating systems. This is particularly true of species such as the Nile crocodile (*Crocodylus niloticus*), which forms large aggregations along riverbanks and lakeshores. Males in some populations defend suitable basking sites as territories all year, but males in other populations are territorial only during the breeding season. Females tend to aggregate on the territories of the largest males (Lang 1989). Other crocodilians have mating systems that are closer to resource defense. For example, male American alligators (*Alligator mississippiensis*) defend territories dispersed in marshy habitats and often out of sight of other males. They rely on long-distance acoustic signals (bellowing) to advertise their territories and attract females. The criteria females use to choose mates are not known.

Resource Defense

Although many anurans form choruses that resemble leks, others have resource-defense mating systems. The two best-studied examples are the North American green frog (*Rana clamitans*) and bullfrog (*Rana*

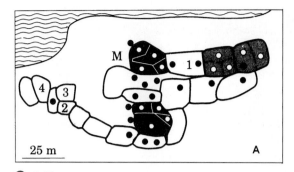

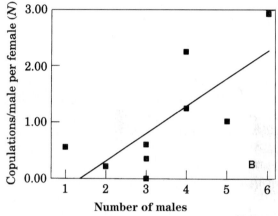

Figure 12–12   **Lekking behavior of the marine iguana (*Amblyrhynchus cristatus*).** (a) Density of iguanas during the breeding season. Open circles indicate territorial males. Solid circles are peripheral males. Territorial males cluster their territories in areas where females aggregate. (b) Number of copulations per male per female as a function of the number of males on a lek. Males on large leks acquired more mates. *(Source: Wikelski et al. 1996.)*

catesbeiana) (Wells 1977b, 1978, Howard 1978a). Both of these species have summer breeding seasons lasting 2 to 3 months. Males set up territories early in the breeding season and defend them with aggressive calls and wrestling against other males (Figure 12–13a). Some individuals maintain their territories for almost 2 months. Territories are situated around suitable oviposition sites, where eggs are laid in surface films (Figure 12–13b). Females sometimes visit the territories of several males before choosing a mate. The quality of the oviposition site defended by a male appears to be more important in mate choice than the size of the male or his behavior, although the largest males typically control the best territories. In his study of bullfrogs, Richard Howard (1978b) found that water temperature was an important component of territory quality. Eggs that were laid in cool water

(a)

(b)

Figure 12–13 **Reproductive behavior of green frogs (*Rana clamitans*).** (a) Two male green frogs wrestling for control of a territory. (b) Male and female green frogs in amplexus in the male's territory. *(Photographs by Kentwood Wells.)*

took longer to develop and were subject to intense predation by leeches, whereas eggs that were laid in water that was too warm developed abnormally. The largest males controlled territories in which eggs developed at the fastest rates and suffered the least predation (Figure 12–14). Territorial males shifted their activities to different parts of the pond as temperatures changed throughout the season.

Resource defense by male salamanders is not common, probably because most species have internal fertilization and mating often is separated from egg laying by weeks or months. This means that males cannot attract females by defending oviposition sites (Verrell 1989, Sullivan et al. 1995). The crypto-

branchid salamanders (*Andrias* and *Cryptobranchus*) have external fertilization and are an exception to this generalization. Males of these species establish territories under rocks on the bottom of streams and rivers. Other males are attacked, but females are allowed to enter the territory, where they are courted by the male. Females lay eggs attached to the undersides of rocks in the males' territories and then leave, while the male remains with the eggs and guards them. Additional females may subsequently enter the nest site and lay their eggs. Mate choice by female cryptobranchids has not been studied, but variation in the quality of nest sites could be an important criterion for choosing mates because the eggs take a long time to develop (5 to 6 months). Observations of the Japanese giant salamander (*Andrias japonicus*) suggest that nonterritorial males may interfere with the courtship of territorial males and perhaps fertilize

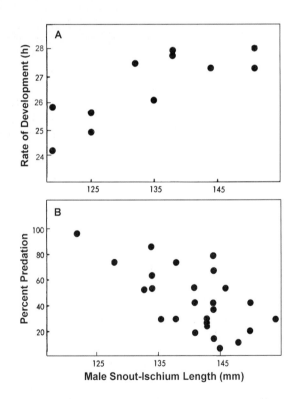

Figure 12–14 **Territory quality among male bullfrogs (*Rana catesbeiana*).** Relationship of male body size to (a) rate of egg development and (b) percentage of eggs eaten by leeches. Large male bullfrogs usually defend the best-quality territories where eggs develop rapidly and suffer the lowest rates of predation. *(Source: Howard 1978b.)*

some eggs, because several males have been observed in a nest site when eggs were being laid (Kuwabara et al. 1989). This type of sexual interference is common in fishes that lay eggs in nests inside male territories.

Extensive studies of the territorial behavior of the red-backed salamander (*Plethodon cinereus*) by Robert Jaeger and his students suggest that this species, and perhaps some other terrestrial plethodontids, has a resource-defense mating system (Jaeger and Forester 1993). Both males and females are aggressive toward other individuals in the laboratory. In the field they defend territories centered on cover objects such as rotting logs that also provide rich food supplies of termites, ants, and other invertebrates. Home ranges of the females sometimes overlap the territories of several males. This gives the females an opportunity to compare males or their territories before choosing a mate. Both large males and males with large territories seem to be preferred. In addition, males that have fed on termites appear to be more attractive to females than those that have fed on ants, presumably because termites are a better-quality, more digestible food. Females use the fecal pellets of males to determine the type of food found in their territories. It is not yet clear how the reproductive success of females is increased by mating with males in high-quality territories.

Defense of resource-based territories is the most common type of mating system for small, insectivorous iguanian lizards. Males typically defend their home ranges as territories throughout the breeding season, spending a lot of time and energy displaying to other males, chasing intruders, and courting females. Food, basking sites, display perches, and retreat sites all are encompassed within their territories. Females of some insectivorous lizards defend territories, and female territories usually overlap territories of males. Females of other species occupy overlapping home ranges but are not aggressive toward other lizards (see Chapter 10). Males with the largest territories overlap the home ranges of several females and probably obtain more mates than those with smaller territories (Stamps 1977, 1983). Because of the advantages of defending a large territory, males of most territorial species have home ranges that are two to four times larger than those of females. The same pattern is seen in tuatara (*Sphenodon punctatus*). Both males and females are territorial, but the territories of males are about twice as large and overlap the territories of as many as three females (Gillingham et al. 1995).

In a resource-based mating system, the ability of a male to monopolize females, and hence his tendency to mate polygynously, depends in part on the size of home ranges occupied by females (Stamps 1983). When females have small home ranges, perhaps because food is very abundant, the density of females increases and males can monopolize several females while defending a relatively small territory (Figure 12–15a). As female home range size increases, males would have to defend larger and larger areas to encompass several female home ranges. Presumably the costs of territory defense increase with increasing territory size. At some point, defending a territory with multiple females would require too much energy, or the territory would become so large that a male could not effectively monitor the movements of intruders. At this point males should concentrate their efforts on defending fewer females more effectively. Hence, the size of male and female home ranges should become more similar as female home range size increases (Figure 12–15b), and males should be less polygynous. At very large female home range sizes, many males are expected to have only a single mate, although they will mate polygynously whenever possible. In general, widely foraging lizards, such as teiids and lacertids, have larger home ranges than do insectivorous iguanians (see Chapter 10). These species usually are nonterritorial, with similar home range sizes in males and females (Figure 12–15b). However, territoriality has been reported in some very dense populations of *Lacerta* and *Cnemidophorus*, where small female home ranges make possible the defense of territories by males.

Males of insectivorous lizards often do not alter their territorial behavior in response to short-term changes in territory quality or the distribution of females. For example, Hews (1993) added food to the territories of some male *Uta palmeri*, but they did not shift the boundaries of their territories even though some females did move from their old home ranges to areas with extra food. Deslippe and M'Closkey (1991) removed all of the females from the territories of some *Sceloporus graciosus* males, but the males continued to defend their territories. This result contrasts with the results of a similar removal experiment with *Urosaurus ornatus* in which males left their territories when females were removed (M'Closkey et al. 1987). *Sceloporus graciosus* is a relatively long-lived species, and males have several breeding seasons in which to find mates. Males adopt a long-term strategy of de-

Figure 12–15 **Home ranges of lizards.** (a) Relationship of home range size of females to density of females in habitat for a variety of species of lizards. (b) Relationship of home range size of females to the ratio of male to female home range size in territorial lizards (solid circles) and nonterritorial lizards (open circles). *(Source: Stamps 1983.)*

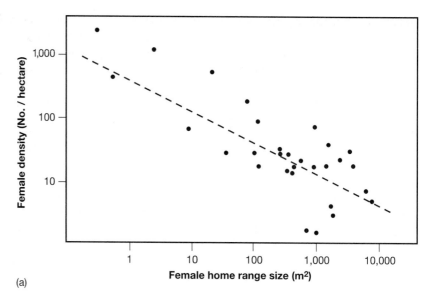

(a)

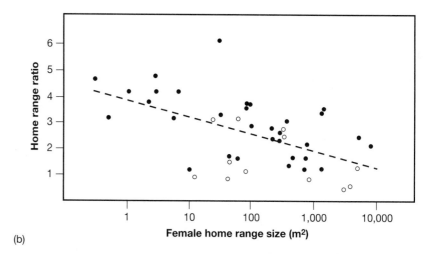

(b)

fending a high-quality territory, even if females are temporarily unavailable there. In contrast, male *Urosaurus ornatus* usually die after one breeding season and presumably cannot afford to waste time defending unprofitable territories.

Variables Affecting Male Reproductive Success

Most studies of sexual selection and mating systems of amphibians and reptiles have focused on determinants of male mating success, because this is what affects the intensity of sexual selection. When males have many opportunities to mate in a single season, some males are likely to mate many times, but many others will not mate at all. This means that the variance in male mating success will be relatively large, and consequently sexual selection on male traits that affect mating success will be strong. On the other hand, if most males can obtain only one mate in a season, variation in male mating success will be smaller, and sexual selection on male traits will be weaker. However, this can be complicated by differences in sex ratios in populations. Even when males have few opportunities to obtain more than one mate, as is often the case in explosive-breeding aggregations, competition among males will be intense if there is an excess of males in the breeding population. In that case, strong sexual selection on male traits is expected.

Exactly which male traits are affected by sexual selection depends on the nature of the mating system and which stage of the mating process has the greatest impact on male mating success. This is illustrated in a flow chart developed by Arnold and Duvall (1994) (Figure 12–16). They divided factors affecting male reproductive success into a hierarchy of male traits. If we begin with total age-specific reproductive success, we see that this is affected by two variables, the number of offspring the male fertilizes and the male's probability of dying. Number of offspring in turn is affected by the number of mates a male acquires and the number of offspring each of those mates produces. Most sexually selected male traits affect male reproductive success through their effect on mating success. One class of traits that affects mating success is anything that affects the time between successive matings. Another is any sort of courtship behavior used by the male to persuade a female to mate with him. Time between matings is a product of the time required to find a female in the first place and the time required to mate and become physiologically ready to mate again. Males also can have a direct influence on the fecundity of their mates if they provide parental care to the offspring, thereby increasing the number that actually survive to reproduce (see Chapter 7).

The mating systems described in the first part of this chapter provide the framework for the operation of sexual selection on male traits because they determine the types of traits that are likely to be important for male mating success. For example, in a species such as the prairie rattlesnake, male mating success probably is limited by the time required to find each mate. Consequently, any traits that can reduce search time will tend to increase male mating success. A capacity to follow chemical trails produced by females might be one such trait that would be strongly favored by sexual selection, because a male that can follow scent trails is more likely to find a mate than a male that cannot follow these trails. Finding females is relatively easy for red-sided garter snakes, and mating success may be limited by how rapidly males can mate with additional females. Males of this species are ready to mate again within 12 hours of mating, about half the time required by *Thamnophis radix*, a species with a much longer breeding season (Whittier et al. 1985). Mating success of green iguanas, marine iguanas, and Nile crocodiles probably is not limited by search time because females aggregate on male territories, but prolonged courtship by males may be important and hence under strong sexual selection. The same may be true for amphibians that form choruses and leks; elaborate vocal signals and courtship are common in these species.

Sexual selection on male traits is also affected by the type of competition that occurs in particular mating systems. In some mating systems, one form of male competition predominates, but in others, males compete for mates in several different ways. For example, because male prairie rattlesnakes search for females but seldom fight with other males, they are essentially engaged in scramble competition to find mates. On the other hand, male toads that form explosive mating aggregations not only scramble to find mates, but also fight over females. Hence, sexual selection should favor traits that enable a male to locate a female quickly as well as traits that enhance his ability to retain possession of a female once he finds her. In territorial species, such as lizards and bullfrogs, males engage in contests for possession of territories, so traits that improve male fighting ability would be favored by sexual selection. However, females also choose mates from among the established territorial males, so traits that increase a male's attractiveness to females or his ability to acquire resources needed by females also could be important.

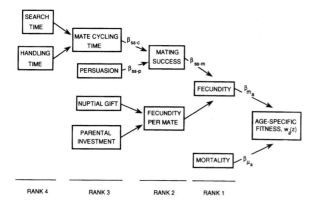

Figure 12–16 **Summary of factors that affect age-specific reproductive success (fitness) of male animals.** Traits that affect mating success are subject to sexual selection. (*Source: Arnold and Duvall 1994.*)

Male Persistence and Allocation of Resources

One factor that often affects male mating success is the ability of a male to continue breeding activities for a long period of time. For example, the amount of time a male frog spends in a chorus often is a major determinant of mating success (Halliday and Tejedo 1995). Yet a common pattern is for most males to be active on only a few nights, perhaps because the energetic costs of calling limit male breeding activity (see Chapter 6). Males often depend on stored energy reserves to make it through the breeding season, and they must reduce their activity as reserves are depleted. Although they sometimes replenish energy reserves by feeding, males often reduce feeding during the breeding season because it conflicts with time needed for territory defense, courtship, and mating. For example, territorial male carpenter frogs (*Rana virgatipes*) call almost all night and feed only during the day; lipid reserves decline throughout the breeding season (Given 1988). Males of the Puerto Rican coquí (*Eleutherodactylus coqui*) do little feeding early in the evening, and total nightly food intake is lower for calling males than for noncalling males or females. Similar patterns have been described for some reptiles. Many territorial male marine iguanas do not feed during the peak breeding period, and those that do feed make very short foraging trips (Figure 12–17). Consequently, territorial males lose up to 26 percent of body mass in about a month (Trillmich 1983).

Even when mating activity itself is not energetically expensive, males may be constrained by limited supplies of sperm, and patterns of allocation of reproductive resources reflect the availability of mates. For example, males of explosive-breeding salamanders, such as *Ambystoma maculatum* and *A. texanum*, apparently use a substantial portion of their annual spermatophore production in a single courtship episode, and only a few spermatophores are actually picked up by a female. Although this may seem like a wasteful strategy, there would be little advantage to a male in conserving his sperm supply if he has few opportunities to encounter additional females. This behavior contrasts with that of newts and plethodontid salamanders, which generally deposit only one spermatophore in each courtship bout. Although the time and energy invested in courting each female may be considerable, there is a much higher probability that each spermatophore will be picked up by a female and

used to fertilize her eggs. Hence, a male's limited supply of spermatophores is conserved for future mating attempts (Arnold 1976, Halliday 1990).

Male Competitive Ability

Explosive-breeding aggregations, mate searching, and mate guarding are systems in which competition among males to locate and hold on to females probably has a greater effect on male mating success than does female choice of mates. Therefore, one would expect sexual selection to favor male traits that enhance a male's competitive ability. Such traits also would be important for territorial males that form leks or have resource-defense mating systems. Competitive ability is most frequently measured as body size, because large males almost always win more fights than small males. Large body size has been shown to enhance the mating success of many amphibians and reptiles (Table 12–1). In explosive-breeding anurans, where struggles among males for possession of females are common, large males are more likely to mate than small males (Figure 12–18). The same is true for territorial species such as green frogs and bullfrogs. In contrast, male body size often is not very important in frogs that form choruses or leks, but only occasionally fight among themselves (Halliday and Tejedo 1995). There are few estimates of variation in male mating success for salamanders in the field, but large male *Desmognathus ochrophaeus* attack and chase away smaller males that attempt to court females in laboratory courtship encounters.

Large body size increases male mating success of snakes that fight over females, such as European adders (Madsen et al. 1993). Large body size is advantageous as well for male European grass snakes (*Natrix natrix*) and northern water snakes (*Nerodia sipedon*), which engage in scramble competition for females but do not fight (Madsen and Shine 1993, Weatherhead et al. 1995, Luiselli 1996). In contrast, large body size did not increase mating success in the explosive mating aggregations of red-sided garter snakes (Joy and Crews 1988).

Large males have greater mating success than small males in most species of territorial lizards that have been studied, including *Iguana iguana* (Rodda 1992), *Amblyrhynchus cristatus* (Trillmich 1983), *Anolis garmani* (Trivers 1976), and *Uta palmeri* (Hews 1990), and in nonterritorial species such as *Lacerta agilis*

Figure 12–17 **Energy cost of breeding for male marine iguanas (*Amblyrhynchus cristatus*).** (a) Percentage of males not feeding on a given day. The mating period is in early January. (b) Amount of time males were away from their territories on foraging trips in different months. *(Source: Trillmich 1983.)*

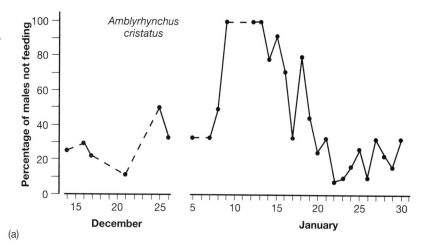

(a)

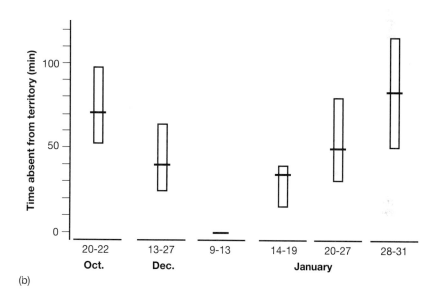

(b)

(Olsson 1993) and *Eumeces laticeps* (Cooper and Vitt 1993). Large male wood turtles (*Clemmys insculpta*) and desert tortoises (*Gopherus agassizii*) dominated smaller males in aggressive interactions and had greater mating success (Kaufmann 1992, Niblick et al. 1994).

Male traits other than large body size are related to male fighting ability of some species and almost certainly have evolved through sexual selection. Male green frogs wrestle while clasping one another with their forelimbs (see Figure 12–13a), and males have much more robust arms than females of the same body size. Sexual dimorphism in forelimb size is even

more pronounced in other frogs. The arms of males of some species of *Leptodactylus* from South and Central America become greatly enlarged during the breeding season (Figure 12–19), and they often have sharp spines on their front feet that are used to jab at opponents in fights (Halliday and Tejedo 1995). Males of *Rana blythi*, a ranid frog from Borneo, are larger than females and have larger heads and stronger jaws. Greatly enlarged fangs on the lower jaw probably are used to bite other males in fights (Figure 12–20a). Similar fanglike projections are present on the lower jaws of male *Pyxicephalus adspersus*, an African ranid frog. Males engage in violent and some-

Table 12–1   **Examples of amphibians and reptiles in which large body size increases male mating success.**

Species	Mating System	Mechanism of Sexual Selection
Anurans		
Bufo americanus	Explosive aggregation	Male competition
Bufo bufo	Explosive aggregation	Male competition
Physalaemus pustulosus	Chorus (lek)	Female choice
Rana catesbeiana	Resource defense	Male competition and female choice
Rana clamitans	Resource defense	Male competition and female choice
Rana sylvatica	Explosive aggregation	Male competition
Caudates		
Desmognathus ochrophaeus	Mate guarding	Male competition
Triturus cristatus	Lek	Female choice
Lizards		
Amblyrhynchus cristatus	Lek	Female choice
Anolis garmani	Resource defense	Male competition and female choice
Eumeces laticeps	Mate guarding	Male competition and female choice
Iguana iguana	Lek	Male competition and female choice
Lacerta agilis	Mate guarding	Male competition
Uta palmeri	Resource defense	Male competition and female choice
Snakes		
Natrix natrix	Mate searching	Male competition
Nerodia sipedon	Mate searching	Male competition
Vipera berus	Mate guarding	Male competition
Turtles		
Clemmys insculpta	Mate guarding	Male competition
Gopherus agassizii	Mate guarding	Male competition

Sources: Andersson 1994, Halliday and Tejedo 1995, and references cited in text.

times fatal fights during the breeding season (Hayes and Licht 1992). The skink *Eumeces laticeps* also fights by biting opponents. Males not only are larger than females, they also have longer and wider heads than do females of the same body length (Figure 12–20b). Similar sexual dimorphism in head size is found in the plethodontid salamander *Aneides aeneus*, and the males of this species are very aggressive toward other males (Staub 1993).

When aggressive encounters between males are frequent, it may be advantageous for males to assess the fighting ability of their opponents and to advertise their own ability. Males of some frogs and toads can judge the size of an opponent by the pitch of its call, because large males have lower-pitched calls than small males (Halliday and Tejedo 1995). Visual threat displays of red-backed salamanders (*Plethodon cinereus*) are probably used by males to advertise their size and willingness to fight other males (Jaeger and Forester 1993). Males of many lizards develop brightly colored patches on the body during the breeding season, and several recent studies have shown that other males use these badges to assess their fighting ability. Large male sand lizards (*Lacerta agilis*) have larger green

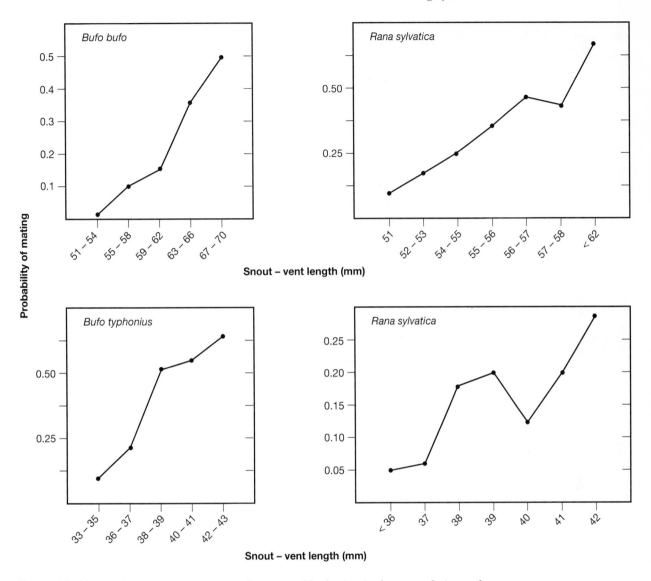

Figure 12–18 **Probability of mating as a function of body size in four populations of explosive-breeding anurans.** *(Sources:* Bufo bufo *data from Davies and Halliday 1978.* Bufo typhonius *data from Wells 1979.* Rana sylvatica *data from Berven 1981 (top) and Howard 1980 (bottom).)*

patches than small males, and large males are more likely to win fights (Olsson 1994a). Males interact repeatedly with the same opponents and learn which can be defeated and which cannot. Subsequently they avoid challenging males of superior fighting ability (Olsson 1994b). The size of the blue patches on the dewlaps of male *Urosaurus ornatus* is used by other males to assess fighting ability (Thompson and Moore 1991). Color is sometimes a better predictor of fighting ability than body size, and when males are similar in size, those with large amounts of blue color on their dewlaps win fights with males that have orange dewlaps. Previous experience in fights also affects the outcome of contests (Carpenter 1995).

The size or quality of a male's territory often influences the choice of mates by females, and sexual selection for traits correlated with male competitive ability may be indirect. That is, females may pay little or no attention to male body size or color per se, but larger or more colorful males may be able to obtain

(a)

(b)

Figure 12–19 **Sexual dimorphism in *Leptodactylus insularum*.** (a) Male, showing hypertrophied forelimbs used for wrestling with other males. (b) Female, showing normal forelimb size. *(Photos by Kentwood Wells.)*

more mates because they control territories that are attractive to females. This situation was documented for the lizard *Uta palmeri* by Diana Hews (1990). She found that males that mated were consistently longer and heavier than males in the population as a whole, and they also had wider, deeper heads with longer jaws. However, females appeared to choose mates mainly on the basis of territory quality rather than morphological traits. All of these traits affected a male's ability to defend a territory, so sexual selection indirectly favored large, heavy males with big heads and jaws. Indirect selection also appears to be operating on green coloration in male sand lizards. Males with more green color obtain more mates, but females do not prefer brighter males (Olsson 1994a).

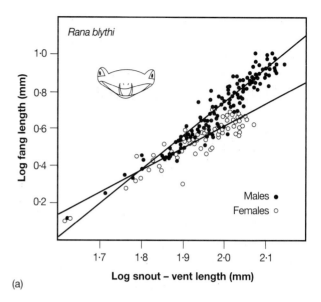

(a)

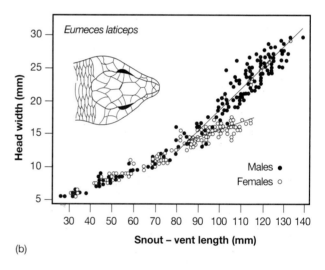

(b)

Figure 12–20 **Sexual dimorphism related to fighting behavior.** (a) Male *Rana blythi* have larger fangs on their lower jaws than females. (b) Male *Eumeces laticeps* have wider heads than females. *(Sources: Emerson and Voris 1992, Emerson 1994, Vitt and Cooper 1985.)*

Female Choice

The ability of females to compare the qualities of potential mates and select the best ones available was the most controversial part of Darwin's theory of sexual selection, and it also is the major focus of much of the recent work on sexual selection (Andersson 1994). The easiest kind of female choice to understand is choice based on resources controlled by males, because such resources can have a direct effect on female reproductive success. Howard's (1978b) study of the effect of male territories on survival of eggs laid by female bullfrogs is a good example.

More problematic is female choice based on morphological or behavioral traits of males. One hypothesis to explain such preferences is that once a preference for some type of male trait arises, it can be maintained, and even increased, through a process of runaway selection. This phenomenon would occur if there were a genetic correlation between the expression of a trait in males and the preference for that trait in females, so that females choosing males with the trait would pass along genes for the trait to their male offspring. Another hypothesis is that male traits chosen by females are indicators of male genetic quality, so that females choosing males with these traits would pass along good genes to their offspring. These could be genes that increase resistance to disease, foraging ability, metabolic efficiency, or any other characteristic not directly related to mating behavior. A third hypothesis is that females choose males with traits that elicit responses from the female's sensory system. For example, if females are particularly good at perceiving red color, then males that evolve displays of red structures might be more successful in obtaining mates than those with display structures of some other color. This hypothesis has been called the sensory exploitation model of sexual selection, because it assumes that sexual selection favors males that exploit a preexisting sensory bias of females. Although many studies have shown correlations between male morphological or behavioral traits with male mating success or female preferences for certain traits, it has been much more difficult to distinguish among these models of sexual selection (Andersson 1994).

The most common approaches to studying female choice are experimental studies that give females a direct choice between male traits and field studies that correlate variation in male traits with variation in mating success. Frogs and toads are particularly suitable for experimental studies because they communicate mostly with acoustic signals, which are more easily manipulated than visual or chemical signals. The usual method is to give a female a choice of two or more call variants in an arena and measure a preference when the female moves toward or contacts one of the speakers playing the calls. Playback experiments have consistently shown that female frogs pre-fer faster, louder, longer, or more complex calls (those with more notes) (Andersson 1994, Sullivan et al. 1995). In other words, females are attracted by calls that require more energy to produce. This may be because these calls provide some indication of differences in the physical condition of males or male quality, or it may be that such calls are more easily detected in a noisy chorus of calling males. In contrast, experiments in which females have been given a choice of calls differing in dominant frequency (pitch), which is negatively correlated with body size, have produced mixed results. Some studies have shown that females choose low-pitched calls produced by large males over high-pitched calls, but others have shown a preference for calls of average dominant frequency over either higher or lower frequencies, indicating that this trait may be subject to stabilizing rather than directional selection (Gerhardt 1991).

Experiments showing a preference for features of calls that might be indicators of male quality suggest that these call features should be under directional selection in natural choruses. It has been surprisingly difficult to document such patterns, however. One problem is that only a few studies have examined individual variation in calling behavior in natural choruses. If variation in calling rate, call intensity, or call complexity within an individual is similar to variation among individuals, then these features will not provide females with reliable information about male quality. In fact, several studies have shown that males do differ in their calling behavior, especially in features such as calling rate and call duration (Runkle et al. 1994, Jennions et al. 1995). Correlations between calling rate and mating success have been found, but in most cases, variation in calling rate explains only a small proportion of the variation in male mating success, because variation in calls is often less important than the amount of time a male spends in the chorus (Wagner and Sullivan 1995). In addition, females often compare only a few males before choosing a mate, so those with the highest calling rates or longest calls are not always selected (Sullivan et al. 1995). Finally, patterns of female choice may be confounded by alternative mating tactics adopted by some males who intercept females moving toward other males.

Much less is known about traits preferred by females of other groups of amphibians and reptiles. In general, female salamanders do not show clear evidence of preferring large males over small males, although in some species large males are more successful because of greater competitive ability (Sullivan et al. 1995). However, in the lek-breeding crested newt (*Triturus cristatus*), females prefer both large males and those with large, conspicuous crests on their backs and tails (Hedlund 1990). There also is evidence from laboratory experiments that female *Triturus vulgaris* prefer large males and males with large crests (Green 1991). One might expect variation in courtship intensity to correlate with male mating success, and in laboratory studies male newts that are persistent or court females several times improve their chances of transferring spermatophores (Halliday 1990). However, nothing is known about variation in courtship intensity among male newts in the field. Most male salamanders use pheromones to increase female receptivity, and preventing males from transferring pheromones reduces their mating success. Unfortunately it is very difficult to quantify the relative effectiveness of pheromones produced by different males, so we have no information about whether females use pheromones to assess differences in male quality.

Even less is known about criteria used by female reptiles to choose mates. Indeed, a recent review provided very few examples of sexually selected male traits in reptiles, and did not list any examples for turtles or crocodilians (Andersson 1994). In most studies in which male body size affected mating success, competition among males rather than female choice was the principal mechanism of selection (see Table 12–1). Only one study has shown an apparent female preference for large males, and in this species, *Eumeces laticeps*, females preferred males with large heads (Cooper and Vitt 1993). Not only did females more frequently associate with such males in the field, they also actively rejected courtship by small males. This species is not territorial, so large male size is not correlated with the quality of defended resources, but it is related to fighting ability. Females did not show a clear preference for the bright orange color that develops on the heads of males in the breeding season, suggesting that this color functions mainly in aggressive interactions among males. This result is consistent with other studies of brightly colored lizards, which have produced little evidence that females use color to choose mates (Tokarz 1995). Olsson and Madsen (1995) suggested that female choice of mates based on male traits such as body size or color may be uncommon among lizards because such traits provide unreliable information about male quality. The same may be true for other groups of reptiles. Another possibility is that female choice of male traits is difficult for investigators to detect because the traits are hard to manipulate experimentally. Clearly more work is needed in this area.

Alternative Mating Tactics

In mating systems in which a few males monopolize most of the available females, less competitive males often adopt alternative mating tactics that enable them to obtain some matings, although usually not as many as those obtained by dominant males. Such tactics can include various forms of sneak matings, male mimicry of female behavior, and sexual interference with the mating attempts of other males (Andersson 1994). One common type of alternative mating tactic is for satellite males to associate with dominant or territorial males and attempt to intercept females. This behavior has been reported for many species of frogs, including those that defend resource-based territories, such as bullfrogs and green frogs, and those that form choruses or leks and defend individual calling perches (Halliday and Tejedo 1995). Satellite male frogs are almost always smaller, younger males that cannot compete for choice territories. However, among species that defend only calling sites, such as tree frogs and toads, satellite males are not always smaller than calling males. In a study of mating behavior in the natterjack toad (*Bufo calamita*), Anthony Arak (1988) found that small males were more likely to adopt a satellite tactic than large males (Figure 12–21a). He also found that the frequency of satellite behavior increased with increasing chorus density (Figure 12–21b), a pattern seen in many other species of frogs as well. For example, in a study of *Eleutherodactylus johnstonei*, a terrestrial frog from Barbados, Ovaska and Hunte (1992) found many more satellite males in a high-density population than at a low-density site. Some satellite males were small, noncalling males that lacked territories, whereas others were callers that switched between calling and satellite tactics and were about the same size as the males with which they associated. Sometimes groups of males interfered with the mating of territorial males by breaking up their courtship attempts.

Several types of alternative mating tactics and sexual interference have been observed in salamanders (Halliday and Tejedo 1995). Males of explosive-breeding ambystomatids commonly cover the spermatophores of other males with their own sper-

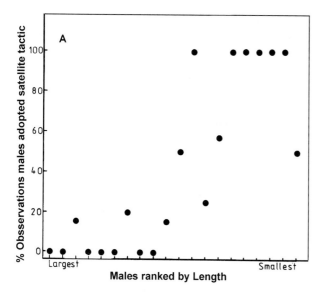

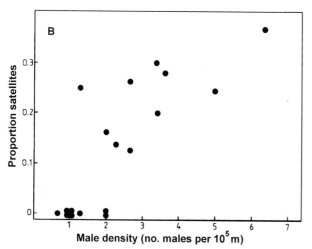

Figure 12–21   **Behavior of satellite males of the natterjack toad (*Bufo calamita*).** (a) Small males are more likely to adopt a satellite tactic than large males. (b) The proportion of males in a population adopting a satellite tactic increases with the density of males in the chorus. (*Source: Arak 1988.*)

gion of the courting male to elicit spermatophore deposition. The interfering male then covers the first male's spermatophore with his own (Figure 12–22). Male plethodontid salamanders lead females in a tail-straddling walk before depositing a spermatophore. Again, interfering males often insert themselves between the courting male and female and elicit spermatophore deposition. However, in this case, the second male does not cover the spermatophore, but he does cause the first male to waste a spermatophore, leaving the female to pick up the second male's spermatophore (Arnold 1976). Similar behavior has been described for several species of newts (*Triturus, Taricha,* and *Notophthalmus*).

Female mimicry of another sort occurs in red-sided garter snakes in explosive-mating aggregations. Some males produce a pheromone in the skin that resembles that of females and is attractive to other males (see Chapter 11). This causes males to waste time and energy courting the so-called she-males while allowing the female mimics to get closer to the female in a mating ball and increasing his chances of successful mating (Mason and Crews 1985). Female mimicry has not been reported among lizards, but satellite behavior is common in some territorial and

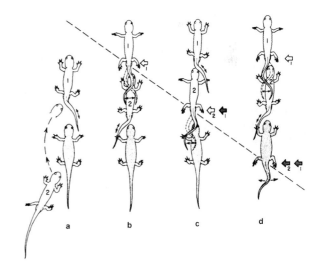

Figure 12–22   **Sexual interference by male tiger salamanders (*Ambystoma tigrinum*).** (a) The female is following male 1 during courtship. Male 2 inserts himself between the courting male and female. (b) Male 2 nudges male 1, a behavior normally used by females to elicit spermatophore deposition. (c) Male 2 deposits a spermatophore on top of the spermatophore previously deposited by male 1. (d) The female moves forward and picks up the spermatophore of male 2. (*Source: Arnold 1976.*)

matophores (see Figure 12–4). When a female encounters one of these multiple spermatophores, she is most likely to pick up the one on top. Male *Ambystoma tigrinum* court females individually, rather than engaging in the sort of scramble competition seen in *A. maculatum* and other explosive breeders. Interfering males sometimes insert themselves between the male and female as the first male leads the female during courtship. The interfering male mimics the behavior of a female by touching the cloacal re-

lekking species. For example, small male green iguanas hang around the periphery of larger males' territories and attempt to copulate with females that are passing through. Most of these attempts are resisted by the female and are unsuccessful, but a few are successful. Both territorial and peripheral males sometimes try to force females to copulate rather than engaging in lengthy courtship (Rodda 1992). Forced copulation has been observed in other lizards as well, including the Lake Eyre dragon (*Ctenophorus maculosus*) from the deserts of southern Australia (Olsson 1995). Forced copulation can be costly to females because they may be injured and because it circumvents their attempts to choose particular males as mates. From a male's perspective, it is a way to mate quickly with as many females as possible in the face of intense competition from other males.

Sperm Competition

When females mate with several different males, either simultaneously in species with external fertilization or sequentially in species with internal fertilization, there is a potential for sperm of different males to compete to fertilize eggs. Sperm competition can have a major impact on male mating success (Devine 1984, Halliday and Verrell 1984). Indeed, measuring male mating success simply by counting the number of matings a male obtains may be inaccurate in situations in which sperm competition is possible. In such cases, genetic studies using DNA fingerprinting or other techniques are needed to measure true reproductive success. Multiple paternity has been documented for several plethodontid salamanders, which sometimes store sperm for long periods before laying eggs (Verrell 1989). Despite vigorous mate guarding by males, female sand lizards (*Lacerta agilis*) often mate with several males, resulting in multiple paternity of egg clutches (Olsson et al. 1996).

Multiple paternity also occurs in snakes and is particularly common when aggregated males engage in direct competition over females. These include northern water snakes (*Nerodia sipedon*), some populations of garter snakes (*Thamnophis sirtalis sirtalis*), and adders (*Vipera berus*) (Schwartz et al. 1989, Barry et al. 1992, Madsen et al. 1992). Female adders often mate with three or four different males in a single season. The sperm of one male certainly is sufficient to fertilize all of a female's eggs, so what is the advantage to

females in mating so frequently? One possibility is that they simply allow males to mate with them to avoid constant harassment by courting males. Madsen et al. (1992) proposed another, more direct benefit to females. They reported that the viability of offspring was higher when females obtained multiple mates. Their interpretation of these results was that multiple mating resulted in direct competition between sperm of different males, with the sperm of genetically superior males being more likely to fertilize a female's eggs and produce high-quality offspring.

Males of frogs with external fertilization usually ensure that they father all of the eggs in a female's clutch by grasping the female in amplexus, with sperm being released while the male holds his cloaca close to that of the female. However, groups of males of some arboreal frogs that form explosive breeding aggregations sometimes participate in simultaneous fertilization of a female's eggs. This behavior has been observed in a number of rhacophorid frogs that build foam nests hanging in trees over rain pools. Once a female begins to lay her eggs, groups of males climb onto her back and participate in beating the mucus around the eggs into a thick foam (Figure 12–23). With this type of mating system, one way for a male to enhance his reproductive success is to release enormous numbers of sperm. Foam-nesting rhacophorid frogs in which group mating is common have larger testes relative to body size than other frogs, with the largest testes being found in *Chiromantis xeramplina* and *Rhacophorus arboreus*, species in which multiple males participate in 80 to 90 percent of matings (Kusano et al. 1991, Jennions and Passmore 1993).

Patterns of Sexual Size Dimorphism

Darwin originally formulated his theory of sexual selection to account for the evolution of sexual differences in morphology and body size. Several examples of sexually selected male traits have already been mentioned—the tail fins and crests of newts, the enlarged heads, jaws, and teeth of some lizards and salamanders, the hypertrophied forelimbs and enlarged fangs and tusks of frogs. We also have seen that large body size increases the mating success of males in many amphibians and reptiles. Yet the typical pattern in most lineages of amphibians and reptiles is for females to be larger than males, and sometimes considerably larger. This is true for most frogs and toads, salamanders, turtles, and snakes. Only among crocodilians and lizards do we find a large proportion of species with males as large as or larger than females. In all of these groups, there are species that depart from the typical pattern. For example, there are frogs, salamanders, snakes, and turtles in which males are larger than females, and lizards and crocodilians in which females are larger than males. To understand these patterns of sexual dimorphism, we need to know something about selective forces acting on body size of both males and females.

For most amphibians and reptiles, the number of offspring produced by a female increases with body size because a larger female can pack more eggs into her body than a small female. Hence, one can expect selection for large clutch size (fecundity selection) to favor large body size in females. However, this is not true for all groups. For example, *Anolis* lizards lay one egg at a time regardless of body size, and geckos typically lay clutches of two eggs, so variation in clutch size cannot explain the evolution of body size of these lizards. Furthermore, increased fecundity of females does not necessarily mean than females are expected to be larger than males. When the reproductive success of females increases faster with increasing body size than the reproductive success of males, selection should favor larger body size in females. On the other hand, if the reproductive success of males is more affected by body size than that of females, selection may favor larger body size in males. This situation is most likely to occur when males fight among themselves for possession of females or territories.

In general, the evolution of unusually large body size in males is associated with male combat in frogs (Wells 1978, Shine 1979, Howard 1981), salamanders (Shine 1979), lizards (Stamps 1983), snakes (Shine 1994), and turtles (Berry and Shine 1980). In snakes, this relationship holds even when differences in phylogenetic history are taken into account. Nevertheless, there are numerous exceptions to this pattern. In frogs, relatively large male body size seems to be associated with resource-defense mating systems, such as those of bullfrogs and green frogs. When males defend only calling sites or fight for possession of females, they usually are considerably smaller than females, even when large males are more likely to mate (Halliday and Tejedo 1995). Males are larger than females in many viperid, colubrid, and elapid snakes with male combat. However, in the European adder (*Vipera berus*) and the European grass snake (*Natrix*

Figure 12–23 **The African gray tree frog, *Chiromantis xeramplina*.** Several males mating with a single female. *(Photograph by K. G. Preston-Mafham.)*

natrix), males are smaller than females, even though large males win more fights and acquire more mates (Madsen 1988, Madsen and Shine 1993).

Females of most aquatic turtles are larger than males, whereas the males of some terrestrial turtles, especially tortoises, are larger than females. Berry and Shine (1980) attributed this relationship to differences in male combat, which is relatively common in male tortoises but rare in most aquatic turtles. In contrast, Gibbons and Lovich (1990) proposed that males of terrestrial species become sexually mature later than males of aquatic turtles, allowing them to put most of their energy into growth and reach a size that protects them from predators.

While patterns of association between behavioral traits such as male combat and relative sizes of males and females suggest a role for sexual selection in the evolution of sexual dimorphism, these correlations alone do not reveal what actually causes differences in body size. In some cases, sexual size dimorphism may result from natural selection favoring different food habits of males and females, or other types of ecological divergence. However, it often is difficult to tell whether such divergence is a cause or an effect of sexual size dimorphism. For example, if males and females of a certain species of snake differ greatly in size and also eat different sizes of prey, this situation could result from selection for ecological divergence, or sexual selection might lead to divergence in size that would enable the two sexes to eat prey of different sizes. Most examples of sexual size dimorphism that have been studied so far have been attributed to sexual selection, but in many cases the data are inadequate to refute the possibility that ecological divergence is important as well (Shine 1989).

Most interspecific comparisons of sexual dimorphism are based on average sizes of males and females with the assumption that any difference is due to natural or sexual selection on body size. However, it is possible for average sizes of males and females in a population to differ even in the absence of differences in selection on body size. For example, average sizes may differ if males and females grow at similar rates and reach similar body sizes but large individuals of one sex are susceptible to predation. This was the case in the population of bullfrogs studied by Howard (1981). He did not find any difference in growth rates of male and female bullfrogs. However, males spent more time in the water during the breeding season than females, and they were more active. Their calling and territorial behavior made them conspicuous, and some of the largest males were eaten by snapping turtles, sometimes while fighting with other males. A similar situation was observed for European adders. Males that were most successful in obtaining mates also suffered the highest mortality, and these generally were the largest males (Madsen and Shine 1994).

A further complication in interpreting patterns of sexual dimorphism among amphibians and reptiles is that these animals often continue to grow after reaching sexual maturity, in contrast to most birds and some mammals. Growth is rapid early in life and gradually levels off, producing an asymptotic growth curve (Figure 12–24). The maximum size of males and females actually provides the best estimate of asymptotic size. Maximum size is a better measure of

sexual dimorphism than average size for *Anolis* lizards, and probably for other amphibians and reptiles. Ideally, the best way to investigate sexual dimorphism is to have data on growth rates of males and females. If these differ, then it is likely that selection acts differently on body size in the two sexes (Stamps 1995).

Amphibians and reptiles exhibit several different patterns of sex-specific growth. In species with little or no sexual dimorphism, males and females grow at similar rates. In other species, the growth rates of males and females diverge early, resulting in different asymptotic sizes. Examples include the African bullfrog (*Pyxicephalus adspersus*), the Neotropical lizard (*Basiliscus basiliscus*), and the American alligator (*Alligator mississippiensis*), all species in which males are considerably larger than females (Figures 12–24a,b,d). The same pattern is seen in the slider turtle (*Trachemys scripta*) and many other aquatic turtles, but in these species, females grow more rapidly and are larger than males (Figure 12–24c). The pattern is somewhat different in wood turtles (*Clemmys insculpta*). Males and females grow at similar rates for the first 20 years of life, but after that growth virtually ceases in females, whereas males continue to grow slowly and reach a larger size (Lovich et al. 1990).

Stamps (1995) investigated patterns of sexual dimorphism in different species of *Anolis* lizards by comparing asymptotic sizes of males and females. Territorial male lizards generally acquire more mates when population densities of females are high because they can defend territories that overlap the home ranges of several females (Stamps 1983). The increased availability of females at high densities also is expected to increase competition among males for access to those females. Because large body size is correlated with fighting ability, sexual selection on male body size should be strongest in species that occur at high densities. Indeed, when Stamps plotted an index of sexual size dimorphism against female population density for a number of species of *Anolis*, she found that sexual dimorphism increased with increasing density (Figure 12–25). This relationship would not necessarily be expected if differences in sizes of males and females were related solely to differences in prey size or some other ecological variable. Although this type of analysis has not been done for other groups of amphibians and reptiles, it seems likely that similar variation in sexual dimorphism will be found in other groups in which densities of females vary among species or populations, such as crocodilians.

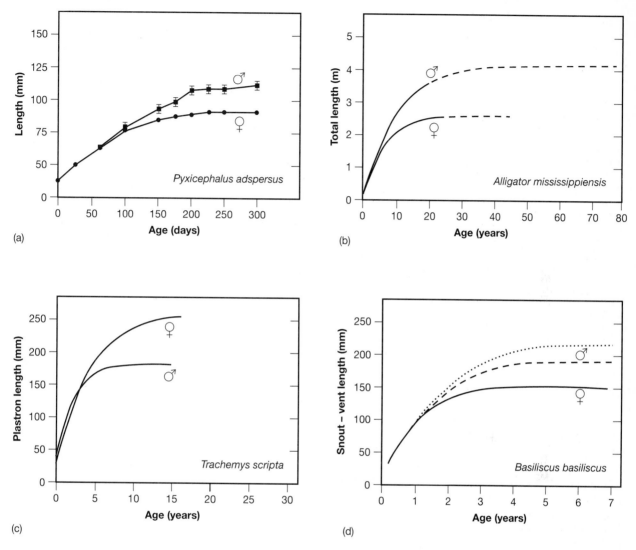

Figure 12–24 **Asymptotic growth resulting in sexual dimorphism in body size of amphibians and reptiles.** (a) Males of the African bullfrog (*Pyxicephalus adspersus*), grow faster than females after reaching sexual maturity. (b) Males and female American alligators (*Alligator mississippiensis*) grow at similar rates for the first 5 years of life, after which growth of females slows. (c) Female slider turtles (*Trachemys scripta*) grow faster than males after the first few years of life and reach a much larger size. (d) Males of the neotropical lizard *Basiliscus basiliscus* grow faster than females after the first year. Growth curves for two different populations of males are shown. (*Sources: (a) Hayes and Licht 1992, (b) Chabreck and Joanen 1979, (c) Dunham and Gibbons 1990, (d) Van Devender 1978.*)

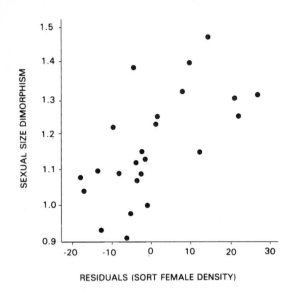

Figure 12–25   **Relationship of sexual dimorphism (the ratio of male to female size) to female density for 25 populations of *Anolis* lizards.** *(Source: Stamps 1995.)*

Summary

The study of mating systems and sexual selection began with Charles Darwin's work in the 1870s, but only in the last 20 or 30 years has there been much serious work on sexual selection in amphibians and reptiles. Most species of amphibians and reptiles are polygynous, with males attempting to acquire multiple mates in a breeding season. Females of some species acquire several different mates as well. Sexual selection generally favors multiple mating by males, because their reproductive success is limited by the number of mates they acquire. Reproductive success in females usually is limited by the energy reserves needed to produce eggs, but multiple matings may provide some genetic benefits to females.

The type of mating system adopted by males of a particular species depends in large part on the spatial and temporal distribution of females. When females are highly aggregated, the resulting male mating system can be either an explosive mating aggregation when the breeding season is short and females are available simultaneously, or a more stable aggregation of males (a lek) when the breeding season is long and females are available asynchronously. When females are moderately aggregated, males often defend resource-based territories in areas that are attractive to females. When females are highly dispersed, males usually are forced to find mates by searching over wide areas and sometimes guard individual females once they are encountered. Explosive mating aggregations are common among species of amphibians and snakes that mate in early spring. Leklike mating aggregations occur in many frogs, some salamanders, large herbivorous lizards, and some crocodilians. Resource-defense mating systems predominate in small insectivorous lizards and are found in some amphibians as well. Mate searching and mate guarding are behaviors held in common by salamanders, widely foraging, nonterritorial lizards, and snakes.

The type of male traits that are favored by sexual selection depend on the nature of the mating system. Explosive mating aggregations generally favor large body size and other traits correlated with male competitive ability, because females have few opportunities to choose mates. Large body size and conspicuous advertisement signals are often favored in species that form leks or defend resource-based territories, situations in which both male–male competition and female choice become important determinants of male mating success. Hence, sexual selection is responsible for the evolution of loud, complex vocal signals in frogs, for bright coloration and elaborate courtship in newts, and for large body size, bright colors, and conspicuous display structures in many lizards. Mate searching will tend to favor traits related to rapid movement and the ability to detect and locate receptive females, such as responsiveness to pheromonal signals. If males also guard females once they are encountered, large size and conspicuous color may be favored as well, even if females do not choose mates based on these traits. Sexual selection also can favor the production of large quantities of sperm, especially in those species in which females are likely to mate with more than one male, leading to direct sperm competition.

13

Foraging Ecology
and Interspecific Interactions

In the first section of this chapter, we describe broad patterns of diet and foraging behavior of amphibians and reptiles. We then consider amphibians and reptiles as the food resources of other animals. Numerous kinds of parasites and predators find their sustenance in amphibians and reptiles. Parasitism and predation have very different effects on individuals and populations. Because parasites are so small compared to their hosts, they are usually difficult or impossible to avoid. The response to parasitism is thus largely physiological, and defensive mechanisms involve the host's immune system. Susceptibility to parasitism is genetic, and selection will alter the interaction between parasites and hosts. In contrast, predators can often be avoided, deterred or fought off, and amphibians and reptiles have evolved a range of defensive mechanisms that enhance their survival.

Foods and Feeding

Most amphibians and reptiles are carnivores, that is, they eat other animals. Some are herbivores, and this category includes both species that feed exclusively on plants and are highly specialized, and species that ingest plant material only occasionally and show no particular specializations. Carnivorous reptiles assimilate about 90 percent of the energy in vertebrate prey. Herbivores have lower assimilation efficiencies than carnivores because vertebrates cannot digest cellulose,

the material that makes up the cell walls of plants. Herbivorous reptiles assimilate only 30 to 60 percent of the energy in plant material, even with the help of microbial symbionts. Carnivores usually have short digestive tracts, whereas herbivores have long digestive tracts that increase the time that foods are exposed to digestive processes. Digestive efficiency is a function of diet, not of phylogenetic lineage, and amphibians and reptiles extract energy and nutrients with the same efficiency as birds and mammals (Karasov and Diamond 1988).

What does differ between ectotherms and endotherms is the rate of food processing; mammals process their food about ten times faster than reptiles, largely because of greater surface area of the mammalian than the reptilian gut (Karasov and Diamond 1985). This difference in rate of digestion is not a handicap for amphibians and reptiles because their energy needs are about ten times lower than those of endotherms.

The expression "we are what we eat" is aptly applied to amphibians and reptiles. Diets are associated with morphological, physiological, and behavioral characteristics that facilitate location, identification, capture, ingestion, and digestion of food items. Morphological specializations associated with feeding were discussed in Chapter 9. In the following section we discuss the diets of carnivorous and herbivorous amphibians and reptiles and some of the ecological and evolutionary factors that affect their diets.

Carnivory

Amphibian and reptilian carnivores typically have diets that are restricted to particular kinds of prey—insects and other invertebrates, fishes and frogs, squamates, or mammals and birds. Within each prey type, carnivores can be categorized by the size of their prey relative to the size of the predator and by specialization on particular types of prey.

In general, insects and other invertebrates are eaten by small amphibians and reptiles and vertebrates are eaten by large species. Given their small body sizes, it is not surprising that the diets of most amphibians and lizards consist mostly of invertebrates and those of snakes and crocodilians consist mostly of vertebrates. Exceptions occur, of course. Large frogs, salamanders, and lizards occasionally eat vertebrates, including members of their own species, and juvenile crocodilians eat insects.

The size of feeding structures is often more closely related to the size of prey than to the size of the predator. For example, all adult anurans swallow their prey whole, and the size of an anuran's mouth is a key to the sizes and types of prey it eats (Emerson 1985b). Anurans that feed on prey that are small relative to the predator (such as mites, termites, and ants) have small heads and narrow mouths, whereas anurans that feed on large prey have broad heads and wide mouths (Figure 13–1). At one extreme are members of the Central American family Rhinophrynidae. These fossorial frogs have tiny heads, and apparently feed on termites by shooting their tongues through a groove in the small opening of the mouth. On the other extreme are the African bullfrog, *Pyxicephalus*, and several South American frogs, including the genera *Ceratophrys* and *Lepidobatrachus*, that feed on anurans (including members of their own species). These frogs have heads so large relative to their bodies that they look like cartoons of hopping mouths.

The evolution of snakelike shapes provided new ecological opportunities for many lizards. The ability to move easily through dense vegetation, litter, sand, and even soil gives access to prey and to hiding places denied to limbed squamates. Like snakes, snake-shaped lizards must feed a large body through a small mouth. This constraint is especially severe for lizards, which have far less skull kinesis than do snakes. Termites are small and abundant in many habitats, and they are frequent items in the diets of snake-shaped lizards. For example, the fossorial and legless *Ty-phlosaurus* skinks in the Kalahari desert of Africa feed largely on termites, even specializing on particular castes within a species of termite (Huey et al. 1974).

Even within adult populations of amphibians and reptiles, body size affects prey size. Sexual differences in the body size of males and females are often associated with differences in foraging behavior and prey size. For example, female file snakes, *Acrochordus arafurae* (an aquatic species from northern Australia), have much larger bodies and heads than males. Females forage in deep water and eat large species of fish, whereas males forage in shallow water and eat smaller species of fish (Shine 1991).

Individuals of the larger sex may eat relatively large prey, and a difference in prey may appear even before the sexes reach adult size. For example, adult males of the lizard *Anolis conspersus* are considerably larger than adult females, and adult males eat larger insects than adult females (Figure 13–2). Moreover, juvenile males eat larger prey than juvenile females, even when both sexes are the same size (Schoener 1967).

Within a prey type, a species may also be characterized by the degree of specialization on particular species or categories of prey. Many species of amphibians and reptiles eat a wide variety of prey taxa within a particular prey type. This generalist diet is characteristic of species that feed on insects and other invertebrates. For example, a dietary analysis revealed that individuals of the North American salamander *Plethodon cinereus* had eaten representatives of nine classes of invertebrates, including ten orders of insects (Jaeger 1990). The total diversity of prey items must have included hundreds of species. Despite the apparently catholic tastes of many insectivorous amphibians and reptiles, a comparison of dietary items with the availability of those items in the habitat indicates that individuals are selective. *Plethodon cinereus*, for example, exhibits selectivity based on digestibility (Jaeger 1990). During wet nights when salamanders are able to forage most successfully, they eat many small soft-bodied insects that can be processed rapidly. In contrast, on dry nights when foraging activity is limited by the risk of dehydration, individuals eat more hard-bodied (more heavily chitinized) insects that take longer to digest. Presumably, hard-bodied prey are not eaten on wet nights because they take so long to process.

Dietary specialization occurs widely within both amphibians and reptiles. Species with diets that are

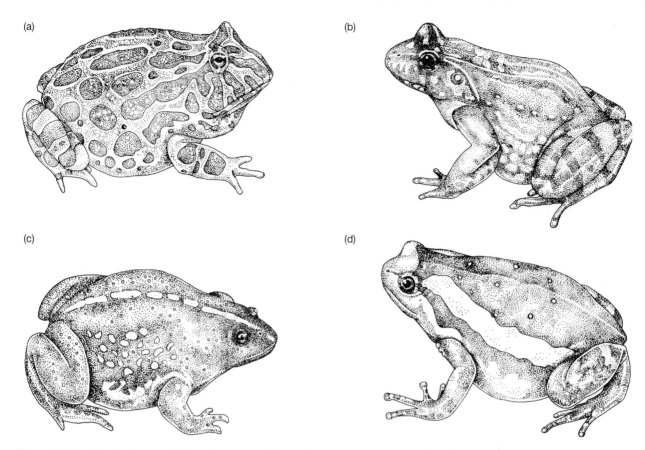

Figure 13–1   **Body form and diet of anurans.** Frogs that eat large prey items have large heads and wide mouths and frogs that eat small prey items have small heads and narrow mouths. Contrast the relative head sizes of two South American bullfrogs *Ceratophrys ornata* (a), and *Leptodactylus pentadactylus* (b), that eat large prey, including vertebrates, with the Mexican burrowing frog *Rhinophrynus dorsalis* (c), and the Asian painted frog *Kaloula pulchra* (d), that eat termites and ants, respectively. Frogs are not drawn to scale. *(Source: Modified from Cogger and Zweifel 1992.)*

limited to one or a few prey types often exhibit morphological, physiological, and behavioral specializations for feeding. For example, dietary specialization on ants (myrmecophagy) is exhibited by some North American horned lizards (*Phrynosoma*: Phrynosomatidae) and the Australian thorny devil (*Moloch horridus*: Agamidae). These lizards have spiny, tanklike bodies, peglike teeth, and cryptic coloration. Their convergence in morphology and behavior may be associated with myrmecophagy. That is, the tanklike bodies of horned lizards and thorny devils may be needed to accommodate the large volumes of small prey they eat, and the stout bodies may make rapid locomotion impossible. Unable to run away from predators, horned

lizards and thorny devils have evolved defensive spines. Moreover, cryptic coloration and limited activity may make these lizards relatively inconspicuous to predators. This hypothesis may be correct, but it must be noted that other species of lizards that subsist largely on ants (*Sceloporus magister* in North America and *Liolaemus monticola* in Chile) have dental specializations similar to those of horned lizards and thorny devils but do not have the other morphological specializations of these species (Greene 1982).

Amphibians and reptiles that eat shelled molluscs cope with the indigestible shell in a variety of ways. Two hyperoliid frogs (*Tomierella*) endemic to the highlands of Ethiopia feed on snails and slugs

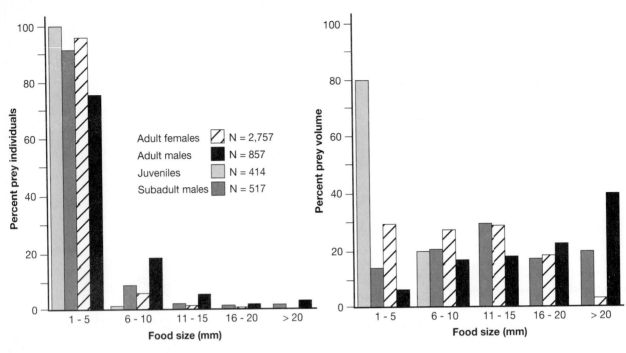

Figure 13–2   **Sex-based difference in diet.** Percentage of prey individuals (left) and prey volume (right) in five categories of prey size eaten by four age and sex classes of *Anolis conspersus*. Subadult males and adult females have the same body sizes (snout–vent lengths of 36–54 millimeters) yet subadult males eat more and greater volumes of large prey than do adult females. This is particularly obvious for the largest prey size category in the right-hand graph. (*Source: Schoener 1967.*)

(Drewes and Roth 1981). Unlike related species that do not eat snails, these frogs have skulls in which the upper jaw is strongly braced against the braincase. The frogs do not crush snails but eat them whole, and their rigid skulls may help them pull snails from their attachment to the substrate. The New World snakes in the genus *Dipsas* also feed on snails (Sazima 1989). Unlike *Tomierella*, however, these snakes remove snails from their shells before eating them. Some map turtles (*Graptemys*) are sexually dimorphic; females have relatively large jaws that are used to crush shelled mollusks, while males have relatively small jaws and feed on smaller and softer prey items than females (Ernst et al. 1994).

Marine turtles and snakes appear particularly prone to dietary specialization. The hawksbill sea turtle (*Eretmochelys imbricata*) feeds almost exclusively on sponges (Meylan 1988). Many sponges are protected from grazing predators because their skeletons are made of silica (glass) spicules. Hawksbills must eat large volumes of silica to gain access to the nutritious portion of their prey. How they do this without abrading their digestive tract is not known. The dietary habits of leatherback turtles (*Dermochelys coriacea*) are peculiar as well: they eat large numbers of jellyfish, which not only contain stinging nematocysts but are mostly water. Judging by one collection of sea snakes from the coast of the Malay Peninsula, radiation within this group has involved extreme dietary specialization (Glodek and Voris 1982). Only one of the 16 species collected could be considered a prey generalist. The others were placed into one of five diet categories: eels, gobys, burrowing gobys, catfishes, or fish eggs. Each of these dietary specializations is associated with sensory, behavioral, and morphological adaptations for locating and capturing prey. For example, sea snakes that capture eels in their burrows have narrow heads and necks, and sea snakes that eat eggs have greatly reduced dentition.

Herbivory

Herbivory is widely but unevenly distributed among amphibians and reptiles. Many amphibians and reptiles occasionally eat plant material, and some eat both plant and animal material regularly. For example, the diet of the North American slider turtle, *Trachemys scripta*, includes plants as well as a wide range of aquatic invertebrates and small vertebrates such as fish and tadpoles and carrion (Gibbons 1990). Opportunistic feeding by sliders extends to neustophagia—the process of skimming small floating items such as pollen and algae (neuston) from the water's surface (Belkin and Gans 1968). This feeding behavior has been reported for other aquatic turtles as well (Figure 13–3).

Fruit and flowers are particularly common food items for many omnivorous species because these structures have large, energy-rich cells that are easily ruptured and their energy and nutrients are readily accessible (Schall and Ressel 1991). In contrast, nutrients are not easily released from within the cell walls of leaves, especially by amphibians and reptiles that do not chew their food. Cellulose, the polysaccharide that makes up the cell wall of plants, is not digested by vertebrates, although it represents a potentially rich source of nutrients. Some herbivores have symbiotic intestinal microbes (bacteria and protozoans) that break down cellulose into simple compounds, such as volatile fatty acids, that are absorbed by the gut.

This specialized association with symbiotic microbes is characteristic of land tortoises (Testudinidae), the green sea turtle (*Chelonia mydas*), all members of the lizard family Iguanidae, the agamid lizards *Uromastyx* and *Hydrosaurus*, and the scincid lizard *Corucia*. These species are herbivorous throughout life, and their diets are largely made up of leaves (Iverson 1982). Microbial processing takes place in an enlarged proximal portion of the large intestine (see Figure 9–42). In general, the digestive efficiencies of these specialized herbivorous turtles and lizards are similar to those of mammalian hind-gut fermenters, such as horses and rabbits (Bjorndal 1985). For example, *Iguana iguana* of Central America digests 54 percent of its plant diet, and microbial fermentation provides at least 38 percent of its caloric intake (McBee and McBee 1982, Troyer 1984). The digestive efficiencies of these herbivorous lizards are thus similar to those of ruminant mammals

Unlike mammals, herbivorous reptiles do not exhibit any parental care, and the gut symbionts needed to digest food are not passed directly from parents to their offspring. How, then, do hatchlings acquire their gut microbes? Katherine Troyer's (1982) observations of *Iguana iguana* in Panama suggest that transferring the gut symbionts between generations is the function of a specialized social interaction that occurs soon after hatching. Iguanas nest communally, and many clutches are deposited in close proximity. Hatchlings swallow soil from the nest chamber and from the

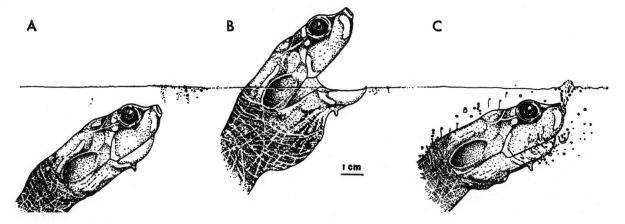

A　　　　　　**B**　　　　　　**C**

Figure 13–3 **Neustophagia.** The pelomedusid turtle *Podocnemis unifilis* feeding on small particles floating on the water surface. (a) Approaching surface, (b) skimming particles, (c) expelling water from pharynx and retaining particles. *(Source: Belkin and Gans 1968.)*

nesting area during the first week after hatching. Soil bacteria form a simple fermentation system in the gut that facilitates digestion of plant material during the second week of life. Hatchlings then disperse as family groups (Burghardt et al. 1977) into the forest canopy, where they associate with adults and acquire their microbial symbionts by seeking out and eating feces of adults (Troyer 1982). This brief association with adults is necessary because the symbionts are obligate anaerobes and do not live long when exposed to oxygen outside a lizard's gut. After inoculation with gut symbionts, juvenile iguanas return to the forest edge vegetation, where they remain for several years while they grow to adult size.

Because specialized herbivorous lizards are large and carnivorous lizards tend to be small, Pough (1973) suggested that herbivory might be incompatible with small body size. He reasoned that plants have low digestibility and protein content compared to animals. Species with small body sizes or small individuals of large-bodied species may not be able to fill their energy and nutrient demands on a strictly plant diet. Recent observations on juveniles of a large-bodied herbivore, however, suggest that herbivory at small body sizes is effective, perhaps because small herbivores select the most highly nutritious plant parts (seeds, flowers, or new tissue). For example, the 12-gram hatchlings of the herbivorous turtle *Pseudemys nelsoni* had a higher assimilation of energy and nitrogen than 3000-gram adults when both fed on hydrilla, an aquatic plant. Because of their smaller heads and mouths, the hatchlings were able to feed selectively on the leaves of hydrilla, which are high in nutrients relative to stems, while the adults fed nonselectively on both leaves and stems (Bjorndal and Bolten 1992).

Cannibalism

Some amphibians and reptiles kill and eat individuals of their own species (Polis and Myers 1985). Cannibalism is particularly widespread among amphibians, especially at the larval stage, and is usually associated with limited resources, such as an ephemeral breeding pond (Crump 1992). Typically, cannibalism is asymmetrical because adults eat larvae or eggs, and larvae eat smaller larvae or eggs. Asymmetry can also be produced if individuals vary greatly in size, either from high variance in growth rate or from a long breeding period. Larvae of tiger salamanders (*Am-bystoma tigrinum*) and several species of spadefoot toads (*Spea*) have cannibalistic morphs with enlarged jaw muscles and predatory foraging behavior (Figure 13–4). The development of these morphs seems to be facultative (i.e., its occurrence depends on conditions in the ponds) and is controlled by a combination of genetic and environmental factors.

In general, eating a conspecific should be selectively disadvantageous because that individual may be a relative. Because cannibalism is so common among amphibians, several researchers have evaluated the responses of potential cannibals to close relatives. When cannibals are given a choice between siblings and unrelated individuals of the same size, they usually eat more unrelated individuals than siblings. However, in a study of the marbled salamander, *Ambystoma opacum*, Susan Walls and Andrew Blaustein (1995) found the opposite: kin were the preferred prey. They suggested that cannibalism of smaller kin may greatly increase the direct fitness of an individual (its own potential reproductive success) without decreasing the individual's indirect fitness (the potential reproductive success of relatives) because smaller relatives could not metamorphose before the pond dried out.

The risk of eating a relative is not the only cost of cannibalism. Another cost is the increased probability of contracting a disease or acquiring a parasite from the prey. Cannibalistic morphs of tiger salamanders have higher rates of infection by nematodes and are more likely to die from bacterial disease than are noncannibalistic morphs (Pfennig et al. 1991). Lakes that are contaminated with bacteria have lower frequencies of the cannibal morphs than lakes that are relatively free of bacteria, probably because of the risk to a cannibal.

Cannibalism also occurs in situations where the victim is clearly not kin and the cannibal presumably benefits both from the meal, and from eliminating a potential competitor for food or mates. Female California newts (*Taricha torosa*) eat the eggs of other females as they are being extruded from the cloaca during oviposition (Kaplan and Sherman 1980). Males of *Eleutherodactylus coqui* raid other males' nests and consume the eggs. This behavior would increase the relative fitness of the cannibal by reducing the reproductive success of a competitor. Thus, a primary function of egg-guarding by the males of this species is thought to be protection from oophagy by conspecific males (see Chapter 7).

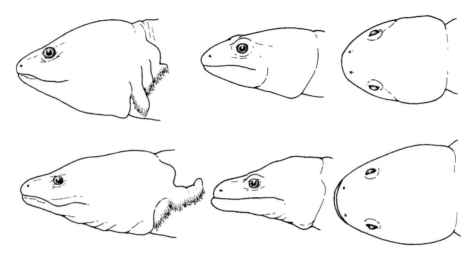

Figure 13–4 **Normal and cannibalistic morphs of the tiger salamander,** *Ambystoma tigrinum.* Top row: larval and adult normal morph. Bottom row: larval and adult cannibalistic morph. *(Duellman and Trueb 1986, after Rose and Armentrout 1976.)*

Cannibalism is not as common in reptiles as in amphibians, and the limited information available supports the idea that eating conspecifics may not be adaptive. Adult *Anolis cuvieri*, the giant anole of Puerto Rico, do not attack conspecific juveniles, although they readily eat similarly sized individuals of other species of *Anolis* (Rand and Andrews 1975). Adults can distinguish conspecific juveniles from other sympatric species of *Anolis* because juvenile *A. cuvieri* have a distinct pattern of broad dark vertical bands.

Some adult curly-tailed lizards (*Leiocephalus schreibersi*) in the Dominican Republic do not eat juveniles, even though cannibalism by adults appears to be a major cause of juvenile mortality. Tom Jenssen and his colleagues (1989) found that only about one third of the adult lizards in their study population were cannibalistic. They presented juvenile *Leiocephalis* and *Anolis* of the same body size to adult curly-tailed lizards. Two thirds of the adult *Leiocephalis* ate the *Anolis*, but only one third of them attacked juvenile *Leiocephalis.* Of the 24 resident juveniles in the study area, all but three had home ranges that were completely within the home ranges of a noncannibal male and a noncannibal female; the other three juveniles had home ranges partially overlapped by a cannibalistic adult. Adults actively defend their territories from intruding adults, so any juvenile within a territory of a noncannibal would gain protection. It may be that a noncannibal pair is necessary to protect juvenile *Leiocephalis* from adult cannibals. Perhaps nonreproductive adults eat conspecific juveniles, but reproductive adults do not eat juveniles within their territories. This hypothesis and the related hypothesis that noncannibalistic adults are protecting their own young require further testing.

Ontogenetic Variation in Diet

In contrast to birds and mammals, in which only individuals of adult size forage, populations of amphibians and reptiles are made up of individuals varying widely in body size. Changes in the diets of carnivores during ontogeny reflect an increasing ability to capture, subdue, and swallow large prey. Both maximum prey size and the range of prey sizes eaten typically increase as individuals grow because large individuals continue to eat small prey while they add large prey to their diets (Figure 13–2). The types of prey eaten may change as well. For example, two species of water snakes, *Nerodia fasciata* and *N. erythrogaster,* that are anuran specialists as adults eat small fish as juveniles (Mushinsky et al. 1982). Their diets shift to frogs and toads when they reach body lengths of between 50 and 70 centimeters (Figure 13–5).

Ontogenetic changes in diet are particularly striking for species that are large as adults because their

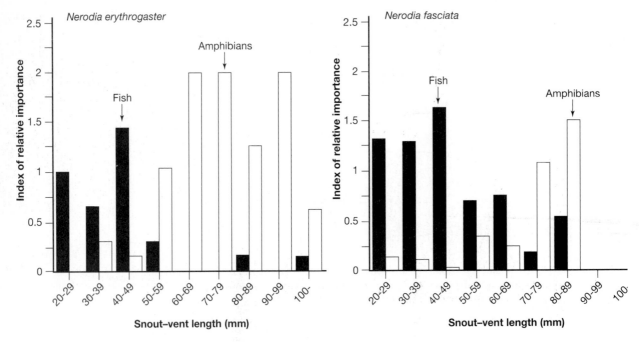

Figure 13–5   **Ontogenetic changes in the diets of the two species of water snakes.** The index of relative importance is a measure of the contribution of either fish or amphibians to the diets of each of the size classes (snout–vent length) of snakes. *Nerodia erythrogaster* and *N. fasciata* eat fish when they are small and switch to anurans when large. (*Source: Adapted from Mushinsky et al. 1982.*)

hatchlings are so much smaller. For example, adult Komodo dragons (*Varanus komodoensis*) in Indonesia weigh as much as 100 kilograms and reach a total length of 3 meters. Adults prey on wild boar, deer, and water buffalo (Figure 13–6). In contrast, at hatching Komodo dragons weigh about 100 grams and eat insects and other small prey (Auffenberg 1981).

A few species of reptiles shift from a largely carnivorous diet as juveniles to a largely herbivorous diet as adults. For example, juvenile slider turtles are largely carnivorous, whereas adults are omnivorous. Juveniles may be able to capture moving prey more readily than adults because juveniles are more agile. Additionally, the carnivorous diets of juveniles provide more energy and nutrients (especially nitrogen) for growth than the herbivorous diets of adults.

Ecological and Evolutionary Determinants of Diets

Why do members of a species eat particular foods or exhibit a particular manner of foraging? At the level of an individual, feeding is the result of genetically de-

termined behaviors that are mediated through the sensory system and are influenced by external factors such as experience and prey availability (Feder and Lauder 1986).

One explanation for a particular diet composition is selectivity by a predator, and the mechanisms of selectivity are diverse. Green tree frogs (*Hyla cinerea*) in Florida eat more larvae of particular moth and beetle species than would be expected from their abundance in the habitat (Freed 1982). Attacks on prey by green tree frogs are elicited by movement. Because these moth and beetle larvae are moving, the frogs eat them and ignore other kinds of insects that are less active.

Herbivorous lizards and tortoises are selective feeders and often concentrate their feeding on a few species of plants or even on a particular age class of leaves of particular plant species. The reasons for this selectivity are well known. The leaves of many plants have unpalatable or toxic chemicals, and these chemicals are more concentrated in older than in younger leaves. Furthermore, older leaves usually have more fiber and are harder to digest than are younger leaves from the same plant. Nonetheless, adult iguanid lizards eat mostly mature leaves, perhaps because

Figure 13–6 **Komodo dragons feeding on a carcass.** *(Photograph by Philippa Scott/NHPA.)*

younger leaves are produced seasonally, and iguanids cannot depend on them for year-round sustenance (Rand et al. 1990).

Selectivity is often exhibited before an individual has had any exposure to potential prey, showing that preferences for particular prey are genetically based (innate). Newborn snakes flick their tongues and strike at cotton swabs that have been dipped in water extracts of potential prey (Burghardt 1967). Moreover, newborns respond most strongly to odors of the prey normally eaten by members of their own population. For example, smooth green snakes (*Opheodrys vernalis*) eat insects, and newborn green snakes respond most strongly to insect extracts, while brown snakes (*Storeria dekayi*) eat earthworms, and newborn brown snakes respond most strongly to earthworm extracts. Thus, the specialization often seen in diets of snakes is produced, in part, by innate responses to the chemical cues of particular prey types.

In contrast to the stimulation of innate prey preferences by olfaction among snakes, the innate feeding responses of some lizards are stimulated by visual cues. *Anolis lineatopus*, a small Jamaican lizard, is insectivorous. At hatching each individual has its own menu of acceptable and unacceptable prey types (von Brockhusen 1977). Some individuals will not eat crickets unless they have been starved, whereas other individuals readily eat crickets. Individual preferences change slowly as individuals gain experience so that as adults, *A. lineatopus* individuals eat a greater diversity of prey types than they do as hatchlings.

Because diet preferences are often exhibited by naive individuals, we assume that these preferences have a genetic basis, but this assumption is seldom tested. One of the few well-analyzed examples is Stevan Arnold's (1981a,b) study of the genetic basis of feeding preferences of the North American garter snake, *Thamnophis elegans*. Individuals from coastal areas where slugs are common feed mostly on slugs, whereas individuals from inland areas that do not have slugs feed mostly on frogs and fish. Arnold tested newborn snakes from the two areas by presenting them with small pieces of slugs on 10 consecutive days and scoring the snakes on the basis of the number of pieces eaten. Snakes were classified as slug eaters if they ate five or more pieces or slug refusers if they ate four or fewer. Of the snakes from coastal areas, 73 percent were slug eaters, whereas only 17 percent of snakes from inland areas were slug eaters. Newborn snakes from matings between individuals from the two areas had intermediate preferences—about 28 percent were slug eaters. These observations indicate that geographic variation in prey preference has a genetic basis, and that the slug-refusing trait exhibits at least partial dominance. Arnold concluded that preferences of garter snakes for different types of prey are adaptive, reflecting geographic variation in prey distribution.

Availability of prey is another factor that affects diets. Diets vary from season to season or from year to year in any one locality. This variation reflects the dynamic nature of natural communities; physical conditions change and affect the biological community. For example, terrestrial amphibians can use a wider range of microhabitats under wet than under dry conditions, and they may encounter more prey on wet days when they can forage more widely. During dry periods, red-backed salamanders (*Plethodon cinereus*) are restricted to relatively moist microhabitats such as beneath logs and underground, where they encounter few prey. In contrast, during rainy periods they not only forage on the ground, they also climb on plants. Salamanders foraging on plants on wet nights ate a mean volume of prey of 1.2 mm^3 per night, while salamanders foraging on the forest floor ate a mean volume of prey of 0.6 mm^3 per night (Jaeger 1978).

Substantial year-to-year variation occurs in diet (Figure 13–7). At Eagle Lake in California, annual variation in rainfall determines water level and whether or not the toad *Bufo boreas* is able to breed. In wet years, garter snakes (*T. elegans*) fed in meadows near the lake and ate newly metamorphosed toads, whereas in dry years snakes fed on fish and leeches that they caught in shallow water at the edge of the lake. Fish and leeches are always available, but they are eaten infrequently in years when toads are available (Kephart and Arnold 1982).

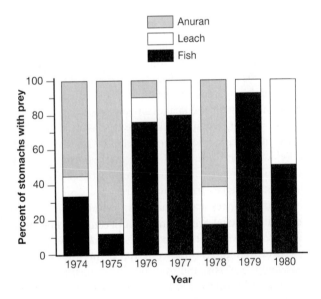

Figure 13–7 **Annual variation in the summer diets of *Thamnophis elegans*.** The histograms show the proportion of snakes at Eagle Lake, California that had eaten anurans, leeches, and fish in each of 7 years. (*Source: Kephart and Arnold 1982.*)

In an evolutionary context, the size and taxonomic diversity of food items eaten by a species are related to the fitness of individuals that exhibit particular feeding strategies. Complex interactions of morphology, physiology, and behavior set the framework for individual fitness. The evolutionary origins of these interactions can be assessed if phylogenetic relationships among species are known. Catherine Toft (1995) and Janalee Caldwell (1996) made such an assessment in their studies of dendrobatid frogs. Dendrobatids are common terrestrial and semiarboreal frogs in the rain forests of Central and South America. When the diets of dendrobatids are superimposed on a cladogram depicting their phylogenetic relationships, the result is a picture of the evolution of feeding behavior within this group (Figure 13–8).

The genus *Colostethus* is an ancestral group within the family. These frogs are cryptically colored, have nontoxic skin, and feed on insects that are large relative to the size of the frog. *Colostethus* do not have to move often because a few large prey fulfill their daily energy requirements. Their feeding behavior is thus compatible with crypsis. The common ancestor of the genera *Epipedobates*, *Phyllobates*, and *Dendrobates* made a critical shift in diet and physiology. These frogs began to feed on relatively small prey, including ants. The dietary shift to small prey was associated with the uptake of toxic alkaloids from these prey and the incorporation of those alkaloids into the frogs' skin. *Epipedobates*, *Phyllobates*, and *Dendrobates* are characterized by highly toxic skin secretions and bright or aposematic colors that warn potential predators that they are poisonous. Because they feed on small prey, these frogs must eat large numbers of prey to fulfill their daily energy needs. As a consequence they must move frequently, but because of their toxic skins and aposematic coloration they can do so without increasing their risk of predation.

Experimental studies support this evolutionary scenario. Members of the genus *Colostethus* do not accumulate alkaloids in their skins even when these chemicals are provided experimentally. Moreover, the dietary origin for the toxic alkaloids in the skin of *Dendrobates*, at least, is supported by the absence of these chemicals from the skins of individuals raised in the laboratory on insects without alkaloids, and by the immediate incorporation of alkaloids when these chemicals are provided experimentally (Daly, Garraffo, et al. 1994; Daly, Secunda, et al. 1994).

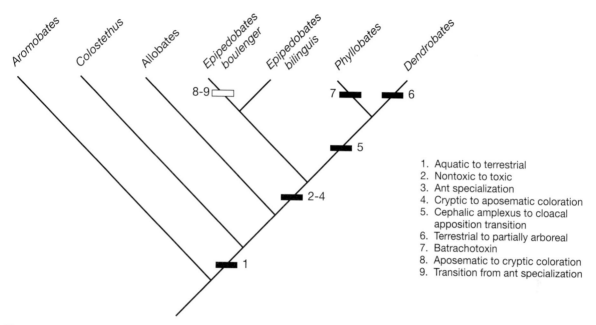

Figure 13–8   **Evolution of myrmecophagy in the frog family Dendrobatidae.** Derived characters are superimposed on the cladogram at the points where they presumably arose during the evolution of this group. Note particularly the association among feeding on ants, toxic skins, and aposematic coloration. The open bar represents the reversal of derived characters within the dendrobatids. *(Source: Adapted from Caldwell 1996.)*

Foraging Strategies

The behaviors that amphibians and reptiles use to encounter food items are more limited than one would expect from the diversity in their diets. For this reason, foraging modes lend themselves to broad conceptual generalization. The foraging modes of small insectivorous frogs and lizards are better known than those of other groups, but generalizations based on frogs and lizards appear relevant to other taxa, including fishes, birds, and mammals.

Foraging Modes

Two extreme types of foraging behavior by amphibians and reptiles have been identified—sit-and-wait foraging and active foraging. Sit-and-wait foragers search for prey from a fixed site until a potential prey animal comes close enough to capture. Capture may follow a brief pursuit, or the predator may wait until the prey can be ambushed. Active foragers move frequently, search large areas, and may dig or probe actively for concealed prey. Prey are usually detected at close range and captured with little or no pursuit.

Sit-and-Wait Predators.   Many frogs in tropical forests are sit-and-wait predators of insects in the litter of the forest floor. These frogs are cryptically colored and eat relatively large prey items. In the New World tropics, sit-and-wait anurans include many species of *Eleutherodactylus* (Leptodactylidae). In Africa, species in the genera *Arthroleptis* and *Arthroleptella* (Ranidae) are sit-and-wait predators. The similarities in feeding behavior and reproductive biology (direct development) of these forest floor frogs reflect convergent evolution.

Sit-and-wait lizards typically make forays from a perch site to capture prey, and large prey are pursued farther than small prey. In the New World, *Sceloporus* and *Anolis* are classic sit-and-wait predators. In the Old World, this foraging mode is exhibited by agamids and chamaeleonids. Some pit vipers maximize the chance of prey capture by waiting next to established trails of small mammals. After striking and injecting its prey with venom, a pit viper uses its olfactory senses to follow the scent trail of the dying victim.

Some amphibians and reptiles do not merely sit and wait but use parts of their bodies as lures to attract prey (Murphy 1976, Sazima 1991). Tongues, tails, and toes that resemble worms are used to bring prey within striking range (Figure 13–9). Alligator snapping turtles (*Macroclemys*) display a wormlike appendage on their tongue that attracts fish. Juveniles of some Old and New World vipers and of the pygopodid, *Lialis burtonis* in Australia attract small frogs and lizards by waving or twitching their tails. Their tails are often lighter in color than the rest of the body, a feature that probably makes the lure conspicuous to prey. The South American frog *Ceratophrys* uses its toes as lures. Depending on the location of the potential prey, these frogs hold a hind foot out from their body or raise it over their head and twitch several toes to attract smaller frogs.

Actively Foraging Predators.   Active foraging (also referred to as wide foraging) lies at the opposite end of the spectrum of predatory behaviors. Active foragers expend energy as they search but use little energy in pursuit of individual prey items. As a result, active foragers appear to eat all of the suitable items that they encounter, and the individual prey are small. In the New World, teiids (*Cnemidophorus* and *Ameiva*) and the colubrid *Masticophis* are classic active foragers, and this foraging mode is paralleled in Eurasia by many lacertids and in Australia by scincids in the genus *Ctenotus*, by monitor lizards (*Varanus*), and by the elapid *Demansia*.

Figure 13–9 **Luring with tongues and tails.** (a) Alligator snapping turtle (*Macroclemys temminckii*) displaying wormlike appendage on tongue. (b) Young copperhead (*Agkistrodon contortrix*) luring frog into striking range by wiggling its yellow tail. (*Photographs by (a) F. Harvey Pough, (b) Zig Leszcznski/Animals Animals.*)

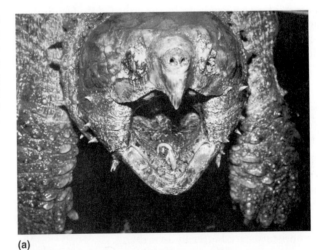

(a)

(b)

Association of Traits.   The correlation of foraging modes, diet, and defense in the family Dendrobatidae emphasizes the point that foraging modes cannot be viewed in isolation from other traits. In general, the behaviors that allow amphibians and reptiles to find and capture food must be compatible with their morphology and physiology as well as with other important functions such as reproduction and the avoidance of predators or stressful physical conditions. It is thus not surprising that the type of foraging mode used by a particular species is consistently associated with other traits (Table 13–1).

Some morphological and physiological traits are clearly linked to the mechanics and energetics of locomotion (Taigen and Pough 1983). Sit-and-wait foragers have high sprint speed but low stamina, appropriate traits for sedentary animals that seldom move but must move quickly when they do. Active foragers have low sprint speed and high stamina, desirable characteristics for animals that are in motion for long periods. Sit-and-wait foragers capture a smaller volume of prey per day than active species do and have lower daily energy requirements.

Other characteristics associated with these foraging strategies are related to predator avoidance (Toft 1981). Sit-and-wait foragers usually match their background, and crypsis is enhanced by low rates of movement. In contrast, actively foraging species often have patterns and colors that are conspicuous, at least to a human observer. Because their frequent movements make them conspicuous, being cryptic would have little value, and other selective forces act on color and patterns.

Sit-and-wait foragers capture prey infrequently and eat relatively large items. Prey tend to be mobile, because moving prey are easily seen by a stationary predator. For example, the Central American casque-headed lizard, *Corytophanes cristatus*, is an extreme sit-and-wait forager (Andrews 1979). Individuals of *Corytophanes* rely on crypsis for protection from predators in their rain forest habitat, feed once a day or less, and eat insects that average one-half of their body length (Figure 13–10). Actively foraging predators are conspicuous and may be captured by sit-and-wait predators, whereas sit-and-wait predators are most likely to be discovered by actively foraging predators (Huey and Pianka 1981).

Among squamates, foraging and the sensory mode used to detect and identify prey are tightly correlated. Sit-and-wait foragers rely on vision to detect moving

Table 13–1   **Correlates of foraging mode based on studies of small insectivorous lizards and frogs. Except where noted, patterns are probably general and can be applied to other amphibians and reptiles. Relative clutch mass is the ratio of clutch mass to female body mass; this entry applies only to lizards.**

Correlates	Sit-and-Wait Foragers	Active Foragers
Prey types	Mobile, large	Sendentary, small, often clumped
Prey mass captured/day	Low	High
Daily energy requirement	Low	High
Physiological correlates	Limited endurance, high sprint speed, low aerobic capacity	High endurance, low sprint speed, high aerobic capacity
Chemosensory tongue flicking (squamates)	No	Yes
Types of predators	Active foraging	Sit-and-wait and active foragers
Mode of escape	Camouflage, speed, saltation	Speed (squamates) skin toxins/aposematic (frogs)
Morphology	Stocky, wide-mouthed	Streamlined, narrow-mouthed
Relative clutch mass	High	Low

Source: Adapted from Huey and Pianka (1981), Toft (1981), Taigen and Pough (1983), Huey et al. (1984), and Cooper (1994).

Figure 13–10   **Typical prey item of the Central American lizard** *Corytophanes cristatus.* The juvenile lizard had a head-and-body length of 53 millimeters and the beetle larva removed from its stomach had a total length of 28 millimeters. *(Photograph by Robin Andrews.)*

prey at a distance, they tongue-flick infrequently, and they do not use scent to detect or identify prey. In contrast, active foragers often search for hidden prey. Their frequent movements bring them close to potential prey, they sample the environment by tongue-flicking, and they use chemosensation to detect and identify prey (Cooper 1994, Schwenk 1995).

Reproductive traits also are associated with foraging mode. Most sit-and-wait foragers have chunky bodies and large clutch masses relative to body size, presumably because a large clutch mass is not a handicap to females that do not move far or often. In contrast, actively foraging species have streamlined bodies and small clutch masses. Among lizards the annual reproductive output of sit-and-wait foragers is not necessarily greater than that of actively foraging species, however, because actively foraging species may produce several small clutches per season while sit-and-wait foragers usually produce just one.

Foraging modes are evolutionarily conservative traits that characterize particular groups, often at fairly high taxonomic levels. For example, iguanians and scleroglossans represent two monophyletic groups that diverged early in squamate history. The iguanians retained the ancestral sit-and-wait foraging mode while the scleroglossans developed an active foraging mode (Cooper 1994). Because of the tight link between phylogeny and foraging mode, deciding which attributes of sit-and-wait and actively foraging species are the direct consequences of foraging mode

per se and which are the result of phylogeny is difficult. To resolve this problem, Raymond Huey and his colleagues (Huey and Pianka 1981, Huey et al. 1984) looked at morphological, physiological, and ecological traits of sit-and-wait and actively foraging lacertid lizards that are both closely related and sympatric. They found that the broad spectrum of differences suggested by comparisons of distantly related taxa (Table 13–1) are applicable to closely related taxa as well.

Observations of squamates that have foraging modes unlike the ancestral mode for their lineage suggest that shifts in foraging mode are associated with shifts in the ability to detect prey chemicals. For example, iguanids use active foraging to locate their food plants. This is a derived foraging behavior, because the ancestral condition was sit-and-wait predation on insects. The derived foraging mode is accompanied by a derived means of food detection—at least one iguanid, the North American *Dipsosaurus,* uses its tongue to detect food chemicals as do other actively foraging squamates (Cooper 1994).

Viewing foraging modes as a dichotomy has provided useful insights into complex adaptations like foraging behavior. But is the dichotomy between sit-and-wait and active species real, or are these two foraging modes simply the ends of a continuum of foraging behaviors? Robert McLaughlin's (1989) synthesis of movement patterns of foraging lizards suggests that the dichotomy is real. The distribution of the number of moves per minute for lizards is bimodal, with mean numbers of moves per hour of 14 and 145 for sit-and-wait foragers and active foragers, respectively (Figure 13–11). However, the distinction between active foraging and sit-and-wait predation depends on the scale of comparison. For example, all the lacertids studied by Huey and Pianka (1981) were classified by McLauglin as active foragers. In other words, all lacertids move frequently when they are compared with lizards such as *Anolis* and *Sceloporus,* which are sit-and-wait foragers, whereas Huey and Pianka were able to show distinctions among lacertids between active foragers and sit-and-wait predators.

Optimal Diets

We assume, probably reasonably, that specialized structures used for feeding or prey capture promote dietary specialization by increasing the efficiency of

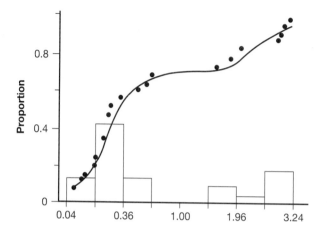

Figure 13–11   **Movement frequencies of 22 species of lizards.** The observed (dots) and predicted (line) cumulative distributions do not differ significantly (Kolmogorov-Smirnov one-sample test, D = 0.08, *P* > 0.20). *(Source: Based on McLaughlin 1989.)*

capture of certain kinds of prey. Measuring changes in efficiency as morphological characters evolve may be difficult or even impossible. In contrast, because foraging behavior can be observed directly, we can ask, Do animals forage optimally? To be an optimal forager, an individual should maximize its net energy or nutrient intake while foraging. This definition of optimal foraging assumes that efficiency is related to fitness, an assumption that seems reasonable in many cases. Presumably an individual that can rapidly fulfill its energy requirements for a certain time period will have greater reproductive success than individuals that must spend more time foraging. The efficient individual will have additional time for other activities, such as searching for mates or defending a territory. Another benefit of efficient foraging is reducing the time an individual is exposed to predators that it encounters during foraging.

What is an optimal diet? Several mathematical models predict optimal diets on the basis of the relative profitability of each prey type, the probability of encountering each prey type, and the energy required to subdue and swallow the prey. Schoener's (1971) model was designed for predators that expend little energy searching for prey, that is, they are sit-and-wait foragers, such as *Anolis* lizards. Profitability (e_i/t_i) is assessed from the net energy value of prey type i (e_i) divided by the time (t_i) required to capture, subdue, and swallow prey type i. Profitability thus represents

the net amount of energy gain per unit time spent subduing and swallowing the prey. The potential energy gained from each prey type is a function of the number of prey that a predator encounters and captures during a foraging bout. We will call this the encounter rate (ER_i) for prey type i. Finally, we need to know the energy requirement of the predator (M) in order to determine what prey type or types should be eaten.

Prey types are ranked in order of highest to the lowest profitability. Then the total energy gain (E) for the first-ranked prey type is calculated by multiplying e_i by ER_i, and the total handling time (T) for that prey type is calculated by multiplying t_i by ER_i. The ratio E/T is the total net energy acquired per unit time spent foraging. If the predator has not fulfilled its energy requirement by eating only individuals of the highest-ranking prey type, then we need to include the second-ranked prey type in the optimal diet. To do this, we simply calculate E/T for this prey type and add it to the E/T for the first-ranked prey type to give the total net energy acquired per unit time for the two highest-ranked prey types. If the predator still has not fulfilled its energy requirements, the calculations are extended to additional prey types until the predator has fulfilled its energy requirement. The set of prey types that fulfill the predator's energy requirement makes up the optimal diet.

This model makes three predictions. First, predators should prefer the most profitable prey. Second, predators should become increasingly specialized (eliminate lower-ranked items) when profitable prey become more common. Third, predators should not eat prey that are not part of the optimal diet.

The prediction that is most generally supported by studies of amphibians and reptiles (and other taxa) is that increasing prey abundance is associated with increasing specialization on the most profitable prey (those with the highest e_i/t_i). Robert Jaeger and Debra Barnard (1981) tested optimal diet theory under controlled laboratory conditions (Figure 13–12) using red-backed salamanders (*Plethodon cinereus*) as predators and two sizes of fruit flies (*Drosophila*) as prey. As prey density increased, the salamanders ate relatively more of the larger flies and usually ignored the smaller flies, even when small flies could have been readily captured. Both results are in accord with predictions of optimal foraging theory. In contrast to theoretical predictions, however, some small flies were eaten at all densities.

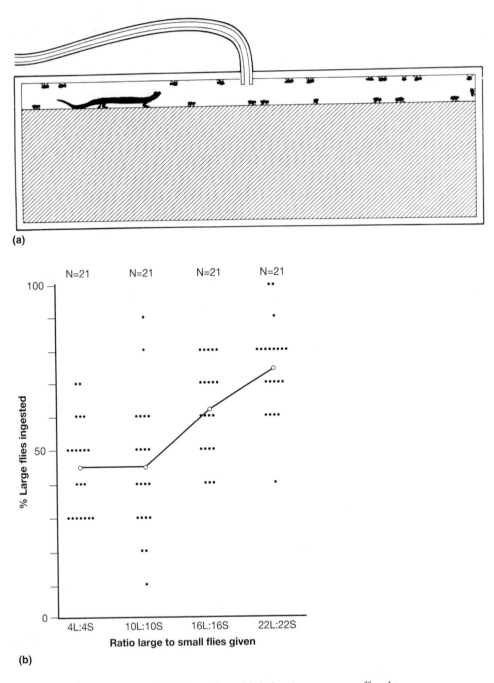

(a)

(b)

Figure 13–12   **Optimal diets.** Red-backed salamanders, *Plethodon cinereus*, were offered two sizes of fruit flies (*Drosophila*). (a) Diagram of testing chamber. The salamander is shown in a position typically adopted just before striking at a prey while pursuing it. (b) The percentage of large flies ingested as a function of the density of large (L) plus small (S) flies provided. Open circles are mean values and solid circles are the responses of individual salamanders. Note that as the density of flies increases, the salamanders eat more of the larger and more profitable prey type. *(Source: Jaeger and Barnard 1981, Jaeger and Rubin 1982.)*

Early models of optimal diets assumed that the abundance and energy value of their prey were the only factors of importance. Because the behavior of real predators does not match the predictions of these models (e.g., red-backed salamanders continue to eat smaller prey even when larger prey are abundant), more sophisticated models have been developed. More realistic assumptions (for example, that learning to recognize new prey may require several encounters, that choice may be affected by the degree of satiety of the predator, and that predator experience may affect diet) has led to predictions that are more consistent with the results of experiments. Red-backed salamanders, for example, forage optimally on larger flies only when they have had previous experience with both large and small flies (Jaeger and Rubin 1982). Another improvement in optimal foraging models was recognition that nutrients could be the currency of fitness rather than energy (Westoby 1984).

Parasite–Host Interactions

The parasites of amphibians and reptiles are virtually the same as those that plague other groups of vertebrates. The biology of parasite–host relationships is best known for humans because of the social and economic impact of diseases such as tuberculosis, AIDS, malaria, and schistosomiasis. For similar reasons, the effects of parasitism are considerably better known for birds and nonhuman mammals than for amphibians and reptiles. There is no reason to suspect, however, that parasitism has qualitatively different effects on amphibians and reptiles than it has on other vertebrates. Crowley (1992) provides a synthesis of this complex topic.

Microparasites

Microparasites consist of viruses, bacteria, and protozoans. Microparasites multiply rapidly within their hosts, and rapid reproduction is facilitated by small size and short generation time. Because of the speed with which microparasites multiply in a host, infection typically has acute effects and high levels of mortality. The duration of infection in an individual may be short, however, because infection stimulates the host's immune system. If the host lives, immunity may

protect it from further infection. Protozoans in particular have evolved elaborate mechanisms to escape the host's immune response and may persist despite an immune response. *Lacerta vivipara* in France, for example, that had high numbers of the haemogregaraine protozoans in their blood cells in 1992 also had high numbers in 1993 (Sorci 1995). These observations suggested that individual lizards have limited ability to recover from infection.

Amphibians and reptiles are associated with many viruses and bacteria, but most studies have been carried out in laboratories, and the results cannot necessarily be extrapolated to natural populations. Natural infections by viruses and bacteria are poorly known and draw attention only when epidemics cause conspicuous die-offs of common species. For example, the bacterium *Aeromonas hydophila* is associated with epidemics of red-leg, a disease that affects both larval and postmetamorphic amphibians (Carey 1993). This disease is discussed in Chapter 15. Amphibians and reptiles are infected by protozoans, including *Plasmodium* (malaria), as well as amoebae, flagellates, coccidians, and ciliates that are intestinal parasites.

Microparasites are often transmitted directly from one host to another. The protozoan *Sarcocystis* is a parasite of the giant agamid lizard (*Gallotia stehlini*) on the island of Gran Canaria (Matuschka and Bannert 1989). Infective stages of *Sarcocystis* encyst in the muscles of an individual's tail. These lizards are extremely aggressive, and they break off their opponents' tails during fights. Transmission of the parasite to a new host occurs when a lizard eats the tail of an infected individual.

For many parasites with direct transmission, the reproductive cycle of the parasite is synchronized with that of the host. For example, the reproductive cycle of *Opalina*, the most abundant and widespread of the intestinal flagellate protozoans in amphibians, is controlled by the reproductive hormones of its host. Infective *Opalina* cysts are released along with feces when the adult frogs enter water to breed, and tadpoles are infected when they consume the *Opalina* cysts.

Direct transmission is effective when infected and uninfected hosts encounter each other frequently, but when hosts are widely distributed the probability that a parasite will encounter a suitable host is very low. This ecological problem may explain the evolution of life cycles that include one or more intermediate hosts. For example, sand flies are the intermediate host and the transmission vector for malarial parasites of the western fence lizard, *Sceloporus occidentalis*. A sand fly becomes infected by feeding on the blood of an infected lizard. The parasite goes though a series of developmental stages in the sand fly, and only then can the bite of an infected sand fly transfer the parasite to another lizard.

The protozoan *Plasmodium* is one of the few parasites of reptiles for which the effect of parasitism on the host is known under field conditions. Many biologists believe that lizard malaria is an evolutionarily old parasite–host relationship because about half of the 120 species of *Plasmodium* that infect vertebrates are parasites of lizards. (Only four species of *Plasmodium* infect humans.) Lizard malaria can exist in a stable equilibrium in which the host is no longer affected adversely. Studies in Panama of the lizard *Anolis limifrons* infected with *Plasmodium* show that survival, growth, reproduction, and the mass of infected individuals are no different from those of noninfected individuals (Rand et al. 1983).

In contrast to this apparently benign relationship between *A. limifrons* and *Plasmodium*, malarial parasites of western fence lizards (*Sceloporus occidentalis*) have a strong negative impact on their hosts. Malarial infection reduces the maximum rate of oxygen consumption and locomotor endurance, and the diminished performance that infected lizards exhibit in the laboratory is also evident in the field. During the breeding season, infected males spend less time interacting with other males than do noninfected males. Infected males are presumed to have reduced reproductive success because social interactions with other males determine access to mates (Schall and Sarni 1987). Similarly, female fence lizards infected with malaria store 17 to 32 percent less fat and produce one to two fewer eggs than uninfected females. Perhaps the difference in the effect of malarial parasitism on *A. limifrons* and *S. occidentalis* is historical; populations of lizards that experience strong impacts from malaria may be those in which the host–parasite relationship is relatively recent.

Macroparasites

Macroparasites typically do not multiply within their hosts. Life cycles of macroparasites involve the release of eggs or larvae that must encounter a new definitive host (the host in which reproduction of the parasite occurs) or an intermediate host or hosts (the host in which larval stages of the parasite occur). Many characteristics of macroparasites are related to the low probability of the transmission from one host to another. For example, macroparasites have long reproductive life spans, which increases the probability of successful transmission. Also, macroparasites typically have sublethal effects on their hosts because the host must live long enough for the parasite to reproduce successfully. If the host dies too soon, the macroparasite will fail to reproduce, providing strong selection for parasites with nonlethal effects. The effect of macroparasites on their hosts is a function of their numbers. Typically, most hosts have low to moderate levels of parasites, and a few individuals have very heavy infestations. Because the number of parasites within one host is usually a function of the infection rate, individuals tend to accumulate parasites throughout their lives. The number of parasites is also affected by the interaction between infection rates and immune responses of the host, and the immune response has a strong genetic component.

Macroparasites of amphibians and reptiles may be internal or external. Internal macroparasites include various kinds of worms. Parasitic flatworms (Platyhelminthes) are common. Tapeworms inhabit the intestinal tract where they absorb nutrients directly through their integument, and flukes inhabit several organs, feeding on body tissue and blood. Parasitic roundworms, such as filarial worms, threadworms, hookworms, and pinworms (Nematoda), are particularly abundant in the intestinal tract, lungs, and blood vessels. Thorny-headed worms (Acanthocephala) are common intestinal parasites that have no guts and absorb nutrients directly from their hosts. Lungworms (Pentastomida) attach themselves to the lungs or respiratory passages of squamates and crocodilians, and feed on blood. Common external parasites of amphibians and reptiles include mites, ticks, mosquitoes, and leeches, all of which feed on lymph or on blood.

Macroparasites have evolved many different types of life cycles and ways of finding and infecting new hosts. Some parasites have simple life cycles in which infective stages are directly transmitted to another host. The ticks that infest Bengal monitors (*Varanus bengalensis*) use the monitors' burrows for refuges during the dry season and for egg laying (Auffenberg 1994). Monogenean flukes that parasitize spadefoot toads (*Scaphiopus* and *Spea*) in the arid regions of western North America are infective only during the 1 to

3 days each year when the toads enter temporary pools to breed. The aquatic fluke larvae invade a spadefoot toad through its nostrils, and the larvae migrate into the lungs and then into the urinary bladder of the toad, where they mature and reproduce while the toad estivates underground. A new cycle of infection is generated when the host enters water to breed and the fluke larvae that were encapsulated within the ovaries of mature female flukes are released into the water (Tinsley and Earle 1983). Other flukes that infect frogs have complex life cycles that require snails as a first intermediate host. If they are eaten by another invertebrate such as a copepod or a dragonfly larva when they leave the snail, they encyst and wait. The life cycle of the fluke is completed if the second intermediate host is eaten by either a tadpole or a frog. In this case, successful transmission of the parasite is facilitated by the use of another species that has a specific interaction with the host.

Individual amphibians and reptiles play host to diverse communities of parasites. For example, the lacertid *Takydromus tachydromoides* at one locality in Japan is host to eight species of worms, six species of intestinal protozoans, and five species of blood parasites, including protozoans, a bacterium, a virus, and a tick (Telford 1997). This community of parasites included species (mostly worms) that were also found in a wide variety of other vertebrates at that site and other species (mostly protozoans) that were largely specific to *Takydromus*.

Macroparasites impose a drain on the host's energy resources that may reduce survivorship, growth, and fecundity and perhaps also weaken individuals and increase their susceptibility to other causes of death. Juvenile *Bufo bufo* infected by the nematode *Rhabdias bufonis*, a lung parasite of European toads and frogs, illustrate some of these problems (Goater and Ward 1992). *Rhabdias bufonis* has a short generation time (10 to 12 days) and direct transmission. Larvae are expelled in the feces of the host and penetrate the skin of a new host after a short developmental period in the soil. In the laboratory, growth and survival of juvenile toads were reduced by infection, and the greater the number of worms in the lungs, the greater the reduction in growth and survival. The mechanism for this detrimental effect was simple; parasitized toads ate as much as 50 percent less than uninfected toads. At least two mechanisms could account for the reduction in feeding by infected toads: the worms may diminish energy stores of the host or affect specific activity patterns such as foraging. In either case, lethargic toads capture less prey than normal toads. Alternatively, reduced feeding might be an adaptive response by an infected toad. Because the life span of the worms is limited and toads recover if they are not reinfected, reduced feeding and activity may allow toads to wait out an infection.

Observations of plethodontid salamanders and the trombiculid mite *Hannemania dunni* suggest that parasitism can have social ramifications (Anthony et al. 1994). Larval mites burrow into and become embedded in the skin of their host, where they feed on lymph and blood for as long as 6 months. Most individuals in populations of five species of salamanders, including *Plethodon ouachitae*, which are endemic to the Ouachita Mountains, are infected with larval mites. Males are more likely to be infected than females or juveniles. For example, all male *P. ouachitae* examined were infected, whereas only 79 percent of females were infected. When mites embed in the snout, they occlude the nasolabial groove, which carries chemicals from the substrate to the vomeronasal organ. Because chemical reception is important to foraging success and to the recognition of other *Plethodon*, such damage might affect the behavior and reproductive success of infected males.

The cumulative effects of a parasite community on its host are likely to be different from the effects of a single parasite species. For example, the body condition (body mass relative to snout–vent length) of male western fence lizards infected either by malaria or by ticks did not differ from the body condition of uninfected individuals, whereas males infected by both parasites were about 10 percent lighter than uninfected individuals (Dunlap and Mathies 1993). Surprisingly, however, measures of body size, fat body size, gonad size, blood counts, and liver masses of *Takydromus* were not correlated with numbers and kinds of parasites present (Telford 1997). Either these static measures of health do not necessarily assess the impact of parasitism, or parasites have little effect on these lizards.

Predator–Prey Interactions

In contrast to parasitism, which involves a prolonged interaction between the host individual and one or more parasitic organisms, predation involves relatively rapid capture and consumption. Predators of amphibians and reptiles include many invertebrates, fish, other amphibians and reptiles, birds, and mammals (Figure 13–13).

Predators

Survivorship curves provide a quantitative assessment of mortality. If most deaths are due to predation, the intensity of predation is mirrored by the rate of decline in the number of survivors as a function of time. The exact shapes of survivorship curves are difficult to determine because of year to year variation in survival and the need to follow cohorts (groups of individuals of the same age) until the last member of the cohort dies. The results of long-term studies of a population of slider turtles (*Trachemys scripta*) at Ellenton Bay, South Carolina, do show, however, typical features of survivorship curves (Frazer et al. 1990). The initial precipitous drop in the number of survivors is associated with intense predation on eggs in nests; over 5 years, only 10.5% of eggs laid produced hatchlings that entered the water (Figure 13–14). Hatchlings that survive the winter (which is spent in the nest) and the subsequent migration to Ellenton Bay have comparatively high survival thereafter, and annual survival of adults is about 80 percent. Some individuals reach ages of 20 or 30 years, but these are a tiny fraction of any cohort because so many individuals die in the nest. Most amphibians and reptiles do not live as long as the slider turtle, but the pattern of high initial mortality for eggs and juveniles and lower mortality for mature individuals is characteristic of most species.

Both generalized and specialized predators feed on the eggs of amphibians and reptiles. Amphibian eggs deposited in water are eaten by other amphibians, invertebrates, fish, turtles, and snakes. Fish are important predators of both eggs and larvae of amphibians in many habitats, and the presence of fish may prevent amphibians from breeding altogether, or limit breeding to species that have defensive mechanisms such as eggs that are deposited in water too shallow for fish, larvae that live in microhabitats fish seldom use, or larvae that are distasteful (Holomuzki 1995).

The eggs of reptiles are eaten by many predators. The majority of eggs of *Anolis limifrons* in Panama that are killed by predators are eaten by ants in the genus *Solenopsis* (Andrews 1988), and ants are important predators on eggs of other species as well. The

highly synchronized nesting of sea turtles is usually considered an evolutionary response to high nest predation. Because nest construction and egg laying are such conspicuous activities, a female is more likely to produce surviving hatchlings if she nests with other females than if she nests by herself. Nonetheless, predation on sea turtles eggs is intense. Hundreds of thousands of eggs may be deposited within just a few days, and the majority may be destroyed by crabs, wild pigs, raccoons, and vultures. Not all mortality is directly due to predation—many eggs are tossed out on the beach by females that dig through the nests of other females while constructing their own nest.

The vulnerability of amphibians and reptiles to predation after hatching varies with the size of the prey and the type of predator. The vulnerability of tadpoles to attack by size-limited predators such as aquatic insect larvae decreases with increasing tadpole size (Formanowicz 1986). Larger tadpoles may be less vulnerable because they swim more efficiently or because size-limited predators have more trouble handling them. In contrast, vertebrate predators such as birds may prefer large tadpoles, and thus vulnerability to vertebrate predators may increase with increasing tadpole size (Crump and Vaira 1991). Metamorphosing tadpoles are more vulnerable to both invertebrate and vertebrate predators than are individuals in younger or older developmental stages. When a tadpole's front limbs erupt and tail resorption begins it neither swims as well as a tadpole nor hops as well as a frog. The stomachs of garter snakes collected near ponds contained large numbers of transforming anurans, even though the most abundant stages in the ponds were premetamorphic (Figure 13–15). In laboratory studies garter snakes were able to capture transforming individuals more readily than either premetamorphic tadpoles or newly metamorphosed frogs (Wassersug and Sperry 1977).

Generally, small species of amphibians and reptiles are vulnerable to invertebrate and small vertebrate predators, whereas larger species are vulnerable only to larger vertebrate predators. Spiders are important predators of small frogs, and tadpoles are eaten by aquatic insects such as dragonfly larvae and dytiscid beetles. Birds are important predators of medium-sized amphibians and reptiles. Some birds, such as the secretary bird in Africa and the laughing falcon in South America, feed largely on snakes. Large reptiles have large predators. Adult leatherback sea turtles are attacked by killer whales, and dead individuals have

Figure 13–13 **Predation on amphibians and reptiles.** (a) Spider eating a frog at Barro Colorado Island, Panama. (b) Snake eating frog eggs, Panama. (c) Australian kookaburra eating a skink. (d) Frog-eating bat about to seize a frog, Panama. *(Photographs by (a) Robin Andrews, (b) Michael and Patricia Fogden/DRK, (c) John Cancalosi/DRK, (d) Merlin Tuttle/Photo Researchers.)*

417

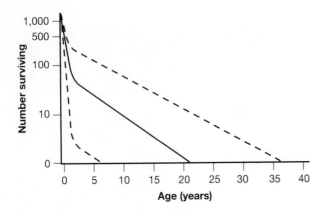

Figure 13–14 **Survivorship curves for slider turtles,** *Trachemys scripta,* **in Ellenton Bay, South Carolina.** Each line represents a theoretical cohort of 1,000 individuals. The solid line represents survivorship based on mean estimates. The upper and lower dotted lines represent best-case and worst-case scenarios, respectively. *(Source: Frazer et al. 1990.)*

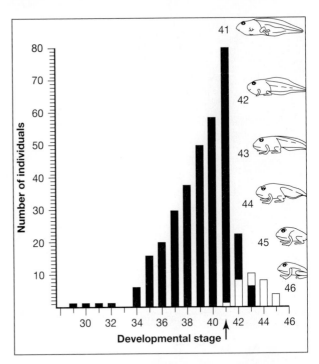

Figure 13–15 **Differential predation on metamorphosing tadpoles.** The frequencies of developmental stages of *Pseudacris regilla* from a pond sample and from the stomachs of *Thamnophis.* The pond sample is indicated by solid bars and the stomach sample is indicated by open bars. The arrow indicates the last premetamorphic stage of the frogs. *(Source: Arnold and Wassersug 1978.)*

been found with missing parts that match the gapes of white sharks (Long 1996).

Defensive Mechanisms

Searching for food and avoiding being eaten are closely related activities. Animals expose themselves to predators while they are foraging, and the mode of foraging, the habitats used, and the times when foraging occurs determine the kinds of predators a species will encounter. Moreover, feeding strategies and defensive mechanisms are functionally related. For example, amphibians and reptiles that subdue and kill their prey using strong jaws or venom-injecting fangs are likely to use those structures in their own defense. Amphibians and reptiles have evolved a wide variety of antipredator mechanisms. Some mechanisms protect individuals against many kinds of predators, whereas others have evolved in response to a specific predator. Predator-specific defenses may continue to evolve in response to the evolution of counteroffenses by the predator. Coevolution is the term used to describe such an evolutionary arms race. Defensive mechanisms have been reviewed by Brodie (1983), Greene (1988), and Pough (1988).

Most amphibians and reptiles have to deal with many predators, each with different visual, auditory, and olfactory capabilities. Thus, no species can rely on a single defense. Any defensive mechanism is a part of a suite of defenses used by an individual during a single encounter with a predator or during its lifetime. Defenses usually vary ontogenetically, and a defense employed by an individual when it is young and small may not be used when it is older and larger. Suites of defenses are usually employed in a hierarchical fashion. For example, an individual may first avoid detection by a predator because it is cryptic. If it is detected, it uses other defensive mechanisms such as rapid flight, exposure of a startling color or pattern, or release of noxious secretions. Several mechanisms may be used simultaneously. The particular defensive mechanism that is exhibited may be specific to the predator. Horned lizards (*Phrynosoma*) squirt blood when they are threatened by dogs or foxes (Figure 13–16) but not when they are attacked by birds or rodents (Middendorf and Sherbrooke 1992). Defensive behaviors that are elicited only by specific predators are probably common, but the responses of amphibians and reptiles to real predators have seldom been recorded.

Figure 13–16 **Projection of blood from the eye of a horned lizard, *Phrynosoma cornutum*.** This defense is elicited only by attacks by canid predators. *(Photograph by Raymond Mendez.)*

Phylogenetic Considerations.

Defensive mechanisms are part of the evolutionary history of an organism. Thus, species from the same lineage tend to share patterns of defense. All amphibians, for example, have both mucus and poison glands in their skin, and both types of glands are used for defense. Mucus glands, which normally serve to keep the skin moist, may secrete adhesive materials that hinder a predator. Poison glands are commonly used for defense, and many lineages of amphibians are characterized by particular defensive chemicals. The potent secretions of some dendrobatid frogs (batrachotoxin) and some salamanders (tetrodotoxin) are examples.

Crypsis, locomotor escape, struggling, defecating, hissing, and biting may be ancestral defensive mechanisms of reptiles (Greene 1988). Crocodilians and sphenodontians exhibit these mechanisms with little modification. Turtles have added specialized forms of plastron hinging and noxious secretions to their defensive repertoire. The greatest diversity of defensive mechanisms is found among squamates. Within the iguanian lineage, defensive mechanisms parallel the ancestral sit-and-wait foraging mode. Crypsis is commonly the first line of defense. Sit-and-wait predators

are rarely far from a retreat site, and they use brief sprints to reach shelter if they are detected. In contrast, active foraging is ancestral in the scleroglossian lineage, and scleroglossans may be far from safety when they are attacked. Defensive behaviors include struggling, defecating, biting with powerful jaws, or exhibiting ritualized displays. The presence of post-cloacal scent glands is an ancestral feature of snakes, and snakes may add noxious secretions from these glands to the urine and feces that all squamates void during encounters with predators.

Specific Defensive Mechanisms.

We know virtually nothing about defensive mechanisms against predators that use sensory modalities other than vision. This is an important bias, because many predators of amphibians and reptiles use scent or hearing to detect prey. These predators may be particularly important for inactive amphibians and reptiles. For example, nocturnal snakes may use chemosensory cues to locate diurnal prey in nighttime refuges such as burrows, beneath rocks, or hidden in the leaf litter. We know little about the defenses used against such predators, or about how successfully a diurnal lizard can escape a nocturnal snake in the dark.

Avoiding Detection.

The sequence of events in a successful predation event are detection, identification, approach, subjugation, and consumption (Endler 1986). The first line of defense for many species is to reduce the probability of detection by the predator. Many amphibians and reptiles exhibit crypsis, which is defined as a general or specific resemblance to features of the habitat (Figure 13–17). Animals that exhibit general resemblance blend in with their background so that predators have difficulty distinguishing them from their surroundings. Familiar examples of general resemblance are the brown color of squamates that rest on tree trunks or litter on the ground and the green color of squamates that rest on leaves. General resemblance may be enhanced by morphology. The Asian frog *Megophrys nasuta* is flattened and has a light line running down its back that resembles the midrib of a leaf. Its appearance thus makes it difficult to distinguish a frog from leaf litter on the forest floor. Other amphibians and reptiles have disruptive patterns or structures that break up or obscure the outline of the body. For example, fleshy protuberances above the eyes, on the snouts, and on the ankles

(a)

(b)

Figure 13–17   **Crypsis.** (a) General resemblance to background: *Hyla arenicolor* in Arizona. (b) Specific resemblance: *Phrynosoma modestum*, a stone mimic. *(Photographs by (a) R. M. Andrews, (b) Wade Sherbrooke.)*

of some frogs may increase crypsis by disrupting the body outline of the animal.

Alternatively, an animal may specifically resemble an object in the environment that the predator does not eat. The smallest species of horned lizard in North America, *Phrynosoma modestum*, looks like a pebble in its desert environment (Sherbrooke and Montanucci 1988). What appears to be a bird dropping on a leaf may, after close inspection, turn into a black and white hylid tree frog. Fecal mimicry is common among insects but less so in vertebrates, probably because they are generally too large to make the resemblance credible.

Cryptic animals usually maximize the protection they receive by selecting appropriate backgrounds. Do individuals of species with polymorphic color patterns also select appropriate backgrounds? Stephen Morey conducted experiments with green and brown morphs of the Pacific tree frog *Pseudacris regilla* to see if these frogs selected matching substrates more often than contrasting substrates, and found that they did (Morey 1990). To further test whether substrate matching actually confers protection on the frogs, Morey placed each color morph on both green and brown substrates and exposed them to garter snakes, *Thamnophis elegans*, which are diurnally active, visual predators. (We do not know that diurnal colubrids can see color, although the structure of the retina suggests that they do have color vision.) He found that substrate matching greatly increased the survival of nonmoving frogs. When frogs moved, however, the

snakes pursued and ate matching and nonmatching frogs equally.

We cannot assume that animals that appear cryptic to humans necessarily appear cryptic to other animals, because humans see only wavelengths of 400 to 700 nanometers, whereas some other animals see shorter or longer wavelengths. For example, many tree frogs are green and appear cryptic to humans when they rest on green leaves. Leaves, however, are strongly reflective in the near infrared (wavelengths of 700 to 900 nanometers). The green tree frog *Hyla cinerea* of North America absorbs infrared radiation and when it rests on leaves it would be conspicuous to snakes and birds that can see near-infrared radiation. In the tropics, however, other species of green tree frogs are reflective in the near-infrared radiation and are cryptic on leaves in both parts of the spectrum (Schwalm et al. 1977).

Fitness, in the evolutionary sense, requires that individuals not only survive but grow to adult size and reproduce, and do so in a timely fashion. Some activities that promote survival are incompatible with activities that promote growth and reproduction. For example, immobility may promote survival, but an individual that does not move cannot eat or encounter mates. Predator avoidance is always a compromise.

Studies of amphibian larvae suggest that predator avoidance has a high cost in terms of activities a larva cannot engage in when predators are present. For example, larvae of *Hyla chrysoscelis* and *Eurycea bislineata* are palatable to sunfish and live in habitats where these fish may be present. Larvae that were exposed to the chemical odors of sunfish spent more time hiding in a refuge than did larvae in clean water (Kats et al. 1988). Larvae reduce the risk of predation by hiding instead of feeding, but they also reduce food intake, and they make this compromise only when predators are actually present. The evolutionary lability of predator detection capabilities is emphasized by comparison of pond and stream populations of *Ambystoma texanum*. Only individuals from stream populations normally encounter fish, and only individuals in these populations reduce their activity in the presence of fish.

The work of Michael Ryan (1985) and his associates in Panama on the tungara frog *Physalaemus pustulosus* provides another example of how individuals make compromises between conflicting activities. By vocalizing, male túngara frogs increase their attractiveness to females, but the particular calls most attractive to females are also the calls that are the most attractive to bats that eat frogs. Ryan and his colleagues tested the responses of calling frogs to bats by using models of bats. This experimental approach allowed them to vary the height of the bats over the pond and also to control the level of illumination. Their results suggested that frogs detected bats visually. The intensity of calling was reduced when bat models were flown overhead on nights with a full moon, whereas frogs continued to call in the presence of bat models on nights that were completely dark. Moreover, when the models of bats were close to the frogs, the frogs deflated their vocal sacs, dived under the water, and swam away. In contrast, when the models of bats were high over the pond, the frogs stopped calling but kept their vocal sacs inflated, and thus were able to resume calling quickly. These behaviors suggested that male frogs were adjusting calling behavior to the risk of predation.

Signaling Inedibility. Once the prey has been detected, it must be identified as edible or inedible. Recognition of edibility can include assessment of size, chance of capture, or noxious qualities. Amphibians and reptiles use morphology, sound, and color to signal inedibility to predators. Morphological warning devices include the hoods of cobras and perhaps the spines of *Phrynosoma* and *Moloch*.

Animals that are unpalatable (taste bad), poisonous (contain a toxic substance that harms a predator when the prey is bitten or eaten), or venomous (can inject a toxin into a predator) are frequently brightly colored. This form of advertisement is called warning, or aposematic, coloration. Species that display aposematic coloration include many dart-poison frogs (Dendrobatidae), the newts *Salamandra* and *Pseudotriton* (Salamandridae), the venomous Gila monster and Mexican beaded lizards (Helodermatidae), and the venomous coral and sea snakes (Elapidae). The message conveyed in their appearance is "Watch out! Don't attack me!"

Potential predators of warningly colored species can learn to associate the conspicuous coloration with an unpleasant experience. These individuals subsequently avoid similarly colored prey. Birds and mammals can learn to avoid warningly colored prey by observing the bad experience of another individual, but observational learning has not been demonstrated for amphibians or reptiles. If the consequences of attacking warningly colored prey are serious enough to reduce the fitness of a predator, innate avoidance could evolve.

Many species exhibit warning coloration but lack the noxious qualities to back up the warning. This phenomenon is called Batesian mimicry, in recognition of the naturalist H. W. Bates, who first described mimicry from his observations of butterflies in South America. Edible species dupe predators by mimicking the appearance of an inedible or noxious model. For example, the resemblance of the erythristic morph (uniform red, rather than brown with a red stripe) of the edible red-backed salamander *Plethodon cinereus* to the toxic eft stage (bright reddish orange) of *Notophthalmus viridescens* in eastern North America has been interpreted as Batesian mimicry. This hypothesis was tested by offering wild birds red efts, the erythristic morph of the red-backed salamander, the normal striped morph of the red-backed salamander, and another edible species of salamander. The survival of the erythristic red-backed salamanders was considerably greater than that of the normal morph, indicating that birds learned to avoid the mimetic *Plethodon* after attacking the noxious efts (Brodie and Brodie 1980).

Müllerian mimicry, named after F. Müller, another naturalist who studied butterflies in South America, describes the situation in which different species, all with noxious properties, exhibit the same warning color. Predators generalize a bad experience with one species to all similar species. The eft stage of *Notophthalmus viridescens* and the red salamander *Pseudotriton ruber* occur in the same habitats in eastern North America. Because both of these salamanders are toxic, they are considered Müllerian mimics.

Venomous snakes are the most common models for mimetic systems involving reptiles (Pough 1988). Venomous snakes are particularly dangerous to potential predators because of their large size. If a mistake by a predator can be lethal, predators should generalize the characteristics of the model broadly, and the protection conveyed by mimicry should be substantial even if the mimic is not very similar to the model. Neotropical coral snakes (*Micrurus*) and the similarly patterned false coral snakes in several colubrid genera represent both Batesian and Müllerian mimicry. Snakes in this mimicry complex resemble each other in having red, black, and white or yellow rings around the body. Two families are represented, Elapidae and Colubridae. The colubrids include rear-fanged species that are Müllerian mimics of coral snakes and nonvenomous species that are Batesian mimics of coral snakes and of the rear-fanged colubrids.

Parallel geographic variation in the color patterns of some models and mimics provides strong indirect evidence for mimicry of coral snakes (*Micrurus*) by colubrids (*Pliocercus*) (Greene and McDiarmid 1981). *Pliocercus* are brightly colored, rear-fanged colubrids found in tropical forests. Throughout most of Mexico and northern Central America, *Micrurus* and *Pliocercus* have red bands alternating with yellow-bordered black bands. In southern Central America and northwestern South America two species of coral snakes are bicolored, with broad black bands alternating with red, pink, or white rings; *Pliocercus* has this bicolored pattern only where it is sympatric with the bicolored coral snakes.

Observations by Edmund Brodie III (1993) in Costa Rica provide direct evidence that coral snake patterns are actually aposematic and that coral snakes are avoided by predators in the wild. He placed plasticine replicas of coral snakes and of unicolored brown snakes on the forest floor. He also placed replicas of these snakes on sheets of white paper where neither would be cryptic. Attacks by predators were recorded by impressions of beaks and claws in the plasticine material (Figure 13–18). Most attacks on the replicas were made by birds, and the plain brown replicas were attacked far more often than the coral snake replicas, irrespective of the background (Figure 13–19). This means that coral snakes are aposematic to birds and are avoided. In another series of experiments, Brodie placed replicas of the six major types of coral snake models and mimics and a unicolored brown snake replica on the forest floor. Again, the plain brown replica was attacked the most. The number of attacks on the types of coral snake replicas varied, suggesting that some patterns are more effective than others at deterring birds. This result supports the prediction that resemblance to coral snakes protects a range of imprecise mimics (Pough 1988).

Coral snakes are highly venomous, and predators that are bitten may die. Thus, predators may not have the opportunity to learn by experience to avoid coral snakes. Instead of learning by individuals, generations of selection favoring birds that did not attack prey with coral snake patterns may have produced a genetically based innate avoidance of those patterns. Susan Smith (1975) raised motmots, a moderately sized Neotropical bird, from hatching so that they would be naive about the warning coloration of coral snakes and their mimics. These naive birds fled to the back of their cage and gave alarm calls when they were shown models with the color patterns of coral snakes, whereas they readily approached and attacked similarly shaped models that were colored like palatable prey. For these naive birds, the avoidance of the coral snake pattern was innate. The avoidance genotype should be favored by selection, and such innate avoidance of coral snake patterns would protect coral snake mimics as well as their deadly models.

Avoiding Capture.   Once it has identified an appropriate target, the predator approaches the prey. The approach may be an immediate rush or a slow stalk followed by attack. At this point, prey exhibit defensive mechanisms that reduce the probability of capture.

For many animals, rapid flight is the option of choice when they have been detected by a predator. Real speedsters among reptiles include lizards in the genus *Callisaurus* in the deserts of North America and some *Ctenotus* in the deserts of Australia. Rapid flight is often toward a hiding place, and treetops, crevices, holes, or dense vegetation may baffle a predator. Some arboreal species take flight more literally. In southeast Asia, some *Rhacophorus* frogs have enlarged webbed feet, and the agamid *Draco* and the gecko *Ptychozoon* have flaps of skin extending along both sides of the body. These species leap from a perch and glide to a distant location. Other species simply

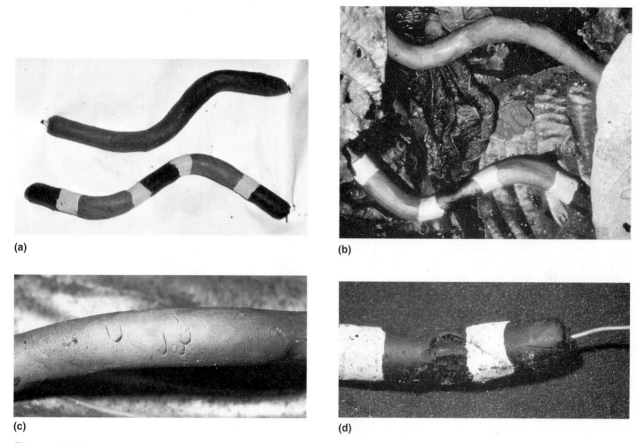

(a)

(b)

(c)

(d)

Figure 13–18 **Experimental tests of coral snake mimicry.** Tricolor coral and unmarked brown snake replicas. (a) On white paper, and (b) a natural background. Replicas were widely spaced under experimental conditions. Imprints left by attacks on snake replicas in the field. (c) U-shaped bill marks from birds, and (d) serrated bill marks from a bird (motmot) attack. *(Source: Brodie 1993.)*

change their media when discovered. Preferred basking sites for aquatic turtles are logs and rocks from which they can drop into the water for quick escape from terrestrial predators, and aquatic snakes and some large tropical lizards, such as *Iguana, Basiliscus, Hydrosaurus,* and *Physignathus,* rest in branches overhanging water. At the approach of a predator, the individual drops from the tree into the water and dives to the bottom, where it can remain for a considerable period.

Because most amphibians and reptiles have low oxidative metabolic capacities, they cannot run far before they are exhausted. Thus, they usually seek cover rapidly or try to baffle a predator rather than try to

outrun it. Salamanders and amphisbaenians flip from side to side, a behavior that may confuse a predator so that it cannot predict the animal's next move. Escape by some anurans is accompanied by the display of bright, contrasting colors on the sides or hind legs when the frog jumps. If the contrasting color defines the predator's image of the prey, the predator may fail to detect the frog after it lands and resumes its cryptic resting posture. This behavior is effective for anurans because most species can jump short distances quickly, although they lack the stamina for long-distance escape.

Behavior and color patterns of snakes are coordinated in ways that facilitate escape from predators.

Figure 13–19 **Results of experimental tests of coral snake mimicry.** (a) The number of avian attacks on tricolor coral and unmarked brown replicas placed against natural or plain white backgrounds. Note that regardless of background, the coral snake replicas were attacked less than the brown replicas. (b) The number of avian attacks on unmarked brown replicas and on six common patterns of coral snakes and their mimics. Solid horizontal lines above the histogram indicate groups that differ significantly from one another. The codes and their patterns are shown below the histogram. Note that unmarked brown replicas were attacked more frequently than any of the banded patterns, but that some banded patterns convey more protection than others. *(Source: Brodie 1993.)*

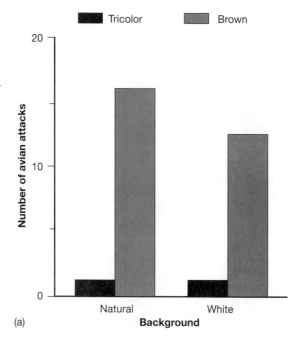

(a)

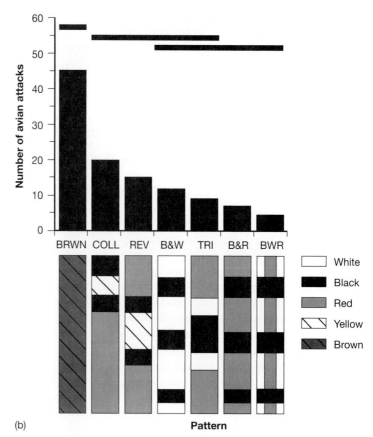

(b)

Snakes that are longitudinally striped are usually diurnal and use rapid flight as a primary mode of escape from predators. Striping makes motion difficult to detect and speed difficult to judge. On the other hand, snakes that are blotched or banded usually have secretive habits and rely on crypsis. Blotches or bands make fixed reference points for the eye and enhance detection when the snake moves. However, rapid movement of a small blotched snake blurs the bands so that the snake appears to be unicolored (Pough 1976b). When movement is abruptly followed by stillness, the snake becomes cryptic, thus appearing to disappear (Figure 13–20).

Studies of a North American garter snake, *Thamnophis ordinoides*, support the hypothesis that particular combinations of dorsal pattern and behavior of juvenile snakes are associated with higher fitness than other combinations. Individuals of this species can be striped, spotted, or solid-colored, and they also exhibit an array of defensive behaviors including flight and sudden reversals of direction followed by stillness. By releasing and recapturing hundreds of juvenile snakes that had been previously scored for pattern and behavior, Brodie (1992) found that individuals with the highest survival exhibited one of two combinations: striping and rapid flight, or spotting and frequent reversals. He concluded that natural selection favoring these two combinations was probably the result of the optical properties of color patterns during movement producing differential predation by birds.

Anurans often void the contents of their bladders when approached or captured. This behavior may startle a predator or deter it from further attack. In addition, emptying the bladder reduces the weight of a frog and may enhance jumping. Squirrel tree frogs with full bladders jumped an average of 31 centimeters, whereas frogs with empty bladders jumped 36 centimeters (Buchanan and Taylor 1996).

Birds and mammals tend to attack the head of their prey, and the tail displays of some squamates and amphisbaenians may divert the attention of a predator toward this less vulnerable part of the body. Many lizards and some salamanders shed their tails readily when attacked (caudal autotomy), sometimes even before they are touched (E. N. Arnold 1988). This behavior facilitates escape because the wiggling tail that remains behind may divert the attention of the predator. The tail is often eaten by the thwarted predator, thus providing it with a meal, albeit smaller than anticipated. Tail au-

totomy is effective, but it does have a cost—the energy requirements of tail regeneration can result in a missed reproductive season (Chapter 7), and tailless individuals may have reduced status within a social hierarchy (Fox and Rostker 1982). Tailless individuals are also easier for predators to capture than individuals with intact tails because they cannot deflect attack and also because they may run more slowly. Ground skinks, *Scincella lateralis*, that have lost their tails compensate for their increased risk. These individuals become more cryptic by reducing their activity and by not fleeing until potential predators are relatively close (Formanowicz et al. 1990).

Rather than attempting to flee, many species of amphibians and reptiles stand their ground and exhibit threatening behaviors. These behaviors may be genuine warnings that the animal can harm an attacker, or the behavior may be a bluff. Threatening behavior includes postures that enhance size, lunging, gaping, hissing, and tail lashing. The Australian agamid *Chlamydosaurus kingii* suddenly spreads the large frill around its neck, gapes its mouth, and lunges at a predator (Figure 13–21). By spreading the mandibular joints laterally, some colubrid snakes make their heads appear larger and more similar to those of vipers. Many salamanders arch their tails over their backs as a defensive behavior, thereby increasing their apparent size. Sudden noises may startle or frighten predators, and many salamanders, anurans, crocodilians, and squamates vocalize when threatened or grasped.

On the other hand, many species make threats that can be followed by formidable defenses, at least against some predators. The sounds made by rattlesnakes are warnings of potential bites and the injection of venom (Greene 1992). Snapping turtles threaten by lowering the front of the shell and extending their rear legs to raise the posterior end. The mouth may be kept open and the threatening posture may be accompanied by hissing, lunging, and biting. Frogs of the genus *Hemiphractus* gape widely at potential predators, displaying a bright yellow buccal cavity. If the predator persists, the frog bites and inflicts a surprisingly painful puncture wound using sharp, fanglike projections on the lower jaw.

Some anurans (e.g., some *Pleurodema* and *Physalaemus nattereri*) display eyespots on their rumps when provoked. These large eyespots may give the impression of a large animal and thus may frighten away the potential predator. If this defense does not work and

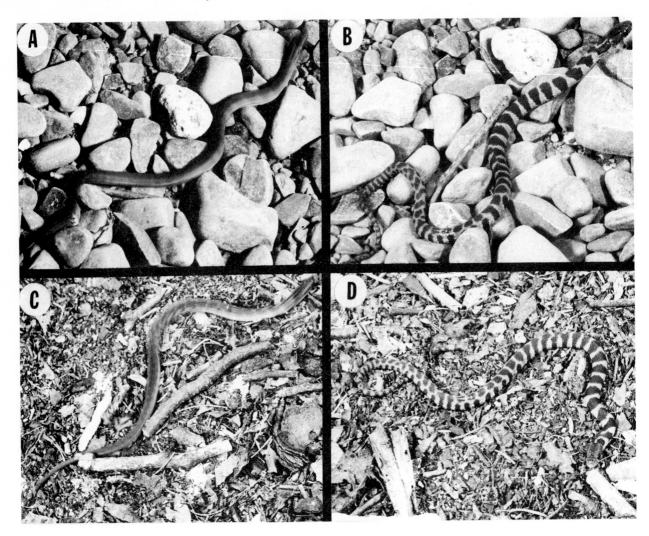

Figure 13–20   **Optical illusion as a defence mechanism.** Appearance of a newborn water snake, *Nerodia sipedon*, while crawling and while motionless. Photographs were made at an exposure of 1/30th of a second and show the snake as it would appear to a predator with a critical flicker frequency (the frequency at which two images blur) about twice that of a human. When the snakes are in motion, they appear unicolored (a, c). The motionless snake is less cryptic on gravel (b) than on plant matter (d). *(Source: Pough 1976b.)*

the predator continues to attack, the frog produces a noxious secretion from glands located near the eye-spots. Ambystomatid and plethodontid salamanders that have concentrations of caudal glands with noxious secretions use the tail aggressively, often lashing at a predator. Salamandrid, ambystomatid, and plethodontid salamanders that have noxious secretions concentrated in parotoid glands on the head may butt the potential predator. Fire salamanders, *Salamandra salamandra*, can spray their neurotoxic skin secretions,

even directing the spray toward the perceived direction of threat (Brodie and Smatresk 1990; Figure 13–22).

Response to a predator may depend on the body temperature of the potential prey. Stanley Rand (1964) found that when the ambient temperature was low, *Anolis lineatopus* fled at greater distances when he approached than when the ambient temperature was high. Because their running speed is greater at high than at low body temperatures, the lizards presumably gauged the probability of escape by their body temper-

Figure 13–21   **Threat display of the Australian agamid *Chlamydosaurus kingii.** (Photograph by Michael and Patricia Fogden.)*

Figure 13–22   **The European fire salamander, *Salamandra salamandra.** The salamander is spraying secretions from dorsal skin glands. (Photograph by Edmund Brodie, Jr.)*

ature. Paul Hertz and colleagues (1982) found that the agamids *Agama savignyi* and *A. pallida* fled or fought when threatened as a function of their body temperature. At body temperatures of 34 to 42°C, immediate flight was a common response to simulated predation. At body temperatures of 20°C, when the lizards could not run well, all individuals fought, and their defensive repertoire included gaping, lunging, and attempting to bite. Other factors, such as reproductive status, energy levels, and distance to hiding places, also affect how individuals attempt to avoid capture.

Preventing Consumption.   An attack is successful only if, once captured, the predator can subdue and kill the prey and then eat it. Amphibians and reptiles have several mechanisms that reduce the probability that they will be killed and eaten even after they are in the clutches of a predator.

Some species gain protection through strictly physical means. The armor and relatively large adult size of most species of turtles and crocodilians are defensive traits. All turtles have shells, and many species have hinged plastrons so that both the head and tail ends can be tightly closed. Crocodilians have thick dorsal osteoderms. Other smaller reptiles, and even some amphibians follow this model. Squamates in the family Cordylidae have skin that is heavily armored with interlocking osteoderms, and all skinks have a plate of dermal bone within each scale. Spines scattered over the bodies of the Australian *Moloch* and the bony cranial protuberances of the frog *Hemiphractus* may make them difficult to handle and eat. However, such spines can be the undoing of both predator and prey. Dead snakes with spiny prey such as horned lizards and catfish lodged in their throats have been found by many naturalists.

Many species inhibit ingestion by becoming hard to handle. Frogs and toads inflate their lungs with air when provoked, swelling up and making themselves difficult for a predator to manipulate. Many lizards such as chuckwallas (*Sauromalus obesus*) of western North America wedge themselves into crevices in and under rocks by inflating their bodies with air so that they are hard to remove.

Immobility (remaining still, but not unresponsive) or death-feigning behavior (completely unresponsive) may protect individuals from some predators. Although this behavior might only hasten ingestion by reptilian predators, it may be effective against predators that rely on movement to elicit prey-killing behavior. Birds that carry a death-feigning lizard or snake back to their nest may lose it when they arrive and relax their grip. Immobility or death feigning may be the last of a sequence of defenses. Hognose snakes (*Heterodon*) of North America, for example, first hiss loudly, and then strike if approached more closely. If harassment continues, they writhe violently and void feces and urine; the terminal behavior is to turn on their backs, gape their mouths, and feign death.

Chemicals are part of the repertoire of defenses of many amphibians and reptiles that come into play when a predator seizes its prey. The most toxic defensive secretions are produced by poison dart frogs (*Phyllobates*), bufonids (*Atelopus*), and newts (*Taricha, Salamandra, Notophthalmus*). Many species that have noxious skin secretions couple this defense with aposematic coloration and postural displays. Other species use crypsis as the first line of defense, but once discovered, warn potential predators by exposing areas of aposematic coloration. For example, the Asian warty newts (*Paramesotriton*), which are often seen in pet stores, and the frogs *Bombina* and *Melanophryniscus* arch their backs and raise the palms and soles to expose bright ventral coloration. This posture is referred to as an unken reflex. Müllerian mimicry by the salamander *Pseudotriton ruber* and the red eft stage of *Notophthalmus viridescens* is enhanced by their modified unken displays; both species have lateral and vertical tail displays that expose their bright venters. The salamander *Echinotriton* combines potent toxins with an injection system. Their ribs have sharp points and are forced through the body wall of the salamander when a predator bites it. As the ribs penetrate the salamander's skin, they pierce warts composed of poison glands and carry toxins from the glands into the predator's mouth.

Skin secretions of some lizards and many amphibians function as adhesives rather than toxins. Australian geckos in the genus *Diplodactylus* have secretory glands in their tails. A liquid is expelled through hollow spines when the lizard is attacked, and it can be sprayed as far as a meter. This liquid turns into sticky threads that adhere to the head of a predator, probably distracting it while the gecko escapes.

Many amphibians produce sticky secretions. Christine Evans and Edmund Brodie, Jr. (1994) measured the adhesive strength of skin secretions of some amphibians by determining the force necessary to separate two disks stuck together with the secretions. Microhylid frogs in the genus *Dyscophus* had the stickiest secretions. Disks stuck together with their secretions took slightly more than 1.0 kilogram of mass to separate, while disks stuck together with rubber cement separated with 0.4 kilogram of mass. Other amphibians, including ambystomatid and plethodontid salamanders and the caecilian *Dermophis*, had secretions approximately as strong as rubber cement. Plethodontid salamanders studied by S. J. Arnold (1982) were able to delay ingestion and even to escape from garter snakes by immobilizing the snakes with their skin secretions. In one case loops of a snake's body were glued for several days in the position initially used to grasp the salamander.

Even after the prey has been eaten, defensive mechanisms can come into play and the prey can be disgorged. Despite the fact that hatchlings of aquatic turtles and predatory fish are found in the same bodies of water, the bite-sized hatchlings are seldom eaten by fish. When Carol Britson and William Gutzke (1993) fed live and dead hatchling red-eared sliders and painted turtles to largemouth bass, the live hatchlings were immediately ejected unharmed, whereas dead hatchlings were consumed. These observations suggested that the hatchlings deterred ingestion by scratching and biting, rather than by a chemical secretion.

Coevolution of Predators and Prey

Coevolution of predators and their prey is an attractive idea, but the evidence for reciprocal and continued adaptations of predators and prey is quite limited. The presence of elaborate defensive mechanisms does not necessarily indicate coevolution. To constitute coevolution, the behavior or other features of the predator that invoked defensive mechanisms by the prey must subsequently have been altered as well. For example, the sudden display of eyespots by *Physalaemus nattereri* may frighten away birds or other predators, but the evolution of eyespots has not necessary selected for better visual discrimination by these predators. Inter- and intraspecific variation in the resistance of garter snakes to the toxins of the newt *Taricha granulosa* provides one of the best-documented examples of adaptive responses of predators to toxic prey (Brodie and Brodie 1990). The question of whether *Taricha* have become more toxic as a result of increased resistance of garter snakes has yet to be answered, however.

Summary

The foods and feeding habits of amphibians and reptiles reflect the evolutionary origins and histories of these vertebrate taxa. Ancestral tetrapods were carnivores, and this feeding mode is typical of most living amphibians and reptiles. The major exceptions are the larvae of anurans, which are largely suspension feeders on phytoplankton and bacteria, and the specialized herbivores, land tortoises and iguanid lizards. Prey sizes are generally correlated with predator size. As a consequence, small species of amphibians and reptiles tend to feed on invertebrates and large species of amphibians and reptiles usually feed on vertebrates. Amphibians and reptiles exhibit two different types of foraging behavior: sit-and-wait and active foraging. Members of each of these groups display distinctive suites of characteristics that integrate foraging behavior with morphology, physiology, defensive mechanisms, and reproductive mode. Individuals also exhibit a behavioral flexibility that facilitates immediate responses to both short- and long-term variation in the abundance of particular prey types.

Because amphibians and reptiles are components of communities, they serve as hosts for parasites and as prey for other predators. The immediate effect of parasitism is a reduction in the fitness of the host, which loses nutrients or suffers injury. In the long term, parasitism involves reciprocal interactions between the immune system of the host and the antigens of the parasite. The immediate effect of predation is the death of the prey. A long-term consequence of predation is the evolution of defensive mechanisms, and perhaps the subsequent evolution of counterstrategies by predators.

Amphibians and reptiles exhibit a wide diversity of defensive mechanisms. Some mechanisms, such as the secretion of noxious chemicals by amphibians and fleeing, biting, and discharging cloacal products by reptiles, are used by many species and reflect their common evolutionary history. Other defenses are innovations that characterize particular lineages. The projection of toxins by fire salamanders, sticky substances by geckos, and venoms by spitting cobras are parallel defensive adaptations in unrelated groups. The defenses exhibited by different individuals may vary considerably. These responses may be adaptive, or simply reflect inescapable constraints imposed by their size, nutritional status, or the habitat where they confront a predator.

CHAPTER

14

Species Assemblages

The noise coming from a frog pond on a rainy spring evening can be deafening. With males of eight or ten species all calling at once, identifying individual frog calls can be a challenge. Why are all these particular species—not fewer, not more, not other species with different traits—coexisting at this spot? What has determined the particular composition and set of traits of this assemblage of species? Why are some species common and others rare? In what ways do the different species of frogs interact with each other? Do the species compete with each other for calling sites or oviposition sites? These are some of the questions that can be asked at the level of community ecology. Community ecology as a discipline focuses on the multispecies patterns (e.g., number of species, absolute and relative densities of those species, patterns of resource use) that occur within a site or across the landscape, and the processes and mechanisms (e.g., competition, predation, environmental tolerances, historical factors) that generate those patterns.

In the broad sense, a community consists of the populations of all species—plants and animals—existing at a particular point in the landscape, and the interactions among these species. Usually, though, an ecologist studies only a subset of the community. A herpetologist, for example, might choose the subset consisting of all the arboreal frogs and lizards in that rain forest. The choice is a matter of convenience: it is impossible to study every species, so each study focuses on the species that seem to be appropriate for the question being asked. Thus, we will refer to the specific set of species being considered as an assemblage.

Communities and their component species assemblages are variable and dynamic. Because the geographic range of many species is restricted, the composition of local assemblages changes from place to place. Furthermore, the abundance of the different species that compose an assemblage varies through time even in the short term. Over longer time frames, species assemblages change dramatically as new species immigrate and old ones vanish.

One of the central questions in community ecology is whether immigrations, emigrations, and extinctions are haphazard or whether existing communities are instead groups of interacting species that fit particularly well together and thus have structure. If the community is a nonrandom sample from the species pool available in the landscape, what forces determine and maintain this structure? Beginning in the nineteenth century, with renewed interest in the 1940s, and with a burst of enthusiasm in the 1960s and 1970s, many animal ecologists inferred that communities were truly nonrandom samples, and they undertook searches for possible organizing forces. Ecologists eventually reached a tentative consensus on interspecific competition as the main organizing force. Resources such as food and space were shown (or assumed) to be limited. Detailed observations or experiments often showed that individuals of different species could negatively affect each other's access to these resources. These findings and a great body of theory led to the inference that competition for limited resources might determine how many species could coexist and what sorts of traits they could possess. The next step was a proliferation of field studies detailing the differences in resource use among coexisting species, with the assumption that such resource partitioning permitted coexistence (MacArthur 1972, Cody and Diamond 1975).

Over the past two decades, however, increasing numbers of ecologists have questioned the importance of competition in structuring communities, and some have questioned whether communities are truly organized at all (Strong et al. 1984, Diamond and Case 1986). This debate has led to different ways of studying communities. For example, experimental manipulations and simulations can be designed to determine what sorts of structure, if any, exist, and if they exist, to identify the processes responsible without necessarily hypothesizing any one factor beforehand.

During both the early and recent phases of community ecology, amphibians and reptiles have been crucial to testing predictions and formulating new hypotheses. The home ranges of most amphibians and reptiles are small enough to be encompassed by a study plot, population densities are often large enough to provide good samples, and many assemblages contain several potentially interacting species. Furthermore, amphibians and reptiles are suitable for methods of investigation ranging from field observations through manipulative field studies to controlled laboratory experiments. In this chapter we will highlight some of the areas in which amphibians and reptiles have provided insights into our understanding of community ecology.

Species Diversity Gradients

Often one of the first exercises in the study of an assemblage is the practical task of enumerating the species within the assemblage. The count of species at a given site, or species richness, is the simplest means of expressing the concept of species diversity and is the most common meaning of the all-inclusive buzzword biodiversity. Species richness data are obtained from long-term sampling in the field. The more person-hours spent on the survey, the more reliable and complete are the data.

Comparative studies over broad geographic scales reveal that the species richness of amphibians and reptiles varies greatly from place to place (Figure 14–1). One identifiable pattern involves latitudinal variation (Table 14–1). As many as 150 to 200 species of amphibians and reptiles can be found in relatively small areas of lowland rain forest at the equator. At midlatitudes in Europe, North America, or South America the number of species is less than half that, and farther north or south the number is even lower. On closer examination, variations in this pattern emerge. Some groups, such as frogs, crocodilians, and snakes, follow the overall trend, but salamanders have their highest species richness in eastern North America and the highlands of southern Mexico and Guatemala, turtles are most speciose in eastern North America and southeastern Asia (including India), and lizards display their greatest diversity in the deserts of Australia.

Species richness also varies with elevation. For example, the number of species of anurans decreases with increasing altitude in a transect along the equator from lowland rain forest (340 meters) in eastern Ecuador to the paramo vegetation that occurs above 3,500 meters at the eastern crest of the Andes Mountains (Figure 14–2). Species richness doesn't always follow this trend, however. Elevational transects in Central America reveal that plethodontid salamanders are most diverse in intermediate-altitude cloud forest habitats, with fewer species in lowland rain forest and in habitats above the cloud forest. Changes in temperature, moisture, and habitat structure associated with differences in elevation are probably responsible for elevational differences in species richness for both amphibians and reptiles.

Why are more total species found near the equator than farther away? Why does the number of species vary with elevation? Why does eastern North America have so many species of salamanders and turtles? Answers to these and other questions about species richness are complex. Single-factor explanations for geographic variation in species richness have proved inadequate, and ecologists now recognize a hierarchy of factors that affect diversity. These factors range from events that took place in the distant past to ongoing interactions among species.

For the remainder of this section we will focus on two present-day factors that are related to latitudinal variation in species richness: climate and resource richness. We do not mean to imply that these two factors are the only determinants of latitudinal variation in species richness. Rather, they are two important

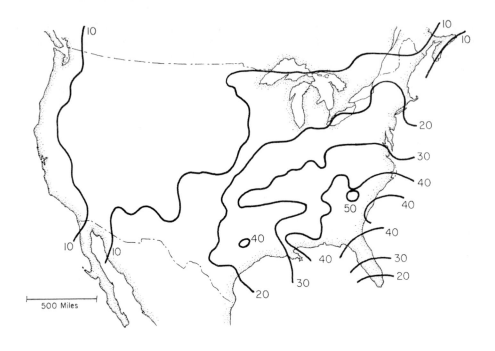

(a)

(b)

Figure 14–1   **Amphibian and reptile species-density contour maps for the continental United States and southern Canada.** Note the latitudinal gradient in number of species of amphibians (a) and reptiles (b). *(Source: Kiester 1971.)*

Table 14–1 **Species richness of amphibians and reptiles at lowland forested sites in the New World. This table illustrates a latitudinal gradient in species richness; the total number of species increases from temperate to equatorial regions. Note that salamanders and turtles show the opposite pattern. Because the sites vary in size and because not all species present may be known, especially for the tropical sites, the number of species at each site should be considered approximate.**

Site (degrees N latitude)	Number of Species							
	Total	Frogs	Caecilians	Salamanders	Crocodilians	Turtles	Lizards	Snakes
George Reserve, MI (42)	39	10	0	8	0	5	1	15
Savannah River, GA (32)	99	25	0	16	1	12	9	36
El Petén, Guatemala (16)	91	20	1	2	1	8	20	39
La Selva, Costa Rica (10)	134	44	1	3	1	4	25	56
Santa Cecelia, Ecuador (0)	185	86	5	2	2	6	31	53

Source: Data from Duellman 1963, 1990, Collins and Wilbur 1979, Gibbons 1990, Conant and Collins 1991.

contributors to observed patterns. We focus on latitudinal gradients for illustrative purposes only. Many sorts of species-richness gradients exist, and these also may be correlated with variation in climate, resource richness, and other factors (Rosenzweig 1995).

Many climatic features vary with increases in latitude. The mean temperature decreases and the magni-

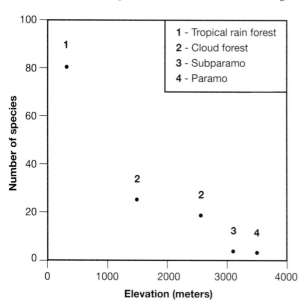

Figure 14–2 **Altitudinal variation in species density.** Number of species of anurans found in assemblages along an altitudinal transect along the equator in Ecuador, from lowland rain forest (340 meters) to the eastern crest of the Andes mountains (paramo habitat above 3500 meters). *(Data from Duellman and Trueb 1986.)*

tude of change in weather from one season to another increases. Moisture usually decreases with increasing latitude, although not in a regular fashion. Many groups are associated with particular climatic regimes, which suggests that species richness within a particular landscape may be directly related to climate. This idea was tested by Schall and Pianka (1978), who reasoned that if the species richness of herpetofaunas was causally related to climate, the relationship between species richness and local climatic features (e.g., rainfall and temperature) should display the same sort of pattern on different continents. They compared the species richness of frogs, lizards, turtles, and snakes in the United States and Australia. Despite differences in latitude (the United States is farther from the equator) and climate (Australia is drier), the statistical association between climate and species richness on these two continents was similar (Table 14–2).

Associations between climate and species richness make biological sense. On both continents, species richnesses of frogs and turtles are highly and positively correlated with mean annual rainfall and negatively correlated with mean annual sunshine. The species richness of lizards is highly and positively correlated with mean annual sunshine and negatively correlated with mean annual precipitation, and the species richness of snakes is positively related to both mean annual rainfall and mean annual temperature. These patterns make intuitive sense: conspicuously aquatic forms (frogs and turtles) are found in wet areas, and strongly heliothermic animals (lizards) are found in sunny regions.

Table 14–2   **Correlation coefficients (*r*) between three climatic variables and species richness of frogs, turtles, lizards, and snakes in the United States and Australia. Positive values of *r* indicate a positive relationship between two variables, and negative values of r indicate an inverse relationship between two variables. (*r* = +1 when the two variables are perfectly correlated, 0 when the two variables are unrelated, and –1 when the two variables are perfectly and negatively correlated.) For the United States, species richnesses and climate are for quadrats of about 10,500 km^2, and for Australia, species richnesses and climate are for quadrats of about 57,500 km^2.**

	Correlation Coefficients by Climatic Variable		
	Mean Annual Hours of Sunshine	Mean Annual Precipitation	Mean Annual Temperature
Frogs			
US	–0.30	0.80	0.52
Australia	–0.56	0.73	0.04
Turtles			
US	–0.20	0.73	0.60
Australia	–0.45	0.77	0.30
Lizards			
US	0.74	–0.25	0.71
Australia	0.44	–0.37	0.09
Snakes			
US	0.22	0.46	0.78
Australia	–0.18	0.57	0.43

Source: Data from Schall and Pianka 1978.

The examples in the preceding paragraph show some relatively direct effects of climate on species richness, but climate has indirect effects on geographic gradients as well. For example, forests in the lowland tropics are generally taller than forests in temperate regions, and tropical forests have greater structural diversity (e.g., an understory, a canopy, vines that extend from the floor to the canopy, epiphytic plants living on tree trunks and branches, stems and twigs of various shapes and sizes, great diversity of leaf size and shape). This structural diversity is accompanied by a great diversity in potential prey species of insects and small vertebrates. Not surprisingly, arboreal frogs, lizards, and snakes are more diverse in tropical lowland forests than in temperate forests.

Determinants of Assemblage Structure and Composition

Why do some species occur in a given assemblage while others are absent? Do interactions among species favor some species over others? Is the physical environment (temperature and humidity) more or less significant than the biological environment (competition and predation)? Are the assemblages we see now determined by current-day events, or are they the results of events that occurred tens of thousands of years ago? For many years such questions constituted the core of research in animal community ecology, and they are still generating some surprising and important answers.

Pattern, Process, and Mechanism

Ecological interactions among species can be viewed on three levels (Feinsinger and Tiebout 1991).

At the most basic level is the mechanism by which individual organisms interact. For example, the mechanism of interspecific competition is an interaction in which an individual of one species either hinders access to resources by an individual of another species (interference competition) or uses those resources more effectively (exploitative competition).

Process is the next level of interaction. Processes are seen at the population level and result from all of the mechanisms that act on individuals. For example, if species A and B are competing, their populations might show reciprocal variation in density: where there are many individuals of species A there are few of species B, and vice versa. This process is called reciprocal negative density dependence.

Pattern, the final level of interaction, is seen at the community or regional level. Patterns include observations of which species are present, where they are in the habitat, what they eat, and so on.

Community ecologists have long assumed that cause-and-effect relationships exist among the three levels of this hierarchy. For example, a researcher might observe reciprocal negative density dependence in populations of two species of lizards. If both species use the same limited resource (prey or shelter sites, for example), and if experiments show that aggression (a competitive mechanism) occurs between individuals of the two species of lizards, the researcher might propose that competition is the cause of the process that is observed.

Alternatively, cause-and-effect interpretations can be based on negative results. For example, if patterns expected to result from interspecific competition cannot be found, a researcher might dismiss competition as an important mechanism of species interaction.

The problem with such interpretations is that different mechanisms can lead to indistinguishable processes. For example, reciprocating negative density dependence is commonly interpreted as a result of competition, but it can also occur when two non-competing species are sensitive to each other's parasites. Likewise, differences in diet among coexisting species are often interpreted as the result of avoiding competition, but they could be simply the result of independently evolved differences in morphology or behavior. Conversely, a study that failed to find the patterns that competition is expected to produce cannot conclude that competition never occurs between the species in question. Other factors (e.g., species differences in heat tolerance) might have overwhelmed competition in the particular site studied.

Furthermore, as we noted in Chapter 13, the life history characteristics of amphibians and reptiles make the simultaneous occurrence of multiple mechanisms of interaction quite likely. Adults of a large species of salamander may prey on adults of a smaller species, but juveniles of both species are the same size,

and juveniles may compete for insect prey. In short, demonstrating that a mechanism exists is neither necessary nor sufficient to explain patterns in herpetofaunal assemblages. Furthermore, demonstrating that a pattern exists is only part of a community-level analysis; studies at the levels of mechanism and process are also needed.

Competition.   Three patterns often attributed to competition are resource partitioning, geographic displacement of similar species, and morphological differentiation of sympatric species. Resource partitioning will be discussed here as the example of pattern.

Resource partitioning refers to the differential use of resources by species within an assemblage. Differences among species may be slight or substantial, and they may occur on one resource axis (e.g., food, time of activity, microhabitat) or on several. Reviews of resource partitioning in amphibians and reptiles reveal considerable variation among assemblages in terms of which resource axis is partitioned the most (Toft 1985). For example, food is the major resource that is partitioned by lizards in North American deserts, but in the Kalahari Desert of southern Africa, lizards eat mainly termites, and there is only slight partitioning of food resources. Lizards in the Kalahari exhibit considerable differences in microhabitat use and in time of activity. Australian desert lizards exhibit strong separation along all three niche dimensions: food, microhabitat, and time of activity (Pianka 1986).

Two decades ago, most ecologists attributed resource partitioning to the influence of past or ongoing interspecific competition. Differences among sympatric species in morphology (especially body size and traits related to feeding), timing of activity, preferred habitat, and prey selection were thought to permit species to coexist by minimizing competition for limited resources. Recently, however, ecologists have questioned the relationship between resource partitioning and competition. First, it makes sense that different species use resources somewhat differently. Second, competition is certainly not the only mechanism that can lead to partitioning of resources. Patterns of resource partitioning can also result from predation and from factors that operate independently of interspecific interactions, such as physiological and morphological constraints.

In a classic study of pattern, Ernest Williams (1983) divided the nine species of *Anolis* lizards on

Puerto Rico into two groups based on their occurrence in sunny or shaded habitats. Within the sun or shade group, species differ in the heights and diameters of their perches (Figure 14–3). Presumably, partitioning of space in this manner provides more or less exclusive perch sites. The location of perch sites is important because these are the places from which prey is detected. The lizards also have different periods of activity, and they feed on different sizes or types of prey. Competition may be (or may have been) an important mechanism in establishing this pattern, but neither the mechanism nor the process has been demonstrated, and other mechanisms are plausible. For example, the species of *Anolis* on Puerto Rico differ in thermoregulatory behavior (see Figure 5–20) and in resistance to evaporative water loss.

Demonstrating the importance of competition usually requires experimental manipulations in the field. For example, by removing individuals of one species, a researcher can analyze the effect of that species on another. Arthur Dunham (1980) assessed competitive interactions between two small insectivorous lizards, the tree lizard (*Urosaurus ornatus*) and the canyon lizard (*Sceloporus merriami*), in an area of the Chihuahuan desert in Texas. He found that population density, foraging success, growth rate, body mass before hibernation, and lipid levels of *U. ornatus* were significantly greater in plots from which *S. merriami* had been removed than in plots where the two species co-occurred. These differences occurred only during the 2 driest years of his 4-year study. The abundance of food (arthropods) was positively correlated with rainfall, and this observation suggests that these lizards compete only during periods of drought-induced food scarcity. The observation that the intensity of interspecific competition varied with environmental conditions indicates the importance of long-term studies in determining the impact of competition, and reveals that processes are not always consistent.

Figure 14–3 **Perch and climatic preferences of *Anolis* lizards from Puerto Rico.** Numbers indicate the maximum adult male snout–vent lengths in millimeters. *Anolis roosevelti* is indicated in brackets because it is only known from the island of Culebra, and in fact the species may now be extinct. *(Source: Williams 1983.)*

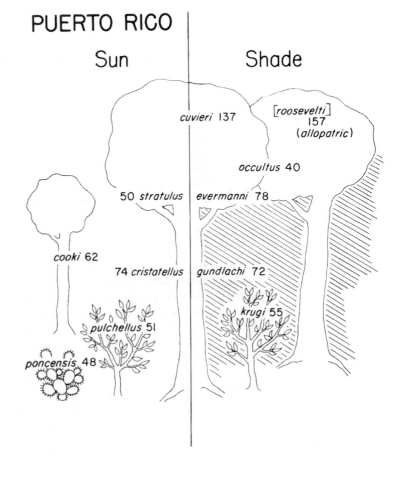

Predation and Parasitism. Predation (a mechanism) occurs when one individual captures and eats another. Predation is generally easier to observe and quantify than is competition. Most predators feed on more than one prey species within a community, and the choice of prey at any given moment often depends on the relative abundance of different prey species. As a given prey species increases in abundance, it may become a primary diet item for increasing numbers of predator species. Eventually increased predation pressure will reduce the density of the prey species, thus making it a less preferred prey item. In this way the feedback system of predation can shape community structure by limiting the abundance of prey populations.

From an evolutionary standpoint, one of the ways a species can respond to an increase in predation pressure is submergent behavior—that is, avoiding predation by reducing activity. An individual that reduces its foraging activity also reduces its food intake, which in turn affects the amount of energy it can devote to reproduction. The effects of submergent behavior can be felt at the community level (Maiorana 1976). For example, as the availability of the species exhibiting submergent behavior is reduced, a predator may switch to a different prey whose availability is now proportionately higher, and community-wide interactions of competition and predation may change.

Predation and competition can act together in structuring communities. For example, predation can reduce population density of amphibian larvae to such low levels that food resources are no longer limited. Several experimental studies have demonstrated that surviving tadpoles grow faster in the presence of predators than in their absence because competition has been reduced. The interaction of competition and predation can also produce indirect effects within communities by changing normal foraging behavior, habitat use, timing of activity, and life history characteristics.

Like competition and predation, parasites operate in a density-dependent manner and may regulate the density of species and alter the composition and normal functioning of a community (Dobson and Hudson 1986). Perhaps half of all animal species are parasites. A parasite invading a community could affect some species and not others, causing major shifts in competitive and predator–prey interactions. Such changes could completely restructure the community, by reducing the number of species present and by affecting densities of the constituent species. Although ecologists acknowledge that parasitism may be extremely important in structuring communities, few empirical studies have documented this phenomenon. Work with a malarial parasite of *Anolis* lizards, discussed in the last section of this chapter, is one of these few studies (Schall 1992).

Pathogens likewise can have serious consequences at the population and community levels. Cynthia Carey (1993) reported on the disappearance of 11 populations of boreal toads (*Bufo boreas boreas*) from Colorado. The apparent cause of local extinction was a disease called red-leg, which is caused by the bacterium *Aeromonas hydrophila*. This bacterium is present on the skin and in the digestive tract of healthy fish and amphibians. Why did these populations of toads succumb to infection? Carey speculated that an environmental factor or combination of factors caused sublethal stress in the toads. This stress then caused suppression of the immune system, so that the toads were unable to fight off infection by *Aeromonas*.

In this example the effect of *Aeromonas* on the population was local extinction. What about consequences at the community level? Presumably, different species have different levels of tolerance to stress, pathogens, and the interactions among these factors. If a pathogen becomes established within a community, the composition, relative abundance of species, and interactions among species will change in unpredictable ways. How much of the apparent structure we see in assemblages today is due to the past and present influence of pathogens? At any point in time the species of amphibians present in a given assemblage may be there because they can tolerate *Aeromonas* or other pathogens at present. Twenty years from now we might find a different assemblage of species. Some of the species currently present may succumb to pathogens, and new species might immigrate.

Habitat Complexity. Community structure can be strongly influenced by the heterogeneity of the environment. For example, Eric Pianka (1967, 1986) documented the pattern of increasing species richness of desert lizards with increasing plant structural diversity and overall microhabitat diversity. Several studies of *Anolis* lizards have emphasized the importance of morphological and behavioral traits and their relationship to habitat structure as an important factor in explaining pattern and mechanism of organization in these lizard assemblages (e.g., Moermond 1979, 1986, Pounds 1988, Losos 1990).

Basically, lizards that jump from branch to branch are not found in a habitat where there aren't any branches. On the other hand, the availability of a very complex matrix of limb diameters, angles, and heights provides the resource diversity to support a great diversity of lizards with different jumping modes. For frogs, the number of species that can coexist generally depends on the number of different kinds of sites available for hiding from predators, laying eggs, perching when locating food resources, or attracting females by calling. Rocky outcrops in grassland habitats have species of snakes not found in the surrounding grassy areas. Community ecologists have often assumed that the division of spatial resources reflects current or historical competition or predation. The patterns we see, however, may simply reflect the fact that the added habitat complexity means that more species with their individual preferences and tolerances are attracted to the area.

Physiological Tolerances. Because each species has a unique set of physiological attributes, each species exhibits unique preferences and tolerances to its physical environment. These preferences and tolerances explain in part why species are found where they are, their observed abundances, and why they use the resources they do.

The microhabitat an animal lives in can influence its ecological performance (Huey 1991). Thus, the physiological requirements of species within a given assemblage should influence their use of the habitat. These requirements vary seasonally in response to climatic variables, reproductive state, and nutritional status. Furthermore, animals in different life history stages differ in their needs and thus in their use of the environment. Once again, we see that an assemblage is not static but is a dynamic group of interacting species because conditions and physiological requirements are constantly changing.

The influence of physiological tolerances is illustrated by the ecological relations of four species of *Eleutherodactylus* frogs from Jamaica (Pough et al. 1977). *Eleutherodactylus cundalli* and *E. gossei* are native to Jamaica and are usually found in moister sites with dense vegetation. In contrast, *E. planirostris* and *E. johnstonei* are colonist species, introduced from elsewhere; the colonists are not found in places where the native species are abundant. Instead they are usually found in drier, disturbed areas such as in pastures, lawns, along road banks, and in coconut and banana groves. Do the native species compete more successfully and exclude the invaders from their preferred habitats, or do the invaders simply have different moisture and temperature tolerances that allow them to exploit disturbed habitats that are not suitable for the two native species? Physiological studies revealed that the two native species were more sensitive to high temperatures than were the colonist species. Although all four species lost water at the same rate, the introduced species were less sensitive to dehydration than were the native species; in fact, they lost half again as much water as the native species before they were incapacitated. It seems, then, that the differences in microhabitat use can be explained in large part by differences in physiological tolerance. The two invaders are likely successful colonizers on Jamaica and in other areas because they can tolerate hot, dry conditions.

Environmental Variation Imposed by Weather

The structure of a community can vary considerably through time as a function of unpredictable or random events such as weather conditions. The effect of weather on the relative abundance and composition of an assemblage of lizards from the Chihuahuan desert in New Mexico was documented over a 5-year period of time (Whitford and Creusere 1977). The amount and seasonality of rainfall is unpredictable, and spring annuals grow only in years when rainfall in late fall or winter exceeds 75 millimeters. Heavy rains in midsummer produce a larger biomass of grasses and forbs than normal. Primary productivity (the amount of carbon fixed by photosynthesis) strongly affects the number and activity of the arthropod prey the lizards eat. Insect abundance was increased in years of above-average rainfall, resident species of lizards increased either their clutch sizes or their frequency of egg laying, and immigrant species from wetter habitats became established in the study sites. In contrast, during years of below-average rainfall, insect abundance was reduced, resident species reduced either their clutch size or the frequency of egg laying, and immigrant species exhibited reduced population sizes or disappeared entirely. Thus, the effect of weather is a strong force in structuring the assemblages of lizards in the

Chihuahuan desert, and these assemblages can change on a yearly basis.

This example shows that the dynamics of interactions within an assemblage will change drastically because each species is affected differently. The same situation results when a temporary pond dries. The length of time a site contains water during a year affects the number of species and the reproductive success of amphibians in a given assemblage. Joseph Pechmann and colleagues monitored three wetland sites in South Carolina for 3 to 8 years (Pechmann et al. 1989). During this time the sites filled with water each winter and dried during the spring or summer, but the actual dates of filling and drying varied considerably among sites and years. Five species of salamanders and 11 species of anurans deposited eggs at these three sites. The number of species that successfully produced juveniles was positively correlated with the number of days the site contained water (Figure 14–4). Likewise, the total number of metamorphosed juveniles (all species combined) was positively correlated with pond duration.

Prolonged drought can strongly modify assemblage composition and species interactions for terrestrial amphibians. Drought will affect short-lived and long-lived species differently. Short-lived species may be locally eliminated, at least temporarily, until colonization from adjacent areas occurs. In contrast, long-lived species may merely experience a drop in reproduction and thus decreased population density for a while. An extremely dry El Niño year may have been responsible for the extinction of a population of golden toads (*Bufo periglenes*) from the Monteverde Cloud Forest Reserve in Costa Rica (Pounds and Crump 1994). Many other anurans also declined or disappeared from the area during this period (1987–1988). The assemblage of amphibian species at Monteverde in the mid-1990s is very different from the assemblage that was present 10 years earlier. This example shows that the short-term effects of weather may be the most important proximate determinant of the composition of an assemblage.

Effects of Humans

Many of the geographic distributions and community compositions we see today may reflect past activities of humans. There are probably few habitats in the world that haven't been modified at least to some extent by humans. When Europeans began exploring

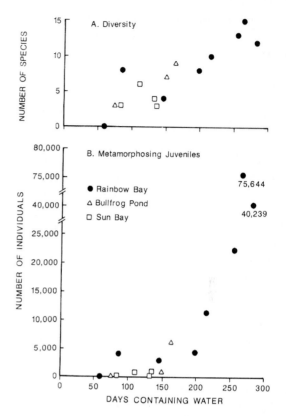

Figure 14–4 **Effect of hydroperiod on anuran reproduction.** The length of time a site contains water (its hydroperiod) can strongly affect the species diversity and reproductive success of species within a given assemblage of aquatic-breeding amphibians. Panel (a) shows that the number of species that successfully produced juveniles increased with the number of days the site contained water. Panel (b) shows that the total number of metamorphosed juveniles (all species combined) also increased with hydroperiod. Each data point represents one site for 1 year. Notice the extreme range of breeding success at Rainbow Bay: from none to 75,644 metamorphosed juveniles, representing from zero to 15 species. (*Source: Pechmann et al. 1989.*)

the New World, they thought they had found vast, undisturbed areas of wilderness. We know now, however, that Native Americans managed many of these landscapes and in the process vastly altered population densities of animals and caused extinctions (Nabhan 1995). For many thousands of years, indigenous cultures of humans worldwide have directly and indirectly influenced the structure of communities of plants and animals.

The Maya culture is one of the oldest in Middle America, and the agricultural practices of the Maya were quite sophisticated. They created terraced fields and devised effective irrigation systems, which meant they could cultivate large areas of land. In fact, during the Classic period (300–900 A.D.) the Maya may have deforested much of the Yucatán Peninsula and converted the land to crops. What effect did this have on the amphibians and reptiles? Julian Lee (1980) proposed that the present disjunct distributions of certain species of amphibians and reptiles on the peninsula resulted in part from shifting land use patterns from Mayan times to the present. He hypothesized that during the time when much of the peninsula was under cultivation, nonforest species expanded their ranges. Later, after the collapse of the Mayan civilization, the forest regenerated, and as it did so nonforest species became restricted to savanna and other non-forested areas of the peninsula.

The reptile fauna on tropical Pacific islands consists of a mixture of native and introduced species. Polynesians and Melanesians inadvertently introduced geckos and skinks throughout much of the Pacific during their extensive travels about 4,000 years ago. More recent settlement by Europeans has further modified community composition. For example, the common house gecko (*Hemidactylus frenatus*) has been inadvertently introduced to many tropical Pacific islands since the 1930s. The density of mourning geckos (*Lepidodactylus lugubris*) is often depressed where house geckos are present. Kenneth Petren and Ted Case (1996) interpreted this decline as a result of exploitative competition. House geckos are larger, can run faster, and are more efficient at harvesting insects that are clumped around light sources. Petren and Case suggested that the construction of European-style buildings with flat, open walls and lights that attract high densities of insects was probably responsible for shifts in interactions, abundances, and composition of species within these lizard assemblages.

Salamanders used as live bait by fishermen are routinely transported out of their normal range of habitats into other areas where they are raised in farm ponds and later harvested. Some of these animals escape. Often the nearby habitats are unfavorable, so that wild populations cannot become established, but in some cases introduced species have successfully become established. One such example is the black-bellied salamander, *Desmognathus quadramaculatus*, commonly called spring lizards. The existence of populations of black-bellied salamanders in Piedmont habitat in northeastern Georgia, far from the species' natural habitat, has been attributed to introduction for the fishing bait industry (Martof 1953). Who knows how many other present-day salamander distributions are the result of accidental or intentional transport by nineteenth-century century anglers? In habitats that have experienced human intervention for hundreds or thousands of years what proportion of the present-day members of herpetological assemblages is a result of conscious (or unconscious) introduction by humans?

Other Historical Events

Historical factors such as colonization and extinction, degree of isolation, the existence of barriers, and available biotic stocks may be responsible for much of the organization apparent in some communities. In order to understand community structure more fully we need to consider historical events and factors such as the biogeographic histories of the species in the communities being studied and the climatic histories of the areas being studied.

Another historical contingency that is increasingly being considered is phylogeny. For example, differences in the number of species in two assemblages in different regions might come about simply because one assemblage is composed of lineages that tend to diversify more than the other (Losos 1996). The difference may have nothing to do with vegetation heterogeneity, competition, or predation. John Cadle and Harry Greene (1993) argued that phylogenetic information about the species found within an assemblage can provide insight into the historical basis of observed patterns of resource use. Using phylogenetic information, they reinterpreted patterns of food resource (Vitt and Vangilder 1983) for a Brazilian assemblage of colubrid snakes. Whereas the earlier study explained observed patterns in terms of present-day ecological factors, Cadle and Greene pointed out that the specific lineages involved provide answers. They argued that every lineage carries with it some evolutionary baggage that permits only certain options in terms of what an animal eats. Thus, there is a strong association between dietary preferences and phylogenetic relationships, and this association affects community composition.

For example, there are no snakes that specialize on invertebrates within the Brazilian assemblage. Vitt and Vangilder suggested that this absence is the result of competition with insectivorous mammals, the presence of predators on small snakes, and lack of suitable microhabitats. In contrast, Cadle and Greene pointed out that lineages of invertebrate-eating snakes are rare or absent in most of the Neotropics, so we wouldn't expect to find them in the Brazilian assemblage. Furthermore, the assemblage consists of a large number of species of snakes that feed on anurans. Vitt and Vangilder attributed this pattern to convergence and to the year-round abundance of frogs at the site. Cadle and Greene argued that it is not surprising that so many of the species feed on frogs since a large proportion of the snakes belong to a clade that feeds primarily on frogs. Finally, there are a few snakes that specialize on mammals. Vitt and Vangilder attributed this pattern to the unpredictability of mammals as a food resource. Once again, Cadle and Greene argued that the lack of mammal-eating snakes is explained more simply by phylogenetic considerations: few Neotropical snake assemblages studied to date have many mammal-eating specialists.

Case Studies of Amphibian and Reptile Assemblages

The remainder of this chapter will highlight studies of amphibians and reptiles that illustrate the importance of the determinants of community structure. It is important to realize that determinants work simultaneously, and their actions can change in time and space. We focus on four types of assemblages here: *Anolis* lizard assemblages in the Caribbean, assemblages of North American salamanders, aquatic larval amphibian assemblages, and tropical anuran assemblages.

Anolis Assemblages in the Caribbean

Anolis lizards in the Caribbean islands have been the focus of numerous community-level studies. Anoles are relatively easy to observe and manipulate, and they are frequently extremely abundant. Many of the smaller islands have two sympatric species of *Anolis*, but at some sites on larger islands there are as many as nine coexisting species. Some islands have few species of birds and no native mammals. With few predators, population densities of anoles may be very high, and

food becomes a limiting resource. *Anolis* on islands in the eastern Caribbean provide classic examples of assemblages where patterns of resource partitioning suggest that competition among ecologically similar species may be reduced by division of resources. Co-occurring species differ in perch site location, prey size, and microhabitat use. In some cases aggressive interactions may be the mechanism that serves to exclude one species from a particular microhabitat. For example, when individuals of *Anolis lineatopus* were removed from their perches on tree trunks near Kingston, Jamaica, individuals of *Anolis opalinus* immediately shifted from perches higher on tree trunks and in the canopy to the low perches previously occupied by *A. lineatopus* (Jenssen 1973). These two species may be competitors for limited food resources. Through interspecific aggression, the behaviorally dominant *A. lineatopus* does not allow *A. opalinus* to perch near the ground where arthropod prey is presumably abundant.

Competition may be expressed less overtly than in the previous example. *Anolis wattsi* is a small species of *Anolis* that occurs on the islands of St. Maarten and St. Eustatius. On each island it coexists with a large species of *Anolis*—*A. gingivinus* on St. Maarten and *A. bimaculatus* on St. Eustatius. The co-occurring species share the same habitat (forested ravines) and have similar diets. The co-occurring species on St. Maarten are more similar in perch height and microhabitat selection than the species on St. Eustatius. Because of the greater ecological overlap on St. Maarten, competition for food was hypothesized to be greater than on St. Eustatius, where the species are more distinct (Pacala and Roughgarden 1985). To test this hypothesis, Pacala and Roughgarden constructed four 12 × 12 meter enclosures on each island. On St. Maarten, two enclosures were stocked with natural densities of *A. gingivinus* and *A. wattsi* and two were stocked only with *A. gingivinus*. On St. Eustatius, two enclosures were stocked with natural densities of *A. bimaculatus* and *A. wattsi* and two were stocked only with *A. bimaculatus*. This experiment was designed to determine if and how the presence of *A. wattsi* affected the co-occurring species. On St. Maarten, *A. gingivinus* ate fewer and smaller prey, grew more slowly, produced fewer eggs, and perched higher in the enclosures with *A. wattsi* than in the enclosures where they were alone (Figure 14–5). In contrast, on St. Eustatius, *A. bimaculatus* exhibited no consistent differences between enclosures with and without *A.*

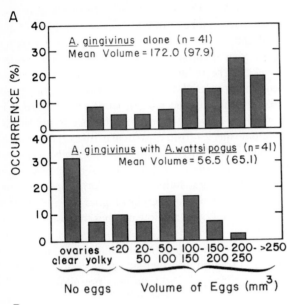

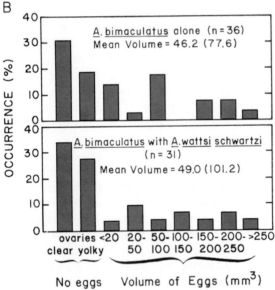

Figure 14–5   **Effect of *Anolis wattsi* on the reproductive condition of *Anolis gingivinus* and *Anolis bimaculatus*.** *Anolis gingivinus* housed in the absence of *A. wattsi* contained on the average from two to four times the volume of eggs found in individuals housed with *A. wattsi* (A). In contrast, there was no significant difference in the volume of eggs in female *A. bimaculatus* housed with or without *A. wattsi* (B). (*Source: Pacala and Roughgarden 1985.*)

wattsi. These results support the hypothesis that competition for food is more intense for ecologically similar than for ecologically dissimilar species.

Additional work with these same two species of *Anolis* on St. Maarten reveals another twist to the story (Schall 1992). Schall suggested that a malarial parasite mediates competition between *A. gingivinus* and *A. wattsi* and in large part determines the present distribution of the two species on the island. *Anolis gingivinus* occurs throughout the island, but *A. wattsi* occurs only in the central hills. The malarial parasite *Plasmodium azurophilum* is very common in *A. gingivinus* in some areas, but it rarely infects *A. wattsi*. The parasite is unusual in that it infects both red and white blood cells. It causes anemia and reduced hemoglobin concentration in red cells, and alters the function of the white blood cells (less acid phosphatase is produced) and presumably reduces immune function. The distribution of the parasite relative to the lizards is intriguing. *Anolis gingivinus* occurs alone in areas where the parasite is absent, but *A. wattsi* is present wherever the parasite infects *A. gingivinus*. Apparently *A. wattsi* can compete successfully with *A. gingivinus* only in areas where *A. gingivinus* is weakened by malaria.

One line of evidence used to support the importance of competition in structuring communities is the phenomenon of niche expansion, which is the response to reduced levels of competition. The classic example of niche expansion, also called ecological release, is found in a comparison between island and mainland species. Because islands have fewer species than mainland areas of similar size, island species theoretically coexist with fewer competitors. One would predict that insular species should have higher population densities, eat a wider variety of prey, and utilize a wider diversity of microhabitats—all examples of niche expansion.

The mainland–island dichotomy can be refined by comparing islands with different numbers of species. Lister (1976) examined three aspects of resource use by West Indian anoles: structural niche (distribution of perch heights), thermal niche (range of body temperatures over which a population is active within its habitat), and food niche (distribution of prey sizes and the diversity of prey species eaten). Results of this study provided evidence of niche expansion in areas where the number of congeners (species in the same genus) was low for all the niche dimensions except for prey diversity, which was fairly constant. For example, both male and female *Anolis sagrei* exhibit greater diversity in the height of perches used on islands where they are the only species of anole present than on islands where there are three or four species of co-occurring *Anolis* (Figure 14–6).

Not all community structure of Caribbean anoles, however, can be attributed to competition. Caribbean *Anolis* also provide a good example of the effect of physiological tolerances on microhabitat distribution. In Puerto Rico, *A. gundlachi* is found within the rain forest, whereas *A. cristatellus* is found at the edge of the rain forest or in open areas. These two species are closely related, are similar in body size, and are both generalized insectivores. One might assume that the habitat partitioning exhibited by these two species is the result of present-day competition. A more likely explanation, however, is that differences in habitat use reflect different physiological tolerances. *Anolis gundlachi* is less heat tolerant than is *A. cristatellus*, a physiological difference corresponding to the microhabitats they occupy. In September 1989, Hurricane Hugo provided a test of the hypothesis that habitat differences are the result of physiological tolerances (Reagan 1991). Strong winds removed the canopy of forest trees in many places. When this happened, individuals of *A. gundlachi* left their former perches on tree trunks and retreated to lower perches that were shaded by fallen debris. Thus, it was not the presence of *A. cristatellus* that prevented *A. gundlachi* from using sunny perch sites in open areas. The differences in habitat use exhibited by *A. gundlachi* and *A. cristatellus* can be explained by physiological tolerances and preferences.

Assemblages of North American Salamanders

Salamanders have formed the basis of classic studies of community structure. Streamside salamanders (mostly plethodontid salamanders in the genus *Desmognathus*) are especially suitable for field experimental manipulations, and some of the most rigorous tests of hypotheses in salamander community ecology have focused on these species. Terrestrial salamanders of the genus *Plethodon* are also suitable for field experimental manipulations, and have been the focus of classic studies that have dealt with the importance of interspecific competition as a population-level process.

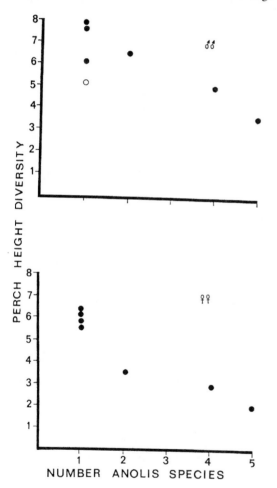

Figure 14–6   **Structural niche of *Anolis sagrei*.** Diversity of perch heights used by male (top) and female (bottom) *Anolis sagrei* on several islands in the West Indies, plotted against the total number of *Anolis* species occurring in the study areas on these islands. The data points for the sites where there is only a single species of *Anolis* are where *A. sagrei* occurs alone. These data (except for the population indicated by o) suggest expansion in the structural niche for *A. sagrei*. (*Source: Lister 1976.*)

Streamside Salamanders. Habitat differences among species of *Desmognathus* have long been recognized (Dunn 1917). Over the past 20 years investigators have studied and documented the patterns of microhabitat preferences in detail. A typical assemblage of *Desmognathus* in the deciduous forests of the southeastern United States consists of the black-bellied salamander (*D. quadramaculatus*), a largely aquatic species that weighs about 12 grams as an adult; the seal salamander (*D. monticola*), a semiterrestrial stream-bank species that grows to 9 grams; and the mountain dusky salamander (*D. ochrophaeus*), the most terrestrial species and, at an adult mass of 3 grams, the smallest (Figure 14–7). Because the larvae of all three species are aquatic, movements between aquatic and terrestrial habitats are a normal part of the ecology of the two terrestrial species. Parallel differences in habitat and body size of the three species were initially attributed to competition-based resource partitioning. Experimental manipulations, however, tell quite a different story.

When Nelson Hairston Sr. and his colleagues removed individuals of the smallest species, *D. ochrophaeus*, from streamside plots, the numbers of the two larger species decreased. This decrease suggested that the small species is an important prey item for the two larger species. If the interaction had been competition for food, the numbers of the two larger species should have increased because more prey would have been made available when *D. ochrophaeus* was removed. In contrast, when the intermediate-sized species, *D. monticola*, was removed, the numbers of both the largest species and the smallest species increased. The increase of the largest species suggested that the population-level interaction between the two larger species is competition. The inferred mechanism is that removing the intermediate-sized species increased food resources for the largest species. The interaction between the intermediate-sized species and the smallest species could be either competition or predation, because the smallest species should respond positively to the removal of either a competitor or a predator. Hairston's conclusion from experimental manipulations of density is that ecological differences among species of *Desmognathus* are the result of both predation and competition (Hairston 1986, 1987). These relationships presumably began when the species assemblage formed, and they continue today. The smallest and most terrestrial species, *D. ochrophaeus*, has shifted its microhabitat away from streams, thereby avoiding predation by *D. quadramaculatus*. The medium-sized species, *D. monticola*, has shifted its microhabitat from the stream to the streambank. The inferred mechanism is that competition with adult *D. quadramaculatus* is avoided, and predation by *D. quadramaculatus* on juvenile *D. monticola* is avoided.

Habitat complexity is another factor that can be important in structuring assemblages of *Desmognathus* salamanders. Adding rocks to the stream or stream-

(a)

(b)

(c)

Figure 14–7   **Three species of *Desmognathus* salamanders that occur in the deciduous forests of the southeastern United States.** (a) *Desmognathus quadramaculatus*, (b) *Desmognathus monticola*, (c) *Desmognathus ochrophaeus*. (*Photographs by R. Wayne Van Devender.*)

edge habitat at a site in southwestern North Carolina increased the density of *D. monticola*, suggesting that the population is limited by availability of cover objects (Kleeberger 1984). Likewise, adding rocks and pieces of wood increased population densities of *D. monticola* along streambanks in North Carolina (Figure 14–8). The additional retreat sites did not increase the number of prey, but they allowed juvenile *D. monticola* to find refuges from the predatory *D. quadramaculatus* and the larger *D. monticola*. Thus, the amount of streambank cover affected the abundance of *D. monticola* through its effect on survival of juveniles (Southerland 1986).

Terrestrial Salamanders. Nearly 50 years ago, Nelson Hairston Sr. interpreted the fact that the slimy salamander (*Plethodon glutinosus*) and Jordan's salamander (*Plethodon jordani*) have altitudinal distributions that overlap very little (70 to 120 meters) in the Black Mountains as being the result of intense interspecific competition (Hairston 1949). In other words, he inferred that the pattern of observed distribution is one that most likely resulted from competition. It was later discovered that whereas these two species overlapped only little in altitudinal distribution in the nearby Great Smoky Mountains, they overlapped by as much as 1,220 meters in the Balsam Mountains (Figure 14–9). Hairston pointed out that if his original interpretation of mutual competitive exclusion were correct, then competition as a population-level process must be weak in the Balsam Mountains.

In field experiments designed to address the hypothesis of strong versus weak competition, Hairston removed each species of salamander separately from different study plots in the different areas, and then compared the densities of salamanders with those of unmanipulated control plots (Hairston 1980a,b, 1987). Data from these long-term field experiments support the prediction that the response of each species of salamander to the removal of the other species is greater in the Great Smoky Mountains than in the Balsam Mountains. When *P. jordani* was removed from plots in the Great Smoky Mountains, the density of *P. glutinosus* increased significantly, demonstrating a negative effect of the former on the latter species, but the reciprocal experiment—removal of *P. glutinosus*—did not result in a significant increase in *P. jordani*. Interestingly, however, the proportion of the two youngest age classes of *P. jordani* was significantly greater in the plots where *P. glutinosus* had been removed, suggesting that *P. jordani* reproduced more successfully in the absence of *P. glutinosus*.

Even in the Balsam Mountains there was a significant increase in the density of *P. glutinosus* in plots from which *P. jordani* had been removed, but the increase was observed only at the end of the fourth year of the experiment, suggesting much less intense competition than in the Great Smoky Mountains. There was also an increase in the proportion of juvenile *P. jordani* in plots from which *P. glutinosus* had been removed, but again the difference was less dramatic than that observed in the Great Smoky Mountains.

Hairston's experiments provide evidence for the importance of competition as a process influencing the distribution of these salamanders, but they do not explain the mechanism. Subsequent studies by two of Hairston's students, Sarah Stenhouse and Kiisa Nishikawa, focused on the mechanism of competition. Stenhouse examined the possibility that the two species partition food resources or foraging microhabitats more extensively in the Balsam Mountains than in the Black Mountains, but found this not to be the case (see Hairston 1987). Nishikawa carried out a series of laboratory experiments with both species from both mountain ranges to determine the extent and importance of intraspecific and interspecific aggression related to space. She found that both species respond aggressively to the other species and to mem-

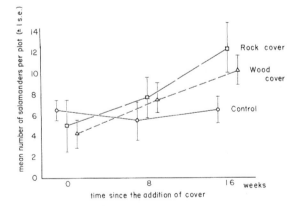

Figure 14–8 **Effect of cover objects on population size.** Addition of rock and wood cover objects to stream habitats in North Carolina resulted in an increase in the number of salamanders found during subsequent searches. (*Source: Southerland 1986.*)

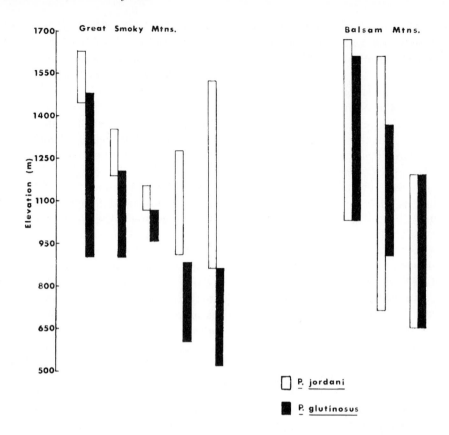

Figure 14–9   **Altitudinal distribution of the salamanders *Plethodon jordani* and *Pletho-don glutinosus*.** Five vertical transects are shown for the Great Smoky Mountains and three for the Balsam Mountains. The distributions of salamanders are nearly mutually exclusive in the Great Smoky Mountains but are broadly overlapping in the Balsam Mountains. *(Source: Hairston 1980b.)*

bers of their own species. This result suggests that the salamanders may be territorial and that space is the resource for which the salamanders compete (Nishikawa 1985). *Plethodon glutinosus* from the Great Smoky Mountains is more aggressive toward intruders than is *P. glutinosus* from the Balsam Mountains. In contrast, there is no significant difference in aggressive behavior between *P. jordani* from the two mountain ranges. Nishikawa concluded that the high intensity of aggression of *P. glutinosus* in the Great Smoky Mountains is the reason for the small overlap in altitudinal distribution of *P. glutinosus* and *P. jordani*.

Another important study of *Plethodon* was carried out by Robert Jaeger in Virginia. The Shenandoah salamander (*P. shenandoah*) is known from only three mountains in Virginia, where it occurs primarily on dry, rocky talus slopes. The red-backed salamander (*P. cinereus*) occurs in soil habitats outside the talus. Jaeger interpreted this nearly exclusive pattern of distribution as being the likely result of competition between the two species. Through field enclosure experiments, he found that *P. cinereus* inhabits areas of deep soil because it requires a moist substrate; it cannot live in the talus slopes because that habitat is too dry. Although *P. shenandoah* does not prefer the rocky, dry conditions of the talus slope it is able to survive there because it can withstand the stress of dehydration better than *P. cinereus* can (Jaeger 1971a,b). At the level of the population, *P. cinereus* appears to be competitively excluding *P. shenandoah* from the more optimal soil habitat. Through a series of elegantly designed experiments, Jaeger and his colleagues identified aggressive interference on the part of *P. cinereus* as the mechanism of competition (Wrobel et al. 1980, Jaeger 1981). *Plethodon cinereus* is territorial and bites intruding salamanders.

Larval Amphibian Assemblages

Assemblages of larval amphibians are extremely dynamic. Most species that lay their eggs in water metamorphose within a few weeks, and the composition and biotic interactions of assemblages change constantly. As larvae grow they become less (or sometimes more) vulnerable to predators, and more (or sometimes less) competitive with larvae of other species. There is continuous input and output as additional species lay their eggs in the pond and metamorphosing individuals leave the pond, and new opportunities for competition and predation arise. The abundance of food resources can change rapidly as algae bloom in response to pulses of nutrients washed in by storms. Unpredictable events such as pond drying and extreme water temperatures affect species differently, thus modifying biotic interactions.

Although resources for larval amphibians are difficult to measure, some investigators have attempted to identify patterns of resource partitioning in these assemblages. Space resources were partitioned to a much greater extent than food resources by assemblages of tadpoles at a seasonal tropical site in Thailand (Heyer 1973, 1974). For example, different species of tadpoles feed in different positions in the water column (Figure 14–10). The time axis is also partitioned. Different species of anurans deposit their eggs at different times of the year, resulting in temporal partitioning of a breeding site for developing larvae. The assumption is that such partitioning reduces the potential competition for food resources.

Assemblages of larval amphibians have provided excellent models for experimental studies of the effects of competition, predation, and interactions between these factors (Wilbur 1972, 1989). Amphibian assemblages are particularly amenable to controlled manipulation because they are often found in a well-defined space (e.g., a small pond) that is relatively easy to simulate artificially. Furthermore, larval amphibian assemblages include both competitors (filter-feeding tadpoles) and predators (salamander larvae) that prey on each other as well as on tadpoles. Henry Wilbur

and his colleagues have used replicated enclosures in natural ponds or replications of artificial ponds (cattle watering tanks) to simulate natural habitats (Figure 14–11). Cattle tanks can be stocked with predetermined quantities of food, tadpoles of competing species, predatory salamander larvae, and predatory insects. Because all variables can be controlled, powerful parametric statistics can be used to detect the direct effects of predation and competition as well as the interactions among the variables. These studies suggest that intra- and interspecific competition and predation are extremely important in structuring assemblages, and that these factors interact in complex ways.

Assemblages of six species of anuran tadpoles (*Scaphiopus holbrooki, Bufo terrestris, Hyla chrysoscelis, Hyla gratiosa, Pseudacris crucifer,* and *Rana sphenocephala*) were used to test the effect of several densities of two species of predaceous salamanders (newts, *Notophthalmus viridescens,* and tiger salamanders, *Ambystoma tigrinum*) (Morin 1983). Tiger salamanders were such effective predators that they eliminated most tadpoles of all six species. In the presence of newts, the four competitively superior species of anurans (*S. holbrooki, B. terrestris, H. chrysoscelis,* and *R. sphenocephala*) had reduced densities, but the two competitively inferior tadpoles (*P. crucifer* and *H. gratiosa*) survived at greater densities in tanks containing newts than in tanks without newts (Figure 14–12).

A positive correlation between predator density and wet body mass of tadpoles at metamorphosis suggested that the effect of interspecific competition for food decreased as predation pressure from newts increased (Figure 14–13). Furthermore, three species of anurans exhibited significantly shorter developmental time to metamorphosis in tanks with high newt densities. This experiment shows that predation on competitively dominant species allows competitively inferior species to persist because of reduced competition for food.

Predators do not have to kill their prey to have an impact on community structure. For example, the presence of predators can modify the behavior of their prey. Laboratory experiments by Earl Werner (1991) suggest that competitive interactions between tadpoles of the bullfrog (*Rana catesbeiana*) and the green frog (*Rana clamitans*) in permanent ponds of eastern North America are altered by the presence of predaceous larvae of a dragonfly (*Anax junius*). In the

Figure 14–10   **Characteristic positions of tadpoles found in ponds in Thailand.** Some species of tadpoles normally are found in only one of these positions; other species are found in two or more positions. *(Source: Heyer 1973.)*

absence of the predator, tadpoles of both species grew at the same rates when housed together with a limited food supply. In contrast, bullfrog larvae grew faster and green frog larvae grew slower when caged predators were present. In the presence of the predator, both species reduced their activity levels (Figure 14–14), but because bullfrog tadpoles were relatively more active they were able to gather more of the limited food provided and thus enhance their growth rates, while green frogs ate so little that their growth was reduced. A critical aspect of this experiment was that the predaceous dragonfly larvae were physically isolated so that the response of the tadpoles was induced visually, mechanically, or indirectly from chemical cues in the water, but not from direct attacks.

Nonlethal effects of this same dragonfly predator, *Anax*, on the competitive interactions of *Rana catesbeiana* and *R. clamitans* tadpoles were examined in an

Figure 14–11   **Cattle tanks, being used for controlled experiments with amphibian larvae.** *(Photograph by Joseph Pechmann.)*

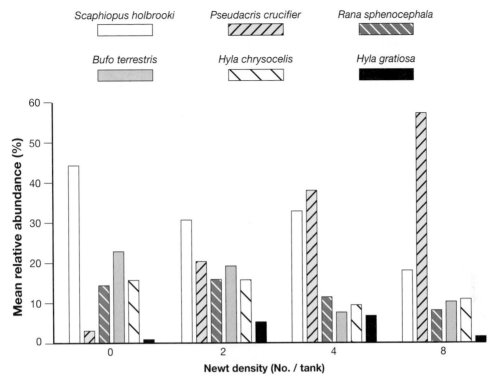

Figure 14–12   **Effect of density of newts (*Notophthalmus viridescens*) on the mean relative abundances of tadpoles and metamorphs from experimental tanks.** Note that *Pseudacris crucifer* had low survivorship in the absence of newts but increasingly higher survivorship with higher newt density. *Hyla gratiosa* tadpoles survived best at intermediate newt densities. The other four species of anuran larvae exhibited depressed densities in tanks where newts were abundant. *(Source: Morin 1983.)*

outdoor experiment using cattle watering tanks (Werner and Anholt 1996). Again, the *Anax* were housed in cages so they were unable to attack the tadpoles. Three size classes of tadpoles were used (large and small bullfrogs and small green frogs) in different density combinations, in the presence or absence of caged dragonfly predators. It was found that the presence of the predator greatly influenced the nature of competitive interactions among the tadpoles. Werner and Anholt suggested that the mechanisms responsible for these effects were probably based on behavioral responses by the tadpoles. By reducing foraging behavior in the presence of predators (an example of submergent behavior), both small bullfrog and green frog tadpoles experienced decreased growth rates. In contrast, the large bullfrog tadpoles experienced increased growth rate and size at metamorphosis in the presence of caged *Anax*, presumably because they re-

mained active and therefore consumed the additional resources made available by the decreased foraging activity of the small size classes of tadpoles (Figure 14–15).

Unpredictable factors such as weather can alter species densities and relative abundances of amphibian larvae. Marbled salamanders (*Ambystoma opacum*) lay eggs in the fall, and their larvae spend the winter in ponds. With this head start, marbled salamander larvae are normally the dominant predator on other larval amphibians in permanent ponds in Indiana. During severe winters, marbled salamander larvae die. In the following summer the larvae of both Jefferson salamanders (*A. jeffersonianum*) and wood frogs (*Rana sylvatica*) are highly successful because they are not eaten by the larger marbled salamander larvae. When larvae of both marbled and Jefferson's salamanders are killed by severe weather, the larvae of spotted

Figure 14–13 **Mass of metamorphosed *Pseudacris crucifer* from experimental tanks housing different densities of newts (*Notophthalmus viridescens*).** These data suggest that as predation increased, the effects of intraspecific competition decreased. *(Source: Morin 1983.)*

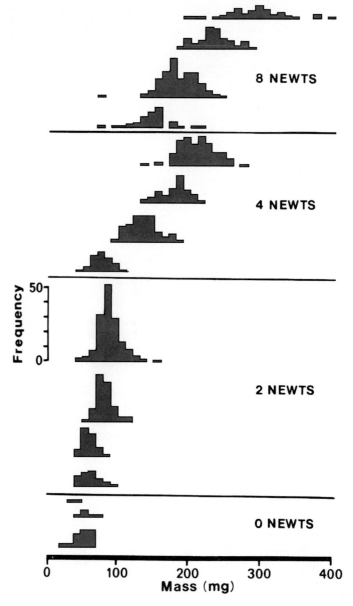

salamanders (*Ambystoma maculatum*) metamorphose successfully because their larvae are not eaten by the larger larvae of the two other species of *Ambystoma* (Cortwright and Nelson 1990).

The structure of larval amphibian assemblages is also influenced by the order in which species deposit their eggs in a pond (referred to as priority effects). All else being equal, species that deposit their eggs early are more likely to become dominant as predators or as competitors because they are the largest lar-

vae in the pond. Competitors or predators already present at a pond can influence the later composition of species at the site. For example, the gray tree frog (*Hyla chrysoscelis*) avoids artificial oviposition sites with certain predators (fish and *Ambystoma*) and with *H. chrysoscelis* larvae (Resetarits and Wilbur 1989). Likewise, *Hyla pseudopuma* avoids artificial oviposition sites with tadpoles of its own species, which are extremely cannibalistic on eggs and tadpoles (Crump 1991). The timing of oviposition may also affect interactions be-

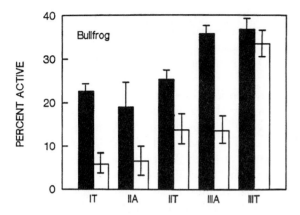

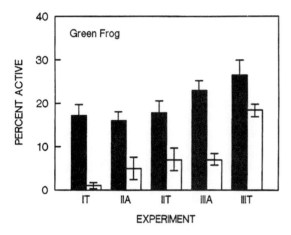

ern leopard frogs (*Rana sphenocephala*) grew and developed better (as measured by mean mass at metamorphosis, length of the larval period, and number of surviving individuals) when raised alone than when raised together. When raised together, both species did best when *Bufo* larvae were introduced to the experimental pond first, and *Rana* larvae were added later. What are the implications of these results for the species in nature? Since both species did best when raised alone, the ideal situation would seem to be habitat segregation. If the species are forced to coexist (e.g., if suitable breeding sites are limited), then the order of oviposition optimal for both species is expected to be *Bufo* first, then *Rana*. Thus, a female *Bufo* should prefer not to lay eggs in a site that already has *Rana* eggs or hatchlings. The decision may be more

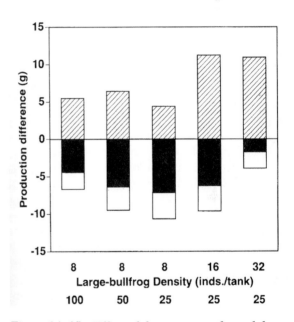

Figure 14–14 **Effect of the presence of caged dragonfly larvae (*Anax junius*) on activity of bullfrog and green frog tadpoles.** Three experiments are indicated as I, II, and III. A, the species was housed alone; T, the two species of tadpoles were housed together. Open bars indicate the presence of dragonfly larvae, solid bars indicate the absence of dragonfly larvae. Fewer tadpoles of both species were active when the caged predators were present. *(Source: Werner 1991.)*

Figure 14–15 **Effect of the presence of caged dragonfly larvae (*Anax junius*) on growth of tadpoles.** Comparison of biomass production among large bullfrog tadpoles (hatched), small bullfrog tadpoles (black), and green frog tadpoles (white) in the presence of caged dragonfly predators (*Anax junius*). The data are average differences in production by each species class between predator treatments (presence of dragonfly predators minus absence of dragonfly predators) for each of five tadpole density combinations. Each treatment had 30 green frog tadpoles. Large bullfrog tadpoles increased biomass production in the presence of the predators relative to the decline in production of small bullfrog and green frog tadpoles. *(Source: Werner and Anholt 1996.)*

tween species, determining levels of competition and predation. For example, adult newts (*Notophthalmus viridescens*) eat the eggs of tiger salamanders (*Ambystoma tigrinum*), but tiger salamander larvae grow quickly and can eat larval newts. Depending on the body size differential, larval *Ambystoma opacum* may either compete with or prey on *Ambystoma maculatum*.

The order of hatching can affect the outcome of larval competition (Alford and Wilbur 1985). Tadpoles of American toads (*Bufo americanus*) and south-

complex for a female *Rana* since she cannot assess whether a *Bufo* will lay eggs in the pond after she has deposited her eggs. The possibility of priority effects suggests that the timing of oviposition and the choice of a breeding site may involve a complex set of decisions for a female anuran.

Tropical Anuran Assemblages

Of the four types of assemblages discussed in this section, tropical anuran assemblages are the least well understood, in part because they are so species-rich and in part because so few long-term ecological studies have been carried out in tropical areas. Species richness of anurans is much greater in the tropics than in temperate regions. One might find as many as 80 species of anurans at an Amazonian lowland site in Peru or Ecuador. Whereas it is relatively straightforward to do a long-term survey and compile a list of species, it is much more difficult to study the processes responsible for apparent organization within the assemblage. We will discuss two aspects of tropical anuran assemblages: first, the high diversity of modes of reproduction and the high proportion of species that deposit their eggs out of water, and second, acoustic resource partitioning.

Assemblages of tropical frogs exhibit high diversity in modes of reproduction. For example, Martha Crump (1974) studied an assemblage of 78 species of anurans in an area of 3 square kilometers in eastern Ecuador, at Santa Cecilia. Included in this assemblage are species that lay their eggs in water, carry their eggs on their backs (both aquatic and terrestrial species), deposit terrestrial eggs and subsequently carry their tadpoles to water, deposit eggs on leaves overhanging water, with the tadpoles later falling into the water below, and deposit terrestrial eggs that undergo direct development. Thirty-six of these species (46 percent) lay their eggs in water and have aquatic tadpoles. None of the other 42 species deposits eggs in water, although 25 of these species have aquatic tadpoles. Thus, 54 percent of the species within this assemblage deposit their eggs out of water. This diversity of reproductive modes results in extensive partitioning of breeding sites and may help explain how so many species of frogs can coexist in such a small area. But what is the selective force that favored the evolution of terrestrial oviposition in such a large proportion of anurans within this assemblage, and what maintains this mode of reproduction?

A high frequency of terrestrial oviposition is a general characteristic of assemblages of tropical frogs as well (Duellman 1990). For example, of the 44 species of anurans at La Selva, Costa Rica, 26 of them (59 percent) lay eggs out of water (Donnelly 1994). At Manaus, Brazil, 16 of 42 species (38 percent) lay eggs out of water, whereas at Manu, Peru, 31 of 75 species (41 percent) lay eggs out of water, and at Barro Colorado Island in Panama, 24 of 50 species (48 percent) lay eggs out of water. What is it about water as a place to lay eggs that tropical anurans are avoiding?

William Magnusson and Jean-Marc Hero (1991) sought to answer this question by studying anurans that have aquatic tadpoles and breed in the Amazonian rain forest near Manaus, Brazil. Seventeen species used the aquatic sites (for oviposition, larval development, or both) included in the study. Eleven of these 17 species (65 percent) deposit eggs that develop on land. Magnusson and Hero examined desiccation, predation on eggs by fish, predation on eggs by tadpoles and aquatic invertebrates, competition with other species of anurans, and water quality. Through a series of experiments and field observations they were able to identify heavy predation on eggs by tadpoles and aquatic invertebrates as the most likely selective force favoring terrestrial eggs. The results of the study revealed a significant positive relationship between egg predation in each pond and the proportion of frogs that laid their eggs on land.

Let's return now to the phenomenon of a pond resonating with the raucous sounds of calling male frogs. Such breeding sites in the tropics may be composed of 15 or more species of anurans at any moment. Is there evidence of structure to such an assemblage?

Just as with larval anuran assemblages, assemblages of adult frogs at a breeding site are unpredictable and extremely variable in time. Different species have different distances to migrate, and thus some will appear earlier than others once the pond has filled with water. The composition of an assemblage at a particular site may vary daily because of species-specific responses to rainfall, humidity, and temperature conditions. Thus, different species may use the pond at different times throughout the breeding season. There is also partitioning of space at a pond throughout a 24-hour period, with some species active primarily during early evening, some in late evening, some in early morning, and so forth. For these reasons, an investigator studying a pond from

8:00 P.M. to 1:00 A.M. for 1 month would have a very biased impression of composition, relative abundances, patterns of resource use, and interspecific interactions at the pond. Unfortunately, many studies are biased in just this way.

One aspect of tropical pond-breeding anuran assemblages that has been studied in some detail is the partitioning of acoustic resources. Over the past three decades numerous studies have shown that within a given assemblage, different species use different microhabitats for calling—in water, on the ground, perched on submerged vegetation, or perched on tree trunks, stems, or leaves of bushes or trees near the pond. William Duellman and Rebecca Pyles (1983) analyzed acoustic characteristics (notes per call group, note repetition rate, number of secondary notes, duration of note, pulse rate, fundamental frequency, and dominant frequency) of hylid frogs from anuran assemblages in rain forest habitat in Brazil, Costa Rica, and Ecuador to answer three questions, as follows.

First, are advertisement calls of closely related species more similar if the species are allopatric (their geographic ranges do not overlap) than if they are sympatric (the ranges overlap)? Duellman and Pyles addressed this question by subjecting the data for the seven call characteristics of 39 species of tree frogs in three sites (Belém, Brazil; Puerto Viejo, Costa Rica; Santa Cecilia, Ecuador) to a cluster analysis, which provides a visual description of which species are the most similar. Four major clusters resulted, indicated as I–IV in Figure 14–16. Within any one cluster, species from all three geographic areas are represented. Examination of pairs of closely related species reveals that allopatric pairs have more similar call characteristics than do sympatric pairs. For example, *Phyllomedusa hypocondrialis* (33) and *Phyllomedusa palliata* (34) are allopatric, found in Brazil and Ecuador, respectively. Notice that their calls are fairly similar, as indicated in the cluster analysis. In contrast, *Phyllomedusa tarsius* (35) and *Phyllomedusa palliata* (34) are sympatric at Santa Cecilia. Notice how different their calls are—they appear in clusters I and II. This is just one example of the overall pattern found, suggesting that calls of closely related species are more similar if the species occur in different geographic areas.

Second, is there an inverse relationship between the similarity of call characteristics and ecological overlap within a given geographic area (that is, do species with similar calls exhibit minimal overlap in

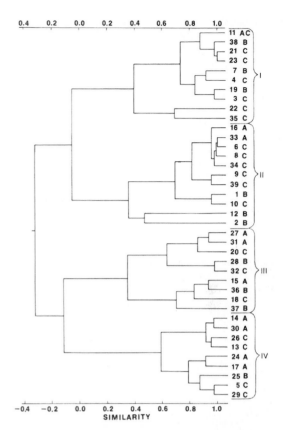

Figure 14–16 **Cluster diagram showing similarity of 39 species of hylid frogs based on seven parameters of their mating calls.** Each Arabic number represents a different species, and letters represent geographic localities: A, Belém, Brazil; B, Puerto Viejo, Costa Rica; C, Santa Cecilia, Ecuador. Roman numerals identify the four major clusters. See text for discussion. *(Source: Duellman and Pyles 1983.)*

use of habitat)? The answer to this question seems to be yes. A good example is a group of small tree frogs in the *Hyla leucophyllata* group, all located in cluster I. *Hyla ebraccata* (7) is found in Costa Rica, and the other four occur at Santa Cecilia, Ecuador (Figure 14–17). *Hyla leucophyllata* (11) and *H. triangulum* (23) have the most similar calls, but they do not overlap in habitat. *Hyla leucophyllata* calls from forest pools, whereas *H. triangulum* calls from ponds and puddles in open, disturbed areas. *Hyla bifurca* (3) calls from the same sites as *H. triangulum*, and their calls sort out very differently in the cluster analysis. *Hyla saryacuensis* (22) also breeds in forest pools, but notice how different its call characteristics are from those of the other three species at Santa Cecilia.

Figure 14–17 **Four species of the *Hyla leucophyllata* group, all found at Santa Cecilia, Ecuador.** (a) *Hyla triangulum*, (b) *Hyla bifurca*, (c) *Hyla saryacuensis*, (d) *Hyla leucophyllata*. *(Photographs by Martha Crump.)*

Third, do different assemblages of hylid frogs from similar habitats have similar acoustic patterns? The answer to this question is also yes. In fact, the means for six of the seven call parameters analyzed did not differ significantly among assemblages. The only parameter that differed was fundamental frequency, which was significantly higher at Santa Cecilia than at Belém and Puerto Viejo. Thus, even though the species composition differed among assemblages, the groups of frogs from similar but disjunct habitats had similar acoustic patterns overall. The similarity perhaps is explained because the physical factors that affect the acoustic environment (e.g., dense vegetation that attenuates sound) are the same in similar habitats.

Summary

Species evolve against an ever-shifting background of geographic, physical, and biological environments, but we observe them at a particular time. We can't assume that the moment captured in that one frame represents the evolutionary history of any one species or, even less, of all the species making up an assemblage. An assemblage is composed of species, some of which may have existed and even coexisted there for some time, some of which may be recent arrivals, some of which are long-term members, and some that are just spilling over from places nearby. There probably is no such thing as a static assemblage of amphibians or reptiles. Instead, the composition of assemblages and the interactions of the constituent species change through time. With this potential complexity, how can we interpret assemblage composition and apparent structure?

We need to consider how historical factors, habitat complexity, chance events, and physiological tolerances may have influenced the patterns we see now. Processes and mechanisms such as competition and predation may be going on all the time, and they can have subtle influences on pattern even though they may not by themselves be responsible for the overall pattern.

The cryptic and little-studied effects of parasites and pathogens could profoundly influence interspecific interactions and assemblage composition. The relative importance of competition and predation (and interactions between these factors) depends on the kinds of organisms that make up the community and on the nature of the environment itself. Competition is a dominant interaction among lizards on small tropical islands, whereas competition and predation are both dominant interac-

tions among aquatic amphibian larvae. The outcome of interactions in any assemblage can be changed or modified by such unpredictable events as extreme weather conditions.

We have focused on interactions among species in assemblages of amphibians and among species in assemblages of reptiles because these groupings are readily defined. Interactions between amphibians and reptiles, however, may also be relevant. Reptiles and amphibians may compete for the same space and food resources and may prey on each other. Other groups of animals may also strongly influence patterns of assemblage organization. Because interactions among phylogenetically distant groups are difficult to study, their consequences are less well known than, say, interactions among larval amphibians. Nonetheless, such interactions can be important determinants of assemblage structure.

What Are Their Prospects for Survival?

15

Conservation and the Future of Amphibians and Reptiles

One of the greatest all-time champions of amphibians and reptiles was Archie Carr (1909–1987)—professor of zoology, naturalist, conservationist, writer of popular books and articles, and a world authority on sea turtles. Over four decades ago, in his delightful book *The Windward Road* (1955), Carr warned of the impending decline of one of his favorite animals, the green sea turtle (*Chelonia mydas*):

Where twenty years ago most Caribbean shore was wilderness or lonesome cocal, aluminum roofing now shines in new clearings in the seaside scrub. The people are breeding too fast for the turtles. The drain on nesting grounds is increasing by jumps. It is this drain that is hard to control, and it is this that will finish Chelonia.

Extinction is a natural phenomenon. At present, however, most extinctions result from human activities. Estimates up to thousands of extinctions per year do not seem unrealistic when one considers all of the plants, fungi, vertebrates, and invertebrates that are affected adversely by humans. How have amphibians and reptiles fared? While many are exploited or are experiencing severe loss of habitat, amphibians seem to be especially affected.

Declining Amphibians: A Model Issue

In the past two decades, many herpetologists worldwide have reported that populations of frogs, toads, and salamanders are declining (Phillips 1994). The extent of decline varies from region to region and varies within and among species. In North America, declines have been reported most frequently from the West (Corn 1994). Population declines have been reported from Canada to South America and from northern Europe to Australia. They have been documented from montane areas as well as from lowland sites. Some of the declines, especially those at low elevations, are associated with human disturbance and habitat destruction. Other declines, including many of those in montane regions, are not obviously associated with habitat destruction. Some of the reported declines have involved single species, but others have involved declines at the community level. For example, Drost and Fellers (1996) reported that at least five of the seven species of frogs and toads from a transect of the Sierra Nevada mountains in the Yosemite area of California have suffered serious declines. Likewise, multiple declines have been reported from California's Great Central Valley (Fisher and Shaffer 1996). At least 14 species of endemic stream-dwelling frogs (including the gastric-brooding frog, *Rheobatrachus silus*) have disappeared or declined sharply from montane rain forests of eastern Australia (Laurance et al. 1996).

What's happening? Are these declines a new and ominous phenomenon, perhaps the result of global climate change or widespread pollution? Or have they always happened at the same rate, and biologists are simply becoming more attuned to population fluctuations? Long-term field studies are critically needed in order to understand and evaluate the significance of fluctuating populations.

Natural population fluctuations probably explain some of the declines, especially in marginal habitat, but they are unlikely to explain synchronous worldwide declines of amphibians. At present there is no evidence for a single causal factor, and most scientists suggest that local effects and global factors probably interact to affect population densities. Some of the proposed local effects include introduction of predators and competitors, pesticides and other forms of pollution, disease, habitat destruction (especially destruction of breeding sites), and overexploitation by humans. Global factors include higher temperatures due to global warming, acidified rain, and changes in regional precipitation patterns (especially drought) as a consequence of global warming and large-scale deforestation. Exposure to increases in ultraviolet radiation may have detrimental effects for some species (Blaustein, Hoffman, et al. 1994). Several factors may act together, such that although one in itself would not be lethal, together they overwhelm the immune system and the animal eventually dies. For example, stressed amphibians are more likely than nonstressed animals to die from red-leg disease, which is caused by a bacterium ubiquitous in freshwater. This combination of stress and bacterial infection has been suggested as a hypothesis to explain local population extinctions of the toad *Bufo boreas* in Colorado (Carey 1993). Another lethal combination could be drought and airborne pesticides or other contaminants (Pounds and Crump 1994). Toxic organic residues can concentrate in the soil, and an amphibian that is rehydrating from that soil will absorb the toxins. Increased exposure to excessive levels of ultraviolet radiation may increase susceptibility of amphibian eggs to *Saprolegnia* fungal infections, leading to death (Blaustein, Hokit, et al. 1994; Blaustein and Wake 1995).

Amphibians and their declining populations have received a lot of attention in recent years, but populations of reptiles are declining as well; both are part of the worldwide decline in biodiversity. Although humans are willing to spend millions of dollars each year to protect giant pandas, whales, elephants, gorillas, or chimpanzees, we seem unwilling to make a comparable commitment to the more than 10,000 species of amphibians and reptiles. These two groups represent about 25 percent of all known living species of vertebrates, but they are overlooked in terms of conservation efforts. Most conservation organizations allocate less than 3 percent of their annual budgets to the study and protection of amphibians and reptiles; most of the rest goes to birds and mammals. Conservation efforts are unbalanced, but fortunately, conservation of birds and large mammals requires large reserves that simultaneously protect other organisms. As a result, amphibians and reptiles benefit indirectly from these conservation efforts.

Major Themes

We will emphasize three major themes in this chapter. First, if a conservation program is to be effective, it must involve the local people. As an initial step, the cultural perceptions, values, and biases of local people must be considered. How do they view the animals in question? If a threatened species is feared or hated, effort must be made to change that perception, or at least to understand and overcome (or work around) that bias. Furthermore, local people must be given an economic incentive to preserve a habitat and its flora and fauna. Long-term conservation will not be accomplished simply because it is judged to be morally right by conservationists; it must be perceived as being beneficial to the local people.

Second, the single most important negative effect that humans have on amphibians and reptiles is destruction of their habitat. More than 80 percent of all species of amphibians and reptiles occur in tropical areas, many of which are rapidly being destroyed by humans. Significant habitat destruction occurs in our own backyards as well, as bulldozers destroy natural areas to prepare for construction of parking lots and shopping malls, or wetlands are drained and filled to be converted to residential areas. The ultimate problem is uncontrolled human population growth and the demands that more and more people put on the environment. Therefore, the most important action we can take to protect amphibians and reptiles (outside of human population control) is protection of their habitat. A focus on habitat preservation, rather than on individual charismatic animals, offers the most protection for communities of animals.

Finally, more research is essential for successful conservation. We simply don't know enough about most species to be certain that we are protecting the right habitats, the right resources, or the right life history stages. Without knowledge of habitat requirements, reproductive biology, life history characteristics, dietary needs, and movement patterns, conservation efforts may be ineffective.

Human Perceptions of Amphibians and Reptiles

Human behavior toward animals is influenced by cultural perceptions—animals that are held in awe are protected, animals associated with evil are often killed. For example, on the island of New Caledonia, children are warned not to kill lizards because they might be the child's own ancestors. In contrast, lizards are often persecuted in Iran because they are believed to carry the devil's soul. Turtles seem well designed to carry burdens on their backs, and perhaps it is for this reason that they feature prominently in myths of creation and are revered. In various cultures in India, China, Japan, and North and South America huge turtles are believed to support mountains and even entire continents. In Hindu cosmology a turtle supports the entire universe. In contrast, turtles are despised and thus persecuted in some parts of the Amazon Basin because they are believed to be associated with human sin. Crocodilians are worshipped in parts of Madagascar, where they are considered to be supernatural beasts; the spirits of chiefs are believed to pass into crocodiles after death. The ancient Egyptians believed crocodiles to be divine and in fact built the holy city of Crocodilopolis in their honor. On the other hand, in many cultures of both the Old and the New World people fear crocodilians because of their aggressive nature and their attacks on livestock, pets, and people; they believe the only good crocodilian is a dead one.

In some parts of the world frogs are revered because they are thought to possess supernatural powers. According to legends in both the Old and New World, lunar eclipses occur when a great frog swallows the moon. In India, China, and Siberia, myths hold that the world rests on the back of a frog and that earthquakes shake the world whenever the frog moves. The alternating appearance and disappearance of populations of frogs and their seemingly magical metamorphosis have led to the worship of frogs as symbols of fertility, resurrection, and creation. Amulets (charms or fetishes) of frogs are possessed by people worldwide to bring luck, ward off the evil eye, or bring rain (Figure 15–1). In contrast, some folklore suggests that toads are evil and should be avoided—

Figure 15–1 **Native American frog fetish.** This frog fetish was carved by a member of the Zuni tribe of New Mexico. The frog represents a common bond among all life forms because of its associations with water. The frog is also associated with the coming of rains that promote growth of crops. *(Photograph by Martha Crump.)*

for example "a toad's breath will cause convulsions in children." Toads frequently symbolize ugliness; in Shakespeare's play, Richard III, the king is called "a poisonous hunch-back'd toad."

This dichotomy of perception is especially strong regarding snakes. All over the world and throughout recorded history, snakes have been a source of fascination and fear for humans; snakes are both worshipped and despised. Snakes symbolize love or hate, procreation or death, health or disease. Snakes hold a focal position in mythology, and in many cultures they are the most honored of all mythical supernatural beings. Snakes are frequently associated with rejuvenation and immortality because of their ability to throw off their old skin and acquire a fresh one. This ability, no doubt coupled with the power of venom that some snakes possess, has led to the prominence of snake cults and ophiolatry (snake worship) throughout the world.

Snakes served as important symbols for early cultures throughout the Americas. The Mayas, Aztecs, and Incas all had abundant images of snakes and mythologies about them. Rattlesnakes were respected, honored, and protected by most Native American tribes in North America; some groups did not kill rattlesnakes for fear the snake's relatives would seek vengeance. Snakes still play a prominent role for the Hopis of northern Arizona, who look upon snakes as messengers to their gods of rain. Each August they gather up bull snakes (*Pituophis*), desert striped racers (*Masticophis*), and rattlesnakes (*Crotalus*) from their fields for an elaborate 9-day ceremony. The ceremony is climaxed by a ritual Snake Dance, during which the Snake priests dance while holding live snakes in their mouths. The snakes are entrusted with the prayers of the Hopi; according to legend, the snakes relay the prayers for rain and adequate crops to the gods.

Rattlesnakes are not universally respected in North America, however. They are persecuted through roundups, events during which rattlesnakes are collected from their natural habitats and killed (Weir 1992). Dozens of rattlesnake roundups, superficially legitimized by civic or charitable organizations as fund-raisers, are held in the United States every year. Ostensibly the excuse is to rid the vicinity of dangerous snakes, but some of these events have turned into community extravaganzas that feature sensationalism, capitalizing on the public's fear of rattlesnakes. The animals are often severely mistreated. Funds are raised through sale of spectator tickets and sale of the snakes for their hides, rattles, and meat (Figure 15–2). The impact of intense collecting efforts on local populations is largely unknown, but the numbers of snakes killed each year are staggering. For example, analysis of data from 28 years (1959–1986) of the Sweetwater, Texas, roundup reveals a range of 1,900 to 15,680 rattlesnakes killed annually, with a mean of 5,563 rattlesnakes killed per year (Campbell et al. 1989).

Thus, depending on the cultural bias to which a person is exposed, he or she could have a positive or negative attitude about amphibians and reptiles that would influence his or her likelihood of believing that these animals are worth conserving. Conservation efforts must work within the culture of the region if they are to be successful.

Impact of Humans on Amphibians and Reptiles

The following sections summarize some of the major ways that humans negatively affect amphibians and reptiles, and thus represent areas where conservation problems exist. A major factor responsible for population declines of amphibians and reptiles is habitat modification and destruction. Some populations are on the verge of extinction because of predators and

competitors that humans have introduced onto islands. Pollution seems to be responsible for some declines. The underlying problem with much of the direct exploitation of amphibians and reptiles is that too many animals are removed from a given population, resulting in declines or even extirpations (local population extinctions); if fewer animals were removed per population, populations could recover more easily.

Habitat modification and destruction

The single most important negative impact of humans on amphibians and reptiles is habitat modification and destruction. Many habitats are shrinking or disappearing at an accelerating pace due to pressures of human population growth and economic development. Increasing numbers of people require more land and increase the global demand for natural products. Unfortunately, in many places in the world habitat destruction goes hand-in-hand with the social problems of poverty, lack of education, and economic disruption. Tropical forests are some of the most species-rich habitats in the world and are particularly vulnerable to human destruction. In 1991 the U.N. Food and Agriculture Organization reported that the world's tropical forests were being destroyed 50 percent faster than they were a decade ago. Tropical forests contain a high diversity and abundance of amphibians and reptiles. At the current rate of deforestation, within 30 years there will remain neither extensive tropical forests nor their endemic amphibian and reptile fauna. As indicated in Table 15–1, extensive nontropical forest habitat loss is occurring worldwide as well.

Humans convert and develop the land in different ways, depending on the needs of the community, accessibility of areas, and the potential productivity of different sites. For example, a large tract of forest may be separated into a patchwork of forest remnants within a matrix of human-modified landscape consisting of towns, roads, cornfields, and golf courses. Some human-modified habitats may be acceptable for wildlife, others are not. Remnants of the original forest may be left in various sizes, shapes, and distances from other forest areas. Wildlife may or may not be able to disperse among isolated patches of habitat. Populations restricted to isolated habitat fragments may be vulnerable to local extinction through chance

Figure 15–2 **Two negative aspects of rattlesnake roundups.** (a) mistreatment—a pit full of rattlesnakes; (b) rattles made into keychains for sale. *(Photographs by Lee Fitzgerald.)*

(a)

(b)

environmental and demographic catastrophes and loss of genetic heterozygosity.

Within the United States, southern Florida has been a focus of rapid human population growth for the past three decades. The major impacts of humans on the amphibians and reptiles in south Florida are destruction of natural cover and manipulation of the hydrologic cycle (Wilson and Porras 1983). For example, currently the biggest threat to alligators in Florida is not uncontrolled or illegal harvest (although it was at one time), but rather destruction of

wetland habitat by humans. Nearly 1,000 people move to Florida every day, creating a serious imbalance in the state's environment. Habitats are severely modified or destroyed to make room for condominiums and retirement villages, bringing alligators and humans into closer contact. During wet years, flooding opens up new land for the alligators; often the ideal habitat happens to be in neighborhood ponds and backyards. The Florida Game and Fresh Water Fish Commission receives over 15,000 requests each year to remove nuisance alligators, resulting in the

Table 15–1 **Examples of worldwide loss of habitat. Percentages represent the amount that had already been lost in each country, as of 1991.**

Forest		Wetlands/Marshes	
Argentina	50%	United States	54%
Guatemala	60%	Cameroon	80%
Mexico	66%	New Zealand	90%
Greece	70%	Italy	94%
India	78%	Australia	95%
Philippines	79%	Thailand	96%
Ethiopia	86%	Vietnam	Almost 100%
Bangladesh	96%		
Savannah/Grassland		Desert/Scrub	
Madagascar	78%	Pakistan	69%
Nigeria	80%	Sri Lanka	75%
New Zealand	90%	India	88%
United States	Almost 100%		

Data from World Resources Institute 1990.

deaths of many thousands of animals. People encourage alligators to hang around their boat docks by feeding marshmallows to the alligators; then, when their pet toy poodle is eaten, they demand to have the so-called nuisance alligator killed.

Draining and filling of wetlands to convert natural ecosystems into human habitation also destroys breeding habitat for numerous species of amphibians that depend on ephemeral bodies of water for reproduction. A population of salamanders or frogs may be totally dependent on a single quarter-acre pond for its existence. Small isolated wetlands are particularly vulnerable to human modification because they are viewed as less valuable than larger habitats (Moler and Franz 1987). Rather than being simply dwarf versions of large wetlands, small wetlands (especially those free of fishes) often provide unique habitat for wildlife that prefer smaller bodies of water. Protection of small wetland sites is a constant battle as developers fight for the right to dredge and fill habitat for building opportunities.

Habitat for amphibians and reptiles is destroyed through various methods of harvesting timber. For example, clear-cutting has reduced or eliminated populations of at least seven species of salamanders in forests in the Pacific Northwest of the United States and at least 16 species of salamanders from mesic hardwood forests of the southern Appalachian Mountains in the southeastern United States, and it has reduced populations of terrestrial amphibians by up to 70 percent in coastal forests in Canada. Bruce Means and colleagues monitored the largest known breeding migration of the flatwoods salamander, *Ambystoma cingulatum*, for 22 years in Florida. They watched the population dwindle over the years, from 200 to 300 adult salamanders per night crossing the highway in 1970–1972 to less than one per night on average in 1990–1992 (Means et al. 1996). The practice of converting native longleaf pine savanna to bedded slash pine plantation may have been responsible for the species' decline. Bedding is a silvicultural practice whereby the topsoil is plowed into long parallel ridges to elevate the newly planted trees so that their roots are raised above the water level. This conversion is thought to have interfered with many aspects of the salamanders' biology, including migration to breeding sites, successful hatching, feeding, and finding suitable retreat sites after metamorphosis.

Modification of habitat by humans doesn't always affect native species adversely. For some species, certain types of habitat modification actually improve conditions for their existence. Population sizes of the Florida king snake (*Lampropeltis getula floridana*) have increased in some areas where native habitat has been converted to sugarcane fields. The high density of rodents associated with the cane fields provides additional food, and the banks of limestone dredge material along the irrigation canals provide increased shelter for the snakes. Particularly in arid regions, agricultural practices that make more standing water available (e.g., irrigation ditches, stock ponds, and flooding of fields) have benefited resident amphibians and allowed other species to expand their ranges. Some lizards (especially geckos) and snakes are more commonly found around human dwellings than in more natural habitat because of the abundance of insect and rodent prey.

Introduction of Exotic Species

Human-induced introductions of exotic animals to islands provide revealing insights into the impact of predation on species that have not evolved with these predators. For example, introduced domestic dogs and cats have had devastating impacts on populations of rock iguanas (*Cyclura carinata*) and on smaller lizards on Pine Cay in the Caicos Islands (Iverson 1978). Mongooses were introduced to Jamaica from India in 1872 to kill rats in the sugarcane fields. Diurnal mongooses are not specialists on nocturnal rats, however, and they do not remain in cane fields; they prey heavily on birds and reptiles. The introduced mongooses are thought to be responsible for the elimination or drastic reduction of several lizard species, including the iguanid *Cyclura collei* and the colubrid snake *Alsophis ater*.

South Pacific iguanas of the genus *Brachylophus* have likewise been particularly affected by human introduction of domestic mammals. These lizards evolved in an environment free of ground-dwelling predators. Their clutch sizes are small (three to six eggs per year), and egg incubation time is long (18 to 35 weeks). Drastic declines of banded iguanas (*Brachylophus fasciatus*) and crested iguanas (*Brachylophus vitiensis*) in the South Pacific during the twentieth century are linked to the introduction of predators, particularly cats (Gibbons and Watkins 1982). Iguanas are now scarce or absent on islands where feral cats are abundant. Furthermore, introduced goats and pigs

have destroyed the understory vegetation, and with this loss of cover the lizards are more vulnerable both to their natural predators and to cats.

Likewise, mammals introduced onto New Zealand have caused declines of amphibians and reptiles. Tuatara (*Sphenodon*) have become extinct on North Island and South Island and on some of the smaller islands of New Zealand, in some cases due to competition and predation associated with sheep, goats, and rats introduced by the early settlers. Alison Cree and colleagues studied the reproduction of tuatara on rat-free and rat-inhabited islands of New Zealand and found that the introduced Pacific rats inhibited recruitment of young tuatara into the population (Figure 15–3). Rats most likely exert negative effects by direct predation on eggs and juveniles and by indirect competition for food. Introduced rats have apparently also caused the extinction of several species of *Leiopelma* frogs in New Zealand. Only three species of the genus exist today; the largest of these, *Leiopelma hamiltoni*, is found only on two rat-free islands.

Amphibians and reptiles themselves have been implicated in the declines of other species. For example, the brown tree snake (*Boiga irregularis*), which was unintentionally introduced onto the island of Guam, has drastically reduced or extirpated not only populations of endemic birds but also several species of lizards, especially the native geckos (Rodda and

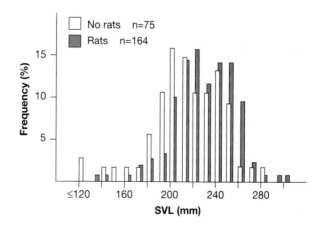

Figure 15–3 **Tuatara and rats.** Comparison of size class distribution (SVL, snout–vent length) of tuatara (*Sphenodon punctatus*) on rat-free islands and on rat-inhabited islands. *(Source: Cree et al. 1995.)*

Fritts 1992). The African clawed frog, *Xenopus laevis*, is now expanding its range and density in southern Africa because it breeds successfully in farm ponds and irrigation systems. The problem is that *X. laevis* is competitively superior to a less common species, *Xenopus gilli*. In some areas where only *X. gilli* occurs naturally but where *X. laevis* has entered as a result of human modification of the environment, *X. gilli* has experienced severe population declines or even local extinctions (Picker 1985). Furthermore, the genetic identity of *X. gilli* is threatened because the two species may hybridize in areas where they overlap. Iberian waterfrogs (*Rana perezi*) are in danger of decline owing to introduced species of *Rana*. Again, the fear is that these introduced *Rana* may be competitively superior or may hybridize with *R. perezi*, causing changes that would modify the genetic structure of the populations on the Iberian peninsula (Arano et al. 1995).

Introduced fishes have caused the local extinction of various populations of amphibians, presumably by eating the tadpoles. For example, introduced trout are thought to be responsible for extirpations of various populations of harlequin frogs (*Atelopus*) in Costa Rica. In the Sierra Nevada mountains of California, several species of introduced salmonid fishes caused the extinction of populations of mountain yellow-legged frogs, *Rana muscosa*, from many lakes and streams earlier in this century. Bradford et al. (1993) suggested that the introduced fishes have also indirectly caused population declines and disappearances of *R. muscosa* in more recent years by isolating the remaining populations. Because the frogs can survive only in waters lacking fish, their habitat has become fragmented, making reestablishment of populations more difficult. Introduced mosquitofish and crayfish are the predators most likely to be responsible for localized declines of California newts (*Taricha torosa*) from streams in the Santa Monica Mountains of southern California (Gamradt and Kats 1996).

Sometimes introduced species are assumed to have negative impacts on endemic fauna, but once the evidence is examined the species turns out to be benign. Introduced populations of *Xenopus laevis* appear to be such a case in southern California. Anecdotal descriptions portrayed the species as a voracious predator, but these reports were based on small sample sizes from the wild and on observations in captivity. McCoid and Fritts (1980) found that the frogs in their study sites in southern California ate almost ex-

clusively slow-moving invertebrates. The most common vertebrate prey found in the stomachs was conspecific eggs and tadpoles, but these represented only 1.5 percent of the food items by number. They concluded that *X. laevis* most likely is not a major threat to native vertebrate fauna in southern California.

Pollution of the Environment

Environmental pollution is a possible cause for the decline of some populations of amphibians. Amphibians may serve as sensitive biological indicators of environmental deterioration because their highly permeable skin rapidly absorbs toxic substances. Examples of pollutants include pesticides, high concentrations of heavy metals washed into aquatic breeding sites, and poisoning resulting from mining and logging operations.

A major form of pollution in some areas is atmospheric acid deposition (acid rain). Pure, unpolluted rain is slightly acidic, with a pH of about 5.6. Much of the rain that currently falls in the eastern United States and in other parts of the world is much more acidic, however, with a pH below 4.5. Acid rain lowers the pH of water below the tolerance level of eggs and larvae of many species of amphibians. Amphibians are vulnerable to acid deposition for several reasons (Wyman 1990). First, many species breed in small, ephemeral ponds that have little buffering capacity. These sites are often filled with water directly from rains, before the water has any contact with soil that can buffer the low pH. Second, in temperate areas some amphibians breed in the early spring, following snow melt. In areas that receive strongly acidic precipitation, snow melt forms acidic ponds. Third, high-elevation forest soils in areas of acid precipitation become acidic because the underlying bedrock has low acid-neutralizing capacity. Forest soils become increasingly acidic when they dry out, causing pH problems for amphibians in contact with the soil.

The toxic effects of low pH on amphibian development have been well documented (Pierce 1985). Acidic conditions reduce sperm motility and may cause the sperm to disintegrate, reducing fertilization success. For those eggs that are fertilized, acidity may cause developmental abnormalities. Some species are amazingly tolerant of low pH conditions (3.7 to 4.0), whereas other species are extremely sensitive to a pH below 5.0. Acid precipitation has been implicated in the declines of some populations of salamanders (e.g., *Ambystoma tigrinum* in the Rocky Mountains of Colorado, Harte and Hoffman 1989) and anurans (e.g., *Bufo calamita* from the lowland heath areas in Britain, Beebee et al. 1990).

Another type of pollution is solid waste. An estimated 24,000 metric tons of plastic packaging material is dumped into the world's oceans each year. The material may be out of sight for us, but some sea turtles encounter and inadvertently eat it (National Research Council 1990). Green turtles (*Chelonia mydas*) ingest plastic bags as they feed on plant material from the substrate, and leatherbacks (*Dermochelys coriacea*), whose primary diet is jellyfish, may mistake plastic bags for prey. Half of the sea turtles examined in some areas have plastic debris in their intestines. Plastic debris may interfere with the turtles' digestive processes, respiration, and buoyancy, and some plastics are toxic.

Some forms of chemical contamination interfere with the endocrine system of animals. For example, in humans certain contaminants may cause an increased incidence of breast cancer and endometriosis in females and testicular cancer and lowered sperm count in males. Reptiles are affected also, and perhaps are more susceptible to the effects of such contaminants because of their lability in sex determination (Guillette and Crain 1996). Polychlorinated biphenyls (PCBs), industrial chemicals such as those used in fire retardants and adhesives, persist and bioaccumulate (build up through the food chain) in the environment. They readily vaporize, and thus are transported long distances through the atmosphere. Some PCB compounds have a molecular structure so similar to that of estrogen that they act as estrogen when they enter an animal's body (they are called EDCs—endocrine-disrupting contaminants) and thus can alter sexual differentiation. For example, PCBs can reverse gonadal sex in the red-eared slider (*Trachemys scripta*), a species that exhibits temperature-dependent sex determination (Bergeron et al. 1994). PCBs counteract the effects of cool temperatures that produce males and instead induce ovarian development, creating females. Obviously, such unnatural altering of sex determination can have disastrous effects on populations.

Louis Guillette and his colleagues have studied possible causes of reproductive failure in alligators from a contaminated lake, Lake Apopka, in central Florida (Guillette et al. 1994, Figure 15–4). The alligators from this lake are exposed to the pesticide dicofol and to DDT and its metabolites that originated from a major chemical spill at a nearby pesticide plant. Results indicated that clutch viability (percentage of eggs in a clutch that produce viable hatchlings) was significantly lower in animals from Apopka compared to the control site. Of the eggs that did hatch at Apopka, 41 percent died within 10 days, compared to less than 1 percent at the control site. At Apopka, 6-month-old females had significantly higher plasma estradiol-17β concentrations than females at the control site and abnormal ovarian morphology; juvenile males had only about one-fourth the concentration of plasma testosterone as animals from the control site, and their penises were abnormally small. These data support the hypothesis that environmental contamination has detrimental effects on endocrine and reproductive functions and thus depresses reproductive success in alligators. The implications of the widespread effects of environmental contamination for all organisms, including humans, are frightening.

Commercial Exploitation for Food

Many people eat amphibians and reptiles because they are readily available and provide a good source of protein. Unfortunately, modern commercialization of amphibians and reptiles for the world's luxury food market is generally done without regard to population dynamics and has often led to depletion of wild populations. Most of the frogs that are killed for human consumption end up not as a critical component of local peoples' diet, but in distant lands as gourmet dishes such as stir-fried frog legs smothered with oyster sauce, frog legs au gratin, frog legs teriyaki, and giant bullfrog chop suey. The same is true for reptiles, whose meat is served either as an oddity or as a delicacy in such forms as steaks, soups, stews, pies, creoles, burgers, and even spaghetti.

Commerce in frog legs is substantial. The most commonly eaten frogs are *Rana catesbeiana* (North America), *Rana esculenta* (Europe), *Rana tigrina* (southern Asia), and *Pyxicephalus adspersus* (Africa). During the first half of the twentieth century the demand for frog legs became so great in the United States and the prices were so high that hunters earned up to $500 a day hunting frogs in Florida. In 1976 alone the United States imported over 2.5 million kilograms of frog legs, mostly from Japan and India. Annual consumption of frog legs in France is estimated at 3,000 to 4,000 tons, mostly imported from

Figure 15–4 **Research with alligators at polluted Lake Apopka, Florida.** (a) sampling blood from female alligator; (b) measuring phallus in male alligator. *(Photographs by Howard Suzuki.)*

(a)

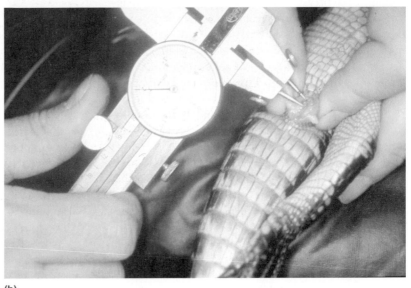

(b)

Bangladesh and Indonesia. In recent years, an estimated 200 million pairs of frog legs have been exported annually from Asia, destined for the United States, Europe, and Australia. Since 1987, however, India has banned exportation of frog legs because densities of insect pests increased dramatically in agricultural areas where frog densities had declined.

Commerce in reptile meat is also substantial. Each year between 1979 and 1987, approximately 45,000 kilograms of alligator meat was sold from the regu-

lated harvests in Louisiana. Currently, several species of turtles are also heavily exploited as food, particularly snapping turtles, red-eared sliders, sea turtles, and softshell turtles (Figure 15–5). Green iguanas (*Iguana iguana*) and spiny-tailed iguanas (*Ctenosaura similis*) have been eaten by humans in Central America for centuries. In the past 30 years, the combination of extensive habitat destruction and overhunting associated with increased human population growth has caused drastic declines of both species (Fitch et al.

Figure 15–5 **Green sea turtle being butchered on Masirah Island, Oman.** Sea turtles provide a valuable source of protein to people on the southern coast of Oman during the southwestern monsoon season when fishing is difficult. Harvest is carefully regulated and is maintained at around 500 turtles per year. Nesting turtles are protected. *(Photograph by Perran Ross.)*

1982). Central America has a highly organized industry of professional iguana hunters who harvest enormous numbers of lizards and ship them to markets in major cities (Figure 15–6). During the late 1960s an estimated 150,000 iguanas were eaten annually in Nicaragua alone. Flesh from adult lizards is in great demand not only because it is delicious, but because of its supposed medicinal properties (including its use as a cure for impotence). Fat from iguanas is valued as a salve for burns, cuts, and sore throats. Females are exposed to especially heavy hunting pressures because their unlaid eggs are considered a delicacy, and because they supposedly increase sexual potency.

Commercial Exploitation for Skins, Art, and Souvenirs

Reptile skins have long been used for making shoes, boots, purses, belts, buttons, wallets, lamp shades, and other accessories (Figure 15–7). During the American Civil War (1860 to 1865) many thousands of alligators were killed and their skins used for shoe leather at a time when there was no free commerce in cow leather. Populations of many species of boas, pythons, crocodilians, and varanid lizards are now declining because of heavy hunting pressure for their skins. During 1981 alone, 304,189 pairs of shoes made from *Boa constrictor* and 176,204 pairs of shoes made from *Python reticulatus* were imported legally into the United States. Because commercial breeding programs for these species do not exist, the skins must have come from wild populations. Between 1980 and 1985 more than 1 million *Caiman crocodilus* skins were traded per year. Over 12 million tegu (*Tupinambis*) skins were imported to consumer nations during this same time period. Some 1 to 2 million monitor lizard skins are exported annually from Africa and Asia. According to the World Resources Institute (1990), during 1986 alone nearly 10.5 million reptile skins (snakes, lizards, and crocodilians combined) were traded legally; if illegal and domestic commerce are added in, this total figure would be much greater. About 56 percent of these skins were lizards, 26 percent were snakes, and 18 percent were crocodilians. Major exporting countries of these 10.5 million reptile skins included Indonesia, Singapore, Thailand, and Argentina; major importers included Singapore, the United States, Italy, and Spain.

Frog skins are used in the manufacture of shoes, purses, belts, key cases, and other novelties; frog leather is used for binding small books. The skins of frogs are also used in making glue and for coverings of artificial fishing lures. Toad skin is used to make change purses, slippers, and shoes. Many amphibians and reptiles are killed each year and made into cheap souvenirs. Teeth and claws from crocodilians are sold as curios, and their feet are made into key rings. Rattlesnake rattles, fangs, and freeze-dried heads are popular souvenirs. Toads (particularly *Bufo marinus*), iguanas, and turtles are stuffed, fitted with glass eyes, and varnished (Figure 15–8). Exploitation of horned lizards from southern California for the curio trade used to be popular. At least 115,000 horned lizards are thought to have been harvested and stuffed over a 45-year period, especially between 1890 and 1910.

Traditional and Modern Medicine

Products from amphibians and reptiles are common constituents of traditional medicines. Iranians use broths made from snakes (particularly vipers) and tor-

Figure 15–6   **Exploitation of iguanas for food in Central America.** Live green iguanas, *Iguana iguana*, in market at Masaya, Nicaragua. Each lizard has its limbs tied together over the back for immobilization. *(Photograph by Robert W. Henderson.)*

toises in the belief that these combat a variety of diseases. In Asia, the abdominal fat of monitor lizards (*Varanus*) is used as a salve for bacterial infections of the skin, and oil extracted from monitor fat is sold on street corners in Pakistan as an aphrodisiac. Sea turtle eggs are used as aphrodisiacs, and turtle oil is used for lubricants and in making cosmetics. Gonads, musk, and urine from crocodilians are used to make perfume. Fat, and oil derived from the fat, from crocodilians are used in Madagascar to treat burns, skin ulcers, and cancer. In the Dominican Republic and Haiti, fat from crocodilians is used as a remedy for asthma. Pliny, in the first century A.D., wrote in his book *Natural History* the following concerning early Roman use of crocodilians: "Romans carried stones from a crocodile's belly as charms against aching joints, bound crocodile teeth to their arms as aphrodisiacs, treated whooping cough in their children with

doses of crocodile meat, and trustfully administered burned crocodile skin mixed with vinegar as an anesthetic to patients about to undergo surgery. Women used crocodile dung in a lint tampon as a contraceptive" (Minton and Minton 1973).

Anthropologists have speculated that ancient cultures of Mesoamerica may have used toad secretions as hallucinogens during religious ceremonies. Numerous small, toad-shaped bowls have been found in archaeological sites in Veracruz and adjacent areas of southeastern Mexico; a prominent feature of toad images on the bowls is the parotoid glands. Davis and Weil (1992) speculated that the toad used by pre-Columbian people was *Bufo alvarius*. This species is unique within the genus (and within the animal kingdom, so far as is known) in possessing a specific enzyme that converts the alkaloid bufotenine to one of the most powerful hallucinogens known in nature, 5-methoxy-*N,N*-

Figure 15–7 **Commercial exploitation of reptile skins.** (a) Snakeskin-tanning shop in Madras, India; (b) rattlesnake boots. *((a) photograph by Romulus Whitaker; (b) photograph by Lee Fitzgerald.)*

(a)

(b)

dimethyltryptamine. Huge amounts of this hallucinogen (up to 15 percent of the dry weight of the gland) accumulate in the parotoid glands. The authors corroborated informants' reports (through personal experience) that smoking the dried parotoid secretion of *B. alvarius* results in hallucinations.

The Mayoruna men of Brazil use skin secretions from the tree frog, *Phyllomedusa bicolor*, as a drug for hunting magic (Daly et al. 1992). Frogs are harassed until they release defensive secretions. The secretion is dried and then applied to the hunter's arm or chest area, on fresh burns to the skin. The secretion enters

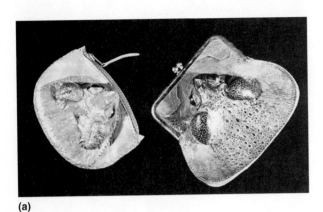

(a)

(b)

Figure 15–8   **Commercial exploitation of marine toads, *Bufo marinus*.** (a) change purses; (b) stuffed toads playing pool. *(Photographs by (a) Martha Crump; (b) Arthur Echternacht.)*

the bloodstream rapidly through the open burn wounds, causing repeated vomiting. The person eventually falls into a condition described as a feeling of being very drunk. The Mayoruna believe that this secretion improves the hunter's aim, makes him more powerful, and sharpens his senses. Occasionally women take the drug in the belief that it allows them to work harder.

In modern medicine, antivenin serum made from snake venom neutalizes the effect of snakebite. Venom is extracted by procedures that are not harmful to the snake (Figure 15–9). The liquid venom is frozen and then dried under strong vacuum; this process yields a powder that is easy to store and ship. Antivenin is made by injecting venom into horses or sheep in increasing sublethal doses until immunity is achieved. Blood is then drawn from the horse or sheep, and the serum containing antibodies against the venom is purified. When this serum is injected into a person who has been snakebitten, the venom is neutralized. Numerous venom extraction facilities operate around the world to provide antidotes for venomous snakebite.

Over 200 pharmacologically active alkaloids have been extracted and identified from the skin of various anurans. These alkaloids are used by the frogs as chemical defense against predators. Since investigators have learned how these alkaloids affect nerve and muscle tissue of a target victim, considerable effort has been made to synthesize and utilize these alkaloids as research tools in neurobiology (Grenard

1994). For example, batrachotoxin (found in the dendrobatid genus *Phyllobates*) prevents the closing of sodium ion channels in the surface membranes of nerve and muscle cells. The result to a victim that has been exposed to batrachotoxin is that an influx of sodium ions electrically depolarizes the cell mem-

Figure 15–9   **Venom extraction.** Venom is being extracted from a Russell's viper, *Vipera russelli*, to be used for making antivenin serum. *(Photograph by Romulus Whitaker.)*

branes; thus the nerve cells cannot transmit impulses and the muscle cells remain in a contracted state and cannot function. The end result is heart failure. Neurobiologists currently use batrachotoxin as a research probe for voltage-sensitive sodium channels. In a radiolabeled form, it is being used to study the interaction of local anesthetics and anticonvulsants.

Epibatidine, a unique class of alkaloids, has so far been isolated only from the dendrobatid genus *Epipedobates*. Epibatidine is a powerful painkiller, and experiments with rats suggest that it is many times more potent than the plant alkaloid morphine. Perhaps someday humans will use a form of synthetic epibatidine in place of morphine. The advantages of epibatidine over morphine are that the former is nonsedating and probably is nonaddictive.

Pets and Zoo Exhibits

Within the United States there is considerable commercial exploitation of amphibians and reptiles for the pet trade. For example, in 1990 the Florida Game and Fresh Water Fish Commission began to collect information on the magnitude of commercial trade in amphibians and reptiles. Everyone in Florida who sells live, wild-caught native amphibians or reptiles now must report his or her collecting activities. The first 2 years' worth of data (June 1990 to June 1992) are astounding: 119,831 animals were removed from the wild (Buck 1997). Of these, 49,240 were snakes (most common was the corn snake, *Elaphe guttata*, with 13,827 individuals), and 41,493 were frogs and toads (most common was the green tree frog, *Hyla cinerea*, with 13,166 individuals); the least popular group was salamanders, at 1,050 individuals (most common was the three-lined salamander, *Eurycea longicauda guttolineata*, with 265 individuals). These figures underestimate the actual number of animals that were removed from the wild for the pet trade because many people doubtlessly failed to report their transactions, nonresidents who collected animals were not included, and amphibians and reptiles collected for personal use were not included.

North American box turtles (genus *Terrapene*) are popular pets, not just in this country but in Europe and other places as well. According to United States Fish and Wildlife Service figures, approximately 74,000 Gulf Coast box turtles were exported from the United States during 1992–1994 (Buhlmann 1996). As a result, populations of box turtles are declining. Box turtles are an excellent example of how important

public opinion is for conservation. Several years ago the U.S. Fish and Wildlife Service began soliciting information from scientists and from the public on population sizes, levels of trade, and the effect of harvesting on populations of box turtles. Based on the input received, box turtles were afforded some protection in 1994. Subsequently, the Office of Scientific Authority issued a statement that the 1996 export quota for box turtles from anywhere in the United States would be zero. This decision was based largely on the input from numerous scientists and from the public, who argued that the risk of continued population decline was great and that we just don't know enough about the turtles' population biology to determine a sustainable level of harvest.

Research and Teaching

Collections of amphibians and reptiles are often made for research purposes. These collections range from very small (one or two vouchers of one species) to very large (vouchers for community-wide inventories). After the study is finished, specimens should be deposited in public museums for use by other investigators.

Amphibians and reptiles are widely used in medical and biological teaching for dissection and demonstration purposes. Collectors for biological supply houses in North America have long been aware that many local populations of ranid frogs are declining. In the early 1970s almost all leopard frogs used in teaching (13 million) and research (2 million) were captured from wild populations (Nace and Rosen 1979). One commercial supplier reported that in 1970–1971 they collected over 10 tons of leopard frogs from one western state alone; 4 years later fewer than 250 pounds of leopard frogs were collected from that same state. One company's volume declined from an average of 30 tons of frogs per year (about 1 million individuals) in the late 1960s to 5 tons in 1973, not because of decreased demand but because of difficulty in finding the frogs. Nature can no longer meet the demand for ranid frogs needed for teaching and research. The problem has been dealt with in three ways. First, a greater percentage of frogs currently used in teaching and research are laboratory-bred and -reared than was the case previously. Second, laboratory instructors increasingly use demonstrations or have groups of students work with one specimen (a trend motivated as much by economics as by conservation) rather than one animal per student, as used to be the case. Third, fetal pigs are being used more often now as an alternative to frogs for classroom dissections.

Patterns of Species Extinction and Extirpation

Humans are responsible for most current extinctions, extirpations, and population declines of amphibians and reptiles. Can we identify species that are most likely to be affected by humans? Species hunted for human consumption, leather, and the pet trade are in danger of overexploitation by humans. Beyond these characteristics we can identify several broadly overlapping categories of species most likely to become extinct or extirpated.

Long-Lived Species

Species that live a long time (for example, many turtles) exhibit a suite of life history characteristics (delayed sexual maturity, low fecundity, and high adult survival rates) that constrain the population's ability to respond to increased mortality (Congdon et al. 1993, 1994). If adults are harvested, the population may not be able to build up its numbers again. For example, a 10 percent annual increase in mortality of adult common snapping turtles older than 15 years of age (adults are often heavily exploited for food) could result in a 50 percent reduction in the population size within 20 years (Figure 15–10).

Species with Low Reproductive Rates

Some species of amphibians and reptiles reproduce only every other year, or even less frequently. Species with low reproductive rates are less likely to recover quickly from population declines. Two species of primitive wart snakes (Acrochordidae) from the Indoaustralian region are highly prized for their skins because, unlike most snakes, their scales are nonoverlapping. In some years as many as 300,000 individuals are killed and their skins tanned for shoes and handbags. Shine et al. (1995) have argued that one of these species, *Acrochordus arafurae*, has such a low reproductive rate that populations cannot withstand much commercial harvesting.

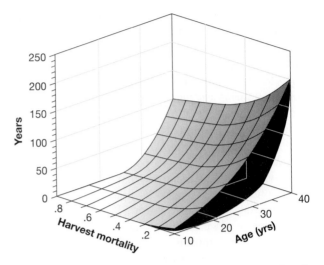

Figure 15–10   **Harvesting snapping turtles.** Results of a computer simulation showing effects of increased harvesting mortality on the rate of population decline for common snapping turtles (*Chelydra serpentina*). Years = number of years taken to remove half the population. Postponing the onset of the same level of harvesting mortality results in a much longer time until the population is reduced by 50 percent. This relationship argues for protection of adult and older juvenile snapping turtles, not just eggs and hatchlings. *(Source: Congdon et al. 1994.)*

Species That Have Poor Dispersal and Colonization Abilities

Many amphibians, especially salamanders, move only small distances on a daily basis. If the habitat of a given population is destroyed or modified, and if the species has poor dispersal abilities, the population may be doomed to extinction, with few opportunities for natural reinvasion.

Continental Endemics

This category includes species that have unusually restricted distributions and require specialized habitats. These are often referred to as relict populations. Endemics may be abundant in the restricted areas where they occur, but they often have rigid habitat requirements. Often it takes only a small alteration of the environment to endanger a species that is restricted geographically. The vulnerability of such species stems from the fact that if local populations are extirpated, there is no chance for recolonization. Examples of such species include the golden toad (*Bufo periglenes;*

Figure 15–11), found only in one small mountain range in Costa Rica, and two species of gastric brooding frogs (*Rheobatrachus*), found in restricted stream habitats in rain forest of southeastern Queensland, Australia. All three species have apparently become extinct within the past decade.

Oceanic Island Endemics

Communities on islands seem to be extraordinarily fragile, and most extinctions of amphibians, reptiles, birds, and mammals within the past several hundred years have occurred on islands. Island species are often extremely vulnerable to predators such as cats, dogs, mongoose, and rats introduced by humans. Associated with a long history of living in a predator-free habitat, these species do not have efficient antipredator defenses. Other adverse factors include severe habitat destruction and excessive hunting pressure from humans. Examples in this category range from small frogs to giant tortoises.

Species with Colonial Nesting Habits

When large numbers of animals gather for breeding activities, they are extremely vulnerable to exploitation by humans and to predation by mammals associated with humans. The classic example is sea turtles, particularly species such as Kemp's ridley (*Lepidochelys*

kempi), which engages in mass nesting, a phenomenon called an *arribada* (arrival, in Spanish). In the 1940s, on one day alone, an *arribada* estimated at more than 40,000 Kemp's ridleys nested on a remote area of beach called Rancho Nuevo, between Tampico, Mexico, and Brownsville, Texas. By the 1960s the huge arribadas had vanished, and only small nesting groups of females were observed. The main reason for this decline is thought to be commercial exploitation of the eggs and nesting females. For decades, people gathered the eggs and transported them to markets in Mexico City and elsewhere, where they were sold for food and as aphrodisiacs. By the mid-1970s, the estimated number of nesting females had dropped to under 1,000, making this species critically endangered. Fewer than 500 females nested on Rancho Nuevo in 1992. The future may be brighter for the turtles as a result of international protection, however, and in 1994 1,566 nests were found at Rancho Nuevo.

Migratory Species

This category, which includes sea turtles, is vulnerable because the animals often migrate between countries so that even if a species is protected by one country, it may be severely exploited in the country to which it migrates. This is the case for green turtles (*Chelonia mydas*); they are protected in Australia but exploited when they migrate to Indonesia. Furthermore, the

Figure 15–11   **The golden toad, *Bufo periglenes*.** This species, once endemic to a small mountain range in Costa Rica, is now apparently extinct. (*Photograph by Martha Crump.*)

often extensive migration routes of sea turtles may take them through polluted waters or human-modified landscapes and otherwise subject them to a greater range of environmental problems than are experienced by more sedentary species. On a more local scale, populations of ambystomatid salamanders and many anurans such as tree frogs, spadefoot toads, and toads that migrate to breeding ponds often suffer high losses due to traffic on roads and highways.

Conservation Options

What can be done to reverse population declines of amphibians and reptiles? Many options are available. Generally, conservation programs must be multifaceted, involving field and laboratory research to understand the animals, education to inform the public, legislation to protect the endangered species and their habitat, and, if appropriate, captive breeding and management programs. Currently considerable emphasis is placed on protection of individual endangered species, perhaps because a colorful or cute animal is a way of grabbing the public's attention. As we have already emphasized, however, long-term conservation efforts should involve habitat protection to maintain evolutionary potential on a community-wide scale.

Considerable effort is now being spent in many areas of the world to involve local people in conservation efforts. Biologists in developed countries are realizing that many people in developing countries view habitat and wildlife protection as a luxury they simply cannot afford. Rural peoples need to provide for their families. Too often they hunt-out or farm-out an area, and then move on to a different site for the next few years until that area is no longer productive. Local people must be given economic incentives to conserve natural resources.

Ultimately the most successful conservation programs are those that identify and deal with the reason a species is endangered and at the same time provide economic benefits to local people. The Irula venom industry in southern India is an excellent example of a project that accommodates both snakes and people (Whitaker 1989). At least as far back as the 1930s many of the rural Irula tribal people in southern India made their living catching snakes and selling the skins to tanneries. During a boom period between the early 1950s and the 1970s, an average of 5,000 skins per day were processed at a single tannery in Madras. On average, 5 million snake skins were exported from India per year. By the 1970s the densities of pythons and certain other species had become so reduced that it was difficult to find the snakes. The density of rats increased, associated with the decline of their main predators—the snakes. Agriculture was severely affected because rats destroyed massive quantities of grain. One estimate suggests that rats destroyed 30 to 50 percent of India's grain every year. In 1976, the government of India banned the export of snakeskins. Although this legislation was obviously good for both the snakes and the local farmers, the impact was devastating for the Irula people, who had lost their primary means of earning a living.

Romulus Whitaker offered an alternative that demonstrated that preservation and exploitation can be compatible. He formed the Irula Snake Catchers' Cooperative for venom extraction (Figure 15–12). The members of the cooperative catch cobras (*Naja*), Russell's vipers (*Vipera russelli*), saw-scaled vipers (*Echis*), and kraits (*Bungarus*), extract their venom three times each, and then release the snakes back into the wild (Whitaker 1989). The cooperative not only provides a livelihood for the Irula, it also increases the limited supply of antivenin serum for the country. A bonus for the Irula is that their income is many times greater than what they could earn by selling snakeskins. Snake venom is one of the most expensive natural resources on Earth. Depending on the species, venom is worth 5 to 1,000 times as much as gold! About 50 kraits are needed to produce 1 gram of dried venom; 50 snakes for venom extraction are worth about 3,000 rupees, whereas 50 skins would be worth a few hundred rupees. The Irula cooperative is now the largest producer of snake venom in India, and the Irula take pride in what they have to offer their country and the world.

Habitat Protection

The most important action we can take for amphibians and reptiles is to protect habitat. Although we lack data on the basic biology of most species, and thus considerable effort must be spent on research, our ignorance should not be used as an excuse for delaying habitat protection. Because of the intense pressures on the environment caused by the ever-growing human population, preservation of land that benefits

Figure 15–12   **Chockalingam, one of the most famous of the Irula snake catchers, and his wife.** He is shown holding a rat snake (*Ptyas mucosa*). *(Photograph by Harvey Lillywhite.)*

all plants and animals is a critical priority. Habitats for multispecies assemblages may be protected by establishing parks, reserves, and conservation easements with private landowners.

If a reserve is to be established, how should it be designed? This question has generated considerable debate and discussion. Basically, we need to consider three aspects: biological considerations, the culture of indigenous peoples, and political and economic constraints and realities (Meffe and Carroll 1994).

The biological considerations include decisions regarding location, size, and shape of the area to be protected. We must determine if the reserve can be connected to other natural areas, and how the surrounding land is used and whether this land use pre-

sents a threat to the reserve. In planning nature reserves, we need to minimize the degree of habitat fragmentation in order to minimize extinction rates. Where the habitat is already fragmented, outside and inside of reserves, corridors connecting the fragments can serve as a way of increasing available habitat by allowing the animals to travel among the remnants.

As mentioned previously, if conservation efforts are to be successful, the needs and cultures of indigenous peoples must be respected—both for ethical and for pragmatic reasons. A reserve and its flora and fauna have a much greater chance of being protected if local peoples support existence of the reserve.

Too often, instead of biological considerations, political and economic constraints dictate the design of a reserve. Whereas it is important to recognize these constraints, ideally with public education programs trade-offs can be instituted so that everyone benefits.

One innovative idea concerning habitat preservation has been the debt-for-nature swap program whereby millions of hectares of land have been set aside as reserves in exchange for release of national governments from international loan debts (Ayres 1989). Developing countries collectively owe about $1.3 trillion to international financial institutions. Because of economic problems, however, many of these countries are unable to repay their loans. As a result, financial institutions are often willing to sell the debts at huge discounts. For example, secondary market prices for Costa Rican debt are as low as 14 percent of the original debt face value. For Peru the figure is 5 percent, and for Nicaragua it is 4 percent. Debt-for-nature swaps work as follows. First, an international, nongovernmental conservation organization (such as Conservation International) works with the debtor country in developing a conservation project; the project often involves land protection, but it could be some other worthwhile conservation endeavor, such as environmental education. The international organization then purchases part of the loan at a discounted price, which frees the debtor country from future payments on that part of the loan. In return, the debtor country agrees to fund the chosen conservation project. The first debt-for-nature swap was begun in 1987, in Bolivia; Conservation International bought $650,000 of Bolivian debt for $100,000, and in exchange, Bolivia put funds into managing and protecting the Beni Biosphere Reserve (135,000 hectares) and three large buffer zones around the reserve. Debt-for-nature swaps have already taken place in several parts of the world, including Costa Rica, Ecuador, Mexico, the Philippines, and Madagascar. Although the idea is wonderful in theory, unfortunately there have been numerous sociological problems, and there is often little or no follow-up vigilance to ensure that the habitat is really protected. If the problems associated with the debt-for-nature swap program can be resolved, the program may prove to be a valuable way of protecting habitats worldwide.

Not all species need to have reserves set aside for their protection. Whereas some species require complete habitat protection, others have lifestyles compatible with human habitation. The following discussion provides some examples of ways in which, by altering use of the environment, humans can coexist with amphibians and reptiles.

Coexistence with Humans

Travel by road has become the dominant mode of transportation for people throughout most of the world. New roads are continually being built for the convenience of the ever-growing and increasingly mobile human population. Roads fragment habitat for wildlife, and with the growing network of roads, animals are increasingly forced to cross roads during their daily activities and are often killed doing so. A conservative estimate of the average number of amphibians and reptiles killed on paved roads each year in Australia is almost 5.5 million (Ehmann and Cogger 1985). In particular, amphibians migrating to breeding sites are slaughtered on roads in areas where traffic is heavy. Over 40 percent of the breeding adults within a population of amphibians may be killed each year by vehicles on the roads (Langton 1989). Public concern has resulted in a variety of measures designed to decrease this source of mortality.

Since the late 1960s road signs have been used to warn motorists about migrations of toads in Germany, Switzerland, and the Netherlands, and for over a decade road signs have been used in the United Kingdom (Figure 15–13). Motorists are advised to avoid use of certain roads during the peak migration periods, and if the roads must be used motorists are asked to drive slowly enough to be able to steer around the toads. In the 1960s Europeans began experimenting with corridors that could link crucial habitats for amphibians, such as a wooded area on one side of a road

Figure 15–13   **Toad crossing sign.** These signs are used in the United Kingdom to warn motorists about migrations of toads crossing the roads. *(Photograph by John Baker.)*

and a breeding pond on the other side. The primary design involves drift fences (upright fencelike structures usually made of metal or plastic). An animal cannot cross the drift fence, so it moves along the fence to a tunnel that provides an underpass beneath the road (Figure 15–14). Considerable engineering research has focused on designing tunnel systems that allow safe crossing for amphibians. The engineering design must consider ideal temperature, air circulation, humidity, and light level conditions, or amphibians will not use the structures. Furthermore, tunnels work only if the associated drift fences are maintained. Tunnels have also been used for spotted salamanders (*Ambystoma maculatum*), pine snakes (*Pituophis melanoleucus*), turtles and tortoises, and for other reptiles and amphibians.

In Europe, as part of a campaign to rescue amphibians from roads, volunteers form toad patrols; toads are gathered up in buckets and carried safely across the roads. In some places drift fences are built along road edges to channel toads to collecting points. In the United Kingdom, at more than 400 sites, an estimated 500,000 animals (mostly the common toad, *Bufo bufo*) are saved each year (Langton 1989). Volunteers are also involved with protecting and restoring breeding sites. New breeding ponds are

Figure 15–14 **A tunnel system and accompanying drift fence.** This system channels animals toward an underpass that allows amphibians, reptiles, and other small animals to avoid crossing road surfaces.

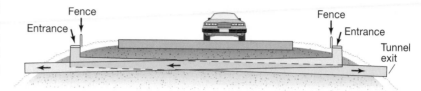

Cross section of one-way tunnel system, showing the direction of movement in each tunnel.

constructed and existing ponds are fenced to exclude cattle.

Another example of how humans can modify their behavior to share critical habitat concerns nesting beaches of sea turtles. As hatchling turtles emerge at night they instinctively head for the brightest horizon, which is normally moonlight or starlight reflected off the ocean's surface. On residential beaches with artificial lighting, hatchlings confuse electrical lights for moonlight. Instead of heading for the ocean, they become disoriented and head toward the residential area. Many hatchlings ultimately desiccate or are run over by cars. Many coastal communities in the United States now have beachfront light ordinances that prohibit lights during designated time periods. For example, ordinances in Florida for some nesting beaches generally permit lights only until 11 P.M. The absence of artificial light for the rest of the evening allows emerging turtles to orient correctly toward the ocean, but not all turtles wait until 11 P.M. Approximately 31 percent of loggerheads emerging from their nests at Melbourne Beach, Florida, do so before 11 P.M. on any given night (Witherington et al. 1990). Low-pressure sodium vapor street lights have proved to be less of a problem for both adult female turtles and hatchlings than normal incandescent lighting. Thus, in areas along nesting beaches where artificial lighting cannot be completely eliminated, low-pressure sodium vapor lights may be a partial solution to the problem.

In the United States, more sea turtles die as a result of drowning in shrimp trawls than from all other human-induced sources of mortality combined. The turtles are unable to come to the surface for air, and they often drown in less than 1 hour. Up to 50,000 loggerheads (*Caretta caretta*) and 5,000 Kemp's ridleys (*Lepidochelys kempi*) drown annually in shrimp trawls in U.S. waters (National Research Council 1990). Progress is being made to reduce this source of mortality. Turtle excluder devices (TEDs) have been designed that can be attached to shrimp-trawling gear. A TED is a small net or metal grid inside the shrimp net

that allows shrimp to pass to the back, but allows turtles to escape. The most effective TED designs exclude 97 percent of the sea turtles that would have been caught in nets without the devices. Other benefits are that the shrimp fishermen don't have to deal with handling the heavy turtles, damage to the shrimp catch is eliminated, and unwanted bycatch such as jellyfish and horseshoe crabs is reduced. Although all U.S. shrimpers are required to install TEDs and conservation organizations offer to pay for them, compliance has been poor. Furthermore, use of TEDs by shrimpers in other countries is rare at present.

Desert tortoises (*Gopherus agassizii*) and domesticated livestock compete for habitat throughout the tortoise's range in the western United States. In California, about 70 percent of tortoise habitat exists on public land, the other 30 percent on private land. About 56 percent of tortoise habitat is also grazed by sheep and cattle. These mammals can have a heavy impact on the tortoises (Berry 1986). For example, in 1 day of grazing, a herd of sheep can consume 60 percent of the biomass of annual plants, the same food the tortoises eat. Heavy grazing by sheep reduces (up to 68 percent by volume, 29 percent by area) the perennial shrubs that serve as protective cover for tortoises. Tortoise burrows are damaged or destroyed when trampled by sheep. Can humans and their livestock share habitat with the desert tortoise? The Bureau of Land Management has instituted various restrictions on grazing in an attempt to minimize competition between tortoises and sheep (Foreman et al. 1986). For example, sheep may not graze in areas designated as crucial tortoise habitat unless at least 37 kilograms per hectare (dry mass) of forage is present in the habitat. In areas identified as highly crucial to tortoises, the minimum forage level is raised to 64 kilograms per hectare, and sheep owners are allowed only one grazing pass through the area. One could argue that sheep should be kept out of tortoise habitat entirely, but as we have emphasized throughout this chapter, long-term conservation won't happen simply because conservationists believe it is morally right.

The livelihood of local people must be considered. With compromises, sheep owners and desert tortoises can better coexist.

Research

One constraint on effective conservation is a lack of information on the basic biology of the species in question. We simply don't know enough about the basic requirements of most species of amphibians and reptiles to be sure that we are protecting the right habitat, the right resources, or the right life history stages. Ideally, basic research should include examination of population size and structure, age-specific survivorship and sources of mortality, habitat preference, spatial requirements and activity patterns (including migrations to feeding and breeding sites), reproductive patterns and frequency of breeding activity, life history traits (including age at maturity and longevity), social behavior, feeding ecology, and genetic variability.

Furthermore, we must be careful not to employ conservation solutions that are merely halfway technology—that is, the types of things done after the problem has already occurred but that do not address the underlying causes of the problem (Frazer 1992). An example of halfway technology is releasing 1,000 toads into an area where a population has recently become extinct without investigating the causes of extinction or whether the environment is still appropriate for that species. In order to practice effective conservation, we must understand the causes behind the problem, and to do this we need more basic research. Following are two examples of how specific types of data are used and why they are critical for successful conservation measures.

Before convincing arguments can be made for habitat protection for a given species, habitat and spatial requirements during all of its life history stages and during both the breeding and non-breeding seasons should be identified. If a frog species migrates from wooded areas to aquatic sites to breed, not only must both areas be protected, but corridors between the sites must also be protected, especially if the habitat is fragmented by roads. It is useless to protect the breeding sites and woods if most of the population gets killed on roads during breeding migration. Vincent Burke and Whitfield Gibbons (1995) studied three species of semiaquatic turtles that live in a wet-

land area in South Carolina to determine the effectiveness of current wetland policies. Federal statutes protect wetlands larger than 0.4 hectare by requiring delineation of the wetland–upland border and then preventing development from occurring within the wetland area. Results of the study showed that all of the turtles' nesting sites and terrestrial hibernation burrows occurred outside the federally delineated boundary, and that critical habitats extended 275 meters beyond the wetland boundary (Figure 15–15). In this case, obviously, current wetland statutes do not adequately protect the habitats that these turtles require throughout their life cycle. Burke and Gibbons convincingly demonstrated the critical need for terrestrial buffer zones. The key to making a strong argument for the turtles' future was gathering extensive data on habitat used for foraging, mating activity, nesting sites, and hibernation burrows.

Data concerning life history traits, age-specific mortality, and the causes of mortality are crucial for identifying which life history stages most need protection. As an example, sea turtles are long-lived, highly fecund animals that have extremely high natural mortality at the egg and hatchling stages. In some species survivorship to reproductive maturity has been estimated to be less than 1 percent. This combination of traits suggests that conservation measures must be directed primarily at the subadult and adult stages rather than at the egg or hatchling stages (Figure 15–16). A good example is Kemp's ridley (*Lepidochelys kempi*), the most endangered of all sea turtles. Over 14 years, more than $4 million has been spent protecting eggs on the nesting beach in Mexico and airlifting thousands of eggs from Mexico to the United States where they have been laboratory-raised and then released into the ocean at 9 to 12 months of age. This effort turned out to be halfway technology. After all these years of protection the population has not increased in size. Although nesting females were protected, biologists eventually realized that these efforts were wasted because so many adults were drowning in shrimp nets. With the use of TEDs, instituted in 1992, the adult nesting population appears to be recovering slowly. The most effective way to protect these turtles is to reduce the mortality of adults and subadults by protecting them from shrimp trawlers. If all conservation measures were aimed at increasing the survivorship of hatchlings and no protection were afforded to older stages, sea turtles would likely soon become extinct. Clearly, some pro-

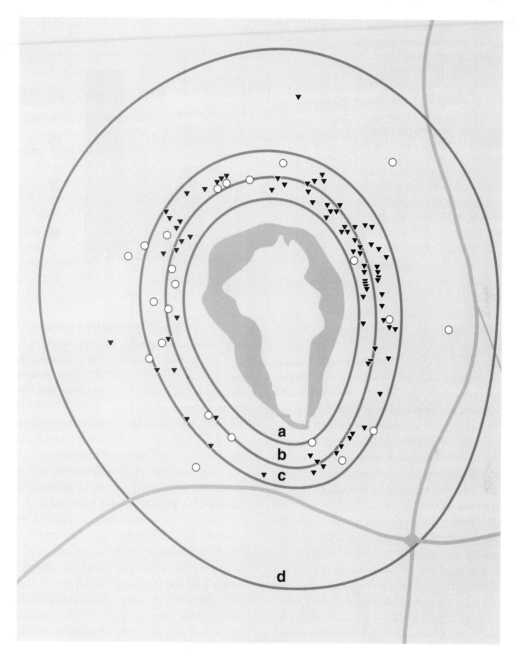

Figure 15–15   **Discrepancy between habitats used by turtles and the wetland area protected.** Shown here is Ellenton Bay, a 10 hectare isolated wetland area in the Atlantic coastal plain in South Carolina. The maximum dimension of area d is approximately 1 kilometer. The area enclosed by ring a is protected by U.S. wetland statutes. Habitats within ring b would be protected by the strictest state statutes. Ring c encloses 90 percent of the turtle nesting sites (triangles) and hibernation sites (circles), but protection of 100 percent of these sites is not attained until ring d. Obviously, current wetland statutes do not adequately protect the habitats of these semi-aquatic turtles throughout their life cycles; terrestrial buffer zones are needed. *(Source: modified from Burke and Gibbons 1995.)*

Figure 15–16 **Annual survivorship of loggerhead turtles (*Caretta caretta*) in five life history stages.** Large juveniles (age class indicated by arrow) offer the greatest management potential for increasing future population growth for the turtles. Stages younger than large juveniles have less reproductive potential because most individuals will not reach reproductive maturity. Although older individuals have greater reproductive potential than do large juveniles, there are few of them remaining in the population. (*Source: Crouse et al. 1987, based on data from Frazer 1983.*)

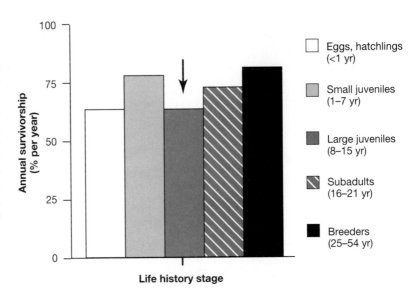

tection is needed at all life history stages for long-lived organisms.

One very positive research-oriented step has been taken by establishing a task force to address declining amphibian issues. In 1991 the Species Survival Commission of the International Union for the Conservation of Nature formed a Declining Amphibians Populations Task Force, designed to be a coordinating center for investigators and agencies from all over the world involved in investigating amphibian declines. The task force identifies priorities for research, offers grants for start-up projects, recommends uniform field methods so that different species and habitats can be compared, and maintains a computerized database and a global monitoring program. Various regional working groups are gathering data on species at local study sites in an effort to document declines and investigate the causes of observed declines. Over 1,200 collaborators are currently organized as working groups in 91 regions or countries. Symposia and workshops are held where scientists interact to share and discuss their data and evaluate protocols. A newsletter, *FROGLOG*, communicates information and activities of the task force to members and to other interested people.

Education

Education is urgently needed at all levels if we hope to maintain viable populations of amphibians and reptiles. Educators must teach the local people who share

the environment with species of concern, but they must also teach policymakers, managers, the media, and the general public. The communication gap between the biologist and the lay person must be narrowed through direct interactions and hands-on experience. Training in basic areas of habitat protection, wildlife management, and conservation biology is needed, especially in tropical countries, where most species of amphibians and reptiles are found.

The success of conservation and management programs ultimately depends on how well the programs are tailored to the interests and needs of the people on whose land the threatened or endangered animals live. Citizens must be taught not only the need to conserve wildlife but also the rationale for doing so. Local people should be encouraged to participate in the teaching as well as the learning part of the education process. Effective education methods that involve local people include construction of museum and zoo displays, distribution of pamphlets and newsletters, production of radio and television programs, and participation as nature guides and park guards. Educators and curriculum developers should be encouraged to produce materials for use in the schools, as conservation-minded children are vital for the future. Children are usually receptive to new ideas, are naturally curious and enthusiastic, and are the ones who will be making policy decisions in the future.

EARTHWATCH is one of several nonprofit organizations that bring together lay persons interested in wildlife with scientists who both need help in their field studies and wish to educate the interested public. The program is immensely successful, each year involving over 3,000 participants. Recent herpetological projects have included studies of sea turtles on nesting beaches in St. Croix and the Yucatan, population studies of diamondback terrapins in South Carolina, and surveys of amphibians and reptiles in Brazilian rain forest, Madagascar, and in the South China Sea islands. A similar educational opportunity is a program cosponsored by the Massachusetts Audubon Society and the Caribbean Conservation Corporation in which participants assist in tagging green sea turtles and in nesting and hatchling research on leatherbacks at Tortuguero in Costa Rica.

Education can and should be a vital component of ecotourism (tourism based on natural history). Each year millions of people spend their vacations viewing wildlife in the animals' native habitats. Ecotourism provides an opportunity to educate the public about the value of both wildlife conservation and habitat protection. Ideally a long-term benefit is a change in attitudes toward nature. Ecotourism also represents a valuable source of income (Figure 15–17). For example, ecotourism is now the leading source of foreign exchange and nongovernmental employment for Costa Rica (bananas rank second place and coffee is third).

Loggerhead turtles (*Caretta caretta*) are an excellent species for ecotourism because they are fascinating, impressive, and easily watched when they come ashore to lay their eggs. Many organizations conduct turtle watches in Florida; approximately 10,000 people participated in organized turtle watches during the 1993 nesting season in the state. These watches provide an ideal opportunity for educating the public about sea turtle biology and the need for their conservation. The concern has been raised, however, that these watches might be detrimental for the turtles even though there are rigid guidelines that must be strictly adhered to by the participants. Recently the effects of organized watches were evaluated by comparing nesting behavior and hatchling success between two groups of loggerheads: experimental females that were observed by an organized turtle watch group and control females that were not observed by a turtle watch group (Johnson et al. 1996). The results were good news. Although the experimental turtles spent significantly less time camouflaging their nests than did the control turtles, hatchling success and hatchling emergence success were not significantly different between the two groups of turtles. These results should encourage other countries to capitalize on sea turtles for ecotourism, providing that guidelines are enforced so that disturbance to nesting females is minimal.

National Legislation

Countries vary widely in the level of national protection of amphibians and reptiles. Some countries provide no protection at all. At the other extreme are countries such as Belgium, where all amphibians and

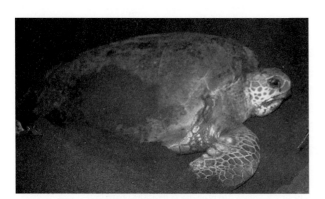

Figure 15–17 **Female green sea turtle (*Chelonia mydas*) laying eggs on beach at Tortuguero, Costa Rica.** For nearly 20 years the beach at Tortuguero, the largest green sea turtle rookery in the Caribbean, has been protected as a national park. Each year more than 15,000 tourists come to see the turtles. *(Photograph by Martha Crump.)*

reptiles with the exception of two common species of ranid frogs have been protected since 1973. Protection is especially good in the Flanders region of Belgium, where the law forbids the capture, killing, taking of eggs or young, and even habitat disturbance of all species except for the two ranids, and then only on private land.

In 1973 the United States enacted the Endangered Species Act, a law that provides protection for both domestic and foreign species that are considered actually or potentially in danger of becoming extinct. Such species are classified as either endangered or threatened. Endangered species are those that are in danger of extinction throughout all or a significant portion of their range; threatened species are those that are identified as likely to become endangered within the foreseeable future throughout all or a significant portion of their range. Once a species is listed as either endangered or threatened, activities detrimental to the species (e.g., collection and habitat destruction) are restricted, and a recovery plan is developed by a team of experts. The objective of the recovery plan is protection so that ultimately the species will recover from the threat of extinction; the goal is eventual removal of species from the list. The status of listed species is reviewed every 5 years, and recommendations for delisting or reclassification are made as warranted. As of August 1994, there were 23 species and subspecies of amphibians and 112 species and subspecies of reptiles listed under the U.S. Endangered Species Act. Of these 135 taxa, 99 are considered endangered and 36 are considered threatened (U.S. Fish and Wildlife Service 1994; Table 15–2).

Control of International Trade

During the 1960s there was increased awareness worldwide of environmental issues ranging from pollution to extinction. With realization of the alarming rate of extinctions, governments around the world instituted national legislation to protect their wildlife. By the early 1970s, however, it became clear that international laws were needed to control trade in wildlife. In 1972, the United Nations Conference on the Human Environment adopted the idea of a convention on trade in endangered species. The following year, a conference was held in Washington, D.C., with the goal of drafting an international endangered species treaty. The result was the Convention on In-

Table 15–2 **Number of endangered and threatened species of amphibians and reptiles listed worldwide under the endangered species act.**

Taxa	Endangered	No. of Species and Subspecies*	
		Threatened	Total
Crocodilians	19	2	21
Snakes	9	7	16
Lizards	23	14	37
Turtles and tortoises	29	8	37
Tuatara [†]	1	—	1
Salamanders	6	3	9
Frogs and toads	12	2	14
Total	**99**	**36**	**135**

*In several cases some populations of a given species are listed as endangered whereas other populations of the species are listed as threatened; in these cases the species is included here as endangered.
[†]Tuatara is still listed as one species in the 20 August 1994 update, but both recognized species are considered endangered.
Source: U.S. Fish and Wildlife Service, 1994.

ternational Trade in Endangered Species of Wild Fauna and Flora (CITES), drafted by 81 nations. The aim of CITES is not to stop trade of wildlife and their products, but instead to regulate trade based on assessment of the status of each species of concern. The CITES treaty mandates that international trade in species (and products thereof) listed by the convention is unlawful unless authorized by permit. An important aspect of CITES is that consumer nations agree to share responsibility for international trade in plants and animals with producer nations by forbidding the importation of illegal wildlife and their products.

At present, four species of salamanders (two species of *Ambystoma* and both species of *Andrias*) are protected by CITES. Within the Anura, three species of *Bufo*, two *Rana*, one *Atelopus*, one *Dyscophus*, *Mantella aurentiaca*, both species of *Rheobatrachus*, all species of *Nectophrynoides*, and many dendrobatids are protected. All sea turtles, all land tortoises (Testudinidae), and many other turtles and tortoises are protected. Both species of tuatara are protected, as are all boas, pythons, and numerous other snakes. All iguanas of the genera *Iguana* and *Brachylophus* are protected, as are all chameleons of the genus *Chamaeleo* and all ground iguanas of the genus *Cyclura*. Tegus (*Tupinambis* spp.), monitor lizards (*Varanus* spp.), and day geckos (*Phelsuma* spp.) are among some of the other lizards protected by CITES. Nearly 600 species of reptiles and amphibians are currently protected by CITES (U. S. Fish and Wildlife Service 1995).

Reestablishing Populations

Increasing effort is being focused on reestablishing self-sustaining populations of threatened or endangered species in the wild; animals are moved from one place to another, or raised in captivity and then released into the wild. Obviously, extreme care must be taken not to introduce disease into wild populations. Although such efforts seem laudable and are extremely popular with the public, unfortunately many of the projects carried out so far with amphibians and reptiles have failed as effective conservation measures because viable populations have not been reestablished (Dodd and Seigel 1991). In some cases the animals have died, in others they migrated from the site, and in most other cases no long-term follow-up studies have been done so we don't know the outcome.

For example, efforts to reintroduce the Houston toad (*Bufo houstonensis*) to ten sites at the Attwater Prairie Chicken National Wildlife Refuge have been unsuccessful. Since 1982 half a million adults, juveniles, and tadpoles have been introduced, but not one new population has been successfully established in the area. Unfortunately, because the causes of the species' decline were not understood, the problems could not be eliminated, and the reintroductions turned out to be halfway technology.

Reestablishment projects are attempted in several ways. Repatriations involve releasing animals into areas that were formerly or are currently occupied by that species. For example, a species of lizard might be released into a habitat that experienced a human-caused forest fire that killed all the individuals formerly living there. Tadpoles may be released into an existing population of frogs that is experiencing serious declines. Translocations involve moving animals into areas that were not historically occupied by that species. Sometimes translocations are necessary because, for example, all the habitat previously occupied by the species may now be under concrete. Extreme care must be taken with translocation to avoid disrupting the biology of the native fauna. If the introduced species hybridizes with native species, as has been documented for some ranid frogs, the hybrid may be competitively superior to the native species; the result may be an eventual decline of the native species. Relocations involve moving animals from areas where they are threatened (e.g., by impending deforestation) to more protected areas (e.g., where

Figure 15–18   **Gopher tortoise (*Gopherus polyphemus*) with upper respiratory tract disease.** *(Photograph by Grace McLaughlin.)*

they would be less vulnerable to loss of habitat). Ideally, animals should be relocated to areas historically occupied by that species. Captive propagation involves maintaining adults in captivity and raising their offspring. The ultimate goal of captive propagation is reintroduction into the wild, assuming that habitat is still available. In all of these population reestablishment activities, animals ideally are released into protected habitat, and then the population should be monitored to determine success or failure of the program.

Head-starting is an experimental practice frequently used in reestablishment projects. Eggs are hatched in captivity, and then the hatchlings are reared to a size such that when they are released into the wild they will be less vulnerable to predators. Head-starting is being used for tuataras, turtles and tortoises, crocodilians, lizards, and amphibians. An example of a success story is the work being done at the Charles Darwin Research Station in the Galápagos Islands, where giant tortoises (*Geochelone*) and land iguanas (*Conolophus*) are captive-reared. Once they are past the size of greatest vulnerability to predators, they are released onto various islands where predators and competitors are controlled. So far, it looks as though populations of giant tortoises and land iguanas have a better outlook for the future, thanks to these efforts.

The value of head-starting for sea turtles, however, is controversial. A reduction in mortality may be offset by major developmental problems. The turtles might experience a nutritional deficiency at a critical stage because of their artificial diet during captivity. Early captivity may interfere with imprinting mechanisms that later guide turtles to their nesting beaches. Hatchlings that spend their first few months in captivity may also develop behavioral abnormalities that could affect later social interactions and reproductive success. Curiously, no head-started sea turtle has ever been shown to have survived to adulthood. A major problem in evaluating this observation, however, is that no tag put onto a hatchling sea turtle has ever been recovered on an adult turtle. Obviously there is a critical need for more permanent tags. Bowen et al. (1994) suggested that perhaps a few Kemp's ridleys found to be nesting on the East Coast of the United States in 1989 and 1992—outside their historical nesting range—may actually be head-started turtles from a restoration project begun in the 1970s. Eggs were transported to Texas from the nesting beach in Tamaulipas, Mexico. The eggs were incubated indoors, then the hatchlings were allowed to run down to the surf zone at Padre Island, Texas. Afterward, the hatchlings were kept in captivity in Galveston for 9 to 12 months before they were released. The suggestion is that these turtles did not imprint correctly to the beach in Texas. Instead, abnormal migra-

Figure 15–19 **Collection of crocodile eggs, Papua New Guinea.** In Papua New Guinea, all wildlife resources are the property of traditional tribal land owners, who have exclusive rights to their use. One program developed as an incentive for protecting habitat and regulating use of crocodiles at sustainable levels, involves traditional land owners locating crocodile nests and selling the eggs to a centralized crocodile ranch. Sellers are paid in cash and also receive one hen's egg for every crocodile egg (to avoid protein deficiency). Cash is vital to these largely subsistence villagers, who use the money to pay school fees and to purchase school books. As a result of this program, adult female crocodiles are protected as a valuable resource for future economic gains. *(Photograph by Perran Ross.)*

Figure 15–20   **Farming of caimans in Colombia.** (a) Grow-out pens for juvenile caimans at the Monterrey Forestal farm in Zambrano, Colombia. (b) Preparation of food for the caimans involves passing fish through a grinder, adding mineral and vitamin supplements, and then packing the mixture into sausages. *(Photographs by John Thorbjarnarson.)*

(a)

tory behavior resulted in their ending up on the East Coast of the United States.

Another problem that must be faced when moving animals from one area to another, and especially if they have spent time in captivity, is the spread of disease (Figure 15–18). Some biologists speculate that an upper respiratory tract disease, mycoplasmosis (caused by *Mycoplasma* bacteria), that is proving fatal to many desert tortoises (*Gopherus agassizii*) in the western Mojave Desert may have been spread by captive individuals released into wild populations. Alternatively, the disease may have always been there, but outbreaks may be caused by stress. We don't know. A similar upper respiratory tract disease has been reported from several wild populations of gopher tortoises (*Gopherus polyphemus*) in Florida. Captive tortoises are known to have been released into at least one of these populations, suggesting that the disease may have been introduced by a captive individual. Now that the disease occurs in wild populations, great care must be taken not to subject healthy animals to the disease. The fear is that as developers are increasingly being allowed to build on land if they relocate tortoises, more tortoises will be exposed to and will succumb to the disease.

Farming and Ranching

One way to reduce the number of wild animals that are harvested for food, skins, or the pet trade is to

(b)

(a)

(b)

Figure 15–21   **Exploitation of tegu lizards (*Tupinambis*) in Argentina.** (a) Belly skin of *Tupinambis rufescens* (most valued part of the skin). (b) Environmental education poster aimed at tegu hunters, informing that small skins are illegal and that small individuals should be left alone. *(Photographs by Lee Fitzgerald.)*

raise species of economic value in captivity for future harvest or trade. Two types of operations exist: farms and ranches. Farming represents a closed system where the operations breed their own stock; other than the initial animals, none is taken from the wild.

The negative aspect of farming is that because all the animals marketed are captive-bred, there is often no incentive to protect wild populations or their natural habitat. In contrast, ranching involves taking eggs or hatchlings from the wild and raising them for market

(Figure 15–19). Because ranching depends on sustainable harvest, considerable emphasis is placed on protection of both the habitat and wild populations.

Several species of crocodilians are being raised successfully in captivity. The Monterrey Forestal caiman farm in Zambrano, Colombia, is one of the largest and best run of the farms in South America (Figure 15–20). Hundreds of thousands of skins are exported from this farm each year. High-quality skins of Nile crocodiles are being produced by farms and ranches in Africa, and the hope is that skins from these centers will eventually replace harvesting crocodiles from the wild completely. Many other countries also have successful operations that may take pressure off wild populations.

A good example of a project in which local people have been given economic incentives to participate in the conservation and management of a reptile and its habitat is work that was initiated in Panama by Dagmar Werner and the Smithsonian Tropical Research Institute and is currently continuing in Costa Rica with green iguanas (*Iguana iguana*). As mentioned earlier, iguanas are an important source of food in some areas of Central America, and unfortunately, populations are declining because of overexploitation. The iguana project involves a combination of farming and sustainable harvesting (Ocana et al. 1988). Local people farm iguanas, and then captive-raised juveniles are released into rural communities to repopulate areas where iguanas have been hunted out. Local people can then harvest the iguanas when they are large enough to eat (usually 2 to 3 years later). A large iguana can weigh more than 10 pounds, and customers are willing to pay more for iguana meat than for fish, chicken, pork, or beef. This project is helping to conserve declining populations of iguanas and is providing an incentive for local people to protect their tropical forests. If iguana farmers want to be successful, they have to provide food for their livestock—leaves in the treetops of the forest.

Captive breeding of many exotic species of amphibians and reptiles for the pet trade is taking pressure off wild populations. For example, most of the leopard geckos (*Eublepharis macularius*), green iguanas (*Iguana iguana*), and horned frogs (*Ceratophrys*) currently sold in pet stores are captive-bred. The same is true for many dart-poison frogs, chameleons, bearded dragon lizards, boas and pythons, and king snakes.

We have discussed the positive side of captive breeding, but there can be problems also. Mass breeding operations may generate health problems. Hatchling red-eared sliders (*Trachemys scripta*), raised on turtle farms in Louisiana, Mississippi, and Arkansas, used to be sold by the millions in pet stores across the United States. On the farms, the hatchlings were fed raw chicken that was often infected with *Salmonella* bacteria. Sale of the hatchlings in pet stores was banned in 1975 by the federal government because of health concerns: the hatchling turtles were infecting humans with *Salmonella*. Many children who handled their baby turtles, or touched contaminated water from the turtle bowl and then put their fingers into their mouths, infected themselves with salmonellosis. In humans, salmonellosis causes diarrhea. Most healthy adults recover from salmonellosis, but small children and babies are likely to become severely dehydrated, and sometimes die from the infection. Since the national ban took effect in 1975, the turtle farms have continued to export millions of hatchling red-eared sliders to Europe, Asia, and Latin America, where they are popular as pets. In the 5-year period between 1989 and 1994, an estimated 26 million hatchlings were exported from turtle farms in the United States to other countries. Although personnel at the farms no longer feed contaminated chicken to the turtles, the *Salmonella* bacteria are still present in the water and soil from so many decades of high levels of infestation—and turtles still become infected. Even treatment of the eggs and hatchlings with antibiotics hasn't eliminated the problem, as antibiotic-resistant strains of the bacteria have developed.

Sustainable Harvesting

To many people the goals of conservation and utilization of resources in need of protection seem mutually exclusive. Conservation, however, has long been closely tied to the value and utilization of resources. Unfortunately, too often harvesting is not done on a sustainable basis. Sustainable harvesting means removing individuals from a population in such a fashion that the resource is renewable—the population will continue indefinitely. Often primarily adult stages are harvested because they provide the most meat or leather. The life history characteristics of many species of amphibians and reptiles (especially snakes, turtles, and crocodilians), however, make exploitation of adults problematic. Many species have a relatively late age at first reproduction, a high egg and neonate

(a)

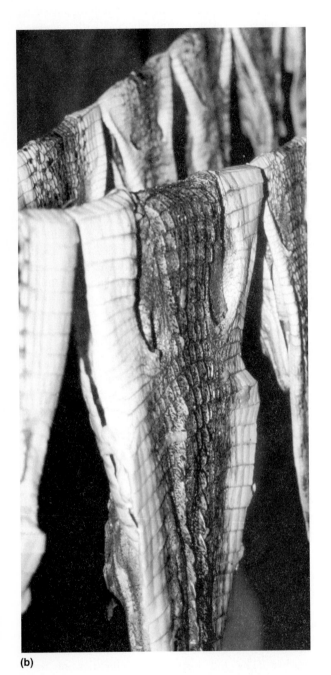

(b)

(c)

Figure 15–22   **Farming and captive breeding of crocodilians in Venezuela.** (a) Skinning ranched *Caiman crocodilus.* (b) Drying caiman skins after they have been pickled in a brine solution. (c) Concerned ranchers participating in Orinoco crocodile captive breeding program. *(Photographs by John Thorbjarnarson.)*

mortality coupled with high adult annual survival and longevity (thus a long reproductive life span), and infrequent breeding (annual, biennial, or even less frequent intervals). Therefore, harvest of adults will have a much larger negative impact on population size than will harvest of juveniles or subadults.

The sustainable use of wildlife can be an extremely valuable approach to conservation by providing economic incentives to local people and by protecting habitat. If the economic value of species is great enough (whether for tourism, subsistence use, or commercial trade), that fact alone will justify preservation of both the species and the habitats they occupy (Fitzgerald 1994). The CITES trade convention, the International Union for the Conservation of Nature, the World Wide Fund for Nature, and other conservation organizations endorse the idea that some species of wildlife can be harvested as renewable, commercial resources. The high price of legal reptile skins, for example, provides incentive for producer nations to manage their local populations for sustainable harvest. To do this requires information about the age-class structure and dynamics of each species' population so that individuals can be removed without adversely affecting the productivity of the population. Adequate enforcement of regulations and control of trade are also needed. In this way the resource can be sustained indefinitely without leading to extinction, and tanners can prepare skins for those people who desire reptilian handbags. Another benefit of regulated harvest is that for many programs, the income generated in the form of license fees and taxes is allocated directly to conservation.

Wild populations of two species of tegu lizards (*Tupinambis*) are currently heavily exploited in Argentina, Paraguay, and parts of Brazil and Bolivia (Figure 15–21). Each year more than 1.2 million skins are exported to the United States, Canada, Mexico, Asia, and Europe; many of these skins are destined to become cowboy boots (Fitzgerald et al. 1991). Trade in tegu skins is important to the economy of Paraguay and Argentina, with a net export value worth millions of dollars annually. Currently the governments of Paraguay and Argentina are involved in attempts to convert exploitation of tegu lizards into a system of sustainable use (Fitzgerald 1994). Laws in both countries prohibit the sale of skins from subadult tegus. Both countries are committed to monitoring populations through harvest monitoring programs, perma-

nently funded by a legally binding tax that is paid by the tanning industry.

Venezuela is such a success story in the conservation of crocodilians that it serves as a model for other countries (King 1989). During the 1950s and early 1960s American crocodile (*Crocodylus acutus*), Orinoco crocodile (*Crocodylus intermedius*), and common caiman (*Caiman crocodilus*) populations in the country were severely exploited. In response to population declines, in 1972 the government of Venezuela prohibited commercial hunting of these species. This prohibition was enforced for 10 years while wildlife biologists studied the ecology and population biology of these species. The only species that recovered was the caiman. Now caiman populations are carefully managed; quotas and size limits are set, and wildlife officials reassess the program annually (Figure 15–22). One unexpected outcome is that the five licensed tanners in the country formed a professional association to promote conservation of the caiman; they voluntarily pay $0.50 per hide they sell. The major ranch owners also formed an association of caiman producers, and they voluntarily pay $0.50 per hide harvested on their ranches. The combined money collected goes to a foundation that monitors the status and ecology of wild caiman populations and underwrites studies of other wildlife species. This program is a superb example of a successful synthesis of research, education, economic development, habitat preservation, and conservation. The concept of value-added conservation (the idea that part of the return from commercial harvest is funneled back into the conservation and management of the species and its habitat) is being used successfully in other parts of the world as well.

Summary

The world is experiencing a widespread decline in biodiversity, in large part due to human activity—habitat modification and destruction, pollution of the environment, and direct exploitation of flora and fauna. We have presented examples of the many ways that humans directly and indirectly affect populations of amphibians and reptiles. When we look at the numbers of local populations and entire species that have gone extinct in the past few decades, it is easy to become discouraged. On the other hand, when we realize all the efforts being made by scientists and nonscientists alike toward the conservation of amphibians and reptiles, we have reason to be hopeful.

The most successful conservation programs are those that involve the local people—programs that are aware of the cultural perceptions, values, and biases regarding flora and fauna, that take the needs of local people into consideration, and that provide economic incentives for preserving habitat and sustainable populations of amphibians and reptiles. The most important action we can take to protect amphibians and reptiles is protection of their habitats. Whenever possible, conservation efforts need to focus on ecosystems instead of individual species. In this way we can protect entire communities of organisms. We need to come up with innovative ways of protecting habitats that are in the best interests of local people. In addition, much more basic research is needed to understand the habitat requirements, reproductive biology, life history characteristics, dietary needs, and genetic diversity of threatened and endangered species.

Much is being done, but there is much more to do. Everyone can make an important contribution, from keeping only captive-bred animals as pets or helping toads cross roads, to becoming directly involved in scientific research and conservation efforts. We need to find ways to more effectively translate results of scientific research into law and policy. Any interested person can become involved in habitat preservation, in helping to draft local wetland protection laws, or in providing input on threatened or endangered species. If we all do our little bit to help, we can all continue to enjoy and better understand amphibians and reptiles.

Bibliography

Adams, M. J. 1993. Summer nests of the tailed frog (*Ascaphus truei*) from the Oregon coast range. *Northwestern Naturalist* 74:15–18.

Adler, K. 1970. The role of extraoptic photoreceptors in amphibian rhythms and orientation: a review. *Journal of Herpetology* 4:99–112.

Adler, K. 1976. Extraocular photoreception in amphibians. *Photochemistry and Photobiology* 23:275–298.

Adler, K. 1989. *Contributions to the History of Herpetology.* Society for the Study of Amphibians and Reptiles, Oxford, OH.

Ahlberg, P. E., and A. R. Milner. 1994. The origin and early diversification of tetrapods. *Nature* 368:507–514.

Alberch, P. 1981. Convergence and parallelism in foot morphology in the neotropical salamander genus *Bolitoglossa* I. Function. *Evolution* 35:84–100.

Alberch, P. 1983. Morphological variation in the neotropical salamander genus *Bolitoglossa. Evolution* 37:906–919.

Alberch, P., and J. Alberch. 1981. Heterochronic mechanisms of morphological diversification and evolutionary change in the neotropical salamander, *Bolitoglossa occidentalis* (Amphibia: Plethodontidae). *Journal of Morphology* 167:249–264.

Alberch, P., S. J. Gould, G. Oster, and D. B. Wake. 1979. Size and shape in ontogeny and phylogeny. *Paleobiology* 5:296–317.

Alberts, A. C. 1989. Ultraviolet visual sensitivity in desert iguanas: implications for pheromone detection. *Animal Behaviour* 38:129–137.

Alberts, A. C. 1992. Constraints on the design of chemical communication systems in terrestrial vertebrates. *American Naturalist* 139:S62–S89.

Alberts, A. C., J. A. Phillips, and D. I. Werner. 1993. Sources of intraspecific variability in the protein composition of lizard femoral gland secretions. *Copeia* 1993:775–781.

Alberts, A. C., D. C. Rostal, and V. A. Lance. 1994. Studies on the chemistry and social significance of chin gland secretions in the desert tortoise, *Gopherus agassizii. Herpetological Monographs* 8:116–124.

Albino, A. M. 1996. The South American fossil Squamata (Reptilia: Lepidosauria). *Münchner Geowissenschaftliche Abhandlungen* 30 A (*Geologie und Paläontologie*):185–202.

Alcala, A. C., and W. C. Brown. 1982. Reproductive biology of some species of *Philautus* (Rhacophoridae) and other Philippine anurans. *Kalikasan, Philippine Journal of Biology* 11:203–226.

Alford, R. A., and H. M. Wilbur. 1985. Priority effects in experimental pond communities: competition between *Bufo* and *Rana. Ecology* 66:1097–1105.

Altig, R. (1979): Toads Are Nice People. Gates House/MANCO. Columbia and Eldon, MO.

Altig, R., and G. F. Johnston. 1989. Guilds of anuran larvae: relationships among developmental modes, morphologies, and habitats. *Herpetological Monographs* 3:81–109.

Anderson, D. J., M. C. Stoyan, and R. E. Ricklefs. 1987. Why are there no viviparous birds? A comment. *American Naturalist* 130:941–947.

Anderson, R. A. 1993. An analysis of foraging in the lizard, *Cnemidophorus tigris.* Pages 83–116 in *Biology of Whiptail Lizards (genus Cnemidophorus),* edited by J. W. Wright and L. J. Vitt. Oklahoma Museum of Natural History, Norman, OK.

Andersson, M. 1994. *Sexual Selection.* Princeton University Press, Princeton, NJ.

Andrén, C. 1982. The role of the vomeronasal organs in the reproductive behavior of the adder *Vipera berus. Copeia* 1982:148–157.

Andrén, C., G. Nilson, M. Höggren, and H. Tegelström. 1997. Reproductive strategies and sperm competition in the adder, *Vipera berus*. Pages 129–141 in *Venomous Snakes: Ecology, Evolution and Snakebite*, edited by R. S. Thorpe, W. Wüster, and A. Malhotra. Clarendon Press, Oxford, UK.

Andrews, R. M. 1979. The lizard *Corytophanes cristatus*: an extreme "sit-and-wait" predator. *Biotropica* 11:136–139.

Andrews, R. M. 1988. Demographic correlates of variable egg survival for a tropical lizard. *Oecologia* 76:376–382.

Andrews, R. M., F. R. Mendez de la Cruz, and M. V. Santa Cruz. 1997. Body temperatures of female *Sceloporus grammicus:* thermal stress or impaired mobility? *Copeia* 1997:108–115.

Andrews, R. M., and F. H. Pough. 1985. Metabolism of squamate reptiles: allometric and ecological relationships. *Physiological Zoology* 58:214–231.

Andrews, R. M., F. H. Pough, A. Collazo, and A. de Quieroz. 1987. The ecological cost of morphological specialization: feeding by a fossorial lizard. *Oecologia* 73:139–145.

Andrews, R. M., and A. S. Rand. 1974. Reproductive effort in anoline lizards. *Ecology* 55:1317–1327.

Andrews, R. M., and A. S. Rand. 1983. Limited dispersal of juvenile *Anolis limifrons*. *Copeia* 1983:429–434.

Anthony, C. D., J. R. Mendelson III, and R. R. Simons. 1994. Differential parasitism by sex on plethodontid salamanders and histological evidence for structural damage to the nasolabial groove. *American Midland Naturalist* 132:302–307.

Anthony, J. 1955. Essai sur l'evolution anatomique de l'appareil venimeux des ophidiens. *Annales des Sciences Naturelles, Zoologie, Paris* 11:7–53.

Anthony, J., and J. Guibé. 1952. Les affinités anatomiques de *Bolyeria* et de *Casarea* (Boidés). *Mémoires de l'Institut Scientifique de Madagascar* ser. A, 12:189–201.

Arak, A. 1988. Callers and satellites in the natterjack toad: evolutionarily stable decision rules. *Animal Behaviour* 36:416–432.

Arano, B., G. Llorente, M. Garcia-Paris, and P. Herrero. 1995. Species translocation menaces Iberian waterfrogs. *Conservation Biology* 9:196–198.

Arnold, E. N. 1979. Indian Ocean giant tortoises: their systematics and island adaptations. *Philosophical Transactions of the Royal Society of London* B 286:127–145.

Arnold, E. N. 1984. Evolutionary aspects of tail shedding in lizards and their relatives. *Journal of Natural History* 18:127–169.

Arnold, E. N. 1984. Variation in the cloacal and hemipenial muscles of lizards and its bearing on their relationships. *Symposia of the Zoological Society of London* 52:47–85.

Arnold, E. N. 1988. Caudal autotomy as a defense. Pages 235–273 in *Biology of the Reptilia, Vol. 16 (Ecology B, Defense and Life History)*, edited by C. Gans and R. B. Huey. Alan R. Liss, Inc., New York, NY.

Arnold, E. N. 1989. Systematics and adaptive radiation of Equatorial African lizards assigned to the genera *Adolfus, Bedriagaia, Gastropholis, Holaspis,* and *Lacerta* (Reptilia: Lacertidae). *Journal of Natural History* 23:525–555.

Arnold, E. N. 1989. Towards a phylogeny and biogeography of the Lacertidae: relationships within an Old-World family of lizards derived from morphology. *Bulletin of the British Museum (Natural History)* 55:209–257.

Arnold, E. N. 1991. Relationships of the South African lizards assigned to *Aporosaura, Meroles* and *Pedioplanis* (Reptilia: Lacertidae). *Journal of Natural History* 25:783–807.

Arnold, E. N. 1995. Identifying the effects of history on adaptation: origins of different sand-diving techniques in lizards. *Journal of Zoology* 235:351–388.

Arnold, S. J. 1977. The evolution of courtship behavior in New World salamanders with some comments on Old World salamandrids. Pages 141–183 in *The reproductive biology of amphibians*, edited by D. H. Taylor and S. I. Gutman. Plenum press, New York, NY.

Arnold, S. J. 1981a. Behavioral variation in natural populations. I. Phenotypic, genetic and environmental correlations between chemoreceptive responses to prey in the garter snake, *Thamnophis elegans*. *Evolution* 35:489–509.

Arnold, S. J. 1981b. Behavioral variation in natural populations. II. The inheritance of a feeding response in crosses between geographic races of the garter snake, *Thamnophis elegans*. *Evolution* 35:510–515.

Arnold, S. J. 1982. A quantitative approach to antipredator performance: salamander defense against snake attack. *Copeia* 1982:247–253.

Arnold, S. J. 1983. Sexual selection: the interface of theory and empiricism. Pages 67–107 in *Mate Choice*, edited by P. Bateson. Cambridge University Press, New York, NY.

Arnold, S. J. 1987. Genetic correlation and the evolution of physiology. Pages 189–215 in *New Directions in Ecological Physiology*, edited by M. E. Feder, A. F. Bennett, W. W. Burggren, and R. B. Huey. Cambridge University Press, Cambridge, UK.

Arnold, S. J., and D. Duvall. 1994. Animal mating systems: a synthesis based on selection theory. *American Naturalist* 143:317–348.

Arnold, S. J., and R. J. Wassersug. 1978. Differential predation on metamorphic anurans of garter snakes (*Thamnophis*): social behavior as a possible defense. *Ecology* 59:1014–1022.

Arntzen, J. W., and M. Sparreboom. 1989. A phylogeny for the Old World newts, genus *Triturus*: biochemical and behavioural data. *Journal of Zoology, London* 219:645–664.

Arntzen, J. W., and J. M. Szymura. 1984. Genetic differentiation between African and European midwife toads (*Alytes*, Discoglossidae). *Bijdragen tot de Dierkunde* 54:157–162.

Ashley-Ross, M. A. 1994. Hindlimb kinematics during terrestrial locomotion in a salamander (*Dicamptodon tenebrosus*). *Journal of Experimental Biology* 193:255–283.

Auffenberg, W. 1963. A note on the drinking habits of some land tortoises. *Animal Behaviour* 11:72–73.

Auffenberg, W. 1965. Sex and species discrimination in two sympatric South American tortoises. *Copeia* 1965: 335–342.

Auffenberg, W. 1966. On the courtship of *Gopherus polyphemus*. *Herpetologica* 22:113–117.

Auffenberg, W. 1977. Display behavior in tortoises. *American Zoologist* 17:241–250.

Auffenberg, W. 1981. *Behavioral ecology of the Komodo dragon*. University Presses of Florida, Gainesville, FL.

Auffenberg, W. 1982. Feeding strategy of the Caicos ground iguana, *Cyclura carinata*. Pages 84–116 in *Iguanas of the World*, edited by G. M. Burghardt and A. S. Rand. Noyes Publications, Park Ridge, NJ.

Auffenberg, W. 1988. *Gray's Monitor Lizard*. University Presses of Florida, Gainesville, FL.

Auffenberg, W. 1994. *The Bengal Monitor*. University Presses of Florida, Gainesville, FL.

Austin, C. C., and H. B. Shaffer. 1992. Short-, medium-, and long-term repeatability of locomotor performance in the tiger salamander *Ambystoma californiense*. *Functional Ecology* 6:145–153.

Autumn, K., R. B. Weinstein, and R. J. Full. 1994. Low cost of locomotion increases performance at low temperature in a nocturnal lizard. *Physiological Zoology* 67: 238–262.

Avery, R. A. 1976. Thermoregulation, metabolism, and social behaviour in Lacertidae. Pages 245–260 in *Morphology and Biology of Reptiles (Linnean Society Symposium Series 3)*, edited by A. d'A. Bellairs and C. B. Cox. Academic Press, London, UK.

Avery, R. A. 1982. Field studies of body temperatures and thermoregulation. Pages 93–166 in *Biology of the Reptilia, Vol. 12 (Physiology C, Physiological Ecology)*, edited by C. Gans and F. H. Pough, Academic Press, London, UK.

Avila-Pires, T. C. S. 1995. Lizards of Brazilian Amazonia (Reptilia: Squamata). *Zoologische Verhandelingen, Leiden* 299:1–706.

Ayres, J. M. 1989. Dept-for-equity swaps and the conservation of tropical rain forests. *Trends in Ecology and Evolution* 4:331–332.

Bachmeyer, H., H. Michl, and B. Roos. 1967. Chemistry of cytotoxic substances in amphibian toxins. Pages 395–399 in *Animal Toxins*, edited by F. E. Russell and P. R. Saunders. Pergamon Press, Oxford.

Badenhorst, A. 1978. The development and the phylogeny of the organ of Jacobson and the tentacular apparatus of *Ichthyophis glutinosus* (Linné). *Annale Universiteit van Stellenbosch* 1:1–26.

Baéz, A. M. 1996. The fossil record of the Pipidae. Pages 329–347 in *The Biology of Xenopus*, edited by R. C. Tinsley and H. R. Kobel. Clarendon Press, Oxford.

Baéz, A. M., and N. G. Basso. 1996. The earliest known frogs of the Jurassic of South America: review and cladistic appraisal of their relationships. *Münchner Geowissenschaftliche Abhandlungen* 30 A (Geologie und Paläontologie):131–158.

Baker, R. E., and J. C. Gillingham. 1983. An analysis of courtship behavior in Blanding's turtle, *Emydoidea blandingi*. *Herpetologica* 39:166–173.

Bakken, G. S. 1992. Measurement and application of operative and standard operative temperatures in ecology. *American Zoologist* 32:194–216.

Banks, B., and T. J. C. Beebee. 1987. Factors influencing breeding site choice by the pioneering amphibian *Bufo calamita*. *Holarctic Ecology* 10:14–21.

Banks, B., T. J. C. Beebee, and J. S. Denton. 1993. Long-term management of a natterjack toad (*Bufo calamita*) population in southern Britain. *Amphibia Reptilia* 14:155–168.

Barrett, R. 1970. The pit organs of snakes. Pages 277–300 in *Biology of the Reptilia, Vol. 2 (Morphology B)*, edited by C. Gans. Academic Press, New York, NY.

Barry, F. E., P. J. Weatherhead, and D. P. Philipp. 1992. Multiple paternity in a wild population of northern water snakes, *Nerodia sipedon*. *Behavioral Ecology and Sociobiology* 30:193–199.

Bartholomew, G. A. 1982. Physiological control of body temperature. Pages 167–211 in *Biology of the Reptilia, Vol. 12 (Physiology C, Physiological Ecology)*, edited by C. Gans and F. H. Pough, Academic Press, London, UK.

Bartholomew, G. A., and V. A. Tucker. 1963. Control of changes in body temperature, metabolism, and circulation by the agamid lizard, *Amphibolurus barbatus*. *Physiological Zoology* 36:199–218.

Bartram, W. 1791. *Travels Through North and South Carolina, Georgia, East and West Florida*. James and Johnson, Philadelphia, PA.

Bates, H. W. 1876. *The Naturalist on the River Amazons*. John Murray, London, UK.

Bauer, A. M. 1992. Lizards. Pages 126–173 in *Reptiles and Amphibians*, edited by H. G. Cogger and R. G. Zweifel. Smithmark, New York, NY.

Bauer, A. M., and A. P. Russell. 1991. Pedal specialisations in dune-dwelling geckos. *Journal of Arid Environments* 20:43–62.

Bauwens, D., and R. Díaz-Uriarte. 1997. Covariation of life-history traits in lacertid lizards: a comparative study. *American Naturalist* 149:91–111.

Bauwens, D., and C. Thoen. 1981. Escape tactics and vulnerability to predation associated with reproduction in the lizard *Lacerta vivipara*. *Journal of Animal Ecology* 50:733–744.

Baverstock, P. R., D. King, M. King, J. Birrell, and M. Krieg. 1993. The evolution of species of the Varanidae: microcomplement fixation analysis of serum albumins. *Australian Journal of Zoology* 41:621–638.

Beachy, C. K., and R. C. Bruce. 1992. Lunglessness in plethodontid salamanders is consistent with the hypothesis of a mountain stream origin: a response to Ruben and Boucot. *American Naturalist* 139:839–847.

Beaupre, S. J. 1995. Comparative ecology of the mottled rock rattlesnake, *Crotalus lepidus*, in Big Bend National Park. *Herpetologica* 51:45–56.

Beaupre, S. J. 1996. Field metabolic rate, water flux, and energy budgets of mottled rock rattlesnakes, *Crotalus lepidus*, from two populations. *Copeia* 1996:319–329.

Beavers, S. C., and E. R. Cassano. 1996. Movements and dive behavior of a male sea turtle (*Lepidochelys olivacea*) in the eastern tropical Pacific. *Journal of Herpetology* 30:97–104.

Beck, D. D. 1990. Ecology and behavior of the Gila monster in southwestern Utah. *Journal of Herpetology* 24:54–68.

Beck, D. D., and C. H. Lowe. 1991. Ecology of the beaded lizard, *Heloderma horridum*, in a tropical dry forest in Jalisco, Mexico. *Journal of Herpetology* 25:395–406.

Beebee, T. J. C., R. J. Flower, A. C. Stevenson, S. T. Patrick, P. G. Appleby, C. Fletcher, C. Marsh, J. Natkanski, B. Rippey, and R. W. Battarbee. 1990. Decline of the natterjack toad *Bufo calamita* in Britain: palaeoecological, documentary and experimental evidence for breeding site acidification. *Biological Conservation* 53:1–20.

Belkin, D. A., and C. Gans. 1968. An unusual chelonian feeding niche. *Ecology* 49:768–769.

Bell, B. D. 1978. Observations on the ecology and reproduction of the New Zealand leiopelmid frogs. *Herpetologica* 34:340–354.

Bell, D. 1989. Functional anatomy of the chameleon tongue. *Zoologische Jarbuch Anatomie* 119:313–336.

Bellairs, A. 1970. *The Life of Reptiles, 2 vols.* Universe Books, New York, NY.

Bellairs, A., and G. Underwood. 1951. The origin of snakes. *Biological Reviews of the Cambridge Philosophical Society* 26:193–237.

Bels, V. L., and Y. J.-M. Crama. 1994. Quantitative analysis of the courtship and mating behavior in the loggerhead musk turtle *Sternotherus minor* (Reptilia: Kinosternidae) with comments on courtship behavior in turtles. *Copeia* 1994:676–684.

Belvedere, P., L. Colombo, C.Giacoma, C.Malacarne, and G. E. Andreoletti. 1988. Comparative ethological and biochemical aspects of courtship pheromones in European newts. *Monitore Zoologico Italiano* 22:397–403.

Bemis, W. E., K. Schwenk, and M. H. Wake. 1983. Morphology and function of the feeding apparatus in *Dermophis mexicanus* (Amphibia: Gymnophiona). *Zoological Journal of the Linnean Society* 77:75–96.

Bennett, A. F. 1987. Interindividual variability: an underutilized resource. Pages 147–169 in *New Directions in Ecological Physiology*, edited by M. E. Feder, A. F. Bennett, W. W. Burggren, and R. B. Huey. Cambridge University Press, Cambridge, UK.

Bennett, A. F., T. T. Gleeson, and G. C. Gorman. 1981. Anaerobic metabolism in a lizard (*Anolis bonairensis*) under natural conditions. *Physiological Zoology* 54:237–241.

Bennett, A. F., and R. B. Huey. 1990. Studying the evolution of physiological performance. Pages 251–284 in *Oxford Surveys in Evolutionary Biology*, edited by D. J. Futuyma and J. Antonovics. Oxford University Press, Oxford, UK.

Benton, M. J. 1990. Phylogeny of the major tetrapod groups: morphological data and divergence dates. *Journal of Molecular Evolution* 30:409–424.

Benton, M. J., and J. M. Clark. 1988. Archosaur phylogeny and the relationships of the Crocodylia. Pages 295–338 in *The Phylogeny and Classification of the Tetrapods, Volume 1: Amphibians, Reptiles, Birds. Systematics Association Special Volume no. 35A*, edited by M. J. Benton. Clarendon Press, Oxford, UK.

Berger, L. 1977. Systematics and hybridization in the *Rana esculenta* complex. Pages 367–410 in *The Reproductive Biology of Amphibians*, edited by D. H. Taylor and S. I. Guttman. Plenum Press, New York, NY.

Bergeron, J. M., D. Crews, and J. A. McLachlan. 1994. PCBs as environmental estrogens: turtle sex determination as a biomarker of environmental contamination. *Environmental Health Perspectives* 102:780–781.

Berry, J. F., and R. Shine. 1980. Sexual size dimorphism and sexual selection in turtles (Order Testudines). *Oecologia* 44:185–191.

Berry, K. H. 1986. Desert tortoise (*Gopherus agassizii*) research in California, 1976–1985. *Herpetologica* 42:62–67.

Berven, K. A. 1981. Mate choice in the wood frog, *Rana sylvatica*. *Evolution* 35:707–722.

Berven, K. A., and T. A. Grudzien. 1990. Dispersal in the wood frog (*Rana sylvatica*): implications for genetic population structure. *Evolution* 44:2047–2056.

Bevier, C. R. 1995. Biochemical correlates of calling activity in Neotropical frogs. *Physiological Zoology* 68:1118–1142.

Bezy, R. L. 1988. The natural history of the night lizards, Family Xantusiidae. Pages 1–12 in *Proceedings of the Conference on California Herpetology*, edited by H. F. De Lisle, P. R. Brown, and B. M. McGurty. Southwestern Herpetologists Society.

Bezy, R. L. 1989a. Morphological differentiation in unisexual and bisexual xantusiid lizards of the genus *Lepidophyma* in Central America. *Herpetological Monographs* 3:61–80.

Bezy, R. L. 1989b. Night lizards: the evolution of habitat specialists. *Terra* 28:29–34.

Bickham, J. W., and J. L. Carr. 1983. Taxonomy and phylogeny of the higher categories of cryptodiran turtles based on a cladistic analysis of chromosomal data. *Copeia* 1983:918–932.

Bickham, J. W., P. K. Tucker, and J. M. Legler. 1985. Diploid-triploid mosaicism: an unusual phenomenon in side-necked turtles (*Platemys platycephala*). *Science* 227: 1591–1593.

Billo, R., and M. H. Wake. 1987. Tentacle development in *Dermophis mexicanus* (Amphibia, Gymnophiona) with an hypothesis of tentacle origin. *Journal of Morphology* 192:101–111.

Bjorndal, K. A. 1985. Nutritional ecology of sea turtles. *Copeia* 1985:736–751.

Bjorndal, K. A., and A. B. Bolten. 1992. Body size and digestive efficiency in a herbivorous freshwater turtle: advantages of small bite size. *Physiological Zoology* 65: 1028–1039.

Blackburn, D. G. 1982. Evolutionary origins of viviparity in the reptilia. I. Sauria. *Amphibia Reptilia* 3:185–205.

Blackburn, D. G. 1992. Convergent evolution of viviparity, matrotrophy, and specializations for fetal nutrition in reptiles and other vertebrates. *American Zoologist* 32: 313–321.

Blackburn, D. G. 1993. Chorioallantoic placentation in squamate reptiles: structure, function, development, and evolution. *Journal of Experimental Zoology* 266:414–430.

Blackburn, D. G., and H. E. Evans. 1986. Why are there no viviparous birds? *American Naturalist* 128:165–190.

Blackburn, D. G., L. G. Vitt, and C. A. Beuchat. 1984. Eutherian-like reproductive specializations in a viviparous reptile. *Proceedings of the National Academy of Sciences, USA* 81:4860–4863.

Blair, W. F. 1963. Acoustic behaviour of amphibians. Pages 694–708 in *Acoustic Behaviour of Animals*, edited by R. G. Busnel. Elsevier, New York, NY.

Blair, W. F. 1972. *Evolution in the genus Bufo*. University of Texas Press, Austin, TX.

Blake, A. J. D. 1973. Taxonomy and relationships of myobatrachine frogs (Leptodactylidae): a numerical approach. *Australian Journal of Zoology* 21:119–149.

Blaustein, A. R., P. D. Hoffman, D. G. Hokit, J. M. Kiesecker, S. C. Walls, and J. B. Hays. 1994. UV repair and resistance to solar UV-B in amphibian eggs: a link to population declines? *Proceedings of the National Academy of Sciences, USA* 91:1791–1795.

Blaustein, A. R., D. G. Hokit, and R. K. O'Hara. 1994. Pathogenic fungus contributes to amphibian losses in the Pacific Northwest. *Biological Conservation* 67:251–254.

Blaustein, A. R., and D. B. Wake. 1990. Declining amphibian populations: a global phenomenon? *Trends in Ecology and Evolution* 5:203–204.

Blaustein, A. R., and D. B. Wake. 1995. The puzzle of declining amphibian populations. *Scientific American* 272:56–61.

Blaustein, A. R., D. B. Wake, and W. P. Sousa. 1994. Amphibian declines: judging stability, persistence, and susceptibility of populations to local and global extinctions. *Conservation Biology* 8:60–71.

Blommers-Schlosser, R. M. A. 1975. Observations on the larval development of some Malagasy frogs, with notes on their ecology and biology (Anura: Dyscophinae, Scaphiophryninae and Cophylinae). *Beaufortia* 24:7–26.

Boake, C. R. B. (editor). 1994. *Quantitative Genetic Studies of Behavioral Evolution*. University of Chicago Press, Chicago, IL.

Bock, B. C., and A. S. Rand. 1985. Seasonal migration and nesting site fidelity in the green iguana. *Contributions in Marine Science* 27:435–443.

Bogert, C. M. 1943. Dentitional phenomena in cobras and other elapids with notes on adaptive modifications of fangs. *Bulletin of the American Museum of Natural History* 81:285–360.

Bogert, C. M. 1947. A field study of homing in the Carolina toad. *American Museum Novitates Number* 1355:1–24.

Bogert, C. M. 1960. The influence of sound on the behavior of amphibians and reptiles. Pages 137–320 in *Animal Sounds and Communication*, edited by W. E. Lanyon and W. N. Tavolga. American Institute of Biological Sciences, Washington, DC.

Bogert, C. M., and R. Martín del Campo. 1956. The Gila Monster and its allies, the relationships, habits, and behavior of the lizards of the Family Helodermatidae. *Bulletin of the American Museum of Natural History* 109:1–238.

Böhme, W. 1995. Hemiclitoris discovered: a fully differentiated erectile structure in female monitor lizards (*Varanus* spp.) (Reptilia: Varanidae). *Journal of Zoological Systematics and Evoulutionary Research* 33:129–132.

Bolt, J. R. 1977. Dissorophoid relationships and ontogeny, and the origin of the Lissamphibia. *Journal of Paleontology* 51:235–249.

Bolt, J. R. 1979. *Amphibamus grandiceps* as a juvenile dissorophid: evidence and implications. Pages 529–563 in *Mazon Creek Fossils*, edited by M. H. Niteki. Academic Press, New York, NY.

Bolt, J. R. 1991. Lissamphibian origins. Pages 194–222 in *Origins of the higher groups of tetrapods, controversy and consensus*, edited by H.-P. Schultze and L. Trueb. Comstock Publishing Associates, Ithaca, NY.

Bolt, J. R., and R. E. Lombard. 1985. Evolution of the amphibian tympanic ear and the origin of frogs. *Biological Journal of the Linnaean Society* 24:83–99.

Borsuk-Bialynicka, M., and S. M. Moody. 1984. Priscagaminae, a new subfamily of the Agamidae (Sauria) from the late Cretaceous of the Gobi Desert. *Acta Palaeontological Polonica* 29:51–81.

Bosa, M. A., D. Jukofsky, and C. Wille. 1995. Costa Rica is a laboratory, not ecotopia. *Conservation Biology* 9:684–685.

Bossy, K. V. 1976. Morphology, paleoecology, and evolutionary relationships of the Pennsylvanian urocordylid nectrideans (Subclass Lepospondyli, Class Amphibia). Unpublished thesis, Yale University.

Boulenger, G. A. 1902. Further notes on the African batrachians *Trichobatrachus* and *Campsosteonyx*. Proceedings of the Zoological Society (London) 1901:709–710.

Boulenger, G. A. 1921. *Monograph of the Lacertidae, 2 volumes.* British Museum (Natural History), London, UK.

Bourgeois, M. 1965. *Contribution a la morphologie comparée du crâne des ophidiens de L'Afrique Centrale.* Unpubl. Ph.D. Diss., Universite Officielle du Congo.

Boutilier, R. G., D. F. Stiffler, and D. P. Toews. 1992. Exchange of respiratory gases, ions, and water in amphibious and aquatic amphibians. Pages 81–124 in *Environmental Physiology of the Amphibians*, edited by M. E. Feder and W. W. Burggren. University of Chicago Press, Chicago, IL.

Bowen, B., J. C. Avise, J. I. Richardson, A. B. Meylan, D. Margaritoulis, and S. R. Hopkins-Murphy. 1993. Population structure of loggerhead turtles (*Caretta caretta*) in the northwestern Atlantic Ocean and Mediterranean Sea. *Conservation Biology* 7:834–844.

Bowen, B. W., T. A. Conant, and S. R. Hopkins-Murphy. 1994. Where are they now? The Kemp's ridley headstart project. *Conservation Biology* 8:853–856.

Bowen, B. W., and S. A. Karl. 1997. Population genetics, phylogeography, and molecular evolution. Pages 29–50 in *The Biology of Sea Turtles*, edited by P. L. Lutz and J. A. Musick. CRC Press, Boca Raton, FL.

Bowen, B. W., A. B. Meylan, and J. C. Avise. 1989. An odyssey of the green sea turtle: Ascension Island revisited. *Proceedings of the National Academy of Sciences, USA* 86:573–576.

Bowen, B. W., A. B. Meylan, J. P. Ross, C. J. Limpus, G. H. Balazs, and J. C. Avise. 1992. Global population structure and natural history of the green turtle (*Chelonia mydas*) in terms of matriarchal phylogeny. *Evolution* 46:865–881.

Bowen, B. W., W. S. Nelson, and J. C. Avise. 1993. A molecular phylogeny for marine turtles: trait mapping, rate assessment, and conservation relevance. *Proceedings of the National Academy of Sciences, USA* 90:5574–5577.

Bowler, J. K. (1977): 1977. Longevity of reptiles and amphibians in North American collections. Society for the Study of Amphibians and Reptiles, Herpetological circular no. 6. Oxford, OH.

Boy, J. A. 1971. Zur Problematik der Branchiosaurier (Amphibia, Karbon-Perm). *Palaontologische Zeitschrift* 45: 107–119.

Boycott, R. C. 1988. Description of a new species of *Heleophryne* Sclater, 1899 from the Cape Province, South Africa (Anura: Heleophrynidae). *Annals of the Cape Provincial Museums (Natural History)* 16:309–319.

Bradford, D. F. 1990. Incubation time and rate of embryonic development in amphibians: the influence of ovum size, temperature, and reproductive mode. *Physiological Zoology* 63:1157–1180.

Bradford, D. F., F. Tabatabai, and D. M. Graber. 1993. Isolation of remaining populations of the native frog, *Rana mucosa*, by introduced fishes in Sequoia and Kings Canyon National Parks, California. *Conservation Biology* 7:882–888.

Bradshaw, S. D. 1986. *Ecophysiology of Desert Reptiles.* Academic Press, Sydney, Australia.

Bramble, D. M. 1974a. Emydid shell kinesis: biomechanics and evolution. *Copeia* 1974:707–727.

Bramble, D. M. 1974b. Occurrence and significance of the os transiliens in gopher tortoises. *Copeia* 1974: 102–109.

Bramble, D. M., and D. B. Wake. 1985. Feeding mechanisms of lower tetrapods. Pages 230–261 in *Functional Vertebrate Morphology*, edited by M. Hildebrand, D. M. Bramble, K. F. Liem, and D. B. Wake. Belknap Press, Cambridge, MA.

Brattstrom, B. H. 1979. Amphibian temperature regulation studies in the field and laboratory. *American Zoologist* 19:345–156.

Breckenridge, R., and M. J. Dufton. 1987. The structural evolution of cobra venom cytotoxins. *Journal of Molecular Evolution* 26:274–283.

Breden, F. 1987. The effect of post-metamorphic dispersal on the population genetic structure of Fowler's toad, *Bufo woodhousei fowleri. Copeia* 1987:386–395.

Breidenbach, C. H. 1990. Thermal cues influence strikes in pitless vipers. *Journal of Herpetology* 24:448–450.

Brillet, C., and M. Paillette 1991: Acoustic signals of the nocturnal lizard *Gekko gecko*: analysis of the long complex sequence. Bioacoustics 3:33–44.

Brinkman, D. 1980a. The hind limb step cycle of *Caiman sclerops* and the mechanics of the crocodile tarsus and metatarsus. *Canadian Journal of Zoology* 58:2187–2200.

Brinkman, D. 1980b. Structural correlates of tarsal and metatarsal functioning in *Iguana* (Lacertiilia; Iguanidae) and other lizards. *Canadian Journal of Zoology* 58: 277–289.

Brinkman, D. 1981. The hind limb step cycle of *Iguana* and primitive reptiles. *Journal of Zoology*, 193:91–103.

Britson, C. A., and W. H. N. Gutzke. 1993. Antipredator mechanisms of hatchling freshwater turtles. *Copeia* 1993:435–440.

Broadley, D. G., and C. Gans. 1978. Southern forms of *Chirindia* (Amphisbaenia, Reptilia). *Annals of the Carnegie Museum of Natural History* 47:29–51.

Brodie, E. D., Jr. 1977. Salamander antipredator postures. *Copeia* 1977:523–535.

Brodie, E. D., Jr. 1983. Antipredator adaptations of salamanders: evolution and convergence among terrestrial species. Pages 109–133 in *Plant, Animal, and Microbial Adaptations to Terrestrial Environment*, edited by N. S.

Margaris, M. Arianoutsour-Faraggitaki, and R. J. Reiter. Plenum Press, New York, NY.

Brodie, E. D., Jr., and E. D. Brodie III. 1980. Differential avoidance of mimetic salamanders by free-ranging birds. *Science* 208:181–183.

Brodie, E. D. Jr., and R. R. Howard. 1973. Experimental study of Batesian mimicry in the salamanders *Plethodon jordani* and *Desmognathus ochrophaeus*. *American Midland Naturalist* 90:38–46.

Brodie, E. D., Jr., and N. J. Smatresk. 1990. The antipredator arsenal of fire salamanders: spraying of secretions from highly pressurized dorsal skin glands. *Herpetologica* 46:1–7.

Brodie, E. D., III. 1992. Correlational selection for color pattern and antipredator behavior in the garter snake *Thamnophis ordinoides*. *Evolution* 46:1284–1298.

Brodie, E. D., III. 1993. Differential avoidance of coral snake banded patterns by free-ranging avian predators in Costa Rica. *Evolution* 47:227–235.

Brodie, E. D., III, and E. D. Brodie, Jr. 1990. Tetrodotoxin resistance in garter snakes: an evolutionary response of predators to dangerous prey. *Evolution* 44:651–659.

Brown, H. A. 1989. Developmental anatomy of the tailed frog (*Ascaphus truei*): a primitive frog with large eggs and slow development. *Journal of Zoology, London* 217:525–537.

Brown, W. S., and W. S. Parker. 1976. Movement ecology of *Coluber constrictor* near communal hibernacula. *Copeia* 1976:225–242.

Brust, D. G. 1993. Maternal brood care by *Dendrobates pumilio*: a frog that feeds its young. *Journal of Herpetology* 27:96–98.

Buchanan, B. W., and R. C. Taylor. 1996. Lightening the load: micturition enhances jumping performance of squirrel treefrogs. *Journal of Herpetology* 30:410–413.

Buck, S. V. 1997. Florida's herp trade: a collector's paradise and . . . a land exploited. *Reptile & Amphibian Magazine*. Jan/Feb 1997:72–81.

Buffetaut, E. 1985. Zoogeographical history of African crocodilians since the Triassic. Pages 453–469 in *Proceedings of the International Symposium on African Vertebrates; Systematics, Phylogeny and Evolutionary Ecology*, edited by K.-L. Schuchmann. Museum Alexander Koenig, Bonn, Germany.

Buhlmann, K. A. 1996. Legislation and Conservation — USA: U.S. Fish and Wildlife Service. *Herpetological Review* 27:54–55.

Bull, C. M. 1994. Population dynamics and pair fidelity in sleepy lizards. Pages 159–174 in *Lizard Ecology. Histori-cal and Experimental Perspectives*, edited by L. J. Vitt and E. R. Pianka. Princeton University Press, Princeton, NJ.

Bull, J. J. 1983. *Evolution of Sex Determining Mechanisms*. Benjamin/Cummings Publishing Co., Menlo Park, CA.

Bullock, D. J. 1986. The ecology and conservation of reptiles on Round Island and Gunner's Quoin, Mauritius. *Biological Conservation* 37:135–156.

Burchell, W. J. 1822. *Travels in the Interior of Southern Africa. Robert MacLehose and Company, Glasgow, Scotland (1953 reprint)*. Batchworth Press, London, UK.

Burggren, W. W. 1988. Role of the central circulation in regulation of cutaneous gas exchange. *American Zoologist* 28:985–998.

Burggren, W. W. 1989: Lung structure and function: Amphibians. Pages 153–192 in *Comparative Pulmonary Physiology: Current Concepts vol 39 (Lung Biology in Health and Disease)*, edited by S. C. Wood, Marcel Dekker, New York, NY.

Burggren, W. W., and J. J. Just. 1992. Developmental changes in physiological systems. Pages 467–530 in *Environmental Physiology of the Amphibians*, edited by M. E. Feder and W. W. Burggren. University of Chicago Press, Chicago, IL.

Burghardt, G. M. 1967. Chemical-cue preferences of inexperienced snakes: comparative aspect. *Science* 157:718–721.

Burghardt, G. M. 1977. Of iguanas and dinosaurs: social behavior and communication in neonate reptiles. *American Zoologist* 17:177–190.

Burghardt, G. M., H. W. Greene, and A. S. Rand. 1977. Social behavior in hatchling green iguanas: life at a reptile rookery. *Science* 195:689–691.

Burke, A. C. 1989. Development of the turtle carapace: implications for the evolution of a novel bauplan. *Journal of Morphology* 199:363–378.

Burke, A. C. 1991. The development and evolution of the turtle body plan: inferring intrinsic aspects of the evolutionary process from experimental embryology. *American Zoologist* 31:616–627.

Burke, V. J., and J. W. Gibbons. 1995. Terrestrial buffer zones and wetland conservation: a case study of freshwater turtles in a Carolina Bay. *Conservation Biology* 9:1365–1369.

Burkey, T. V. 1989. Extinction in nature reserves: the effect of fragmentation and the importance of migration between reserve fragments. *Oikos* 55:75–81.

Burkey, T. V. 1995. Extinction rates in archipelagoes: implications for populations in fragmented habitats. *Conservation Biology* 9:527–541.

Burrage, B. R. 1974. Population structure in Agama agama and Cordylus cordylus cordylus in the vicinity of De Kelders, Cape Province. *Annals of the South African Museum* 66:1–23.

Burton, T. M., and G. E. Likens. 1975. Energy flow and nutrient cycling in salamander populations in the Hubbard Brook Experimental Forest. *Ecology* 56:1068–1080.

Busack, S. D. 1986. Biochemical and morphological differentiation in Spanish and Moroccan populations of *Discoglossus* and the description of a new species from southern Spain (Amphibia: Anura: Discoglossidae). *Annals of the Carnegie Museum of Natural History* 55:41–61.

Busby, A. B., III. 1989. Form and function of the feeding apparatus of *Alligator mississippiensis*. *Journal of Morphology* 202:99–127.

Buttemer, W. A. 1990. Effect of temperature on evaporative water loss of the Australian tree frogs *Litoria caerulea* and *Litoria chloris*. *Physiological Zoology* 63:1043–1057.

Cadle, J. E. 1983. Problems and approaches in the interpretation of the evolutionary history of venomous snakes. *Memorias do Instituto Butantan* 46 (1982):255–274.

Cadle, J. E. 1984. Molecular systematics of neotropical xenodontine snakes. III. Overview of xenodontine phylogeny and the history of New World snakes. *Copeia* 1984:641–652.

Cadle, J. E. 1987. The geographic distribution of snakes: problems in phylogeny and zoogeography. Pages 77–105 in *Snakes: Ecology and Evolutionary Biology*, edited by R. A. Seigel, J. T. Collins, and S. S. Novak. Macmillan Publ. Co., New York, NY.

Cadle, J. E. 1988. Phylogenetic relationships among advanced snakes: a molecular perspective. *University of California Publications in Zoology* 119:1–70.

Cadle, J. E. 1992. Phylogenetic relationships among vipers: immunological evidence. Pages 41–48 in *Biology of Pit Vipers*, edited by J. A. Campbell and E. D. Brodie Jr. Selva Natural History Book Publishers, Tyler, TX.

Cadle, J. E. 1994. The colubrid radiation in Africa (Serpentes: Colubridae): phylogenetic relationships and evolutionary patterns based on immunological data. *Zoological Journal of the Linnean Society* 110:103–140.

Cadle, J. E., H. C. Dessauer, C. Gans, and D. F. Gartside. 1990. Phylogenetic relationships and molecular evolution in uropeltid snakes (Serpentes: Uropeltidae): allozymes and albumin immunology. *Biological Journal of the Linnaean Society* 40:293–320.

Cadle, J. E., and H. W. Greene. 1993. Phylogenetic patterns, biogeography, and the composition of Neotropical snake assemblages. Pages 281–293 in *Species Diversity in Ecological Communities: Historical and Geographical Perspectives*, edited by R. E. Ricklefs and D. Schluter. University of Chicago Press, Chicago, IL.

Caldwell, J. P. 1996. The evolution of myrmecophagy and its correlates in poison frogs (Family Dendrobatidae). *Journal of Zoology, London* 240:75–101.

Calow, L. J., and R. McN. Alexander. 1973. A mechanical analysis of a hind leg of a frog (*Rana temporaria*). *Journal of Zoology* 171:293–321.

Camp, C. L. 1923. Classification of the lizards. *Bulletin of the American Museum of Natural History* 48:289–435.

Campbell, J. A., and J. L. Camarillo. 1992. The Oaxacan dwarf boa, *Exiliboa placata* (Serpentes: Tropidophiidae): descriptive notes and life history. *Caribbean Journal of Science* 28:17–20.

Campbell, J. A., and J. L. Camarillo R. 1994. A new lizard of the genus *Diploglossus* (Anguidae: Diploglossi-nae) from Mexico, with a review of the Mexican and northern Central American species. *Herpetologica* 50:193–209.

Campbell, J. A., D. R. Formanowicz, Jr., and E. D. Brodie, Jr. 1989. Potential impact of rattlesnake roundups on natural populations. *Texas Journal of Science* 41:301–317.

Campbell, J. A., and D. R. Frost. 1993. Anguid lizards of the genus *Abronia*: revisionary notes, descriptions of four new species, a phylogenetic analysis, and key. *Bulletin of the American Museum of Natural History* 216:1–121.

Campbell, J. A., and W. W. Lamar. 1989. *The Venomous Reptiles of Latin America*. Comstock Publishing Associates, Ithaca, NY.

Cannatella, D. C., and R. O. De Sá. 1993. *Xenopus laevis* as a model organism. *Systematic Biology* 42:476–507.

Cannatella, D. C., and D. M. Hillis. 1993. Amphibian relationships: phylogenetic analysis of morphology and molecules. *Herpetological Monographs* 7:1–7.

Cannatella, D. C., and K. de Queiroz. 1989. Phylogenetic systematics of the anoles: is a new taxonomy warranted? *Systematic Zoology* 38:57–69.

Cannatella, D. C., and L. Trueb. 1988a. Evolution of pipoid frogs: intergeneric relationships of the aquatic frog family Pipidae (Anura). *Zoological Journal of the Linnean Society* 94:1–38.

Cannatella, D. C., and L. Trueb. 1988b. Evolution of pipoid frogs: morphology and phylogenetic relationships of *Pseudhymenochirus*. *Journal of Herpetology* 22:439–456.

Capranica, R. R. 1965. *The Evoked Vocal Response of the Bullfrog: a Study of Communication by Sound*. MIT Press, Cambridge, MA.

Card, W., and A. G. Kluge. 1995. Hemipeneal skeleton and varanid lizard systematics. *Journal of Herpetology* 29:275–280.

Carey, C. 1993. Hypothesis concerning the causes of the disappearance of boreal toads from the mountains of Colorado. *Conservation Biology* 7:355–362.

Carpenter, C. C. 1967. Aggression and social structure of iguanid lizards. Pages 87–105 in *Lizard Ecology: A Symposium*, edited by W. S. Milstead. University of Missouri Press, Columbia, MO.

Carpenter, C. C. 1977. Communication and displays of snakes. *American Zoologist* 17:217–223.

Carpenter, C. C. 1978. Ritualistic social behaviors of lizards. Pages 253–267 in *Behavior and Neurology of Lizards*, edited by N. Greenberg and P. D. MacLean. National Institute of Mental Health, Bethesda, MD.

Carpenter, C. C., and G. W. Ferguson. 1977. Variation and evolution of stereotyped behavior in reptiles. Pages 335–554 in *Biology of the Reptilia, Vol. 7 (Ecology and Behaviour A)*, edited by C. Gans and D. W. Tinkle. Academic Press, New York, NY.

Carpenter, G. C. 1995. Modeling dominance: the influence of size, coloration, and experience on dominance relations in tree lizards (*Urosaurus ornatus*). *Herpetological Monographs* 9:88–101.

Carr, A. F. 1952. *Handbook of Turtles, the Turtles of the United States and Canada*. Cornell University Press, Ithaca, NY.

Carr, A. 1955. *The Windward Road*. Alfred A. Knopf, Inc. New York, NY.

Carr, A., M. H. Carr, and A. B. Meylan. 1978. The ecology and migrations of sea turtles & the west Caribbean green turtle colony. *Bulletin of the American Museum of Natural History* 162:1–46.

Carr, A., and P. J. Coleman. 1974. Seafloor spreading theory and the odyssey of the green turtle from Brazil to Ascension Island, Central Atlantic. *Nature* 249:128–130.

Carr, J. L., and J. W. Bickham. 1986. Phylogenetic implications of karyotypic variation in the Batagurinae (Testudines: Emydidae). *Genetica* 70:89–106.

Carr, S. M., A. J. Brothers, and A. C. Wilson. 1987. Evolutionary inferences from restriction maps of mitochondrial DNA from nine taxa of *Xenopus* frogs. *Evolution* 41:176–188.

Carrier, D. R. 1986. Lung ventilation during walking and running in four species of lizards. *Experimental Biology* 47:33–42.

Carrier, D. R. 1987. The evolution of locomotor stamina in tetrapods: circumventing a mechanical constraint. *Paleobiology* 13:326–341.

Carroll, R. L. 1988. *Vertebrate Paleontology and Evolution*. W. H. Freeman Co., New York, NY.

Carroll, R. L. and D. Baird. 1972. Carboniferous stem-reptiles of the family Romeriidae. *Bulletin of the Museum of Comparative Zoology* 143:321–363.

Carroll, R. L., and P. J. Currie. 1975. Microsaurs as possible apodan ancestors. *Zoological Journal of the Linnean Society* 57:229–247.

Carroll, R. L., and P. Gaskill. 1978. The order Microsauria. *Memoires of the American Philosophical Society* 126: 1–211.

Case, S. M., and M. H. Wake. 1977. Immunological comparisons of caecilian albumins (Amphibia: Gymnophiona). *Herpetologica* 33:94–98.

Castilla, A. M., and R. Van Damme. 1996. Cannibalistic propensities in the lizard *Podarcis hispanica atrata*. *Copeia* 1996:991–994.

Cei, J. M. 1962. *Batracios de Chile*. Universidad de Chile, Santiago, Chile.

Cei, J. M. 1980. Amphibians of Argentina. *Monitore Zoologico Italiano n.s. Monografia* 2:xii + 609.

Chabreck, R. H., and T. Joanen. 1979. Growth rates of American alligators in Louisiana. *Herpetologica* 35:51–57.

Chang, M.-M. 1991. "Rhipidistians," dipnoans, and tetrapods. Pages 3–28 in *Origins of the Higher Groups of Tetrapods: Controversy and Consensus*, edited by H.-P.

Schultze and L. Trueb. Comstock Publishing Associates, Ithaca, NY.

Channing, A. 1978. A new bufonid genus (Amphibia: Anura) from Rhodesia. *Herpetologica* 34:394–397.

Channing, A. 1989. A re-evaluation of the phylogeny of Old World treefrogs. *South African Journal of Science* 24: 116–131.

Charland, M. B. 1993. Thermal consequences of reptilian viviparity: thermoregulation in gravid and nongravid garter snakes (*Thamnophis*). *Journal of* Herpetology 29:383–390.

Charnov, E. L., and J. J. Bull. 1977. When is sex environmentally determined? *Nature* 266:828–830.

Chelazzi, C. 1992. Reptiles. Pages 235–261 in *Animal Homing*, edited by F. Papi. Chapman and Hall, New York, NY.

Chelazzi, G., and G. Delfino. 1986. A field test on the use of olfaction in homing by *Testudo hermanni* (Reptilia: Testudinidae). *Journal of Herpetology* 20:451–455.

Chiszar, D. R., K. K. Lee, C. W. Radcliffe, and H. M. Smith. 1992. Searching behaviors by rattlesnakes following predatory strikes. Pages 369–388 in *Biology of Pitvipers*, edited by J. A. Campbell and E. D. Brodie, Jr., Selva, Tyler, TX.

Christian, K. A., and C. R. Tracy. 1981. The effect of thermal environment on the ability of hatchling Galapagos land iguanas to avoid predation during dispersal. *Oecologia* 49:218–223.

Ciofi, C., and G. Chelazzi. 1994. Analysis of homing pattern in the colubrid snake *Coluber viridiflavus*. *Journal of Herpetology* 28:477–484.

Clack, J. A. 1992. The stapes of *Acanthostega gunnari* and the role of the stapes in early tetrapods. Pages 405–420 in *The Evolutionary Biology of Hearing*, edited by D. B. Webster, R. R. Fay, and A. N. Popper. Springer-Verlag, New York, NY.

Clack, J. A. 1994. Earliest known tetrapod braincase and the evolution of the stapes and fenestra ovalis. *Nature* 369:392–394.

Clarke, B. T. 1981. Comparative osteology and evolutionary relationships in the African Raninae (Anura, Ranidae). *Monitore Zoologico Italiano n.s., Suppl.* 15:285–331.

Clarke, B. T. 1987. A description of the skeletal morphology of *Barbourula* (Anura: Discoglossidae), with comments on its relationships. *Journal of Natural History* 21:879–891.

Clobert, J., M. Massot, J. Lecomte, G. Sorci, M. de Fraipont, and R. Barbault. 1994. Determinants of dispersal behavior: the common lizard as a case study. Pages 183–206 in *Lizard Ecology: Historical and Experimental Perspectives*, edited by L. J. Vitt and E. R. Pianka. Princeton University Press, Princeton, NJ.

Coates, M. I., and J. A. Clack. 1990. Polydactyly in the earliest-known tetrapod limbs. *Nature* 347:66–69.

Coates, M. I., and J. A. Clack. 1991. Fish-like gills and breathing in the earliest known tetrapod. *Nature* 352:234–236.

Cock Buning, T. J. de. 1985. Qualitative and quantitative explanation of the forms of heat sensitive organs in snakes. *Acta Biotheoretica* 34:193–206.

Cocroft, R. B. 1994. A cladistic analysis of chorus frog phylogeny (Hylidae: *Pseudacris*). *Herpetologica* 50:420–437.

Cocroft, R. B., and K. Hambler. 1989. Observations on a commensal relationship of the microhylid frog *Chiasmocleis ventrimaculata* and the burrowing theraphosid spider *Xenesthis immanis* in southeastern Peru. *Biotropica* 21:2–8.

Cody, M. L., and J. M. Diamond (editors). 1975. *Ecology and Evolution of Communities*. The Belknap Press of Harvard University Press, Cambridge, MA.

Cogalniceanu, D. 1994. The relative importance of vision and olfaction in mate recognition in male newts (genus *Triturus*). *Herpetologica* 50:344–349.

Cogger, H. G. and R. G. Zweifel (editors). 1992. *Reptiles and Amphibians*. Smithmark, New York, NY.

Collins, J. P., and H. M. Wilbur. 1979. Breeding habits and habitats of the amphibians of the Edwin S. George Reserve, Michigan, with notes on the local distribution of fishes. *Occasional Papers of the Museum of Zoology, University of Michigan* 686 pp 1–34.

Conant, R., and J. T. Collins. 1991. *Reptiles and Amphibians: Eastern and Central North America, 3rd edition*. Houghton Mifflin Co., Boston, MA.

Congdon, J. D. 1989. Proximate and evolutionary constraints on energy relations of reptiles. *Physiological Zoology* 62:356–373.

Congdon, J. D., A. E. Dunham, and R. C. Van Loben Sels. 1993. Delayed sexual maturity and demographics of Blanding's turtles (*Emydoidea blandingii*): implications for conservation and management of long-lived organisms. *Conservation Biology* 7:826–833.

Congdon, J. D., A. E. Dunham, and R. C. Van Loben Sels. 1994. Demographics of common snapping turtles (*Chelydra serpentina*): implications for conservation and management of long-lived organisms. *American Zoologist* 34:397–408.

Congdon, J. D., and J. W. Gibbons. 1987. Morphological constraint on egg size: a challenge to optimal egg size theory? *Proceedings of the National Academy of Sciences, USA* 84:4145–4147.

Congdon, J. D., J. W. Gibbons, and J. L. Breene. 1983. Parental investment in the chicken turtle, (*Deirochelys reticularia*). *Ecology* 64:419–425.

Conner, J., and D. Crews. 1980. Sperm transfer and storage in the lizard, *Anolis carolinensis*. *Journal of Morphology* 163:331–348.

Cooper, W. E., Jr. 1994: Prey chemical discrimination, foraging mode, and phylogeny. Pages 95–116 in *Lizard Ecology: Historical and Experimental Perspectives*, edited by L. J. Vitt and E. R. Pianka. Princeton University Press, Princeton, NJ.

Cooper, W. E., Jr., and N. Greenberg 1992: Reptilian coloration and behavior. Pages 298–422 in *Biology of the Reptilia, Vol. 18 (Physiology E, Hormones, Brain, and Behavior)*, edited by C. Gans and D. Crews. University of Chicago Press, Chicago, IL.

Cooper, W. E., Jr., P. Lopez, and A. Salvador. 1994. Pheromone detection by an amphisbaenian. *Animal Behaviour* 47:1401–1411.

Cooper, W. E., Jr., and L. J. Vitt. 1986. Lizard pheromones: behavioral responses and adaptive significance in skinks of the genus *Eumeces*. Pages 323–340 in *Chemical Signals in Vertebrates 4 (Ecology, Evolution, and Comparative Biology)*, edited by D. Duvall, D. Muller-Schwarze, and R. M. Silverstein. Plenum Press, New York, NY.

Cooper, W. E., Jr., and L. J. Vitt. 1993. Female mate choice of large male broad-headed skinks. *Animal Behaviour* 45:683–693.

Corn, P. S. 1994: What we know and don't know about amphibian declines in the West. Pages 59–67 in *Sustainable Ecological Systems: Implementing an Ecological Approach to Land Management*, edited by W. W. Covington and L. F. DeBano. USDA Forest Service, Rocky Mountain Forest and Range Experiment Station, Ft. Collins, Colorado. General Technical Report RM-247. May 1994.

Cortwright, S. A., and C. E. Nelson. 1990. An examination of multiple factors affecting community structure in an aquatic amphibian community. *Oecologia* 83:123–131.

Cowles, R. B., and C. M. Bogert. 1944. A preliminary study of the thermal requirements of desert reptiles. *Bulletin of the American Museum of Natural History* 83:261–296.

Cree, A., and D. Butler. 1993. *Tuatara Recovery Plan (Sphenodon ssp). Threatened Species Recovery Plan no 9*. New Zealand Department of Conservation, Wellington, NZ.

Cree, A., C. H. Daugherty, and J. M. Hay. 1995. Reproduction of a rare New Zealand reptile, the tuatara *Sphenodon punctatus*, on rat-free and rat-inhabited islands. *Conservation Biology* 9:373–383.

Cree, A., M. B. Thompson, and C. H. Daugherty. 1995. Tuatara sex determination. *Nature* 375:543.

Crews, D. 1980. Interrelationships among ecological, behavioral, and neuroendocrine processes in the reproductive cycle of *Anolis carolinensis* and other reptiles. *Advances in the Study of Behavior.* 11:1–74.

Crews, D., J. M. Bergeron, J. J. Bull, D. Flores, A. Tousignant, J. K. Skipper, and T. Wibbels. 1994. Temperature-dependent sex determination in reptiles: proximate mechanisms, ultimate outcomes, and practical applications. *Developmental Genetics* 15:297–312.

Crews, D., and W. R. Garstka. 1982. The ecological physiology of a garter snake. *Scientific American* 247:158–168.

Crews, D., and M. C. Moore. 1993. Psychobiology of reproduction of unisexual whiptail lizards. Pages 257–282 in *Biology of Whiptail Lizards (genus Cnemidophorus)*, edited by J. W. Wright and L. J. Vitt. Oklahoma Museum of Natural History, Norman, OK.

Crother, B. I., and W. F. Presch. 1992. The phylogeny of xantusiid lizards: the concern for analysis in the search for a best estimate of phylogeny. *Molecular Phylogenetics and Evolution* 1:289–294.

Crouse, D. T., L. B. Crowder, and H. Caswell. 1987. A stage-based population model for loggerhead sea turtles and implications for conservation. *Ecology* 68:1412–1423.

Crowell Comuzzie, D. K., and D. W. Owens. 1990. A quantitative analysis of courtship behavior in captive green sea turtles (*Chelonia mydas*). *Herpetologica* 46: 195–202.

Crowley, M. J. (editor). 1992. *Natural Enemies: The Population Biology of Predators, Parasites, and Diseases*. Blackwell Scientific Publications, Oxford.

Crumly, C. R. 1982. A cladistic analysis of *Geochelone* using cranial osteology. *Journal of Herpetology* 16:215–234.

Crumly, C. R. 1994. Phylogenetic Systematics of North American Tortoises (Genus *Gopherus*): Evidence for Their Classification. *Fish and Wildlife Research* 13:7–32.

Crump, M. L. 1974. Reproductive strategies in a tropical anuran community. *Miscellaneous Publication Museum of Natural History, University of Kansas* 61 pp 1–68.

Crump, M. L. 1986. Homing and site fidelity in a neotropical frog, *Atelopus varius* (Bufonidae). *Copeia* 1986: 438–444.

Crump, M. L. 1988. Aggression in harlequin frogs: male-male competition and a possible conflict of interest between the sexes. *Animal Behaviour* 36:1064–1077.

Crump, M. L. 1989. Life history consequences of feeding versus non-feeding in a facultatively non-feeding toad larva. *Oecologia* 78:486–489.

Crump, M. L. 1991. Choice of oviposition site and egg load assessment by a treefrog. *Herpetologica* 47:308–315.

Crump, M. L. 1992. Cannibalism in amphibians. Pages 256–276 in *Cannibalism: Ecology and Evolution Among Diverse Taxa*, edited by M. A. Elgar and B. J. Crespi. Oxford University Press, Oxford, UK.

Crump, M. L. 1996. Parental care among the Amphibia. *Advances in the Study of Behavior* 25:109–144.

Crump, M. L., and M. Vaira. 1991. Vulnerability of *Pleurodema borelli* tadpoles to an avian predator: effect of body size and density. *Herpetologica* 47:316–321.

Cundall, D. 1983. Activity of head muscles during feeding by snakes: A comparative study. *American Zoologist* 23:383–396.

Cundall, D. 1987. Functional morphology. Pages106–140 in *Snakes: Ecology and Evolutionary Biology*, edited by R. A. Seigel, J. T. Collins, and S. S. Novak. Macmillan Publishing Co., New York, NY.

Cundall, D., and C. Gans. 1979. Feeding in water snakes: An electromyographic study. *Journal of Experimental Zoology* 209:189–207.

Cundall, D., and F. J. Irish. 1989. The function of the intramaxillary joint in the Round Island boa, *Casarea dussumieri. Journal of Zoology, London* 217:569–598.

Cundall, D., J. Lorenz-Elwood, and J. D. Groves. 1987. Asymmetric suction feeding in primitive salamanders. *Experientia* 43:1229–1231.

Cundall, D., and D. A. Rossman. 1993. Cephalic anatomy of the rare Indonesian snake *Anomochilus weberi. Zoological Journal of the Linnean Society* 109:235–273.

Cundall, D., and J. Shardo. 1995. Rhinokinetic snout of thamnophiine snakes. *Journal of Morphology* 225: 31–50.

Cundall, D., V. Wallach, and D. A. Rossman. 1993. The systematic relationships of the snake genus *Anomochilus. Zoological Journal of the Linnean Society* 109: 275–299.

Cuny, G., J. J. Jaeger, Mahboubi, and J. C. Rage. 1990. Les plus anciens serpents (Reptilia, Squamata) connus. Mise au point sur l'âge géologique des serpents de la partie moyenne du Crétacé. *Comptes Rendus de l'Academie des Sciences, Paris* 311, ser. II:1267–1272.

Dalrymple, G. H. 1979. On the jaw mechanism of the snail-crushing lizards, *Dracaena* Daudin 1802 (Reptilia, Lacertilia, Teiidae). *Journal of Herpetology* 13:303–311.

Daltry, J. C., W. Wüster, and R. S. Thorpe. 1996. Diet and snake evolution. *Nature* 379:537–540.

Daly, J. W. 1986. Effects of alkaloids on ion transport. Pages 121–133 in *Proceedings of the NATO Symposium on Natural Products and Biological Activity*. Tokyo Press, Toyko.

Daly, J. W. 1989. Structure activity relationships for alkaloid modulators of ion channels. Pages 385–412 in *New Aspects of Organic Chemistry I*. Kodansha Ltd., Tokyo, Japan.

Daly, J. W. 1995. The chemistry of poisons in amphibian skin. *Proceedings of the National Academy of Sciences, USA* 92:9–13.

Daly, J. W., N. R. Andriamaharavo, M. Andriantsiferana, and C. W. Myers. 1996. Madagascan poison frogs (*Mantella*) and their skin alkaloids. *American Museum Novitates* 3177:1–34.

Daly, J. W., J. Caceres, R. W. Moni, F. Gusovsky, M. Moos Jr., K. B. Seamon, K. Milton, and C. W. Myers. 1992. Frog secretions and hunting magic in the upper Amazon: identification of a peptide that interacts with an adenosine receptor. *Proceedings of the National Academy of Sciences, USA* 89:10960–10963.

Daly, J. W., H. M. Garraffo, T. F. Spande, C. Jaramillo, and A. S. Rand. 1994. Dietary source for skin alkaloids of poison frogs (Dendrobatidae)? *Journal of Chemical Ecology* 20:943–955.

Daly, J. W., C. W. Myers, and N. Whittaker. 1987. Further classification of skin alkaloids from neotropical poison frogs (Dendrobatidae), with a general survey of toxic/noxious substances in the amphibia. *Toxicon* 25:1023–1095.

Daly, J. W., S. I. Secunda, H. M. Garraffo, T. F. Spande, A. Wisnieski, and J. F. Cover. 1994. An uptake system for dietary alkaloids in poison frogs (Dendrobatidae). *Toxicon* 32:657–663.

Darda, D. M. 1994. Allozyme variation and morphological evolution among Mexican salamanders of the genus *Chiropterotriton* (Caudata: Plethodontidae). *Herpetologica* 50:164–187.

Dart, R. C., and H. F. Gomez. 1996. Reptile bites and scorpion stings. Pages 864–867 in *Emergency Medicine: A Comprehensive Study Guide*, 4th Ed., edited by J. E. Tintinalli, E. Ruiz, and R. L. Krome. McGraw-Hill, New York, NY.

Darwin, C. 1871. *The Descent of Man and Selection in Relation to Sex. 2 vols.* D. Appleton, New York, NY.

Daugherty, C. H., A. Cree, J. M. Hay, and M. B. Thompson. 1990. Neglected taxonomy and continuing extinctions of tuatara (*Sphenodon*). *Nature* 347:177–179.

Daugherty, C. H., L. R. Maxson, and B. D. Bell. 1982. Phylogenetic relationships within the New Zealand frog genus *Leiopelma*: immunological evidence. *New Zealand Journal of Zoology* 8:543–550.

Davenport, J., A. Munks, and P. J. Oxford. 1984. A comparison of the swimming of marine and freshwater turtles. *Proceedings of the Royal Society of London* 220B: 447–475.

Davies, N. B., and T. R. Halliday. 1979. Competitive mate searching in male common toads, *Bufo bufo*. *Animal Behaviour* 27:1268–1276.

Davis, M. S. 1987. Acoustically mediated neighbor recognition in the North American bullfrog, *Rana catesbeiana*. *Behavioral Ecology and Sociobiology* 21:185–190.

Davis, W., and A. T. Weil. 1992. The identity of a New World psychoactive toad. *Ancient Mesoamerica* 3:51–59.

Dawley, E. M. 1984a. Identification of sex through odors by male red-spotted newts, *Notophthalmus viridescens*. *Herpetologica* 40:101–105.

Dawley, E. M. 1984b. Recognition of individual, sex, and species odours by salamanders of the *Plethodon glutinosus-P. jordani* complex. *Animal Behaviour* 32:353–361.

Dawley, E. M. 1987. Species discrimination between hybridizing and non-hybridizing terrestrial salamanders. *Copeia* 1987:924–931.

Dawley, E. M. 1992. Sexual dimorphism in a chemosensory system: the role of the vomeronasal organ in salamander reproductive behavior. *Copeia* 1992:113–120.

Dawley, E. M., and A. H. Bass. 1989. Chemical access to the vomeronasal organs of a plethodontid salamander. *Journal of Morphology* 200:163–174.

Dawley, R. M. and J. P. Bogart (editors). 1989. *Evolution and Ecology of Unisexual Vertebrates* (Bulletin of the New York State Museum no. 466). New York State Museum, Albany, New York.

de Jongh, H. J. 1968: Functional morphology of the jaw apparatus of larval and metamorphosing *Rana temporaria* L. *Netherlands Journal of Zoology* 18:1–103, pl. 1–25.

de Jongh, H. J., and C. Gans. 1969. On the mechanism of respiration in the bullfrog, *Rana catesbeiana*: a reassessment. *Journal of Morphology* 127:259–290.

Del Pino, E. M. 1989. Marsupial frogs. *Scientific American* 260:110–118.

DeMarco, V. G. 1993. Estimating egg retention times in sceloporine lizards. *Journal of Herpetology* 27:453–458.

Densmore, L. D., and P. S. White. 1991. The systematics and evolution of the Crocodilia as suggested by restriction endonuclease analysis of mitochondrial and nuclear ribosomal DNA. *Copeia* 1991:602–615.

De Queiroz, K. 1987. Phylogenetic systematics of iguanine lizards, a comparative osteological study. *University of California Publications in Zoology* 118:1–203.

Derr, J. N., J. W. Bickham, I. F. Greenbaum, A. G. J. Rhodin, and R. A. Mittermeier. 1987. Biochemical systematics and evolution in the South American turtle genus *Platemys* (Pleurodira: Chelidae). *Copeia* 1987:370–375.

Deslippe, R. J., and R. T. M'Closkey. 1991. An experimental test of mate defense in an iguanid lizard (*Sceloporus graciosus*). *Ecology* 72:1218–1224.

Dessauer, H. C., J. E. Cadle, and R. Lawson. 1987. Patterns of evolution in snakes suggested by their proteins. *Fieldiana, Zoology* new series, 34:1–34.

Devine, M. C. 1984. Potential for sperm competition in reptiles: behavioral and physiological consequences. Pages 509–521 in *Sperm Competition and the Evolution of Animal Mating Systems*, edited by R. L. Smith. Academic Press, New York, NY.

Dewitt, C. B. 1967. Precision of thermoregulation and its relation to environmental factors in the desert iguana, *Dipsosaurus dorsalis*. *Physiological Zoology* 40:67–82.

Dial, B. E., and L. L. Grismer. 1992. A phylogenetic analysis of physiological-ecological character evolution in the lizard genus *Coleonyx* and its implications for historical biogeographic reconstruction. *Systematic Biology* 41: 178–195.

Diamond, J., and T. J. Case (editors). 1986. *Community Ecology*. Harper and Row Publishers, Inc., New York, NY.

Diemer, J. E. 1992. Home range and movements of the tortoise *Gopherus polyphemus* in northern Florida. *Journal of Herpetology* 26:158–165.

Dingle, H. 1996. *Migration: The Biology of Life on the Move*. Oxford University Press, New York, NY.

Dixon, J. R. 1973. A systematic review of the teiid lizards, genus *Bachia*, with remarks on *Heterodactylus* and *Anotosaura*. *University of Kansas Museum of Natural History, Miscellaneous Publications* no. 57:1–47.

Dixon, J. R., and F. S. Hendricks. 1979. The wormsnakes (Family Typhlopidae) of the neotropics, exclusive of the Antilles. *Zoologische Verhandelingen, Leiden* 173:1–39.

Dixon, J. R., and C. P. Kofron. 1983. The Central and South American anomalepid snakes of the genus *Liotyphlops. Amphibia Reptilia* 4:241–264.

Dobson, A. P., and P. J. Hudson. 1986. Parasites, disease and the structure of ecological communities. *Trends in Ecology and Evolution* 1:11–15.

Dodd, C. K., and E. D. Brodie, Jr. 1975. Notes on the defensive behavior of the snapping turtle, *Chelydra serpentina. Herpetologica* 31:286–288.

Dodd, C. K., Jr. 1986. Importation of live snakes and snake products into the United States, 1977–1983. *Herpetological Review* 17:76–79.

Dodd, C. K., Jr. 1987. Status, conservation, and management. Pages 478–513 in *Snakes: Ecology and Evolutionary Biology*, edited by R. A. Seigel, J. T. Collins, and S. S. Novak. MacMillan Publication Co., New York, NY.

Dodd, C. K., Jr., and R. A Seigel. 1991. Relocation, repatriation, and translocation of amphibians and reptiles: are they conservation strategies that work? *Herpetologica* 47:336–350.

Dole, J. W. 1965. Summer movements of adult leopard frogs, *Rana pipiens* Schreber, in northern Michigan. *Ecology* 46:236–255.

Donnelly, M. A. 1994. Amphibian diversity and natural history. Pages 199–209 in *La Selva. Ecology and Natural History of a Neotropical Rain Forest*, edited by L. A. McDade, K. S. Bawa, H. A. Hespenheide, and G. S. Hartshorn. University of Chicago Press, Chicago, IL.

Donnelly, M. A., R. O. de Sa, and C. Guyer. 1990. Description of the tadpoles of *Gastrophryne pictiventris* and *Nelsonophryne aterrima* (Anura: Microhylidae), with a review of morphological variation in free-swimming microhylid larvae. *American Museum Novitates* 2976:1–19.

Donoghue, M. J., J. Doyle, J. Gauthier, A. G. Kluge, and T. Rowe. 1989. The importance of fossils in phylogeny reconstruction. *Annual Review of Ecology and Systematics* 20:431–460.

Dowling, H. G., C. A. Hass, S. B. Hedges, and R. Highton. 1996. Snake relationships revealed by slow-evolving proteins: a preliminary survey. *Journal of Zoology, London* 240:1–28.

Dowling, H. G., R. Highton, G. C. Maha, and L. R. Maxson. 1983. Biochemical evaluation of colubrid snake phylogeny. *Journal of Zoology, London* 201:309–329.

Dowling, H. G., and J. M. Savage. 1960. A guide to the snake hemipenis: a survey of basic structure and systematic characteristics. *Zoologica* 45:17–31.

Drewes, R. C. 1984. A phylogenetic analysis of the Hyperoliidae (Anura): treefrogs of Africa, Madagascar, and the Seychelles Islands. *Occasional Papers of the California Academy of Sciences* 139:1–70.

Drewes, R. C., and R. Altig. 1996. Anuran egg predation and heterocannibalism in a breeding community of East African frogs. *Tropical Zoology* 9:333–347.

Drewes, R., and B. Roth. 1981. Snail eating frogs from the Ethiopian highlands: a new anuran specialization. *Zoological Journal of the Linnean Society* 73:267–287.

Drost, C. A., and G. M. Fellers. 1996. Collapse of a regional frog fauna in the Yosemite area of the Califor-nia Sierra Nevada, USA. *Conservation Biology* 10:414–425.

Dubach, J., A. Sajewicz, and R. Pawley. 1997. Parthenogenesis in the Arafuran filesnake *(Achrochordus arafurae). Herpetological Natural History* In Press.

Dubois, A. 1981. Liste des genres et sous-genres nominaux de Ranoidea (Amphibiens anoures) du monde, avec identification de leurs espèces-types: conséquences nomenclaturales. *Monitore Zoologico Italiano* n.s., Suppl. 15: 225–284.

Duellman, W. E. 1963. Amphibians and reptiles of the rainforest of southern El Petén, Guatemala. *University of Kansas Publications, Museum of Natural History* 15: 205–249.

Duellman, W. E. 1970. *The Hylid Frogs of Middle America, 2 vols.* Museum of Natural History, The University of Kansas, Kansas.

Duellman, W. E. 1975. On the classification of frogs. *Occasional Papers of the Museum of Natural History, University of Kansas* no. 42:1–14.

Duellman, W. E. 1990. Herpetofaunas in Neotropical rainforests: comparative composition, history, and resource use. Pages 455–505 in *Four Neotropical Rainforests*, edited by A. H. Gentry. Yale University Press, New Haven, CT.

Duellman, W. E. 1993. *Amphibian Species of the World: Additions and Corrections.* University of Kansas, Museum of Natural History (Spec. Publ. No. 21), Lawrence, KS.

Duellman, W. E., and M. Lizana. 1994. Biology of a sit-and-wait predator, the leptodactylid frog *Ceratophrys cornuta. Herpetologica* 50:51–64.

Duellman, W. E., L. R. Maxson, and C. A. Jesiolowski. 1988. Evolution of marsupial frogs (Hylidae: Hemiphractinae): Immunological evidence. *Copeia* 1988: 527–543.

Duellman, W. E., and R. A. Pyles. 1983. Acoustic resource partitioning in anuran communities. *Copeia* 1983: 639–649.

Duellman, W. E., and L. Trueb. 1986. *Biology of Amphibians.* McGraw-Hill Book Co., New York, NY.

Dufton, M. J. 1984. Classification of elapid snake neurotoxins and cytotoxins according to chain length: evolutionary implications. *Journal of Molecular Evolution* 20: 128–134.

Dugan, B. 1982. The mating behavior of the green iguana, *Iguana iguana*. Pages 320–341 in *Iguanas of the World*, edited by G. M. Burghardt and A. S. Rand. Noyes Publications, Park Ridge, NJ.

Dundee, H. A. 1971. *Cryptobranchus, Cryptobranchus alleganiensis. Catalogue of American Amphibians and Reptiles* 101:1–4.

Dunham, A. E. 1980. An experimental study of interspecific competition between the iguanid lizards *Sceloporus merriami* and *Urosaurus ornatus*. *Ecological Monographs* 50: 309–330.

Dunham, A. E., and J. W. Gibbons. 1990. Growth of the slider turtle. Pages 135–145 in *Life History and Ecology of the Slider Turtle*, edited by J. W. Gibbons. Smithsonian Institution Press, Washington, D. C.

Dunham, A. E., B. W. Grant, and K. L. Overall. 1989. Interfaces between biophysical and physiological ecology and the population ecology of terrestrial vertebrate ectotherms. *Physiological Zoology* 62:335–355.

Dunham, A. E., D. B. Miles, and D. N. Reznick. 1988. Life history patterns in squamate reptiles. Pages 442–511 in *Biology of the Reptilia, Vol. 16 (Ecology B, Defense and Life History)*, edited by C. Gans and R. B. Huey. Alan R. Liss, Inc., New York, NY.

Dunlap, K. D., and T. Mathies. 1993. Effects of nymphal ticks and their interaction with malaria on the physiology of male fence lizards. *Copeia* 1993:1045–1048.

Dunn, E. R. 1917. The salamanders of the genera *Desmognathus* and *Leurognathus*. *Proceedings of the U. S. National Museum* 53:393–433.

Dunn, E. R. 1923. The salamanders of the family Hynobiidae. *Proceedings of the American Academy of Arts and Sciences* 58:445–523.

Dunn, E. R. 1926. *The Salamanders of the Family Plethodontidae*. Smith College, Northampton, MA.

Dunson, W. A. 1976. Salt glands in reptiles. Pages 413–445 in *Biology of the Reptilia, Vol 5 (Physiology A)*, edited by C. Gans and W. R. Dawson, Academic Press, London, UK.

Duvall, D., S. J. Arnold, and G. W. Schuett. 1992. Pitviper mating systems: ecological potential, sexual selection, and microevolution. Pages 321–336 in *Biology of the Pitvipers*, edited by J. A. Campbell and E. D. Brodie, Jr., Selva, Tyler.

Duvall, D., M. J. Goode, W. K. Hayes, J. K. Leonhardt, and D. G. Brown. 1990. Prairie rattlesnake vernal migration: field experimental analysis and survival value. *National Geographic Research* 6:457–469.

Duvall, D., G. W. Schuett, and S. J. Arnold. 1993. Ecology and evolution of snake mating systems. Pages 165–200 in *Snakes: Ecology and Behavior*, edited by R. A. Seigel and J. T. Collins. McGraw Hill, New York, NY.

Echelle, A. A., A. F. Echelle, and H. S. Fitch. 1971. A comparative analysis of aggressive display in nine species of Costa Rican Anolis. *Herpetologica* 27:271–288.

Edwards, J. L. 1976. Spinal nerves and their bearing on salamander phylogeny. *Journal of Morphology* 148: 305–328.

Edwards, J. L. 1977. The evolution of terrestrial locomotion. Pages 553–577 in *Major Patterns in Vertebrate Evolution*, edited by M. K. Hecht, P. C. Goody, and B. M. Hecht. Plenum Press, New York, NY.

Edwards, J. L. 1985. Terrestrial locomotion without appendages. Pages 159–172 in *Functional Vertebrate Morphology*, edited by M. Hildebrand, D. M. Bramble, K. F. Liem, and D. B. Wake. Belknap Press, Cambridge, MA.

Eernisse, D. J., and A. G. Kluge. 1993. Taxonomic congruence versus total evidence, and the phylogeny of amniotes inferred from fossils, molecules and morphology. *Molecular Biology and Evolution* 10:1170–1195.

Ehmann, H., and H. Cogger. 1985. Australia's endangered herpetofauna: a review of criteria and policies. Pages 435–448 in *The Biology of Australasian Frogs and Reptiles*, edited by G. Grigg, R. Shine, and H. Ehmann. Surrey Beatty and Sons, Chipping Norton, Australia.

Elepfandt, A. 1996. Underwater acoustics and hearing in the clawed frog, *Xenopus*. Pages 177–193 in *The Biology of Xenopus*, edited by R. C. Tinsley and H. R. Kobel. Clarendon Press, Oxford.

Ellis, T. M., and M. A. Chappell. 1987. Metabolism, temperature relations, maternal behavior, and reproductive energetics in the ball python (*Python regius*). *Journal of Comparative Physiology B* 157:393–402.

Ellis-Quinn, B. A., and C. A. Simon. 1989. Homing behavior of the lizard *Sceloporus jarrovi*. *Journal of Herpetology* 3:146–152.

Elwood, J. R. L., and D. Cundall. 1994. Morphology and behavior of the feeding apparatus in *Cryptobranchus alleganiensis* (Amphibia: Caudata). *Journal of Morphology* 220:47–70.

Emerson, S. B. 1976. Burrowing in frogs. *Journal of Morphology* 149:437–458.

Emerson, S. B. 1982: Frog postcranial morphology: Identification of a functional complex. *Copeia* 1982:603–613.

Emerson, S. B. 1983. Functional analysis of frog pectoral girdles. The epicoracoid cartilages. *Journal of Zoology, London* 201:293–308.

Emerson, S. B. 1984. Morphological variation in frog pectoral girdles: testing alternatives to a traditional adaptive explanation. *Evolution* 38:376–388.

Emerson, S. B. 1985a. Jumping and leaping. Pages 58–72 in *Functional Vertebrate Morphology*, edited by M. Hildebrand, D. M. Bramble, K. F. Liem, and D. B. Wake. Belknap Press, Cambridge, MA.

Emerson, S. B. 1985b. Skull shape in frogsÑcorrelations with diet. *Herpetologica* 41:177–188.

Emerson, S. B. 1988a. The giant tadpole of *Pseudis paradoxa. Biological Journal of the Linnaean Society* 34: 93–104.

Emerson, S. B. 1988b. Convergence and morphological constraint in frogs: variation in postcranial morphology. *Fieldiana* 43:1–19.

Emerson, S. B. 1988c. Testing for historical patterns of change: a case study with frog pectoral girdles. *Paleobiology* 14:174–186.

Emerson, S. B. 1994. Testing predictions of sexual selection: a frog example. *American Naturalist* 143:848–869.

Emerson, S. B., and M. A. R. Koehl. 1990. The interaction of behavioral and morphological change in the evolution of a novel locomotor type: "flying" frogs. *Evolution* 44:1931–1946.

Emerson, S. B., J. Travis, and M. A. R. Koehl. 1990. Functional complexes and additivity in performance: a test case with "flying" frogs. *Evolution* 44:2153–2157.

Emerson, S. B., and H. Voris. 1992. Competing explanations for sexual dimorphism in a voiceless Bornean frog. *Functional Ecology* 6:654–660.

Emmons, L. H. 1989. Jaguar predation on chelonians. *Journal of Herpetology* 23:311–314.

Endler, J. A. 1986. Defense against predators. Pages 109–134 in *Predator-prey Relationships: Perspectives and Approaches from the Study of Lower Vertebrates*, edited by M. E. Feder and G. V. Lauder. University of Chicago Press, Chicago, IL.

Erdman, S., and D. Cundall. 1984. The feeding apparatus of the salamander *Amphiuma tridactylum*: morphology and behavior. *Journal of Morphology* 181:175–204.

Ernst, C. H., and R. W. Barbour. 1989. *Turtles of the World*. Smithsonian Institution Press, Washington, D. C.

Ernst, C. H., J. E. Lovich, and R. W. Barbour. 1994. *Turtles of the United States and Canada*. Smithsonian Institution Press, Washington, D. C..

Estes, R. 1981. Gymnophiona, Caudata. Pages 1–115 in *Handbuch der Paläoherpetologie, vol. 2*, edited by P. Wellnhofer. Gustav Fischer Verlag, Stuttgart, Germany.

Estes, R. 1983. The fossil record and early distribution of lizards. Pages 365–398 in *Advances in Herpetology and Evolutionary Biology*, edited by A. G. Rhodin and K. Miyata. Museum of Comparative Zoology, Harvard University, Cambridge, MA.

Estes, R., K. De Queiroz, and J. A. Gauthier. 1988. Phylogenetic relationships within Squamata. Pages 119–281 in *Phylogenetic Relationships of the Lizard Families, Essays Commemorating Charles L. Camp*, edited by R. Estes and G. Pregill. Stanford University Press, Stanford, CA.

Estes, R., T. H. Frazzetta, and E. E. Williams. 1970. Studies on the fossil snake *Dinilysia patagonica* Woodward: Part I. Cranial morphology. *Bulletin of the Museum of Comparative Zoology* 140:25–74.

Estes, R., and G. Pregill. 1988. *Phylogenetic Relationships of the Lizard Families, Essays Commemorating Charles L. Camp*. Stanford University Press, Stanford, CA.

Estes, R., Z. V. Spinar, and E. Nevo. 1978. Early Cretaceous pipid tadpoles from Israel (Amphibia: Anura). *Herpetologica* 34:374–393.

Etheridge, R. 1995. Redescription of *Ctenoblepharys adspersa* Tschudi, 1845, and the taxonomy of Liolaeminae (Reptilia: Squamata: Tropiduridae). *American Museum Novitates* 3142:1–34.

Etheridge, R., and K. De Queiroz. 1988. A phylogeny of Iguanidae. Pages 283–367 in *Phylogenetic Relationships of the Lizard Families, Essays Commemorating Charles L. Camp*, edited by R. Estes and G. Pregill. Stanford University Press, Stanford, CA.

Etheridge, R., and E. E. Williams. 1985. Notes on *Pristidactylus* (Squamata: Iguanidae). *Breviora* 483:1–18.

Etheridge, R., and E. E. Williams. 1991. A review of the South American lizard genera *Urostrophus* and *Anisolepis* (Squamata: Iguania: Polychridae). *Bulletin of the Museum of Comparative Zoology* 152:317–361.

Evans, C. M., and E. D. Brodie, Jr. 1994. Adhesive strength of amphibian skin secretions. *Journal of Herpetology* 28:499–502.

Evans, S. E. 1993. Jurassic lizard assemblages. *Revue de Paléobiologie* vol. spéc.:55–65.

Evans, S. E. 1995. Lizards: evolution, early radiation and biogeography. Pages 51–55 in *Sixth Symposium on Mesozoic Terrestrial Ecosystems and Biota, Short Papers*, edited by A. Sun and Y. Wang. China Ocean Press, Beijing.

Evans, S. E., and A. R. Milner. 1993. Frogs and salamanders from the Upper Jurassic Morrison Formation (Quarry Nine, Como Bluff) of North America. *Journal of Vertebrate Paleontology* 13:24–30.

Evans, S. E., and A. R. Milner. 1996. A metamorphosed salamander from the early Cretaceous of Las Hoyas, Spain. *Philosophical Transactions of the Royal Society of London* B 351:627–646.

Evans, S. E., A. R. Milner, and F. Mussett. 1988. The earliest known salamanders (Amphibia: Caudata): a record from the Middle Jurassic of England. *Geobios* 21:539–552.

Evans, S. E., A. R. Milner, and F. Mussett. 1990. A discoglossid frog from the Middle Jurassic of England. *Palaeontology* 33:298–311.

Evans, S. E., A. R. Milner, and C. Werner. 1996. Sirenid salamanders and a gymnophionan amphibian from the Cretaceous of the Sudan. *Palaeontology* 39:77–95.

Ewert, M. A., D. R. Jackson, and C. E. Nelson. 1994. Patterns of temperature-dependent sex determination in turtles. *Journal of Experimental Zoology* 270:3–15.

Fahrig, L., J. H. Pedlar, S. E. Pope, P. D. Taylor, and J. F. Wegner. 1995. Effect of road traffic on amphibian density. *Biological Conservation* 73:177–182.

Farris, J. S., A. G. Kluge, and M. F. Mickevich. 1982. Phylogenetic analysis, the monothetic group method, and myobatrachid frogs. *Systematic Zoology* 31:317–327.

Feder, M. E. 1982. Thermal ecology of Neotropical lungless salamanders (Amphibia: Plethodontidae). Environmental temperatures and behavioral responses. *Ecology* 63:1665–1674.

Feder, M. E., and S. J. Arnold. 1982. Anaerobic metabolism and behavior during predatory encounters between snakes (*Thamnophis elegans*) and salamanders (*Plethodon jordani*). *Oecologia* 53:93–97.

Feder, M. E., and W. W. Burggren. 1985. Cutaneous gas exchange in vertebrates: Designs, patterns, control and implications. *Biological Reviews* 60:1–45.

Feder, M. E., and G. V. Lauder (editors). 1986. *Predator-prey Relationships: Perspectives and Approaches from the Study of Lower Vertebrates.* University of Chicago Press, Chicago, IL.

Feinsinger, P., and H. M. Tiebout III. 1991. Competition among plants sharing hummingbird pollinators: laboratory experiments on a mechanism. *Ecology* 72:1946–1952.

Ferguson, M. W. J., and T. Joanen. 1982. Temperature of egg incubation determines sex in *Alligator mississippiensis. Nature* 296:850–853.

Firth, B. T., and J. S. Turner 1982. Sensory, neural, and hormonal aspects of thermoregulation. Pages 213–274 in *Biology of the Reptilia, Vol. 12 (Physiology C, Physiological Ecology),* edited by C. Gans and F. H. Pough, Academic Press, London, UK.

Fisher, R. N., and H. B. Shaffer. 1996. The decline of amphibians in California's Great Central Valley. *Conservation Biology* 10:1387–1397.

Fitch, H. S., R. W. Henderson, and D. M. Hillis. 1982. Exploitation of iguanas in Central America. Pages 397–417 in *Iguanas of the World,* edited by G. M. Burghardt and A. S. Rand. Noyes Publications, Park Ridge, NJ.

Fitzgerald, L. A. 1994a. The interplay between life history and environmental stochasticity: implications for the management of exploited lizard populations. *American Zoologist* 34:371–381.

Fitzgerald, L. A. 1994b. *Tupinambis* lizards and people: a sustainable use approach to conservation and development. *Conservation Biology* 8:12–16.

Fitzgerald, L. A., J. M. Chani, and O. E. Donadío. 1991. *Tupinambis* lizards in Argentina: implementing management of a traditionally exploited resource. Pages 303–316 in *Neotropical Wildlife Use and Conservation,* edited by J. G. Robinson and K. H. Redford. The University of Chicago Press, Chicago, IL.

Fitzgerald, S. 1989. *International wildlife trade: whose business is it?* World Wildlife Fund, Washington, D. C..

Fleischman, L. J. 1992. The influence of the sensory system and the environment on motion patterns in the visual displays of anoline lizards and other vertebrates. *American Naturalist* 139:S36–S61.

Ford, L. S. 1993. The phylogenetic position of the dart-poison frogs (Dendrobatidae) among anurans: an examination of the competing hypotheses and their characters. *Ethology, Ecology, and Evolution* 5:219–231.

Ford, L. S., and D. C. Cannatella. 1993. The major clades of frogs. *Herpetological Monographs* 7:94–117.

Ford, N. B., and G. M. Burghardt. 1993. Perceptual mechanisms and the behavioral ecology of snakes. Pages 117–164 in *Snakes: Ecology and Behavior,* edited by R. A. Seigel and J. T. Collins. McGraw Hill, New York, NY.

Foreman, L. D., J. M. Brode, R. Haussler, and K. Kramer. 1986. The responsibilities of federal and state agencies for protection of the desert tortoise in California. *Herpetologica* 42:59–62.

Forey, P. L., B. G. Gardiner, and C. Patterson. 1991. The lungfish, the coelacanth, and the cow revisited. Pages 145–172 in *Origins of the Higher Groups of Tetrapods: Controversy and Consensus,* edited by H.-P. Schultze and L. Trueb. Comstock Publishing Associates, Ithaca, NY.

Formanowicz, D. R., Jr. 1986. Anuran tadpole/aquatic insect predator-prey interactions: tadpole size and predator capture success. *Herpetologica* 42:367–373.

Formanowicz, D. R., Jr., E. D. Brodie, Jr., and P. J. Bradley. 1990. Behavioral compensation for tail loss in the ground skink, *Scincella lateralis. Animal Behavior* 40:782–784.

Formanowicz, D. R., Jr., E. D. Brodie, Jr., and S. C. Wise. 1989. Foraging behavior of matamata turtles: the effects of prey density and the presence of a conspecific. *Herpetologica* 45:61–67.

Fouquette, M. J. 1969. Rhinophrynidae, *Rhinophrynus dorsalis. Catalogue of American Amphibians and Reptiles* 78:1–2.

Fox, S. F., and M. A. Rostker. 1982. Social cost of tail loss in *Uta stansburiana. Science* 218:692–693.

Frazer, N. B. 1983. Demography and life history evolution of the Atlantic loggerhead sea turtle, *Caretta caretta.* Ph. D. Dissertation, University of Georgia, Athens, GA.

Frazer, N. B. 1992. Sea turtle conservation and halfway technology. *Conservation Biology* 6:179–184.

Frazer, N. B., J. W. Gibbons, and J. L. Greene. 1990. Life tables of a slider turtle population. Pages 183–200 in *Life History and Ecology of the Slider Turtle,* edited by J. W. Gibbons. Smithsonian Institution Press, Washington, D. C..

Frazzetta, T. 1983. Adaptation and function of cranial kinesis in reptiles: A time-motion analysis of feeding in alligator lizards. Pages 222–244 in *Advances in Herpetology and Evolutionary Biology,* edited by A. G. J. Rhodin and K. Miyata. Museum of Comparative Zoology, Harvard University, Cambridge, MA.

Freda, J. 1986. The influence of acidic pond water on amphibians: a review. *Water, Air and Soil Pollution* 30:439–450.

Freed, A. N. 1982. A treefrog's menu: selection for an evening's meal. *Oecologia* 53:20–26.

Freeland, W. J., and K. C. Martin. 1985. The rate of range expansion by *Bufo marinus* in Northern Australia 1980–1984. *Australian Wildlife Research* 12:555–559.

Frey, E., J. Riess, and S. F. Tarsitano. 1989. The axial tail musculature of recent crocodiles and its phyletic implications. *American Zoologist* 29:857–862.

Frost, D. R. 1985. *Amphibian Species of the World, a Taxonomic and Geographical Reference.* Allen Press, Inc. and

The Association of Systematics Collections, Lawrence, KS.

Frost, D. R. 1992. Phylogenetic analysis and taxonomy of the *Tropidurus* group of lizards (Iguania: Tropiduridae). *American Museum Novitates* 3033:1–68.

Frost, D. R. and R. Etheridge. 1989. A phylogenetic analysis and taxonomy of Iguanian lizards (Reptilia: Squamata). *Miscellaneous Publications of the Museum of Natural History, University of Kansas* 81:1–65.

Gaffney, E. S. 1975. A phylogeny and classification of the higher categories of turtles. *Bulletin of the American Museum of Natural History* 155:387–436.

Gaffney, E. S. 1977. The side-necked turtle family Chelidae: a theory of relationships using shared derived characters. *American Museum Novitates* 2620:1–28.

Gaffney, E. S. 1979. Comparative cranial morphology of recent and fossil turtles. *Bulletin of the American Museum of Natural History* 164:65–376.

Gaffney, E. S. 1984. Historical analysis of theories of chelonian relationship. *Systematic Zoology* 33:283–301.

Gaffney, E. S. 1990. The comparative osteology of the Triassic turtle *Proganochelys*. *Bulletin of the American Museum of Natural History* 194:1–263.

Gaffney, E. S., J. H. Hutchinson, F. A. Jenkins, and L. J. Meeker. 1987. Modern turtle origins: the oldest known cryptodire. *Science* 237:289–291.

Gaffney, E. S., and J. W. Kitching. 1994. The most ancient African turtle. *Nature* 369:55–58.

Gaffney, E. S., and P. A. Meylan. 1988. A phylogeny of turtles. Pages 157–219 in *The Phylogeny and Classification of the Tetrapods, Volume 1: Amphibians, Reptiles, Birds*, edited by M. J. Benton. Clarendon Press, Oxford, UK.

Gaffney, E. S., P. A. Meylan, and A. R. Wyss. 1991. A computer assisted analysis of the relationships of the higher categories of turtles. *Cladistics* 7:313–335.

Gallardo, J. M. 1961. On the species of Pseudidae (Amphibia, Anura). *Bulletin of the Museum of Comparative Zoology* 125:109–134.

Gamradt, S. C., and L. B. Kats. 1996. Effect of introduced crayfish and mosquitofish on California newts. *Conservation Biology* 10:1155–1162.

Gans, C. 1961. The feeding mechanism of snakes and its possible evolution. *American Zoologist* 1:217–227.

Gans, C. (senior editor). 1969–1992. *Biology of the Reptilia*. A total of 18 volumes. (vol. 1–13) Academic Press, London, UK; (vol. 14,15) Wiley, New York, NY; (vol. 16) Liss, New York, NY; (vol. 17,18) University of Chicago Press, Chicago, IL.

Gans, C. 1974. *Biomechanics: An Approach to Vertebrate Biology*. The University of Michigan Press, Ann Arbor, MI.

Gans, C. 1975. Tetrapod limblessness: evolution and functional corollaries. *American Zoologist* 15:455–467.

Gans, C. 1976. Aspects of the biology of uropeltid snakes. Pages 191–204 in *Morphology and Biology of Reptiles. Lin-naean Society Symposium Series* No. 3, edited by Ad'A. Bellairs and C. B. Cox. Academic Press, London, UK.

Gans, C. 1978a. The characteristics and affinities of the Amphisbaenia. *Transactions of the Zoological Society of London* 34:347–416.

Gans, C. 1978b. Reptilian venoms: Some evolutionary considerations. Pages 1–42 in *Biology of the Reptilia, Vol. 8 (Physiology B)*, edited by C. Gans and K. A. Gans. Academic Press, New York, NY.

Gans, C. 1984. Slide-pushing—a transitional locomotor method of elongate squamates. Pages 13–26 in *The Structure Development and Evolution of Reptiles. Symp. Zool. Soc. London, No. 52*, edited by M. W. J. Ferguson. Academic Press, New York, NY.

Gans, C. 1987. Studies on amphisbaenians (reptilia). 7. The small round-headed species (*cynisca*) from Western Africa. *American Museum Novitates* 2896 pp 1–84.

Gans, C. 1990. Patterns in amphisbaenian biogeography: a preliminary analysis. Pages 133–143 in *Vertebrates in the Tropics*, edited by G. Peters and R. Hutterer. Museum Alexander Koenig, Bonn, Germany.

Gans, C. 1992. Amphisbaenians. Pages 212–217 in *Reptiles and Amphibians*, edited by H. G. Cogger and R. G. Zweifel. Smithmark, New York, NY.

Gans, C. 1996. An overview of parental care among the Reptilia. *Advances in the Study of Behavior.* 25:145–157.

Gans, C., H. C. Dessauer, and D. Baic. 1978. Axial differences in the musculature of uropeltid snakes: the freight-train approach to burrowing. *Science* 199:189–192.

Gans, C., and G. C. Gorniak. 1982a. Functional morphology of lingual protrusion in marine toads (*Bufo marinus*). *American Journal of Anatomy* 163:195–222.

Gans, C., and G. C. Gorniak. 1982b. How does the toad flip its tongue? Test of two hypotheses. *Science* 216:1335–1337.

Gans, C., and G. M. Hughes. 1967. The mechanism of lung ventilation in the tortoise *Testudo graeca*. *Journal of Experimental Biology* 47:1–20.

Gans, C., and F. H. Pough 1982. Physiological ecology: its debt to reptilian studies, its value to studies of reptiles. Pages 1–13 in *Biology of the Reptilia, Vol 12 (Physiology C, Physiological Ecology)*, edited by C. Gans and F. H. Pough, Academic Press, London, UK.

Gao, K., and L. Hou. 1996. Systematics and taxonomic diversity of squamates from the Upper Cretaceous Djadochta Formation, Bayan Mandahu, Gobi Desert, People's Republic of China. *Canadian Journal of Earth Sciences* 33:578–598.

Gao, K., and H. Lianhai. 1995. Iguanians from the Upper Cretaceous Djadochta Formation. Gobi desert, China. *Journal of Vertebrate Paleontology* 15:57–78.

Gardiner, B. G. 1993. Haematothermia: Warm-Blooded Amniotes. *Cladistics* 9:369–395.

Garland, T., Jr. 1984. Physiological correlates of locomotory performance in a lizard: an allometric approach. *American Journal of Physiology* 247:R806–R815.

Garland, T., Jr., and J. B. Losos. 1994. Ecological morphology of locomotor performance in squamate reptiles. Pages 240–302 in *Ecological Morphology, Integrative Organismal Biology*, edited by P. C. Wainwright and S. M. Reilly. University of Chicago Press, Chicago, IL.

Garland, T., Jr., and P. A. Carter. 1994. Evolutionary physiology. *Annual Review of Physiology* 56:579–621.

Garrick, L. D., and J. W. Lang. 1977. Social signals and behaviors of adult alligators and crocodiles. *American Zoologist* 17:225–239.

Garstka, W. R., and D. Crews. 1985. Mate preference in garter snakes. *Herpetologica* 41:9–19.

Gasc, J.-P. 1981. Axial musculature. Pages 355–435 in *Biology of the Reptilia, Vol. 11 (Morphology F)*, edited by C. Gans and T. S. Parsons. Academic Press, New York, NY.

Gasparini, Z. 1996. Biogeographic evolution of the South American crocodilians. *Münchner Geowissenschaftliche Abhandlungen* 30 A (Geologie und Paläontologie):159–184.

Gatesy, J., and G. D. Amato. 1992. Sequence similarity of 12s ribosomal segment of mitochondrial DNAs of gharial and false gharial. *Copeia* 1992:241–243.

Gatesy, S. M. 1991. Hind limb movements of the American alligator (*Alligator mississippiensis*) and postural grades. *Journal of Zoology*, 224:577–588.

Gatten, R. E., Jr., K. Miller, and R. J. Full. 1992. Energetics at rest and during locomotion. Pages 314–377 in *Environmental Physiology of the Amphibians*, edited by M. E.Feder and W. W. Burggren, University of Chicago Press, Chicago, IL.

Gauthier, J. A. 1982. Fossil xenosaurid and anguid lizards from the early Eocene Wasatch Formation, southeast Wyoming, and a revision of the Anguioidea. *Contributions to Geology, University of Wyoming* 21:7–54.

Gauthier, J. A., D. Cannatella, K. De Queiroz, A. Kluge, and T. Rowe. 1989. Tetrapod phylogeny. Pages 337–353 in *The Hierachy of Life*, edited by B. Fernholm, K. Bremer, and H. Jornvall. Elsevier Science Publishers, Biomedical Division, Amsterdam.

Gauthier, J. A., A. G. Kluge, and T. Rowe. 1988. Amniote phylogeny and the importance of fossils. *Cladistics* 4:105–209.

Gauthier, J. A., and K. Padian. 1985. Phylogenetic, functional, and aerodynamic analyses of the origin of birds and their flight. Pages 185–197 in *The Beginnings of Birds. Proceedings of the International Archaeopteryx Conference, Eichstatt*, edited by M. K. Hecht, J. H. Ostrom, G. Viohl, and P. Wellnhofer. Eichstatt, Germany.

Gaymer, R. 1971: New method of locomotion in limbless terrestrial vertebrates. *Nature* 234:150–151.

Gehlbach, F. R., and R. S. Baldridge. 1987. Live blind snakes (*Leptotyphlops dulcis*) in eastern screech owl (*Otus asio*) nests: a novel commensalism. *Oecologia* 71:560–563.

Gehlbach, F. R., R. Gordon, and J. B. Jordan. 1973. Aestivation of the salamander, *Siren intermedia. American Midland Naturalist* 89:455–463.

Georges, A., and M. Adams. 1992. A phylogeny for Australian chelid turtles based on allozyme electrophoresis. *Australian Journal of Zoology* 40:453–476.

Gerhardt, H. C. 1978. Temperature coupling in the vocal communication system of the gray treefrog *Hyla versicolor. Science* 199:992–994.

Gerhardt, H. C. 1988. Acoustic properties used in call recognition by frogs and toads. Pages 455–483 in *The Evolution of the Amphibian Auditory System*, edited by B. Fritzsch, M. J. Ryan, W. Wilczynski, T. E. Hetherington, and W. Walkowiak. John Wiley and Sons, New York, NY.

Gerhardt, H. C. 1991. Female mate choice in treefrogs: static and dynamic acoustic criteria. *Animal Behaviour* 42:615–635.

Gerhardt, H. C. 1994. The evolution of vocalization in frogs and toads. *Annual Review of Ecology and Systematics* 25:293–324.

Gerhardt, H. C., and J. J. Schwartz (1995): Interspecific interactions in anuran courtship. Pages 603–632 in *Amphibian Biology Vol 2: Social Behaviour*, edited by H. Heatwole and B. K. Sullivan. Surrey Beatty and Sons, Chipping Norton, New South Wales, Australia.

Gibbons, J. R. H., and I. F. Watkins. 1982. Behavior, ecology, and conservation of South Pacific banded iguanas, *Brachlophus*, including a newly discovered species. Pages 418–441 in *Iguanas of the World*, edited by G. M. Burghardt and A. S. Rand. Noyes Publications, Park Ridge, NJ.

Gibbons, J. W. (editor). 1990. *Life History and Ecology of the Slider Turtle*. Smithsonian Institution Press, Washington, D. C.

Gibbons, J. W., J. L. Greene, and J. D. Congdon. 1990. Temporal and spatial movement patterns of sliders and other turtles. Pages 201–215 in *Life History and Ecology of the Slider Turtle*, edited by J. W. Gibbons. Smithsonian Institution Press, Washington, D. C..

Gibbons, J. W., and J. E. Lovich. 1990. Sexual dimorphism in turtles with emphasis on the slider turtle (*Trachemys scripta*). *Herpetological Monographs* 4:1–29.

Gibbons, J. W., and R. D. Semlitsch. 1982. Survivorship and longevity of a long-lived vertebrate species: How long do turtles live? *Journal of Animal Ecology* 51:523–527.

Gill, D. E. 1978. The metapopulation ecology of the red-spotted newt, *Notophthalmus viridescens* (Rafinesque). *Ecological Monographs* 48:145–166.

Gillingham J. C. 1987. Social behavior. Pages 184–209 in *Snakes: Ecology and Evolutionary Biology*, edited by R. A. Seigel, J. T. Collins, and S. S. Novak. Macmillan, New York, NY.

Gillingham, J. C., C. Carmichael, and T. Miller. 1995. Social behavior of the tuatara, *Sphenodon punctatus. Herpetological Monographs* 9:5–16.

Given, M. F. 1988. Growth rate and the cost of calling activity in male carpenter frogs, *Rana virgatipes. Behavioral Ecology and Sociobiology* 22:153–160.

Given, M. F. 1993. Male responses to female vocalizations in the carpenter frog, *Rana virgatipes*. *Animal Behaviour* 46:1139–1149.

Glasheen, J. W., and T. A. McMahon 1996. A hydrodynamic model of locomotion in the basilisk lizard. Nature 380:340–342.

Gleeson, T. T. 1996. Post-exercise lactate metabolism: a comparative review of sites, pathways, and regulation. *Annual Review of Physiology* 58:565–581.

Gleeson, T. T., and P. M. Dalessio. 1990. Lactate: a substrate for reptilian muscle gluconeogenesis following exhaustive exercise. *Journal of Comparative Physiology B* 169:331–338.

Glodek, G. S., and H. K. Voris. 1982. Marine snake diets: prey composition, diversity and overlap. *Copeia* 1982:661–666.

Gloyd, H. K., and R. Conant. 1990. *Snakes of the Agkistrodon Complex, a Monographic Review*. Society for the Study of Amphibians and Reptiles, Oxford, OH.

Goater, C. P., and P. I. Ward. 1992. Negative effects of *Rhabdias bufonis* (Nematoda) on the growth and survival of toads (*Bufo bufo*). *Oecologia* 89:161–165.

Godley, J. S. 1983. Observations on the courtship, nests and young of *Siren intermedia* in southern Florida. *American Midland Naturalist* 110:215–219.

Gollman, G., L. J. Borkin, and P. Roth. 1993. Genic and morphological variation in the fire-bellied toad, *Bombina bombina* (Anura, Discoglossidae). *Zoologische Jahrbuch fur Systematik* 120:129–136.

Gonser, R. A., and L. L. Woolbright. 1995. Homing behavior of the Puerto Rican frog, *Eleutherodactylus coqui*. *Journal of Herpetology* 29:481–484.

Good, D. A. 1987a. An allozyme analysis of anguid subfamilial relationships (Lacertilia: Anguidae). *Copeia* 1987: 696–701.

Good, D. A. 1987b. A phylogenetic analysis of cranial osteology in the gerrhonotine lizards. *Journal of Herpetology* 21:285–297.

Good, D. A. 1988. Phylogenetic relationships among Gerrhonotine lizards, an analysis of external morphology. *University of California Publications in Zoology* 121:1–139.

Good, D. A. 1989. Hybridization and cryptic species in *Dicamptodon* (Caudata: Dicamptodontidae). *Evolution* 43:728–744.

Good, D. A. 1994. Species limits in the genus *Gerrhonotus* (Squamata: Anguidae). *Herpetological Monographs* 8:180–202.

Good, D. A., and D. B. Wake. 1992. Geographic variation and speciation in the torrent salamanders of the genus *Rhyacotriton* (Caudata: Rhyacotritonidae). *University of California Publications in Zoology* 126:1–91.

Good, D. A., and D. B. Wake. 1993. Systematic studies of the Costa Rican moss salamanders, genus *Nototriton*, with descriptions of three new species. *Herpetological Monographs* 7:131–159.

Goodrich, E. S. 1930. *Structure and Development of Vertebrates*. Macmillan, London, UK.

Gorman, G. C. 1973. The chromosomes of the Reptilia, a cytotaxonomic interpretation. Pages 349–424 in *Cytotaxonomy and Vertebrate Evolution*, edited by A. B. Chiarelli. Academic Press, New York, NY.

Gorniak, G. C., H. I. Rosenberg, and C. Gans. 1982. Mastication in the tuatara, *Sphenodon punctatus* (Reptilia: Rhynchocephalia): structure and activity of the motor system. *Journal of Morphology* 171:321–353.

Gorzula, S. J. 1978. An ecological study of *Caiman crocodilus crocodilus* inhabiting savannah lagoons in the Venezuelan Guayana. *Oecologia* 35:21–34.

Gradwell, N. 1971. *Ascaphus* tadpole: Experiments on the suction and gill irrigation mechanisms. *Canadian Journal of Zoology* 49:307–332.

Gradwell, N. 1972a. Gill irrigation in *Rana catesbiana*. Part I. On the anatomical basis. *Canadian Journal of Zoology* 50:481–499.

Gradwell, N. 1972b. Gill irrigation in *Rana catesbiana*. Part II. On the musculoskeletal mechanism. *Canadian Journal of Zoology* 50:501–521.

Graf, J.-D. 1996. Molecular approaches to the phylogeny of *Xenopus*. Pages 379–389 in *The Biology of Xenopus*, edited by R. C. Tinsley and H. R. Kobel. Clarendon Press, Oxford, UK.

Graf, J.-D., and M. Polls Pelaz. 1989. Evolutionary genetics of the *Rana esculenta* complex. Pages 289–302 in *Evolution and Ecology of Unisexual Vertebrates (New York State Museum Bulletin 466)*, edited by R. M. Dawley and J. P. Bogart. State University of New York, Albany, NY.

Graham, T., A. Georges, and N. McElhinney. 1996. Terrestrial orientation by the eastern long-necked turtle, *Chelodina longicollis*, from Australia. *Journal of Herpetology* 30:467–477.

Grandison, A. G. C. 1980. Aspects of breeding morphology in *Mertensophryne micranotis* (Anura: Bufonidae): secondary sexual characters, eggs and tadpole. *Bulletin of the British Museum (Natural History)* 39:299–304.

Gray, J. 1968. *Animal Locomotion*. W. W. Norton & Co., New York, NY.

Graybeal, A. 1993. The phylogenetic utility of cytochrome b: lessons from bufonid frogs. *Molecular Phylogenetics and Evolution* 2:256–269.

Graybeal, A., and D. C. Cannatella. 1995. A new taxon of Bufonidae from Peru, with descriptions of two new species and a review of the phylogenetic status of supraspecific bufonid taxa. *Herpetologica* 51:105–131.

Green, A. J. 1991. Large male crests, an honest indicator of condition, are preferred by female smooth newts, *Triturus vulgaris* (Salamandridae) at the spermatophore transfer stage. *Animal Behaviour* 41:367–369.

Green, D. 1984. Long-distance movements of Galapagos green turtles. *Journal of Herpetology* 18:121–130.

Green, D. M. 1988a. Cytogenetics of the endemic New Zealand frog, *Leiopelma hochstetteri*: extraordinary supernumerary chromosome variation and a unique sex chromosome. *Chromosoma* 97:55–70.

Green, D. M. 1988b. Heteromorphic sex-chromosomes in the rare and primitive frog *Leiopelma hamiltoni* from New Zealand. *Journal of Heredity* 79:165–169.

Green, D. M., and D. C. Cannatella. 1993. Phylogenetic significance of the amphicoelous frogs, Ascaphidae and Leiopelmatidae. *Ethology, Ecology, and Evolution* 5:233–245.

Green, D. M., and S. K. Sessions (editors). 1991. *Amphibian Cytogenetics and Evolution*. Academic Press, Inc., San Diego.

Green, D. M., and T. F. Sharbel. 1988. Comparative cytogenetics of the primitive frog, *Leiopelma archeyi* (Anura; Leiopelmatidae). *Cytogenetics and Cell Genetics* 47:212–216.

Greenberg, N. and D. Crews (editors). 1977. *Social behavior of reptiles* (a symposium). American Zoologist 17: 151–286.

Greene, H. W. 1982. Dietary and phenotypic diversity in lizards: why are some organisms specialized? Pages 107–128 in *Environmental Adaptation and Evolution: A Theoretical and Empirical Approach*, edited by D. Mossakowski and G. Roth. Fischer Verlag, Stuttgart, Germany.

Greene, H. W. 1983. Dietary correlates of the origin and radiation of snakes. *American Zoologist* 23:431–441.

Greene, H. W. 1988. Antipredator mechanisms in reptiles. Pages 1–152 in *Biology of the Reptilia, Vol. 16 (Ecology B, Defense and Life History)*, edited by C. Gans and R. B. Huey. Alan R. Liss, NY.

Greene, H. W. 1989. Ecological, evolutionary, and conservation implications of feeding biology in Old World cat snakes, genus *Boiga* (Colubridae). *Proceedings of the California Academy of Science* 46:193–207.

Greene, H. W. 1992. The ecological and behavioral context for pitviper evolution. Pages 107–117 in *Biology of Pit Vipers*, edited by J. A. Campbell and E. D. Brodie Jr. Selva Natural History Book Publishers, Tyler, TX.

Greene, H. W., and G. M. Burghardt. 1978. Behavior and phylogeny: constriction in ancient and modern snakes. *Science* 200:74–77.

Greene, H. W., and R. W. McDiarmid. 1981. Coral snake mimicry: does it occur? *Science* 213:1207–1212.

Greer, A. E. 1970. A subfamilial classification of scincid lizards. *Bulletin of the Museum of Comparative Zoology* 139:151–184.

Greer, A. E. 1985. The relationships of the lizard genera *Anelytropsis* and *Dibamus*. *Journal of Herpetology* 19: 116–156.

Greer, A. E. 1990. Limb reduction in the scincid lizard genus *Lerista*. 2. Variation in the bone complements of the front and rear limbs and the number of postsacral vertebrae. *Journal of Herpetology* 24:142–150.

Greer, A. E. 1991. Limb reduction in squamates: identification of the lineages and discussion of the trends. *Journal of Herpetology* 25:166–173.

Gregory, P. T. 1982. Reptilian hibernation. Pages 53–154 in *Biology of the Reptilia, Vol 13 (Physiology D, Physiological Ecology)*, edited by C. Gans and F. H. Pough, Academic Press, London, UK.

Gregory, P. T., J. M. Macartney, and K. W. Larsen. 1987. Spatial patterns and movements. Pages 366–395 in *Snakes: Ecology and Evolutionary Biology*, edited by R. A. Seigel, J. T. Collins, and S. S. Novak. MacMillan, New York, NY.

Grenard, S. 1994. *Medical Herpetology*. Reptile & Amphibian Magazine. A Division of NG Publishing Inc. Pottsville, PA.

Grimmond, N. M., M. R. Preest, and F. H. Pough. 1994. Energetic cost of feeding on different kinds of prey for the lizard *Chalcides ocellatus*. *Functional Ecology* 8:17–21.

Grismer, L. L. 1988. Phylogeny, taxonomy, classification, and biogeography of eublepharid geckos. Pages 369–469 in *Phylogenetic Relationships of the Lizard Families, Essays Commemorating Charles L. Camp*, edited by R. Estes and G. K. Pregill. Stanford University Press, Stanford, CA.

Groombridge, B. C. 1979a. On the vomer in Acrochordidae (Reptilia: Serpentes), and its cladistic significance. *Journal of Zoology, London* 189:559–567.

Groombridge, B. C. 1979b. Variations in morphology of the superficial palate of henophidian snakes and some possible systematic implications. *Journal of Natural History* 13:447–475.

Groombridge, B. C. 1984. The facial carotid artery in snakes (Reptilia, Serpentes): variations and possible cladistic significance. *Amphibia Reptilia* 5:145–155.

Gross, M. R., and R. Shine. 1981. Parental care and mode of fertilization in ectothermic vertebrates. *Evolution* 35: 775–793.

Guillette, L. J., Jr. 1982. The evolution of viviparity and placentation in the high elevation, Mexican lizard *Sceloporus aeneus*. *Herpetologica* 38:94–103.

Guillette, L. J., Jr. 1987. The evolution of viviparity in fishes, amphibians and reptiles: an endocrine approach. Pages 523–560 in *Hormones and Reproduction in Fishes, Amphibians, and Reptiles*, edited by D. O. Norris and R. E. Jones. Plenum, New York, NY.

Guillette, L. J., Jr., and D. A. Crain. 1996. Endocrine-disrupting contaminants and reproductive abnormalities in reptiles. *Comments Toxicology* 5:381–399.

Guillette, L. J., Jr., T. S. Gross, G. R. Masson, J. M. Matter, H. F. Percival, and A. R. Woodward. 1994. Developmental abnormalities of the gonad and abnormal sex hormone concentrations in juvenile alligators from contaminated and control lakes in Florida. *Environmental Health Perspectives* 102:680–688.

Guimond, R. W., and V. H. Hutchison. 1973. Aquatic respiration: an unusual strategy in the hellbender *Crypto-*

branchus alleganiensis alleganiensis (Daudin). *Science* 182: 1263–1265.

Gullison, R. E., and E. C. Losos. 1993. The role of foreign debt in deforestation in Latin America. *Conservation Biology* 7:140–147.

Guyer, C. 1991. Orientation and homing behavior as a measure of affinity for the home range in two species of iguanid lizards. *Amphibia Reptilia* 12:373–384.

Guyer, C., and J. M. Savage. 1986. Cladistic relationships among anoles (Sauria: Iguanidae). *Systematic Zoology* 35: 509–531.

Guyer, C., and J. M. Savage. 1992. Anole systematics revisited. *Systematic Biology* 41:89–110.

Gyi, K. K. 1970. A revision of colubrid snakes of the subfamily Homalopsinae. *University of Kansas Publications, Museum of Natural History* 20:47–223.

Haacke, W. D. 1969. The call of the barking geckos (Gekkonidae: Reptilia). *Scientific Papers of the Namib Desert Research Station* 46:83–93.

Haas, A. 1995. Cranial features of dendrobatid larvae (Amphibia: Anura: Dendrobatidae). *Journal of Morphology* 224:241–264.

Haas, G. 1955. The systematic position of *Loxocemus bicolor* Cope (Ophidia). *American Museum Novitates* 1748: 1–8.

Haas, G. 1960. On the trigeminus muscles of the lizards *Xenosaurus grandis* and *Shinisaurus crocodilurus*. *American Museum Novitates* 2017:1–54.

Haas, G. 1973. Muscles of the jaws and associated structures in the Rhynchocephalia and Squamata. Pages 285–490 in *Biology of the Reptilia, Vol. 4 (Morphology D)*, edited by C. Gans and T. S. Parsons. Academic Press, New York, NY.

Hahn, D. E. 1978. A brief review of the genus *Leptotyphlops* (Reptilia, Serpentes, Leptotyphlopidae) of Asia, with description of a new species. *Journal of Herpetology* 12: 477–489.

Haiduk, M. W., and J. W. Bickham. 1982. Chromosomal homologies and evolution of testudinoid turtles with emphasis on the systematic placement of *Platysternon*. *Copeia* 1982:60–66.

Hailey, A., and P. M. C. Davies. 1988. Activity and thermoregulation of the snake *Natrix maura*. 2. A synoptic model of thermal biology and the physiological ecology of performance. *Journal of Zoology, London* 214:325–342.

Hairston, N. G. 1949. The local distribution and ecology of the plethodontid salamanders of the southern Appalachians. *Ecological Monographs* 19:47–73.

Hairston, N. G. 1980a. Evolution under interspecific competition: field experiments on terrestrial salamanders. *Evolution* 34:409–420.

Hairston, N. G. 1980b. The experimental test of an analysis of field distributions: competition in terrestrial salamanders. *Ecology* 61:817–826.

Hairston, N. G., Sr. 1986. Species packing in *Desmognathus* salamanders: experimental demonstration of predation and competition. *American Naturalist* 127:266–291.

Hairston, N. G., Sr. 1987. *Community Ecology and Salamander Guilds*. Cambridge University Press, Cambridge, UK.

Hall, J. C. 1994. Central processing of communication sounds in the anuran auditory system. *American Zoologist* 34:670–684.

Halliday, T. R. 1977. The courtship of European newts: an evolutionary perspective. Pages 185–232 in *The Reproductive Biology of Amphibians*, edited by D. H. Taylor and S. I. Guttman. Plenum Press, New York, NY.

Halliday, T. R. 1983. Information and communication. Pages 43–81 in *Animal Behaviour*. Volume 2. *Communication*, edited by T. R. Halliday and P. J. B. Slater. Freeman, New York, NY.

Halliday, T. R. 1990. The evolution of courtship behavior in newts and salamanders. *Advances in the Study of Behavior.* 19:137–169.

Halliday, T., and B. Arano. 1991. Resolving the phylogeny of European newts. *Trends in Ecology and Evolution* 6:113–117.

Halliday, T. R., and M. Tejedo (1995): Intrasexual selection and alternative mating behaviour. Pages 419–468 in *Amphibian Biology. Vol. 2. Social Behaviour*, edited by H. Heatwole and B. K. Sullivan. Surrey Beatty and Sons, Chipping Norton, New South Wales, Australia.

Halliday, T. R., and P. A. Verrell. 1984. Sperm competition in amphibians. Pages 487–508 in *Sperm Competition and the Evolution of Animal Mating Systems*, edited by R. L. Smith. Academic Press, New York, NY.

Halpern, M. T. 1992. Nasal chemical senses in reptiles: structure and function. Pages 423–523 in *Biology of the Reptilia, Vol. 18 (Physiology E, Hormones, Brain, and Behavior)*, edited by C. Gans and D. Crews. Alan R. Liss, Inc., New York, NY.

Hanken, J. 1983. Miniaturization and its effects on cranial morphology in plethodontid salamanders, genus *Thorius* (Amphibia: Plethodontidae): II. The fate of the brain and sense organs and their role in skull morphogenesis and evolution. *Journal of Morphology* 177: 255–268.

Hanken, J. 1984. Miniaturization and its effects on cranial morphology in plethodontid salamanders, genus *Thorius* (Amphibia: Plethodontidae). I. Osteological variation. *Biological Journal of the Linnaean Society* 23:55–75.

Hanken, J. 1986. Developmental evidence for amphibian origins. *Evolutionary Biology* 20:389–417.

Hanken, J. 1993. Model systems versus outgroups: alternative approaches to the study of head development and evolution. *American Zoologist* 33:448–456.

Hanken, J., and D. B. Wake. 1994. Five new species of minute salamanders, genus *Thorius* (Caudata: Plethodontidae), from northern Oaxaca, Mexico. *Copeia* 1994: 573–590.

Hanna, G., and W. J. P. Barnes. 1991. Adhesion and detachment of the toe pads of tree frogs. *Journal of Experimental Biology* 155:103–125.

Hansen, S. 1989. Debt for nature swaps: overview and discussion of key issues. *Ecological Economics* 1:77–93.

Hardaway, T. E., and K. L. Williams. 1976. Costal cartilages in snakes and their phylogenetic significance. *Herpetologica* 32:378–387.

Harding, K. A. 1982. Courtship display in a Bornean frog. *Proceedings of the Biological Society of Washington* 95:621–624.

Hardy, D. L. 1994. A re-evaluation of suffocation as the cause of death during constriction by snakes. *Herpetological Review* 25:45–47.

Harkness, L. 1977. Chameleons use accommodation cues to judge distance. *Nature* 267:346–349.

Harris, D. M. 1982. The *Sphaerodactylus* (Sauria: Gekkonidae) of South America. *Occasional Papers of the Museum of Zoology, University of Michigan* 704:1–31.

Harris, D. M. 1985. Infralingual plicae: support for Boulenger's Teiidae (Sauria). *Copeia* 1985:560–565.

Harris, D. M. 1994. Review of the teiid lizard genus *Ptychoglossus*. *Herpetological Monographs* 8:226–275.

Harris, D. M., and A. G. Kluge. 1984. The *Sphaerodactylus* (Sauria: Gekkonidae) of Middle America. *Occasional Papers of the Museum of Zoology, University of Michigan* 706:1–59.

Harris, J. B. 1997. Toxic phospholipases in snake venom: an introductory review. Pages 235–250 in *Venomous Snakes: Ecology, Evolution and Snakebite*, edited by R. S. Thorpe, W. Wüster, and A. Malhotra. Clarendon Press, Oxford.

Harris, L. D. 1984. *The Fragmented Forest*. The University of Chicago Press, Chicago, IL.

Harte, J., and E. Hoffman. 1989. Possible effects of acidic deposition on a Rocky Mountain population of the tiger salamander *Ambystoma tigrinum*. *Conservation Biology* 3:149–158.

Hartline, P. H., L. Kass, and M. S. Loop. 1978. Merging of modalities in the optic tectum: Infrared and visual integration in rattlesnakes. *Science* 199:1225–1229.

Hass, C. A., J. F. Dunski, L. R. Maxson, and M. S. Hoogmoed. 1995. Divergent lineages within the *Bufo margaritifera* complex (Amphibia: Anura: Bufonidae) revealed by albumin immunology. *Biotropica* 27:238–249.

Hass, C. A., M. A. Hoffman, L. D. Densmore, and L. R. Maxson. 1992. Crocodilian evolution: insights from immunological data. *Molecular Phylogenetics and Evolution* 1:193–201.

Hass, C. A., R. A. Nussbaum, and L. R. Maxson. 1993. Immunological insights into the evolutionary history of caecilians (Amphibia: Gymnophiona): relationships of the Seychellean caecilians and a preliminary report on

family-level relationships. *Herpetological Monographs* 7:56–63.

Hasumi, M. 1994. Reproductive behavior of the salamander *Hynobius nigrescens*: monopoly of egg sacs during scramble competition. *Journal of Herpetology* 28:264–267.

Hay, J. M., I. Ruvinsky, S. B. Hedges, and L. R. Maxson. 1995. Phylogenetic relationships of amphibian families inferred from DNA sequences of mitochondrial 12S and 16S ribosomal RNA genes. *Molecular Biology and Evolution* 12:928–937.

Hayes, T., and P. Licht. 1992. Gonadal involvement in sexual size dimorphism in the African bullfrog (*Pyxicephalus adspersus*). *Journal of Experimental Zoology* 264:130–135.

Hayes, W. K., I. I. Kaiser, and D. Duvall. 1992. The mass of venom expended by prairie rattlesnakes when feeding on rodent prey. Pages 383–388 in *Biology of Pitvipers*, edited by J. A. Campbell and E. D. Brodie, Jr. Selva, Tyler, TX.

Hayes, W. K., P. Lavín-Murcio, and K. V. Kardong. 1993. Delivery of Duvernoy's secretion into prey by the brown tree snake, *Boiga irregularis* (Serpentes: Colubridae). *Toxicon* 31:881–887.

Hayes, W. K., P. Lavín-Murcio, and K. V. Kardong. 1995. Northern Pacific rattlesnakes (*Crotalus viridis oreganus*) meter venom when feeding on prey of different sizes. *Copeia* 1995:337–343.

Healy, W. R. 1974. Population consequences of alternative life histories in *Notophthalmus viridescens*. *Copeia* 1974: 221–229.

Heath, J. E. 1965. Temperature regulation and diurnal activity in horned lizards. *University of California Publications in Zoology* 64:97–136.

Heatwole, H. (editor) 1994, 1995. *Amphibian Biology* (2 vol.). Surrey Beatty, Chipping Norton, NSW, Australia.

Heatwole, H., and N. S. Poran. 1995. Resistances of sympatric and allopatric eels to sea snake venoms. *Copeia* 1995:136–147.

Hecht, M. K., and J. L. Edwards. 1976. The determination of parallel or monophyletic relationships: the proteid salamanders—a test case. *American Naturalist* 110: 653–677.

Hecht, M. K., V. Walters, and G. Ramm. 1955. Observations on the natural history of the Bahaman pigmy boa, *Tropidophis pardalis*, with notes on autohemorrhage. *Copeia* 1955:249–251.

Hedges, S. B. 1986. An electrophoretic analysis of Holarctic hylid frog evolution. *Systematic Zoology* 35:1–21.

Hedges, S. B., and R. L. Bezy. 1993. Phylogeny of xantusiid lizards: concern for data and analysis. *Molecular Phylogenetics and Evolution* 2:76–87.

Hedges, S. B., J. P. Bogart, and L. R. Maxson. 1992. Ancestry of unisexual salamanders. *Nature* 356:708–710.

Hedges, S. B., and O. H. Garrido. 1992. A new species of *Tropidophis* from Cuba (Serpentes: Tropidophiidae). *Copeia* 1992:820–825.

Hedges, S. B., and L. R. Maxson. 1993. A molecular perspective on Lissamphibian phylogeny. *Herpetological Monographs* 7:27–42.

Hedges, S. B., K. Moberg, and L. R. Maxson. 1990. Tetrapod phylogeny inferred from 18S and 28S ribosomal RNA sequences and a review of the evidence for amniote relationships. *Molecular Biology and Evolution* 7:607–633.

Hedges, S. B., R. A. Nussbaum, and L. R. Maxson. 1993. Caecilian phylogeny and biogeography inferred from mitochondrial DNA sequences of the 12s rRNA and 16s rRNA genes (Amphibia: Gymnophiona). *Herpetological Monographs* 7:64–76.

Hedlund, L. 1990. Factors affecting differential mating success in male crested newts, *Triturus cristatus*. *Journal of Zoology, London* 220:33–40.

Hedlund, L., and J. G. M. Robertson. 1989. Lekking behaviour in crested newts, *Triturus cristatus*. *Ethology* 80: 111–119.

Heise, P. J., L. R. Maxson, H. G. Dowling, and S. B. Hedges. 1995. Higher-level snake phylogeny inferred from mitochondrial DNA sequences of 12S rRNA and 16s rRNA genes. *Molecular Biology and Evolution* 12: 259–265.

Heisler, N., P. Neumann, and G. M. O. Maloiy. 1983. The mechanism of intra-cardiac shunting in the lizard *Varanus exanthematicus*. *Journal of Experimental Blogy* 105: 15–32.

Hennig, W. 1966. *Phylogenetic Systematics*. University of Illinois Press, Urbana, IL.

Henrici, A. C. 1994. *Tephrodytes brassicarvalis*, new genus and species (Anura: Pelodytidae), from the Arikareean Cabbage Patch Geds of Montana, USA, and pelodytid-pelobatid relationships. *Annals of the Carnegie Museum of Natural History* 63:155–183.

Hertz, P. E., R. B. Huey, and E. Nevo. 1982. Flight versus flight: body temperature influences defensive responses of lizards. *Animal Behavior* 30:676–679.

Hertz, P. E., R. B. Huey, and R. D. Stevenson. 1993. Evaluating temperature regulation by field-active ectotherms: the fallacy of the inappropriate question. *American Naturalist* 142:796–818.

Hetherington, T. E. 1989. Use of vibratory cues for detection of insect prey by the sandswimming lizard *Scincus scincus*. *Animal* Behavior, 37:290–297.

Hetherington, T. E. 1992. The effects of body size on the evolution of the amphibian middle ear. Pages 421–437 in *The Evolutionary Biology of Hearing*, edited by D. B. Webstrer, R. R. Fay, and A. N. Popper. Springer-Verlag. New York, NY.

Heulin, B., C. P. Guillaume, A. Bea, and M. J. Arrayago. 1993. Interprétation biogéographique de la bimodalité de reproduction du lézard *Lacerta vivipara* Jacquin (Sauria, Lacertidae): un modèle pour l'étude de l'évolution de las viviparité. *Biogeographica* 69:3–13.

Heusser, H. 1968. Die Lebensweise der Erdkröte *Bufo bufo* (L.) Wanderungen und Sommerquartiere. *Revue Suisse de Zoologie* 75:927–982.

Hews, D. K. 1990. Examining hypotheses generated by field measures of sexual selection on male lizards, *Uta palmeri*. *Evolution* 44:1956–1966.

Hews, D. K. 1993. Food resource affect female distribution and male mating opportunities in the iguanian lizard Uta palmeri. *Animal Behaviour* 46:279–291.

Heyer, W. R. 1973. Ecological interactions of frog larvae at a seasonal tropical location in Thailand. *Journal of Herpetology* 7:337–361.

Heyer, W. R. 1974. Niche measurements of frog larvae from a seasonal tropical location in Thailand. *Ecology* 55:651–656.

Heyer, W. R. 1975. A preliminary analysis of the intergeneric relationships of the frog family Leptodactylidae. *Smithsonian Contributions to Zoology* 199:1–55.

Heyer, W. R. 1976. Studies in larval amphibian habitat partitioning. *Smithsonian Contributions to Zoology* 242:1–27.

Heyer, W. R. 1994. Variation within the *Leptodactylus podicipinus-wagneri* complex of frogs (Amphibia: Leptodactylidae). *Smithsonian Contributions to Zoology* 546:1–124.

Heyer, W. R., and D. S. Liem. 1976. Analysis of the intergeneric relationships of the Australian frog Family Myobatrachidae. *Smithsonian Contributions to Zoology* 233: 1–29.

Heyer, W. R., and L. R. Maxson. 1983. Relationships, zoogeography, and speciation mechanisms of frogs of the genus *Cycloramphus* (Amphibia, Leptodactylidae). *Arquivos de Zoologia, Sao Paulo* 30:341–373.

Hicks, J. W., A. Ishimatsu, S. Molloi, A. Erskin, and N. Heisler. 1996. The mechanism of cardiac shunting in reptiles: a new synthesis. *Journal of Experimental Biology* 199:1435–1446.

Hicks, J. W., and S. C. Wood 1989: Oxygen homeostasis in lower vertebrates: The impact of external and internal hypoxia. Pages 311–341 in *Comparative Pulmonary Physiology: Current Concepts, vol 39 (Lung Biology in Health and Disease)*, edited by S. C. Wood, Marcel Dekker, New York, NY.

Hiestand, P., and R. R. Hiestand. 1979. *Dispholidus typus* (boomslang) snake venom: purification and properties of the coagulant principle. *Toxicon* 17:489–498.

Highton, R. 1962. Revision of the North American salamanders of the genus *Plethodon*. *Bulletin of the Florida State Museum (Biological Sciences)* 6:235–267.

Highton, R. 1991. Molecular phylogeny of plethodonine salamanders and hylid frogs: statistical analysis of protein comparisons. *Molecular Biology and Evolution* 8:796–818.

Highton, R., G. C. Maha, and L. R. Maxson. 1989. Biochemical evolution in the slimy salamanders of the *Plethodon glutinosus* complex in the eastern United States. *Illinois Biological Monographs* 57:1–153.

Hildebrand, M. 1985. Walking and running. Pages 38–57 in *Functional Vertebrate Morphology*, edited by M. Hildebrand, D. M. Bramble, K. F. Liem, and D. B. Wake. Belknap Press, Cambridge, MA.

Hildebrand, M. 1995. *Analysis of Vertebrate Structure, Fourth Ed.* John Wiley & Sons, New York, NY.

Hillenius, D. 1986. The relationship of *Brookesia, Rhampholeon* and *Chamaeleo* (Chamaeleonidae, Reptilia). *Bijdragen tot de Dierkunde* 56:29–38.

Hillenius, D. 1988. The skull of *Chamaeleo nasutus* adds more information to the relationship of *Chamaeleo* with *Rhampholeon* and *Brookesia* (Chamaeleonidae, Reptilia). *Bijdragen tot de Dierkunde* 58:7–11.

Hillis, D. M. 1991. The phylogeny of amphibians: current knowledge and the role of cytogenetics. Pages 7–31 in *Amphibian Cytogenetics and Evolution*, edited by D. M. Green and S. K. Sessions. Academic Press, San Diego.

Hillis, D. M., L. K. Ammerman, M. T. Dixon, and R. O. de Sá. 1993. Ribosomal DNA and the phylogeny of frogs. *Herpetological Monographs* 7:118–131.

Hillis, D. M., and D. M. Green. 1990. Evolutionary changes of heterogametic sex in the phylogenetic history of amphibians. *Journal of Evolutionary Biology* 3:49–64.

Hirayama, R. 1984. Cladistic analysis of batagurine turtles (Batagurinae: Emydidae: Testudinoidea); a preliminary result. *Studia Geologica Salmanticiensia* Spec. vol. 1:141–157.

Hofman, A., L. R. Maxson, and J. W. Arntzen. 1991. Biochemical evidence pertaining to the taxonomic relationships within the family Chamaeleonidae. *Amphibia Reptilia* 12:245–265.

Höglund, J., and R. V. Alatalo. 1995. *Leks*. Princeton University Press, Princeton, NJ.

Holman, J. A. 1984. *Texasophis galbreathi*, new species, the earliest New World colubrid snake. *Journal of Vertebrate Paleontology* 3:223–225.

Holman, J. A. 1995. *Pleistocene Amphibians and Reptiles in North America*. Oxford University Press, New York, NY.

Holomuzki, J. R. 1995. Oviposition sites and fish-deterrent mechanisms of two stream anurans. *Copeia* 1995:607–613.

Hoogmoed, M. S. 1969. Notes on the herpetology of Surinam II. On the occurrence of *Allophryne ruthveni* Gaige (Amphibia, Salientia, Hylidae) in Surinam. *Zoologische Mededelingen* 44:75–81.

Houck, L. D. 1977. Life history patterns and reproductive biology of neotropical salamanders. Pages 43–72 in *The Reproductive Biology of Amphibians*, edited by D. H. Taylor and S. I. Guttman. Plenum Publ. Co., New York, NY.

Houck, L. D., and N. L. Reagan. 1990. Male courtship pheromones increase female receptivity in a plethodontid salamander. *Animal Behaviour* 39:729–734.

Houck, L. D., and D. M. Sever 1994. Role of the skin in reproduction and behavior. Pages 351–381 in *Amphibian Biology Vol 1: The Integument*, edited by H. Heatwole and G. T. Barthalmus. Surrey Beatty and Sons, Chipping Norton, New South Wales, Australia.

Houck, L. D., and S. K. Woodley. 1995. Field studies of steroid hormones and male reproductive behaviour in amphibians. Pages 677–703 in *Amphibian Biology*, edited by H. Heatwole and B. K. Sullivan. Surrey Beatty and Sons, NSW, Australia.

Hourdry, J., A. L'Hermite, and R. Ferrand. 1996. Changes in the digestive tract and feeding behavior of anuran amphibians during metamorphosis. *Physiological Zoology* 69:219–251.

Howard, R. D. 1978a. The evolution of mating strategies in bullfrogs, *Rana catesbeiana*. *Evolution* 32:850–871.

Howard, R. D. 1978b. The influence of male-defended oviposition sites on early embryo mortality in bullfrogs. *Ecology* 59:789–798.

Howard, R. D. 1980. Mating behviour and mating success in wood frogs, *Rana sylvatica*. *Animal Behaviour* 28:705–716.

Howard, R. D. 1981. Sexual dimorphism in bullfrogs. *Ecology* 62:303–310.

Howard, R. D. 1988. Sexual selection on male body size and mating behaviour in American toads, *Bufo americanus*. *Animal Behaviour* 36:1796–1808.

Howard, R. D., and A. G. Kluge. 1985. Proximate mechanisms of sexual selection in wood frogs. *Evolution* 39:260–277.

Huey, R. B. 1982: Temperature, physiology, and the ecology of reptiles. Pages 25–91 in *Biology of the Reptilia, Vol. 12 (Physiology C, Physiological Ecology)*, edited by C. Gans and F. H. Pough, Academic Press, London, UK.

Huey, R. B. 1987. Phylogeny, history, and the comparative method. Pages 76–101 in *New Directions in Ecological Physiology*, edited by M. E. Feder, A. F. Bennett, W. W. Burggren, and R. B. Huey. Cambridge University Press, New York, NY.

Huey, R. B. 1991. Physiological consequences of habitat selection. *American Naturalist* 137, Supplement: S91–S115.

Huey, R. B., and A. F. Bennett. 1986. A comparative approach to field and laboratory studies in evolutionary biology. Pages 82–98 in *Predator-Prey Relationships*, edited by M. E. Feder and G. V. Lauder. University of Chicago Press, Chicago, IL.

Huey, R. B., A. E. Dunham, K. L. Overall, and R. A. Newman. 1990. Variation in locomotor performance in demographically known populations of the lizard *Sceloporus merriami*. *Physiological Zoology* 63:845–872.

Huey, R. B., H. John-Alder, and K. A. Nagy. 1984. Locomotor capacity and foraging behavior of Kalahari lizards. *Animal Behavior* 32:41–50.

Huey, R. B., and E. R. Pianka. 1977. Natural selection for juvenile lizards mimicking noxious beetles. *Science* 195:201–203.

Huey, R. B., and E. R. Pianka. 1981. Ecological consequences of foraging mode. *Ecology* 62:991–999.

Huey, R. B., E. R. Pianka, M. E. Egan, and L. W. Coons.
1974. Ecological shifts in sympatry: Kalahari fossorial
lizards (*Typhlosaurus*). *Ecology* 55:304–316.

Huey, R. B., E. R. Pianka, and T. W. Schoener (editors).
1983. *Lizard Ecology: Studies of a Model Organism*. Harvard University Press, Cambridge, MA.

Huey, R. B., and R. D. Stevenson. 1979. Integrating thermal physiology and ecology of ectotherms: A discussion
of approaches. *American Zoologist* 19:357–366.

Hutchison, J. H. 1991. Early Kinosterninae (Reptilia: Testudines) and their phylogenetic significance. *Journal of
Vertebrate Paleontology* 11:145–167.

Hutchison, V. H., and R. K. Dupr. 1992. Thermoregula-
tion. Pages 206–249 in *Environmental Physiology of the
Amphibians*, edited by M. E. Feder and W. W. Burggren,
University of Chicago Press, Chicago, IL.

Hutchison, V. H., H. B. Haines, and G. Engbretson. 1976.
Aquatic life at high altitude: respiratory adaptations in
the Lake Titicaca frog, *Telmatobius culeus*. *Respiration
Physiology* 27:115–129.

Inger, R. F. 1967. The development of a phylogeny of frogs.
Evolution 21:369–384.

Ingram, G. J., M. Anstis, and C. J. Corben. 1975. Observations on the Australian leptodactylid frog, *Assa darlingtoni*. *Herpetologica* 31:425–429.

Ivachnenko, M. F. 1978. Urodelans from the Triassic and
Jurassic of Soviet Central Asia. *Paleontologie Zhurnal*
1978:84–89.

Iverson, J. B. 1978. The impact of feral cats and dogs on
populations of the West Indian rock iguana, *Cyclura carinata*. *Biological Conservation* 14:63–73.

Iverson, J. B. 1979. Behavior and ecology of the rock
iguana, *Cyclura carinata*. *Bulletin of the Florida State Museum, Biological Series* 24:175–358.

Iverson, J. B. 1980. Colic modifications in iguanine lizards.
Journal of Morphology 163:79–93.

Iverson, J. B. 1982. Adaptations to herbivory in iguanine
lizards. Pages 60–76 in *Iguanas of the World: Their
Behavior, Ecology, and Conservation*, edited by G. M.
Burghardt and A. S. Rand. Noyes Publications, Park
Ridge, NJ.

Iverson, J. B. 1986. Notes on the natural history of the
Caicos Islands Dwarf boa, *Tropidophis greenwayi*. *Caribbean Journal of Science* 22:191–198.

Iverson, J. B. 1991. Phylogenetic hypotheses for the evolution of modern kinosternine turtles. *Herpetological Monographs* 5:1–27.

Iverson, J. B. 1992. *A Revised Checklist with Distribution Maps
of the Turtles of the World*. Privately Printed, Richmond,
IN.

Izecksohn, E. 1971. Novo genero e nova especie de Brachycephalidae do Estado do Rio de Janeiro, Brasil (Amphibia, Anura). *Boletim do Museu Nacional, Rio de Janeiro*
280:1–12.

Jackman, T. R., and D. B. Wake. 1994. Evolutionary and
historical analysis of protein variation in the blotched
forms of salamanders of the *Ensatina* complex (Amphibia: Plethodontidae). *Evolution* 48:876–897.

Jackson, C. G., and J. D. Davis. 1972. Courtship display behavior in *Chrysemys concinna suwanniensis*. *Copeia* 1972:
385–387.

Jackson, C. G., and J. D. Davis. 1972. A quantitative study
of the courtship display of the red-eared turtle, *Chrysemys scripta elegans* (Wied). *Herpetologica* 28:58–64.

Jackson, K., D. G. Butler, and J. H. Youson. 1996. Morphology and ultrastructure of possible integumentary
sense organs in the estuarine crocodile (*Crocodylus
porosus*). *Journal of Morphology* 229:315–324.

Jackson, K., and T. H. Fritts. 1995. Evidence from tooth
surface morphology for a posterior maxillary origin of
the proteroglyph fang. *Amphibia Reptilia* 16:273–288.

Jacobs, J. S., J. J. Greenhaw, J. M. Goy, and M. V. Plummer.
1985. Pectoral glands of *Scaphiopus* and *Megophrys*. *Journal of Herpetology* 19:419–420.

Jaeger, R. G. 1971a. Competitive exclusion as a factor influencing the distributions of two species of terrestrial salamanders. *Ecology* 52:632–637.

Jaeger, R. G. 1971b. Moisture as a factor influencing the
distributions of two species of terrestrial salamanders.
Oecologia 6:191–207.

Jaeger, R. G. 1978. Plant climbing by salamanders: periodic
availability of plant-dwelling prey. *Copeia* 1978:686–
691.

Jaeger, R. G. 1981. Dear enemy recognition and the costs of
aggression between salamanders. *American Naturalist*
117:962–974.

Jaeger, R. G. 1986. Pheromonal markers as territorial advertisement in terrestrial salamanders. Pages 191–203 in
*Chemical Signals in Vertebrates 4. Ecology, Evolution, and Com-
parative Biology*, edited by D. Duvall, D. Muller-Schwarze,
and R. M. Silverstein. Plenum Press, New York, NY.

Jaeger, R. G. 1990. Territorial salamanders evaluate size and
chitinous content of arthropod prey. pages 111–126 in
Behavioral Mechanisms of Food Selection, edited by R. N.
Hughes. NATO ASI Series, Subseries G: Ecological Sciences. Springer-Verlag, Heidelberg, Germany.

Jaeger, R. G., and D. E. Barnard. 1981. Foraging tactics of a
terrestrial salamander: choice of diet in structurally simple environments. *American Naturalist* 117:639–664.

Jaeger, R. G., and D. C. Forester. 1993. Social behavior of
plethodontid salamanders. *Herpetologica* 49:163–175.

Jaeger, R. G., and A. M. Rubin. 1982. Foraging tactics of a
terrestrial salamander: judging prey profitability. *Journal
of Animal Ecology* 51:167–176.

Jansen, W. D., and R. C. Foehring. 1983. The mechanism
of venom secretion from Duvernoy's gland of the snake

Thamnophis sirtalis (Serpentes: Colubridae). *Journal of Morphology* 175:271–277.

Janzen, F. J. 1995. Experimental evidence for the evolutionary significance of temperature-dependent sex determination. *Evolution* 49:864–873.

Janzen, F. J., and G. L. Paukstis. 1988. Environmental sex determination in reptiles. *Nature* 332:790.

Janzen, F. J., and G. L. Paukstis. 1991. Environmental sex determination in reptiles: Ecology, evolution, and experimental design. *Quarterly Review of Biology* 66: 149–179.

Janzen, R. J. 1994. Climate change and temperature-dependent sex determination in reptiles. *Proceedings of the National Academy of Sciences, USA* 91:7487–7490.

Jayes, A. S., and R. McN. Alexander. 1980. The gaits of chelonians: walking techniques for very low speeds. *Journal of Zoology* 191:353–378.

Jayne, B. C. 1985. Swimming in constricting (*Elaphe g. guttata*) and nonconstricting (*Nerodia fasciata pictiventris*) colubrid snakes. *Copeia* 1985:195–208.

Jayne, B. C. 1988a. Muscular mechanisms of snake locomotion: An electromyographic study of lateral undulation of the Florida banded water snake (*Nerodia fasciata*) and the yellow rat snake (*Elaphe obsoleta*). *Journal of Morphology* 197:159–181.

Jayne, B. C. 1988b. Muscular mechanisms of snake locomotion: An electromyographic study of the sidewinding and concertina modes of *Crotalus cerastes*, *Nerodia fasciata*, and *Elaphe obsoleta*. *Journal of Experimental Biology* 140: 1–33.

Jayne, B. C., and A. F. Bennett. 1989. The effect of tail morphology on locomotor performance of snakes: A comparison of experimental and correlative methods. *Journal of Experimental Zoology* 252:126–133.

Jenkins, F. A., Jr., and G. E. Goslow. 1983. The functional anatomy of the shoulder of the savannah monitor lizard (*Varanus exanthematicus*). *Journal of Morphology* 175: 195–216.

Jenkins, F. A., Jr., and D. M. Walsh. 1993. An early Jurassic caecilian with limbs. *Nature* 365:246–250.

Jennings, M. R. 1987. Impact of the curio trade for San Diego horned lizards (*Phrynosoma coronatum blainvillii*) in the Los Angeles Basin, California: 1885–1930. *Journal of Herpetology* 21:356–358.

Jennions, M. D., P. J. Bishop, P. R. Y. Backwell, and N. I. Passmore. 1995. Call rate variability and female choice in the African frog *Hyperolius marmoratus*. *Behaviour* 132:709–720.

Jennions, M. D., and N. I. Passmore. 1993. Sperm competition in frogs: testis size and a 'sterile male' experiment on *Chiromantis xerampelina* (Rhacophoridae). *Biological Journal of the Linnean Society* 50:211–220.

Jenssen, T. A. 1970. Female response to filmed displays of *Anolis nebulosus* (Sauria, Iguanidae). *Animal Behaviour* 18:640–647.

Jenssen, T. A. 1973. Shift in the structural habitat of *Anolis opalinus* due to congeneric competition. *Ecology* 54: 863–869.

Jenssen, T. A. 1977. Evolution of anoline lizard display behavior. *American Zoologist* 17:203–215.

Jenssen, T. A., D. L. Marcellini, K. A. Buhlmann, and P. H. Goforth. 1989. Differential infanticide by adult curly-tailed lizards, *Leiocephalus schreibersi*. *Animal Behavior* 38: 1054–1061.

Joger, U. 1991. A molecular phylogeny of agamid lizards. *Copeia* 1991:616–622.

Johnson, S. A., K. A. Bjorndal, and A. B. Bolten. 1996. Effects of organized turtle watches on loggerhead (*Caretta caretta*) nesting behavior and hatchling production in Florida. *Conservation Biology* 10:570–577.

Joly, J. 1968. Données ecologiques sur las salamandre tachetée *Salamandra salamandra* (L.). *Annales des Sciences Naturelles, Zoologie (12th series)* 10:301–366.

Jones, T., A. G. Kluge, and A. J. Wolf. 1993. When theories and methodologies clash: a phylogenetic reanalysis of the North American ambystomatid salamanders (Caudata: Ambystomatidae). *Systematic Biology* 42: 92–102.

Joy, J. E., and D. Crews. 1988. Male mating success in red-sided garter snakes: size is not important. *Animal Behaviour* 36:1839–1841.

Juncá, F. A., R. Altig, and C. Gascon. 1994. Breeding biology of *Colostethus stepheni*, a dendrobatid frog with a non-transported nidicolous tadpole. *Copeia* 1994:747–750.

Kam, Y.-C., Z.-S. Chuang, and C.-F. Yen. 1996. Reproduction, oviposition-site selection, and tadpole oophagy of an arboreal nester, *Chirixalus eiffingeri* (Rhacophoridae), from Taiwan. *Journal of Herpetology* 30:52–59.

Kamel, L. T., S. E. Peters, and D. P. Bashor. 1996. Hopping and swimming in the leopard frog, *Rana pipiens*: II. A comparison of muscle activities. *Journal of Morphology* 230:17–31.

Kaplan, R. H. 1985. Maternal influences on offspring development in the California newt, *Taricha torosa*. *Copeia* 1985:1028–1035.

Kaplan, R. H., and S. N. Salthe. 1979. The allometry of reproduction: an empirical view in salamanders. *American Naturalist* 113:671–689.

Kaplan, R. H., and P. W. Sherman. 1980. Intraspecific oophagy in California newts. *Journal of Herpetology* 14: 183–185.

Karasov, W. H., and J. M. Diamond. 1985. Digestive adaptations for fueling the cost of endothermy. *Science* 228:202–204.

Karasov, W. H., and J. M. Diamond. 1988. Interplay between physiology and ecology in digestion. *Bioscience* 38:602–611.

Kardong, K. V. 1995. *Vertebrates: Comparative Anatomy, Function, Evolution*. Wm. C. Brown Publishers, Dubuque, IA.

Kardong, K. V., and P. A. Lavin-Murcio. 1993. Venom delivery of snakes as high-pressure and low-pressure systems. *Copeia* 1993:644–650.

Kats, L. B., J. W. Petranka, and A. Sih. 1988. Antipredator defenses and the persistence of amphibian larvae with fishes. *Ecology* 69:1865–1870.

Kaufmann, J. H. 1992. The social behavior of wood turtles, *Clemmys insculpta*, in central Pennsylvania. *Herpetological Monographs* 6:1–25.

Kaufmann, J. H. 1995. Home ranges and movements of wood turtles, *Clemmys insculpta*, in central Pennsylvania. *Copeia* 1995:22–27.

Keinath, J. A., and J. A. Musick. 1993. Movements and diving behavior of a leatherback turtle, *Dermochelys coriacea*. *Copeia* 1993:1010–1017.

Kelleher, K. E., and J. R. Tester. 1969. Homing and survival in the Manitoba toad, *Bufo hemiophrys*, in Minnesota. *Ecology* 50:1040–1048.

Kephart, D. G., and S. J. Arnold. 1982. Garter snake diets in a fluctuating environment: a seven year study. *Ecology* 63:1232–1236.

Kiester, A. R. 1971. Species density of North American amphibians and reptiles. *Systematic Zoology* 20:127–137.

King, D. and B. Green. 1993. *Goanna: The Biology of the Varanid Lizards*. New South Wales University Press, Kensington, New South Wales, Australia.

King, F. W. 1989. Conservation and management. Pages 216–229 in *Crocodiles and Alligators*, edited by C. A. Ross. Facts on File, New York, NY.

King, F. W., and R. L. Burke (editors). 1989. *Crocodilian, Tuatara, and Turtle Species of the World, A Taxonomic and Geographic Reference*. Association of Systematics Collections, Washington, D. C.

King, F. W., and R. L. Burke (editors). 1989. *Crocodilian, Tuatara, and Turtle Species of the World, a Taxonomic and Geographic Reference*. Association of Systematics Collections, Washington, D. C.

King, M. B., and D. Duvall. 1990. Prairie rattlesnake seasonal migrations: episodes of movement, vernal foraging and sex differences. *Animal Behaviour* 39:924–935.

King, W., and F. G. Thompson. 1968. A review of the American lizards of the genus *Xenosaurus* Peters. *Bulletin of the Florida State Museum (Biological Sciences)* 12:93–123.

Kizirian, D. A. 1996. A review of Ecuadorian *Proctoporus* (Squamata: Gymnophthalmidae) with descriptions of nine new species. *Herpetological Monographs* 10:85–155.

Klauber, L. M. 1940. The worm snakes of the genus *Leptotyphlops* in the United States and northern Mexico. *Transactions of the San Diego Society of Natural History* 9:87–162.

Klauber, L. M. 1972. *Rattlesnakes, Their Habits, Life Histories, and Influence on Mankind, second edition, 2 volumes*. University of California Press, Berkeley.

Klaver, C., and W. Böhme. 1986. Phylogeny and classification of the Chamaeleonidae (Sauria) with special reference to hemipenis morphology. *Bonner Zoologische Monographien* 22:1–64.

Kleeberger, S. R. 1984. A test of competition in two sympatric populations of desmognathine salamanders. *Ecology* 65:1846–1856.

Kluge, A. G. 1967. Higher taxonomic categories of Gekkonid lizards and their evolution. *Bulletin of the American Museum of Natural History* 135:1–60.

Kluge, A. G. 1974. A taxonomic revision of the lizard family Pygopodidae. *Miscellaneous Publications of the Museum of Zoology, University of Michigan* 147:vi+221.

Kluge, A. G. 1976. Phylogenetic relationships in the lizard family Pygopodidae: an evaluation of theory, methods and data. *Miscellaneous Publications of the Museum of Zoology, University of Michigan* no. 152:1–72.

Kluge, A. G. 1977. *Chordate Structure and Function*. Macmillan, New York, NY.

Kluge, A. G. 1987. Cladistic relationships in the Gekkonoidea (Squamata, Sauria). *Miscellaneous Publications of the Museum of Zoology, University of Michigan* 173:1–54.

Kluge, A. G. 1991. Boine snake phylogeny and research cycles. *Miscellaneous Publications of the Museum of Zoology, University of Michigan* 178:1–58.

Kluge, A. G. 1993a. *Aspidites* and the phylogeny of pythonine snakes. *Records of the Australian Museum* Suppl. 19:1–77.

Kluge, A. G. 1993b. *Calabaria* and the phylogeny of erycine snakes. *Zoological Journal of the Linnean Society* 107: 293–351.

Kluge, A. G., and J. S. Farris. 1969. Quantitative phyletics and the evolution of anurans. *Systematic Zoology* 18:1–32.

Kluger, M. J. 1979. Fever in ectotherms: evolutionary implications. *American Zoologist* 19:295–304.

Knight, A., and D. P. Mindell. 1993. Substitution bias, weighting of DNA sequence evolution, and the phylogenetic position of Fea's viper. *Systematic Biology* 42:18–31.

Kochva, E., M. Shayer-Wollberg, and R. Sobol. 1967. The special pattern of the venom gland in *Atractaspis* and its bearing on the taxonomic status of the genus. *Copeia* 1967:763–772.

Kochva, E., and M. Wollberg. 1970. The salivary glands of Aparallactinae (Colubridae) and the venom glands of *Elaps* (Elapidae) in relation to the taxonomic status of this genus. *Zoological Journal of the Linnean Society* 49:217–224.

Kofron, C. 1988. The central and South American blindsnakes of the genus *Anomalepis*. *Amphibia Reptilia* 9:7–14.

Kok, D., L. H. du Preez, and A. Channing. 1989. Channel construction by the African bullfrog: another anuran parental care strategy. *Journal of Herpetology* 23:435–437.

Kraus, F. 1989. Constraints on the evolutionary history of the unisexual salamanders of the *Ambystoma laterale-texanum* complex as revealed by mitochondrial DNA analysis. Pages 218–227 in *Evolution and Ecology of Unisexual*

Vertebrates, Bulletin, New York State Museum, Albany, edited by R. M. Dawley and J. P. Bogart. New York State Museum, Albany, NY.

Kusano, T., M. Toda, and K. Fukuyama. 1991. Testes size and breeding systems in Japanese anurans with special reference to large testes in the treefrog, *Rhacophorus arboreus* (Amphibia: Rhacophoridae). *Behavioral Ecology and Sociobiology* 29:27–31.

Kuwabara, K., N. Suzuki, F. Wakabayashi, H. Ashikaga, T. Inoue, and J. Kobara. 1989. Breeding the Japanese giant salamander *Andrias japonicus*. *International Zoo Yearbook* 28:22–31.

Laan, R., and B. Verboom. 1990. Effects of pool size and isolation on amphibian communities. *Biological Conservation* 54:251–262.

Lahanas, P. N., and J. M. Savage. 1992. A new species of caecilian from the Península de Osa of Costa Rica. *Copeia* 1992:703–709.

Lamb, T., and C. Lydeard. 1994. A molecular phylogeny of the gopher tortoises, with comments on familial relationships within the Testudinoidea. *Molecular Phylogenetics and Evolution* 3:283–291.

Lamb, T., C. Lydeard, R. B. Walker, and J. W. Gibbons. 1994. Molecular systematics of map turtles (*Graptemys*): a comparison of mitochondrial restriction site versus sequence data. *Systematic Biology* 43:543–559.

Landwer, A. J. 1994. Manipulation of egg production reveals costs of reproduction in the tree lizard (*Urosaurus ornatus*). *Oecologia* 100:243–249.

Lang, J. W. 1989. Social behavior. Pages 102–117 in *Crocodiles and Alligators*, edited by C. A. Ross. Facts on File, New York, NY.

Lang, J. W., and H. V. Andrews. 1994. Temperature-dependent sex determination in crocodilians. *Journal of Experimental Zoology* 270:28–44.

Lang, M. 1988. Notes on the genus *Bombina* Oken (Anura: Bombinatoridae): I. Recognized species, distribution, characteristics and use in laboratory. *British Herpetological Society Bulletin* 26:.

Lang, M. 1989. Phylogenetic and biogeographic patterns of basiliscine iguanians (Reptilia: Squamata: "Iguanidae"). *Bonner Zoologische Monographien* 28:1–171.

Lang, M. 1990. Phylogenetic analysis of the genus group *Tracheloptychus-Zonosaurus* (Reptilia: Gerrhosauridae), with a hypothesis of biogeographic unit relationships in Madagascar. Pages 261–274 in *Vertebrates in the Tropics*, edited by G. Peters and R. Hutterer. Museum Alexander Koenig, Bonn, Germany.

Lang, M. 1991. Generic relationships within Cordyliformes (Reptilia: Squamata). *Bulletin de l'Institute Royal des Sciences Naturelles de Belgique* 61:121–188.

Langton, T. E. S. (editor). 1989. *Amphibians and Roads*. ACO Polymer Products, Shefford, Bedfordshire, UK.

Langton, T. E. S. 1989. Reasons for preventing amphibian mortality on roads. Pages 75–80 in *Amphibians on Roads*, edited by T. E. S. Langton. ACO Polymer Products Ltd., Bedfordshire, UK.

Lannoo, M. J., D. S. Townsend, and R. J. Wassersug. 1987. Larval life in the leaves: arboreal tadpole types, with special attention to the morphology, ecology, and behavior of the oophagous *Osteopilus brunneus* (Hylidae) larva. *Fieldiana, Zoology* n.s., no. 38:1–31.

Laporta-Ferreira, I. L., and M. G. Salomo. 1991. Morphology, physiology and toxicology of the oral glands of a tropical cochleophagous snake, *Sibynomorphus neuwiedi* (Colubridae—Dipsadinae). *Zoologischer Anzeiger* 1991: 198–208.

Larsen, J. H., Jr., and D. J. Guthrie. 1975. The feeding mechanism of terrestrial tiger salamanders (*Ambystoma tigrinum melanostictum* Baird). *Journal of Morphology* 147: 137–153.

Larson, A. 1983. A molecular phylogenetic perspective on the origins of a lowland tropical salamander fauna. I. Phylogenetic inferences from protein comparisons. *Herpetologica* 39:85–99.

Larson, A. 1984. Neontological inferences of evolutionary pattern and process in the salamander family Plethodontidae. *Evolutionary Biology* 17:119–217.

Larson, A. 1991. Evolutionary analysis of length-variable sequences: divergent domains of ribosomal RNA. Pages 221–248 in *Phylogenetic Analysis of DNA Sequences*, edited by M. M. Miyamoto and J. Cracraft. Oxford University Press, Oxford.

Larson, A., and W. W. Dimmick. 1993. Phylogenetic relationships of the salamander families: an analysis of congruence among morphological and molecular characters. *Herpetological Monographs* 7:77–93.

Larson, A., D. B. Wake, L. R. Maxson, and R. Highton. 1981. A molecular phylogenetic perspective on the origins of morphological novelties in the salamanders of the tribe Plethodontini (Amphibia, Plethodontidae). *Evolution* 35:405–422.

Larson, A., and A. C. Wilson. 1989. Patterns of ribosomal RNA evolution in salamanders. *Molecular Biology and Evolution* 6:131–154.

Lauder, G. V. 1985. Aquatic feeding in lower vertebrates. Pages 210–229 in *Functional Vertebrate Morphology*, edited by M. Hildebrand, D. M. Bramble, K. F. Liem, and D. B. Wake. Belknap Press, Cambridge, MA.

Lauder, G. V., and H. B. Shaffer. 1985. Functional morphology of the feeding mechanism in aquatic ambystomatid salamanders. *Journal of Morphology* 185:297–326.

Lauder, G. V., and H. B. Shaffer. 1988. Ontogeny of functional design in tiger salamanders (*Ambystoma tigrinum*): Are motor patterns conserved during major morphological transformations? *Journal of Morphology* 197:249–268.

Lauder, G. V., and H. B. Shaffer. 1993. Design of feeding systems in aquatic vertebrates: Major patterns and their evolutionary interpretations. Pages 113–149 in *The*

Skull, volume 3, edited by J. Hanken and B. K. Hall. University of Chicago Press, Chicago, IL.

Lauff, R. F., A. P. Russell, and A. M. Bauer. 1993. Topography of the digital cutaneous sensilla of the tokay gecko, *Gekko gecko* (Reptilia, Gekkonidae), and their potential role in locomotion. *Canadian Journal of Zoology* 71:2462–2472.

Laurance, W. F., K. R. McDonald, and R. Speare. 1996. Epidemic disease and catastropic decline of Australian rain forest frogs. *Conservation Biology* 10:406–413.

Laurent, R. 1950. Revision du genre *Atractaspis* A. Smith. *Memoires de l'Institute Royale des Sciences Naturelles de Belgique* ser. 2, 38:1–49.

Laurent, R. F. 1964. A revision of the *punctatus* group of African *Typhlops* (Reptilia: Serpentes). *Bulletin of the Museum of Comparative Zoology* 130:387–444.

Laurent, R. F. 1972. Tentative revision of the genus *Hemisus* Günther. *Annales Musee Royal de l'Afrique Centrale, Tervuren, Belgique* 194:1–67.

Laurin, M., and R. R. Reisz. 1995. A re-evaluation of early amniote phylogeny. *Zoological Journal of the Linnean Society* 113:165–223.

Lawson, P. A. 1994. Orientation abilities and mechanisms in nonmigratory populations of garter snakes (*Thamnophis sirtalis* and *T. ordinoides*). *Copeia* 1994:263–274.

Lee, J. C. 1980. An ecogeographic analysis of the herpetofauna of the Yucatán Peninsula. *Miscellaneous Publication Museum of Natural History, University of Kansas* 67:1–75.

Lee, M. S. Y. 1993. The origin of the turtle body plan: bridging a famous morphological gap. *Science* 261:1716–1720.

Lee, M. S. Y. 1995. Historical burden in systematics and the interrelationships of 'Parareptiles'. *Biological Reviews of the Cambridge Philosophical Society* 70:459–547.

Lee, M. S. Y. 1996. Correlated progression and the origin of turtles. *Nature* 379:812–815.

Lee, M. S. Y., and B. G. M. Jamieson. 1992. The ultrastructure of the spermatozoa of three species of myobatrachid frogs (Anura, Amphibia) with phylogenetic considerations. *Acta Zoologica Stockholm* 73:213–222.

Lee, M. S. Y., and B. G. M. Jamieson. 1993. The ultrastructure of the spermatozoa of bufonid and hylid frogs (Anura, Amphibia): implications for phylogeny and fertilization biology. *Zoologica Scripta* 22:309–323.

Li, W.-H., and D. Graur. 1991. *Fundamentals of Molecular Evolution*. Sinauer Associates, Inc., Sunderland, MA.

Liem, K. F., H. Marx, and G. B. Rabb. 1971. The viperid snake *Azemiops*: its comparative cephalic anatomy and phylogenetic position in relation to viperinae and crotalinae. *Fieldiana, Zoology* 59:65–126.

Liem, S. S. 1970. The morphology, systematics, and evolution of the Old World Treefrogs (Rhacophoridae and Hyperoliidae). *Fieldiana, Zoology* 57:1–145.

Lillywhite, H. B., and P. F. A. Maderson 1982: Skin structure and permeability. Pages 397–442 in *Biology of the Reptilia, Vol. 12 (Physiology C, Physiological Ecology)*, edited by C. Gans and F. H. Pough, Academic Press, London, UK.

Linnaeus, C. 1758. *Systema Naturae, 10th edition*. Stockholm, Sweden.

Lister, B. C. 1976. The nature of niche expansion in West Indian *Anolis* lizards I: ecological consequences of reduced competition. *Evolution* 30:659–676.

Lofts, B. (editor) 1974, 1976. *Physiology of the Amphibia*. Vol. 2, 3. Academic Press, New York, NY.

Lohmann, K. J. 1992. How sea turtles navigate. *Scientific American* 200:100–106.

Lohmann, K. J., B. E. Witherington, C. M. F. Lohmann, and M. Salmon. 1997. Orientation, navigation, and natal beach homing in sea turtles. Pages 107–135 in *The Biology of Sea Turtles*, edited by P. L. Lutz and J. A. Musick. CRC Press, Boca Raton, FL.

Lombard, R. E., and J. R. Bolt. 1979. Evolution of the tetrapod ear: an analysis and reinterpretation. *Biological Journal of the Linnaean Society* 11:19–76.

Lombard, R. E., and D. B. Wake. 1976. Tongue evolution in the lungless salamanders, Family Plethodontidae. I. Introduction, theory and a general model of dynamics. *Journal of Morphology* 148:265–286.

Lombard, R. E., and D. B. Wake. 1977. Tongue evolution in the lungless salamanders, Family Plethodontidae. II. Function and evolutionary diversity. *Journal of Morphology* 153:39–79.

Lombard, R. E., and D. B. Wake. 1986. Tongue evolution in the lungless salamanders, Family Plethodontidae IV. Phylogeny of plethodontid salamanders and the evolution of feeding dynamics. *Systematic Zoology* 35:532–551.

Long, D. J. 1996. Records of white shark-bitten leatherback sea turtles along the central California coast. Pages 317–319 in *Great White Sharks: The Biology of Carcharodon carcharias*, edited by A. P. Klimley and D. G. Ainley. Academic Press, NY.

Lopez, T. J., and L. R. Maxson. 1995. Mitochondrial DNA sequence variation and genetic differentiation among colubrine snakes (Reptilia: Colubridae: Colubrinae). *Biochemical Systematics and Ecology* 23:487–505.

Lopez, T. J., and L. R. Maxson. 1996. Albumin and mitochondrial DNA evolution: phylogenetic implications for colubrine snakes (Colubridae: Colubrinae). *Amphibia Reptilia* 17:247–259.

Losos, J. B. 1985. An experimental demonstration of the species-recognition role of *Anolis* dewlap color. *Copeia* 1985:905–910.

Losos, J. B. 1990. Ecomorphology, performance capability, and scaling of West Indian *Anolis* lizards: an evolutionary analysis. *Ecological Monographs* 60:369–388.

Losos, J. B. 1996. Phylogenetic perspectives on community ecology. *Ecology* 77:1344–1354.

Louw, G. N. 1972. The role of advective fog in the water economy of certain Namib Desert animals. Pages 297–314 in *Comparative Physiology of Desert Animals (Symposium of the Zoological Society of London No. 31)*, edited by G. M. O Maloy. Academic Press, London, UK.

Loveridge, A., and E. E. Williams. 1957. Revision of the African tortoises of the suborder Cryptodira. *Bulletin of the Museum of Comparative Zoology* 115:1–557.

Lovich, J. 1993. *Macroclemys. Catalogue of American Amphibians and Reptiles* 562:1–4.

Lovich, J. E., C. H. Ernst, and J. F. McBreen. 1990. Growth, maturity, and sexual dimorphism in the wood turtle, *Clemmys insculpta. Canadian Journal of Zoology* 68:672–677.

Lovich, J. E., and C. J. McCoy. 1992. Review of the *Graptemys pulchra* group (Reptilia: Testudines: Emydidae), with descriptions of two new species. *Annals of the Carnegie Museum of Natural History* 61:293–315.

Lovich, J. E., D. W. Herman, and K. M. Fahey. 1992. Seasonal activity and movements of bog turtles (*Clemmys muhlenbergii*) in North Carolina. *Copeia* 1992:1107–1111.

Luiselli, L. 1996. Individual success in mating balls of the grass snake, *Natrix natrix*: size is important. *Journal of Zoology, London* 239:731–740.

Luke, C. 1986. Convergent evolution of lizard toe fringes. *Biological Journal of the Linnean Society* 27:1–16.

Luschi, P., F. Papi, H. C. Liew, E. H. Chan, and F. Bonadonna. 1996. Long-distance migration and homing after displacement in the green turtle (*Chelonia mydas*): a satellite tracking study. *Journal of Comparative Physiology A* 178:447–452.

Lynch, J. D. 1971. Evolutionary relationships, osteology, and zoogeography of leptodactyloid frogs. *University of Kansas Museum of Natural History, Miscellaneous Publications* no. 53:1–238.

Lynch, J. D. 1973. The transition from archaic to advanced frogs. Pages 133–182 in *Evolutionary Biology of the Anurans: Contemporary Research on Major Problems*, edited by J. L. Vial. University of Missouri Press, Columbia, MO.

Lynch, J. D. 1978. A re-assessment of the telmatobiine leptodactylid frogs of Patagonia. *Occasional Papers of the Museum of Natural History, University of Kansas* no. 72:1–57.

Lynch, J. D. 1982. Relationships of the frogs of the genus *Ceratophrys* (Leptodactylidae) and their bearing on hypotheses of Pleistocene forest refugia in South American and punctuated equilibria. *Systematic Zoology* 31:166–179.

Lynch, J. D. 1986. The definition of the Middle American clade of *Eleutherodactylus* based on jaw musculature (Amphibia: Leptodactylidae). *Herpetologica* 42:248–258.

Lynch, J. D. 1993. The value of the *M. Depressor Mandibulae* in phylogenetic hypotheses for *Eleutherodactylus* and its allies (Amphibia: Leptodactylidae). *Herpetologica* 49: 32–41.

Lynch, J. D., P. M. Ruiz-Carranza, and J. Rueda V. 1983. Notes on the distribution and reproductive biology of *Centrolene geckoideum* Jimenez de la Espada in Colombia and Ecuador (Amphibia: Centrolenidae). *Studies on Neotropical Fauna and Environment* 18:239–243.

MacArthur, R. H. 1972. *Geographical Ecology*. Harper & Row Publishers, New York, NY.

Macartney, J. M., P. T. Gregory, and K. W. Larsen. 1988. A tabular survey of data on movements and home ranges of snakes. *Journal of Herpetology* 22:61–73.

MacDonald, D. W., and G. M. Carr. 1989. Food security and the rewards of tolerance. Pages 75–99 in *Comparative Socioecology. The Behavioural Ecology of Humans and Other Mammals*, edited by V. Standen and R. A. Foley. Blackwell Scientific Publications, Oxford.

Macedonia, J. M., C. S. Evans, and J. B. Losos. 1994. Male *Anolis* lizards discriminate video-recorded conspecific and heterospecific displays. *Animal Behaviour* 47: 1220–1223.

Macedonia, J. M., and J. A. Stamps. 1994. Species recognition in *Anolis grahami* (Sauria: Iguanidae): evidence from responses to video playbacks of conspecific and heterospecific displays. *Ethology* 98:246–264.

Macey, J. R., A. Larson, N. B. Ananjeva, and T. J. Papenfuss. 1997. Evolutionary shifts in three major structural features of the mitochondrial genome among iguanian lizards. *Journal of Molecular Evolution* in press.

Madsen, T. 1984. Movements, home range size and habitat use of radio-tracked grass snakes (*Natrix natrix*) in southern Sweden. *Copeia* 1984:707–713.

Madsen, T. 1988. Reproductive success, mortality, and sexual size dimorphism in the adder, *Vipera berus. Holarctic Ecology.* 11:77–80.

Madsen, T., and J. Loman. 1987. On the role of colour display in the social and spatial organization of male rainbow lizards (*Agama agama*). *Amphibia Reptilia* 8:365–372.

Madsen, T., and R. Shine. 1993. Male mating success and body size in European grass snakes. *Copeia* 1993: 561–564.

Madsen, T., and R. Shine. 1994. Components of lifetime reproductive success in adders, *Vipera berus. Journal of Animal Ecology* 63:561–568.

Madsen, T., and R. Shine. 1996. Seasonal migration of predators and prey: a study of pythons and rats in tropical Australia. *Ecology* 77:149–156.

Madsen, T., R. Shine, J. Loman, and T. Håkansson. 1992. Why do female adders copulate so frequently? *Nature* 355:440–441.

Madsen, T., R. Shine, J. Loman, and T. Håkansson. 1993. Determinants of mating success in male adders, *Vipera berus. Animal Behaviour* 45:491–499.

Magnusson, W. E., and J.-M. Hero. 1991. Predation and the evolution of complex oviposition behaviour in Amazon rainforest frogs. *Oecologia* 86:310–318.

Magnusson, W. E., A. P. Lima, and R. M. Sampaio. 1985. Sources of heat for nests of *Paleosuchus trigonatus* and a review of crocodilian nest temperatures. *Journal of Herpetology* 19:199–207.

Magnusson, W. E., E. Vieira da Silva, and A. P. Lima. 1987. Diets of Amazonian crocodilians. *Journal of Herpetology* 21:85–95.

Maiorana, V. C. 1976. Predation, submergent behavior, and tropical diversity. *Evolutionary Theory* 1:157–177.

Marcellini, D. L. 1977. Acoustic and visual display behavior of gekkonid lizards. *American Zoologist* 17:251–260.

Márquez, R., and J. Bosch. 1995. Advertisement calls of the midwife toads *Alytes* (Amphibia, Anura, Discoglossidae) in continental Spain. *Journal of Zoological Systematics and Evolutionary Research* 33:185–192.

Martin, J., and A. Wolfe. 1992. *Chameleons: Nature's Masters of Disguise*. Blandford, London, UK.

Martins, E. P. 1993. Contextual use of the push-up display by the sagebrush lizard, *Sceloporus graciosus*. *Animal Behaviour* 45:25–36.

Martins, E. P. 1994. Phylogenetic perspectives on the evolution of lizard territoriality. Pages 117–144 in *Lizard Ecology: Historical and Experimental Perspectives*, edited by L. J. Vitt and E. R. Pianka. Princeton University Press, Princeton, NJ.

Martof, B. 1953. The "spring lizard" industry: a factor in salamander distribution and genetics. *Ecology* 34:436–437.

Martof, B. S. 1974. Sirenidae. *Catalogue of American Amphibians and Reptiles* :151.1–151.2.

Mason, R. T. (1992): Reptilian pheromones. Pages 114–228 in *Biology of the Reptilia, Vol. 18 (Physiology E, Hormones, Brain, and Behavior)*, edited by C. Gans and D. Crews. University of Chicago Press, Chicago, IL.

Mathis, A., R. G. Jaeger, W. H. Keen, P. K. Ducey, S. C. Walls, and B. W. Buchanan. 1995. Aggression and territoriality in salamanders and a comparison with the territorial behaviour of frogs. Pages 633–676 in *Amphibian Biology Vol 2: Social Behaviour*, edited by H. Heatwole and B. K. Sullivan. Surrey Beatty and Sons, Chipping Norton, New South Wales, Australia.

Mason, R. T. 1992. Reptilian pheromones. Pages 114–228 in *Biology of the Reptilia, Vol. 18 (Physiology E, Hormones, Brain, and Behavior)*, edited by C. Gans and D. Crews. University of Chicago Press, Chicago, IL.

Mason, T. T., and D. Crews. 1985. Female mimicry in garter snakes. *Nature* 316:59–60.

Massey, A. 1988. Sexual interactions in red-spotted newt populations. *Animal Behaviour* 36:205–210.

Mathis, A., R. G. Jaeger, W. H. Keen, P. K. Ducey, S. C. Walls, and B. W. Buchanan 1995: Aggression and territoriality by salamanders and a comparison with the territorial behaviour of frogs. Pages 633–676 in *Amphibian Biology. Vol. 2. Social Behaviour*, edited by H. Heatwole and B. K. Sullivan. Surrey Beatty and Sons, Chipping Norton, New South Wales, Australia.

Matuschka, F. R., and B. Bannert. 1989. Recognition of cyclic transmission of *Sarcocystis stehlinii* N. Sp. in the Gran Canarian giant lizard. *Journal of Parasitology* 75:383–387.

Maxson, L. R. 1976. The phylogenetic status of phyllomedusine frogs (Hylidae) as evidenced from immunological studies of their serum albumins. *Experientia* 32:1149–1150.

Maxson, L. R. 1984. Molecular probes of phylogeny and biogeography in toads of the widespread genus *Bufo*. *Molecular Biology and Evolution* 1:345–356.

Maxson, L. R. 1992. Tempo and pattern in anuran speciation and phylogeny: an albumin perspective. Pages 41–57 in *Herpetology: Current Research on the Biology of Amphibians and Reptiles. Proceedings of the First World Congress of Herpetology*, edited by K. Adler. Society for the Study of Amphibians and Reptiles, Oxford, OH.

Maxson, L. R., and W. R. Heyer. 1982. Leptodactylid frogs and the Brasilian shield: an old and continuing adaptive radiation. *Biotropica* 14:10–15.

Maxson, L. R., and W. R. Heyer. 1988. Molecular systematics of the frog genus *Leptodactylus* (Amphibia: Leptodactylidae). *Fieldiana, Zoology* new ser., no. 41:1–13.

Maxson, L. R., P. E. Moler, and B. W. Mansell. 1988. Albumin evolution in salamanders of the genus *Necturus*. *Journal of Herpetology* 22:231–235.

Maxson, L. R., and C. W. Myers. 1985. Albumin evolution in tropical poison frogs (Dendrobatidae): a preliminary report. *Biotropica* 17:50–56.

Maxson, L. R., V. M. Sarich, and A. C. Wilson. 1975. Continental drift and the use of albumin as an evolutionary clock. *Nature* 255:397–400.

Maxson, L. R., and J. M. Szymura. 1984. Relationships among discoglossid frogs: an albumin perspective. *Amphibia Reptilia* 5:245–252.

Maxson, L. R., and A. C. Wilson. 1975. Albumin evolution and organismal evolution in tree frogs (Hylidae). *Systematic Zoology* 24:1–15.

McBee, R. H., and V. H. McBee. 1982. The hindgut fermentation in the green iguana, *Iguana iguana*. Pages 77–83 in *Iguanas of the World: Their Behavior, Ecology, and Conservation*, edited by G. M. Burghardt and A. S. Rand. Noyes Publications, Park Ridge, NJ.

McCarthy, C. J. 1985. Monophyly of elapid snakes (Serpentes: Elapidae). An assessment of the evidence. *Zoological Journal of the Linnean Society* 83:79–93.

McCarthy, C. J. 1986. Relationships of the laticaudine sea snakes (Serpentes: Elapidae: Laticaudinae). *Bulletin of the British Museum (Natural History)* 50:127–161.

McCoid, M. J., and T. H. Fritts. 1980b. Observations of feral populations of *Xenopus laevis* (Pipidae) in southern California. *Bulletin of the Southern California Academy of Sciences* 79:82–86.

McDiarmid, R. W. 1971. Comparative morphology and evolution of frogs of the neotropical genera *Atelopus, Dendrophryniscus, Melanophryniscus,* and *Oreophrynella*. *Bulletin of the Los Angeles County Museum of Natural History* Science, no. 12:1–66.

McDowell, S. B. 1964. Partition of the genus *Clemmys* and related problems in the taxonomy of the aquatic Testudinidae. *Proceedings of the Zoological Society of London* 143:239–279.

McDowell, S. B. 1967. *Aspidomorphus*, a genus of New Guinea snakes of the Family Elapidae, with notes on related genera. *Journal of Zoology, London* 151:497–543.

McDowell, S. B. 1968. Affinities of the snakes usually called *Elaps lacteus* and *E. dorsalis*. *Journal of the Linnean Society (Zoology)* 47:561–578.

McDowell, S. B. 1969. Notes on the Australian sea-snake *Ephalophis greyi* M. Smith (Serpentes: Elapidae, Hydrophiinae) and the origin and classification of sea snakes. *Zoological Journal of the Linnean Society* 48:333–349.

McDowell, S. B. 1972. The genera of sea-snakes of the *Hydrophis* group (Serpentes: Elapidae). *Transactions of the Zoological Society of London* 32:189–247.

McDowell, S. B. 1974. A catalogue of the snake of New Guinea and the Solomons, with special reference to those in the Bernice P. Bishop Museum, Part I. Scolecophidia. *Journal of Herpetology* 8:1–57.

McDowell, S. B. 1975. A catalogue of the snakes of New Guinea and the Solomons, with special reference to those in the Bernice P. Bishop Museum. Part II. Anilioidea and Pythoninae. *Journal of Herpetology* 9:1–80.

McDowell, S. B. 1979. A catalogue of the snakes of New Guinea and the Solomons, with special reference to those in the Bernice P. Bishop Museum. Part III. Boinae and Acrochordoidea (Reptilia, Serpentes). *Journal of Herpetology* 13:1–92.

McDowell, S. B. 1983. The genus *Emydura* (Testudines: Chelidae) in New Guinea with notes on the penial morphology of Pleurodira. Pages 169–189 in *Advances in Herpetology and Evolutionary Biology*, edited by A. J. G. Rhodin and K. Miyata. Museum of Comparative Zoology, Cambridge, MA.

McDowell, S. B. 1986. The architecture of the corner of the mouth in colubroid snakes. *Journal of Herpetology* 20:353–407.

McDowell, S. B. 1987. Systematics. Pages 3–50 in *Snakes: Ecology and Evolutionary Biology*, edited by R. A. Seigel, J. T. Collins, and S. S. Novak. Macmillan Publishing Co., New York, NY.

McDowell, S. B., and C. M. Bogert. 1954. The systematic position of *Lanthanotus* and the affinities of the Anguinomorphan lizards. *Bulletin of the American Museum of Natural History* 105:1–142.

McGuire, J. 1996. Phylogenetic systematics of crotaphytid lizards (Reptilia: Iguania: Crotaphytidae). *Bulletin of the Carnegie Museum of Natural History* 32:1–143.

McLaughlin, R. L. 1989. Search modes of birds and lizards: evidence for alternative movement patterns. *American Naturalist* 133:654–670.

M'Closkey, R. T., K. A. Baia, and R. W. Russell. 1987. Tree lizard (*Urosaurus ornatus*) territories: experimental perturbation of the sex ratio. *Ecology* 68:2059–2062.

Means, D. B., J. G. Palis, and M. Baggett. 1996. Effects of slash pine silviculture on a Florida population of flatwoods salamander. *Conservation Biology* 10:426–437.

Mebs, D. 1978. Pharmacology of reptilian venoms. Pages 437–560 in *Biology of the Reptilia, Vol. 8 (Physiology B)*, edited by C. Gans and K. A. Gans. Academic Press, New York, NY.

Medem, F. 1958. The crocodilian genus *Paleosuchus*. *Fieldiana, Zoology* 39:227–247.

Medem, F. 1963. Osteología craneal, distribución geográfica y ecología de *Melanosuchus niger* (Spix) (Crocodylia, Alligatoridae). *Revista de la Academia Colombiana de Ciencias Exactas, Físicas y Naturales* 12:5–19.

Medem, F. M. 1981. *Los Crocodylia de Sur America, Volumen I. Los Crocodylia de Colombia*. Ministerio de Educación Nacional y Fondo Colombiano de Investigaciones Científicas y Proyectos Especiales "Francisco José de Caldas", Bogotá, Colombia.

Medem, F. M. 1983. *Los Crocodylia de Sur America, Volumen II. Venezuela - Trinidad - Tobago - Guyana - Suriname - Guayana Francesa - Ecuador - Perú - Bolivia - Brasil - Paraguay - Argentina - Uruguay*. Universidad Nacional de Colombia y Fondo Colombiano de Investigaciones Cientificas Y Proyectos Especiales "Francisco José de Caldas", Bogotá, Colombia.

Medem, F., and H. Marx. 1955. An artificial key to the New World species of crocodilians. *Copeia* 1955:1–2.

Meffe, G. K., and C. R. Carroll. 1994. *Principles of Conservation Biology*. Sinauer Associates, Inc., Sunderland, MA.

Mendonça, M. T. 1983. Movements and feeding ecology of immature green turtles (*Chelonia mydas*) in a Florida lagoon. *Copeia* 1983:1013–1023.

Mengden, G. A. 1983. The taxonomy of Australian elapid snakes: a review. *Records of the Australian Museum* 35: 195–222.

Metter, D. E. 1968. *Ascaphus, Ascaphus truei*. Catalogue of American Amphibians and Reptiles 69:1–2.

Meyer, A. 1995. Molecular evidence on the origin of tetrapods and the relationships of the coelacanth. *Trends in Ecology and Evolution* 10:111–116.

Meylan, A. 1982. Sea turtle migration—evidence from tag returns. Pages 91–100 in *Biology and Conservation of Sea Turtles*, edited by K. A. Bjorndal. Smithsonian Institution Press, Washington, D. C.

Meylan, A. B. 1988. Spongivory in hawksbill turtles: a diet of glass. *Science* 239:393–395.

Meylan, P. A. 1987. The phylogenetic relationships of soft-shelled turtles (Family Trionychidae). *Bulletin of the American Museum of Natural History* 186:1–101.

Middendorf, G. A., III, and W. C. Sherbrooke. 1992. Canid elicitation of blood-squirting in a horned lizard (*Phrynosoma cornutum*). *Copeia* 1992:519–527.

Middlebrook, J. L. 1992. Molecular cloning of snake toxins and other venom components. Pages 281–295 in *Hand-*

book of Natural Toxins, Vol. 5: Reptile Venoms and Toxins, edited by A. T. Tu. Marcel Dekker, Inc., New York, NY.

Miller, J. D. 1997. Reproduction in sea turtles. Pages 51–81 in The Biology of Sea Turtles, edited by P. L. Lutz and J. A. Musick. CRC Press, Boca Raton, FL.

Miller, K. 1993. The improved performance of snapping turtles (Chelydra serpentina) hatched from eggs incubated on a wet substrate persists through the neonatal period. Journal of Herpetology 27:228–233.

Miller, K., G. C. Packard, and M. J. Packard. 1987. Hydric conditions during incubation influence locomotor performance of hatchling snapping turtles. Journal of Experimental Biology 127:401–412.

Milner, A. R. 1983. The biogeography of salamanders in the Mesozoic and early Caenozoic: a cladistic-vicariance model. Pages 431–468 in Evolution, time and space: the emergence of the biosphere (Syst. Assoc. Spec. Vol. 23), edited by R. W. Sims, J. H. Price, and P. E. S. Whalley. Academic Press, London, UK.

Milner, A. R. 1988. The relationships and origin of living amphibians. Pages 59–102 in The Phylogeny and Classification of the Tetrapods. Systematics Association Special Vol. no. 35A, edited by M. J. Benton. Clarendon Press, Oxford, UK.

Milner, A. R. 1993. The Paleozoic relatives of Lissamphibians. Herpetological Monographs 7:8–27.

Milstead, W. W. (editor) 1967. Lizard Ecology: A Symposium. University of Missouri, Columbia, MO.

Milstead, W. W. 1969. Studies on the evolution of box turtles (genus Terrapene). Bulletin of the Florida State Museum (Biological Sciences) 14:1–113.

Milton, K. 1994. No pain, no game. Natural History (9):44–51.

Mink, D. G., and J. W. Sites, Jr. 1996. Species limits, phylogenetic relationships, and origins of viviparity in the scalaris complex of the lizard genus Sceloporus (Phrynosomatidae: Sauria). Herpetologica 52:551–571.

Minnich, J. E. 1982. The use of water. Pages 325–395 in Biology of the Reptilia, Vol. 12 (Physiology C, Physiological Ecology), edited by C. Gans and F. H. Pough, Academic Press, London, UK.

Minton, S. A., Jr. 1990. Venomous bites by nonvenomous snakes: an annotated bibliography of colubrid envenomation. Journal of Wilderness Medicine 1:119–127.

Minton, S. A., Jr., and M. R. Minton 1973. Giant Reptiles. Scribner's New York, NY.

Mittermeier, R. A., J. L. Carr, I. R. Swingland, T. B. Werner, and R. B. Mast. 1992. Conservation of amphibians and reptiles. Pages 59–80 in Herpetology: Current Research on the Biology of Amphibians and Reptiles, edited by K. Adler. Proceedings of the First World Congress of Herpetology. Society for the Study of Amphibians and Reptiles, Oxford, OH.

Moermond, T. C. 1979. Habitat constraints on the behavior, morphology, and community structure of Anolis lizards. Ecology 60:152–164.

Moermond, T. C. 1986. A mechanistic approach to the structure of animal communities: Anolis lizards and birds. American Zoologist 26:23–37.

Molenaar, G. J. 1992. Anatomy and physiology of infrared sensitivity of snakes. Pages 367–453 in Biology of the Reptilia, Vol. 17 (Neurology C, Sensorimotor Integration), edited by C. Gans and P. S. Ulinski. University of Chicago Press, Chicago, IL.

Moler, P. E., and R. Franz. 1987. Wildlife values of small, isolated wetlands in the Southeastern Coastal Plain. Pages 234–241 in Proceedings 3rd S. E. Nongame and Endangered Wildlife Symposium, edited by R. R. Odom, K. A. Riddleberger, and J. C. Ozier. Georgia Department of Natural Resources, Atlanta, GA.

Moler, P. E., and J. Kezer. 1993. Karyology and systematics of the salamander genus Pseudobranchus (Sirenidae). Copeia 1993:39–47.

Moore, F. L. 1987. Regulation of reproductive behaviors. Pages 505–522 in Hormones and Reproduction in Fishes, Amphibians, and Reptiles, edited by D. O. Norris and R. E. Jones, Plenum Press, New York, NY.

Moore, F. R., and R. E. Gatten Jr. 1989. Locomotor performance of hydrated, dehydrated, and osmotically stressed anuran amphibians. Herpetologica 45:101–110.

Moore, J. A. (editor) 1964. Physiology of the Amphibia, Vol. 1. Academic Press, New York, NY.

Moore, M. C., and J. Lindzey. 1992. Physiological regulation of sexual behavior in male reptiles. Pages 70–113 in Biology of the Reptilia, Vol. 18 (Physiology E, Hormones, Brain, and Behavior), edited by C. Gans and D. Crews, The University of Chicago Press, Chicago, IL.

Morescalchi, A. 1979. New developments in vertebrate cytotaxonomy I. Cytotaxonomy of the amphibians. Genetica 50:179–193.

Morey, S. R. 1990. Microhabitat selection and predation in the Pacific treefrog, Pseudacris regilla. Journal of Herpetology 24:292–296.

Morin, P. J. 1983. Predation, competition, and the composition of larval anuran guilds. Ecological Monographs 53:119–138.

Moritz, C., C. J. Schneider, and D. B. Wake. 1992. Evolutionary relationships within the Ensatina eschscholtzii complex confirm the ring species interpretation. Systematic Biology 41:273–291.

Mortimer, J. A., and A. Carr. 1987. Reproduction and migration of the Ascension Island green turtle (Chelonia mydas). Copeia 1987:103–113.

Moura-da-Silva, A. M., R. D. G. Theakston, and J. M. Crampton. 1997. Molecular evolution of phospholipase A2s and metalloproteinase/disintegrins from venoms of vipers. Pages 173–187 in Venomous Snakes: Ecology, Evolu-

tion and Snakebite,edited by R. S. Thorpe, W. Wüster, and A. Malhotra. Clarendon Press, Oxford, UK.

Mrosovsky, N. 1983. *Conserving Sea Turtles*. British Herpetological Society. London, UK.

Mrosovsky, N. 1994. Sex ratios in sea turtles. *Journal of Experimental Zoology* 270:16–27.

Munger, J. C. 1984. Home ranges of horned lizards (*Phrynosoma*): circumscribed and exclusive? *Oecologia* 62:351–360.

Murphy, C. G. 1994a. Chorus tenure of male barking treefrogs, *Hyla gratiosa. Animal Behaviour* 48:763–777.

Murphy, C. G. 1994b. Determinants of chorus tenure in barking treefrogs (*Hyla gratiosa*). *Behavioral Ecology and Sociobiology* 34:285–294.

Murphy, J. B. 1976. Pedal luring in the leptodactylid frog, *Ceratophrys calcarata* Boulenger. *Herpetologica* 32:339–341.

Murphy, J. B., and W. E. Lamoreaux. 1978. Mating behavior in three Australian chelid turtles (Testudines: Pleurodira: Chelidae). *Herpetologica* 34:398–405.

Mushinsky, H. R., J. J. Hebrard, and D. S. Vodopich. 1982. Ontogeny of water snake foraging ecology. *Ecology* 63:1624–1629.

Musick, J. A., and C. J. Limpus. 1997. Habitat utilization and migration in juvenile sea turtles. Pages 137–163 in *The Biology of Sea Turtles*, edited by P. L. Lutz and J. A. Musick. CRC Press, Boca Raton, FL.

Myers, C. W. 1967. The familial position of *Typhlophis* Fitzinger (Serpentes). *Herpetologica* 23:75–77.

Myers, C. W. 1987. New generic names for some neotropical poison frogs (Dendrobatidae). *Papeis Avulsos de Zoologia, Sao Paulo* 36:301–306.

Myers, C. W., and J. W. Daly. 1976. Preliminary evaluation of skin toxins and vocalizations in taxonomic and evolutionary studies of poison-dart frogs (Dendrobatidae). *Bulletin of the American Museum of Natural History* 157:173–262.

Myers, C. W., J. W. Daly, and B. Malkin. 1978. A dangerously toxic new frog (*Phyllobates*) used by Embera Indians of western Colombia, with discussion of blowgun fabrication and dart poisoning. *Bulletin of the American Museum of Natural History* 161:307–366.

Myers, C. W., A. Paolillo, and J. W. Daly. 1991. Discovery of a defensively malodorous and nocturnal frog in the Family Dendrobatidae: phylogenetic significance of a new genus and species from the Venezuelan Andes. *American Museum Novitates* 3002:1–33.

Nabhan, G. P. 1995. Cultural parallax in viewing North American habitats. Pages 87–101 in *Reinventing Nature? Responses to Postmodern Deconstruction*, edited by M. E. Soulé and G. Lease. Island Press, Washington, D. C..

Nace, G. W., and J. K. Rosen. 1979. Source of amphibians for research II. *Herpetological Review* 10:8–15.

Nagy, K. A. 1992. The doubly labeled water method in ecological energetics studies of terrestrial vertebrates. *Bulletin of the Society of Ecophysiology, Supplement T* 17:9–14.

Nagy, K. A., and P. A. Medica. 1986. Physiological ecology of desert tortoises in southern Nevada. *Herpetologica* 42:73–92.

Narins, P. M. 1990. Seismic communication in anuran amphibians. *Bioscience* 40:268–274.

Narins, P. M. 1992. Evolution of anuran chorus behavior: neural and behavioral constraints. *American Naturalist* 139:S90–S104.

Narins, P. M., and R. R. Capranica. 1978. Communicative significance of the two-note call of the treefrog *Eleutherodactylus coqui. Journal of Comparative Physiology* 127:1–9.

Narins, P. M., and R. Zelick. 1988. The effects of noise on auditory processing and behavior in anurans. Pages 511–536 in *The Evolution of the Amphibian Auditory System*, edited by B. Fritzsch, M. J. Ryan, W. Wilczynski, T. E. Hetherington, and W. Walkowiak. John Wiley and Sons, New York, NY.

National Research Council. 1990. *Decline of the Sea Turtles*. National Academy Press, Washington, DC.

Naylor, B. G. 1978. The earliest known *Necturus* (Amphibia, Urodela), from the Paleocene Ravenscrag Formation of Saskatchewan. *Journal of Herpetology* 12: 565–569.

Neill, W. T. 1974. *Reptiles and Amphibians in the Service of Man*. Pegasus, Bobbs-Merrill, IN.

Nelson, C. E., and J. R. Meyer. 1967. Variation and distribution of the Middle American snake genus, *Loxocemus* Cope (Boidae?). *Southwestern Naturalist* 12:439–453.

Nessov, L. A. 1988. Late Mesozoic amphibians and lizards of Soviet Middle Asia. *Acta Zoologica Cracoviensia* 31: 475–486.

Nevo, E., and R. Estes. 1969. *Ramonellus longispinus*, an early Cretaceous salamander from Israel. *Copeia* 1969:540–547.

Newman, E. A., and P. H. Hartline. 1982. The infrared "vision" of snakes. *Scientific American* 246:116–127.

Newman, R. A. 1992. Adaptive plasticity in amphibian metamorphosis. *Bioscience* 42:671–678.

Niblick, H. A., D. C. Rostal, and T. Classen. 1994. Role of male-male interactions and female choice in the mating system of the desert tortoise, *Gopherus agassizii. Herpetological Monographs* 8:124–132.

Nickerson, M. A., and C. E. Mays. 1973. *The hellbenders: North American "Giant Salamanders"*. Milwaukee Public Museum, Milwaukee, WI.

Nilson, G., and C. Andren. 1986. The mountain vipers of the Middle East—the *Vipera xanthina* complex (Reptilia, Viperidae). *Bonner Zoologische Monographien* 20:1–90.

Nishikawa, K. C. 1985. Competition and the evolution of aggressive behavior in two species of terrestrial salamanders. *Evolution* 39:1282–1294.

Nishikawa, K. C. 1990. Intraspecific spatial relationships of two species of terrestrial salamanders. *Copeia* 1990: 418–426.

Nishikawa, K. C., and D. C. Cannatella. 1991. Kinematics of prey capture in the tailed frog *Ascaphus truei* (Anura: Ascaphidae). *Zoological Journal of the Linnean Society* 103:289–307.

Noble, G. K. 1925. The integumentary, pulmonary, and cardiac modifications correlated with increased cutaneous respiration in the Amphibia: a solution of the "hairy frog" problem. *Journal of Morphology and Physiology* 40:341–416.

Noble, G. K. 1931. *The Biology of the Amphibia*. McGraw-Hill Book Co., New York, NY.

Noble, G. K., and H. T. Bradley. 1933. The mating behavior of lizards: its bearing on the theory of sexual selection. *Annals of the New York Academy of Sciences* 35:25–100.

Noble, G. K., and R. C. Noble. 1923. The Anderson tree frog (*Hyla andersonii* Baird). Observations on its habits and life history. *Zoologica* 2:414–455.

Norell, M. A. 1989. The higher level relationships of the extant Crocodylia. *Journal of Herpetology* 23:325–335.

Norris, K. S. 1967. Color adaptation in desert reptiles and its thermal relationships. Pages 162–229 in *Lizard Ecology, A Symposium*, edited by W. W. Milstead. University of Missouri Press, Columbia, MO.

Nürnberger, B., N. Barton, C. MacCallum, J. Gilchrist, and M. Appleby. 1995. Natural selection on quantitative traits in the *Bombina* hybrid zone. *Evolution* 49:1224–1238.

Nussbaum, R. A. 1976. Geographic variation and systematics of salamanders of the genus *Dicamptodon* Strauch (Ambystomatidae). *Miscellaneous Publications of the Museum of Zoology, University of Michigan* 149:1–94.

Nussbaum, R. A. 1977. Rhinatrematidae: a new family of caecilians (Amphibia: Gymnophiona). *Occasional Papers of the Museum of Zoology, University of Michigan* 682:1–30.

Nussbaum, R. A. 1979a. Mitotic chromosomes of Sooglossidae (Amphibia: Anura). *Caryologia* 32:279–298.

Nussbaum, R. A. 1979b. The taxonomic status of the caecilian genus *Uraeotyphlus* Peters. *Occasional Papers of the Museum of Zoology, University of Michigan* 687:1–20.

Nussbaum, R. A. 1980. Phylogenetic implications of amplectant behavior in sooglossid frogs. *Herpetologica* 36:1–5.

Nussbaum, R. A. 1983. The evolution of a unique dual jaw-closing mechanism in caecilians (Amphibia: Gymnophiona) and its bearing on caecilian ancestry. *Journal of Zoology, London* 199:545–554.

Nussbaum, R. A. 1985a. The evolution of parental care in salamanders. *Miscellaneous Publications of the Museum of Zoology, University of Michigan* 169:1–50.

Nussbaum, R. A. 1985b. Systematics of caecilians (Amphibia: Gymnophiona) of the family Scolecomorphidae. *Occasional Papers of the Museum of Zoology, University of Michigan* no. 713:1–49.

Nussbaum, R. A. 1991. Cytotaxonomy of caecilians. Pages 33–66 in *Amphibian Cytogenetics and Evolution*, edited by D. M. Green and S. K. Sessions. Academic Press, San Diego.

Nussbaum, R. A. 1992. Caecilians. Pages 52–59 in *Reptiles and Amphibians*, edited by H. G. Cogger and R. G. Zweifel. Smithmark, New York, NY.

Nussbaum, R. A., and H. Hinkel. 1994. Revision of East African caecilians of the genera *Afrocaecilia* Taylor and *Boulengerula* Tornier (Amphibia: Gymnophiona: Caeciliaidae). *Copeia* 1994:750–760.

Nussbaum, R. A., and C. K. Tait. 1977. Aspects of the life history and ecology of the Olympic salamander, *Rhyacotriton olympicus* (Gaige). *American Midland Naturalist* 98:176–199.

Nussbaum, R. A., and M. Wilkinson. 1989. On the classification and phylogeny of caecilians (Amphibia: Gymnophiona), a critical review. *Herpetological Monographs* 3:1–42.

Nussbaum, R. A., and M. Wilkinson. 1996. A new genus of lungless tetrapod: a radically divergent caecilian (Amphibia: Gymnophiona). *Proceedings of the Royal Society of London* B 261:331–335.

Ocana, G., I. Rubinoff, N. Smythe, and D. Werner. 1988. Alternatives to destruction: research in Panama. Pages 370–376 in *Biodiversity*, edited by E. O. Wilson and F. M. Peter. National Academy Press, Washington, D. C..

O'Connor, M. P., and J. R. Spotila. 1992. Consider a spherical lizard: animals, models, and approximations. *American Zoologist* 32:19–193.

O'Connor, M. P., L. C. Zimmerman, D. E. Ruby, S. J. Bulova, and J. R. Spotila. 1994. Home range size and movements by desert tortoises, *Gopherus agassizii*, in the eastern Mojave Desert. *Herpetological Monographs* 8:60–71.

Oftedal, O. T., M. E. Allen, A. L. Chung, R. C. Reed, and D. E. Ullrey. 1994. Nutrition, urates, and desert survival: potassium and the desert tortoise (*Gopherus agassizii*). Pages 308–313 in *Proceedings of the Association of Reptilian and Amphibian Veterinarians and American Association of Zoo Veterinarians*, edited by ., Pittsburgh, PA.

Oliver, J. A. 1955. *The Natural History of North American Amphibians and Reptiles*. Van Nostrand-Reinhold, Princeton, NJ.

Olsson, M. 1993. Contest success and mate guarding in male sand lizards, *Lacerta agilis*. *Animal Behaviour* 46:408–409.

Olsson, M. 1994a. Nuptial coloration in the sand lizard, *Lacerta agilis*: an intra-sexually selected cue to fighting ability. *Animal Behaviour* 48:607–613.

Olsson, M. 1994b. Rival recognition affects male contest behavior in sand lizards (*Lacerta agilis*). *Behavioral Ecology and Sociobiology* 35:249–252.

Olsson, M. 1995. Forced copulations and costly female resistance behavior in the Lake Eyre dragon, *Ctenophorus maculosus*. *Herpetologica* 51:19–24.

Olsson, M., A. Gullberg, and H. Tegelstrom. 1994. Sperm competition in the sand lizard, *Lacerta agilis*. *Animal Behaviour* 48:193–200.

Olsson, M., A. Gullberg, and H. Tegelstrom. 1996. Mate guarding in male sand lizards (*Lacerta agilis*). *Behaviour* 133:367–386.

Olsson, M., and T. Madsen. 1995. Female choice on male quantitative traits in lizards—why is it so rare? *Behavioral Ecology and Sociobiology* 36:179–184.

Olsson, M., and R. Shine. 1997. The limits to reproductive output: offspring size versus number in the sand lizard (*Lacerta agilis*). *American Naturalist* 149:179–188.

Omland, K. S. 1996. Slope orientation as a mechanism of red-spotted newt (*Notophthalmus viridescens*) migratory guidance. Unpublished M. S. Thesis, University of Vermont, Burlington, VT.

Orton, G. 1953. The systematics of vertebrate larvae. *Systematic Zoology* 2:63–75.

Orton, G. 1957. The bearing of larval evolution on some problems in frog classification. *Systematic Zoology* 6:79–86.

Ott, M., and F. Schaeffel. 1995. A negatively powered lens in the chameleon. *Nature* 373:692–694.

Ouboter, P. E., and L. M. R. Nanhoe. 1988. Habitat selection and migration of *Caiman crocodilus crocodilus* in a swamp and swamp-forest habitat in northern Suriname. *Journal of Herpetology* 22:283–294.

Ovaska, K., and W. Hunte. 1992. Male mating behavior of the frog *Eleutherodactylus johnstonei* (Leptodactylidae) in Barbados, West Indies. *Herpetologica* 48:40–49.

Özeti, N., and D. B. Wake. 1969. The morphology and evolution of the tongue and associated structures in salamanders and newts (Family Salamandridae). *Copeia* 1969:91–123.

Pacala, S. W., and J. Roughgarden. 1985. Population experiments with the *Anolis* lizards of St. Maarten and St. Eustatius. *Ecology* 66:129–141.

Packard, G. C., and M. J. Packard. 1988. The physiological ecology of reptilian eggs and embryos. Pages 523–605 in *Biology of the Reptilia, Vol. 16 (Ecology B: Defense and Life History)*, edited by C. Gans and R. B. Huey. Alan R. Liss, Inc., New York, NY.

Packard, G. C., and M. J. Packard. 1995. The basis for cold tolerance in hatchling painted turtles (*Chrysemys picta*). *Physiological Zoology* 68:129–148.

Packard, G. C., C. R. Tracy, and J. J. Roth. 1977. The physiological ecology of reptilian eggs and embryos, and the evolution of viviparity within the class Reptilia. *Biological Reviews of the Cambridge Philosophical Society* 52:71–105.

Panchen, A. L., and T. R. Smithson. 1987. Character diagnosis, fossils, and the origin of tetrapods. *Biological Reviews of the Cambridge Philosophical Society* 62:341–438.

Papenfuss, T. J. 1982. The ecology and systematics of the amphisbaenian genus *Bipes*. *Occasional Papers of the California Academy of Sciences* 136:1–42.

Parker, H. W. 1934. *A Monograph of the Frogs of the Family Microhylidae*. British Museum (Natural History), London, UK.

Parker, H. W. 1956. Viviparous caecilians and amphibian phylogeny. *Nature* 178:250–252.

Parker, H. W., and A. G. C. Grandison. 1977. *Snakes: A Natural History, 2nd Ed*. Cornell University Press, Ithaca, NY.

Parrish, J. M. 1986. Locomotor adaptations in the hindlimb and pelvis of Thecodontia. *Hunteria* 1:1–35.

Parsons, T. S., and E. E. Williams. 1963. The relationships of the modern amphibia: a re-examination. *Quarterly Review of Biology* :26–53.

Patchell, F. C., and R. Shine. 1986a. Feeding mechanisms in pygopodid lizards: how can *Lialis* swallow such large prey? *Journal of Herpetology* 20:59–64.

Patchell, F. C., and R. Shine. 1986b. Hinged teeth for hard-bodied prey: a case of convergent evolution between snakes and legless lizards. *Journal of Zoology, London* 208:269–275.

Pechmann, J. H. K., D. E. Scott, J. W. Gibbons, and R. D. Semlitsch. 1989. Influence of wetland hydroperiod on diversity and abundance of metamorphosing juvenile amphibians. *Wetlands Ecology and Management* 1:3–11.

Pechmann, J. H. K., D. E. Scott, R. D. Semlitsch, J. P. Caldwell, L. J. Vitt, and J. W. Gibbons. 1991. Declining amphibian populations: the problem of separating human impacts from natural fluctuations. *Science* 253:892–895.

Perret, J.-L. 1966. Les amphibiens du Cameroun. *Zoologische Jahrbucher Abteilung für Systematik* 93:289–464.

Persky, B., H. M. Smith, and K. L. Williams. 1976. Additional observations on ophidian costal cartilages. *Herpetologica* 32:399–401.

Peterson, C. C. 1996. Anhomeostasis: seasonal water and solute relations in two populations of the desert tortoise (*Gopherus agassizii*) during chronic drought. *Physiological Zoology* 69:1324–1358.

Peterson, J. A. 1984. The locomotion of *Chamaeleo* (Reptilia: Sauria) with particular reference to the forelimb. *Journal of Zoology* 202:1–42.

Petranka, J. W., and D. A. G. Thomas. 1995. Explosive breeding reduces egg and tadpole cannibalism in the wood frog, *Rana sylvatica*. *Animal Behaviour* 50:731–739.

Petren, K., and T. J. Case. 1996. An experimental demonstration of exploitation competition in an ongoing invasion. *Ecology* 77:118–132.

Pfennig, D. W., M. L. G. Laeb, and J. P. Collins. 1991. Pathogens as a factor limiting the spread of cannibalism in tiger salamanders. *Oecologia* 88:161–166.

Phillips, J. A. 1995. Rhythms of a desert lizard. *Natural History* 104 (10):51–55.

Phillips, J. A. 1995. Movement patterns and density of *Varanus albigularis*. *Journal of Herpetology* 29:407–416.

Phillips, J. B. 1987. Laboratory studies of homing orientation in the eastern red-spotted newt *Notophthalmus viridescens*. *Journal of Experimental Biology* 131:215–229.

Phillips, J. B., K. Adler, and S. C. Borland. 1995. True navigation by an amphibian. *Animal Behaviour* 50: 855–858.

Phillips, J. B., and S. C. Borland. 1994. Use of a specialized magnetoreception system for homing by the eastern red-spotted newt *Notophthalmus viridescens*. *Journal of Experimental Biology* 188:275–291.

Phillips, K. 1994. *Tracking the Vanishing Frogs: An Ecological Mystery*. St. Martin's Press, New York, NY.

Pianka, E. R. 1967. On lizard species diversity: North American flatland deserts. *Ecology* 48:333–351.

Pianka, E. R. 1986. *Ecology and Natural History of Desert Lizards*. Princeton University Press, Princeton, NJ.

Picker, M. D. 1985. Hybridization and habitat selection in *Xenopus gilli* and *Xenopus laevis* in the South-western Cape Province. *Copeia* 1985:574–580.

Pierce, B. A. 1985. Acid tolerance in amphibians. *BioScience* 35:239–243.

Pinder, A. W., K. B. Storey, and G. R. Ultsch. 1992. Estivation and hibernation. Pages 250–274 in *Environmental Physiology of the Amphibians*, edited by M. E. Feder and W. W. Burggren, University of Chicago Press, Chicago, IL.

Plasa, L. 1979. Heimfindeverhalten bei *Salamandra salamandra* (L.). *Zeitschrift für Tierpsychologie* 51:113–125.

Plummer, M. V., and J. D. Congdon. 1994. Radiotelemetric study of activity and movements of racers (*Coluber constrictor*) associated with a Carolina Bay in South Carolina. *Copeia* 1994:20–26.

Poe, S. 1996. Data set incongruence and the phylogeny of crocodilians. *Systematic Biology* 45:393–414.

Polis, G. A., and C. A. Myers. 1985. A survey of intraspecific predation among reptiles. *Journal of Herpetology* 19:99–107.

Pombal, J. P., I. Sazima, and C. F. B. Haddad. 1994. Breeding behavior of the pumpkin toadlet, *Brachycephalus ephippium* (Brachycephalidae). *Journal of Herpetology* 28:516–519.

Pope, C. H. 1958. Fatal bite of captive African rear-fanged snake (*Dispholidus*). *Copeia* 1958:280–282.

Pough, F. H. 1973. Lizard energetics and diet. *Ecology* 54:837–844.

Pough, F. H. 1976a. Acid precipitation and embryonic mortality of spotted salamanders, *Ambystoma maculatum*. *Science* 192:68–70.

Pough, F. H. 1976b. Multiple cryptic effects of cross-banded and ringed patterns of snakes. *Copeia* 1976:834–836.

Pough, F. H. 1980. The advantages of ectothermy for tetrapods. *American Naturalist* 115:92–112.

Pough, F. H. 1983. Amphibians and reptiles as low-energy systems. Pages 141–188 in *Behavioral Energetics: The Cost of Survival in Vertebrates*. Ohio State University Press, Columbus.

Pough, F. H. 1988. Mimicry and related phenomena. Pages 153–234 in *Biology of the Reptilia, Vol 16 (Ecology B: Defense and life history)*, edited by C. Gans and R. B. Huey. Alan R. Liss, NY.

Pough, F. H. 1989. Organismal performance and Darwinian fitness: Approaches and Interpretations. *Physiological Zoology* 62:199–236.

Pough, F. H. 1993. Zoo-academic research collaborations: how close are we? *Herpetologica* 49:500–508.

Pough, F. H., and R. M. Andrews. 1985a. Energy costs of subduing and swallowing prey for a lizard. *Ecology* 66:1525–1531.

Pough, F. H., and R. M. Andrews. 1985b. Use of anaerobic metabolism by free-ranging lizards. *Physiological Zoology* 58:205–213.

Pough, F. H., and J. D. Groves. 1983. Specializations of the body form and food habits of snakes. *American Zoologist* 23:443–454.

Pough, F. H., J. B. Heiser, and W. N. McFarland. 1996. *Vertebrate Life*, 4th ed. Prentice Hall, New York, NY.

Pough, F. H., W. E. Magnusson, M. J. Ryan, K. D. Wells, and T. L. Taigen. 1992. Behavioral energetics. Pages 395–436 in *Environmental Physiology of the Amphibians*, edited by M. E. Feder and W. W. Burggren, University of Chicago Press, Chicago, IL.

Pough, F. H., M. M. Stewart, and R. G. Thomas. 1977. Physiological basis of habitat partitioning in Jamaican *Eleutherodactylus*. *Oecologia* 27:285–293.

Pough, F. H., T. L. Taigen, M. M. Stewart, and P. F. Brussard. 1983. Behavioral modification of evaporative water loss by a Puerto Rican frog. *Ecology* 64:244–252.

Pounds, J. A. 1988. Ecomorphology, locomotion, and microhabitat structure: patterns in a tropical mainland *Anolis* community. *Ecological Monographs* 58: 299–320.

Pounds, J. A., and M. L. Crump. 1994. Amphibian declines and climate disturbance: the case of the golden toad and the harlequin frog. *Conservation Biology* 8:72–85.

Poynton, J. C. 1964. The amphibia of southern Africa: a faunal study. *Annals of the Natal Museum* 17:1–334.

Poynton, J. C., and D. G. Broadley. 1985. Amphibia Zambesiaca 1. Scolecomorphidae, Pipidae, Microhylidae, Hemisidae, Arthroleptidae. *Annals of the Natal Museum* 26:503–553.

Preest, M. R. 1991. Energetic costs of prey ingestion in a scincid lizard, *Scincella lateralis*. *Journal of Comparative Physiology B* 161:327–332.

Preest, M. R., and F. H. Pough. 1989. Interaction of temperature and hydration on locomotion of toads. *Functional Ecology* 3:693–699.

Pregill, G. K. 1992. Systematics of the West Indian lizard genus *Leiocephalus* (Squamata: Iguania: Tropiduridae).

Miscellaneous Publications, Museum of Natural History, University of Kansas 84:1–69.

Pregill, G. K., J. A. Gauthier, and H. W. Greene. 1986. The evolution of helodermatid squamates, with description of a new taxon and an overview of Varanoidea. *Transactions of the San Diego Society of Natural History* 21:167–202.

Presch, W. 1974. Evolutionary relationships and biogeography of the macroteiid lizards (Family Teiidae, subfamily Teiinae). *Bulletin of the Southern California Academy of Sciences* 73:23–32.

Presch, W. 1980. Evolutionary history of the South American microteiid lizards (Teiidae: Gymnophthalminae). *Copeia* 1980:36–56.

Prestwich, K. N. 1994. The energetics of acoustic signaling in anurans and insects. *American Zoologist* 34:625–643.

Pritchard, P. C. H., and P. Trebbau. 1984. *The Turtles of Venezuela*. Society for the Study of Amphibians and Reptiles, Oxford, OH.

Qualls, C. P., and R. Shine. 1995. The evolution of vivparity within the Australian scincid lizard *Lerista bougainvillii*. *Journal of Zoology, London* 237:13–26.

Rabb, G. B., and M. S. Rabb. 1963a. Additional observations on breeding behavior of the Surinam toad, *Pipa pipa*. *Copeia* 1963:636–642.

Rabb, G. B., and M. S. Rabb. 1963b. On the behavior and breeding biology of the African pipid frog *Hymenochirus boettgeri*. *Zeitschrift fur Tierpsychologie* 20:215–240.

Rage, J.-C. 1987. Fossil history. Pages 51–76 in *Snakes: Ecology and Evolutionary Biology*, edited by R. A. Seigel, J. T. Collins, and S. S. Novak. Macmillan Publishing Company, New York, NY.

Rage, J.-C. 1988. The oldest known colubrid snakes. The state of the art. *Acta Zoologica Cracoviensia* 31:457–474.

Rage, J.-C., E. Buffetaut, H. Buffetaut-Tong, Y. Chaimanee, S. Ducrocq, J.-J. Jaeger, and V. Suteethorn. 1992. A colubrid snake in the late Eocene of Thailand: the oldest known Colubridae (Reptilia, Serpentes). *Comptes Rendus de l'Academie des Sciences, Paris* 314, ser. II: 1085–1089.

Rage, J. C., and G. V. R. Prasad. 1992. New snakes from the late Cretaceous (Maastrichtian) of Naskal, India. *Neue Jarbuch Geologiches Paläontologische Abhandlungen* 187: 83–97.

Rage, J.-C., and Z. Rocek. 1989. Redescription of *Triadobatrachus massinoti* (Piveteau, 1936), an anuran amphibian from the early Triassic. *Palaeontographica Abt. A* 206: 1–16.

Rage, J.-C., and G. Wouters. 1979. Decouverte du plus ancien palaeopheide (Reptilia, Serpentes) dans le Maestrichtien du Maroc. *Geobios* 12:293–296.

Rand, A. S. 1964. Inverse relationship between temperature and shyness in the lizard *Anolis lineatopus*. *Ecology* 45: 863–864.

Rand, A. S. 1967. The adaptive significance of territoriality in iguanid lizards. Pages 106–115 in *Lizard Ecology: A Symposium*, edited by W. S. Milstead. University of Missouri Press, Columbia, MO.

Rand, A. S., and R. M. Andrews. 1975. Adult color dimorphism and juvenile pattern in *Anolis cuvieri*. *Journal of Herpetology* 9:257–260.

Rand, A. S., B. A. Dugan, H. Monteza, and D. Vianda. 1990. The diet of a generalized folivore: *Iguana iguana* in Panama. *Journal of Herpetology* 24:211–214.

Rand, A. S., S. Guerrero, and R. M. Andrews. 1983. The ecological effects of malaria on populations of the lizard *Anolis limifrons* on Barro Colorado Island Panama. Pages 455–471 in *Advances in Herpetology and Evolutionary Biology: Essays in Honor of Ernest E. Williams*, edited by A. G. J. Rhodin and K. Miyata. Harvard University Press, Cambridge, MA.

Rasmussen, A. R. 1997. Systematics of sea snakes: a critical review. Pages 15–30 in *Venomous Snakes: Ecology, Evolution and Snakebite*, edited by R. S. Thorpe, W. Wüster, and A. Malhotra. Clarendon Press, Oxford.

Raxworthy, C. J., and R. A. Nussbaum. 1995. Systematics, speciation and biogeography of the dwarf chamaeleons (*Brookesia*; Reptilia, Squamata, Chamaeleontidae) of northern Madagascar. *Journal of Zoology, London* 235: 525–558.

Raynaud, A. 1985. Development of limbs and embryonic limb reduction. Pages 59–148 in *Biology of the Reptilia, Vol. 15 (Development B)*, edited by C. Gans and F. Billett. John Wiley & Sons, New York, NY.

Reagan, D. P. 1991. The response of *Anolis* lizards to hurricane-induced habitat changes in a Puerto Rican rain forest. *Biotropica* 23:468–474.

Reed, K. M., I. F. Greenbaum, and J. W. Sites. 1995a. Cytogenetic analysis of chromosomal intermediates from a hybrid zone between two chromosome races of the *Sceloporus grammicus* complex (Sauria, Phrynosomatidae). *Evolution* 49:37–47.

Reed, K. M., I. F. Greenbaum, and J. W. Sites. 1995b. Dynamics of a novel chromosomal polymorphism within a hybrid zone between two chromosome races of the *Sceloporus grammicus* complex (Sauria, Phrynosomatidae). *Evolution* 49:48–60.

Reeder, T. W., and J. J. Wiens. 1996. Evolution of the lizard family Phrynosomatidae as inferred from diverse types of data. *Herpetological Monographs* 10:43–84.

Regal, P. J. 1966. Thermophilic responses following feeding in certain reptiles. *Copeia* 1966:588–590.

Reh, W., and A. Seitz. 1990. The influence of land use on the genetic structure of populations of the common frog *Rana temporaria*. *Biological Conservation* 54:239–249.

Reilly, S. M. 1983. The biology of the high altitude salamander *Batrachuperus mustersi* from Afghanistan. *Journal of Herpetology* 17:1–9.

Reilly, S. M. 1994. The ecological morphology of metamorphosis: heterochrony and the evolution of feeding mechanisms in salamanders. Pages 319–338 in *Ecological Morphology, Integrative Organismal Biology*, edited by P. C. Wainwright and S. M. Reilly. University of Chicago Press, Chicago, IL.

Reilly, S. M. 1995. Quantitative electromyography and muscle function of the hind limb during quadrupedal running in the lizard *Sceloporus clarki*. *Zoology* 98:263–277.

Reilly, S. M., and R. A. Brandon. 1994. Partial paedomorphosis in the Mexican stream ambystomatids and the taxonomic status of the genus *Rhyacosiredon* Dunn. *Copeia* 1994:656–662.

Reilly, S. M., and M. J. DeLancey. 1997. Sprawling locomotion in the lizard *Sceloporus clarkii*: Quantitative kinematics of a walking trot. *Journal of Experimental Biology* 200:753–765.

Reilly, S. M., and G. V. Lauder. 1992. Morphology, behavior, and evolution: Comparative kinematics of aquatic feeding in salamanders. *Brain, Behavior, and Evolution* 40:182–196.

Reilly, S. M., E. O. Wiley, and D. J. Meinhardt. 1997. An integrative approach to heterochrony: the distinction between interspecific and intraspecific phenomena. *Biological Journal of the Linnaen Society* 60:119–143.

Resetarits, W. J., Jr., and H. M. Wilbur. 1989. Choice of oviposition site by *Hyla chrysoscelis*: role of predators and competitors. *Ecology* 70:220–228.

Ressel, S. J. 1996. Ultrastructural properties of muscles used for calling in neotropical frogs. *Physiological Zoology* 69:952–973.

Rewcastle, S. C. 1980. Form and function in lacertilian knee and mesotarsal joints; a contribution to the analysis of sprawling locomotion. *Journal of Zoology* 191:147–170.

Richards, S. J., U. Sinsch, and R. A. Alford. 1994. Radio tracking. Pages 155–158 in *Measuring and Monitoring Biological Diversity: Standard Methods for Amphibians*, edited by W. R. Heyer, M. A. Donnelly, R. W. McDiarmid, L. C. Hayek, and M. S. Foster. Smithsonian Institution Press, Washington, D. C.

Rieppel, O. 1977. Studies on the skull of the Henophidia (Reptilia: Serpentes). *Journal of Zoology, London* 181:145–173.

Rieppel, O. 1978. A functional and phylogenetic interpretation of the skull of the Erycinae (Reptilia, Serpentes). *Journal of Zoology, London* 186:185–208.

Rieppel, O. 1980. The trigeminal jaw adductor musculature of *Tupinambis*, with comments on the phylogenetic relationships of the Teiidae (Reptilia, Lacertilia). *Zoological Journal of the Linnean Society* 69:1–29.

Rieppel, O. 1987. The phylogenetic relationships within the Chamaeleontidae, with comments on some aspects of cladistic analysis. *Zoological Journal of the Linnean Society* 89:41–62.

Rieppel, O. 1988. A review of the origin of snakes. *Evolutionary Biology* 22:37–130.

Rieppel, O., and M. deBraga. 1996. Turtles as diapsid reptiles. *Nature* 384:453–455.

Robb, J. 1966. The generic status of the Australasian typhlopids (Reptilia: Squamata). *Annals and Magazine of Natural History* (13) 9:675–679.

Roberts, W. E. 1994. Explosive breeding aggregations and parachuting in a neotropical frog, *Agalychnis saltator* (Hylidae). *Journal of Herpetology* 28:193–199.

Rocek, Z., and L. A. Nessov. 1993. Cretaceous anurans from central Asia. *Palaeontographica* 226:1–54.

Rodda, G. H. 1984. Orientation and navigation of juvenile alligators: evidence for magnetic sensitivity. *Journal of Comparative Physiology A* 154:649–658.

Rodda, G. H. 1985. Navigation in juvenile alligators. *Zeitschrift für Tierpsychologie* 68:65–77.

Rodda, G. H. 1992. The mating behavior of *Iguana iguana*. *Smithsonian Contributions to Zoology, number* 534:1–40.

Rodda, G. H., and T. H. Fritts. 1992. The impact of the introduction of the colubrid snake *Boiga irregularis* on Guam's lizards. *Journal of Herpetology* 26:166–174.

Rodriguez-Robles, J. A., and R. Thomas. 1992. Venom function in the Puerto Rican racer, *Alsophis portoricensis* (Serpentes: Colubridae). *Copeia* 1992:62–68.

Roessler, M. K., H. M. Smith, and D. Chiszar. 1990. Bidder's organ: a bufonid by-product of the evolutionary loss of hyperfecundity. *Amphibia Reptilia* 11:225–235.

Rose, F. L., and D. Armentrout. 1976. Adaptive strategies of *Ambystoma tigrinus* (Green) inhabiting the Llano Estacado of west Texas. *Journal of Animal Ecology* 45:713–729.

Rosen, D. E., P. L. Forey, B. G. Gardiner, and C. Patterson. 1981. Lungfishes, tetrapods, paleontology and plesiomorphy. *Bulletin of the American Museum of Natural History* 167:159–276.

Rosenberg, H. I., A. Bdolah, and E. Kochva. 1985. Lethal factors and enzymes in the secretion from Duvernoy's gland of three colubrid snakes. *Journal of Experimental Zoology* 233:5–14.

Rosenzweig, M. L. 1995. *Species Diversity in Space and Time*. Cambridge University Press, New York, NY.

Ross, C. A., D. K. Blake, and J. T. V. Onions. 1989. Farming and Ranching. Pages 202–213 in *Crocodiles and Alligators*, edited by C. A. Ross. Facts on File, New York, NY.

Ross, D. A. and R. K. Anderson. 1990. Habitat use, movements, and nesting of *Emydoidea blandingi* in central Wisconsin. *Journal of Herpetology* 24:6–12.

Ross, P., and D. Crews. 1977. Influence of the seminal plug on mating behavior in the garter snake. *Nature* 267:344–345.

Roth, G., U. Dicke, and K. Nishikawa. 1992. How do ontogeny, mophology, and physiology of sensory systems constrain and direct the evolution of amphibians? *American Naturalist* 139:S105–S124.

Roth, G., K. C. Nishikawa, C. Naujoks-Manteuffel, A. Schmidt, and D. B. Wake. 1993. Paedomorphosis and simplification in the nervous system of salamanders. *Brain, Behavior and Evolution* 42:137–170.

Roth, G., and A. Schmidt. 1993. The nervous system of plethodontid salamanders: insight into the interplay between genome, organism, behavior, and ecology. *Herpetologica* 49:185–194.

Rougier, G. W., M. S. de la Fuente, and A. B. Arcucci. 1995. Late Triassic turtles from South America. *Science* 268:855–858.

Routman, E., R. Wu, and A. R. Templeton. 1994. Parsimony, molecular evolution, and biogeography: the case of the North American giant salamander. *Evolution* 48:1799–1809.

Roux-Estève, R. 1974. Révision systématique des Typhlopidae d'Afrique (Reptilia - Serpentes). *Mémoires du Muséum National d'Histoire Naturelle, Paris* ser. A (Zoologie) 87:1–313.

Roux-Estève, R. 1975. Recherches sur la morphologie, la biogéographie et la phylogénie des Typhlopidae d'Afrique. *Bulletin de l'Institute Fondamentale d'Afrique Noire* ser. A, 36:428–508.

Rowe, J. W. and E. O. Moll. 1991. A radiotelemetric study of activity and movements of the Blanding's turtle (*Emydoidea blandingi*) in northeastern Illinois. *Journal of Herpetology* 25:178–185.

Roze, J. A. 1996. *Coral Snakes: Biology, Identification, Venoms.* Krieger Publishing Co., Malabar, FL.

Ruben, J. A., and D. E. Battalia. 1979. Aerobic and anaerobic metabolism during activity in small rodents. *Journal of Experimental Zoology* 208:73–76.

Ruben, J. A., and A. Boucot. 1989. The origin of the lungless salamanders (Amphibia: Plethodontidae). *American Naturalist* 134:161–169.

Ruben, J. A., N. L. Reagan, P. A. Verrell, and A. J. Boucot. 1993. Plethodontid salamander origins: a response to Beachy and Bruce. *American Naturalist* 142:1038–1051.

Ruby, D. E., and H. A. Niblick. 1994. A behavioral inventory of the desert tortoise: development of an ethogram. *Herpetological Monographs* 8:88–102.

Ruby, D. E., J. R. Spotila, S. K. Martin, and S. J. Kemp. 1994. Behavioral responses to barriers by desert tortoises: implications for wildlife management. *Herpetological Monographs* 8:144–160.

Ruibal, R., and V. Ernst. 1965. The structure of the digital setae of lizards. *Journal of Morphology* 117:271–294.

Ruiz-Carranza, P. M., and J. D. Lynch. 1991. Ranas Centrolenidae de Colombia I. Propuesta de una nueva clasificacion generica. *Lozania* 57:1–30.

Runkle, L. S., K. D. Wells, C. C. Robb, and S. L. Lance. 1994. Individual, nightly, and seasonal variation in calling behavior of the gray treefrog, *Hyla versicolor*: implications for energy expenditure. *Behavioral Ecology* 5:318–325.

Russell, A. P. 1975. A contribution to the functional analysis of the foot of the Tokay, *Gekko gecko* (Reptilia: Gekkonidae). *Journal of Zoology* 176:437–476.

Russell, A. P. 1979a. Parallelism and integrated design in the foot structure of gekkonine and diplodactyline geckos. *Copeia* 1979:1–21.

Russell, A. P. 1979b. The origin of parachuting locomotion in gekkonid lizards (Reptilia: Gekkonidae). *Zoological Journal of the Linnean Society* 65:233–249.

Russell, A. P. 1981. Descriptive and functional anatomy of the digital vascular system of the tokay, *Gekko gecko*. *Journal of Morphology* 169:293–323.

Russell, A. P. 1986. The morphological basis of weight-bearing in the scansors of the tokay gecko (Reptilia: Sauria). *Canadian Journal of Zoology* 64:948–955.

Russell, A. P., and A. M. Bauer. 1990. Digit I in pad-bearing gekkonine geckos: Alternate designs and the potential constraints of phalangeal number. *Memoirs of the Queensland Museum* 29:453–472.

Russell, F. E. 1980. *Snake Venom Poisoning*. J. B. Lippincott Co., Philadelphia, PA.

Ryan, M. J. 1985. *The Túngara Frog*. The University of Chicago Press, Chicago, IL.

Ryan, M. J. 1988. Energy, calling and selection. *American Zoologist* 28:885–898.

Ryser, J. 1989. The breeding migration and mating system of a Swiss population of the common frog *Rana temporaria*. *Amphibia-Reptilia* 10:13–21.

de Sa, R. O., and D. M. Hillis. 1990. Phylogenetic relationships of the pipid frogs *Xenopus* and *Silurana*: an integration of ribosomal DNA and morphology. *Molecular Biology and Evolution* 7:365–376.

Sage, R. D., E. M. Prager, and D. B. Wake. 1983. A Cretaceous divergence time between pelobatid frogs (*Pelobates* and *Scaphiopus*): immunological studies of serum albumin. *Journal of Zoology, London* 198:481–494.

Saint Girons, H. 1985. Comparative data on Lepidosaurian reproduction and some time tables. Pages 35–58 in *Biology of the Reptilia, Vol. 15 (Development B)*, edited by C. Gans and F. Billett. John Wiley and Sons, New York, NY.

Salmon, M., M. G. Tolbert, D. P. Painter, M. Goff, and R. Reiners. 1995. Behavior of loggerhead sea turtles on an urban beach. II. Hatchling orientation. *Journal of Herpetology* 19:568–576.

Salthe, S. N. 1969. Reproductive modes and the numbers and sizes of ova in the urodeles. *American Midland Naturalist* 81:467–490.

Salthe, S. N. 1973. Amphiumidae, *Amphiuma*. *Catalogue of American Amphibians and Reptiles* 147:1–4.

Sanchiz, B., and Z. Rocek. 1996. An overview of the anuran fossil record. Pages 317–328 in *The Biology of Xenopus*, edited by R. C. Tinsley and H. R. Kobel. Clarendon Press, Oxford.

Sanderson, I. T. 1937. *Animal Treasure*. Viking Press, New York, NY.

Sanderson, S. L., and R. Wassersug. 1993. Convergent and alternative designs for vertebrate suspension feeding. Pages 37–112 in *The Skull*, volume 3, edited by J. Hanken and B. K. Hall. University of Chicago Press, Chicago, IL.

Satrawaha, R., and C. M. Bull. 1981. The area occupied by an omnivorous lizard, *Trachydosaurus rugosus*. *Australian Wildlife Research* 8:435–442.

Savage, J. M., and K. R. Lips. 1993. A review of the status and biogeography of the lizard genera *Celestus* and *Diploglossus* (Squamata: Anguidae), with description of two new species from Costa Rica. *Revista de Biologia Tropical* 41:817–842.

Savage, J. M., and M. H. Wake. 1972. Geographic variation and systematics of the Middle American caecilians, genera *Dermophis* and *Gymnopis*. *Copeia* 1972:680–695.

Savitzky, A. H. 1980. The role of venom delivery strategies in snake evolution. *Evolution* 34:1194–1204.

Savitzky, A. H. 1981. Hinged teeth in snakes: An adaptation for swallowing hard-bodied prey. *Science* 212:346–349.

Savitzky, A. H. 1983. Coadapted character complexes among snakes: Fossoriality, piscivory, and durophagy. *American Zoologist* 23:397–409.

Sazima, I. 1989. Feeding behavior of the snail-eating snake, *Dipsas indica*. *Journal of Herpetology* 23:464–468.

Sazima, I. 1991. Caudal luring in two Neotropical pit vipers, *Bothrops jararaca* and *B. jararacussa*. *Copeia* 1991:245–248.

Scanlon, J. D. 1993. Madtsoiid snakes from the Eocene Tingamarra Fauna of Eastern Queensland. *Kaupia. Darmstädter Beiträge zur Naturgeschichte* 3:3–8.

Schafer, S. F., and C. O. Krekorian. 1983. Agonistic behavior of the Galapagos tortoise, *Geochelone elaphantopus*, with emphasis on its relationship to saddle-backed shell shape. *Herpetologica* 39:448–456.

Schall, J. H., and S. Ressel. 1991. Toxic plant compounds and the diet of the predominantly herbivorous whiptail lizard, *Cnemidophorus arubensis*. *Copeia* 1991:111–119.

Schall, J. J. 1992. Parasite-mediated competition in *Anolis* lizards. *Oecologia* 92:58–64.

Schall, J. J., and E. R. Pianka. 1978. Geographical trends in numbers of species. *Science* 201:679–686.

Schall, J. J., and G. A. Sarni. 1987. Malarial parasitism and the behavior of the lizard, *Sceloporus occidentalis*. *Copeia* 1987:84–93.

Schmidt, B. R. 1993. Are hybridogenetic frogs cyclical parthenogens? *Trends in Ecology and Evolution* 8:271–272.

Schneider, H. 1988. Peripheral and central mechanisms of vocalization. Pages 537–558 in *The Evolution of the Amphibian Auditory System*, edited by B. Fritzsch, M. J. Ryan, W. Wilczynski, T. E. Hetherington, and W. Walkowiak. John Wiley and Sons, New York, NY.

Schoener, T. W. 1967. The ecological significance of sexual dimorphism in size in the lizard *Anolis conspersus*. *Science* 155:474–477.

Schoener, T. W. 1969. Models of optimal prey size for solitary predators. *American Naturalist* 103:277–313.

Schoener, T. W. 1971. The theory of feeding strategies. *Annual Review of Ecology and Systematics* 2:369–404.

Schubauer, J. P., J. W. Gibbons, and J. R. Spotila. 1990. Home range and movement patterns of slider turtles inhabiting Par Pond. Pages 223–232 in *Life History and Ecology of the Slider Turtle*, edited by J. W. Gibbons. Smithsonian Institution Press, Washington.

Schubert, C., T. Steffen, and E. Christophers. 1990. Weitere Beobachtungen zur "dermolytischen Schreckhäutung" bei *Geckolepis typica* (Reptilia, Gekkonidae). *Zoologischer Anzeiger* 224:175–192.

Schuett, G. W., F. J. Fernandez, W. F. Gergits, N. J. Casna, H. M. Smith, J. B. Mitton, S. P. Mackessy, R. A. Odum, and M. J. Demlong. 1997. Production of offspring in the absence of males: evidence for facultative parthenogenesis in bisexual snakes. *Herpetological Natural History* In Press.

Schultze, H.-P. 1994. Comparison of hypotheses on the relationships of sarcopterygians. *Systematic Biology* 43: 155–173.

Schultze, H.-P., and L. Trueb (editors). 1991. *Origins of the Higher Groups of Tetrapods*. Comstock Publishing Associates, Ithaca, NY.

Schumacher, G.-H. 1973. The head muscles and hyolaryngeal skeleton of turtles and crocodilians. Pages.101–199 in *Biology of the Reptilia, Vol. 4 (Morphology D)*, edited by C. Gans and T. S. Parsons. Academic Press, London, UK.

Schwalm, P. A., P. H. Starrett, and R. W. McDiarmid. 1977. Infrared reflectance in leaf-sitting Neotropical frogs. *Science* 196:1225–1227.

Schwartz, A., and R. W. Henderson. 1991. *Amphibians and Reptiles of the West Indies: Descriptions, distributions, and natural history*. University of Florida Press, Gainesville, FL.

Schwartz, A., and R. J. Marsh. 1960. A review of the *pardalis-maculatus* complex of the boid genus *Tropidophis* of the West Indies. *Bulletin of the Museum of Comparative Zoology* 123:49–84.

Schwartz, J. J. 1987. The function of call alternation in anuran amphibians: a test of three hypotheses. *Evolution* 41:461–471.

Schwartz, J. J. 1989. Graded aggressive calls of the spring peeper, *Pseudacris crucifer*. *Herpetologica* 45:172–181.

Schwartz, J. J., and K. D. Wells. 1983. An experimental study of acoustic interference between two species of neotropical treefrogs. *Animal Behaviour* 31:181–190.

Schwartz, J. J., and K. D. Wells. 1985. Intra- and interspecific vocal behavior of the neotropical treefrog *Hyla microcephala*. *Copeia* 1985:27–38.

Schwartz, J. M., G. F. McCracken, and G. M. Burghardt. 1989. Multiple paternity in wild populations of the garter snake, *Thamnophis sirtalis*. *Behavioral Ecology and Sociobiology* 25:269–273.

Schwarzkopf, L. 1994. Measuring trade-offs: a review of studies of costs of reproduction in lizards. Pages 7–29 in

Lizard Ecology: Historical and Experimental Perspectives, edited by L. J. Vitt and E. R. Pianka. Princeton University Press, Princeton, NJ.

Schwenk, K. 1985. Occurrence, distribution, and functional significance of taste buds in lizards. *Copeia* 1985:91–101.

Schwenk, K. 1986. Morphology of the tongue in the tuatara, *Sphenodon punctatus* (Reptilia: Lepidosaura), with comments on function and phylogeny. *Journal of Morphology* 188:129–156.

Schwenk, K. 1993. The evolution of chemoreception in squamate reptiles: a phylogenetic approach. *Brain, Behavior and Evolution* 41:124–137.

Schwenk, K. 1994. Why snakes have forked tongues. *Science* 263:1573–1577.

Schwenk, K. 1995. Of tongues and noses: chemoreception in lizards and snakes. *Trends in Ecology and Evolution* 10:7–12.

Schwenk, K., and G. S. Throckmorton. 1989. Functional and evolutionary morphology of lingual feeding in squamate reptiles: phylogenetics and kinematics. *Journal of Zoology* 219:153–175.

Schwenk, K., and D. B. Wake. 1993. Prey processing in *Leurognathus marmoratus* and the evolution of form and function in desmognathine salamanders (Plethodontidae). *Biological Journal of the Linnaean Society* 49:141–162.

Scott, N. J., Jr., and R. A. Seigel. 1992. The management of amphibian and reptile populations: species priorities and methodological and theoretical constraints. Pages 343–368 in *Wildlife 2001: Populations*, edited by D. R. McCullough and R. H. Barrett. Elsevier Science Publishers, New York, NY.

Secor, S. M. 1995. Ecological aspects of foraging mode for the snakes *Crotalus cerastes* and *Masticophis flagellum*. *Herpetological Monographs* 9:169–186.

Secor, S. M., B. C. Jayne, and A. F. Bennett. 1992. Locomotor performance and energetic cost of sidewinding by the snake *Crotalus cerastes*. *Journal of Experimental Biology* 163:1–14.

Seidel, M. E. 1988. Revision of the West Indian emydid turtles (Testudines). *American Museum Novitates* 2918:1–41.

Seigel, R. A., and J. T. Collins (editors). 1993. *Snakes, Ecology and Behavior*. McGraw-Hill, Inc., New York, NY.

Seigel, R. A., J. T. Collins, and S. S. Novak (editors). 1987. *Snakes: Ecology and Evolutionary Biology*. Macmillan Publishing Co., New York, NY.

Semlitsch, R. D. 1985. Reproductive strategy of a facultatively paedomorphic salamander *Ambystoma talpoideum*. *Oecologia* 65:305–313.

Sergeyev, A. 1939. The body temperature of reptiles in natural surroundings. *Doklady Akademia Nauk SSSR* 22:49–52.

Sessions, S. K. 1982. Cytogenetics of diploid and triploid salamanders of the *Ambystoma jeffersonianum* complex. *Chromosoma* 84:599–621.

Sessions, S. K., and A. Larson. 1987. Developmental correlates of genome size in plethodontid salamanders and their implications for genome evolution. *Evolution* 41:1239–1251.

Sessions, S. K., P. Leon, and J. Kezer. 1982. Cytogenetics of the Chinese giant salamander, *Andrias davidianus* (Blanchard): the evolutionary significance of cryptobranchoid karyotypes. *Chromosoma* 86:341–357.

Sever, D. M. 1988. Male *Rhyacotriton olympicus* (Dicamptodontidae: Urodela) has a unique cloacal vent gland. *Herpetologica* 44:274–280.

Sever, D. M. 1991. Comparative anatomy and phylogeny of the cloacae of salamanders (Amphibia: Caudata). II. Cryptobranchidae, Hynobiidae and Sirenidae. *Journal of Morphology* 207:283–301.

Sever, D. M., L. C. Rania, and J. D. Krenz. 1996. Reproduction of the salamander *Siren intermedia* Le Conte with especial reference to oviducal anatomy and mode of fertilization. *Journal of Morphology* 227:335–348.

Seymour, R. S. and D. Bradford. 1995. Respiration of amphibian eggs. *Physiological Zoology* 68:1–25.

Seymour, R. S., and J. P. Loveridge. 1994. Embryonic and larval respiration in the arboreal foam nests of the African frog *Chiromantis xerampelina*. *Journal of Experimental Biology* 197:31–46.

Seymour, R. S., and J. D. Roberts. 1991. Embryonic respiration and oxygen distribution in foamy and nonfoamy egg masses of the frog *Limnodynastes tasmaniensis*. *Physiological Zoology* 64:1322–1340.

Shaffer, H. B. 1984. Evolution of a paedomorphic lineage. I. An electrophoretic analysis of the Mexican ambystomatid salamanders. *Evolution* 38:1194–1206.

Shaffer, H. B. 1993. Phylogenetics of model organisms: the laboratory axolotl, *Ambystoma mexicanum*. *Systematic Biology* 42:508–522.

Shaffer, H. B., J. M. Clark, and F. Kraus. 1991. When molecules and morphology clash: phylogenetic relationships among North American *Ambystoma* (Caudata: Ambystomatidae). *Systematic Zoology* 40:284–303.

Shaffer, H. B., and G. V. Lauder. 1988. The ontogeny of functional design: metamorphosis of feeding behaviour in the tiger salamander (*Ambystoma tigrinum*). *Journal of Zoology* 216:437–454.

Shaffer, H. B., and M. L. McKnight. 1996. The polytypic species revisited: genetic differentiation and molecular phylogenetics of the tiger salamander *Ambystoma tigrinum* (Amphibia: Caudata) complex. *Evolution* 50:417–433.

Shelton, G., and R. G. Boutilier. 1982. Apnoea in amphibians and reptiles. *Journal of Experimental Biology* 100:245–273.

Sheppard, L., and A. d'A. Bellairs. 1972. The mechanism of autotomy in *Lacerta*. *British Journal of Herpetology* 4:276–286.

Sherbrooke, W. C., and R. R. Montanucci. 1988. Stone mimicry in the round-tailed horned lizard, *Phrynosoma modestum* (Sauria: Iguanidae). *Journal of Arid Environments* 14:275–284.

Shine, R. 1979. Sexual selection and sexual dimorphism in the Amphibia. *Copeia* 1979:297–306.

Shine, R. 1980a. "Costs" of reproduction in reptiles. *Oecologia* 46:92–100.

Shine, R. 1980b. Ecology of the Australian death adder *Acanthophis antarcticus* (Elapidae): Evidence for convergence with the Viperidae. *Herpetologica* 36:281–289.

Shine, R. 1983a. Reptilian reproductive modes: the oviparity-viviparity continuum. *Herpetologica* 39:1–8.

Shine, R. 1983b. Reptilian viviparity in cold climates: testing the assumptions of an evolutionary hypothesis. *Oecologia* 57:397–405.

Shine, R. 1985. The evolution of viviparity in reptiles: an ecological analysis. Pages 605–694 in *Biology of the Reptilia, Vol. 15 (Development B)*, edited by C. Gans and F. Billett. John Wiley and Sons, New York, NY.

Shine, R. 1986. Ecology of a low-energy specialist: food habits and reproductive biology of the Arafura filesnake (Acrochordidae). *Copeia* 1986:424–437.

Shine, R. 1987. Intraspecific variation in thermoregulation, movements and habitat use by Australian blacksnakes, *Pseudechis porphyriacus* (Elapidae). *Journal of Herpetology* 21:165–177.

Shine, R. 1988. Parental care in reptiles. Pages 275–329 in *Biology of the Reptilia, Vol. 16 (Ecology B, Defense and Life History)*, edited by C. Gans and R. B. Huey. Alan R. Liss, Inc., New York, NY.

Shine, R. 1989. Ecological causes for the evolution of sexual dimorphism: a review of the evidence. *Quarterly Review of Biology* 64:419–461.

Shine, R. 1990. Function and evolution of the frill of the frillneck lizard, *Chlamydosaurus kingii* (Sauria: Agamidae). *Biological Journal of the Linnean Society* 40:11–20.

Shine, R. 1991. *Australian Snakes: A Natural History*. Reed Books, NSW, Australia.

Shine, R. 1992. Relative clutch mass and body shape in lizards and snakes: is reproductive investment constrained or optimized? *Evolution* 46:828–833.

Shine, R. 1993. Sexual dimorphism in snakes. Pages 49–86 in *Snakes: Ecology and Behavior*, edited by R. A. Seigel and J. T. Collins. McGraw Hill, Inc., New York, NY.

Shine, R. 1994. Sexual size dimorphism in snakes revisted. *Copeia* 1994:326–346.

Shine, R., and J. J. Bull. 1979. The evolution of livebearing in lizards and snakes. *American Naturalist* 113:905–923.

Shine, R., and M. Fitzgerald. 1995. Variation in mating systems and sexual size dimorphism between populations of the Australian python *Morelia spilota* (Serpentes: Pythonidae). *Oecologia* 103:490–498.

Shine, R., P. Harlow, J. S. Keogh, and Boeadi. 1995. Biology and commercial utilization of acrochordid snakes, with special reference to karung (*Acrochordus javanicus*). *Journal of Herpetology* 29:352–360.

Shine, R., and R. Lambeck. 1985. A radiotelemetric study of movements, thermoregulation and habitat utilization of Arafura filesnakes (Serpentes: Acrochordidae). *Herpetologica* 41:351–361.

Shine, R., and T. Madsen. 1996. Is thermoregulation unimportant for most reptiles? An example using water pythons (*Liasis fuscus*) in tropical Australia. *Physiological Zoology* 69:252–269.

Shine, R., and J. K. Webb. 1990. Natural history of Australian typhlopid snakes. *Journal of Herpetology* 24:357–363.

Shoemaker, V. H., S. Hillman, S. D. Hillyard, D. C. Jackson, L. L. McClanahan, P. C. Withers, and M. L. Wygoda. 1992. Exchanges of water, ions, and respiratory gases in terrestrial amphibians. Pages 125–150 in *Environmental Physiology of the Amphibians*, edited by M. E. Feder and W. W. Burggren, University of Chicago Press, Chicago, IL.

Shoemaker, V. H., and C. Sigurdson. 1989. Brain cooling via evaporation from the eyes in a waterproof treefrog. *American Zoologist* 29:106A.

Shubin, N., and P. Alberch. 1986. A morphogenetic approach to the origin and basic organization of the tetrapod limb. *Evolutionary Biology* 20:319–387.

Shubin, N. H., and F. A. Jenkins. 1995. An early Jurassic jumping frog. *Nature* 377:49–52.

Sinervo, B. 1993. The effect of offspring size on physiology and life history: manipulation of size using allometric engineering. *Bioscience* 43:210–218.

Sinervo, B. 1994. Manipulations of clutch and offspring size in lizards: mechanistic, evolutionary, and conservation considerations. Pages 183–193 in *Captive Management and Conservation of Amphibians and Reptiles,* edited by J. B. Murphy, K. Adler, and J. T. Collins. Society for the Study of Amphibians and Reptiles, Oxford.

Sinervo, B., and P. Licht. 1991. Proximate constraints on the evolution of egg size, number, and total clutch mass in lizards. *Science* 252:1300–1302.

Sinsch, U. 1988. Seasonal changes in the migratory behaviour of the toad *Bufo bufo*: direction and magnitude of movements. *Oecologia* 76:390–398.

Sinsch, U. 1990. Migration and orientation in anuran amphibians. *Ethology, Ecology, and Evolution* 2:65–79.

Sinsch, U. 1992. Amphibians. Pages 213–233 in *Animal Homing*, edited by F. Papi. Chapman and Hall, New York, NY.

Sinsch, U., and D. Seidel. 1995. Dynamics of local and temporal breeding assemblages in a *Bufo calamita* metapopulation. *Australian Journal of Ecology* 20:351–361.

Sites, J. W., J. W. Archie, C. J. Cole, and O. Flores Villela. 1992. A review of phylogenetic hypotheses for lizards of the genus *Sceloporus* (Phrynosomatidae): implications for ecological and evolutionary studies. *Bulletin of the American Museum of Natural History* 213:1–110.

Sites, J. W., J. W. Bickham, B. A. Pytel, I. F. Greenbaum, and B. B. Bates. 1984. Biochemical characters and the reconstruction of turtle phylogenies: relationships among batagurine genera. *Systematic Zoology* 33:137–158.

Sjogren, P. 1991. Extinction and isolation gradients in metapopulations: the case of the pool frog (*Rana lessonae*). *Biological Journal of the Linnean Society* 42:135–147.

Skoczylas, R. 1978. Physiology of the digestive tract. Pages 589–717 in *Biology of the Reptilia, volume 8 (Physiology B)*, edited by C. Gans and K. A. Gans. Academic Press, New York, NY.

Slip, D. J., and R. Shine. 1988. The reproductive biology and mating system of diamond pythons, *Morelia spilota* (Serpentes: Boidae). *Herpetologica 44:396–404*.

Slowinski, J. B. 1989. The interrelationships of laticaudine sea snakes based on the amino acid sequences of short-chain neurotoxins. *Copeia* 1989:783–788.

Slowinski, J. B. 1994. A phylogenetic analysis of *Bungarus* (Elapidae) based on morphological characters. *Journal of Herpetology* 28:440–446.

Slowinski, J. B. 1995. A phylogenetic analysis of the New World coral snakes (Elapidae: *Leptomicrurus, Micruroides*, and *Micrurus*). *Journal of Herpetology* 29:325–338.

Smith, K. K. 1982. An electromyographic study of the function of the jaw adducting muscles in *Varanus exanthematicus* (Varanidae). *Journal of Morphology* 173:137–158.

Smith, K. K., and W. L. Hylander. 1985. Strain gauge measurement of mesokinetic movement in the lizard *Varanus exanthematicus*. *Journal of Experimental Biology* 114:53–70.

Smith, L. A., M. A. Olson, P. J. Lafaye, and J. O. Dolly. 1995. Cloning and expression of mamba toxins. *Toxicon* 33:459–474.

Smith, M. 1969. *The British Amphibians and Reptiles, 4th edition*. Collins, London, UK.

Smith, M. A. 1926. *Monograph of the sea-snakes (Hydrophiidae)*. British Museum (Natural History), London, UK.

Smith, M. H., and K. T. Scribner. 1990. Population genetics of the slider turtle. Pages 74–81 in *Life History and Ecology of the Slider Turtle*, edited by J. W. Gibbons. Smithsonian Institution Press, Washington, DC.

Smith, S. 1975. Innate responses of coral snake pattern by a possible avian predator. *Science* 187:759–760.

Smith, W. J. 1977. *The Behavior of Communicating*. Harvard University Press, Cambridge, MA.

Snyder, G. K., and G. A. Hammerson. 1993. Interrelationships between water economy and thermoregulation in the Canyon tree-frog *Hyla arenicolor*. *Journal of Arid Environments* 25:321–329.

Soini, P. 1984. Ecología y situación de la charapa (*Podocnemis expansa*): informe preliminar. Pages 177–183 in *Reporte Pacaya-Samiria, Investigaciones en la Estación Biológica Cahuana, 1979–1994 (Published 1995)*, edited by P. Soini, A. Tovar, and U. Valdez. Universidad Nacional Agraria La Molina, Lima, Peru.

Sokol, O. M. 1969. Feeding in the pipid frog *Hymenochirus boettgeri* (Tornier). *Herpetologica* 25:9–24.

Sokol, O. M. 1975. The phylogeny of anuran larvae: a new look. *Copeia* 1975:1–24.

Sorci, G. 1995. Repeated measurements of blood parasite levels reveal limited ability for host recovery in the common lizard (*Lacerta vivipara*). *Journal of Parasitology* 81:825–827.

Southerland, M. T. 1986. The effects of variation in streamside habitats on the composition of mountain salamander communities. *Copeia* 1986:731–741.

Spande, T. F., H. M. Garraffo, M. W. Edwards, H. J. C. Yeh, L. Pannell, and J. W. Daly. 1992. Epibatidine: A novel (chloropyridyl)azabicycloheptane with potent analgesic activity from an Ecuadorian poison frog. *Journal of the American Chemical Society* 114:3475–3478.

Spinar, Z. V. 1952. Revision of some Moravian Discosauriscidae. *Rozpravy Ustredniho Ustadu Geologickeho* 15:1–159.

Spolsky, C. M., C. A. Philips, and T. Uzzell. 1992. Antiquity of clonal salamander lineages revealed by mitochondrial DNA. *Nature* 356:706–708.

Spotila, J. R., M. P. O'Connor, and G. S. Bakken. 1992. Biophysics of heat and mass transfer. Pages 59–80 in *Environmental Physiology of the Amphibians*, edited by M. E. Feder and W. W. Burggren, University of Chicago Press, Chicago, IL.

Stamps, J. A. 1983. Sexual selection, sexual dimorphism, and territoriality. Pages 169–204 in *Lizard Ecology: Studies of a Model Organism*, edited by R. B. Huey, E. R. Pianka, and T. W. Schoener. Harvard University Press, Cambridge, MA.

Stamps, J. A. 1990. The effect of contender pressure on territory size and overlap in seasonally territorial species. *American Naturalist* 135:614–632.

Stamps, J. A. 1994. Territorial behavior: testing the assumptions. *Advances in the Study of Behavior.* 23:173–232.

Stamps, J. A. 1995. Using growth-based models to study behavioral factors influencing sexual size dimorphism. *Herpetological Monographs* 9:75–87.

Starrett, P. H. 1973. Evolutionary patterns in larval morphology. Pages 251–271 in *Evolutionary Biology of the Anurans*, edited by J. L. Vial. University of Missouri Press, Columbia, MO.

Staub, N. L. 1993. Intraspecific agonistic behavior of the salamander *Aneides flavipunctatus* (Amphibia: Plethodontidae) with comparison to other plethodontid species. *Herpetologica* 49:271–282.

Stephenson, E. M., and N. G. Stephenson. 1957. Field observations on the New Zealand frog, *Leiopelma* Fitzinger.

Transactions of the Royal Society of New Zealand 84: 867–882.

Stevenson, R. D. 1985a. Body size and limits to the daily range of body temperature in terrestrial ectotherms. *American Naturalist* 125:102–117.

Stevenson, R. D. 1985b. The relative importance of behavioral and physiological adjustments controlling for body temperature in terrestrial ectotherms. *American Naturalist* 126:362–386.

Stewart, J. R. 1993. Yolk sac placentation in reptiles: structural innovation in a fundamental vertebrate fetal nutritional system. *Journal of Experimental Zoology* 266: 431–449.

Stewart, M. M. 1985. Arboreal habitat use and parachuting by a subtropical forest frog. *Journal of Herpetology* 19:391–401.

Stewart, M. M., and A. S. Rand. 1991. Vocalizations and the defense of retreat sites by male and female frogs, *Eleutherodactylus coqui. Copeia* 1991:1013–1024.

Stickel, L. F. 1950. Populations and home range relationships of the box turtle *Terrepene c. carolina* (Linnaeus). *Ecological Monographs* 20:351–378.

Stickel, L. F. 1989. Home range behavior among box turtles (*Terrepene c. carolina*) of a bottomland forest in Maryland. *Journal of Herpetology* 23:40–44.

Storey, K. B., and J. M. Storey. 1996. Natural freezing survival in animals. *Annual Review of Ecology and Systematics* 27:365–386.

Strong, D. R., Jr., D. Simberloff, L. G. Abele, and A. B. Thistle (editors). 1984. *Ecological Communities: Conceptual Issues and the Evidence.* Princeton University Press, Princeton, NJ.

Strydom, D. J. 1979. The evolution of toxins found in snake venoms. Pages 258–275 in *Snake Venoms*, edited by C.-Y. Lee. Springer Verlag, Berlin, Germany.

Sues, H.-D. 1989. The place of crocodilians in the living world. Pages 14–25 in *Crocodiles and Alligators*, edited by C. A. Ross. Facts on File, New York, NY.

Sullivan, B. K. 1989. Desert environments and the structure of anuran mating systems. *Journal of Arid Environments* 17:175–183.

Sullivan, B. K., M. J. Ryan, and P. A. Verrell. 1995. Female choice and mating system structure. Pages 469–517 in *Amphibian Biology. Vol. 2. Social Behaviour*, edited by H. Heatwole and B. K. Sullivan. Surrey Beatty and Sons, Chipping Norton, New South Wales, Australia.

Swingland, I. R., and C. M. Lessells. 1979. The natural regulation of giant tortoise populations on Aldabra Atoll, movement polymorphism, reproductive success and mortality. *Journal of Animal Ecology* 48:639–654.

Szymura, J. M. 1993. Analysis of hybrid zones with *Bombina*. Pages 261–289 in *Hybrid Zones and the Evolutionary Process*, edited by R. G. Harrison. Oxford University Press, New York, NY.

Szyndlar, Z. 1991a. A review of Neogene and Quaternary snakes of central and eastern Europe. Part I: Scole-

cophidia, Boidae, Colubrinae. *Estudios Geologicos (Museo Nacional de Ciencias Naturales, Madrid)* 47:103–126.

Szyndlar, Z. 1991b. A review of Neogene and Quaternary snakes of central and eastern Europe. Part II: Natricinae, Elapidae, Viperidae. *Estudios Geologicos (Museo Nacional de Ciencias Naturales, Madrid)* 47:237–266.

Szyndlar, Z., and W. Böhme. 1993. The fossil snakes of Germany: history of the faunas and of their exploration. *Mertensiella* 3:381–431.

Szyndlar, Z., and J.-C. Rage. 1990. West Palearctic cobras of the genus *Naja* (Serpentes: Elapidae): interrelationships among extinct and extant species. *Amphibia Reptilia* 11:385–400.

Taigen, T. L., and F. H. Pough. 1983. Prey preference, foragaing behavior, and metabolic characteritics of frogs. *American Naturalist* 122:509–520.

Taigen, T. L., F. H. Pough, and M. M. Stewart. 1984. Water balance of terrestrial anuran (*Eleutherodactylus coqui*) eggs: importance of parental care. *Ecology* 65:248–255.

Taigen, T. L., and K. D. Wells. 1985. Energetics of vocalizations by an anuran amphibian. *Journal of Comparative Physiology* 155:163–170.

Tan, A.-M., and D. B. Wake. 1995. MtDNA phylogeography of the California newt, *Taricha torosa* (Caudata, Salamandridae). *Molecular Phylogenetics and Evolution* 4: 383–394.

Tarsitano, S. F., E. Frey, and J. Riess. 1989. The evolution of the crocodilia: a conflict between morphological and biochemical data. *American Zoologist* 29:843–856.

Taubes, G. 1992. A dubious battle to save the Kemp's ridley sea turtle. *Science* 256:614–616.

Taylor, D. H., and K. Adler. 1973. Spatial orientation by salamanders using plane-polarized light. *Science* 181:285–287.

Taylor, E. H. 1968. *The Caecilians of the World, a Taxonomic Review*. University of Kansas Press, Lawrence, KS.

Telford, S. R., Jr. 1997. *The ecology of a symbiotic community. Vol 1: Population Biology of the Japanese Lizard Takydromus tachydromoides (Schlegel)(Lacertidae)*. Krieger Publishing Co., Malabar, FL.

Thomas, R. G., and F. H. Pough. 1979. The effect of rattlesnake venom on digestion of prey. *Toxicon* 17:221–228.

Thompson, C. W., and M. C. Moore. 1991. Throat color reliably signals status in male tree lizards, *Urosaurus ornatus. Animal Behaviour* 42:745–754.

Thompson, R. L. 1996. Larval habitat, ecology, and parental investment of *Osteopilus brunneus* (Hylidae). Pages 259–269 in *Contributions to West Indian Herpetology: a Tribute to Albert Schwartz*, edited by R. Powell and R. W. Henderson. Society for the Study of Amphibians and Reptiles, Ithaca, NY.

Thorn, R. 1968. *Les salamandres d'Europe, d'Asie, et d'Afrique du Nord*. Ed. P. Lechevalier, Paris.

Thorpe, R. S., W. Wüster, and A. Malhotra. 1997. *Venomous Snakes: Ecology, Evolution and Snakebite*. Clarendon Press, Oxford.

Tiebout, H. M., and J. R. Carey. 1987. Dynamic spatial ecology of the water snake, *Nerodia sipedon. Copeia* 1987:1–18.

Tilley, S. C., and M. J. Mahoney. 1996. Patterns of genetic differentiation in salamanders of the *Desmognathus ochrophaeus* complex (Amphibia: Plethodontidae). *Herpetological Monographs* 10:1–42.

Tinkle, D. W. 1969. The concept of reproductive effort and its relation to the evolution of life histories of lizards. *American Naturalist* 103:501–516.

Tinkle, D. W., and A. E. Dunham. 1986. Comparative life histories of two syntopic sceloporine lizards. *Copeia* 1986:1–18.

Tinkle, D. W., and J. W. Gibbons. 1977. The distribution and evolution of viviparity in reptiles. *Miscellaneous Publications of the Museum of Zoology, University of Michigan* 154:1–55.

Tinsley, R. C., and C. M. Earle. 1983. Invasion of vertebrate lungs by the polystomatid monogeneans *Pseudodiplorchis americanus* and *Neodiplorchis scaphiopodis. Parasitology* 86:501–517.

Tinsley, R. C., and H. R. Kobel. 1996. *The Biology of Xenopus*. Clarendon Press, Oxford, UK.

Titus, T. A., and D. R. Frost. 1996. Molecular homology assessment and phylogeny in the lizard family Opluridae (Squamata: Iguania). *Molecular Phylogenetics and Evolution* 6:49–62.

Titus, T. A., and A. Larson. 1995. A molecular phylogenetic perspective on the evolutionary radiation of the salamander family Salamandridae. *Systematic Biology* 44:125–151.

Titus, T. A. and A. Larson. 1996. Molecular phylogenetics of desmognathine salamanders (Caudata: Plethodontidae): a reevaluation of evolution in ecology, life history, and morphology. *Systematic Biology* 45:451–472.

Toft, C. A. 1981. Feeding ecology of Panamanian litter anurans: patterns in diet and foraging mode. *Journal of Herpetology* 15:139–144.

Toft, C. A. 1985. Resource partitioning in amphibians and reptiles. *Copeia* 1985:1–21.

Toft, C. A. 1995. Evolution of diet specialization in poison-dart frogs (Dendrobatidae). *Herpetologica* 51:202–216.

Tokar, A. A. 1995. Taxonomic revision of the genus *Gongylophis* Wagler 1830: *G. conicus* (Schneider 1801) and *G. muelleri* Boulenger 1892 (Serpentes Boidae). *Tropical Zoology* 8:347–360.

Tokarz, R. R. 1995. Mate choice in lizards: a review. *Herpetological Monographs* 9:17–40.

Townsend, D. S. 1996. Patterns of parental care in frogs of the genus *Eleutherodactylus*. Pages 229–239 in *Contributions to West Indian Herpetology: a Tribute to Albert Schwartz*. Society for the Study of Amphibians and Reptiles, Contributions to Herpetology 12, edited by R. Powell and R. W. Henderson. Ithaca, NY.

Townsend, D. S., M. M. Stewart, and F. H. Pough. 1984. Male parental care and its adaptive significance in a neotropical frog. *Animal Behavior* 32:421–431.

Townsend, D. S., M. M. Stewart, F. H. Pough, and P. F. Brussard. 1981. Internal fertilization in an oviparous frog. *Science* 212:469–470.

Tracy. C. R. 1982. Biophysical modeling in reptilian physiology. Pages 275–321 in *Biology of the Reptilia, Vol. 12 (Physiology C, Physiological Ecology)*, edited by C. Gans and F. H. Pough, Academic Press, London, UK.

Trillmich, K. G. K. 1983. The mating system of the marine iguana (*Amblyrhynchus cristatus*). *Zeitschrift für Tierpsychologie* 63:141–172.

Trivers, R. L. 1976. Sexual selection and resource-accruing abilities in *Anolis garmani. Evolution* 30:253–269.

Troyer, K. 1982. Transfer of fermentative microbes between generations in a herbivorous lizard. *Science* 216:540–542.

Troyer, K. 1984. Structure and function of the digestive tract of a herbivorous lizard *Iguana iguana. Physiological Zoology* 57:1–8.

Trueb, L. 1973. Bones, frogs, and evolution. Pages 65–132 in *Evolutionary Biology of the Anurans: Contemporary Research on Major Problems*, edited by J. L. Vial. University of Missouri Press, Columbia, MO.

Trueb, L. 1996. Historical constraints and morphological novelties in the evolution of the skeletal system of pipid frogs (Anura: Pipidae). Pages 349–377 in *The Biology of Xenopus*, edited by R. C. Tinsley and H. R. Kobel. Clarendon Press, Oxford.

Trueb, L., and D. C. Cannatella. 1986. Systematics, morphology, and phylogeny of genus *Pipa* (Anura: Pipidae). *Herpetologica* 42:412–449.

Trueb, L., and R. Cloutier. 1991. A phylogenetic investigation of the inter- and intrarelationships of the Lissamphibia (Amphibia: Temnospondyli). Pages 223–313 in *Origins of the Higher Groups of Tetrapods: Controversy and Consensus*, edited by H. P. Schultze and L. Trueb. Cornell University Press, Ithaca, NY.

Trueb, L., and C. Gans. 1983. Feeding specializations of the Mexican burrowing toad, *Rhinophrynus dorsalis* (Anura: Rhinophrynidae). *Journal of Zoology, London* 199:189–208.

Tunner, H. G. 1992: Locomotory behaviour in water frogs from Neusiedlersee (Austria, Hungary). 15 km migration of *Rana lessonae* and its hybridogenetic associate *Rana esculenta*. Proceedings of the 6th Ordinary General Meeting of the Societas Europas Herpetologicae, Budapest, pp. 449–452.

Turner, F. B., R. I. Jennrich, and J. D. Weintraub. 1969. Home ranges and the body sizes of lizards. *Ecology* 50:1076–1081.

Twitty, V. C. 1966. *Of Scientists and Salamanders*. Freeman, San Francisco.

Tyler, M. J. 1983. *The Gastric Brooding Frog*. Croon Helm, London, UK.

Tyler, M. J. 1991. Australian fossil frogs. Pages 591–604 in *Vertebrate Palaeontology of Australasia*, edited by P. Vickers-Rich, J. M. Monaghan, R. F. Baird, and T. H. Rich. Pioneer Design Studio, Lilydale, Victoria, Australia.

Underwood, G. 1967. *A Contribution to the Classification of Snakes*. British Museum (Natural History), London, UK.

Underwood, G. 1976. A systematic analysis of boid snakes. Pages 151–175 in *Morphology and Biology of Reptiles, Linnean Society Symposium Series 3*, edited by A. d'A. Bellairs and C. B. Cox. Academic Press, London, UK.

Underwood, G., and E. Kochva. 1993. On the affinities of the burrowing asps *Atractaspis* (Serpentes: Atractaspididae). *Zoological Journal of the Linnean Society* 107:3–64.

Underwood, G., and A. F. Stimson. 1990. A classification of pythons (Serpentes, Pythoninae). *Journal of Zoology, London* 221:565–603.

Underwood, H. 1992. Endogenous rhythms. Pages 229–297 in *Biology of the Reptilia, Vol 18 (Physiology E, Hormones, Brain, and Behavior)*, edited by C. Gans and D. Crews. University of Chicago Press, Chicago, IL.

U S Fish and Wildlife Service. 1994. *Endangered and threatened wildlife and plants*. 50 CFR 17.11 & 17.12. U. S. Fish and Wildlife Service, Department of the Interior. August 20, 1994.

U S Fish and Wildlife Service. 1995. CITES. Appendices I, II, and III to the Convention on International Trade in Endangered Species of Wild Fauna and Flora. U. S. Fish and Wildlife Service, Department of the Interior. February 16, 1995.

Uzzell, T., and I. S. Darevsky. 1975. Biochemical evidence for the hybrid origin of the parthenogenetic species of the *Lacerta saxicola* complex (Sauria: Lacertidae), with a discussion of some ecological and evolutionary implications. *Copeia* 1975:204–222.

van Berkum, F. H., R. B. Huey, and B. A. Adams. 1986. Physiological consequences of thermoregulation in a tropical lizard (*Ameiva festiva*). *Physiological Zoology* 59: 464–472.

van Berkum, F. H., and J. S. Tsuji. 1987. Inter-familiar differences in sprint speed of hatchling *Sceloporus occidentalis* (Reptilia: Iguanidae). *Journal of Zoology, London* 212: 511–519.

Van Devender, R. W. 1978. Growth ecology of a tropical lizard, *Basiliscus basiliscus*. *Ecology* 59:1031–1038.

Van Devender, R. W. 1982. Growth and ecology of spiny-tailed and green iguanas in Costa Rica, with comments on the evolution of herbivory and large body size. Pages 162–183 in *Iguanas of the World: Their Behavior, Ecology, and Conservation*, edited by G. M. Burghardt and A. S. Rand. Noyes Publications, Park Ridge, NJ.

Van Mierop, L. H. S., and S. M. Barnard. 1978. Further observations on thermoregulation in the brooding female *Python molurus bivittatus* (Serpentes: Boidae). *Copeia* 1978:615–621.

Verrell, P. A. 1989a. An experimental study of the behavioral basis of sexual isolation between two sympatric plethodontid salamanders, *Desmognathus imitator* and *D. ochrophaeus*. *Ethology* 80:274–282.

Verrell, P. A. 1989b. The sexual strategies of natural populations of newts and salamanders. *Herpetologica* 45: 265–282.

Videler, J. J., and J. T. Jorna. 1985. Functions of the sliding pelvis in *Xenopus laevis*. *Copeia* 1985:251–254.

Viets, B. E., M. A. Ewert, L. G. Talent, and C. E. Nelson. 1994. Sex-determining mechanisms in squamate reptiles. *Journal of Experimental Zoology* 270:45–56.

Villa, J. 1984. Biology of a neotropical glass frog, *Centrolenella fleischmanni* (Boettger), with special reference to its frogfly associates. *Milwaukee Public Museum Contributions in Biology and Geology* 55:1–60.

Villa, J., and C. E. Valerio. 1982. Red, white, and brown. Preliminary observations on the color of the centrolenid tadpole (Amphibia: Anura: Centrolenidae). *Brenesia* 19/20:1–16.

Vitt, L. J., and W. E. Cooper. 1985. The evolution of sexual dimorphism in the skink *Eumeces laticeps*: an example of sexual selection. *Canadian Journal of Zoology* 63: 995–1002.

Vitt, L. J., and E. R. Pianka (editors). 1994. *Lizard Ecology, Historical and Experimental Perspectives*. Princeton University Press, Princeton, NJ.

Vitt, L. J., and H. J. Price. 1982. Ecological and evolutionary determinants of relative clutch mass in lizards. *Herpetologica* 38:237–255.

Vitt, L. J., and L. D. Vangilder. 1983. Ecology of a snake community in northeastern Brazil. *Amphibia-Reptilia* 4:273–296.

Vliet, K. A. 1989. Social displays of the American alligator (*Alligator mississippiensis*). *American Zoologist* 29:1019–1031.

Vogt, R. C. 1993. Systematics of the false map turtles (*Graptemys pseudogeographica* complex: Reptilia, Testudines, Emydidae). *Annals of Carnegie Museum* 62:1–46.

von Brockhusen, F. 1977. Untersuchungen zur individuellen varibilität der beuteannahme von *Anolis lineatopus* (Reptilia, Iguanidae). *Zeitschrift für Tierpsychologie* 44:13–24.

Voris, H. K. 1975. Dermal scale-vertebra relationships in sea snakes (Hydrophiidae). *Copeia* 1975:746–757.

Voris, H. K. 1977. A phylogeny of the sea snakes (Hydrophiidae). *Fieldiana, Zoology* 70:79–169.

Vorobyeva, E., and H.-P. Schultze. 1991. Description and systematics of panderichthyid fishes with comments on their relationship to Tetrapoda. Pages 68–109 in *Origins of the Higher Groups of Tetrapods: Controversy and Consensus*, edited by H.-P. Schultze and L. Trueb. Comstock Publishing Associates, Ithaca, NY.

Wachtel, S. S., and T. R. Tiersch. 1994. The search for the male-determining gene. Pages 1–22. *In Molecular Genetics of Sex Determination*, edited by S. S. Wachtel. Academic Press, New York, NY.

Wagner, W. E., Jr. 1989. Fighting, assessment, and frequency alteration in Blanchard's cricket frog. *Behavioral Ecology and Sociobiology* 25:429–436.

Wagner, W. E., Jr., and B. K. Sullivan. 1995. Sexual selection in the Gulf Coast toad, *Bufo valliceps*: female choice based on variable characters. *Animal Behaviour* 49: 305–319.

Wainwright, P. C. and A. F. Bennett. 1992a. The mechanism of tongue projection in chameleons. I. Electromyographic tests of functional hypotheses. *Journal of Experimental Biology* 168:1–21.

Wainwright, P. C., and A. F. Bennett. 1992b. The mechanism of tongue projection in chameleons. II. Role of shape change in a muscular hydrostat. *Journal of Experimental Biology* 168:23–40.

Wainwright, P. C., D. M. Kraklau, and A. F. Bennett. 1991. Kinematics of tongue projection in *Chamaeleo oustaleti*. *Journal of Experimental Biology* 159:109–133.

Wainwright, P. C., and S. M. Reilly. 1994. *Ecological Morphology, Integrative Organismal Biology*. University of Chicago Press, Chicago, IL.

Wake, D. B. 1966. Comparative osteology and evolution of the lungless salamanders, family Plethodontidae. *Memoires of the southern California Academy of Sciences* 4:1–111.

Wake, D. B. 1982. Functional and developmental constraints and opportunities in the evolution of feeding systems in urodeles. Pages 51–66 in *Environmental Adaptation and Evolution*, edited by D. Mossakowski and G. Roth. Gustav Fischer, New York, NY.

Wake, D. B. 1987. Adaptive radiation of salamanders in Middle American cloud forests. *Annals of the Missouri Botanical Garden* 74:242–264.

Wake, D. B. 1991a. Declining amphibian populations. *Science* 253:860.

Wake, D. B. 1991b. Homoplasy: the result of natural selection, or evidence of design limitations? *American Naturalist* 138:543–567.

Wake, D. B. 1992. An integrated approach to evolutionary studies of salamanders. Pages 163–177 in *Herpetology: Current Research on the Biology of Amphibians and Reptiles*, edited by K. Adler. Society for the Study of Amphibians and Reptiles, Oxford, OH.

Wake, D. B. 1993. Phylogenetic and taxonomic issues relating to salamanders of the family Plethodontidae. *Herpetologica* 49:229–237.

Wake, D. B., and P. Elias. 1983. New genera and a new species of Central American salamanders, with a review of the tropical genera (Amphibia, Caudata, Plethodontidae). *Natural History Museum of Los Angeles County, Contributions in Science* no.345:1–19.

Wake, D. B., and A. Larson. 1987. Multidimensional analysis of an evolving lineage. *Science* 238:42–48.

Wake, D. B., and J. F. Lynch. 1976. The distribution, ecology, and evolutionary history of plethodontid salamanders in tropical America. *Natural History Museum of Los Angeles County, Science Bulletin* no. 25:1–65.

Wake, D. B., L. R. Maxson, and G. Z. Wurst. 1978. Genetic differentiation, albumin evolution, and their biogeographic implications in plethodontid salamanders of California and southern Europe. *Evolution* 32:529–539.

Wake, D. B., T. J. Papenfuss, and J. F. Lynch. 1992. Distribution of salamanders along elevational transects in Mexico and Guatemala. Pages 303–319 in *Biogeography of Mesoamerica, Proceedings of a symposium, Mérida, Yucatán, Mexico, October 26–30, 1984*, edited by S. P. Darwin and A. L. Welden. Tulane University, New Orleans.

Wake, D. B., and G. Roth. 1989. *Complex Organismal Functions: Integration and Evolution in Vertebrates*. John Wiley and Sons, Chichester.

Wake, M. H. 1977. The reproductive biology of caecilians: an evolutionary perspective. Pages 73–101 in *The Reproductive biology of amphibians*, edited by D. H. Taylor and S. I. Gutman. Plenum Publ. Corp., New York, NY.

Wake, M. H. 1978. The reproductive biology of *Eleutherodactylus jasperi* (Amphibia, Anura, Leptodactylidae), with comments on the evolution of live-bearing systems. *Journal of Herpetology* 12:121–133.

Wake, M. H. 1993. Evolution of oviductal gestation in amphibians. *Journal of Experimental Zoology* 266:394–413.

Wake, M. H., and J. Hanken. 1982. Development of the skull of *Dermophis mexicanus* (Amphibia, Gymnophiona), with comments on skull kinesis and amphibian relationships. *Journal of Morphology* 173:203–223.

Walker, W. F., Jr. 1971a. A structural and functional analysis of walking in the turtle, *Chrysemys picta marginata*. *Journal of Morphology* 134:195–213.

Walker, W. F., Jr. 1971b. Swimming in sea turtles of the Family Cheloniidae. *Copeia* 1971:229–233.

Wallach, V. 1993. The supralabial imbrication pattern of the Typhlopoidea (Reptilia: Serpentes). *Journal of Herpetology* 27:214–218.

Walls, S. C., and A. R. Blaustein. 1995. Larval marbled salamanders, *Ambystoma opacum*, eat their kin. *Animimal Behavior* 50:537–545.

Walton, M., B. C. Jayne, and A. F. Bennett. 1990. The energetic cost of limbless locomotion. *Science* 249:524–527.

Waser, P. M., and R. H. Wiley. 1979. *Mechanisms and evolution of spacing in animals. Pages 159–223 in Handbook of Behavioral Neurobiology vol 3. Social Behavior and Communication, edited by P. Marler and J. G. Vandenberg*. Plenum Press, New York, NY.

Wassersug, R. 1980. Internal oral features of larvae from eight anuran families: functional, systematic, evolutionary and ecological considerations. *Miscellaneous Publica-*

tions, *Museum of Natural History, University of Kansas*, No. 68:1–146.

Wassersug, R. J. 1989. Locomotion in amphibian larvae (or "Why aren't tadpoles built like fishes?"). *American Zoologist* 29:65–84.

Wassersug, R. J. 1997. Where the tadpole meets the world —observations and speculations on biomechanical and biochemical factors that influence metamorphosis in anurans. *American Zoologist* 37:124–136.

Wassersug, R. J., K. J. Frogner, and R. F. Inger. 1981. Adaptations for life in tree holes by rhacophorid tadpoles from Thailand. *Journal of Herpetology* 15:41–52.

Wassersug, R. J., and K. Hoff. 1979. A comparative study of the buccal pumping mechanism of tadpoles. *Biological Journal of the Linnean Society* 12:225–259.

Wassersug, R. J., and W. F. Pyburn. 1987. The biology of the Pe-ret' toad, *Otophryne robusta* (Microhylidae), with special consideration of its fossorial larva and systematic relationships. *Zoological Journal of the Linnean Society* 91:137–169.

Wassersug, R. J., and K. Rosenberg. 1979. Surface anatomy of branchial food traps of tadpoles: A comparative study. *Journal of Morphology* 159:393–426.

Wassersug, R. J., and D. G. Sperry. 1977. The relationship of locomotion to differential predation on *Pseudacris triseriata* (Anura: Hylidae). *Ecology* 58:830–839.

Wassersug, R. J., and D. B. Wake. 1995. Fossil tadpoles from the Miocene of Turkey. *Alytes* 12:145–157.

Watt, W. B., P. A. Carter, and K. Donohue. 1986. Females' choice of good genotypes is promoted by an insect mating system. *Science* 233:1187–1190.

Weatherhead, P. J., F. E. Barry, G. P. Brown, and M. R. L. Forbes. 1995. Sex ratios, mating behavior and sexual size dimorphism of the northern water snake, *Nerodia sipedon*. *Behavioral Ecology and Sociobiology* 36:301–311.

Weatherhead, P. J., and D. J. Hoysak. 1989. Spatial and activity patterns of black rat snakes (*Elaphe obsoleta*) from radiotelemetry and recapture data. *Canadian Journal of Zoology* 67:463–468.

Webb, G. J. W., and C. Gans. 1982. Galloping in *Crocodylus johnstoni*—a reflection of terrestrial activity? *Records of the Australian Museum* 34:607–618.

Weintraub, J. D. 1970. Homing in the lizard *Sceloporus orcutti*. *Animal Behaviour* 18:132–137

Weir, J. 1992. The Sweetwater rattlesnake round-up: a case study in environmental ethics. *Conservation Biology* 6:116–127.

Weldon, P. J., and T. L. Leto. 1995. A comparative analysis of proteins in the scent gland secretions of snakes. *Journal of Herpetology* 29:474–476.

Wells, K. D. 1977a. The social behaviour of anuran amphibians. *Animal Behaviour* 25:666–693.

Wells, K. D. 1977b. Territoriality and male mating success in the green frog (*Rana clamitans*). *Ecology* 58:750–762.

Wells, K. D. 1978. Territoriality in the green frog (*Rana clamitans*): vocalizations and agonistic behaviour. *Animal Behaviour* 26:1051–1063.

Wells, K. D. 1979. Reproductive behavior and male mating success in a neotropical toad, *Bufo typhonius*. *Biotropica* 11:301–307.

Wells, K. D. 1980. Social behavior and communication of a dendrobatid frog, *Colostethus trinitatis*. *Herpetologica* 36:189–199.

Wells, K. D. 1988. The effect of social interactions on anuran vocal behavior. Pages 433–454 in *The Evolution of the Amphibian Auditory System*, edited by B. Fritzsch, M. J. Ryan, W. Wilczynski, T. E. Hetherington, and W. Walkowiak. John Wiley and Sons, New York, NY.

Wells, K. D., and C. R. Bevier. 1997. Contrasting patterns of energy substrate use in two species of frogs that breed in cold weather. *Herpetologica* 53:70–80.

Wells, K. D., and T. L. Taigen. 1984. Reproductive behavior and aerobic capacities of male American toads (*Bufo americanus*): Is behavior constrained by physiology? *Herpetologica* 40:292–298.

Wells, K. D., and T. L. Taigen. 1989. Calling energetics of a neotropical treefrog, *Hyla microcephala*. *Behavioral Ecology and Sociobiology* 25:13–22.

Wells, K. D., T. L. Taigen, and J. A. O'Brien. 1996. The effect of temperature on calling energetics of the spring peeper (*Pseudacris crucifer*). *Amphibia Reptilia* 17:149–158.

Werner, D. I. 1982. Social organization and ecology of land iguanas, *Conolophus subcristatus*, on Isla Fernandina, Galapagos. Pages 342–365 in *Iguanas of the World*, edited by G. M. Burghardt and A. S. Rand. Noyes Publications, Park Ridge, NJ.

Werner, D. I. 1983. Reproduction in the iguana *Conolophus subcristatus* on Fernandina Island, Galapagos: clutch size and migration costs. *American Naturalist* 121:757–775.

Werner, E. E. 1991. Nonlethal effects of a predator on competitive interactions between two anuran larvae. *Ecology* 72:1709–1720.

Werner, E. E., and B. R. Anholt. 1996. Predator-induced behavioral indirect effects: consequences to competitive interactions in anuran larvae. *Ecology* 77:157–169.

Westoby, M. 1984. An analysis of diet selection by large generalized herbivores. *American Naturalist* 108:290–304.

Weygoldt, P., and S. Potsch de Carvalho e Silva. 1991. Observations on mating, oviposition, egg sac formation and development in the egg-brooding frog, *Fritziana goeldii*. *Amphibia Reptilia* 12:67–80.

Whetstone, K. N. 1978. A new genus of cryptodiran turtles (Testudinata, Chelydridae) from the Upper Cretaceous

Hell Creek Formation of Montana. *University of Kansas Science Bulletin* 51:539–563.

Whitaker, Z. 1989. *Snakeman*. India Magazine Books, Bombay,India.

White, F. N. 1973. Temperature and the Galapagos marine iguana: insights into reptilian thermoregulation. *Comparative Biochemistry and Physiology* 45A:503–513.

Whiteman, H. H. 1994. Evolution of facultative paedomorphosis in salamanders. *Quarterly Review of Biology* 69:205–221.

Whitfield, P. (editor). 1979. *The Animal Family*. W. W. Norton, New York, NY.

Whitford, W. G., and F. M. Creusere. 1977. Seasonal and yearly fluctuations in Chihuahuan desert lizard communities. *Herpetologica* 33:54–65.

Whittier, J. M., and D. Crews. 1987. Seasonal reproduction: patterns and control. Pages 385–409 in *Hormones and Reproduction in Fishes, Amphibians, and Reptiles*, edited by D. O. Norris and R. E. Jones, Plenum Press, New York, NY.

Whittier, J. M., R. T. Mason, and D. Crews. 1985. Mating in the red-sided garter snake, *Thamnophis sirtalis parietalis*: differential effects on male and female sexual behavior. *Behavioral Ecology and Sociobiology* 16:257–261.

Whittier, J. M., and R. R. Tokarz. 1992. Physiological regulation of sexual behavior in female reptiles. Pages 24–69 in *Biology of the Reptilia, Vol. 18 (Physiology E, Hormones, Brain, and Behavior)*, edited by C. Gans and D. Crews, The University of Chicago Press, Chicago, IL.

Wibbels, T., J. J. Bull, and D. Crews. 1994. Temperature-dependent sex determination: a mechanistic approach. *Journal of Experimental Zoology* 270:71–78.

Wiens, J. J. 1993. Systematics of the leptodactylid frog genus *Telmatobius* in the Andes of northern Peru. *Occasional Papers of the Museum of Natural History, University of Kansas* 162:1–76.

Wiens, J. J., and T. A. Titus. 1991. A phylogenetic analysis of *Spea* (Anura: Pelobatidae). *Herpetologica* 47:21–28.

Wiewandt, T. A. 1969. Vocalization, aggressive behavior, and territoriality in the bullfrog, *Rana catesbeiana*. *Copeia* 1969:276–285.

Wiewandt, T. A. 1982. Evolution of nesting patterns in iguanine lizards. Pages 119–141 in *Iguanas of the World*, edited by G. M. Burghardt and A. S. Rand. Noyes Publications, Park Ridge, NJ.

Wiggers, W., G. Roth, C. Eurich, and A. Straub. 1995. Binocular depth perception mechanisms in tongue-projecting salamanders. *Journal of Comparative Physiology* A176:365–377.

Wikelski, M., C. Carbone, and F. Trillmich. 1996. Lekking in marine iguanas: female grouping and male reproductive strategies. *Animal Behaviour* 52:581–596.

Wikelski, M., and F. Trillmich. 1994. Foraging strategies of the Galapagos marine iguana (*Amblyrhynchus cristatus*):

adapting behavioral rules to ontogenetic size change. *Behaviour* 128:255–279.

Wilbur, H. M. 1972. Competition, predation, and the structure of the *Ambystoma-Rana sylvatica* community. *Ecology* 53:3–21.

Wilbur, H. M. 1989. In defense of tanks. *Herpetologica* 45:122–123.

Wilbur, H. M., P. J. Morin, and R. N. Harris. 1983. Salamander predation and the structure of experimental communities: anuran responses. *Ecology* 64:1423–1429.

Wilczynski, W., and E. A. Brenowitz. 1988. Acoustic cues mediate intermale spacing in a neotropical frog. *Animal Behaviour* 36:1054–1063.

Wilczynski, W., H. H. Zakon, and E. A. Brenowitz. 1984. Acoustic communication in spring peepers: Call characteristics and neurophysiological aspects. *Journal of Comparative Physiology A* 155:577–584.

Wild, E. R. 1995. New genus and species of Amazonian microhylid frog with a phylogenetic analysis of New World genera. *Copeia* 1995:837–849.

Wilkinson, M. 1989. On the status of *Nectocaecilia fasciata* Taylor, with a discussion of the phylogeny of the Typhlonectidae (Amphibia: Gymnophiona). *Herpetologica* 45:23–36.

Wilkinson, M. 1996. Resolution of the taxonomic status of *Nectocaecilia haydee* (Roze) and a revised key to the genera of the Typhlonectidae (Amphibia: Gymnophiona). *Journal of Herpetology* 30:413–415.

Wilkinson, M., and R. A. Nussbaum. 1996. On the phylogenetic position of the Uraeotyphlidae (Amphibia: Gymnophiona). *Copeia* 1996:550–562.

Williams, E. E. 1950. Variation and selection in the cervical central articulations of living turtles. *Bulletin of the American Museum of Natural History* 94:509–561.

Williams, E. E. 1976a. South American anoles: the species groups. *Papeis Avulsos de Zoologia, Sao Paulo* 29:259–268.

Williams, E. E. 1976b. West Indian anoles: a taxonomic and evolutionary summary. 1. Introduction and a species list. *Breviora* 440:1–21.

Williams, E. E. 1983. Ecomorphs, faunas, island size, and diverse end points in island radiations of *Anolis*. Pages 326–370 in *Lizard Ecology, Studies of a Model Organism*, edited by R. B. Huey, E. R. Pianka, and T. W. Schoener. Harvard University Press, Cambridge, MA.

Williams, E. E. 1988. A new look at the Iguania. Pages 429–488 in *Proceedings of a Workshop on Neotropical Distribution Patterns*, edited by W. R. Heyer and P. E. Vanzolini. Academia Brasileira de Ciencias, Rio de Janeiro, Brazil.

Williams, E. E. 1989. A critique of Guyer and Savage (1986): cladistic relationships among anoles (Sauria: Iguanidae): are the data available to reclassify the anoles? Pages 433–478 in *Biogeography of the West Indies*, edited by C. A. Woods. Sandhill Crane Press, Gainesville, FL.

Williams, E. E., G. Orces-V, J. C. Matheus, and R. Bleiweiss. 1996. A new giant phenacosaur from Ecuador. *Breviora* 505:1–32.

Williams, E. E., and J. A. Peterson. 1982. Convergent and alternative designs in the digital adhesive pads of scincid lizards. *Science* 215:1509–1511.

Williston, S. W. 1912. Restoration of *Limnoscelis*, a cytolosaur reptile from New Mexico. *American Journal of Science* 34:457–468.

Wilson, E. O. 1975. *Sociobiology: The New Synthesis*. Harvard University Press, Cambridge, MA.

Wilson, L. D., and L. Porras. 1983. The ecological impact of man on the south Florida herpetofauna. *Museum of Natural History, University of Kansas Special Publication* 9 pp 1–89.

Witherington, B. E., K. A. Bjorndal, and C. M. McCabe. 1990. Temporal pattern of nocturnal emergence of loggerhead turtle hatchlings from natural nests. *Copeia* 1990:1165–1168.

Withers, P. C. 1992. *Comparative Animal Physiology*. Saunders, Forth Worth.

Withers, P. C., and C. R. Dickman. 1995. The role of diet in determining water, energy and salt intake in the thorny devil *Moloch horridus* (Lacertilia: Agamidae). *Journal of the Royal Society of Western Australia* 78:3–11.

Wood, R. C. 1976. *Stupendemys geographicus*, the world's largest turtle. *Breviora* 436:1–31.

Wood, S. C. 1984. Cardiovascular shunts and oxygen transport in lower vertebrates. *American Journal of Physiology* 247:R3–R14.

Wood, S. C., and J. W. Hicks. 1985. Oxygen homeostasis in vertebrates with cardiovascular shunts. Pages 354–366 in *Cardiovascular Shunts: Phylogenetic, Ontogenetic, and Clinical Aspects*, edited by K. Johansen and W. W. Burggren, Munksgaard, Copenhagen, Denmark.

Woolbright, L. L., E. J. Greene, and G. C. Rapp. 1990. Density-dependent mate searching strategies of male woodfrogs. *Animal Behaviour* 40:135–142.

World Resources Institute 1990. *World Resources 1990–91*. Oxford University Press, New York, NY.

Worthy, T. H. 1987a. Osteology of *Leiopelma* (Amphibia: Leiopelmatidae) and descriptions of three new subfossil *Leioplema* species. *Journal of the Royal Society of New Zealand* 17:201–251.

Worthy, T. H. 1987b. Palaeoecological information concerning members of the frog genus *Leiopelma*: Leiopelmatidae in New Zealand. *Journal of the Royal Society of New Zealand* 17:409–420.

Wright, J. W. 1993. Evolution of the lizards of the genus *Cnemidophorus*. Pages 27–81 in *Biology of Whiptail lizards (genus Cnemidophorus)*, edited by J. W. Wright and L. J. Vitt. Oklahoma Museum of Natural History, Norman, OK.

Wrobel, D. J., W. F. Gergits, and R. G. Jaeger. 1980. An experimental study of interference competition among terrestrial salamanders. *Ecology* 61:1034–1039.

Wu, X. 1996. *Sineoamphisbaena hexatabularis*, an amphisbaenian (Diapsida: Squamata) from the Upper Cretaceous redbeds at Bayan Mandahu (Inner Mongolia, People's Republic of China), and comments on the phylogenetic relationships of the Amphisbaenia. *Canadian Journal of Earth Sciences* 33:541–577.

Wüster, W., and R. S. Thorpe. 1989. Population affinities of the asiatic cobra (*Naja naja*) species complex in southeast Asia: reliability and random resampling. *Biological Journal of the Linnaean Society* 36:391–409.

Wüster, W., and R. S. Thorpe. 1992a. Asiatic cobras: population systematics of the *Naja naja* species complex (Serpentes: Elapidae) in India and Central Asia. *Herpetologica* 48:69–85.

Wüster, W., and R. S. Thorpe. 1992b. Dentitional phenomena in cobras revisited: spitting and fang structure in the Asiatic species of *Naja* (Serpentes: Elapidae). *Herpetologica* 48:424–434.

Wüster, W., R. S. Thorpe, M. J. Cox, P. Jintakune, and J. Nabhitabhata. 1995. Population systematics of the snake genus *Naja* (Reptilia: Serpentes: Elapidae) in Indochina: multivariate morphometrics and comparative mitochondrial DNA sequencing (cytochrome oxidase I). *Journal of Evolutionary Biology* 8:493–510.

Wyman, R. L. 1990. What's happening to the amphibians? *Conservation Biology* 4:350–352.

Wynn, A. H., C. J. Cole, and A. L. Gardner. 1987. Apparent triploidy in the unisexual Brahminy Blind Snake, *Rhamphotyphlops braminus*. *American Museum Novitates* 2868: 1–7.

Yager, D. D. 1992a. Underwater acoustic communication in the African pipid frog *Xenopus borealis*. *Bioacoustics* 4:1–24.

Yager, D. D. 1992b. A unique sound producing mechanism in the pipid anuran *Xenopus borealis*. *Zoological Journal of the Linnean Society* 104:351–375.

Young, B. A., and K. V. Kardong. 1996. Dentitional surface features in snakes (Reptilia: Serpentes). *Amphibia Reptilia* 17:261–276.

Young, J. Z. *The Life of Vertebrates*. Oxford University Press, Oxford.

Zaher, H. 1994. Les Tropidopheoidea (Serpentes; Alethinophidia) sont-ils réellement monophylétiques? Arguments en faveur de leur polyphylétisme. *Comptes Rendus de l'Academie des Sciences, Paris* 317:471–478.

Zhao, E., Q. Hu, Y. Jiang, and Y. Yang. 1988. *Studies on Chinese Salamanders*. Society For the Study of Amphibians and Reptiles, Oxford, Ohio.

Zhao, E., and G. Zhao. 1981. Notes on Fea's viper (*Azemiops feae* Boulenger) from China. *Acta Herpetologica Sinica* 5:71–76.

Zucker, N. 1994. A dual status-signalling system: a matter of redundancy or differing roles? Animal Behaviour 47:15–22. Zucker, N. 1994. Social influences on the use of a modifiable status signal. *Animal Behaviour* 48: 1317–1324.

Zug, G. R. 1993. *Herpetology, an Introductory Biology of Amphibians and Reptiles*. Academic Press, San Diego, CA.

Zweifel, R. G. 1962. A systematic review of the microhylid frogs of Australia. *American Museum Novitates* 2113: 1–40.

Zweifel, R. G. 1972. Results of the Archbold Expeditions. No. 97, A revision of the frogs of the subfamily Asterophryinae, Family Microhylidae. *Bulletin of the American Museum of Natural History* 148:411–546.

Zweifel, R. G. 1986. A new genus and species of microhylid frog from the Cerro de la Neblina Region of Venezuela and a discussion of relationships among New World microhylid genera. *American Museum Novitates* 2863: 1–24.

Zylberberg, L., and M. H. Wake. 1990. Structure of the scales of *Dermophis* and *Microcaecilia* (Amphibia: Gymnophiona), and a comparison to dermal ossifications of other vertebrates. *Journal of Morphology* 206:25–43.

Illustration Credits

Chapter 1

1-1 Prof. Gerhard Roth, University of Bremen.

1-2 *Environmental Physiology of the Amphibians*, edited by M.E. Feder and W.W. Burggren, Fig. 14.1, p. 397 University of Chicago Press.

1-6 Data from *Physiological Zoology* 40:49-66. Fig. 9, p. 58 by C.B. Dewitt, 1967. University of Chicago Press.

1-7 *American Naturalist* 115:92-112 by F.H. Pough, Fig. 2, p. 99, 1980. University of Chicago Press.

1-8 *Journal of Zoology*, London 214:325-342, Fig. 3, p. 330 by A. Hailey and P.M.C. Davies. 1988

1-10 Reprinted by permission of the publisher from *Lizard Ecology: Studies of a Model Organism*, edited by R.B. Huey, E.R. Pianka, and T.W. Schoener, Cambridge, Mass.: Harvard University Press, Copyright © 1983 by the Presidents and Fellows of Harvard College.

Chapter 2

2-2a-c From M.E. Coates and J.A. Clack, 1995 Proceedings of the 7th International Symposium on Lower Vertebrates. Bulletin of the Museum of Natural History Nationale, Paris 17:373-388.

2-3 Reprinted from *Origins of the Higher Groups of Tetrapods: Controversy and Consensus*, edited by Hans-Peter Schultze and Linda Trueb. Copyright © 1991 by Cornell University. Used by permission of the publisher, Cornell University Press.

2-4a From Williston 1912.

2-4b From Spinar 1952.

2-4c From K.V. Bossy 1976, unpublished thesis, Yale University.

2-4d-f From *Vertebrate Paleontology and Evolution* by R.L. Carroll 1988. Fig. 9-30a, d and 10-2. W.H. Freeman and Company. Reprinted with permission of R.L. Carroll.

2-5a From *Biology of Amphibians* by W.E. Duellman and L. Trueb 1986. Fig. 15-20b. The McGraw-Hill Companies.

2-5b From *The Evolutionary Biology of Hearing*, by T.E. Hetherington, edited by D.B. Webster, R.R. Fay and A.N. Popper, Fig. 21.2, p. 423. © Springer-Verlag.

2-6a-b From W. E. Duellman and L. Trueb *Biology of Amphibians*, Fig. 13-15a, p. 311 © 1986 The Johns Hopkins University Press.

2-6c From *Vertebrate Palontaology and Evolution* by R.L. Carroll, 1988, Fig. 9-17a. W.H. Freeman and Company.

2-7 From A. Milner *Herpetological Monographs* 7: Fig. 5, p. 16, 1993.

2-8a From J.R. Bolt *Journal of Paleontology* 51:Figs 1 and 2, p. 236-237, 1977.

2-8b From Williston 1910.

2-8c From J.A. Boy, *Paleontologische Zeit* 45: Fig 1A, p. 110, 1971.

2-9 Adapted from Gauthier et al. 1989 and Laurin and Reisz 1995.

Chapter 3

3-1 From Larson and Dimmick *Herpetological Monographs* 7: Fig. 5, p. 86, 1993.

3-6 Redrawn from *Wake Bulletin of the Society of California Academy of Sciences* 4, 1966; *Larsone et al Evolution* 35: Fig. 2, p. 411, 1981; Wake and Elias, *Natural History Museum of Los Angeles County Contr. Science*, 345: Fig. 9, p. 17, 1983.

3-9a From M. Wake and J. Hanken *Journal of Morphology* 173: Fig. 2, p. 209. Copyright 1982. Reprinted by permisson of Wiley-Liss, Inc., a division of John Wiley & Sons, Inc.

3-9b From R.A. Nussbaum, *Occasional Papers of the Museum of Zoology* 682: Fig. 1, p. 4, 1977. University of Michigan, Ann Arbor.

3-11 From D.M. Hillis *Amphibian Cytogenetics and Evolution*, Fig. 3, p. 14, 1991.

3-13a Reprinted from *Evolutionary Biology of the Anurans* edited by James L. Vial, Fig. 201, p. 72 by permission of teh University of Missouri Press. Copyright © 1973 by the Curators of the University of Missouri.

3-13b From J.Z. Young *The Life of Vertebrates, 2/e*, Fig. 2-21, p. 38, 1962. Reprinted by permisison of Oxford University Press.

3-13c From W.E. Duellman *Hylid Frogs of Middle Americas*, vol. 2, Fig 22.9a-b, p. 630, 1970. Reprinted with permission of Museum of Natural History, University of Kansas, Lawrence.

3-14a Reprinted from *Evolutionary Biology of the Anurans* edited by James L. Vial, Fig 2-8, p. 93 by permission of the University of Missouri Press. Copyright © 1973 by the Curators of the University of Missouri.

3-14b Reprinted from *Evolutionary Biology of the Anurans* edited by James L. Vial, Fig. 2-9d, p. 96 by permission of the University of Missouri Press. Copyright © 1973 by the Curators of the University of Missouri.

3-15 Modified from Orton 1953, 1957; Starrett 1973.

3-16 From J.C. Rage and Z. Rocek *Palaeontographica abt.* A206: Text Fig. 5, p. 15, 1989.

3-17 From L. Ford and D. Cannatella *Herpetological Monographs* 7: Figs. 1-2, p. 96-97, 1993.

Chapter 4

4-1 From A. Bellairs *The Life of Reptiles*, Fig. 16, p. 66 © 1970 Universe Books.

4-2 Revised from Pritchard and Trebbau, *Turtles of Venezuela*, Fig. 11, p. 24, © 1984 Society of the Study of Amphibians and Reptiles.

4-3a From E.S. Gaffney *Comparative Cranial Morphology of Recent and Fossil Turtles* © 1979 Bulletin of the American Museum of Natural History 164, Fig. 210, p. 298.

4-3b From E.S. Gaffney *Comparative Cranial Morphology of Recent and Fossil Turtles* © 1979 Bulletin of the American Museum of Natural History 164, Fig. 184, p. 280.

4-4 From S.B. McDowell, The Genus Emydura (Testudines: Chelidae) in New Guinea with notes on the penial morphology of Pleurodira. *Advancements in Herpetology and Evolutionary Biology*, Fig. 2, p. 173 © 1983 Museum of Comparative Zoology, Harvard.

4-5 Gaffney and Meylan 1988.

4-11 Estes et al. 1988.

4-14a-f From H.G. Dowline and J.M. Savage, A guide to the snake hemipenis: a survey of basic structure and systematic characteristics. *Zoologica* 45: Figs. 3-4, text figs. 4-6 © 1960 Wildlife Conservation Society.

4-15 From L. Sheppard and A. Bellairs, The mechanism of fautotomy in Lacerta, *British Journal of Herpetology* 4: Fig. 1, p. 277 © 1972 The British Herpetological Society, London.

4-16 From K. Kardong, *Vertebrates*, 1995, Fig 13.10, p. 499. William C. Brown Co. a division of The McGraw-Hill Companies.

4-26a-g From H. W. Parker and A.G.C. Grandison, *Snakes, A Natural History*, Fig. 21, p. 41; Fig. 22, p. 61; Fig. 23, p. 63; Fig. 24, p. 64; Fig. 26, p. 68; Fig. 27, p. 74; Fig. 28, p. 79, and Fig. 14D, p. 84 © 1977 The Natural History Museum, London.

4-31a-b From L.C. Klauber, *Rattlesnakes*, Figs. 5, 10, 12 © 1972 University of California Press.

4-33 From S. Poe, Data set incongruence and the phylogeny of crocodilians. *Systematic Biology* 45: Fig. 7, p. 403 © 1996.

Chapter 5

5-1 Reprinted from *Comparative Physiology of Desert Animals*, Fig. 6, p. 309 by G.N. Louw, Symposium of the Zoological Society, London, No. 31 © 1972 by permission of the publisher Academic Press Limited, London.

5-2 From A note on the drinking habits of some land tortoises. *Animal Behavior* 11: Fig. 1, p. 73 by W. Auffenberg. Copyright 1963. Reprinted with permission of the publisher Academic Press Limited, London.

5-3 From W.A. Buttemer. The effect of temperature on evaporative water loss of the Australian tree frogs *Litoria caerulea* and *Litoria chloris*. *Physiological Zoology* 63: Fig. 6, p. 1055. Copyright 1990 The University of Chicago Press.

5-6 Courtesy of William A. Dunson. From Dunson et al. *Science* 173:437-441. Copyright 1971.

5-8 From J.E. Heath. Temperature regulation and diurnal activity in horned lizards. *University of California Publications in Zoology* 64: Fig. 2, p. 101. Copyright 1965 University of California Press, Berkeley.

5-11 From L.H.S. Van Mierop and S.M. Barnard, Further observations on thermoaregulation in the brooding female *Python molurus bivittatus*. *Copeia* Fig. 2, p. 618 and Fig. 3, p. 619. © 1978 American Society of Ichthyologists and Herpetologists.

5-13 From P.C. Withers, *Comparative Animal Physiology*, Fig. 5-15, p. 142, copyright 1992 Saunders College Publishing.

5-14 From G.A. Bartholomew and V.A. Tucker. Control of changes in body temperature, metabolism, and circulation by the agamid lizard, *Amphibolurus barbatus*. *Physiological Zoology* 36: Fig. 3, p. 201. Copyright 1963 The University of Chicago Press.

5-15 From G.A. Bartholomew and V.A. Tucker. Control of changes in body temperature, metabolism, and circulation by the agamid lizard, *Amphibolurus barbatus*. *Physiological Zoology* 36: Fig. 4, p. 205. Copyright 1963 The University of Chicago Press.

5-16 From F.N. White, *Comparative Biochemistry & Physiology* 45A: 503-513. Copyright 1973 Pergamon Press, NY.

5-17 From J.E. Heath. Temperature regulation and diurnal activity in horned lizards. *University of California Publications in Zoology* 64: Fig. 18, p. 124. Copyright 1965 University of California Press, Berkeley.

5-19 From R.B. Huey. Temperature, physiology and the ecology of reptiles, *Biology of the Reptilia*, vol. 12, Fig. 9, p. 51. Edited by C. Gans and F.H. Pough. Copyright 1982 Academic Press Limited, London.

5-20 From P.E. Hertz, R.B. Huey and R.D. Steenson. Evaluating temperature regulation by field-active actotherms: the fallacy of the inappropriate question. *American Naturalist* 142: Fig. 1, p. 803. Copyright 1993 The University of Chicago Press.

5-21 From G.C. Packard and M.J. Packard. The basis for cold tolerance in hatchling pointed turtles (*Chrysemys picta*). *Physiological Zoology* 68: Fig. 2, p. 138. Copyright 1995 The University of Chicago Press.

5-22 From F.H. Pough, T.L. Taigen, M.M. Stewart, and P.F. Brussard. Behavioral modification of evaporative water loss by a Puerto Rican frog. *Ecology* 64: Fig. 1, p. 246. Copyright 1983 The Ecology Society of America.

5-23 From C.C. Peterson. Anhomeostasis: seasonal water and solute relations in two populations of the desert tortoise (*Gopherus agassizii*) during chronic drought. *Physiological Zoology* 69: Fig. 11, p. 1346. Copyright 1996 The University of Chicago Press.

5-24 From C.C. Peterson. Anhomeostasis: seasonal water and solute relations in two populations of the desert tortoise (*Gopherus agassizii*) during chronic drought. *Physiological Zoology* 69: Fig. 13, part a, p. 1354. Copyright 1996 The University of Chicago Press.

Chapter 6

6-3a-c From W.E. Duellman and L. Trueb *Biology of Amphibians*, Fig. 5-6 right, 5-7a, and 5-9, p. 118-119. © 1986 The McGraw-Hill Companies. Reproduced with permission of the authors.

6-4 From H.J. Jongh and C. Gans *Journal of Morphology* 127: Fig. 14, p. 279 © 1969 Reprinted by permission of Wiley-Liss, Inc. a subsidiary of John Wiley & Sons, Inc.

6-5 From D.R. Carrier, The evolution of locomotor stamina in tetrapods: circumventing a mechanical constraint. *Paleobiology* 13: 326-341 © 1987.

6-6 From *Comparative Animal Physiology* by Philip C. Withers, copyright © 1992 by Saunders College Publishing. Reproduced by permission of the publisher.

6-7 From C. Gans and G.M. Houghes *Journal of Experimental Biology* 47: 1-20, © 1967 Company of Biologists Ltd., Cambridge, UK.

6-8 From G. Shelton and R.G. Boutilei *Journal of Experimental Biology* 100: Fig. 10a, p. 265 © 1982 Springer-Verlag.

6-9 From N. Heisler et al. *Journal of Experimental Biology* 105: 15-32 © 1983 Company of Biologists Ltd., Cambridge, UK.

6-10 From A.G. Kluge *Chordate Structure and Function*, Fig. 7-50, p. 386 © 1977 Macmillan, New York.

6-11 From *Environmental Physiology of the Amphibians* edited by M.E. Feder and W.W. Burggren, Fig. 12-19, p. 344 © 1992 University of Chicago Press.

6-13 From F.H. Pough and R.M. Andrews *Physiological Zoology* 58: Fig. 1, p. 211 © 1985 University of Chicago Press.

6-14 From T.T. Gleeson and P.M. Dalessio *Journal of Comparative Physiology* B169: 331-338 Copyright 1990 Springer-Verlag.

6-15 Reprinted with permission from *Science* 249:524-527. Article by J. Walton, B.C. Jayne, and A.F. Bennett. Copyright 1990 American Association for the Advancement of Science.

6-17a-b From F.H. Pough asnd R.M. Andrews *Ecology* 66: Figs. 2 and 3, p. 1527-1528. Copyuright 1985 Reprinted with permission of the Ecological Society of America.

6-18 From T.L. Taigen and K.D. Wells, *Journal of Comparative Physiology*, B155: Fig. 1, p. 166. Copyright 1985 Springer-Verlag.

6-19 Reprinted from *Animal Behavior* 48: Fig. 3 (top), p. 770, © 1994 by permission of the publisher Academic Press Limited, London.

6-20 From C.G. Murphy *Behavioral Ecology and Sociobiology* 34: Fig. 1, p. 286. Copyright 1994 Springer-Verlag.

6-21 From S.J. Beaupre *Copeia* 1996. Fig. 2, p. 325. Copyright 1996 American Society of Ichthyologists and Herpetologists.

6-22 From R.B. Huey and R.D. Stevenson *American Zoologist* 19: Fig. 1, p. 358 © 1979.

6-23 From F.H. Van Berkum, R.B. Huey, and B. A. Adams *Physiological Zoology* 59: Fig. 1, p. 466 Copyright 1986 The University of Chicago Press.

6-24 From From F.H. Van Berkum, R.B. Huey, and B. A. Adams *Physiological Zoology* 59: Fig. 2, p. 470 Copyright 1986 The University of Chicago Press.

6-25 From F.R. Moore and R.E. Gatten, Jr. *Herpetologica* 45: Fig. 1, p. 106. Copyright 1989 The Herpetologist's League.

6-26 From M.R. Preest and F.H. Pough *Functional Ecology* 3: Fig. 3, p. 697. Copyright 1989 Blackwell Science, Ltd., Oxford, UK.

6-27 From G.C. Packard and M.J. Packard *The Biology of the Reptilia*, vol 16 , Fig. 14, p. 579, edited by C. Gans and R.B. Huey. Copyright 1988. Reprinted with permission of Wiley-Liss, Inc. a division of John Wiley & Sons, Inc.

6-28 From K. Miller, *Journal of Herpetology* 27:P Fig. 1, p. 230. Copyright 1993 the Society for the Study of Amphibians and Reptiles.

6-29 From K. Miller, *Journal of Herpetology* 27: Fig. 1, p. 230 Copyright 1993 The Society for the Study of Amphibians and Reptiles.

6-30 From T. Garland, Jr. *American Journal of Physiology* 247: Fig. 1, p. R809. Copyright 1984 The American Physiology Society.

Chapter 7

7-1 Reprinted from *Herpetology* by G. Zug, Fig. 7.2 Copyright 1993 by permission of the publisher Academic Press Limited, London.

7-3 Modified from L.D. Houck and S.K. Woodley Field studies of steroid hormones and male reproductive behavior in amphibians. In *Biology* edited by H. Heatwole and B.K. Sullivan copyright 1995 Surrey Beatty and Sons, Chipping Norton, NSW, Australia.

7-4 From J.M. Whittier and D. Crews, *Seasonal Reproduction: Patterns and Control in Hormones and Reproduction in Fishes, Amphibans, and Reptiles*, edited by D.O. Norris and R.E. Jones, copyright 1987 Plenum Publishing Corporation, NY.

7-5 From Cogger and Zweifel *Reptiles and Amphibians* p. 18 Copyright 1992 Weldon Owen Publishing, McMahons Point NSW, Australia.

7-6 From Crews et al. *Developmental Genetics* vol 15, Fig. 2, p. 297-312. Copyright 1994 Reprinted by permission of Wiley-Liss, Inc. a division of John Wiley & Sons, Inc.

7-7 From James J. Bull *Evolution of Sex Determining Mechanisms*, Fig. 9D2 Copyright 1983 Benjamin/Cummings Publishing Company.

7-8 Based on F.J. Janzen Experimenatal evidence for the evolutionary significance of temperature dependent sex determination, *Evolution* 49: Fig. 1, p. 864-873. Copyright 1995.

7-15a From J. R. Stewart, Yolk sac placentation in reptiles: structural innovation in a fundamental vertebrate fetal nutritional system. *Journal of Experimental Zoology* 166: Fig. 2, p. 431-449 Copyright 1993 Reprinted by permission of Wiley-Liss, Inc. a division of John Wiley & Sons, Inc.

7-15b From D.G. Blackburn Histology of the late stage placentae in the matrotrophic skink *Chalcides chalcides* (Lacertilia: Scincidae) *Journal of Morphology* 216: Fig. 4, p. 179-195. Copyright 1993 Reprinted by permission of Wiley-Liss, Inc. a division of John Wiley & Sons, Inc.

7-17 From A.E. Dunham, D.B. Milesn and D.N. Reznick Life history patterns in squamte reptiles. *Biology of the Reptilia, Ecology B (Defense and Life History)*, vol 16, Fig. 1, p. 465 Copyright 1988 Reprinted by permission of Wiley-Liss, Inc. a division of John Wiley & Sons, Inc.

7-18 From D.W. Tinkle *American Naturalist* 103: Fig. 1, p. 508 Copyright 1969 The Univerisity of Chicago Press.

7-19 From Reilly et al. An integrative approach to heterochrony: the distinction between interspecific and intraspecific phenomena. *Biological Journal of the Linnean Society* 60: Fig. 1, p. 119-143 Copyright 1997 Academic Press Limited.

7-21 From Reilly et al. An integrative approach to heterochrony: the distinction between interspecific and intraspecific phenomena. *Biological Journal of the Linnean Society* 60: Fig. 4, p. 119-143 Copyright 1997 Academic Press Limited.

Chapter 8

8-2a Modified from Gray *Animal Locomotion* Fig. 5.6, p. 91. Copyright 1968 W.W. Norton.

8-2b From Gray *Animal Locomotion* Fig. 6.1, p. 120. Copyright 1968 W.W. Norton.

8-2c Reprinted by permission of the publisher from *Vertebrate Functional Morphology* edited by Milton Hildbrand, Denis M. Bramble, Karen F. Liem and David B. Wake, Cambridge, Mass.: Harvard University Press, Copyright © 1985 by the President and Fellows of Harvard.

8-3c From Gatesy *Journal of Zoology* 224(4): Fig. 1, p. 579. Copyright 1991 Zoological Society of London.

8-3d From Savitsky *Evolution* 34(6): Fig. 1, p. 1195, Copyright 1980.

8-5a From Jenkins and Goslow *Journal of Morphology* 176(2): Fig. 7, p. 205. Copyright 1983. Reprinted with permission of Wiley-Liss, Inc. a division of John Wiley & Sons, Inc.

8-5b From Jenkins and Goslow *Journal of Morphology* 176(2): Fig. 12, p. 214. Copyright 1983. Reprinted with permission of Wiley-Liss, Inc. a division of John Wiley & Sons, Inc.

8-6b Modified from Brinkman *Canadian Journal of Zoology* 58(2): Fig. 4 and 9, p. 281 and 285. Copyright 1980a National Research Council of Canada, Ottawa.

8-8 From Webb and Gans, *Records of the Australian Museum* 34(14): Fig. 2, p. 613. Copyright 1982 The Australian Museum.

8-9 From Walker *Journal of Morphology* 134 (2): Fig. 2, p. 199 Copyright 1971a. Reprinted by permission of Wiley-Liss, Inc., a division of John Wiley & Sons, Inc.

8-12a Cundall 1987.

8-12b Modified from Jayne 1985, 1988a.

8-13a From Hildbrand *Analysis of Vertebrate Structure 4/e*, Fig. 25.12, p. 500. Copyright 1995. Reprinted by permission of John Wiley & Sons, Inc.

8-13b From Gans *Biomechanics*, Fig. 3-13, p. 86, Copyright 1984 The University of Michigan Press.

8-14a From Gans, Structure, Development, and Evolution of Reptiles, *Zoological Soceity of London Symposium No. 52*, Fig. 3, p. 18. Copyright 1984 The Zoological Society of London.

8-14b-d From Gans *Biomechanics*, (b) Fig. 3-19, p. 91, (c) Fig. 3-23, p. 96, (d) Fig. 3-33, Copyright 1984 The University of Michigan Press.

8-16 From Jayne *Copeia* 1985 (1), Fig. 5, p. 204, Copyright 1985 The American Society of Ichthyologists and Herpetologists.

8-17 Modifed From Videler and Jorna *Copeia* 1985a, Fig. 1 and 2, p. 252 and 253. Copyright 1985 The American Society of Ichthyologists and Herpetologists.

8-18a From Davenport et al. *Proceedings of the Royal Society of London*, 220B (1221): Fig. 5, p. 453.

8-18b From Walker *Copeia* 1971 (2): Fig. 2, p. 230. Copyright 1984 The American Society of Ichthyologists and Herpetologists.

8-19a From Gans 1976. In Bellaris and Cox *Morphology and Biology of Reptiles*, Linnean Society Symp. Ser., No. 3, Fig. 1, p. 196.

8-19b Reprinted from *Nature* 234(5325): Fig. 2, p. 151. Copyright 1971 Macmillan Magazines Limited.

8-20a From Gans *Biomechanics*, Fig. 4.17, p. 146. Copyright 1974 The University of Michigan Press.

8-20b-d From Gans *Biomechanics*, Fig. 4.17b-d, p. 162. Copyright 1974 The University of Michigan Press.

8-22b-c From Peterson *Journal of Morphology* 202(1: (b) Fig. 1a (left) and Fig. 2b (right), p. 10; (c) Plate 11d (left) and Plate 11f (right), p. 23. Copyright 1984. Reprinted with permisson of Wiley-Liss, Inc., a division of John Wiley & Sons, Inc.

8-24d From Russell *Journal of Zoology* 176: Fig. 15, p. 461. Copyright 1975 Zoological Society of London.

Chapter 9

9-1 From *Zoological Journal of the Linnean Society* 88(3): Fig. 2, p. 280. Copyright © 1996.

9-2 From *Journal of Zoology* 216(3): Plate 1 (left), frames 1, 6, 8, 12. Copyright 1988 Zoological Society of London.

9-3a From Loauder and Shaffert *Journal of Morphology* 197(3): Fig. 4A and 6A, p. 255 and 260. Copyright 1988. Reprinted by permission of Wiley-Liss, Inc., a division of John Wiley & Sons, Inc.

9-3b From *Journal of Morphology* 185 (3): Fig. 6, p. 307. Copyright 1985. Reprinted by permission of Wiley-Liss, Inc., a division of John Wiley & Sons, Inc.

9-5b-c From Cundall et al. *Experientia* 43: Fig. 3, p. 1230. Copyright 1987 Birkhauser Verlag AG, Basel, Switzerland.

9-6a-d, f From de Jongh *Netherlands Journal of Zoology* 18(1): Figs. 6.7, p. 16 and 18. Copyright 1968.

9-6e From Wasswersug, University of Kansas Museum of Natural History, Miscellaneous publication (68): Fig. 20 (top), p. 53. Copyright 1980. Reprinted with permission.

9-8 From Altig and Johnston *Herpetological Monographs* (3): Fig. 7 as follows: (a) fig. 7A; (b) Fig. 7B; (c) Fig. 7D; (d) Fig. 7I. Copyright 1989.

9-9 From Altig and Johnston *Herpetological Monographs* (3): Fig. 8 as follows: (a) Fig. 8A; (b) Fig. 8B; (c) Fig. 8F; (d) Fig. 8G. Copyright 1989.

9-10 From Altig and Johnston *Herpetological Monographs* (3): Figs. 7-8 as follows: (a) Fig. 7E; (b) Fig. 8D. Copyright 1989.

9-11b From Gradwell *Canadian Journal of Zoology* 49(3): Fig. 4, p. 311. Copyright 1971 National Research Council of Canada, Ottawa.

9-12b-c From Wassersug and Pyburn *Zoological Jorunal of the Linnean Society* 91(2): (a) Fig 8B, p. 150; (b) Fig. 2A, p. 139; (c) Fig. 6, p. 148. Copyright 1987 Academic Press.

9-13 From *Zoological Journal of the Linnean Society* 77: Figs. 2-3, p. 83-84. Copyright 1983 Academic Press.

9-14 From Schumacher *Biology of the Reptilia*, vol. 4, Fig. 4, p. 110. Copyright 1973. Academic Press.

9-15 From Gaffney and Meyland *Phylogical Classification of Tetrapods:* Fig. 5.2, p. 163. Copyright 1988 Clarendon Press, a division of Oxford Univeristy Press.

9-16 From Larsen and Guthrie *Journal of Morphology* 178(3): fig. 5, p. 144. Copyright 1975. Reprinted by permission of Wiley-Liss, Inc., a division of John Wiley & Sons, Inc.

9-17 From Larsen and Guthrie *Journal of Morphology* 178(3): Plate 1, Fig. 9 only, p. 151. Copyright 1975. Reprinted by permission of Wiley-Liss, Inc., a division of John Wiley & Sons, Inc.

9-18 Modified from Lombard and Wake *Journal of Morphology* 148(3): Figs. 2-3, p. 270-271. Copyright 1976. Reprinted by permission of Wiley-Liss, Inc., a division of John Wiley & Sons, Inc.

9-19 From Lombard and Wake *Journal of Morphology* 151(1): Fig. 23, p. 39-79. Copyright 1977. Reprinted by permission of Wiley-Liss, Inc., a division of John Wiley & Sons, Inc.

9-20 From Lombard and Wake *Journal of Morphology* 153(1): Fig. 15, p. 57 (two upper left figures only). Copyright 1977. Reprinted by permission of Wiley-Liss, Inc., a division of John Wiley & Sons, Inc.

9-21 From Gans and Gorniak *American Journal of Anatomy* 163: Figs. 1-3, p. 200-202. Copyright 1982. Reprinted by permission of Wiley-Liss, Inc., a division of John Wiley & Sons, Inc.

9-23 Modified from Wainwright, et al. *Journal of Experimental Biology* 159: Fig. 1, p. 114, © 1991 and Wainwright and Bennett *Journal of Experimental Biology* 168: Fig. 1, p. 5, c 1992. Reprinted with permission of Company of Biologists, Ltd., Cambridge, UK.

9-26 Modified from Smith *Journal of Morphology*, Fig. 1, p. 140. Copyright 1982. Reprinted by permission of Wiley-Liss, Inc., a division of John Wiley & Sons, Inc.

9-27b From Dalrympole *Journal of Herpeteology* 13(3): Fig. 21, p. 304. Copyright 1979.

9-28 From Smith *Journal of Morphology* 173: Fig. 4, p. 144. Copyright 1982. Reprinted by permission of Wiley-Liss, Inc. a division of John Wiley & Sons, Inc.

9-31 Based on Parker and Grandison *Snakes: A Natural History*, Fig. 216, p. 74. Copyright 1977 The British Museum of Natural History, London.

9-32 Modified from Cundall and Gans *Journal of Experimental Zoology* 209(2): Fig. 1, p. 191. Copyright 1979 The Company of Biologists, Ltd, Cambridge, UK.

9-33 Modified from Cundall and Gans *Journal of Experimental Zoology* 209(2): Fig. 3, p. 193. Copyright 1979 The Company of Biologists, Ltd, Cambridge, UK.

9-35 Reprinted with permission from Savitzky *Science* 212: (a) Fig. 11 and (b) Fig. 1H, p. 347. Copyright 1971 American Association for the Advancement of Science.

9-37 From Gans *Biomechanics*, Fig. 20-16, p. 39. Copyright 1974 The University of Michigan Press.

9-38 From I. Sazima, Feeding behavior of the snail-eating snake, *Dipsas indica*, *Journal of Herpetology* 23: 464-468. Copyright 1989.

9-39 Based on Kochva *Biology of the Reptilia* 8: Fig. 37, p. 81. Copyright 1978.

9-40 Based on Kochva *Biology of the Reptilia* 8: Fig. 53, p. 105 and Fig. 66, p. 118. Copyright 1978.

9-41 From Parker and Brandison *Snakes: A Natural History*, Fig. 14, p. 41. Copyright 1977 The British Museum of Natural History, London.

9-43 From Iverson *Journal of Morphology* 163(1): Fig. 6, p. 85. Copyright 1980. Reprinted by permission of Wiley-Liss, Inc., a division of John Wiley & Sons.

Chapter 10

10-1 From K. Nishikawa *Copeia* 1990: Fig. 2, p. 422, © 1990 American Society of Ichthyologists and Herpetologists.

10-2 From J.A. Keinath and J.A. Muisick *Copeia* 1993: Fig. 1, p. 1011, © 1993 American Society of Ichthyologists and Herpetologists.

10-3 From T. Madsen *Copeia* 1984: Fig. 2, p. 709, © 1984 American Society of Ichthyologists and Herpetologists.

10-4 From Walter Auffenberg *The Behavioral Ecology of the Komodo Monitor*, Fig. 5.9, p. 99 © 1981. Reprinted with permission of the University of Florida, Gainesville.

10-5 From E.P. Martins, L.J. Vitt and E.R. Pianka, eds. *Lizard Ecology: Historical and Experimental Perspectives*, Fig. 6.2, p. 122 Princeton University Press © E.P. Martins.

10-6 From J.A. Stamps *American Naturalist* 135: Fig. 1, p. 622 © 1990 The University of Chicago Press.

10-8 From B.W. Bowen and S.A. Karl, *The Biology of Sea Turtles*, P.O. Lutz and J.A. Musick, eds., Fig 2.3, p. 39 © 1997 CRC Press Inc., Boca Raton, FL.

10-9 From G.M. Brughardt *American Zoologist* 17: Fig. 6, p. 184 © 1977 American Zoologist.

10-10 From J.D. Miller *The Biology of Sea Turtles*, P.O. Lutz and J.A. Musick, eds, Fig. 3 © 1977 CRC Pres, Inc., Boca Raton, FL.

10-11 From V.C. Twitty *Of Scientists and Salamanders*, Fig. 42, p. 117 © 1996 W.C. Freeman & Co., San Francisco, CA.

10-12 From C. Ciofi and G. Chelazzi *Journal of Herpetology* 28: Fig. 1, p. 479 © 1994. Reprinted with permission.

10-13 From L. Plasa *Z. Tierpsychologie* 51: Fig. 2, p. 116 © 1979 Blackwell Wissenschafts-Verlag, Wien, Austria.

10-14 From J.D. Weintraub *Animal Behavior* 18: Fig. 2, p. 135 © 1970 Academic Press, Ltd., London.

10-15 From K.J. Lohmann et al. *The Biology of Sea Turtles*, P.O. Lutz and J.A. Musick, eds. Fig. 5.8 © 1997 CRC Press Inc., Boca Raton, FL.

10-16 From K. Adler *Journal of Herpetology* 4: Fig. 3, p. 104 © 1970.

10-17 From K. Adler *Photochemistry and Photobiology* 23: Fig. 7, p. 288 © 1976 International Cytokine Society, Boston, MA.

10-18 From K. Adler *Photochemistry and Photobiology* 23: Fig. 1, p. 276 © 1976 International Cytokine Society, Boston, MA.

10-19 Reprinted with permission from D.H. Adler *Science* 181: Fig. 1, p. 285. Copyright © 1973 American Association for the Advancement of Science.

10-20 From K. Adler *Photochemistry and Photobiology* 23: Fig. 11, p. 293 © 1976 International Cytokine Society, Boston, MA.

Chapter 11

11-1 Reprinted with permission from H.C. Gerhardt *Science* 199: Fig. 1, p. 93 Copyright 1978 American Association for the Advancement of Science.

11-3 From P.M. Narins *Bioscience* 40: Fig. 3A, p. 270. Copyright 1990 American Institute of Biological Sciences.

11-4 From E.M. Dawley and A.H. Bass *Journal of Morphology* 200: Fig. 1, p. 164 Copyright 1989. Reprinted by permission of Wiley-Liss, Inc., a division of John Wiley & Sons, Inc.

11-5 From E.M. Dawley *Animal Behavior* 322: Fig. 3, p. 356 Copyright 1984 Academic Press, Ltd., London.

11-6 From S.J. Arnold *Z. Tierpsychologie* 42: Fig. 19, p. 272 Copyright 1976 Blackwell Wissenschafts-Verlag, Wien, Austria.

11-7 From L. Houck and D.M. Sever *Amphibian Biology* vol. 1, edited by H. Heatwold and G. Barthalmus, Figs. 4, 5, 6, p. 363 Copyright 1994 Surrey Beatty and Sons.

11-8 From S.J. Arnold Z. *Tierpsychologie* 42: Fig. 18, p. 270 Copyright 1976 Blackwell Wissenschafts-Verlag, Wien, Austria.

11-9 From L. Houck and D.M. Sever *Amphibian Biology* vol. 1, edited by H. Heatwold and G. Barthalmus, Figs. 7, 8, p. 364 and 365 Copyright 1994 Surrey Beatty & Sons.

11-10 From T.R. Halliday *Advances in the Study of Behavior* 19: Fig. 3, p. 145 Copyright 1990 Academic Press, Ltd., London.

11-11 From J.W. Arntzen and M. Sparreboom *Journal of Zoology* 219: Fig. 2A, p. 650 and Fig. 1A, p. 649 Copyright 1989 Zoological Society of London.

11-12 From W.E. Duellman and L. Trueb *Biology of Amphibians*, Figs. 4.3 and 4.5, pp. 90 and 91 Copyright 1986 The Johns Hopkins University Press.

11-15 From J.J. Schwartz and K.D. Wells *Copeia* 1985: Fig. 7, p. 34 Copyright 1985 American Society of Ichthyologists and Herpetologists.

11-16 From W. Wilczynski, H.J. Zakon and E. Brenowitz *Journal of Comparative Physiology A* 155: (a) Fig. 3A, p. 580, (b) Fig. 5, p. 591 Copyright 1984 Springer-Verlag.

11-18 From J.A. Oliver *Natural History of North American Amphibians and Reptiles* Copyright 1955 Van Nostrand-Reinhold/International Thomson Publishers.

11-19 From Auffenberg *American Zoologist* 17: (a) Fig. 1, p. 243, (b) Fig. 2, p. 244 Copyright 1977 American Zoologist.

11-20 From W. Auffenberg *Herpetologica* 22: Fig. 1, p. 115 Copyright 1966 Herpetologists League.

11-21 From K.A. Vliet *American Zoologist* 29: Fig. 1, p. 1021 and Fig. 6 p. 1025 Copyright 1989 American Zoology.

11-22 From J.C. Gillingham et al. *Herpetological Monographs* 9: Fig. 5, p. 11 Copyright 1995 Reprinted by permission of Wiley-Liss, Inc., a division of John Wiley & Sons, Inc.

11-23 From K. Schwenk *Trends in Ecology and Evolution* 10: Box 1, p. 9 Copyright 1995 Elsevier Trends Journals, Cambridge, UK.

11-24 From K. Schwenk *Trends in Ecology and Evolution* 10: Fig. 2, p. 10 Copyright 1995 Elsevier Trends Journals, Cambridge, UK.

11-25 From A.A. Echelle et al. *Herpetologica* 27: 221-288 Copyright 1971 Herpetologists League.

11-27 From P. Whitfield *The Animal Family*, p. 47 Copyright 1979 Marshall Editions, London.

Chapter 12

12-1 From C. Darwin The Descent of Man and Selection in Relation to Sex, vol. 2, p. 23, 33, 34. A. Appleton, New York, 1871.

12-2 From R.D. Howard and A.G. Kluge Evolution 39: Fig. 1a and b, p. 260-277. Copyright 1985.

12-4 From S.J. Arnold Z. *Tierpsychologie* 42: Fig. 31, p. 289 and Fig. 27, p. 282. Copyright 1976 Blackwell Wissenschafts-Verlag, Wien, Austria.

12-7 From T.R. Halliday *Advances in the Study of Behavior* 19: Fig. 4B, p. 147. Copyright 1990 Academic Press, Ltd, London.

12-8 From M. Smith *The British Amphibians and Reptiles*, Fig. 80, p. 249. Copyright 1969 Harper Collins.

12-9a-d From D.E. Ruby and H.A. Niblick *Herpetological Monographs* 8: Fig. 4, p. 96. Copyright 1994. Reprinted by permission of Wiley-Liss, Inc., a division of John Wiley & Sons, Inc.

12-9e From W. Auffenberg *American Zoologist* 17: Fig. 4, p. 247. Copyright 1977 American Zoologist.

12-11 From T.R. Halliday *The Reproductive Biology of Amphibians*, edited by D. Taylor and S.E. Guttman, Fig. 11, p. 212. Copyright 1977 Plenum Publishing Corporation.

12-12 From M. Wikelski et al. *Animal Behavior* 52: (a) Fig. 1b, p. 583, (b) Fig. 8b, p. 591. Copyright 1996 Academic Press, Ltd., London.

12-14 From R.D. Howard Ecology 59: (a) Fig. 8, p. 795, (b) Fig. 9, p. 796. Copyright 1978b The Ecological Society of America.

12-15 Reprinted by permission of the pubisher from *Lizard Ecology: Studies of a Model Organism* edited by R.B. Huey, E.R. Pianka and T.W. Schoener, Cambridge, MA: Harvard University Press, Copyright © 1983 by the President and Fellows of Harvard College.

12-16 From S.J. Arnold and D. Duvall *American Naturalist* 143: Fig. 1, p. 319. Copyright 1994 The University of Chicago Press.

12-17 From K.G.K. Trillmich Z. *Tierpsychologie* 63: (a) Fig. 10, p. 153, (b) Fig. 12, p. 154. Copyright 1983 Blackwells Wissenschafts-Verlag, Wien, Austria.

12-20a From S.B. Emerson and H. Voris *Functional Ecology* 6: Fig. 2d, p. 657. Copyright 1992.

12-20b From L.J. Vitt and W.E. Cooper *Canadian Journal of Zoology* 63: Fig. 4, p. 998. Copyright 1985 National Research Council of Canada, Ottawa.

12-21 From A. Arak *Animal Behavior* 36: (a) Fig. 1, p. 418, (b) Fig. 2c, p. 421. Copyright 1988 Academic Press, Ltd., London.

12-22 From S.J. Arnold Z. *Tierpsychgologie* 42: Fig. 28, p. 282. Copyright 1976 Blackwell Wissenschafts-Verlag, Wien, Austria.

12-24a From T. Hayes and P. Licht *Journal of Experimental Zoology* 264: Fig. 1, p. 132. Copyright 1992. Reprinted by pemission of Wiley-Liss, Inc., a division of John Wiley & Sons, Inc.

12-24b From R.H.J. Chabreck and T. Joanen *Herpetologica* 35: Fig. 4, p. 56. Copyright 1979 The Herpetologists League.

12-24c From A.E. Dunham and J.W. Gibbons *Life History and Ecology of the Slider Turtle*, Fig. 10.2c, p. 141. J.W. Gibbons, Editor. Copyright 1990 Smithsonian Institution Press.

12-24d From R.W. Van Devender *Ecology* 59: Fig. 4, p. 1034 Copyright 1978 The Ecological Society of America.

12-25 From J.A. Stamps *Herpetological Monographs* 9: Fig. 2, p. 81. Copyright 1995.

Chapter 13

13-1a-d Modified from Cogger and Zweifel *Reptiles and Amphibians* (a and b) p. 89, (c) p. 87, (d) p. 102. Copyright Weldon Owen Publishing, Sydney, Australia.

13-2 Reprinted with permission from The ecological significance of secual dimorphism in size in the lard Anolis conspersus *Science* 155: pp 474-477 by T.W. Schoener Copyright 1967 American Association for the Advancement of Science.

13-3 From D.A. Belk and C. Gans An unusual chelonian feeding niche. *Ecology* 49: 768-769 Copyright 1968 The Ecological Society of America.

13-4 From Rose and Armentrout 45: 713-739 Copyright © 1976.

13-5 From H.R. Musinsky, J.J. Hebrard and D.S. Vodopich Anatomy of water snake foraging ecology, *Ecology* 63: 1624-1629. Copyright 1982   The Ecological Society of America.

13-7 From D.G. Kephart and S.J. Arnold Garter snake diets in a fluctuating environment: a seven year study. *Ecology* 63: Fig. 2, 1232-1236 Copyright 1982 The Ecological Society of America.

13-8 Form J.P. Caldwell The evolution of myrmecophagy and its correlates in poison frogs (family Dendrobatidae). *Journal of Zoology*, London 240: Fig. 9, p. 75-101 Copyright 1996 Zoological Society of London.

13-9a Drawn from photo by David Scott.

13-9b From J.B. Murphy *Herpetologica* 32: Fig. 1, p. 339-341 Copyright 1976 The Herpetologist's League.

13-9c From I. Sazima *Copeia* 1991: 245-248 Copyright 1991 American Society of Ichthyologists and Herpetologists.

13-11 From R.L. McLaughlin *American Naturalist* 133: Fig. 2, p. 654-670. Copyright 1989 The University of Chicago Press.

13-12a From Jaeger and Rubin *Journal of Animal Ecology* 51: Fig. 1, p. 167-176 Copyright 1982 Blackwell Science Ltd., Oxford, UK.

13-12b From Jaeger and Bernard *American Naturalist* 117: Fig. 12, p. 639-6643 Copyright 1981 The University of Chicago Press.

13-14 From Frazer, Gibbons, and Green *Life History and Ecology of the Slider Turtle*, Fig. 15.7, p. 183-200. J.W. Gibbons, Editor. Copyright 1990 The Smithsonian Institution Press.

13-15 From Arnold and Wassersug *Ecology* 59: Fig. 1, p. 1014-1022 Copyright 1978 The Ecological Society of America.

13-19 From E.D. Brodie III *Evolution* 47: Figs. 4 and 5, pp. 227-235. Copyright 1993 Society for the Study of Evolution.

13-20 From Pough *Copeia* 1976: Fig. 1, p. 834-836 Copyright 1976 American Society of Ichthyologists and Herpetologists.

Chapter 14

14-1 From Kiester *Systematic Zoology* 20: Figs. 3 and 4, pp. 130 and 131 Copyright 1971 The Society of Systematic Biologists.

14-3 Reprinted by permission of the publisher from *Lizard Ecology: Studies of a Model Organism* edited by R.B. Huey, E.R. Pianka and T.W. Schoener, Cambridge, MA: Harvard University Press, Copyright © 1983 by the President and Fellows of Harvard College.

14-4 From Pechmann et al. *Wetlands Ecology and Management* 1: Fig. 1, p. 6 Copyright 1989 SPB Academic Publishing, The Netherlands.

14-6 From Pacala and Roughgarden *Ecology* 66: Fig. 4, p. 136 Copyright 1985 The Ecological Society of America.

14-7 From Lister *Evolution* 30: Fig. 3, p. 663 Copyright 1976 The Society for the Study of Evolution.

14-9 From Southerland *Copeia* 1986: Fig. 3, p. 736 Copyright 1986 The American Society of Ichthyologists and Herpetologists.

14-10 From Hariston *Ecology* 61: Fig. 1, p. 818 Copyright 1980b The Ecological Society of America.

14-12 From Heyer *Journal of Herpetology* 7: Fig. 9, p. 356 Copyright 1973 The Society for the Study of Amphibians and Reptiles.

14-14 From Morin *Ecological Monographs* 53: Fig. 1, p. 124 Copyright 1983 The Ecological Society of America.

14-15 From Morin *Ecological Monographs* 53: Fig. 5, p. 129 Copyright 1983 The Ecological Society of America.

14-16 From Werner *Ecology* 72: Fig. 2, p. 1713 The Ecological Society of America.

14-17 From Werner and Anholt *Ecology* 77: Fig. 6, p. 165 The Ecological Society of America.

14-18 From Duellman and Pyles *Copeia* 1983: Fig. 1, p. 644 Copyright 1983 The American Society of Ichthyologists and Herpetologists.

Chapter 15

15-3 From Cree et al. *Conservation Biology* 9: Fig. 2, p. 377 Copyright 1995 Blackwell Science, Ltd. Oxford, UK.

15-10 From Congdon et al. *American Zoologist* 34: Fig. 5, p. 406 Copyright 1994.

15-14 Courtesy ACO Polymer Products, Chardon, OH.

Author Index

Boldface indicates author is cited in figure legend on that page.

Subject Index